Wetlands through Time

edited by

Stephen F. Greb
Kentucky Geological Survey
228 MMRB
University of Kentucky
Lexington, Kentucky 40506
USA

and

William A. DiMichele
Department of Paleobiology
National Museum of Natural History
Smithsonian Institution
Washington, D.C. 20560
USA

THE
GEOLOGICAL
SOCIETY
OF AMERICA

Special Paper 399

3300 Penrose Place, P.O. Box 9140 ▪ Boulder, Colorado 80301-9140 USA

2006

Copyright © 2006, The Geological Society of America, Inc. (GSA). All rights reserved. GSA grants permission to individual scientists to make unlimited photocopies of one or more items from this volume for noncommercial purposes advancing science or education, including classroom use. For permission to make photocopies of any item in this volume for other noncommercial, nonprofit purposes, contact the Geological Society of America. Written permission is required from GSA for all other forms of capture or reproduction of any item in the volume including, but not limited to, all types of electronic or digital scanning or other digital or manual transformation of articles or any portion thereof, such as abstracts, into computer-readable and/or transmittable form for personal or corporate use, either noncommercial or commercial, for-profit or otherwise. Send permission requests to GSA Copyright Permissions, 3300 Penrose Place, P.O. Box 9140, Boulder, Colorado 80301-9140, USA.

Copyright is not claimed on any material prepared wholly by government employees within the scope of their employment.

Published by The Geological Society of America, Inc.
3300 Penrose Place, P.O. Box 9140, Boulder, Colorado 80301-9140, USA
www.geosociety.org

Printed in U.S.A.

GSA Books Science Editor: Abhijit Basu

Library of Congress Cataloging-in-Publication Data

Wetlands through time / edited by Stephen F. Greb and William A. DiMichele
    p. cm. -- (Special paper ; 399)
  Includes bibliographical references, index.
  ISBN: 0-8137-2399-X (pbk.)
    1. Paleoecology--North America. 2. Paleoecology--Paleozoic. 3. Wetlands--North America--History.
  I. Greb, Stephen F. II. DiMichele, William A. Special papers (Geological Society of America) ; 399.

QE720.2.N7 W48 2006
560'.4568097--dc22

2005058232

Cover: *Wetlands through Time*, a painting by Stephen F. Greb.

10 9 8 7 6 5 4 3 2 1

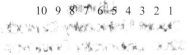

Contents

Preface v

1. **Evolution and importance of wetlands in earth history** 1
   Stephen F. Greb, William A. DiMichele, and Robert A. Gastaldo

2. **Wetlands before tracheophytes: Thalloid terrestrial communities of the
   Early Silurian Passage Creek biota (Virginia)** 41
   Alexandru Mihail Florian Tomescu and Gar W. Rothwell

3. **Sedimentology and taphonomy of the Early to Middle Devonian plant-bearing beds
   of the Trout Valley Formation, Maine** 57
   Jonathan P. Allen and Robert A. Gastaldo

4. **Plant paleoecology of the Late Devonian Red Hill locality, north-central Pennsylvania,
   an Archaeopteris-dominated wetland plant community and early tetrapod site** 79
   Walter L. Cressler III

5. **Tournaisian forested wetlands in the Horton Group of Atlantic Canada** 103
   Michael C. Rygel, John H. Calder, Martin R. Gibling, Murray K. Gingras, and
   Camilla S.A. Melrose

6. **The Fayetteville Flora of Arkansas (USA): A snapshot of terrestrial vegetation patterns
   within a clastic swamp at Late Mississippian time** 127
   Michael T. Dunn, Gar W. Rothwell, and Gene Mapes

7. **A Late Mississippian back-barrier marsh ecosystem in the Black Warrior and
   Appalachian Basins** 139
   Robert A. Gastaldo, Michael A. Gibson, and Allyn Blanton-Hooks

8. **The Hancock County tetrapod locality: A new Mississippian (Chesterian) wetlands fauna
   from western Kentucky (USA)** 155
   William J. Garcia, Glenn W. Storrs, and Stephen F. Greb

9. **A fossil lycopsid forest succession in the classic Joggins section of Nova Scotia:
   Paleoecology of a disturbance-prone Pennsylvanian wetland** 169
   John H. Calder, Martin R. Gibling, Andrew C. Scott, Sarah J. Davies, and Brian L. Hebert

10. **Compositional characteristics and inferred origin of three Late Pennsylvanian coal beds
    from the northern Appalachian Basin** 197
    Cortland F. Eble, William C. Grady, and Brenda S. Pierce

11. *From wetlands to wet spots: Environmental tracking and the fate of Carboniferous elements in Early Permian tropical floras* 223
   William A. DiMichele, Neil J. Tabor, Dan S. Chaney, and W. John Nelson

12. *Carbon isotopic evidence for terminal-Permian methane outbursts and their role in extinctions of animals, plants, coral reefs, and peat swamps* 249
   Gregory J. Retallack and Evelyn S. Krull

13. *Controls on the formation of an anomalously thick Cretaceous-age coal mire* 269
   T.A. Moore, Z. Li, and N.A. Moore

14. *Paleoecology of a late Pleistocene wetland and associated mastodon remains in the Hudson Valley, southeastern New York State* 291
   Norton G. Miller and Peter L. Nester

Index 305

Preface

WHAT ARE WETLANDS?

Wetlands encompass a variety of ecosystems and depositional regimes at the transition between the aqueous and terrestrial realms. When most people think of wetlands, they picture a reedy marsh or a swamp forest with stagnant water, but there are a wide variety of wetland types. Modern wetlands occur from subtidal marine to mountainous environments, in fresh to saline waters. There are many definitions and classifications of wetlands. The Ramsar Convention (2000, Article 1.1, p. 6) defines wetlands as "areas of marsh, fen, peatland or water, whether natural or artificial, permanent or temporary, with water that is static or flowing, fresh, brackish, or salt, including areas of marine water the depth of which at low tide does not exceed 6 m. Wetlands may incorporate riparian and coastal zones adjacent to the wetlands, and islands or bodies of marine water deeper than 6 m at low tide lying within the wetlands." Another commonly used wetland classification system is that of Cowardin et al. (1979, p. 3), which identifies wetlands as "lands transitional between terrestrial and aquatic systems where the water table is usually at or near the surface or the land is covered by shallow water...wetlands must have one or more of the following three attributes: (1) at least periodically, the land supports predominantly hydrophytes (water-adapted plants); (2) the substrate is predominantly undrained hydric soil (waterlogged soils); and (3) the substrate is nonsoil and is saturated with water or covered by shallow water at some time during, the growing season of each year." These circumscriptions show that wetlands encompass a wide range of extant environments, and therefore should be represented in a wide range of depositional facies. Likewise, the definitions show the basic types of criteria needed to interpret ancient wetlands.

Hydric Paleosols and Waterlogged Substrates

Hydric soils are characterized by frequent, prolonged saturation. The root zones in hydric soils are characterized by low oxygen content and anaerobic conditions. Peats, a type of Histosol, are an example of a hydric soil. Peats are preserved in the rock record as dark carbonaceous shales or coals and are the most obvious evidence of ancient wetlands. They are not, however, the only evidence. Anaerobic conditions in extant wetland soils promote gleying and accumulation of minerals such as siderite and pyrite, which can be preserved in ancient wetland paleosols, sometimes in the absence of overlying coals or carbonaceous shales.

Many modern wetlands are only seasonally waterlogged. In floodplain settings, riparian wetlands are commonly only seasonally inundated. These can be preserved as thin clays, stumps, and rooting structures in the floodplain silts and shales. In sedimentological analyses of ancient fluvial complexes, the stumps and rooting are commonly used as evidence for identifying the larger floodplain facies, but each rooting horizon also represents a paleowetland. In dry seasonal wetlands, salt crusts and calcretes can be produced. When found in combination with evidence of rooting or wetland vegetation, these features provide evidence of seasonal wetlands. Rhizocretions can form in calcareous paleosols (similar to extant Entisols), providing clues to seasonal lakes and palustrine wetlands (Retallack, 2001), as well as for inferring groundwater at or near the root zone during the growing season.

Ancient Hydrophytes

Hydrophytes are plants that have adapted to growing in saturated soils or water cover for some part of the growing system. One of the ways to identify modern wetlands is through the identification of hydrophytes. Identifying fossil hydrophytes can also aid in interpretation of ancient wetlands, but such interpretations are not as easy as in the modern. To begin with, most plant fossils are not the remains of the entire plant. Also, many plant fossils represent transported remains. Thus, interpretation of hydrophytic affinities from fossil evidence usually relies on inferences from functional morphology of plant parts or facies association with hydric paleosols. For example, *Lepidocarpon,* which are female reproductive units of some Carboniferous lycopsid trees, are shaped like small canoes, a likely functional adaptation for floating in open water. This morphological feature, in combination with the association of lycopsid trees with coals, carbonaceous shales, and other waterlogged paleosols, suggests that these plants grew in wetland habitats.

THE IMPORTANCE OF WETLANDS, NOW AND THEN

The importance of wetlands to global ecology is increasingly apparent. Wetlands serve a variety of ecological functions including wildlife habitat, food chain support, surface-water retention and detention, groundwater recharge, and nutrient transformation (Mitsch and Gosselink, 2000; National Research Council, 1995; Keddy, 2000). Wetlands are also crucial links in the water, oxygen, nitrogen, carbon, sulfur, and phosphate cycles (Keddy, 2000; Mitsch and Gosselink, 2000; National Research Council, 1995; U.S. Environmental Protection Agency 1995a, 1995b). This is not only true of present wetlands, but has been true for varying degrees of wetlands for at least the past 400 million years. In fact, there has been an evolution of functions and biogeochemical links, which has changed with the evolution of wetland flora and fauna.

Wetlands also harbor a significant amount of the world's biodiversity, with many animals using wetlands during part of their life cycle (Bacon, 1997). Because many wetlands are located in lowland habitats and evidence low-oxygenated substrates, they have the potential for rapid burial without erosion. For this reason, abundant fossil flora and fauna have been found in association with ancient wetlands. In the geologic past, wetlands also served as harbors of biodiversity, particularly in the Paleozoic, sometimes providing critical refuges for both flora and fauna.

The burial of peatlands, a specific type of wetland, leads to the formation of coal. Hence, ancient wetlands are responsible for approximately a third of the world's electric supply (Energy Information Administration, 2004). Not surprisingly, a significant amount of research on ancient wetlands has concentrated on coal and coal-bearing rocks. Interpretations of the specific types or successions of different types of wetlands that formed a coal can aid in understanding and predicting trends in thickness and quality parameters that directly influence the mining and utilization of the coal. Obviously, the burning of coal and other hydrocarbons has also raised environmental concerns, which are in turn, related to changes in the biogeochemical cycles from which the coal originated.

Because modern wetland accumulations generally represent relatively short-term accumulations (10s to 1000s of years) that are sensitive to small amounts of base-level change, ancient wetlands may provide not only a geologic snapshot of the paleoecology of the wetland habitat, but also indications of climate, sea-level, or tectonic change. It can be argued that the initial idea of continental drift was spurred by evidence of belts of coal with similar plant fossils on different continents. Likewise, evidence of coals in areas that are now deserts or covered in glaciers helped scientists understand that climates changed through time.

OVERVIEW

This volume contains fourteen manuscripts representing research on ancient wetlands from several continents spanning the Silurian through the Pleistocene. Its purpose is to highlight the significance of wetlands through time by assembling research that illustrates interpretations of different types of wetlands from different periods of earth history, as well as demonstrating the importance of wetlands as floral and faunal traps or indicators of climate, sea-level, and tectonic change.

The volume begins with a manuscript by Stephen F. Greb, William A. DiMichele, and Robert A. Gastaldo, which analyzes the evolution of wetland types and functions. The manuscript provides background information on changes in wetlands, wetland functions, and wetlands as habitats through time. The paper includes periods of geologic time and significant paleowetland deposits not covered in other papers in the volume, and so may fill in the "temporal gaps" between manuscripts.

The Early Silurian (Llandoverian) Passage Creek biota is discussed in a paper by Alexandru Mihail Florian Tomescu and Gar W. Rothwell. The paper describes material from compression fossils in overbank facies of Silurian braided streams in Virginia. Their research provides a glimpse into the ecology of the oldest known terrestrial wetlands and initial plant colonization of the landscape.

Wetlands representative of early marshes and forest swamps are described in two Devonian manuscripts. The paper by Jonathan P. Allen and Robert A. Gastaldo reports on wetland plant fossils from the early Devonian (Emsian-Eifelian) Trout Valley Formation in Maine, a locality which has been a model for early Devonian land plant communities. This new research integrates the paleobotany with new sedimentological and taphonomic analyses to place the flora in the context of the riverine (riparian) and estuarine wetlands in which they were preserved.

Walter L. Cressler III's report on plant paleoecology of the Late Devonian Red Hill locality in Pennsylvania describes and documents an early swamp (forested, non–peat-accumulating) and marsh (herbaceous, non–peat-accumulating) habitat, in an oxbow lake setting. The Red Hill locality is famous for its vertebrate remains; including tetrapods, and this research helps to place the vertebrates in their ecological setting.

The volume contains four manuscripts on Mississippian (Lower Carboniferous) wetlands. This is significant, because the majority of Carboniferous wetlands research has concentrated on Pennsylvanian (Upper Carboniferous) Euramerican coal-forming wetlands. The four manuscripts presented herein provide useful insight into the non–coal-forming wetlands that were the precursors to the vast Euramerican coal mires, as well as the landscapes through which the Carboniferous radiation of tetrapods took place.

Michael C. Rygel, John H. Calder, Martin R. Gibling, Murray K. Gingras, and Camilla S.A. Melrose describe a Tournaisian forest from the Horton Group of Atlantic Canada, famous for its tetrapod fossils. This research provides a new look at the morphology of *Protostigmaria*, a cormose lycopsid tree, as well as a variety of vegetation-induced sedimentary structures that can be used to identify in situ stumps.

Michael T. Dunn, Gar W. Rothwell, and Gene Mapes describe permineralized and compression flora from the Late Mississippian Fayetteville flora of Arkansas. Two to three community-level ecosystems are interpreted within clastic swamp (non–peat-accumulating forested wetlands) habitats and controls on these habitats are discussed.

Robert A. Gastaldo, Michael A. Gibson, and Allyn Blanton-Hooks also report on cormose lycopsids from siliciclastics of the Late Mississippian of Alabama and carbonates of West Virginia. The broad range of habitats in which these lycopsids grew during the Mississippian is discussed, as is the possibility that some of these marsh plants were adapted to periodic brackish conditions.

The manuscript by William J. Garcia, Glenn W. Storrs, and Stephen F. Greb describes a new tetrapod locality from the Chesterian of Western Kentucky. Vertebrate fossils and flora are interpreted from an oxbow lake and wetland deposit, and compared with other Mississippian vertebrate sites to better understand the radiation and terrestrialization of vertebrates in Carboniferous wetlands.

There are two manuscripts representing Pennsylvanian (Upper Carboniferous) wetlands. The manuscript by John S. Calder, Martin R. Gibling, Andrew C. Scott, Sarah J. Davies, and Brian L. Hebert describes and interprets megaspore, miospore, macroflora, and sedimentological data from the famous fossil stump horizons of Joggins, Nova Scotia. This Early Pennsylvanian locality is where the oldest known amniotes were found within buried stumps of lycopsid trees, and nicely follows the preceding manuscripts on Devonian and Mississippian tetrapod localities.

The continuously disturbed wetlands described at Joggins can be contrasted with the coal-forming peatlands described from the Late Pennsylvanian of Pennsylvania and West Virginia by Cortland F. Eble, William C. Grady, and Brenda S. Pierce. This is the first description of its type for Late Pennsylvanian mineable coals; most previous work concentrated on the extensive Middle Pennsylvanian coals of the region. The research compares palynology, petrography, coal quality, and geometry of three different mined coals to infer how paleogeography, paleoclimate, and mire type may have influenced Late Pennsylvanian coal (and thereby peatland) distribution.

William A. DiMichele, Neil J. Tabor, Dan S. Chaney, and W. John Nelson describe the fate of the Carboniferous Euramerican peat mires and wet floodplains as climates dried in the Permian. Paleobotany, sedimentology, and paleosols are used to demonstrate the effects of drying climates on wetlands from the Late Pennsylvanian into the early Permian in the Midland Basin, Texas.

The final Paleozoic manuscript reports on the terminal mass extinction at the end of the Permian, which, among other things, resulted in a gap in the fossil record of coals (and therefore peat-forming wetlands) in the early Triassic. Gregory J. Retallack and Evelyn S. Krull use carbon isotope data to suggest a model of catastrophic methane outburst and resultant hydrocarbon pollution of the atmosphere to explain Permian marine and terrestrial extinctions. The manuscript is an example of how data from wetland facies (among others) can have implications for global ecology that go well beyond the wetlands themselves.

Mesozoic wetland deposits are represented by the description of an exceptionally thick (>25 m) Cretaceous coal from New Zealand, coauthored by T.A. Moore, Z. Li, and N.A. Moore. Their research shows how thick coals can be interpreted as stacked paleomires, and how changes in coal lithofacies can be used to infer changes in paleoclimate and possibly synsedimentary faulting.

The final manuscript describes a wetland association that aided in the preservation of vertebrate remains. Norton G. Miller and Peter L. Nester report on a Pleistocene oxbow lake and fringing fen that contained a nearly complete mastodon skeleton from the Hudson Valley, New York. Plant macrofossils and sedimentary analyses are used to determine the infilling history of the pond and the type of wetland in which the mastodon perished. Comparisons are made to other known mastodon finds in the region.

ACKNOWLEDGMENTS

We sincerely thank all of the contributors to this volume. We also thank the reviewers of the manuscripts for their efforts and the GSA publications staff for final compilation of the volume.

REFERENCES CITED

Bacon, P.R., 1997, Wetlands and biodiversity, *in* Hails, A.J., ed., Wetlands, Biodiversity and the Ramsar Convention—The Role of the Convention on Wetlands in the Conservation and Wise Use of Biodiversity: IUCN Publications, Cambridge, United Kingdom, Ramsar Convention Library, 196 p.

Cowardin, L.M., Carter, V., Golet, F.C., and LaRoe, E.T., 1979, Classification of wetlands and deepwater Habitats of the United States: Washington, D.C., U.S. Fish and Wildlife Service Publication FWS/OBS79/31, 103 p.

Energy Information Administration, 2004, International energy outlook: U.S. Department of Energy, Office of Integrated Analysis and Forecasting, Washington, D.C., DOE/EIA-0484 (2003), 160 p.

Keddy, P.A., 2000, Wetland ecology—Principles and conservation: Cambridge University Press, Studies in Ecology, 614 p.

Mitsch, W.J., and Gosselink, J.G., 2000, Wetlands: New York, John Wiley & Sons, third edition, 920 p.

National Research Council, 1995, Wetlands: Characteristics and Boundaries: Washington, D.C., National Academies Press, 328 p.

Ramsar Convention Secretariat, 2004, The Ramsar Convention Manual: A Guide to the Convention on Wetlands (Ramsar, Iran, 1971): Gland, Switzerland, third edition, 75 p.

Retallack, G.J., 2001, Soils of the past—An introduction to paleopedology: Oxford, 2nd edition, Blackwell, 404 p.

U.S. Environmental Protection Agency, 1995a, Wetlands fact sheets: Washington, D.C., Office of Water, Office of Wetlands, Oceans and Watersheds, EPA843-F-95-001, varied pagination.

U.S. Environmental Protection Agency, 1995b, America's wetlands: Our vital link between land and water: Washington, D.C., Office of Water, Office of Wetlands, Oceans and Watersheds, EPA843-K-95–001, varied pagination.

Stephen F. Greb
William A. DiMichele

Evolution and importance of wetlands in earth history

Stephen F. Greb
Kentucky Geological Survey, 228 MMRB University of Kentucky, Lexington, Kentucky 40506, USA

William A. DiMichele
Department of Paleobiology, National Museum of Natural History, Smithsonian Institution, Washington, D.C. 20560, USA

Robert A. Gastaldo
Department of Geology, Colby College, Waterville, Maine 04901-8858, USA

ABSTRACT

The fossil record of wetlands documents unique and long-persistent floras and faunas with wetland habitats spawning or at least preserving novel evolutionary characteristics and, at other times, acting as refugia. In addition, there has been an evolution of wetland types since their appearance in the Paleozoic. The first land plants, beginning in the Late Ordovician or Early Silurian, were obligate dwellers of wet substrates. As land plants evolved and diversified, different wetland types began to appear. The first marshes developed in the mid-Devonian, and forest swamps originated in the Late Devonian. Adaptations to low-oxygen, low-nutrient conditions allowed for the evolution of fens (peat marshes) and forest mires (peat forests) in the Late Devonian. The differentiation of wetland habitats created varied niches that influenced the terrestrialization of arthropods in the Silurian and the terrestrialization of tetrapods in the Devonian (and later), and dramatically altered the way sedimentological, hydrological, and various biogeochemical cycles operated globally.

Widespread peatlands evolved in the Carboniferous, with the earliest ombrotrophic tropical mires arising by the early Late Carboniferous. Carboniferous wetland-plant communities were complex, and although the taxonomic composition of these wetlands was vastly different from those of the Mesozoic and Cenozoic, these communities were essentially structurally, and probably dynamically, modern. By the Late Permian, the spread of the *Glossopteris* flora and its adaptations to more temperate or cooler climates allowed the development of mires at higher latitudes, where peats are most common today. Although widespread at the end of the Paleozoic, peat-forming wetlands virtually disappeared following the end-Permian extinction.

The initial associations of crocodylomorphs, mammals, and birds with wetlands are well recorded in the Mesozoic. The radiation of Isoetales in the Early Triassic may have included a submerged lifestyle and hence, the expansion of aquatic wetlands. The evolution of heterosporous ferns introduced a floating vascular habit to aquatic wetlands. The evolution of angiosperms in the Cretaceous led to further expansion of aquatic species and the first true mangroves. Increasing diversification of angiosperms in the Tertiary led to increased floral partitioning in wetlands and a wide variety of specialized wetland subcommunities. During the Tertiary, the spread of grasses, rushes, and sedges into wetlands allowed for the evolution of freshwater and salt-water reed marshes. Additionally, the spread of *Sphagnum* sp. in the Cenozoic allowed bryophytes, an ancient wetland clade, to dominate high-latitude mires, creating some of the most widespread mires of all time. Recognition of the evolution of wetland types and inherent framework positions and niches of both the flora and fauna is critical to understanding both the evolution of wetland functions and food webs and the paleoecology of surrounding ecotones, and is necessary if meaningful analogues are to be made with extant wetland habitats.

Keywords: paleobotany, paleoecology, paleoflora, earth history, wetlands, coal, swamp, mire, marsh, fen, bog.

INTRODUCTION

Modern wetlands are characterized by water at or near the soil surface for some part of the year, soils that are influenced by water saturation all or part of the year, and plants that are adapted to living in conditions of water saturation all or part of the year (National Research Council, 1995; Keddy, 2000; Mitsch and Gosselink, 2000). Many wetlands occupy lowlands and natural depressions, so have a relatively high preservational potential. It is not surprising, then, that a large part of the fossil record of terrestrial flora and fauna (especially in the Paleozoic) is found within wetlands or wetland-associated habitats. These deposits provide windows into ancient biodiversity, but frequently represent a mix of allochthonous and autochthonous material from different ecosystems. In order to examine the importance of wetlands through time, it is important to recognize that there are many different types of wetlands and wetland functions, and that both have changed through time.

Types of Wetlands

Holocene wetlands have been classified variously over the past several decades, with workers on different continents and in different hemispheres using a range of terms to classify wetlands on the basis of hydrology, geography, and flora, among other criteria. Unfortunately, variable definitions and terminology can lead to uncertain or mistaken use of analogues when interpreting the paleoecology of ancient wetlands. For the purposes of this investigation, we use the following general terminology adapted from Keddy (2000): *aquatic* (or *shallow water*) *wetland* for wetlands dominated by submerged vegetation under continually inundated conditions; *marsh* for wetlands dominated by herbaceous, emergent vegetation rooted in mineral (non-peat) substrates; *swamp* for forested wetlands on mineral (non-peat) substrates; *fen* or *nonforested mire* for wetlands dominated by herbaceous or shrub vegetation on peat substrates. Because there is considerable variability in the use of the term *bog* (Keddy, 2000; Mitsch and Gosselink, 2000), the term *forest mire* is used herein for forested peats. These general terms can have a wide array of meanings (Mitsch and Gosselink, 2000) but serve as a starting point for discussion of paleowetlands. The terms are similar to wetland classes in the hierarchical Canadian wetland system (Zoltai and Pollett, 1983). Such hierarchical classifications are commonly used to describe modern wetlands. In the context of characterizing paleowetlands on the basis of standardized wetland classifications, additional modifiers such as marine, estuarine, riverine, palustrine, and lacustrine are used where appropriate for comparison to the U.S. Fish and Wildlife wetland classification (Cowardin et al., 1979). Modifiers such as marine/coastal and inland are used where appropriate to indicate relative equivalence to wetlands in the Ramsar Convention classification. Modifiers also are used to describe wetland forms, types, and varieties as described in the Canadian system or to describe form, hydrology, or nutrient status for peat-producing wetlands (Gore, 1983; Moore, 1989, 1995; Mitsch and Gosselink, 2000). Each of the modern systems is designed for different purposes, so varied modifiers from each are used to describe clearly paleowetlands discussed in this report.

Wetland Functions

Modern wetlands provide many critical functions in global ecology, including providing habitat and food for diverse species, and aiding in groundwater recharge and water retention and detention, which allows for maintenance of high water tables in wetlands as well as reduced flooding in adjacent ecosystems. They also provide erosion and sedimentation controls between adjacent ecosystems, improve water quality through filtering sediment and metals from groundwater, and cycle nutrients to terrestrial and aqueous environments within the wetland and between ecotones (National Research Council, 1995; Keddy, 2000; Mitsch and Gosselink, 2000). Wetlands are also important global sources, sinks, and transformers of various elements in the earth's various biogeochemical cycles (National Research Council, 1995; Keddy, 2000; Mitsch and Gosselink, 2000). As full or part-time habitats, they function as a significant repository of the world's biodiversity (Bacon, 1997; Keddy, 2000; Mitsch and Gosselink, 2000). These functions are important not only within the wetlands themselves, but also to surrounding ecosystems. Not all functions are equally distributed through the different types of wetlands, and many are influenced by particular floras and faunas. Because the floras, faunas, and types of wetlands have evolved through time, wetland functions have changed through time, as well.

Wetland Niches and Associations

The variety of organisms adapted to various wetland habitats is large and includes all major groups of animals and plants (Bacon, 1997). Herein, we examine the evolution of some common wetland faunal and floral associations. Changes in wetland niches and associations have occurred as the various adaptive strategies of plants and animals have evolved. In some cases, the extant wetland biota lives under conditions similar to those of ancient wetland plants and animals. In others, framework positions or habitats have evolved through specialization, resulting in new wetland types and functions.

Analyses of Paleowetlands

There has been extensive research on ancient wetlands, mostly centered on coals because of their economic value. Several papers have specifically examined floral change in coal-forming floras through time, sometimes concentrating on a particular era (e.g., Shearer et al., 1995) or region (e.g., Cross and Phillips, 1990). Some reports also have used various aspects of coal distribution through time to further understanding of global changes in tectonics, climate, and eustasy (e.g., Scotese, 2001). In terms of wetlands, such reviews tend to be focused on peat-forming mires, which represent a subset of wetland types. In fact, coals are often generalized as representing wetlands, which has the unfortunate result of marginalizing the significance of non-coal facies as wetlands of importance. The understanding that coal floras and "roof" shale floras represent different types of wetlands (e.g., Gastaldo, 1987), emphasizes that non-peat producing wetlands are well represented in the fossil record. In some cases, at different times in earth history, these non-peat producing wetlands may have been more important, in terms of their functions and influences on ecotones, than mires.

Numerous botanical and biogeographical studies have demonstrated how changing climate or timing of tectonic movements changed the composition of Tertiary floras (including wetland inhabitants) in different areas (e.g., Aaron et al., 1999). In terms of climate, it also is important to understand the bias imposed by the present global climate on wetlands and wetland floras. Pfefferkorn (1995) noted the need for a reorientation of a perceived north-temperate perspective and search strategy for interpreting ancient mire ecosystems. Likewise, Collinson and Scott (1987) pointed out the importance of understanding differences in a flora through time when attempting to reconstruct ancient mires. Similarly, it is important to understand changes in specific types of wetland ecosystems. Extant floras and faunas occupy specific niches in different types of wetlands, some of which entail unique physiological adaptations and ecological interactions. These adaptations have changed through time. In some cases, novel floral adaptations have led to new types of wetlands, wetland functions, and wetland faunal niches.

Purpose

Herein the evolution of wetland ecosystems through time is analyzed. We focus on the development of new and changing wetland ecosystems, which accompanied the evolution of the terrestrial flora and, in turn, influenced the evolution of numerous animal groups through the evolution of new niche space, food sources, and habitat. The unusual chemistry and sedimentology of wetland systems resulted in a wide variety of traps in which both fauna and flora are preserved. Significant wetland fossil sites that offer snapshots of ancient biodiversity and paleoecology are also highlighted in order to illustrate the importance of wetland ecosystems to our understanding of earth history. Likewise, we examine the origins and changing influences of specific wetland functions through time to illustrate the potential importance of wetland ecosystems on neighboring ecosystems and in some cases, global paleoecology. The fossil record is our best tool for understanding how changes in wetland distribution, type, niches, and functions influence non-wetland ecosystems, which is particularly important when trying to understand potential long-term natural and anthropogenic influences on global ecology.

ORDOVICIAN-SILURIAN

Prevascular Wetlands

The origin of land plants appears to have occurred in the Late Ordovician to Middle Silurian, involving pre-tracheophyte, embrophytic or bryophytic (moss, lichen) plants that were obligate dwellers of wet substrates (Gray et al., 1982; Gensel and Andrews, 1984; Taylor, 1988; Stewart and Rothwell, 1993; Tomescu and Rothwell, this volume). Whether these prevascular plant-vegetated substrates can be considered wetlands depends on the definition used, and Retallack (1992) has proposed a separate terminology for the associated paleosols. If a "wetland" can be defined simply as consisting of vegetation on a wet substrate, then this habitat has its origin with these vascular precursors. Using the classification scheme of Cowardin et al. (1979), these habitats come closest to representing fluvial and paludal moss-lichen wetlands in which mosses or lichens cover a saturated mineral substrate, other than rock, and dominate the vegetation. They obviously would have differed significantly from extant moss-lichen wetlands in not being associated with any vegetation of taller stature. Pre-Devonian moss-like wetlands also were non-peat-accumulating and therefore would not be termed bogs or fens, nor would they be expected to have similar ecology and functions to those of extant *Sphagnum* moss-dominated mires. If wetlands are defined by the presence of hydrophytic vascular plants, then, by definition, wetland origins are tied to the origin of vascular plants in the Middle Silurian.

SILURIAN

The Oldest Vascular Plants in Wetlands

By many accounts, *Cooksonia* (Wenlockian) is considered the oldest vascular plant (Edwards, 1980; Edwards and Fanning, 1985*)*. *Cooksonia* is a rhyniophyte, a group of small, simple, stick-like vascular plants. It is found mostly in autochthonous deposits associated with fluvial sandstones and floodplains. Edwards (1980) inferred *Cooksonia* habitats along large rivers, which might indicate inland fluvial wetlands according to the Cowardin et al. (1979) classification. The term "riparian wetland," which describes wetlands and associated upstream areas influenced by the river, also would apply, although there would be few functional similarities to extant riparian settings because of the small stature of these rhyniophytes. *Cooksonia* only grew to a few centimeters, so was moss-like in stature. The term "marsh" (often used in descriptions of these wetlands) is functionally problematic in its application to *Cooksonia*-dominated wetlands. Marshlands generally are considered to be dominated by deeply rooted herbaceous vegetation (e.g., Keddy, 2000), decimeters to meters in height (Mitsch and Gosselink, 2000; Keddy, 2000). Pre–Late Devonian plants were mostly less than a meter in height and were not deeply rooted (Fig. 1).

Some research has inferred that simple rhyniophytoid plants, like *Cooksonia,* inhabited salt marshes (Jeram et al., 1990; Shear et al., 1989). Modern salt marshes are a special wetland type inhabited by a low diversity of plants adapted to salt stress caused by brackish to marine tidal inundation or sea spray. Late Silurian plants were simple plants lacking morphological features common in modern salt marsh plants, such as deeply buried rhizomes, salt-excluding roots (e.g., pneumatophores), and bark or leaves that might contain salt glands, and do not appear to have any obvious adaptations to varying soil salinities. As a consequence, it is likely that early rhyniophytes grew under freshwater conditions.

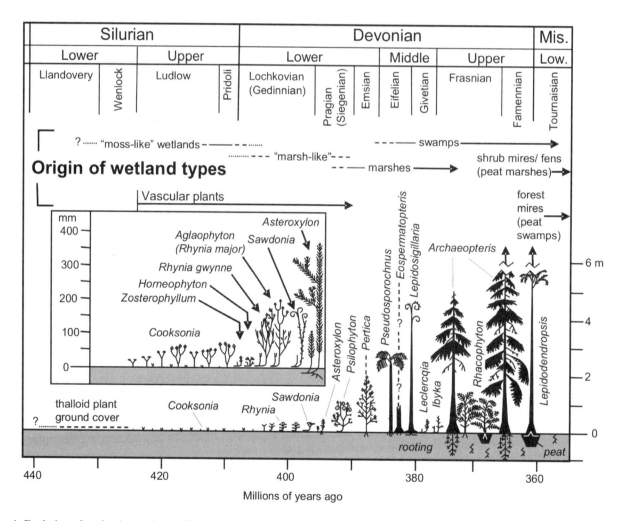

Figure 1. Evolution of wetland types in the Silurian and Devonian. The heights of major floral components are shown as is the inferred depth of rooting. Heights of plants from various sources. Estimates of root depth from Algeo et al. (1995).

LATE SILURIAN–EARLY DEVONIAN

Arthropod Terrestrialization in Wetlands

Arthropods are the oldest terrestrial animals. Putative paleosols and terrestrial arthropod trace fossils are inferred for strata as old as the Ordovician (Retallack and Feakes, 1987; Retallack, 2000; Shear and Selden, 2001), but the oldest undisputed terrestrial land animal, *Pneumodesmus*, is a millipede from the Middle Silurian of Scotland (Wilson and Anderson, 2004). Upper Silurian terrestrial arthropods include trigonotarbids (spider-like arachnids), kampecarids (millipede-like arthropods) and fragments of possible centipedes (Jeram et al. 1990; Rolfe, 1990). Silurian arthropod terrestrialization was linked closely to vascular plant evolution in wetlands (Rolfe, 1980; Jeram et al., 1990). In fact, the transition from an aqueous to a terrestrial habit may have been aided by low-structured vegetation that created humid microclimates near the soil surface (Rolfe, 1985). Most Late Silurian and Early Devonian arthropods are found associated with freshwater marsh-like vegetation in both autochthonous and allochthonous deposits, providing the earliest evidence of habitat function in wetlands. The oldest possible insect is the fragmentary remains of *Rhyniognatha*, from the Lower Devonian (Pragian) Rhynie Chert (Engel and Grimaldi, 2004). The slightly younger and more complete remains of a bristletail from the Emsian (Lower Devonian) of Quebec, Canada, was inferred by Labandeira et al., (1988) to indicate hexapod origins in wet, marsh-like habitats. Similar deposits from the Emsian of Canada have produced millipedes, arthropleurids, and terrestrial scorpions (Shear et al., 1996). The Alken-an-der-Mosel fauna (Emsian), which includes trigonotarbids, arthropleurids, and the oldest non-scorpion arachnid (Størmer, 1976), is preserved along with lycopsids and rhyniophytes (wetland plants) (Jeram et al., 1990; Shear and Selden, 2001). The Middle Devonian (Givettian) Gilboa fauna includes eurypterids and terrestrial arthropods, including arachnids, centipedes, a possible insect, and the oldest spider, and is in association with herbaceous lycopsids and progymnosperms (Shear et al., 1984; Selden et al., 1991).

The spread of kampecarid arthropods (myriapods) is an example of the possible paleoecological significance of wetlands in arthropod evolution. Kampecarids were millipede-like arthropods that were restricted to freshwater aquatic or near-aquatic habitats in which they fed on plant detritus (Almond, 1985). In the Silurian, plant detritus would have been restricted in and around moss-like to marsh-like wetlands. Modern millipedes prefer moist litter horizons and dead wood as habitats, and they are critical agents for nutrient cycling in tropical wetlands and wetland forests as litter-horizon detritivores. The radiation of kampecarids and true diplopods (millipedes) into the earliest wetland communities undoubtedly contributed to increased nutrient cycling, which increased soil quality and contributed to increasingly complex food webs as the terrestrial floral and faunal radiations progressed.

DEVONIAN

The Spread of Wetlands

Most of the Early to Middle Devonian terrestrial fossil record is confined to subtropical-to-tropical wetland habitats, with plants restricted to monotypic stands in freshwater, near-channel, deposits (Edwards, 1980; Beerbower, 1985; Edwards and Fanning, 1985; DiMichele and Hook, 1992). Hence, these assemblages mostly would be classified as paludal or riverine wetlands. Late Silurian rhyniophytes were joined by several new clades in the Early Devonian, including zosterophylls (Gedinnian) and trimerophytes (Siegenian) (Kenrick and Crane, 1997; Bateman et al., 1998), all low-stature (centimeters in height) vegetational types (Fig. 1). Lycopsids also are found in the Early Devonian (Siegenian), and may represent an additional new clade if a Silurian age for *Baragwanathia* is discounted. *Baragwanathia*, a primitive lycopsid from Australia, originally was assigned a Late Silurian (Ludlovian) age (Lang and Cookson, 1935; Garratt et al., 1984), but this determination is controversial. *Baragwanathia* actually may be of Early Devonian age (Edwards et al., 1979).

All Early Devonian vascular plants were small and homosporous, which means that their gametophytes required water-mediated fertilization (Remy, 1982). Likewise, the small rhizoids of these rhyniophytes, trimerophytes, and zosterophylls indicate habitats characterized by nearly continuous moisture (DiMichele and Hook, 1992; Hotton et al., 2001)—in other words, moss-like to at most marsh-like wetlands, but still smaller in height than the flora that typically inhabits extant marshes (Fig. 1).

Geothermal Wetlands

By far the most famous early terrestrial biota is from the Rhynie Chert (Siegenian) of Scotland. Chert in this wetland deposit preserves the three-dimensional remains of fungi, algae, small non-vascular polysporangiophytes, a lycophyte, small vascular plants, arachnids (mites, trigonotarbids), an insect, and freshwater crustaceans (Remy and Remy, 1980; Rolfe, 1980; Trewin, 1996; Rice et al., 2002). Rhyniophytes have been interpreted as "swamp" (e.g., Knoll, 1985), "marsh" (Trewin and Rice, 1992), and "bog" plants (Rice et al., 1995), although the terms have been applied somewhat indiscriminately. Although the term "swamp" is sometimes used informally to describe any type of wetland, formal use in several classification systems requires arborescent vegetation, which were lacking at Rhynie. "Marsh-like" rather than "marsh" might be more appropriate because of the small stature of herbaceous vegetation preserved. The term "bog" is even more problematic because bogs are peat-accumulating wetlands. Although silicified organic laminae have been called "peat mats" at Rhynie (Knoll, 1985), these are not thick (millimeters thick) and much thicker peats would be more typical of the modern peat-forming wetlands classified as bogs.

Recently, the cherts were shown to have been deposited in a fluvio-lacustrine setting within, or on the margin of, a hydrothermal basin (Trewin and Rice, 1992; Trewin, 1994, 1996;

Rice et al., 1995, 2002). In situ plant assemblages accumulated in ambient waters of interfluves and overflow pools between hydrothermal ponds and geysers. Hence, at least some part of the Rhynie Chert biota represents inland geothermal wetlands as defined in the Ramsar classification (Fig. 2).

The association of freshwater crustaceans with the Rhynie biome is interesting because crustaceans are one of the most common groups of modern wetland-inhabiting arthropods. The Rhynie crustaceans (*Lepidocaris*, *Castracollis*) are branchiopods, similar to modern tadpole shrimp (*Triops*) and fairy shrimp (*Artemia*) (Anderson and Trewin, 2003; Fayers and Trewin, 2004). Extant branchiopods are common in wet meadows (vernal ponds) where they are important parts of detritivore-based food webs. Extant wet meadows are ephemeral wetlands dominated by herbaceous grasses and shrubs (Keddy, 2000; Mitsch and Gosselink, 2000). Crustaceans can thrive in ephemeral wetlands because of the lack of fish, which also seems to have been the case in the Rhynie ecosystem. Today, wet meadows (considered by some as a subset of marshes) are dominated by angiosperms (grasses and sedges). In the middle Paleozoic, rhyniophytes may have occupied similar niches, although rhyniophytes were likely less drought resistant than the flora of wet meadows today, and the relationship between their life history pattern and seasonal drought is not understood.

The Oldest Marshes

By the Middle Devonian (Eifelian), several plant groups had evolved shrub or bush morphology (Fig. 1). The lycophyte *Asteroxylon mackiei* (Emsian-Givetian) from the Rhynie Chert may have grown to heights of 50 cm (Gensel and Andrews, 1984; Gensel, 1992). *Pertica quadrifaria*, a trimerophyte from the Trout Valley of Maine (United States), grew to at least a meter in height (Kasper and Andrews, 1972; Allen and Gastaldo, this volume) if not taller. As such, wetlands comprised of these emergent plants formed the earliest marshes (inland shrub-dominated wetlands sensu Ramsar classification). Middle Devonian wetlands began to exhibit floral partitioning (Allen and Gastaldo, this volume), possibly in response to salinity, water chemistry, nutrients, or sedimentation and flooding (duration and periodicity of inundation). This partitioning undoubtedly involved feedback loops with stands of vegetation also influencing flooding and sedimentation as seen in modern freshwater marshes and wet meadows. In the riparian and lake-margin settings in which much of the Middle Devonian flora is found, wet meadows were likely common, as increasing stature, rooting, and floral partitioning allowed for some plants to adapt to seasonal inundation or exposure.

Sphenopsid-like plants are another important shrubby clade that emerged in the Late Devonian. Included among these plants are the Iridopterids (Stein et al., 1984). Calamitalean sphenopsids of the Carboniferous appear to have been adept particularly at colonizing disturbed environments, such as riparian wetlands susceptible to flooding and sedimentation (Scott, 1978; DiMichele and Phillips, 1985; Gastaldo, 1987; Pfefferkorn et al., 2001). In modern coastal, lacustrine, and riverine marsh settings, some emergent, reed-like plants are simplified (reduced) as an adaptation for living in these disturbance-prone areas. Reed-like

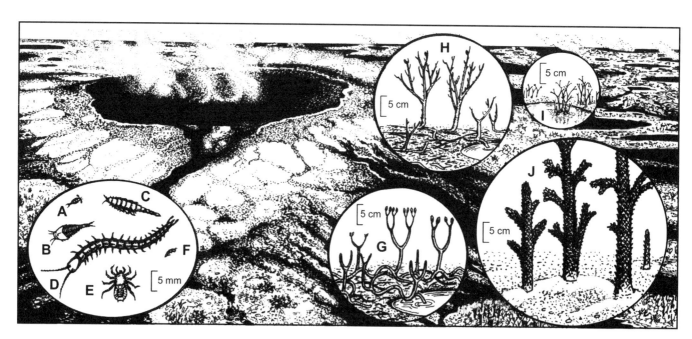

Figure 2. Illustration of the Rhynie geothermal wetlands. Arthropods include the crustaceans (A) *Lepidocaris* and (B) *Castracollis*, (C) a euthycarcinoid, (D) the partial remains of centipede, (E) the trigonotarbid *Palaeocharinus*; and (F) the partial remains of a springtail. Flora include (G) *Aglaophyton*, (H) *Rhynia*, (I) *Horneophyton*, and (J) *Asteroxylon*. All floral members drawn to same scale. Illustrations based on data and reconstructions from the University of Aberdeen, Scotland (www.abdn.ac.uk/rhynie/).

morphologies limit damage from storms and flooding through reduction of surface area, and clonal growth allows reestablishment of aerial shoots if the emergent parts of the plants should be broken (Keddy, 2000) or buried (Gastaldo, 1992). Sphenopsid reed-like morphologies in disturbance-prone Carboniferous environments created a framework similar to that presently created by reeds and rushes. Thick stands of reeds in modern marshes serve important functions in terms of sedimentation control, water filtering, flood control, and habitat, all of which are likely to have originated in Devonian marshes.

The Oldest Swamps

During the Middle to Late Devonian, lycopsids and progymnosperms attained tree-like stature, which led to the evolution of the first true forested wetlands, by definition, swamps (Fig. 1). Lycopsids were the first land plants to develop shallow substrate-penetrating roots (Remy and Remy, 1980), which advanced the process of soil development. Other clades evolved root systems later in the Devonian (Driese and Mora, 2001), altering pedogenic processes. Root systems were essential to the development of an arborescent growth habit because of the centralized growth form of most trees. Arborescence continued the pattern of increasing vegetational zonation, with the development of tiered canopies, including both trees and understory shrubs (Scott, 1980). Zonation contributed directly to the differentiation of swamps and marshes and the development of new niche space (Scheckler, 1986a; Cressler, this volume), and thereby biodiversity.

What may be the oldest swamps (forested wetlands) were reported by Driese et al. (1997) from the Middle Devonian of New York. Large stumps and shallow-penetrating roots, attributable to cf. *Eospermatopteris*, are preserved in a gray-green, gleyed, pyritic mudstone, interpreted as a waterlogged paleosol. Bartholomew and Brett (2003) have redescribed similar in situ stumps of *Eospermatopteris* (possibly a cladoxylalean) from the famous Gilboa locality in New York, from which the genus was described originally (Goldring, 1924). Although the habit of this plant is uncertain, stumps of approximately one meter in diameter have been reported, suggesting large trees adapted to wetland (swamp) conditions.

The progymnosperm *Archaeopteris* sp. is considered the oldest typically woody, tall tree (Figs. 1, 3), growing to heights of 18 m and occupying poorly drained flood plains and coastal areas (Beck, 1962, 1964; Retallack, 1985; Scheckler, 1986; Meyer-Berthaud et al., 1999). As such, they formed true gallery forests in floodplain environments constituting riverine or paludal forested wetlands, riparian forest-wetlands, or swamps (when defined as forested wetlands on mineral substrates). Arborescent progymnosperms had flattened branch systems and leaves, providing for a canopy and the potential for a shaded understory, which, in the Late Devonian, was dominated by the scrambling fern-like plant *Rhacophyton* (Fig. 3). In combination, this plant association would have increased litter input to the swamp floor (DiMichele and Hook, 1992; Algeo et al., 1995; Algeo and Scheckler, 1998), providing increased nutrients to surrounding wetland, fluvial, and upland ecosystems. The result of litterfall detritus in extant wetlands is the formation of a complex detritus-based food web that supports a great diversity of aquatic invertebrates, fish, and amphibians, often with greater biodiversity than in adjacent uplands because of the "edge effects" of ecotones (Bacon, 1997; Keddy, 2000; Mitsch and Gosselink, 2000). Such a food web was likely in place by the Middle Devonian.

The development of deep, extensive roots in Frasnian progymnosperms resulted in increased substrate stabilization (Figs. 1, 3) and a change in the rate at which paleosols formed and sediment was discharged (Algeo and Scheckler, 1998; Algeo et al., 2001). Devonian substrate stabilization also decreased sediment fluxes and reduced catastrophic flooding in wetland habitats (Schumm, 1968; Beerbower et al., 1992). This latter consequence

Figure 3. Devonian lacustrine wetland dominated by the pre-fern *Rhacophyton* and the progymnosperm *Archaeopteris*, whose roots stabilize the banks of the oxbow lake.

is an important function of modern wetlands, where flooding is prevented through the "breaking" action supplied by thick stands of plants against floodwater velocity, as well as through floodwater storage (Mitsch and Gosselink, 2000; Keddy, 2000). It also could lead to reduced runoff and increased precipitation, leading to significant changes in the global hydrological cycle (Algeo and Scheckler, 1998; Algeo et al., 2001).

Roots are central in the process of denitrification, which is important in global nitrogen cycling (e.g., Keddy, 2000). This critical function presumably originated in mid-Devonian marshes but increased with the evolution and spread of true swamps, and the development of upland forests leading to a dramatic increase in vegetative primary productivity. These expansions across the landscape would have increased carbon consumption and atmospheric carbon dioxide (pCO_2) drawdown. In combination with increased nutrient flux and bottom water anoxia and organic carbon fluxes, these perturbations could have led to global cooling, Devonian glaciation, as well as the end-Devonian mass extinction (Berner, 1993, 1997; Algeo et al., 1995; Algeo and Scheckler, 1998).

The Oldest Mires

Late Devonian coals record the evolution of the first peat-accumulating wetlands, indicating when plants had evolved the production and shedding of prolific amounts of biomass, which allowed peat to accumulate under specific chemical conditions. There is a distinction made between modern peat and non-peat-forming wetlands in most discussions (Mitsch and Gosselink, 2000; Keddy, 2000), and many authors differentiate between swamps and mires (bogs, fens; e.g., Gore, 1983). Peats are composed of at least 50% organic (mostly plant) material and accumulate where organic production outpaces decomposition, generally in wet, low-oxygen substrates. Often, the presence of an impervious aquiclude underlying the peat mire allows for the stilting of the water table, promoting litter accumulation (Gastaldo and Staub, 1999). Peat substrates present plants with considerably different challenges than mineral substrates. Most importantly, many peats are relatively nutrient deficient because organic matter chelates mineral nutrients. Stability for rooting also differs from mineral substrates. Finally, pore waters in peat, in some cases, have a lower pH than what most plants experience on other types of substrates (DiMichele et al., 1987; Cross and Phillips, 1990; Gastaldo and Staub, 1999). Peat accumulation in the Devonian resulted in new types of wetlands and new wetland functions associated with mires.

Some of the earliest coals are interpreted as sapropelic "boghead" coals, which form from the accumulation of algae in brackish to freshwater restricted environments (Thiessen, 1925; Sanders, 1968), although most result from the accumulation of terrestrial detritus. Several Late Devonian (Frasnian) coals of eastern North America are dominated by the herbaceous scrambling fern *Rhacophyton* (Scheckler, 1986a; Cross and Phillips, 1990). These sites would be classified as shrub-dominated peat wetlands or "fens" (Fig. 1; Gore, 1983; Keddy, 2000; Mitsch and Gosselink, 2000). Because *Rhacophyton* grew in both mineral and peat substrates, it likely was preadapted to oligotrophic conditions, which allowed this marsh plant to become one of the initial mire creators/occupiers.

Forested mires also appear in the Late Devonian and are composed of lycopsids (Figs. 1, 4). Late Devonian coals of China are dominated by the arborescent lycopsids *Lepidodendropsis, Lepidosigillaria,* and *Cyclostigma* (Xingxue and Xiuyhan, 1996). Arborescent lycopsids originated in non-peat-accumulating Devonian swamps and later expanded their range into peatlands, where they became dominant. It has been inferred that as peatlands expanded, these ecosystems became refugia for relict plants (like the lycopsids), as increasing morphological innovation allowed other clades to expand outside of wetland habitats

Figure 4. Devonian mires were dominated by the pre-fern *Rhacophyton,* but arborescent lycopods with stigmarian roots became increasingly common.

(Knoll, 1985; DiMichele et al., 1987). The stigmarian root systems of lycopods (Fig. 4) permitted growth in wet, oxygen-poor, soft-sediment substrates (Rothwell, 1984; DiMichele and Phillips, 1985; Phillips et al., 1985) and allowed lycopods to become the dominant vegetation of the Carboniferous peatlands.

Late Devonian forested mires may represent the earliest bogs, depending on the use of the term. Bogs generally are differentiated from fens by the accumulation of thicker peat composed of vegetation that is at least partly arborescent. In this respect, Late Devonian forest mires could be termed bogs. Devonian forest mires, however, were not ombrotrophic or dominated by mosses, characteristics implied in some uses of "bog" (Mitsch and Gosselink, 2000; Keddy, 2000). In terms of their ecological functions, these Devonian fens and forest mires mark the initiation of a new carbon sink, contributing to changes in the global carbon cycle and remaining important to this day. Also, the high water-storage capacity of peats means that mires can significantly influence local and regional hydrology (Mitsch and Gosselink, 2000; Keddy, 2000), which likely began in the Devonian but would have greater impact with the spread of mires in the Carboniferous.

Tetrapod Evolution and Wetlands

Tetrapods made landfall in the Late Devonian (Milner et al., 1986; Clack, 2002) from lungfish and lobe-finned fish ancestors. In fact, low-oxygen conditions caused by decaying plant matter in freshwater wetlands and wetland-fringing lakes may have spurred the evolution of tetrapod lungs (Randall et al., 1981; Carroll, 1988; Clack, 2002). Extant lungfish, such as the Australian *Neoceratodos forsteri* and African *Protopterus,* inhabit freshwater rivers, ponds, and marshes. They survive in ephemeral wetlands by burrowing into and estivating within wet substrates, surviving for many months until seasonal rains reflood their habitat (Speight and Blackith, 1983).

Acanthostega is one of the earliest aquatic tetrapods. Its multidigit appendages were preadapted for use on land, having first evolved in water (Gould, 1991; Clack, 1997; Clack and Coates 1995; Coates and Clack 1995). In the fluvial environments in which *Acanthostega* is preserved, it has been hypothesized that digitation was useful in strong currents for grasping onto rocks and water plants (Clack, 1997). Terrestrial mobility may have originated as a preadaptation in these earliest tetrapods that developed in association with maneuvering through vegetation in fluvial (riparian) wetlands dominated by dense stands of *Rhacophyton* in Late Devonian riverine marshes (Fig. 5).

Amphibians are common in many modern riverine/riparian wetlands (Mitsch and Gosselink, 2000) and many extant species require this habitat for part of their life cycle. Modern amphibian distribution is influenced by predation and the stability, light intensity, and temperature of their habitats (Skelly et al., 1999, 2002). Broad wetlands, with distinct microhabitats of overstory, midstory, and shrub, provide different types of food and cover where amphibians generally are abundant (Rudolph and Dickson, 1990). By the Late Devonian, tiering and canopy zonation in marshes, swamps, fens, and forest mires was well established, and created the types of food and cover in which tetrapods could thrive, adding another layer to both freshwater aquatic and terrestrial food webs.

MISSISSIPPIAN

Tetrapod and Wetland Diversification

Tetrapods continued to evolve and diversify into the Carboniferous as exemplified by one of the most famous Lower Carboniferous sites in East Kirkton, England. The fossil-bearing limestone preserves a wide variety of vertebrates, including chondrichthyan and acanthodian fish, lungfish, temnospondyls, anthracosaurs, and a reptiliomorph (reptile-like) animal (Milner and Sequeira, 1994). At one time, the reptiliomorph nicknamed "Lizzie" was interpreted as the oldest amniote (reptile; Smithson, 1989). More recent studies, however, have suggested that it was only a close relative of amniotes (Smithson et al., 1994), and possibly even a stem-tetrapod or an early amphibian, rather than a true amniote (Laurin and Reisz, 1999).

The East Kirkton tetrapod assemblage occurs in an alkaline, freshwater lake rimmed with marshes formed from reed-like calamites and a pteridosperm with *Sphenopteris* foliage (Milner et al., 1986). Volcanogenic rocks preserve several different plant assemblages within hydrothermal hot-spring deposits (Rolfe et al., 1990; Brown et al., 1994; Scott and Rex, 1987; Galtier and Scott, 1994; Scott et al., 1994). The vertical juxtaposition of these assemblages indicates that East Kirkton initially was a

Figure 5. *Acanthostega* maneuvers through stands of the pre-fern *Rhacophyton* and roots of the arborescent progymnosperm *Archaeopteris* in a flooded Devonian riparian marsh.

lake surrounded by drier, pteridosperm-dominated woodlands; these subsequently were altered to wetter substrates in which lycopod-dominated swamps are preserved (Scott et al., 1994). Many Carboniferous tetrapod assemblages accumulated in similar open-water bodies fringed by marshes or forest swamps, and in swamp-filled pools (oxbows, billabongs) (Milner et al., 1986; Hook and Baird, 1986; Garcia et al., this volume).

Tuffs containing fusain at East Kirkton may indicate that volcanic activity ignited wildfires (Brown et al., 1994), which in turn may have driven the vertebrates from this landscape into the lakes where they perished (Scott et al., 1994). Fires are important elements in the ecology of most modern wetlands and influence floral content and community succession in extant wetlands (Keddy, 2000). This probably has been the case since the Late Devonian (Scott, 1989).

Possible Mangal Wetland Origins

The Mississippian provides the first evidence for the expansion of any clade into nearshore and marginal marine sites, those under possible saline influence. Inasmuch as the term "mangrove" often is applied to woody taxa, the term mangal—any salt-tolerant plant—would be applied to these assemblages. Gastaldo (1986) interpreted the stigmarian-rooted lycopsids reported by Pfefferkorn (1972) in the Battleship Wash Formation, Arizona, as representing the first mangal taxon. Gastaldo et al. (this volume) also demonstrate that some Mississippian back-barrier marshes were inhabited by herbaceous, cormose lycopsids. Most arborescent lycopsids are interpreted to have been intolerant of salt water (DiMichele and Phillips, 1985), although smaller, cormose forms, such as *Chaloneria*, have been interpreted as living in coastal marsh-like habitats (DiMichele et al., 1979), as well as fresh-water marshes and peat-forest swamps (Pigg, 1992). It is not a simple proposition to identify morphological features that would support a brackish-habitat interpretation for Paleozoic plants because not all of these adaptations (for example stilt roots) are solely an adaptation to saline tolerance. Transgression (onlap) could result in burial of freshwater, near-coast taxa in marine sediments, confounding interpretations of mangal habit based on sedimentological evidence. In addition, many extant freshwater wetland plants and mangals live on freshwater lenses in the soil, adjacent to brackish or marine waters. Few plants can tolerate the precipitation of salts in marine-water influenced soils. Therefore, interpretation of mangal habit is, in part, a matter of recognizing that the plants did not live directly within fully marine salinities but could tolerate the incursion of salt water, or recognizing physiological features that allow an interpretation of salinity tolerance.

Spreading Mires and Lowland Swamps

Within the coastal plains and continental interiors, extensive swamps and thick peat mires first occur in the Late Mississippian throughout Eurasia including Canada, western Europe, Ukraine, Belarus, Russia, and China (Wagner et al., 1983; Scotese, 2001; Rygel et al., this volume). These are dominated by arborescent lycopsids that range throughout the late Early Carboniferous up to near the Mississippian-Pennsylvanian boundary, where they first are joined by typical Pennsylvanian lycopsid taxa. The lycopsids *Lepidophloios* and *Paralycopodites* remain a component of these mires into the Pennsylvanian, while new species of the lepidodendrid complex replace typical Early Carboniferous forms. A few floristic elements of the Early Mississippian (Visean) persist into the Namurian mires in the Silesian basin, mostly within the sphenopsids (*Archaeocalamites* and *Mesocalamites*) and fern/pteridosperms (Purkynová, 1977; Havlena, 1961). Although much of the global Mississippian is recorded in carbonate ramp deposits, it is these lycopsid-dominant swamps and mires that set the stage for the extensive accumulation of peatlands in the Pennsylvanian.

PENNSYLVANIAN

The Heyday of Tropical Mires

Pennsylvanian (Upper Carboniferous) coals are known from basins in the eastern and central United States, eastern Canada, England, eastern and western Europe as well as parts of China and East Asia (Walker, 2000; Scotese, 2001; Thomas, 2002). These areas straddled the Pennsylvanian equator (Fig. 6), with some coals representing the most widespread tropical mire systems in earth history (Greb et al., 2003).

Much is known about the ecology of Pennsylvanian wetland plants and plant communities, a consequence, in part, of exposures made possible by the mining of economically important coals. The ecologies of the dominant plant groups have been reviewed by DiMichele and Phillips (1994), but recent data from the Early Pennsylvanian (Langsettian) may indicate that the partitioning of ecospace within mires occurred through the Pennsylvanian (Gastaldo et al., 2004). In brief, giant lycopsid trees were restricted mostly to wet, periodically flooded substrates. These trees dominated Early and Middle Pennsylvanian forest mires. Lycopsids were spore producers, although some had seed-like "aquacarps," adapted for aquatic fertilization and dispersal in forested wetlands (Phillips and DiMichele, 1992). They were supported by bark, rather than wood, and had highly specialized rooting systems (*Stigmaria*) that facilitated growth in low-oxygen, soft substrates. There was a variety of lycopsid tree genera with specializations to different levels of disturbance and substrate exposure (Fig. 7A–7E).

Other spore-producing groups coexisted with the lycopsids in these mires. Marattialean tree ferns of the genus *Psaronius* were cheaply constructed plants (Baker and DiMichele, 1997); tree habit was made possible by a thick mantle of adventitious roots (Figs. 7A, 7C, 7F). The calamiteans were another group, closely related to modern scouring rushes and horsetails of the genus *Equisetum*. Extant *Equisetum* is a small, widespread, non-woody plant that grows in moist places and poor soil. Calamiteans appear to have inhabited the same environments (Fig. 7A), although some calamiteans grew to heights in excess of 5 m and, hence, would have served functions more similar to small trees than shrubs (Fig. 7D). The calamiteans were the only major Late Carboniferous tree group to exhibit clonal growth. Aerial stems developed from subterranean rhizomes in most species, a growth form that permitted them to exploit habitats with high rates of sediment aggradation (Fig. 7F) in which the stems could be buried repeatedly by flood-borne siliciclastics and continue to regenerate (Gastaldo, 1992).

Two seed-producing tree groups also were common in peat-substrate mires, the cordaites (Fig. 7F) and the medullosan pteridosperms (Figs. 7C, E). Cordaites were woody trees and shrubs closely related to extant conifers. In the middle Westphalian, cordaitean gymnosperms became abundant in some parts of mire landscapes (Fig. 7F), apparently reflecting areas with periodic extended substrate exposure or disturbance (Phillips et al., 1985). Cordaites were also common components of late Paleozoic Angaran (Asian) wetlands (Oshurkova, 1996). Some forms have been reconstructed as mangroves (Cridland 1964; Raymond and Phillips, 1983), although evidence of stilt-like roots is lacking in preserved *Cordaites* tree trunks (Johnson, 1999). It also has been suggested that they could tolerate brackish conditions (Wartmann, 1969). There is, however, substantial reason to doubt a mangrove interpretation, given that the plants appear to prefer rotted peat, possibly subject to exposure, and that they occur in a complex flora associated with an array of other plants that do not appear to be specifically adapted to salt-water tolerance (Phillips et al., 1985; Raymond et al., 2001).

Medullosan pteridosperms were small trees largely confined to nutrient-enriched substrates. They produced large fronds on which were borne some of the largest seeds known among Carboniferous tropical plants (Gastaldo and Matten, 1978). Medullosans were free standing and formed thickets or tangles of plants that leaned on each other for support (Wnuk and Pfefferkorn, 1984). In addition to these tree forms, representatives within the pteridosperms, ferns, sphenopsids, and lycopsids also displayed ground cover and liana (vine) growth strategies (Fig. 7B–7D, 7F). A liana growth strategy is important in modern tropical wetlands because it allows plants to compete for light amongst tall trees. These forms were systematically diverse and occasionally abundant in the community (Hamer and Rothwell, 1988; DiMichele and Phillips, 1996a; DiMichele and Phillips, 2002).

Non-Peat-Forming Swamps

Non-peat-forming swamps, sometimes referred to as "clastic swamps" (e.g., Gastaldo, 1987; Mapes and Gastaldo, 1986; Gastaldo et al., 1995) also were widespread in Pennsylvanian coal basins. These habitats supported a vegetation much like that of forest mires, although there were many species-level differences and the environments were dominated by different plants. Swamp habitats often were enriched in lycopsids but included pteridosperms as major components (Wnuk and Pfefferkorn,

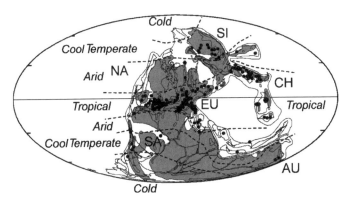

Figure 6. Upper Carboniferous paleogeography and climates showing locations of coal (black dots) and thereby known paleomires (modified from Scotese, 2001). AU = Australia, CH = China, EU = Europe, NA = North America, SA = South America, SI = Siberia.

1984; Scott, 1978; Collinson and Scott, 1987; Gastaldo, 1987). In the late Middle Pennsylvanian, marattialean tree ferns began to increase in abundance in all parts of the wetland landscapes, although the increase in fern abundance can be detected in clastic substrate habitats (marshes and swamps) before it appears in mire habitats (Pfefferkorn and Thomson, 1982). A major extinction at the end of the Middle Pennsylvanian (Westphalian) resulted in a significant reorganization of wetland ecology (Phillips et al., 1974; DiMichele and Phillips, 1996b). Following the extinction, Euramerican Late Pennsylvanian (Stephanian) mire and riparian swamps became more similar in overall patterns of dominance and diversity. *Psaronius* tree ferns were dominant, medullosan pteridosperms were subdominant, and *Sigillaria* (Fig. 7D), a tree lycopsid that may have preferred periodic substrate dryness, was locally common.

The Development of Wetland Successions

Many modern peatlands exhibit a temporal succession of wetland types in response to changing hydrology and nutrients (Gore, 1983; Moore, 1989; Mitsch and Gosselink, 2000; Keddy, 2000). The earliest definitive successions in ancient wetlands are from the Pennsylvanian. Studies of English Pennsylvanian coals by Smith (1957, 1962) noted that many exhibit vertical changes in spore content, which were inferred to represent changes or successions in plant (and wetland) types. Coals also exhibit vertical changes in ash yield, sulfur content, palynology, and petrography, which result from temporal succession of different wetland types (Cecil et al., 1985; Esterle and Ferm, 1986; Eble and Grady, 1990; Greb et al., 1999a, 2002). Successional patterns also have been inferred from coal balls (Raymond, 1988; Pryor, 1993; Greb et al., 1999b).

Many Euramerican coals began in topographic lows as marshes or swamps (Figs. 7A) and then became mires as conditions allowed for the uninterrupted accumulation of biomass. In some cases, successions occurred between different mire types. Thin peats, rooted seat earths (poorly developed soils—e.g., incep-

Figure 7. Pennsylvanian wetlands were diverse and included (A) pioneering topogenous riverine and paludal mires and swamps, (B) flooded swamps and topogenous forest mires, (C) paludal swamps, (D) riverine/riparian-margin marshes and swamps, and (E) ombrogenous mires. Disturbance-prone mires (F) along wetland margins were dominated by disturbance-tolerant flora. Swamps and forest mires were dominated by lycopods including *Paralycopodites* (Lp), *Lepidophloios* (Lls), *Lepidodendron* (Lln), *Sigillaria* (Ls), and *Omphalophloios* (Lo). Juvenile *Lepidodendron* (Llj). Lycopod reconstructions based on DiMichele and Phillips (1985, 1994). Other arborescent flora included tree ferns (Tf), sphenopsids such as *Calamites* (Ca, which ranged from herbaceous to arborescent), and the gymnospermous tree *Cordaites* (Co). Sphenopsids also occurred as vines (lianas). Ground cover was dominated by ferns and sphenopsids.

tisols or entisols), and in situ tree stumps within mineral substrates form in a wide variety of swamp and marsh settings (e.g., Teichmüller, 1990); thick peats can accumulate in fens and forested mires. Extant planar peatlands, also called topogenous mires or low-lying moors, generally occur at or just below the ground-water table and tend to fill in the topography (Gore, 1983; Moore, 1989). Ombrogenous mires, or raised mires in which peat doming may occur, build up above the topography in everwet climates (Gore, 1983; Clymo, 1987; Moore, 1989). Successions from topogenous to ombrogenous mires have been inferred for numerous Euramerican Pennsylvanian coals on the basis of palynological analyses (Cecil et al., 1985; Esterle and Ferm, 1986; Eble and Grady, 1990; Greb et al., 1999a, 1999b). Because modern ombrogenous mires build up above surrounding river levels, they are low-nutrient habitats without standing water cover and are dominated by stunted vegetation in the domed areas (Gore, 1983; Moore, 1989). In the Pennsylvanian, similar conditions are inferred for ombrogenous mires, which appear to have been dominated by stunted lycopsids *(Omphalophloios)* and ferns (Fig. 7E; Esterle and Ferm, 1986; Eble and Grady, 1990; Greb et al., 1999a, 1999b).

Figure 8. Giant arthropods in Pennsylvanian wetlands included the millipede *Arthropleura* and giant mayflies (lower right) here shown in a lycopod swamp.

Giant Arthropods in Wetlands

The record of Carboniferous arthropods is very good, partly because of the many Carboniferous concretion locations that are fossiliferous, including the famous Mazon Creek area of the Illinois Basin and Montceau-les-Mines, France (Darrah, 1969; Gastaldo, 1977; Nitecki, 1979; Baird et al., 1986). Much of the primary plant productivity in Late Carboniferous wetlands continued to reach animal food webs through arthropod detritivores, although a relatively complete trophic web of detritivores, herbivores, and carnivores had developed (DiMichele and Hook, 1992; Labandeira and Eble, 2006). *Arthropleura* was a giant millipede-like arthropod (Fig. 8) that consumed the inside of rotting lycopod trunks on swamp and forest-mire floors (Rolfe, 1980; Hahn et al., 1986; Scott et al., 1992). At 1.8 m in length, *Arthropleura* is the largest terrestrial arthropod of all time (Rolfe, 1985). Their large size suggests that arthropleurids filled a niche that had yet to be shared with tetrapods (DiMichele and Hook, 1992), or that tetrapods were not yet large enough to pose a threat. Millipedes are still important wetland detritivores but are much smaller than *Arthropleura*. Cockroaches are another common extant detritivore and were particularly abundant in Carboniferous wetlands (Durden, 1969; Scott et al., 1992; Easterday, 2003), reaching 8 cm in length.

In addition to their importance as litter-dwelling wetland detritivores, some Carboniferous arthropods also evolved flight (Kukalova-Peck, 1978, 1983; Scott et al., 1992; Labandeira and Eble, 2006). One explanation for the origin of flight is that wings evolved from gills in aquatic stages, and flight evolved through surface-skimming, a process used by extant, wetland-inhabiting stone flies (Plecoptera) and subadult mayflies (Ephemeroptera; Marden and Kramer, 1994). Giant mayflies with wingspans of more than 40 cm are known from Late Carboniferous wetland facies (Fig. 8; Kukalova-Peck, 1983). The most commonly depicted flying insect in Carboniferous illustrations is *Meganeura*, a dragonfly-like hexapod, which had a wingspan of more than 60 cm. The precursors of extant dragonflies, the Protodonata, also evolved in the Carboniferous, and some had wingspans of more than 60 cm (Carpenter, 1960). Extant dragonflies are common predators of wetlands. Because most dragonflies have aquatic nymphs, they require wet habitats for part of their life cycle. In fact, the evolution of metamorphosis in insects appears to have occurred in wetland or wetland-fringing ecosystems (Kukalova-Peck, 1983; Truman and Riddiford, 1999).

Insect flight also may have contributed to the rise of insect herbivory, as flying insects could exploit new food resources (DiMichele and Hook, 1992). Some Carboniferous insects (e.g., megasecopterans and paleodictyopterans) developed mouth parts for sucking and piercing. Evidence for this strategy is found in permineralized swamp-and-mire plants (Scott et al., 1992; Labandeira and Phillips, 1996; Labandeira and Eble, 2006). In fact, most major insect herbivore functional feeding groups on land were established by the late Paleozoic and are preserved in wetland and wetland-fringing estuarine and lacustrine sediments (Labandeira and Eble, 2006). Insect herbivory brought the wetland food web closer to modern trophic systems.

Amniote Evolution and Wetlands

The oldest undisputed amniote, the "protorothyridid" *Hylonomus* from the Middle Pennsylvanian of Joggins, Nova Scotia (Dawson, 1854; Carroll, 1964; DiMichele and Hook, 1992; Calder, et al., 1997; Calder et al., this volume), appears to be a very early member of the lineage that led to diapsids. Although reptiles do not require aqueous conditions for breeding, as do amphibians, many do require wetlands for food and cover (Fig. 9; Clark, 1979). At present, reptile abundance is influenced by the availability of horizontal and vertical habitat (Jones, 1986), as may have been the case in the Carboniferous. In the layered canopies of Pennsylvanian peatlands and forest swamps, there was abundant habitat availability for food and cover. At Joggins, reptiles were found within fossil hollowed lycopsid tree stumps; Dawson (1854) originally thought that the animals had fallen into the stumps and been trapped. More recent investigations interpreted the stumps as possible dens in which the reptiles died during wildfires (Calder et al., 1997; Falcon-Lang, 1999). This interpretation is plausible, given that modern wetlands are susceptible to seasonal wildfires, especially crown fires (Scott, 2001).

PERMIAN

High-Latitude Peatlands in Gondwana

Although Permian coals are sometimes considered part of the first great coal-forming period (Permo-Carboniferous), most, with the exception of some coals from the Permian of China (Xingxue and Xiuyhan, 1996), are geographically and floristically separate from their Carboniferous precursors. Pennsylvanian coals represent mostly tropical to subtropical mires that were widespread in Euramerican basins. By the Early Permian, North American coals were restricted to the northern Appalachian Basin, and these mires represented a holdover of Pennsylvanian floras into the Permian. Tropical coals became restricted to several Asian plates (Scotese, 2001). The most widespread peatlands flourished in the cool-temperate climates of the southern Gondwana supercontinent (Fig. 10). These included the first evidence of peats accumulating under permafrost conditions, similar to modern palsa mires (Krull, 1999). Some of these high-latitude Gondwana mires were the first extensive, nontropical mires in earth history and data suggest that there was latitudinal plant zonation (toward both poles), analogous to the modern latitudinal gradients in Northern Hemisphere wetlands (Retallack, 1980; Archangelski, 1986; Cuneo, 1996; Xingxue and Xiuyhan, 1996).

The majority of coal resources in present-day Australia, India, South Africa, and Antarctica are of Permo-Triassic age (Archangelsky, 1986; Walker, 2000; Thomas, 2002). The floral composition of Gondwana coals is distinctly different from the Carboniferous coals of the Northern Hemisphere. Whereas Carboniferous mires were dominated by lycopods and tree ferns, Permian Gondwana mires were dominated by gymnosperms (Archangelski, 1986; Falcon, 1989; Cross and Phillips, 1990; Shearer et al., 1995). In the Early Permian, *Gangamopteris* was dominant. By the Middle Permian, *Glossopteris* was dominant. Many species are interpreted to have had both herbaceous and arborescent growth strategies (Falcon, 1989; Taylor and Taylor, 1990; White, 1990; Stewart and Rothwell, 1993; Shearer et al., 1995). Arborescent *Glossopteris* taxa were tall, with *Dadoxylon-Araucarioxylon*-type gymnospermous wood and *Vertebraria*-type roots (Fig. 11, Gould and Delevoryas, 1977; Stewart and Rothwell, 1993). The arrangement of secondary xylem and the presence of large air chambers in the roots indicate that these trees were adapted to standing water or waterlogged soils in swamp and forest mire settings (Gould, 1975; Retallack and Dilcher, 1981; White, 1990). The similarities between *Glossopteris* taxa on different Southern Hemisphere continents, and the recognition that *Glossopteris*-rich, coal-bearing strata accumulated under different climatic conditions from those of today, were some of the original data used to support the theory of continental drift.

Glossopteris mires also were composed of abundant horsetails, ferns, herbaceous lycopsids, and bryophytes (Neuburg, 1958; Archangelski, 1986; White, 1990) in a wide array of wetland types, including algal ponds, reed fens dominated by the sphenopsid *Phyllotheca*, wet forest mires, and dry swamp forests (Diessel, 1982). The association of bryophytes with high-latitude mires continues to this day in the world's most widespread peatlands, the *Sphagnum*-dominated peats of West Siberia (Botch and Masing, 1983) and the Hudson Bay Lowlands (Zoltai and Pollett, 1983). Another wetland association that began in the Permian was that with large semiaquatic vertebrates. Today alligators, crocodiles, and gavials play a similar ecological role. In

Figure 9. *Hylonomus*, one of the first reptiles, takes shelter in a hollow lycopod trunk during a Pennsylvanian swamp fire in what is now Nova Scotia.

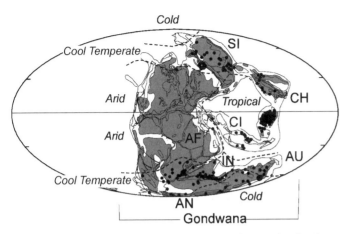

Figure 10. Permian paleogeography and paleoclimates showing locations of coal (black dots) and thereby known paleomires (modified from Scotese, 2001). AF = Africa, AN = Antarctica, AU = Australia, CH = China, CI = Cimmeria, IN = India, SI = Siberia

the Permian, (and Late Carboniferous) large, semi-aquatic labyrinthodont temnospondyls were found in these roles. The 1.8-m long *Eryops* is one of the most common and widespread early Permian labyrinthodonts (Carroll, 1988). Later in the Permian, the rhinesuchids evolved from the eryptoid labyrinthodonts. Rhinesuchids had elongated skulls with eyes on top of their skull similar to extant crocodilians (Fig. 11).

Climatic Changes and Shrinking Wetlands

At the same time that the northern and southern continents were amalgamating to Pangaea, the late Paleozoic ice age was ending, with the last vestiges of Southern Hemisphere ice disappearing in the earliest Permian (Frakes et al., 1992). The termination of ice-age climates, and the sea-level periodicity associated with them, led to an overall climatic warming, which resulted in drying and a dramatic decrease in the scale and extent of wetlands when compared with the Carboniferous. Under these new conditions, some of the previously dominant spore-producing plants were restricted to narrow riparian corridors and lakeside settings (DiMichele and Chaney, this volume). The exception to this pattern occurs on the Chinese microcontinents, which remained climatically wet owing to their proximity to oceanic moisture sources. This region maintained wetland floras similar to those of the Middle Pennsylvanian (lycopsids, cordaites, tree ferns); such floras persisted into the Late Permian (Xingxue and Xiuyhan, 1996; Rees et al., 2002).

Changing climates and flora resulted in distinct global floristic zones (Ziegler, 1990; DiMichele and Hook, 1992; Rees et al., 2002). Today, latitudinal climate distribution results in zonation of different types of wetlands (e.g., extensive *Sphagnum* bogs at high latitudes, marshes in the temperate zone, and mangrove swamps in the Neotropics). Middle to Late Permian coals of the Southern Hemisphere are dominated by wood and leaves of the

Figure 11. During the Permian, forest mires spread across Gondwanaland with gymnosperms replacing lycopods as the dominate wetland trees. The mire is dominated by *Glossopteris* trees and a ground cover of ferns and horsetails. *Rhinesuchus* (in the water) watches a dicynodont on shore. *Glossopteris* tree reconstruction after Gould and Delevoryas (1977).

pteridosperm *Glossopteris* (Fig. 11), whereas coeval peats in Siberia are composed of biomass from ruflorian and voynovskyalean cordaites (e.g., Meyen, 1982; Taylor and Taylor, 1990). Ziegler (1990) discusses latitudinal zonation of Permian biomes. Regional Permian drying resulted in the diversification of seed plants, with the evolution and diversification of ginkgophytes, cycads, peltasperms, and filicalean ferns.

Just as the loss of wetland habitats perturb modern ecosystems, the loss of Permian wetlands had profound influences on terrestrial ecosystems at the close of the Paleozoic. In the Karoo Basin of South Africa, where the most complete terrestrial record occurs across the P-Tr boundary, there is a basinward shift from riparian wetlands to dry uplands through the Permian. This shift is accompanied by a decrease in abundance and ultimate extinction of the *Dicynodon* (a therapsid) assemblage (Smith, 1995), which was replaced by the Early Triassic *Lystrosaurus* assemblage (Rubidge, 1995) soon thereafter. Dicynodonts (Fig. 11) were the most conspicuous terrestrial animals of the Late Permian, and among the first herbivorous vertebrates. They may have used their tusks for digging and slicing horsetail stems and buried rhizomes (Rayner, 1992). Some, like *Lystrosaurus*, were semi-aquatic and inhabited lowland riparian wetlands (Carroll, 1988).

The evolution of vertebrate herbivory opened up a new niche to be exploited in wetland food webs. Modern wetlands support a wide variety of large grazing and browsing mammals including buffalo *(Syncerus caffer)* and hippopotamuses *(Hippopotamus amphibius)* in Africa, moose *(Alces alces)* in North America, water buffalo *(Bubalus bubalis)* in Asia, and the manatee *(Trichecus* sp.*)* in the Neotropics (Bacon, 1997). In extant wetlands, large herbivores modify and reshape wetlands. Their trails become corridors for other animals and may even modify flow paths. Herbivory can lead to increasing diversity of habitat and thereby species, modification of nutrient cycles, as well as expanding resilience and resistance of flora to disturbance (Naiman and Rogers, 1997; Mitsch and Gosselink, 2000).

Effects of the End-Permian Extinction on Wetlands

Reduction in wetland area in the modern world has been shown to decrease biodiversity because so many animals rely on wetlands for at least part of their life cycle (Mitsch and Gosselink, 2000: Bacon, 1997; Keddy, 2000); the reduction of wetland area in the Permian may have caused similar perturbations throughout Gondwana, leading into the end-Permian extinction event. Aside from loss of habitat, food, and nutrients, reductions in wetland area would also have reduced critical hydrological functions provided by wetlands. Decreasing flood storage capacity would have led to increased variability in continental and coastal hydrology, and possibly increased susceptibility of ecotonal areas to flash flooding.

The end-Permian mass extinction caused almost total collapse of the remaining wetland ecosystems (Retallack 1995; Visscher et al., 1996; MacLeod et al., 2000; Rees et al., 2000). This is indicated by the dieback of arborescent vegetation and the high-diversity *Glossopteris* flora (Visscher et al., 1996), as well as the global absence of coal beds in the Early Triassic (Retallack 1995; Retallack and Krull, this volume). In the northern continents, many pteridospermous taxa and most of the arborescent lycopsids that had dominated the vast peatlands of the Carboniferous went extinct (Phillips et al., 1985; DiMichele and Hook, 1992; Stewart and Rothwell, 1993).

TRIASSIC

Wetland Recovery

Postextinction wetland habitat recovery occurred first with the short-term occupation of low-lying areas, by lycopsid isoetalean swamp forests and marshes, presumably from refugia. Isoetaleans were preadapted to oligotrophic conditions, so may have had an advantage in the post-catastrophic environments of the Early Triassic (Looy et al., 1999, 2001). Extant *Isoetes* (quillworts) are terrestrial to submerged aquatic plants with slender, quill-like leaves. Air chambers in the leaves of extant and fossil *Isoetites* support an aquatic ancestry (Taylor and Hickey, 1992). In some modern wetland investigations, submergent and floating vegetation characterizes shallow water or aquatic (e.g., Keddy, 2000) wetlands. Although emergent pteridophytes had been common in wetlands along lake and river margins in the Paleozoic, adaptation to a submerged habit in Triassic *Isoetes* would have allowed for the expansion of wetlands further into the riverine, littoral, and palustrine aquatic realms. Not only did isoetelean lycophytes diversify into freshwater aquatic wetlands, but some genera may also have been salt tolerant. *Pleuromeia* and *Cyclostrobus* have both been interpreted as salt-marsh plants because of their occurrence in coastal lagoon facies (Retallack, 1997).

As the postextinction recovery continued, lycopsid-dominated wetland assemblages were replaced by gymnosperm-dominated assemblages, divided broadly into the *Dicroidium* (pteridosperm) flora of southern Pangaea and the *Sciatophyllum* flora of northern Pangaea (Retallack, 1995; Looy et al., 1999, 2001). By the Middle Triassic, peatlands once again became part of the global ecosystem witnessed by the presence of thin coals in Northern Hemisphere rift basins and more extensive and thick coals in Antarctica (Fig. 12; Visscher et al., 1996; Looy et al., 1999; Retallack, 1995; Walker, 2000; Scotese, 2001; Thomas, 2002). The widespread coals in Antarctica continued the trend of high-latitude peat mires begun in the Permian. In Antarctica, mires were dominated by gymnosperms assigned to the Peltaspermales *(Dicroidium)*, cycadophytes, and ferns (Taylor and Taylor, 1990), whereas tree ferns and rhizomatous ferns, conifers, cycadeoids, gnetaleans, and pentoxylaleans became more common in the Late Triassic and persisted into the Cretaceous (Pigg et al., 1993; Retallack et al., 1996). There is growing evidence that many plant lineages that characterize later Triassic and Jurassic landscapes, including wetlands, originated in the Permian, and thus survived the Permo-Triassic extinction. These include peltasperms (Kerp, 1988), some cycads (DiMichele et al., 2001), and corystosperms (Kerp et al., 2004). As a consequence, the Permo-Triassic event or events that led to massive marine extinctions may have affected terrestrial landscapes mainly by causing ecological restructuring more than mass extinction—this in spite of an apparent global absence of mire habitats in the early Triassic.

Seasonal and Riparian Wetlands

Parts of the famous Petrified Forest of the Chinle Formation in the southwestern United States are examples of the reestablished forest swamps (non-peat-forming wetlands) during wetter Triassic intervals (e.g., Demko et al., 1998; Creber and Ash, 2004). The Chinle represents a paludal complex of streams, lakes, and swamps (Stewart et al., 1972; Blakey and Gubitosa, 1983;

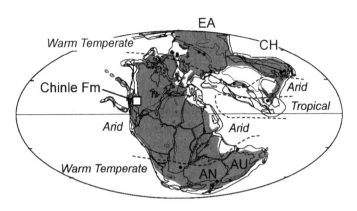

Figure 12. Triassic paleogeography and paleoclimates showing locations of coal (black dots) and thereby known paleomires (modified from Scotese, 2001). AN = Antarctica, AU = Australia, CH = China, EA = Eurasia.

Figure 13. A phytosaur (Ph) maneuvers through a riverine marsh in the Triassic of the southwestern United States. Some of the *Equisetites* (E) horsetails were arborescent. The riparian wetland is also home to the small fish *Semionotus* (S), the large dipnoan *Ceratodus* (Ce), the protosaur *Malerisaurus* (M), large, herding *Placerias* (Pl) dicynodonts, and the small theropod dinosaur *Coelophysis* (Co). The floodplain consists of horsetails and ferns bordered by a riparian forest of giant *Araucarioxylon* (A) conifers.

Long and Padian, 1986). The famous petrified logs are assigned mostly to *Araucarioxylon*-wood, although several new taxa have been recognized (Creber and Ash, 2004). These trees are interpreted as conifers that grew to heights of 56 m with diameters of 3 m (Ash, 2003). The lineages originated in the Southern Hemisphere (Stockey, 1982; Stewart and Rothwell, 1993) and spread northward into riparian settings, including forested wetlands. Common neocalamites, ferns, and lycopsids grew in emergent freshwater marshes within the Chinle paludal complex (Fig. 13), whereas horsetails, cycadeoids, cycads, and ferns occupied floodplains (Demko et al., 1998). Some *Equisetites* were arborescent, similar to their Carboniferous ancestors (Fig. 7D).

Increasing evidence of seasonality in the Chinle complex (Fiorillo et al., 2000; Therrien and Fastovsky, 2000) suggests that wetlands may have been more similar to seasonal riparian marshes and wet meadows than to more continuously wet marshes or bogs. Remains of carnivorous archosaurs, phytosaurs, metoposaurs, and small dinosaurs (such as *Coelophysis*, Fig. 13) are known from the Petrified Forest National Park (Stewart et al., 1972; Long and Padian, 1986; Therrien and Fastovsky, 2000). Dicynodonts, such as *Placerias*, also are found (Fig. 13) and play a role similar to that of large wetland herbivores in Permian wetlands. In modern semiarid to arid areas, riparian wetlands are critical to maintaining vertebrate biodiversity (National Research Council, 1995; Bacon, 1997), and likely were similarly important in ancient semiarid and arid environments (Ashley and Liutkus, 2002).

Phytosaurs and Crocodile Ancestors in Wetlands

Crocodiles, alligators, and gavials are common in modern wetlands especially in estuarine wetlands, coastal marshes, and mangrove swamps. In some wetlands, crocodilians are keystone species and play a crucial role in faunal and floral maintenance as biological "wetland engineers." In the Everglades, for example, the paths and dens of alligators (gator holes) maintain waterways that would otherwise fill with sediment, and may be the only pools remaining in dry seasons. Thus, the alligators' behavior provides crucial habitats for a wide variety of wetland species (Craighead, 1968; Jones et al., 1994). The relationship between crocodylomorphs (crocodile-like and other reptiles) and wetlands began in the mid-Triassic, during the adaptive radiation of archosaurs. In the Triassic, the crocodylomorphs replaced labyrinthodonts as the dominant large, semiaquatic wetland predators. Several archosaur groups with crocodile-like ankles (crurotarsi) evolved in the Triassic, and two taxa are convergent with modern crocodiles in habitat and morphology—the Phytosauria and Suchia. Phytosaurs (Parasuchia) look like modern gavials but had nostrils on top of their heads near their eyes, rather than at the end of the snout (Fig. 13). Phytosaurs were common in the fluvial and riparian marsh and forest wetlands of the Triassic in Virginia and the southwestern United States but were extinct by the end of the Triassic (Chatterjee, 1986; Long and Padian, 1986).

Suchians, the group that includes the Crocodylomorpha and is ancestral to extant crocodilians, originated as small, terrestrial, bipedal reptiles in the Triassic. The evolution of an aquatic habit by eusuchian crocodylomorphs in the Jurassic allowed these semiaquatic archosaurs to replace the phytosaurs. By the Cretaceous, giant crocodile-like eusuchians, such as the 12 m long *Deinosuchus*, were inhabiting estuarine wetlands along the southern coast of North America (Schwimmer, 2002). Also by the Cretaceous, Crocodylia (modern crocodile group) had evolved (Schwimmer, 2002) and represented the only surviving archosaurs (Carroll, 1988).

TRIASSIC-JURASSIC

Frogs, Salamanders, and Turtles in Wetlands

Among the most common animals in extant tropical and temperate wetlands are frogs, salamanders, and turtles. Although the association of amphibians and reptiles with wetlands began in the Paleozoic, extant classes did not evolve until the Mesozoic. The possible ancestor of frogs, *Triadobatrachus*, is reported from the Early Triassic and provides a link between earlier labyrinthodonts and frogs (Carroll, 1988). *Chunerpeton*, the oldest salamander, is known from Triassic lacustrine deposits of Mongolia (Gao and Shuban, 2003). Likewise, *Proganochelys* (=*Triassochelys*), the oldest freshwater turtle, is known from paludal marsh deposits of Germany, Southeast Asia, and North America (Gaffney, 1990). Members of each of these groups are dependent on wetlands for part of their life cycle and serve as important links in the trophic web (Weller, 1994; Mitsch and Gosselink, 2000). For example, tadpoles eat small plants and invertebrates and in turn, are eaten by fish. Later in life, adult frogs eat insects. Similar trophic links between these taxa likely were established by the Jurassic.

JURASSIC

Global Perturbations and Expanding Wetlands

The end-Triassic mass extinction is coincident with greenhouse warming, resulting in global perturbations in the carbon cycle and a near-total species-level turnover of megaflora (McElwain et al., 1999). Throughout the Jurassic, global warming and increased precipitation caused a gradual shift in wetland habitats from narrow riparian, lake-fringing swamps and marshes to more extensive conifer-dominated swamps and mires in the Cretaceous (Cross and Phillips, 1990). Southern Hemisphere swamps and forest mires were dominated by podocarpaceous and araucarian conifers, and Northern Hemisphere swamps and forest mires were dominated by taxodiaceous conifers (Wing and Sues, 1992; Askin and Spicer, 1995). Elements of this zonation remain to this day. An extinct conifer family, the Cheirolepidiaceae, were common in the Tropics, particularly in coastal wetland settings.

Krasilov (1975) interpreted a series of typical Jurassic wetland floral zonations in northern Eurasia. *Ptilophyllum* bennettites are interpreted to have occupied mangrove-like wetlands, while marshes were characterized by monospecific stands of large *Equisetites*. Bogs (forest mires) along lake margins and in riparian settings had a canopy formed from taxodiaceous conifers (*Elatides*) and arborescent ferns *(Dictyophyllum, Todites)*, with an understory composed of ferns and *Ptilophyllum* bennettites. Cycadeoids were the dominant flora of the Middle Jurassic coals of Mexico (Person and Delevoryas, 1982; Cross and Phillips, 1990). Ferns (e.g., *Coniopteris*), with lesser contribution from conifers and ginkophytes, dominated Middle to Late Jurassic mires of western North America (Silverman and Harris, 1967; Miller, 1987). Jurassic coals of China were dominated by tree ferns, dwarf coniferophytes, and secondary cycads (Miao et al., 1989). These examples highlight the increasing variability of floral associations in Jurassic wetlands. By the Late Jurassic, coals also were accumulating in several basins in the former Soviet Union, Mongolia, south China, and Iran (Fig. 14; Scotese, 2001; Walker, 2000; Scotese, 2001; Thomas, 2002).

Wetland Preservation of Early Mammals

Much of our understanding of the early diversification of mammals comes from material collected in a brown coal from the Guimarota coal mine, central Portugal. The mine was worked from 1973 to 1982 exclusively for paleontological purposes (Gloy, 2000; Martin, 2000), providing a detailed insight into the changing seres within the mire. The largest biomass contribution to the Guimarota paleomire was from *Araucariaceae* (conifers) and horsetails *(Equisetites)* with lower biomass contribution from pteridophytes *(Deltoidospora, Dicksoniaceae)*, cycads, and ginkgophytes (Van Erve and Mohr, 1988). Entombed within the Guimarota peat are ostracods, gastropods, freshwater and brackish molluscs, hybodont sharks, amphibians, small reptiles (turtles, crocodiles, lizards), the giant crocodile *Machimosaurus*, small dinosaurs, and mammals. The exceptional mammalian biota consists of Multituburculata, Docodonta, and Holotheria (Martin, 2000). In many modern wetlands, small mammals (especially rodents) are the dominant terrestrial and semiaquatic herbivores (Speight and Blackith, 1983). Although the Guimarota mammals show that small mammals were occupying wetland habitats, expansion into semiaquatic lifestyles may not have occurred until the Tertiary.

JURASSIC-CRETACEOUS

Mangals in Wetlands

Coastal mangals of coniferous affinity are interpreted from Wealden strata across the Late Jurassic–Early Cretaceous of the Northern Hemisphere. This group, informally known as the frenelopsids, are woody trees assigned to the Cheirolepidiaceae that produced *Classopollis*-type pollen (Axsmith et al., 2004).

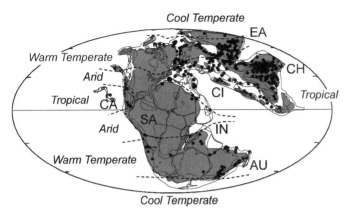

Figure 14. Jurassic paleogeography and paleoclimates showing locations of coal (black dots) and thereby known paleomires (modified from Scotese, 2001). AU = Australia, CA = Central America, CH = China, CI = Cimmeria, EA = Eurasia, IN = India, SA = South America.

Pseudofrenelopsis and related taxa are common components of Early Cretaceous deposits of Africa, England, eastern Europe, and North America, and sedimentological criteria were used by Upchurch and Doyle (1981) to place these trees within a low-diversity, tidally influenced coastal regime. This is similar to the Upper Jurassic Purbeck beds where an in situ forest is preserved within a thin, carbonaceous marl paleosol (a well-drained, immature rendzina) of an intertidal and supratidal sequence (Francis, 1983, 1986). Associated with the Purbeck conifers are a few cycadophyte stems. Although these trees exhibit no evidence of buttressing or mangrove habit, they are encased in an algal stromatolitic limestone that formed in response to a change in base level of saline marine waters. Physiognomic characters of the frenelopsids including shoot morphology, the presence of thick cuticles, reduced leaves, sunken stomata, and succulent appearance, are morphological adaptations to water stress in saline or dry environments (Upchurch and Doyle, 1981; Gomez et al., 2001). Aside from stratigraphic and physiognomic indicators, several isotopic studies of Cretaceous European fossil plant assemblages using isotopic $^{12}C/^{13}C$ analysis indicate that *Frenelopsis* in marginal marine facies has elevated ^{13}C relative to other genera in more distal facies, suggestive of stress and possibly saline influences in salt-water marshes (Nguyen Tu et al., 2002).

JURASSIC-CRETACEOUS

Aquatic Ferns in Wetlands

Marsileaceae and Salviniaceae are heterosporous aquatic ferns whose origins can be traced to the Late Jurassic–Early Cretaceous (Yamada and Kato, 2002) and mid-Cretaceous, respectively (Hall, 1975; Skog and Dilcher, 1992; Pryer, 1999). Extant *Marsilea* are rooted shallow-water ferns, while the Salviniaceae consist of free-floating aquatic ferns. Free-floating habits extended the diversity of vascular macrophytes in wetlands, a

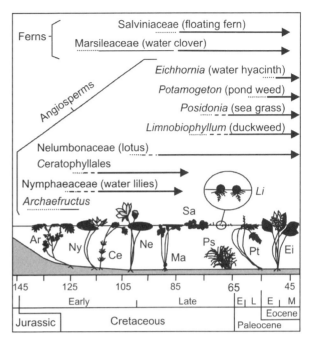

Figure 15. Radiation of aquatic plants and thereby aquatic wetlands in the Cretaceous and early Tertiary periods. Dates and image for *Archaefructus* from Sun et al. (2002); Ceratophyllales from Dilcher et al. (1996); *Eichhornia* from Patil and Singh (1978); *Limnobiophyllum* from Kvaček (1995); Marsileaceae from Lupia (2000); Nelumbonaceae from Dilcher (2000); Nymphaeaceae from Friis et al. (2001); *Potamogeton* from Berry (1937), Bremer (2000), Collinson (2002); ancestral Salviniaceae from Hall (1975); *Posidonia* and *Thalassocharis* (not shown) seagrasses are inferred to have Cretaceous origins in Brasier (1975) and Kuo and den Hartog (2000).

trend that would be duplicated by unrelated angiosperms later in the Cretaceous and in the Tertiary (Fig. 15). Extant *Salvinia* have the ability to grow quickly and can form thick mats that limit sunlight and open water for other wetland plants and aquatic fauna (Julien et al., 2002). By the mid-Cretaceous, water ferns like *Hausmannia* were influencing lacustrine aquatic wetlands, acting as pond colonizers in mires (Spicer, 2002). In extant wetlands, the accumulation of aquatic plant mats and detritus, as well as sediment trapping from rooted aquatic plants, is an important part of pond-filling successions.

CRETACEOUS

Aquatic Angiosperms in Wetlands

Today, with notable exceptions, wetlands are dominated by angiosperms. The timing of origin of this group is subject to considerable debate, but the oldest undisputed fossil angiosperms are from the Early Cretaceous (Hickey and Doyle, 1997; Sun et al., 1998, 2002; Sun and Dilcher, 2002). Angiosperm origins are hotly debated (Scott et al., 1960; Crane, 1993; Crane et al., 1995), with some authors inferring evolution in upland areas (e.g.,

Stebbins 1974, 1976) while others have suggested origination in coastal lowlands (e.g., Retallack and Dilcher, 1981). Regardless of their origin, some of the oldest angiosperms were aquatic plants (Sun et al., 1998). *Archaeofructus,* the oldest known possible angiosperm, is interpreted as a submerged aquatic plant (Sun et al., 2002). Aquatic angiosperms (Fig. 15) developed a series of biochemical, morphological, and physiological specializations that allowed them to diversify into shallow aquatic wetlands (littoral, limnetic). By the Early to mid-Cretaceous, several freshwater families with rooted, floating leaf habits are recorded. These include water lilies (Nymphaeaceae, Cabombaceae), lotus (Nelumbonaceae), plants with affinities to hornworts (Ceratophyllaceae) (Dilcher, 2000; Dilcher et al., 1996; Friis et al., 2001), and possible water milfoils (Halogragaceae) (Hernández-Castillo and Cervallos-Ferriz, 1999). By the Late Cretaceous, the radiation of aquatic angiosperms also included a free-floating habit, with *Lymnobiophyllum* providing a possible ancestral link between duckweeds (Lemnaceae) and the aroids (Araceae) (Stockey et al., 1997). The diversification of aquatic angiosperms and ferns with floating leaves and free-floating morphologies would have provided new habitats and trophic links for fish, amphibians, and aquatic invertebrates in freshwater lacustrine and riverine wetlands, as well as in shallow, open-water wetlands. Likewise, the diversification of various aquatic plant morphologies would have set the stage for increased partitioning of flooded wetlands and hydroseres, more similar to those found in extant limnic and paludal wetlands.

CRETACEOUS

The Return of Extensive Peatlands

The Cretaceous represents the second episode of global coal formation. Extensive Cretaceous coals in western North America, China, the former Soviet Union, Central America, northwestern South America, and New Zealand (Saward, 1992; Walker, 2000; Scotese, 2001; Thomas, 2002) indicate that mires (fens, bogs, forest swamps) once again became widespread (Fig. 16). Northern Hemisphere peatlands continued to be dominated by conifers (*Abietites, Athrotacites, Moriconea, Podozamites, Protophyllocladus, Sequoia, Metasequoia*) with an understory of ferns, *Equisetites,* and less commonly, cycadophytes (Parker, 1975; Knoll, 1985; LaPasha and Miller, 1985; Spicer and Parrish, 1986; Miller, 1987; Cross and Phillips, 1990; Pelzer, et al., 1992; Saward, 1992; Spicer et al., 1992; Shearer et al., 1995; Hickey and Doyle, 1997; Spicer, 2002). In some raised mire successions, ferns and mosses were important (Hickey and Doyle, 1997). In the Southern Hemisphere, the palynology of coals from New Zealand and Australia indicates that podocarps and ferns dominated forest mires (Moore et al., this volume). These trends demonstrate an evolutionary stability and/or longevity in mire settings (as compared with floral changes in upland environments), a pattern of conservatism that has occurred several times in the geologic past and may be explained by incumbency. In essence, there is an ecological asymmetry between swamp environments and terra firma environments (DiMichele et al., 1987); plants adapted to the flooded, often low-nutrient conditions of swamps display physiological specializations that reduce their competitive abilities in terra firma settings. In contrast, the stringent physical conditions of permanently to periodically flooded environments exclude plants from terra firma environments. This results in sharp differences in species richness between these broad environmental categories within any given climatic zone (DiMichele et al., 2001). Hence, although angiosperms dominated many terrestrial ecosystems by the end of the Cretaceous (Lidgard and Crane, 1988; Wing and Boucher, 1998; Graham, 1999), and palms and at least 20 broad-leaved angiosperm taxa, including genera that contain common extant wetland plants such as *Platanus* (sycamore), are preserved in Cretaceous coal-bearing strata (Parker and Balsley, 1977; Tidwell, 1975; Balsey and Parker, 1983; Cross and Phillips, 1990), angiosperms remained only minor components in peat-accumulating wetlands (Pelzer et al., 1992; Saward, 1992; Hickey and Doyle, 1997; Wing and Boucher, 1998; Nguyen Tu et al., 2002). The exception occurs in the Southern Hemisphere, where the coniferous flora began to be replaced by *Nothofagus* (southern beech) in Antarctica and then Australia toward the end of the Cretaceous (Muller, 1984; Saward, 1992; Hill and Dettman, 1996).

Dinosaurs in Paludal Wetlands

Many reptiles inhabited—or traversed—and perished in Mesozoic wetlands. By the early Jurassic, herbivorous dinosaurs had replaced synapsids in terrestrial wetlands. The most famous dinosaurs associated with Cretaceous wetlands are the Bernissart *Iguanodons* from the Luronne coal seam, collected in Belgium in 1878 (Fig. 17). These ornithopods are historically famous for

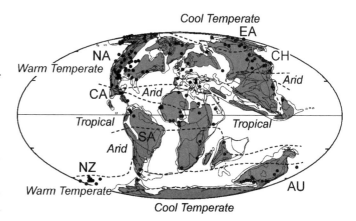

Figure 16. Cretaceous paleogeography and paleoclimates showing locations of coal (black dots) and thereby known paleomires (modified from Scotese, 2001). AU = Australia, CA = Central America, CH = China, EA = Eurasia, NA = North America, NZ = New Zealand, SA = South America.

(1) being the first complete dinosaur skeletons recovered, (2) providing the first evidence that dinosaurs traveled in groups, and (3) proving that some dinosaurs were bipedal (Norman, 1980; Forster, 1997). Although such bones are not preserved commonly in peat, the unusual groundwater chemistries of wetlands can enhance preservation. In North America, the recovery of a putative fossilized four-chambered heart of the ornithischian dinosaur *Thescelosaurus* may owe its preservation to burial in a riparian forest habitat (Fisher et al., 2000).

In many cases, trackways provide evidence of vertebrates in wetlands. Thousands of dinosaur footprints have been found in the roof strata of Cretaceous coal mines (Peterson, 1924; Balsey and Parker, 1983; Parker and Rowley, 1989). Likewise, trackways from the Wessex Formation, Isle of Wight, England, were preserved in coastal floodplain, riparian wetlands. Twenty-two dinosaur species are known from the Isle of Wight, including *Iguanodon* and the fish-eating theropod *Baronyx* (Martill and Naish, 2001). Some beds represent catenas formed in seasonal wetlands, similar to modern tropical and subtropical river systems such as the Pantanal of the Amazon Basin, Brazil (Wright, et al. 2000).

Angiosperm Mangroves

Mangroves are a large group of unrelated, salt-tolerant trees and associated non-woody taxa including ferns (mangals). Although earlier plants have been interpreted as occupying possible mangal habitats, unequivocal salt-tolerant mangroves related to extant species appeared after the angiosperms in the Cretaceous (Muller, 1984; Aaron et al., 1999; Hogarth, 1999; Gee, 2001). *Nypa* palms (Arecaceae) evolved during the Cretaceous and rapidly spread into many wetland and wetland-fringing environments of the Neotropics (Singh, 1999). The Late Cretaceous to Paleocene marks the zenith of systematic diversity in the genus, with only *N. fructicans* constituting monotypic stands of the palm presently. Associated with *N. fructicans* in tidally influenced coastal zones is the mangrove fern *Acrostichum*, which is first reported from the Late Cretaceous (Bonde, 2002), and spread into the Eocene (Collinson, 2002).

Another Cretaceous mangal is *Weichselia reticulata* (Shinaq and Bandel, 1998). This tree fern is found in the Late Cretaceous of Bahariya, North Africa, with bivalves, gastropods, sharks, fish, turtles, crocodyliforms, and at least five genera of dinosaurs. The dinosaur *Paralititan stromeri* is one of the largest herbivores, whereas *Spinosaurus* and *Charcharodontosaurus* are two of the largest carnivores, of all time (Smith et al., 2001; Lacovera et al., 2002). All appear to have lived in or around this Cretaceous coastal swamp.

Modern mangrove swamps serve many important ecological functions, including nutrient cycling, and are net exporters of organic material into adjacent estuaries. They are important habitats for fluvial, estuarine, and coastal ecosystems (Bacon, 1997; Mitsch and Gosselink, 2000). These links lead to high productivity and biodiversity, a possible reason for the diversity and size of the gigantic dinosaurs at the Bahariya site (Smith et al., 2001). Modern mangrove swamps also play an important function in sedimentation and storm surge baffling along tropical coastlines. These functions probably existed in earlier inferred mangal habitats, but the adaptation of extant taxa in the Cretaceous allows for more actualistic comparisons of mangrove functions in the Late Cretaceous through Tertiary, based on the functions of extant genera.

Figure 17. Dinosaurs in a Euramerican Cretaceous mire. *Iguanodon* herd passes through conifer-dominated forest mire. Ground cover consists of abundant ferns, Equisetites, and less common palms.

CRETACEOUS–TERTIARY

Marine Angiosperms: Sea Grasses

Phylogenetic analyses of extant sea grasses suggests that marine angiosperms have evolved in at least three separate lineages (Les et al., 1997). Sea grasses have a relatively poor fossil record, but *Posidonia* (Potamogetonaceae) is known from the Cretaceous (Kuo and den Hartog, 2000) and phylogenetic analyses support a Late Cretaceous origin (Bremer, 2000). Macrofossils of the sea grass genera *Thalasssodendron, Cymodocea* (Potamogetonaceae), and *Thalassia* (Hydrocaritaceae) are known from the middle Eocene of Florida (Lumber et al., 1984). Sea grasses are halophytes, and their evolution involved physical reduction in floral and leaf structures and xylem tissue, changes in reproductive strategies, and a physiological change to bicarbonate utilization in photosynthesis (Brasier, 1975; Stevenson, 1988; Kuo and den Hartog, 2000). Extant sea grasses are completely aquatic, with habitats extending to more than 6 m depth (which is the present limit of wetlands by the Ramsar classification). Hence, the evolution of submerged sea grasses extended the range of wetlands in coastal marine and subtidal estuarine environments, providing new habitats and resources for invertebrates and vertebrates. Sea grasses are particularly important because they dominate some of the most productive habitats on Earth (Stevenson, 1988; Bacon, 1997), and their presence changes local hydrodynamics, thus enhancing sedimentation of fines out of the water column. In fact, there is a recognized facultative successional sequence between mangrove swamps, sea grass meadows, and coral reefs, which may have its origins in the Late Cretaceous with the first appearance of sea grasses and mangroves (Brasier, 1975; McCoy and Heck, 1976). Such a succession and integrated trophic web explain the shared pan-Tethyan distribution of sea grasses with coral reef fish, decapod crustaceans, molluscs (McCoy and Heck, 1976), foraminifera (Brasier, 1975), and even manatees (Domning et al., 1982).

Carnivorous Plants in Wetlands

Low-nutrient fens and bogs support some of the rarest and most diverse plant communities in modern mires, including carnivorous plants (National Research Council, 1995; Bacon, 1997). Modern species of carnivorous plants, including bladderworts *(Utricularia)*, sundews *(Drosera)*, and butterworts *(Pinguicula)*, grow in acidic fens, bogs, and swamps. This relationship may indicate that plant carnivory arose in angiosperms as an adaptation to acidic, low-nutrient conditions of mire habitats. Carnivory arose not just once, but separately in 18 genera among six different plant orders (Juniper et al., 1989; DeGreef, 1997). Seeds of *Paleoadrovanda splendus*, which are similar in appearance to those of the extant carnivorous genus *Aldrovanda* (Droseraceae), a free-floating aquatic plant, are known from the Late Cretaceous (Knobloch and Mai, 1984; DeGreef, 1997). *Aldrovandra* is recognized in the Oligocene (Collinson et al., 1993) and spores of *Utricularia* (Lentibulariaceae) have been identified from the Miocene (Muller, 1984). The fossil history of other carnivorous plants is less certain. In general, their small stature and delicate nature, in combination with alteration due to early and late diagenesis within organic-rich substrates, result in a poor fossil record.

Amber in Wetlands

Most of the world's amber deposits are found in Cretaceous and Tertiary lignites, although amber often is reworked into other sedimentary deposits. Cretaceous ambers are known from England, Alaska, and New Jersey in the United States, Canada, Burma, and the Middle East. More well known are the Tertiary deposits from the Baltic, Dominican Republic, and Mexico (Poinar, 1992; Grimaldi et al., 2002). The New Jersey Cretaceous ambers preserve the most diverse assemblage of plants and animals, including 25 orders comprising 125 families and more than 250 species. New Jersey ambers formed in coastal swamps dominated by the conifer *Pityoxylon* (Pinaceae similar to *Pinus*, *Picea*, or *Larix*). These ambers contain the oldest fossil mushroom, ant, potter wasp, and bee, as well as the only Cretaceous flower preserved in amber (Grimaldi et al., 2000).

Tertiary Baltic amber was produced by *Agathis*-like (Kauri pine) araucariacean trees in conifer-dominated swamps and moist lowland forests. These ambers preserve a diverse assemblage including amphipods, isopods, centipedes, millipedes, dragonflies, roaches, beetles, and the oldest praying mantids (Poinar, 1992). Common wetland forms, including aquatic larvae and nymphs of caddis flies, mayflies, and waterbugs, provide evidence for standing water in some parts of the araucarian swamps (Larsson, 1978).

Blood Suckers in Wetlands

Many people associate black flies (Diptera) and mosquitoes (Culicidae) with wetlands. Although insects have been associated with wetlands since at least the Devonian (e.g., Rolfe, 1980), the oldest undisputed black flies and mosquitoes date from Late Cretaceous amber (Poinar 1992; Grimaldi et al., 2000, 2002). Modern mosquitoes are important transmitters of diseases such as malaria, yellow fever, dengue fever, and encephalitis. The association of these diseases with tropical wetlands is ingrained in our society. In fact, the translation of the word malaria (*mal aria*) means bad air, derived from the disease's association with fetid marshes. When wetland mosquitoes (and other insects) began to transmit diseases is uncertain (Martins-Neto, 2003), although Statz (1994) speculated that Oligocene mosquitoes spread diseases. Insect-borne diseases may have influenced the evolution of our own species, as indicated by the relationship between malaria and sickle cell disease. Although famous as pests and disease vectors, mosquito and black fly larvae are important parts of many wetland food webs (Bacon, 1997).

Effects of the K-T Extinction on Wetlands

The K-T extinction of the dinosaurs and a wide array of vertebrates and invertebrates led to extensive ecological restructuring in wetlands. At the same time, some of the fauna that survived were obligate wetland inhabitants, such as crocodiles, turtles, and frogs, suggesting that wetlands served as a faunal refugium during the K-T event. Wetlands tend to be inhabited by conservative taxa adapted to some aspect of limiting conditions, so wetland fauna may be preadapted to survival of mass extinctions.

The extinction is also associated with global floristic changes (Vajda et al., 2001), although these were mostly concentrated in the Northern Hemisphere, dominantly, North America (Askin, 1988; Johnson et al., 1989; Wolbach et al., 1990; Wing and Sues, 1992; Nichols and Pillmore, 2000). In some parts of western North America, the iridium anomaly occurs within coal beds, which provide a unique glimpse of successive responses to global catastrophe. In these areas, the ejecta cloud from the inferred bolide impact deposited a thin layer of glassy debris in the mires that eventually was altered to kaolinite (Nichols and Pillmore, 2000). This was followed by an increase in ferns, the "fern spike" found at many locations worldwide. The increase in ferns is associated with the elimination of much of the pre-existing swamp flora (especially deciduous dicots), and is interpreted to represent post-catastrophic colonization by pioneering taxa (Tschudy et al., 1984 Askin, 1988; Nichols and Pillmore, 2000). The most significant influences were on the angiosperms; the least were on conifers, ferns, pteridophytes, and mosses (Nichols and Fleming, 1990), the common wetland inhabitants. Likewise, in New Zealand, Vajda et al. (2001) interpreted recolonization of a waterlogged, K-T acidic substrate by a succession of moss and ground ferns, and then tree ferns. These plants would have been preadapted to post-catastrophic acidic environments through adaptations gained in pre-catastrophe mire habitats.

TERTIARY

Thick Peats and Peatland Successions

The Tertiary represents the third major interval during which widespread peat accumulation occurred. Tertiary coals are known from many basins worldwide (Scotese, 2001; Fig. 18), although the greatest resources are in western North America, northwestern and western South America, Germany, and Southeast Asia (Walker, 2000; Thomas, 2002). Tertiary coal beds can be as much as 90 m thick, whereas the thickest modern ombrotrophic mires are generally less than 20 m thick (the peat representing accumulation over the last ~7000 years). In fact, ombrotrophic mires may be limited in their potential thickness by numerous conditions including microbial respiration within the underlying peat (Moore, 1995). Hence, the great thickness of some Tertiary coals suggests that they cannot represent the accumulation of a single peat mire, but rather the accumulation of multiple, stacked mires (Shearer et al., 1994; Moore, 1995).

Even single coal beds may represent a wide variety of mire types. Palynological evidence indicates that Paleogene mire floras initially were dominated by gymnosperms with increasing importance of angiosperms through time (Nichols, 1995). This continued a trend that started in the Late Cretaceous when mires were dominated by conifers (Wing and Boucher, 1998; Graham, 1999). As angiosperms became increasingly important quantitatively, the resultant coals varied significantly in organic facies and in quality (Nichols, 1995), because of the increasing diversity of specialized mire types that could contribute to a single peat and ultimately coal bed. Eocene coals of the U.S. Gulf Coast accumulated from successions of freshwater herbaceous communities enriched in ferns, to freshwater Juglandaceous mire forests codominated with palms and *Nyssa* (tupelo), and, depending upon the sequence stratigraphic relationship of the coal to overlying marine sediments, even to mangrove swamps (Fig. 19A; Raymond et al., 1997). Miocene lignites from central Europe exhibit complex successions of wetlands including limnic to littoral aquatic wetlands with *Potamogeton* (pond weed), reed thickets in freshwater marshes, *Taxodium-Nyssa* forest mires, mixed herbaceous angiosperm fens, palm-dominated fens and forest mires, *Myrica* (bayberry) bogs or fens, riparian emergent wetlands with thickets of *Alnus* and *Cornus* (dogwood), mixed conifer (*Marcoduria, Sequoia*) forest mires, and oligotrophic low-diversity conifer bogs or raised mires (Fig. 19B; Teichmüller, 1958, 1962, 1982; Lancucka-Srodoniowa, 1966; Knobloch, 1970; Schneider, 1992, 1995; Mosbrugger et al., 1994). These examples illustrate the increasing diversity of angiosperms in Tertiary wetlands, as well as resultant wetland partitioning, when compared with those of the Cretaceous and Carboniferous.

Likewise, Tertiary plate tectonics exerted a profound effect on the distribution and biogeography of wetland floras (especially in the Southern Hemisphere), as the Gondwanan continents separated, and in some cases collided with northern continents (Christophel, 1989; Wing and Sues, 1992; Askin and Spicer, 1995; Burnham and Graham, 1999; Graham, 1999).

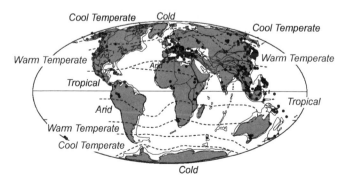

Figure 18. Tertiary (Miocene) paleogeography and paleoclimates showing locations of coal (black dots) and thereby known paleomires (modified from Scotese, 2001).

Cypress Swamps and Mires

Taxodiaceous conifers had dominated Cretaceous Northern Hemisphere wetlands (Stewart and Rothwell, 1993; Shearer et al., 1995), but *Taxodium* sp. (bald and pond cypress) did not become dominant in swamps and forest mires until the early Tertiary (Wing, 1987; Schneider, 1992, 1995; Kvaček, 1998; Collinson, 2000). By the Eocene, angiosperm-dominated wetlands had become more common in temperate riparian and lacustrine-margin settings (Graham, 1999), but taxodiaceous swamps persisted at high latitudes above the Arctic Circle during the Eocene thermal maximum (Francis, 1991; Basinger et al., 1994; Greenwood and Basinger, 1995; Williams et al., 2003a, 2003b). Taxodiaceous swamps on Axel Heiberg Island in the Canadian High Arctic consist of in situ assemblages of mummified tree stumps and forest-floor leaf-litter mats buried at different times over century to millennial time intervals. The picture that emerges in these swamps is one of a vegetational mosaic wherein taxodiaceous conifers *(Metasequoia and Glyptostrobus)* are laterally or stratigraphically adjacent to mixed coniferous forests and angiosperm/fern bogs, with the taxodiaceous swamp phase accounting for peat accumulation. Hence, taxodiaceous swamps were more extensive than at present, with geographic restriction to their present latitudinal distribution occurring during the Paleogene and Neogene. At least by the Oligocene, these wetlands occupied coastal settings of central Europe (Gastaldo et al., 1998), a distribution that continued into the Miocene (Kovar-Eder et al., 2001); along the Atlantic and Gulf coasts of North America, taxodiaceous swamps became well established in the Neogene (Rich et al., 2002). Both peat-accumulating and minerogenic swamps persisted into the Miocene. Taxodiaceous and other coniferous taxa, however, continued to contribute the bulk of biomass to north temperate peat mires, with little contribution from woody angiosperm taxa (Mosbrugger et al., 1994). In the late Cenozoic, access to continuous habitats across latitudinal gradients controlled the distribution of taxodiaceous conifers. *Taxodium* remained in eastern North America because there were continuous habitats it could occupy during late Cenozoic climate changes; *Metasequoia* went extinct in western North America because similar habitats were not present (Potts and Behrensmeyer, 1998).

Tropical Palm Swamps

Angiosperms show marked increase in Tertiary wetlands. Palms (monocots) are found in Tertiary coals from North America, Europe, Asia, and New Zealand (Packnall, 1989; Raymond

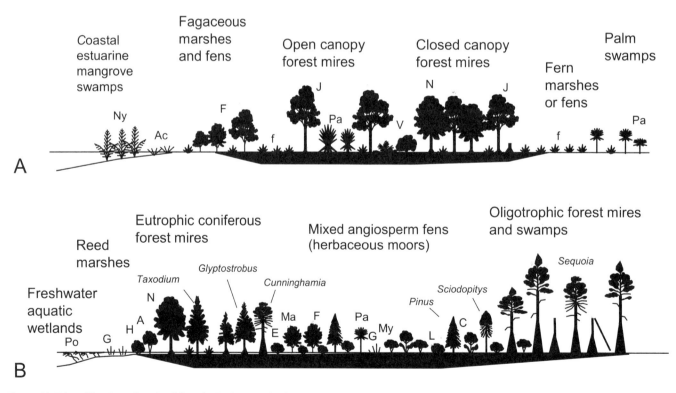

Figure 19. Diversification of wetland flora in Tertiary wetlands. (A) Wetlands interpreted from Eocene Gulf Coast coals (after data from Westgate and Gee, 1990, and Raymond et al., 1995). (B) Wetlands interpreted for Miocene brown coals in Europe (based on data from Teichmuller, 1962, 1982, and Schneider, 1992, 1995). Coniferous trees are labeled. Angiosperms include A = Alnus (alder), C = Cyrilliceae, E = Ericaceae, F = Fagaceae, G = Glumiflorae (reeds), H = Hamamelidaceae (sweet gum), J = Juglandaceae, L = Lauraceae, Ma = Magnoliaceae, My = Myrica (myrtles and bayberry), N = *Nyssa* (Tupelo), Pa = Palm, Po = *Potamogeton* (pond weed), and V = *Viburnum*. Ac = *Acrostichum* (mangrove fern).

et al., 1997; Lenz and Riegel, 2001). Today palms occupy a wide range of habitats including wetlands, but most wetland palms occupy non-peat-producing swamps rather than mires. In the Tertiary, on the other hand, palms inhabited swamps and mires. *Nypa* mangrove palms dominated coastal swamps of the Eocene Gulf Coast of North America, often in close association with tropical woody angiosperms, lycopsids, and ferns, similar to extant genera in coastal mangrove, back-mangrove swamps, and freshwater swamps (Fig. 19A; Fredriksen, 1985; Westgate and Gee, 1990). These estuarine mangrove palm swamps are associated with a diverse fauna including invertebrates, sharks and rays, bony fish, amphibians, turtles, alligators, the giant aquatic snake *Pteroshenus,* and a wide array of mammals including the four-toed horse *Epihippus,* the odd-toed ungulate *Amynodon* (which may have been semiaquatic), and sirenians (Westgate and Gee, 1990).

Although currently confined to a pantropical belt (Uhl and Dransfield, 1987; Myers, 1990), palms extended into midlatitude wetlands during the Eocene global warming event (Uhl and Dransfield, 1987). Palm distribution since the Eocene has been influenced by plate movements and climate changes (Burnham and Graham, 1999). The principal genera in extant palm wetlands are *Mauritia, Raphia,* and *Metroxylon* (Myers, 1990). The oldest *Mauritia* fossils are from the Paleocene and this genus became widespread throughout the Tertiary of South America (Muller, 1984; Junk, 1983; Uhl and Dransfield, 1987; Maraven, 1998). Extant *Mauritia flexurosa* has pneumatophores to cope with inundation in the swamps it inhabits, such as the várzea of the Amazon River in South America. Downriver in the Amazon, pure stands of *Raphia* and *Manicaria* palms are adapted to twice-a-day tidal inundation. *Manicaria* is known from the early Eocene London Clay (Collinson and Hooker, 1987). Further shoreward, mangrove palms dominate the river mouth and coastal estuaries (Junk, 1983; Brinson, 1990). Another palm adaptation can be seen in *Calamus* sp., the rattan palm, spores of which are found from the Paleocene (Muller, 1984). Extant rattan palms are climbing vines and are common in many tropical wetlands. These examples illustrate the wide range of wetland habitats to which palms have adapted and the specialization that typified the radiation of angiosperms in the Tertiary, resulting in a diverse array of wetland types and structural complexity within subcommunities of wetlands.

The Spread of Freshwater Broad-Leaved Wetlands

Aside from Palmae, there is a well-documented latitudinal expansion of angiosperms throughout the Tertiary (see summaries in Wing and Sues, 1992; Potts and Behrensmeyer, 1992; Askin and Spicer, 1995; Wing and Boucher, 1998; Graham, 1999). Although the spread of angiosperms into wetland habitats lagged behind the spread of sister taxa outside of wetlands, partitioning of wetland habitats increased as angiosperms became increasingly specialized, as shown in the examples in Figure 19A and 19B. Higher-latitude wetlands show more floral turnover than tropical and lower latitude wetlands, some elements of which remained from the early Tertiary. In the Southern Hemisphere, one of the most important arborescent angiosperms was *Nothofagus* (southern beech), which originated in Late Cretaceous high latitudes of South America or Antarctica (Muller, 1984; Hill and Dettman, 1996) and dispersed into Tertiary coal-forming mires of Australia and New Zealand (Barlow and Hyland, 1988; Christophel, 1989; Kershaw et al., 1991). Miocene and Oligocene peats of Australia and New Zealand accumulated as coastal and estuarine mires often dominated by *Nothofagus* with Myrtaceae, palms, podocarps, and ferns (Pocknall, 1985; Kershaw et al., 1991; Shearer et al., 1995). Some of these peats may reflect successions from podocarp- and fern-dominated floras to raised bogs with *Nothofagus* (Sluiter et al., 1995).

Among common northern-latitude arborescent genera, *Nyssa* (tupelo, black gum), *Alnus* (alder), *Platanus* (sycamore), *Populus* (poplar), and *Salix* (willow) became increasingly common in Tertiary temperate freshwater wetlands (Berger, 1998; Gastaldo et al., 1998; Kvaček, 1998; Graham, 1999), with many similarities to assemblages in extant North American Gulf Coast swamps (Mosbrugger and Utescher, 1997). Some Eocene forest mires in Germany were dominated by Fagaceae (oak and chestnut) and Betulaceae (beech), (Lenz and Riegel, 2001). Late Eocene to Oligocene Baltic amber swamps included common *Fagus* (chestnut) and *Quercus* (oak) (Poinar, 1992; Stewart and Rothwell, 1993). Following late Miocene cooling, taxonomically diverse broadleaved forests (including *Acer,* Fagaceae, and Juglandaceae) spread into northern-latitude wetlands (Askin and Spicer, 1995; Agar and White, 1997).

Wetland species in these angiosperm groups developed a wide array of adaptations to wet substrates. Tupelo and black gum have pneumatophores and buttressed bases, similar to bald cypress (Mitsch and Gosselink, 2000), an example of parallel evolution in different lineages of plants under the same physical conditions. Willows (*Salix* spp.) and cottonwoods (*Populus deltoides*) have adventitious roots, which permit recovery from periodic flooding. Some modern willow and cottonwood species have seeds that can germinate while submerged (Kozlowski, 1997). Willows also have large lenticels—structures that allow for gas exchange, an advantage in low-oxygen wetland habitats (Mitsch and Gosselink, 2000). These adaptations, and others, resulted in a wide variety of freshwater swamp types (e.g., red maple swamps, bottomland hardwood swamps) that were distinct in terms of dominant tree taxa, climate, frequency of flooding, and flood duration among other factors.

Mangrove and Mangal Wetlands

Mangroves increased in diversity throughout the Cenozoic, with *Rhizophora* (red mangrove) the most common extant genus replacing *Nypa* sp. during the early Tertiary (Plaziat, 1995; Aaron et al., 1999; Graham, 1999). Some of the most recognizable modern genera evolved prior to the Miocene (Fig. 20). All modern mangrove genera, except one, evolved before the close of the eastern Tethys Ocean in the late Miocene, with continental drift and changing climate altering species distributions (Plaziat, 1995; Aaron et al., 1999).

Modern mangroves exhibit a wide variety of adaptations to salinity stress, some being modifications of wetland root types that had previously evolved in other wetland flora in response to inundation and oxygen stress. *Rhizophora* has prop and drop roots, *Bruguiera* has knee roots, and *Avicennia* has pneumatophores (Fig. 20). In mangroves, cell membranes in these root systems exclude salt ions. Some modern mangrove species exhibit new adaptations to salt tolerance among wetland flora, such as salt-secreting glands and the ability to concentrate and then shed salt in bark and old leaves (Kozlowski, 1997; Hogarth, 1999). Viviparity is another important adaptation to salt tolerance in some mangroves (Koslowski, 1997; Hogarth, 1999; Mitsch and Gosselink, 2000). *Rhizophora* propagules germinate on the plant and then fall into the water, where they float until reaching water of appropriate salinity; the propagules then tilt on end and take root. At what point each of these adaptations evolved is uncertain, although fossil evidence of viviparity is known from the early to mid Eocene London Clay (Collinson, et al., 1993; Collinson, 2000).

The biogeographic distribution of mangroves throughout the Cenozoic parallels global climate changes up until the Eocene thermal maximum, with a range contraction of this wetland to its present pantropical distribution thereafter. The timing of mangal expansion toward the polar regions may have differed in the hemispheres; *Nypa* mangrove communities became established in New Zealand (Crouch and Visscher, 2003) and Tasmania (Pole, 1996) prior to the thermal maximum, whereas mangroves related to the genus *Bruguiera* and *Ceriops* are known first from the Eocene London Clay in southern England closer in time to the event (Chandler, 1951; Collinson, 1983). Most localities are identified on the basis of fruits, seeds, and pollen of mangrove taxa. In fact, the preservation and recognition of in situ coastal mangrove paleoswamps is undoubtedly biased because they occupy very narrow coastal habitats, tend to be non-peat producing, and are subject to erosion during sea-level rise (e.g., Liu and Gastaldo, 1992). Marsh-to-swamp transitions may be the result of less than a 30 cm change in elevation (e.g., Gastaldo et al., 1987) and mangrove-to-swamp transitions are similar (e.g., Gastaldo and Huc, 1992).

The onset and zenith of the thermal maximum allowed for the expansion of mangals to higher latitudes but also may have perturbed the tropical wetlands closer to the equator. Rull (1999) documents a stepped and gradual change in the marsh and back-mangrove swamps of the Maracaibo Basin in Venezuela, where Paleocene taxa are interpreted to be of pantropical distribution, whereas Eocene assemblages are more restricted to the Neotropics. Thereafter, there is near-complete replacement of these Middle Eocene forms with typical Oligocene–Recent mangrove taxa, including *Rhizophora,* a trend reported globally (Muller, 1980; Rull, 1998).

Faunal Traps in Wetlands

Tertiary lignites and associated strata contain diverse flora and fauna. Some of the most famous Eocene vertebrates come from German lignites. The Geissel peat was a faunal trap with many fossils found in so-called sinkholes within the accumulation, as well as in lacustrine and fluvial facies. The most common vertebrates are crocodiles, tortoises, and mammals. At least 14 different orders of mammals are recorded as well as fish, amphibians, snakes, lizards, and birds (Franzen et al., 1993). Some component of the famous Messel deposits is also likely related to Eocene wetland inhabitants. Plant fossils in the Messel lake deposit include swamp cypress (Taxodiaceae), water lilies (Nymphaeaceae), sedges (Cyperaceae), club mosses, and ferns (Schaal and Ziegler, 1992), all common Eocene wetland taxa.

In Thailand, claystone interbeds in late Eocene lignites have yielded gastropods, pelecypods, turtles, a crocodile, and an early primate, *Siamopithecus* (Udomkan et al., 2003). A diverse fauna including primates is also known from lignites in Hungary. In fact, Kordos and Begun (2002) suggest that great apes in these wetlands may have migrated to Africa following Miocene climate changes. Some hominoid primates continued to occupy wetlands in the Oligocene of Africa. At the famous Fayum deposits of Egypt, *Aegyptopithecus* and *Propliopithecus* occur with a wide variety of mammals including anthracotheres, arsinöitheres, proboscideans, basilosaurs, and sirenians; reptiles including turtles, crocodiles, and the giant snake *Gigantophis;* and avifauna including storks and herons. The famous vertebrate fauna is associated with coastal mangrove and back-mangrove swamps (Bown et al., 1982), as well as with freshwater marshes and swamps interpreted as similar to modern Ugandan swamps (Olson and Rasmussen, 1986).

Some of the vertebrates associated with these sites have been interpreted as obligate wetland inhabitants similar to modern semiaquatic *Hippopotamus*, including the Eocene perissodactyls *Amynodon* and *Metamynodon* (Wall, 1998), the Eocene pantodont *Coryphodon* (Ashley and Liutkus, 2002), the Eocene–Oligocene

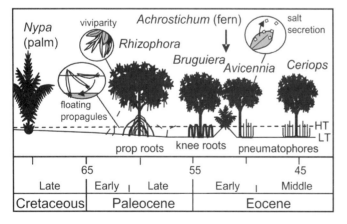

Figure 20. Earliest occurrence of extant mangrove taxon based on data compiled in Aaron et al. (1999). Various adaptations to saline wetland conditions are shown for modern species of genera shown.

ungulate *Moeritherium* (Carroll, 1988; Bown et al., 1982), and the Eocene–Pliocene artiodactyl ungulates of the Anthracotheriidae (Carroll, 1988; Berger, 1998; Kron and Manning, 1998). The latter are a link to true hippopotamuses, which evolved in the Miocene (Carroll, 1988). Likewise, several large proboscideans also may have been adapted to middle Tertiary wetland habitats. The Miocene elephants *Ambelodon* and *Platybelodon* both had broad shovel tusks, commonly interpreted as an adaptation for feeding on aquatic vegetation in marshes and submerged wetlands, although this assumption may be an oversimplification (Janis et al., 1998). All of these large, herbivorous mammals may have relied on wetlands for habitat and food and, in turn, had the potential to exert dramatic influence on the wetlands they inhabited.

Aquatic Mammal Evolution in Wetlands

During the Eocene, some terrestrial mammals evolved morphologic changes that allowed them to become permanent occupants of aquatic environments. This transition from terra firma to a fully aquatic environment occurred within (or at least through) wetlands. The Archaeocetes (ancient whales) had elongate mouths and probably had ecological roles similar to crocodiles in coastal, riverine, and lacustrine habitats (including wetlands) (Thewissen et al., 2001). The small whale *Kutchicetus minimus* was found in Eocene lignites from India and is interpreted to have inhabited backswamp environments (Bajpai and Thewissen, 2002). Likewise, sirenians (manatees, dugongs) are known from Miocene amber-bearing marls of Puerto Rico (Iturralde-Vinent and Hartstein, 1998) and the Oligocene Fayum deposits (Bown et al., 1982). Extant sirenians inhabit estuaries, streams, and coastal areas associated with aquatic and mangrove wetlands, where their primary diet is sea grass (Domning et al., 1982). Since the Eocene, they have played a unique role as large aquatic herbivores. Eocene and Miocene sirenians were mostly restricted to coastal riverine and estuarine aquatic wetlands (Domning, 1982). Evolution through the Tertiary appears to have been driven by the spread of sea grasses and the evolution of new aquatic macrophytes (Domning et al., 1982; Savage et al. 1994).

Birds in Wetlands

Although birds evolved in the Jurassic, transitional shorebirds are not recognized until the Late Cretaceous (e.g., Yang et al., 1994). The adaptive radiation of modern waterfowl lineages did not begin until after the K-T extinction, during the Tertiary radiation of birds (Feduccia, 1995, 1999). This radiation was concurrent with the spread of angiosperm-dominant wetlands and mangrove wetlands, all habitats that are used by birds for food, shelter, and breeding. The spread of aquatic wetlands was likely of particular importance to the diversification of waterfowl, since aquatic plants are a major part of waterfowl diets. Approximately one-third of extant North American bird species use wetlands as habitat and breeding grounds (Kroodsma, 1979; Mitsch and Gosselink, 2000; Stewart, 1996; Keddy, 2000).

The Presbyornithidae were a group of long-legged wading birds that may have originated in the latest Cretaceous and continued into the Tertiary. *Presbyornis* had the body of a flamingo and the head of a duck (Fig. 21; Feduccia, 1999). Mass death accumulations of *Presbyornis* associated with the Eocene Green River shales indicate that these early waterfowl lived in large colonies, similar to modern wetland-inhabiting flamingoes (Olson and Feduccia, 1980). True flamingoes evolved in the Eocene (Feduccia, 1996). A possible charadriiform shorebird has been identified from Eocene subtropical swamp sediments in China (Hou and Ericson, 2002). Long-legged wading birds, such as herons and storks (Ciconiidae), have a limited fossil record but can be traced back to the late Eocene or Miocene (Feduccia, 1996; Miller et al., 1998).

Ducks (Anatidae), the largest group of modern waterfowl and common inhabitants of wetlands around the world, are known from the Oligocene from several places around the world (Olson and Feduccia, 1980). Interestingly, seeds similar to those of modern duckweed (*Lemna* sp.), an aquatic plant favored by many extant duck species, also are known from the Oligocene to the recent (Mai, 1985). Modern duck genera became dominant

Figure 21. *Presbyornis,* a wading bird, that lived in Cretaceous–Tertiary marshes, looks for food in an Eocene lacustrine marsh. *Typha* dominates the lake margin. Marsh inhabitants in the Eocene Green River Formation included the alligator *Procaimanoidea*, the turtle *Baptemys*, the alligator gar *Lepisosteus*, and the famous *Knightia* fish. Sycamores and palms occupy fringing forests.

elements in freshwater marshes by the Pliocene (Carroll, 1988; Feduccia, 1996, 1999). The adaptation of these various bird groups to wetlands introduced a new avian component to wetland trophic systems. Recent studies have shown that increases in waterfowl will cause increases in benthic detritus, macrophytes, and fish in inland lakes (Mitsch and Gosselink, 2000). It is likely that the diversification of birds into wetland habitats in the Cretaceous and into the Tertiary caused similar changes through time.

Freshwater Grass, Reed, Rush, and Sedge Marshes

Modern freshwater marshes are dominated by a mosaic of diverse emergent herbaceous taxa, with some of the most common temperate members belonging to the grasses (Poaceae) such as *Phragmites* (reed grass), reeds (Typhaceae) such as *Typha* (cattails), rushes (Juncaceae) like *Juncus* sp., and sedges (Cyperaceae) such as *Scirpus* (bullrush), *Carex* sp. and *Cyperus* sp. The Cyperaceae alone includes more than 3500 species of grass-like herbs (Plunkett et al., 1995), many of which are common in wetlands. Some species of sedges, such as *Cyperus papyrus* (papyrus reeds), are highly productive wetland plants (Weller, 1994; Mitsch and Gosselink, 2000). Unfortunately, because these monocotyledons have no woody tissue, their preservation potential is limited to fruits and seeds (van der Burgh and Zetter, 1998).

Many extant wetland monocots are thought to have readapted to an emergent marsh habitat after the mid-Tertiary radiation of grasses, but some freshwater monocot wetland taxa may extend back to the Cretaceous. *Typha* (Fig. 21) and partially synonymous *Cyperacites* are reported from the Late Cretaceous (Tidwell, 1975; Muller, 1984; Mai, 1985; Bremer, 2000) but are not common floristic components until the Paleogene (MacGinitie, 1969; Machin, 1971; Muller, 1984; Collinson and Hooker, 1987; Collinson, 2002). Both have aerenchyma in their root tissues, commonly found in plants from mostly wet substrates. Seeds similar to those of the modern rush, *Juncus* sp., are reported from the upper Eocene/lower Oligocene (Collinson, 1983; Collinson et al., 1993) and from the Miocene of Europe (Mai, 1985). Most extant members of the family are freshwater species. Sedges (Cyperaceae) occupy diverse habitats today and are known from Eocene pollen and seed remains (Muller, 1984; Collinson and Hooker, 1987; Cross and Phillips, 1990; Bremer, 2000; Collinson, 2000). Wetland sedges, such as fossil *Phragmites* (Thomasson, 1986), a common constituent of European "reedswamps" and *Scirpus* (Van der Burgh and Zetter, 1998), are known from the latest Oligocene, and sedges were occupying peat mires in Australia by the late Oligocene or early Miocene (Blackburn and Sluiter, 1994), although much of the diversification of the Cyperaceae appears to be post-Miocene (Potts and Behrensmeyer, 1992).

An important aspect of the spread of grass, reed, and rush marshes during the drying climates at the end of the Tertiary is that these wetlands would have been the only sources of water and moist habitat in the vast grasslands that dominated many continental interiors, similar to vernal ponds and prairie potholes today. The diversification of the Anatidae (ducks) beginning in the Miocene and into the Pliocene is coincident with the spread of rushes, sedges, and grasses into freshwater marshes. This parallel expansion perhaps cemented the important wetland-habitat association that exists to this day.

Analyses of extant sedges in freshwater marshes indicate that they use a C_4 pathway for photosynthesis, in which CO_2 is fixed into a four-carbon molecule (Jones, 1988; Keeley, 1998; Ehleringer and Monson, 1993). Most plants use the C_3 pathway (CO_2 fixed into a three-carbon molecule). Molecular phylogenies and fossil evidence suggest that the C_4 pathway has arisen in different families at different times, but the initial appearance of this physiology appears to have been post-Cretaceous (Ehleringer and Monson, 1993; Kellogg, 2001) and in aquatic plants (Sage, 2001). The C_4 pathway has a physiological advantage when atmospheric conditions consist of low CO_2 pressures accompanied by warm, dry climates. Although this would not seem to be an advantage to wetland plants, plants using the C_4 pathway have increased nitrogen efficiency, which is a definite advantage in low-nutrient substrates such as those of oligotrophic wetlands (Jones, 1988; Ehleringer and Monson, 1993).

Salt-water Grass, Rush, and Sedge Marshes

Salt marshes are among the most productive ecosystems on Earth, and are common along tidally influenced coastlines except in the Tropics, where mangroves dominate most coastal wetlands. Although numerous Paleozoic and early Mesozoic wetland deposits have been interpreted as possible salt marshes, the inference is often based on the juxtaposition of overlying transgressive marine deposits, rather than on morphological features of the fossil flora or paleosols. That said, modern salt marshes are dominated by grasses, particularly cordgrass (*Spartina* sp.), rushes (*Juncus* sp.), and sedges (*Carex* sp.), such that the evolution of these wetland habitats postdates the evolution of angiosperms.

The oldest of the extant salt-marsh families is the Juncaceae (rushes), which appears at least by the middle Eocene, similar to the Cyperaceae (sedges) (Graham, 1999; Bremer, 2000). As mentioned previously, most extant members of both families are freshwater species. Thus, it is likely that the evolution of rushes and sedges with salt tolerance postdates the evolution of the families as a whole. Again, analyses of the photosynthetic pathways of saline marsh plants provide insight into the evolution of salt tolerance in emergent marsh monocots in lieu of fossil evidence. In many coastal and inland saline wetlands, C_4 plants replace C_3 plants as salinity increases (Ehleringer and Monson, 1993). The C_4 pathway has better water-use efficiency than the C_3 pathway (Jones, 1988; Keeley, 1998). Although water-use efficiency may not be important in freshwater marshes (except for vernal pools and prairie potholes), it is critical in saline environments (Ehleringer and Monson, 1993). The C_4 pathway appears to have evolved multiple times in monocots since the Cretaceous, but the earliest definite C_4 macrofossils are from the Miocene (Kellogg, 2001). Extant C_4-pathway saline grasses (Poaceae), such as salt grass (*Distichlis* sp.) and cordgrass (*Spartina* sp.), do not have an extensive fossil record and may postdate the Miocene. It is unlikely that there was significant competition from preexisting flora in the niche now occupied by halophytes such as *Spartina*. This is important to consider in terms of paleoecology, because it may indicate that many of the functions provided by extant saltwater marshes (both coastal marine and inland) were nonexistent (or at least diminished) earlier.

TERTIARY-QUATERNARY

Sphagnum-Mire Complexes

The development of cold climate in the Pliocene led to pine dominance of northern conifer forests and to development of lowland tundra, forest tundra, and permafrost (Agar and White, 1997). Two of the world's largest modern wetlands are the high-latitude mire complexes of the West Siberian and Hudson Bay lowlands. Northern-latitude mires are dominated by the peat moss, *Sphagnum* sp., and co-inhabited by a wide variety of plants, including conifers such as black spruce (*Picea mariana*) and tamarack (*Larix laricina*), woody angiosperms such as birch (*Betula*), and groundcover monocotyledons and dicotyledons such as heaths, sedges, and pitcher plants (Botch and Masing, 1983; Zoltai and Pollett, 1983).

Precursors of *Sphagnum*, the Protosphagnales, are reported from the Permian of Russia (Neiburg, 1958), and spores of *Sphagnum* are recorded from Jurassic coals of China (Miao et al., 1989), Cretaceous coals of Alaska (Hickey and Doyle, 1997), and Tertiary coals of North America and Europe (Steere, 1946; Cross and Phillips, 1990). The point at which *Sphagnum* began to dominate oligotrophic mires is uncertain, although it appears to postdate the late Neogene. Obviously, the extent of current *Sphagnum*-dominated wetlands is related to the last Pleistocene glacial retreat. In fact, there is a repetitive expansion and contraction of the northern-latitude coniferous forests (and associated wetlands) with each ice advance and retreat (Agar and White, 1997). In some cases, these wetlands acted as refugia for both flora and fauna during interglacial periods (Speight and Blackith, 1983). Access to continuous habitat across latitudinal gradients was a strong selective criterion in sorting which elements of the wetland flora and fauna survived late Cenozoic climate changes (Potts and Behrensmeyer, 1998).

Sphagnum almost certainly was preadapted to oligotrophic habitats, with the development of extensive aerenchymatous tissues allowing it to grow in low-oxygen environments. This anatomical feature is related to the plant's ability to leak oxygen through its roots to create a local aerobic environment. Its compact growth habit, overlapping leaves, and rolled branch leaves enhance water retention (Mitsch and Gosselink, 2000). Modern *Sphagnum* has the ability to acidify its surroundings, which may aid in retarding bacterial decomposition, allowing peat to accumulate even in an environment of low primary productivity (Mitsch and Gosselink, 2000). Additionally, acidity helps alter mineral substrates on which the peat mosses accumulate, creating and maintaining a clay-rich, impermeable layer that further promotes waterlogging and peat accumulation.

Giant Wetland Mammals

High rates of biomass production and recycling in wetlands support trophic tiers of abundant animals, albeit each of relatively low species diversity. The abundance of food resources in this setting permitted extraordinarily large animals to inhabit these environments. In fact, the largest rodent of all time was a wetland inhabitant. *Phoberomys pattersoni* was more than ten times larger then the largest living rodent, the capybara, and inhabited late Miocene freshwater paludal marshes of Venezuela. *Phoberomys* was semiaquatic or foraged in water on wetland grasses, as do extant capybaras (Sánchez-Villagra et al., 2003).

Beavers belong to the family Castoridae, which appears to have originated in North America during the Oligocene (Kurten and Anderson, 1980; Carroll, 1988). Beavers are not only wetland inhabitants, but also creators of wetland habitat, so-called natural wetland engineers (Jones et al., 1994). Wetlands and ponds created by beavers (Fig. 22) are important habitats for amphibians, mammals, and birds (Keddy, 2000). The oldest beaver, *Dipoides*, is known from the late Neogene (Pliocene) of Eurasia and North America. Fossil beaver dams in the Plio-

Figure 22. Mastodons along the shore of a Pleistocene beaver pond in a boreal fen. The pond is also home to turtles, ducks, and other birds. Aquatic plants include water lilies (lower left) and *Potamogeton* sp. (pond weed) (lower right). Sedges (*Carex* sp.) and mosses are common. The fen is bordered by a black spruce (*Picea mariana*) swamp, with tamarack (*Larix laricina*), balsam fir (*Abies balsamea*), and a few deciduous trees and shrubs such as oak (*Quercus* sp.) and willow (*Salix* sp.).

cene indicate that dam construction was an early part of this animal's behavior (Tedford and Harington, 2003). During the Pleistocene, the giant beaver *(Castoroides ohioensis)* reached lengths of 2.5 m in North America. Remains of ice age giant beavers have been preserved in numerous Eurasian Pleistocene peats and pond-paludal wetland deposits (Kurten and Anderson, 1980; Hansen, 1996).

Many skeletons of ice age mammals, including mammoths, mastodons, ground sloths, and wooly rhinoceroses, have been excavated from fluvial, paludal, and peat permafrost in the northern high latitudes. Some of these finds have included soft-part preservation of hair, skin, and internal organs (e.g., Lister and Bahn, 1994). Numerous mastodons (Fig. 22) also have been found trapped in peat and wetland-fringing pond deposits of eastern North America (Eiseley, 1945; Miller and Nester, this volume). Indeed, the first mastodons to be described were found at Big Bone Lick, Kentucky, along with mammoths, bison, and other mammalian taxa. The Big Bone fauna is interpreted as having accumulated in a "bog" fed by a salt-and-mineral spring (e.g., Jillson, 1968), although a lacustrine marsh (non-peat producing) may be a more appropriate term. These wetland bones were used by Cuvier in the late 1700s and early 1800s to argue for the idea of extinction (Rudwick, 1997; Semonin, 2000).

QUATERNARY

Wetland Archeology

Wetlands have had a profound effect on human civilization and, of course, humans have dramatically influenced wetlands; unfortunately, in modern times the influence mostly has been detrimental. Wetlands were historically used as sources for construction materials, fuels, fishing materials (traps, poisons, dyes), iron, textiles, dyes for cloth, tannin for leather preservation, compost, sugar, vinegar, honey, fermented drinks, medicines, contraceptives, aphrodisiacs, waxes, incense, glues, and as a food resource, through fishing, hunting, and aquaculture (Bacon, 1997).

Shaped stone tools found with the bones of straight-tusked elephants and other mammals in wetland clays of Torralba and Ambrona, Spain, may represent early hominid butchery or scavenging sites on the margins of wetlands (Klein, 1987: Nicholas, 1998). By the late Pleistocene, a hunting-and-gathering lifestyle was firmly established among humans, and several archaeological sites in Europe indicate that wetlands were an important resource base (Nicholas, 1998). Among the European sites are the oldest known hunting spears, found with butchered remains of horses, from a coal mine in Schöningen, Germany (Dennell, 1997; Thieme, 1997). Preservation of the 400,000-year-old spears was aided by tannic acids from the overlying peat.

At Monte Verde, Chile, the oldest human occupation site in the Americas is situated beneath a water-saturated reed-and-shrub bog that covered the site with a layer of peat, isolating the material from oxygen and deterioration. Mastodon bones and meat, wooden lances, planks and stakes, knotted reeds, and animal hides have been recovered from the site, as well as blood hemoglobin from a tool (Tuross and Dillehay, 1995). Not only was the preservation of this site reliant upon wetland chemistry, but it also appears that bogs and freshwater and salt-water marshes provided construction material and food for the Monte Verde culture (Dillehay, 1989). Likewise, in North America, lacustrine aquatic wetlands preserved mastodon intestines filled with sand and gravel, indicating that prehistoric humans filled these organs as "clastic anchors" to keep the bodies on the bottoms of peaty, anoxic ponds for winter meat storage (Fisher, 1995).

The development of modern civilizations around the Fertile Crescent of the Tigris and Euphrates Rivers, as well as the Nile Valley, resulted from their location along rivers with fertile flood plains, marshes, and riparian wetlands. Aside from food and water available within these ecosystems, wetland plants provided the Egyptians with papyrus *(Cyperus papyrus)*. The word "paper" is derived from papyrus, and Egyptians began to use this marsh plant to make paper by 2000 B.C. Many of the classic writings of ancient Egypt, Greece, and Rome were inscribed on the smashed stem piths of these plants. At the same time that paper was being made from wetland plants in Egypt, man-made wetlands were being created to grow rice in lowland deltas and flood plains in Southeast Asia.

One of the most interesting wetland-associated anthropological finds is the "bog bodies" of northern Europe (Glob, 1965; Menon, 1997) and Florida (Doran et al., 1986). These remains date back to 8000 B.C. and are famous because they are mummified with excellent soft-tissue preservation. Additionally, they provide information about social behavior as indicated by a wide range of burial mechanisms, including ritual burial, accidental death, and murder (executions, sacrifices). Possibly the most unusual example of soft-tissue preservation is the Middle Archaic (8000-year-old) brain tissues and DNA recovered from human remains in a pond peat from Windover, Florida (Doran et al., 1986). At Windover, Native Americans buried their dead underwater on the bottom of a pond. Over time, lacustrine peat covered the bodies, promoting exceptional preservation.

Human Impacts on Wetlands

Our civilization's expansion has come at the expense of wetland habitats. More than 70% of the world's population inhabits coastal areas, and increased population growth in the past several hundred years has resulted in a loss of as much as 50% percent of the world's wetlands (Keddy, 2000). This loss has resulted in the deterioration of many wetland functions, such as contributions to wildlife habitat, biodiversity, natural water quality improvement, natural flood mitigation through water storage, as well as shore and bank stabilization (Mitsch and Gosselink, 2000; Keddy, 2000). Bacon (1997) noted that because of their variability, geographic distribution, and biological richness, wetlands contain a significant amount of the world's biodiversity, and thereby a large pool of genetic resources. Loss of wetland habitats may endanger the future genetic resources of plants and animals. The fossil record is our best source of baseline data from which to assess the long-term impacts of environmental perturbations on global ecology and biodiversity. Through better understanding of the fossil record and the evolution of wetland types, wetland functions, and wetland interactions with other ecosystems, we can better understand and possibly mitigate detrimental influences on wetlands and associated ecosystems.

SUMMARY

Modern wetlands are a diverse array of habitats with equally diverse floral and faunal associations, controlled by a host of interacting factors. Most of the functions recognized in extant wetlands originated in the Paleozoic. At some times in the past, certain functions have played a far more significant role than they do today, for example, in natural carbon sequestration or the natural alteration of the world's biogeochemical and hydrological cycles.

The oldest wetlands were similar in stature to moss-lichen communities but were non-peat-accumulating. Floral adaptations and evolution led to the first marshes, swamps, fens, and, eventually, forested mires in the Devonian. The diversification of wetland habitats profoundly influenced the terrestrialization of plants, invertebrates, and vertebrates, as well as sediment stabilization and global biogeochemical cycles. By the Carboniferous, wetlands dominated by trees and other plants were widespread and included the largest tropical mires in earth history. The spread of mires from the late Devonian to the Carboniferous increased the importance of wetlands as global carbon sinks. Within these environments, the flora and fauna greatly diversified through time. Most of the Paleozoic terrestrial fossil record comes from these Carboniferous environments. During the Permian, floral adaptations to cooler climates allowed for the development of the first high-latitude mires, latitudinal zonation of wetland floras, and a switch from mires dominated by lower vascular plants to those dominated by gymnosperms. Large, semiaquatic herbivores and carnivores also made their first appearances in Permo-Carboniferous wetlands. Changes in climate and tectonics at the close of the Paleozoic resulted in dramatic upheavals within wetland habitats, leading to major disruptions of many wetland ecosystems. Decrease in wetland area at the end of the Permian likely was accompanied by a significant decrease in wetland functions.

Recovery in the Mesozoic was slow, with the reconstitution of wetlands ultimately by a different "framework" vegetation than that of similar habitats in the late Paleozoic. During the Mesozoic, continental movements resulted in the physical separation of Northern and Southern Hemisphere landmasses, with resultant evolution of distinct wetland floras in these areas, particularly visible among newly evolved conifer groups; some of these differences persist to this day. In the latter part of the Mesozoic, angiosperm evolution led to dramatic floristic changes in almost all terrestrial environments including wetland habitats. Additionally, angiosperm diversification resulted in novel morphologies that permitted the exploitation of habitats on the margins of existing wetlands. Angiosperm expansion allowed for the development of extensive fresh and saline floating and submerged communities, as well as tidal-estuarine salt marsh and mangrove-forest wetlands fringing today's coastal zones. These wetlands are among the most productive ecosystems on the planet and are tied intricately to food webs in surrounding communities. The evolution of extant wetland groups, such as frogs, salamanders, turtles, and crocodiles occurred during the mid-Mesozoic; genera from these groups would be among the few terrestrial vertebrate survivors of the K-T extinction, illustrating the value of wetlands as refugia. Mid- to late Mesozoic wetlands were also important faunal traps for early mammals and dinosaurs, including some of the largest herbivores and carnivores in earth history. Amber from Mesozoic and Tertiary wetlands provides unique insight into the radiation of insects that accompanied the radiation of angiosperms.

The Cenozoic radiation of angiosperms allowed for abundant floral partitioning and the development of a wide array of specialized subcommunities within wetland ecosystems. This radiation was coeval with the radiations of mammals and birds, both of which developed specialized niches within wetland ecosystems. The development of modern grass- and herb-dominant marshes accompanied mid-Cenozoic cooling and likely was a dramatic influence on extant waterfowl-marsh associations. By the late Cenozoic, most modern biomes had formed including extensive high-latitude *Sphagnum*-dominated mire systems, the most extensive wetlands in the world today. The late Cenozoic has also witnessed the expansion of our own species along the margins of wetlands, with early civilization utilizing a wide array of wetland flora and fauna for food and materials. The expansion of civilization resulted in the infilling and draining of global wetlands. Only recently have the repercussions of wetland loss been realized, with increasing attempts to restore and protect these vital parts of our global ecosystem. As public awareness of the importance of wetlands continues to grow, we will need to understand better how these systems respond to perturbations, how they recover from major environmental disruptions, and how wetland biotas interact with those of surrounding environments. The fossil record is our best source of information on these concerns, and it will become increasingly important as the details of ancient wetlands and the vagaries of their dynamics are investigated and clarified.

ACKNOWLEDGMENTS

The authors thank the reviewers, S.L. Wing and G.W. Rothwell for their helpful comments and suggestions. Additionally, C.F. Eble, A.C. Scott, S. Ash, and T.A. Moore are acknowledged for sharing their expertise during the preparation of this manuscript. Thanks also to C. Rullo, Kentucky Geological Survey, who helped with final figure preparation.

REFERENCES CITED

Aaron, M.E., Farnsworth, E.J., and Merkt, R.E., 1999, Origins of mangrove ecosystems and the mangrove biodiversity anomaly: Global Ecology and Biogeography, v. 8, p. 95–115.

Agar, T., and White, J., 1997, The history of Late Tertiary floras and vegetation change in Beringia based on fossil records of northwestern Canada, Alaska, and northeast Asia, in Elias, S, and Brigham-Grette, J., co-conveners, Beringian Paleoenvironments Workshop, Abstracts and Program: U.S. National Science Foundation, p. 5–6.

Algeo, T.J., Berner, R.A., Maynard, J.B., and Scheckler, S.E., 1995, Late Devonian oceanic anoxic events and biotic crises: "Rooted" in the evolution of vascular land plants?: GSA Today, v. 5, no. 3, p. 45, 64–66.

Algeo, T.J., and Scheckler, S.E., 1998, Terrestrial-marine teleconnections in the Devonian: Links between the evolution of land plants, weathering processes, and marine anoxic events: Royal Society of London Philosophical Transactions, ser. B, v. 353, p. 113–130, doi: 10.1098/rstb.1998.0195.

Algeo, T.J., Scheckler, S.E., and Maynard, J.B., 2001, Effects of early vascular land plants on weathering processes and global chemical fluxes during the Middle and Late Devonian, in Gensel, P., and Edwards, D., eds., Plants invade the land: Evolutionary and environmental perspectives: New York, Columbia University Press, p. 213–236.

Almond, J.E., 1985, The Silurian-Devonian fossil record of the Myriapoda: Royal Society of London Philosophical Transactions, ser. B, v. 309, p. 227–237.

Anderson, L.I., and Trewin, N.H., 2003, An early Devonian arthropod fauna from the Windyfield chert, Aberdeenshire, Scotland: Palaeontology, v. 46, p. 467–510, doi: 10.1111/1475-4983.00308.

Archangelsky, S., 1986, Late Paleozoic floras of the Southern Hemisphere: Distribution, composition and paleoecology, in Broadhead, T.W., ed., Land plants—Notes for a short course: University of Tennessee, Department of Geological Sciences, Studies in Geology, v. 15, p. 128–142.

Ash, S., 2003, The Wolverine Petrified Forest: Utah Geological Survey, Survey Notes, v. 35, no. 3, p. 3–6.

Ashley, G.M., and Liutkus, C.M., 2002, Tracks, trails, and trampling by large vertebrates in a rift valley paleo-wetland, Lowermost Bed 11, Olduvai Gorge, Tanzania: Ichnos, v. 9, p. 23–32, doi: 10.1080/10420940216407.

Askin, R.A., 1988, The palynologic record across the Cretaceous/Tertiary boundary in Seymour Island, Antarctica, in Feldman, R.M., and Woodburn, M.O., eds., Geology and paleontology of Seymour Island, Antarctic Peninsula: Geological Society of America Memoir 169, p. 155–162.

Askin, R.A., and Spicer, R.A., 1995, The Late Cretaceous and Cenozoic history of vegetation and climate at northern and southern high latitudes: A comparison, in Board on Earth Sciences and Resources, eds., Studies in geophysics–Effects of past global change on life: Washington, D.C., National Academies Press, p. 156–173.

Axsmith, B.J., Krings, M., and Waselkov, K., 2004, Conifer pollen cones from the Cretaceous of Arkansas: Implications for diversity and reproduction in the Cheirolepidiaceae: Journal of Paleontology, v. 78, p. 402–409.

Bacon, P.R., 1997, Wetlands and biodiversity, in Hails, A.J., ed., Wetlands, biodiversity and the Ramsar Convention—The role of the convention on wetlands in the conservation and wise use of biodiversity: Cambridge, UK, IUCN Publications, Ramsar Convention Library, 196 p.

Baird, G.C., Sroka, S.D., Shabica, C.W., and Kuecher, G.J., 1986, Taphonomy of Middle Pennsylvanian Mazon Creek area fossil localities, Northeast Illinois: Significance of exceptional fossil preservation in syngenetic concretions: Palaios, v. 1, p. 271–285.

Bajpai, S., and Thewissen, J.G.M., 2002, Vertebrate fauna from Panandhro lignite field (Lower Eocene), District Kachchh, western India: Current Science, v. 82, p. 507–509.

Baker, R., and DiMichele, W.A., 1997, Resource allocation in Late Pennsylvanian coal-swamp plants: Palaios, v. 12, p. 127–132.

Balsey, J.K., and Parker, L.R., 1983, Cretaceous wave-dominated delta, barrier island, and submarine fan depositional systems; Book Cliffs, east central Utah: A field guide: Tulsa, Oklahoma, American Association of Petroleum Geologists, 162 p.

Barlow, B.A., and Hyland, B.P.M., 1988, The origins of the flora of Australia's wet tropics: Proceedings of the Ecological Society of Australia, v. 15, p. 1–17.

Bartholomew, A.J., and Brett, C.E., 2003, The Middle Devonian (Givetian) Gilboa forest: Sequence stratigraphic determination of the world's oldest fossil forest deposit, Schoharie Co., New York State: Geological Society of America Abstracts with Programs, v. 35, no. 3, p. 76.

Basinger, J.F., Greenwood, D.G., and Sweda, T., 1994, Early Tertiary vegetation of Arctic Canada and its relevance to paleoclimatic interpretation: NATO ASI Series, v. 127, p. 175–198.

Bateman, R.M., Crane, P.R., DiMichele, W.A., Kenrick, P.R., Rowe, N.P., Speck, T., and Stein, W.E., 1998, Early evolution of land plants: Phylogeny, physiology, and ecology of the primary terrestrial radiation: Annual Review of Ecology and Systematics, v. 29, p. 263–292, doi: 10.1146/annurev.ecolsys.29.1.263.

Beck, C.B., 1962, Reconstruction of *Archaeopteris* and further consideration of its phylogenetic position: American Journal of Botany, v. 49, p. 373–382.

Beck, C.B., 1964, Predominance of *Archaeopteris* in Upper Devonian flora of western Catskills and adjacent Pennsylvania: Botanical Gazette, v. 125, p. 126–128, doi: 10.1086/336257.

Beerbower, J.R., 1985, Early development of continental ecosystems, in Tiffney, B.H., ed., Geologic factors and the evolution of plants: New Haven, Connecticut, Yale University Press, p. 47–91.

Beerbower, J.R., Boy, J.A., DiMichele, W.A., Gastaldo, R.A., Hook, R., Hotton, N., III, Phillips, T.L., Scheckler, S.E., and Shear, W.A., 1992, Paleozoic terrestrial ecosystems, in Behrensmeyer, A.K., Damuth, J.D., DiMichele, W.A., Potts, R., Sues, H.-D., and Wing, S.L., eds., Terrestrial ecosystems through time: Chicago, University of Chicago Press, p. 205–325.

Berger, J.-P., 1998, 'Rochette' (Upper Oligocene, Swiss Molasse): A strange example of a fossil assemblage: Reviews of Palaeobotany and Palynology, v. 101, p. 95–110.

Berner, R.A., 1993, Paleozoic atmospheric CO_2: Importance of solar radiation and plant evolution: Science, v. 261, p. 68–70.

Berner, R.A., 1997, The rise of plants and their effect on weathering and atmospheric CO_2: Science, v. 276, p. 544–546, doi: 10.1126/science.276.5312.544.

Berry, E.W., 1937, Tertiary floras of eastern North America: New York, The Botanical Review, v. 3, p. 31–46.

Blackburn, D.T., and Sluiter, I.R.K., 1994, The Oligocene-Miocene coal floras of southwestern Australia, in Hill, R.S., ed., History of the Australian vegetation: Cretaceous to Recent: Cambridge, UK, Cambridge University Press, p. 328–367.

Blakey, R.C., and Gubitosa, R., 1983, Late Triassic paleogeography and depositional history of the Chinle Formation, southern Utah and northern Arizona, in Reynolds, M.W., and Dolly, E.D., eds., Mesozoic paleogeography of the west-central United States: Denver, Society of Economic Paleontologists and Mineralogists, Rocky Mountain Section, p. 57–76.

Bonde, S.D., 2002, A permineralized species of mangrove fern *Acrostichum* L. from Deccan Intertrappean Beds of India: Review of Palaeobotany and Palynology, v. 120, p. 285–299, doi: 10.1016/S0034-6667(02)00081-7.

Botch, M.S., and Masing, V.V., 1983, Mire ecosystems in the USSR, in Gore, A.J.P., ed., Mires: Swamp, bog, fen and moor: New York, Elsevier, Ecosystems of the World, v. 4A, p. 95–152.

Bown, T.M., Kraus, M.J., Wing, S.L., Fleagle, J.G., Tiffney, B.H., Simons, E.L., and Vondra, C.F., 1982, The Fayum primate forest revisited: Journal of Human Evolution, v. 11, p. 503–560.

Brasier, M.D., 1975, An outline history of seagrass communities: Palaeontology, v. 18, p. 681–702.

Bremer, K., 2000, Early Cretaceous lineages of monocot flowering plants: Proceedings of the National Academy of Sciences of the United States of America, v. 97, p. 4707–4711, doi: 10.1073/pnas.080421597.

Brinson, M.M., 1990, Riverine forests, in Lugo, A.E., Brinson, M., and Brown, S., eds., Forested wetlands: New York, Elsevier, Ecosystems of the World, v. 15, p. 87–141.

Brown, R., Scott, A.C., and Jones, T.P., 1994, Taphonomy of fossil plants from the Viséan of East Kirkton, West Lothian, Scotland: Transactions of the Royal Society of Edinburgh: Earth Sciences, v. 84, p. 267–274.

Burnham, R.J., and Graham, A., 1999, The history of neotropical vegetation: New developments and status: Annals of the Missouri Botanical Garden, v. 86, p. 546–589.

Calder, J.H., Gibling, M.R., Scott, A.C., and Skilliter, D.M., 1997, The Carboniferous Joggins section reconsidered: Recent paleoecological and sedimentological research: Atlantic Geology, v. 33, p. 54–55.

Carpenter, F.M., 1960, Studies of North American Carboniferous insects: 1, The Protodonata: Psyche, v. 67, p. 98–110.

Carroll, R.L., 1964, The earliest reptiles: Zoological Journal of the Linnean Society, v. 45, p. 61–83.

Carroll, R.L., 1988, Vertebrate paleontology and evolution: New York, W.H. Freeman and Co., 698 p.

Cecil, C.B., Stanton, R.W., Neuzil, S.G., Dulong, F.T., Ruppert, C.F., and Pierce, B.S., 1985, Paleoclimate controls on late Paleozoic sedimentation and peat formation in the Central Appalachian basin (U.S.A.): International Journal of Coal Geology, v. 5, p. 195–230, doi: 10.1016/0166-5162(85)90014-X.

Chandler, M.E.J., 1951, Note on the occurrence of mangroves in the London Clay: Proceedings of the Geologists Association, v. 62, p. 271–272.

Chatterjee, S., 1986, The Late Triassic Dockum vertebrates: Their stratigraphic and paleobiogeographic significance, in Padian, K., ed., The beginning of the Age of Dinosaurs: Faunal change across the Triassic-Jurassic boundary: Cambridge, UK, Cambridge University Press, p. 139–150.

Christophel, D.C., 1989, Evolution of the Australian flora from the Tertiary: Plant Systematics and Evolution, v. 162, p. 63–78, doi: 10.1007/BF00936910.

Clack, J.A., 2002, Gaining ground—The origin of tetrapods: Bloomington, Indiana University Press, 369 p.

Clack, J.A., 1997, Devonian tetrapod trackways and trackmakers: A review of the fossils and footprints: Palaeogeography, Palaeoclimatology, Palaeoecology, v. 130, p. 227–250, doi: 10.1016/S0031-0182(96)00142-3.

Clark, J.E., 1979, Fresh water wetlands: Habitats for aquatic invertebrates, amphibians, reptiles, and fish, in Greeson, P.E., Clark, J.R., and Clark, J.E., eds., Wetland functions and values: The state of our understanding: Minneapolis, American Water Resources Association, p. 330–343.

Clack, J.A., and Coates, M.I., 1995, Acanthostega—a primitive aquatic tetrapod? in Arsenault, M., Lelièvre, H., and Janvier, P., eds., Proceedings of the 7th International Symposium on Lower Vertebrates, Bulletin du Muséum National d'Histoire Naturelle, Paris, p. 359–373.

Clymo, R.S., 1987, Rainwater-fed peats as a precursor of coal, in Scott, A.C., ed., Coal and coal-bearing strata—Recent advances: Geological Society [London] Special Publication 32, p. 7–23.

Coates, M.I., and Clack, J.A., 1995, Romer's gap: Tetrapod origins and terrestriality: Bulletin du Museum National d'Histoire Naturelle, section C, Sciences de la Terre: Paleontologie, Geologie, Mineralogie, v. 17, p. 373–388.

Collinson, M.E., 1983, Fossil plants of the London Clay: Palaeontological Association Field Guides to Fossils, v. 1, 121 p.

Collinson, M.E., 2000, Cenozoic evolution of modern plant communities and vegetation, in Culver, S.J., and Rawson, P.F., eds., Biotic response to global change—The last 145 million years: New York, Cambridge University Press, p. 223–264.

Collinson, M.E., 2002, The ecology of Cainozoic ferns: Review of Palaeobotany and Palynology, v. 119, p. 51–68, doi: 10.1016/S0034-6667(01)00129-4.

Collinson, M.E., and Hooker, J.J., 1987, Vegetational and mammalian faunal changes in the Early Tertiary of southern England, in Friis, E.M., Chaloner, W.G., and Crane, P.R., eds., The origins of angiosperms and their biological consequences: Cambridge, UK, Cambridge University Press, p. 259–304.

Collinson, M.E., and Scott, A.C., 1987, Implications of vegetational change through the geologic record on models for coal-forming environments, in Scott, A.C., ed., Coal and coal-bearing strata—Recent advances: Geological Society [London] Special Publication 32, p. 67–85.

Collinson, M.E., Boulter, M.C., and Holmes, P.L., 1993, Magnoliophyta ("Angiospermae") in Benton, M.J., ed., The fossil record 2: London, Chapman and Hall, p. 809–841.

Cowardin, L.M., Carter, V., Golet, F.C., and LaRoe, E.T., 1979, Classification of wetlands and deepwater habitats of the United States: Washington, D.C., U.S. Fish and Wildlife Service Publication FWS/OBS79/31, 103 p.

Craighead, F.C., 1968, The role of the alligator in shaping plant communities and maintaining wildlife in the southern Everglades: The Florida Naturalist, v. 41, p. 69–74.

Crane, P.R., 1993, Time for the angiosperms: Nature, v. 366, p. 631–632, doi: 10.1038/366631a0.

Crane, P.R., Friis, E.M., and Pederson, K.R., 1995, The origin and early diversification of angiosperms: Nature, v. 374, p. 27–33, doi: 10.1038/374027a0.

Creber, G.T., and Ash, S.R., 2004, The Late Triassic Schilderia adamanica and Woodworthia arizonica trees of the Petrified Forest National Park, Arizona, USA: Palaeontology, v. 47, p. 21–39, doi: 10.1111/j.0031-0239.2004.00345.x.

Cridland, A.A., 1964, Amyelon in American coal balls: Palaeontology, v. 7, p. 189–209.

Cross, A.T., and Phillips, T.L., 1990, Coal-forming plants through time in North America: International Journal of Coal Geology, v. 16, p. 1–46, doi: 10.1016/0166-5162(90)90012-N.

Crouch, E.M., and Visscher, H., 2003, Terrestrial vegetation record across the initial Eocene thermal maximum at the Tawanui marine section, New Zealand, in Wing, S.L., Gingerich, P.D., Schmitz, B., and Thomas, E., eds., Causes and consequences of globally warm climates in the early Paleogene: Geological Society of America Special Paper 369, p. 351–363.

Cuneo, R., 1996, Permian phytogeography in Gondwana: Paleogeography, Paleoclimatology, and Paleoecology, v. 125, p. 75–104.

Darrah, W.C., 1969, Upper Pennsylvanian Floras of North America: Gettysburg, Pennsylvania, privately published, 220 p, 80 pl.

Dawson, J.W., 1854, On the coal Measures of the South Joggins, Nova Scotia: Quarterly Journal of the Geological Society of Canada, v. 10, p. 1–41.

DeGreef, J.D., 1997, Fossil Aldrovanda: Carnivorous Plant Newsletter, v. 26, p. 93–97.

Demko, T.M., Dubiel, R.F., and Parrish, J.T., 1998, Plant taphonomy in incised valleys: Implications for interpreting paleoclimate from fossil plants: Geology, v. 26, p. 1119–1122, doi: 10.1130/0091-7613(1998)026<1119: PTIIVI>2.3.CO;2.

Dennell, R., 1997, The world's oldest spears: Nature, v. 385, p. 767, doi: 10.1038/385767a0.

Diessel, C.F.K., 1982, An appraisal of coal facies based on maceral characteristics: Australian Coal Geology, v. 4, p. 474–483.

Dilcher, D., 2000, Toward a new synthesis: Major evolutionary trends in the angiosperm fossil record: Proceedings of the National Academy of Sciences of the United States of America, v. 97, p. 7030–7036, doi: 10.1073/pnas.97.13.7030.

Dilcher, D.L., Krassilov, V., and Douglas, J., 1996, Angiosperm evolution: Fruits with affinities to Ceratophyllales from the Lower Cretaceous: Fifth Conference of the International Organization of Palaeobotany, Santa Barbara, California, June 30–July 5, Abstracts, p. 23.

Dillehay, T.D., 1989, Monte Verde—A Late Pleistocene settlement in Chile, Volume 1: Palaeoenvironment and site context: Washington, D.C., Smithsonian Institution Press, 306 p.

DiMichele, W.A., and Hook, R.W., 1992, Paleozoic terrestrial ecosystems, in Behrensmeyer, A.K., Damuth, J.D., DiMichele, W.A., Potts, R., Sues, H.-D., and Wing, S.L., eds., Terrestrial ecosystems through time: Chicago, University of Chicago Press, p. 206–325.

DiMichele, W.A., and Phillips, T.L., 1985, Arborescent lycopod reproduction and paleoecology of late Middle Pennsylvanian age (Herrin coal, Illinois, USA): Review of Palaeobotany and Palynology, v. 44, p. 1–26, doi: 10.1016/0034-6667(85)90026-0.

DiMichele, W.A., and Phillips, T.L., 1994, Paleobotanical and paleoecological constraints on models of peat formation in the Late Carboniferous of Euramerica: Palaeogeography, Palaeoclimatology, Palaeoecology, v. 106, p. 39–90, doi: 10.1016/0031-0182(94)90004-3.

DiMichele, W.A., and Phillips, T.L., 1996a, Climate change, plant extinctions, and vegetational recovery during the Middle-Late Pennsylvanian transition: The case of tropical peat-forming environments in North America, in Hart, M.L., ed., Biotic recovery from mass extinctions: Geological Society [London] Special Publication 102, p. 201–221.

DiMichele, W.A., and Phillips, T.L., 1996b, Clades, ecological amplitudes, and ecomorphs: Phylogenetic effects and the persistence of primitive plant communities in the Pennsylvanian-age tropics: Palaeogeography, Palaeoclimatology, Palaeoecology, v. 127, p. 83–106, doi: 10.1016/S0031-0182(96)00089-2.

DiMichele, W.A., and Phillips, T.L., 2002, The ecology of Paleozoic ferns: Review of Palaeobotany and Palynology, v. 119, p. 143–159, doi: 10.1016/S0034-6667(01)00134-8.

DiMichele, W.A., Mahaffy, J.F., and Phillips, T.L., 1979, Lycopods of Pennsylvanian age coals: Polysporia: Canadian Journal of Botany, v. 57, p. 1740–1752.

DiMichele, W.A., Phillips, T.L., and Olmstead, R.G., 1987, Opportunistic evolution—Abiotic environmental stress and the fossil record of plants: Review of Palaeobotany and Palynology, v. 50, p. 151–178, doi: 10.1016/0034-6667(87)90044-3.

DiMichele, W.A., Mammay, S.H., Chaney, D.S., Hook, R.W., and Nelson, W.J., 2001, An Early Permian flora with Late Permian and Mesozoic affinities from north-central Texas: Journal of Paleontology, v. 75, p. 449–460.

DiMichele, W.A., Stein, W.E., and Bateman, R.M., 2001, Ecological sorting during the Paleozoic radiation of vascular plant classes, in Allmon, W.D., and Bottjer, D.J., eds., Evolutionary Paleoecology: New York, Columbia University Press, p. 285–335.

Domning, D.P., 1982, Evolution of manatees: A speculative history: Journal of Paleontology, v. 56, p. 599–619.

Domning, D.P., Morgan, G.S., and Ray, C.E., 1982, North American Eocene sea cows (Mammalia: Sirenia), Washington, D.C., Smithsonian Institution Press, Smithsonian Contributions to Paleobiology, v. 52, 69 p.

Doran, G.H., Dickel, D.N., Ballinger, W.E., Jr., Agee, O.F., Laipis, P.J., and Hauswirth, W.H., 1986, Anatomical, cellular, and molecular analysis of 8,000-year old brain tissue from the Windover archeological site: Nature, v. 323, p. 803–806, doi: 10.1038/323803a0.

Driese, S.G., and Mora, C.I., 2001, Diversification of Siluro-Devonian plant traces in paleosols and influence on estimates of paleoatmospheric CO_2

levels, *in* Gensel, P.G., and Edwards, E., eds., Plants invade the land: New York, Columbia University Press, p. 237–254.

Driese, S.G., Mora, C.I., and Elick, J.M., 1997, Morphology and taphonomy of root and stump casts of the earliest trees (Middle to Late Devonian), Pennsylvania and New York, U.S.A.: Palaios, v. 12, p. 524–537.

Durden, C.J., 1969, Pennsylvanian correlation using blattoid insects: Canadian Journal of Earth Sciences, v. 6, p. 1159–1177.

Easterday, C.R., 2003, Evidence for silk-spinning in trigonotarbid arachnids (Chelicerata: Tetrapulmonata) and other new discoveries from Cemetery Hill (Carboniferous: Desmoinesian-Missourian), Columbiana County, eastern Ohio: Geological Society of America Abstracts with Programs, v. 35, p. 538.

Eble, C.F., and Grady, W.C., 1990, Paleoecological interpretation of a Middle Pennsylvanian coal bed in the Central Appalachian Basin, U.S.A: International Journal of Coal Geology, v. 16, p. 255–286, doi: 10.1016/0166-5162(90)90054-3.

Edwards, D., 1980, Early land floras, *in* Patchen, L., ed., The Terrestrial Environment and the Origin of land vertebrates: New York, Academic Press, p. 55–85.

Edwards, D., Bassett, M.G., and Rogerson, C.W., 1979, The earliest vascular land plants: Continuing the search for proof: Lethaia, v. 12, p. 313–324.

Edwards, D., and Fanning, U., 1985, Evolution and environment in the late Silurian-early Devonian: The rise of the pteridophytes: Royal Society of London Philosophical Transactions, ser. B, v. 309, p. 147–165.

Ehleringer, J.R., and Monson, R.K., 1993, Evolutionary and ecological aspects of photosynthetic pathway variation: Annual Review of Ecology and Systematics, v. 24, p. 411–439, doi: 10.1146/annurev.es.24.110193.002211.

Eiseley, L.C., 1945, The mastodon and early man in America: Science, v. 102, no. 2640, p. 108–110.

Engel, M.S. and Grimaldi, D.A., 2004, New light shed on the oldest insect: Nature, v. 427, p. 627–630.

Esterle, J.S., and Ferm, J.C., 1986, Relationship between petrographic and chemical properties and coal seam geometry, Hance Seam, Breathitt Formation, southeastern Kentucky: International Journal of Coal Geology, v. 6, p. 199–214, doi: 10.1016/0166-5162(86)90001-7.

Falcon, R.M.S., 1989, Macro- and micro-factors affecting coal-seam quality and distribution in southern Africa with particular reference to the No. 2 Seam, Witbank coalfield, South Africa: International Journal of Coal Geology, v. 12, p. 681–731, doi: 10.1016/0166-5162(89)90069-4.

Falcon-Lang, H.J., 1999, Fire ecology of a Late Carboniferous floodplain, Joggins, Nova Scotia: Journal of the Geological Society of London, v. 156, p. 137–148.

Fayers, S.R., and Trewin, N.H., 2004, A new crustacean from the Early Devonian Rhynie Chert, Aberdeenshire, Scotland: Transactions of the Royal Society of Edinburgh: Earth Sciences, v. 93, p. 355–382.

Feduccia, A., 1995, Explosive evolution in Tertiary birds and mammals: Science, v. 267, no. 5198, p. 637–638.

Feduccia, A., 1996, The origin and evolution of birds: New Haven, Connecticut, Yale University Press, 420 p.

Fiorillo, A.R., Padian, K., and Musikasinthorn, C., 2000, Taphonomy and depositional setting of the *Placerius* quarry (Chinle Formation, Late Triassic, Arizona): Palaios, v. 15, p. 373–386.

Fisher, D.C., 1995, Experiments on subaqueous meat caching: Center for the Study of the First Americans, Current Research in the Pleistocene, v. 12, p. 77–80.

Fisher, P.E., Russell, D.A., Stoskopf, M.K., Barrick, R.E., Hammer, M., and Kuzmitz, A.A., 2000, Cardiovascular evidence for an intermediate or higher metabolic rate in an ornithischian dinosaur: Science, v. 288, no. 5465, p. 503–505, doi: 10.1126/science.288.5465.503.

Forster, C.A., 1997, Iguanodontidae, *in* Currie, J., and Padian, K., eds., Encyclopedia of dinosaurs: San Diego, Academic Press, p. 359–361.

Frakes, L.E., Francis, J.E., and Sykta, J.I., 1992, Climate modes of the Phanerozoic: Cambridge, UK, Cambridge University Press, 286 p.

Francis, J.E., 1983, The dominant conifer of the Jurassic Purbeck Formation, England: Palaeontology, v. 26, p. 277–294.

Francis, J.E., 1986, The calcareous paleosols of the basal Purbeck Formation (Upper Jurassic), southern England, *in* Wright, V.P., ed., Paleosols: Their recognition and interpretation: Princeton, New Jersey, Princeton University Press, p. 112–138.

Francis, J.E., 1991, The dynamics of polar fossil forests: Tertiary fossil forests of Axel Heiberg Island, Canadian Arctic Archipelago: Geological Survey of Canada Bulletin, v. 403, p. 29–38.

Franzen, J.L., Haubold, H., and Storch, G., 1993, Relationships of the mammalian faunas from Messel and the Geiseltal: Darmstadt, Germany: Kaupia, v. 3, p. 145–149.

Fredriksen, N.O., 1985, Review of Early Tertiary sporomorph ecology: American Association Stratigraphic Palynologists Contribution Series, v. 15, p. 1–92.

Friis, E.M., Pedersen, K.R., and Crane, P.R., 2001, Fossil evidence of water lilies (*Nymphaeales*) in the Early Cretaceous: Nature, v. 410, p. 357–360, doi: 10.1038/35066557.

Gaffney, E.S., 1990, The comparative osteology of the Triassic turtle *Proganchelys*: Bulletin of the American Museum of Natural History, v. 194, 263 p.

Galtier, J., and Scott, A.C., 1994, Arborescent gymnosperms from the Viséan of East Kirkton, West Lothian, Scotland: Transactions of the Royal Society of Edinburgh: Earth Sciences, v. 84, p. 261–266.

Gao, K.-Q., and Shuban, N.H., 2003, Earliest known crown-group salamanders: Nature, v. 422, no. 6930, p. 424–428, doi: 10.1038/nature01491.

Garratt, M.J., Tims, J.D., Rickards, R.B., Chambers, T.C., and Douglas, J.G., 1984, The appearance of *Baragwanathia (Lycophytina)* in the Silurian: Botanical Journal of the Linnean Society, v. 89, p. 355–358.

Gastaldo, R.A., 1977, A Middle Pennsylvanian Nodule Flora from Carterville, Illinois, *in* Romans, R.C. ed., Geobotany: New York, Plenum Press, p. 133–156.

Gastaldo, R.A., 1986, Implications on the paleoecology autochthonous Carboniferous lycopods in clastic sedimentary environments of the Early Pennsylvanian of Alabama: Palaeogeography, Palaeoclimatology, Palaeoecology, v. 53, p. 191–212, doi: 10.1016/0031-0182(86)90044-1.

Gastaldo, R.A., 1987, Confirmation of Carboniferous clastic swamp communities: Nature, v. 326, p. 871–896.

Gastaldo, R.A., 1992, Regenerative growth in fossil horsetails following burial by alluvium: Historical Biology, v. 6, p. 203–219.

Gastaldo, R.A., and Huc, A.Y., 1992, Sediment facies, depositional environments, and distribution of phytoclasts in the Recent Mahakam River delta, Kalimantan, Indonesia: Palaios, v. 7, p. 574–591.

Gastaldo, R.A., and Matten, L.C., 1978, *Trigonocarpus leeanus*, a new species from the Middle Pennsylvanian of southern Illinois: American Journal of Botany, v. 65, p. 882–890.

Gastaldo, R.A., and Staub, J.R., 1999, A mechanism to explain the preservation of leaf litter lenses in coals derived from raised mires: Palaeogeography, Palaeoclimatology, Palaeoecology, v. 149, p. 1–14, doi: 10.1016/S0031-0182(98)00188-6.

Gastaldo, R.A., Douglass, D.P., and McCarroll, S.M., 1987, Origin, characteristics and provenance of plant macrodetritus in a Holocene crevasse splay, Mobile delta, Alabama: Palaios, v. 2, p. 229–240.

Gastaldo, R.A., Riegel, W., Püttmann, W., Linnemann, U.H., and Zetter, R., 1998, A multidisciplinary approach to reconstruct the Late Oligocene vegetation in central Europe: Review of Palaeobotany and Palynology, v. 101, p. 71–94, doi: 10.1016/S0034-6667(97)00070-5.

Gastaldo, R.A., Stevanović-Walls, I.M., Ware, W.N., and Greb, S.F., 2004, Community heterogeneity of Early Pennsylvanian peat mires: Geology, v. 32, p. 693–696, doi: 10.1130/G20515.1.

Gee, C.T., 2001, The mangrove palm *Nypa* in the geologic past of the New World: Wetlands Ecology and Management, v. 9, p. 181–203, doi: 10.1023/A:1011148522181.

Gensel, P.G., 1992, Phylogenetic relationships of the zosterophylls and lycopsids: Evidence from morphology, paleoecology and cladistic methods of inference: Annals of the Missouri Botanical Garden, v. 79, p. 450–473.

Gensel, P.G., and Andrews, H.N., 1984, Plant life in the Devonian: New York, Praeger, 380 p.

Glob, P.V., 1965, The Bog People, Iron Age Man preserved: New York, Ballantine Books, 200 p.

Gloy, U., 2000, Taphonomy of the fossil lagerstätte Guimarota, *in* Martin, T., and Krebs, B., eds., Guimarota—A Jurassic ecosystem: Munich, Verlag Dr. Friedrich Pfeil, p. 129–136.

Goldring, W., 1924, The Upper Devonian forest of seed ferns in eastern New York: Bulletin of the New York Museum, v. 251, p. 50–72.

Gomez, B., Martin-Closas, C., Meon, H., Thevenard, F., and Barale, G., 2001, Plant taphonomy and palaeoecology in the lacustrine Una Delta (late Barremian, Iberian Ranges, Spain): Palaeogeography, Palaeoclimatology, Palaeoecology, v. 170, p. 133–148, doi: 10.1016/S0031-0182(01)00232-2.

Gore, A.J.P., 1983, Introduction, *in* Gore, A.J.P., ed., Mires: Swamp, bog, fen and moor: New York, Elsevier, Ecosystems of the World, v. 4A, p. 1–34.

Gould, R.E., 1975, A preliminary report on petrified axes of Vertebraria from the Permian of eastern Australia, *in* Campbell, K.S.W., ed., Gondwana Geology: Papers presented at the Third Gondwana Symposium: Canberra, Australian National University Press, p. 109–115.

Gould, R.E., and Delevoryas, T., 1977, The biology of *Glossopteris*: Evidence from petrified seed-bearing and pollen-bearing organs: Alcheringa, v. 1, p. 387–399.

Gould, S.J., 1991, Eight (or fewer) little piggies: Natural History, v. 1991, p. 22–29.

Graham, A., 1999, Late Cretaceous and Cenozoic history of North American vegetation: Oxford, Oxford University Press, 350 pp.

Gray, J., Massa, D., and Boucot, A.J., 1982, Caradocian land plant microfossils from Libya: Geology, v. 10, p. 197–201, doi: 10.1130/0091-7613(1982)10<197:CLPMFL>2.0.CO;2.

Greb, S.F., Eble, C.F., Hower, J.C., and Andrews, W.M., 2002, Multiple-bench architecture and interpretations of original mire phases in Middle Pennsylvanian coal seams—Examples from the Eastern Kentucky Coal Field: International Journal of Coal Geology, v. 49, p. 147–175, doi: 10.1016/S0166-5162(01)00075-1.

Greb, S.F., Eble, C.F., and Hower, J., 1999a, Depositional history of the Fire Clay coal bed (Late Duckmantian), eastern Kentucky, USA: International Journal of Coal Geology, v. 40, p. 255–280, doi: 10.1016/S0166-5162(99)00004-X.

Greb, S.F., Eble, C.F., Chesnut, D.R., Jr., Phillips, T.L., and Hower, J.C., 1999b, An in situ occurrence of coal balls in the Amburgy coal bed, Pikeville Formation (Duckmantian), Central Appalachian Basin, U.S.A.: Palaios, v. 14, p. 433–451.

Greb, S.F., Andrews, W.M., Eble, C.F., DiMichele, W., Cecil, C.B., and Hower, J.C., 2003, Desmoinesian coal beds of the Eastern Interior and surrounding basins: The largest tropical peat mires in earth history, in Chan, M.A., and Archer, A.W., eds., Extreme depositional environments: Mega–end members in geologic time: Geological Society of America Special Paper 370, p.127–150.

Greenwood, D.R., and Basinger, J.F., 1994, The paleoecology of high-latitude Eocene swamp forests from Axel Heiberg Island, Canadian High Arctic: Review of Palaeobotany and Palynology, v. 81, p. 83–97, doi: 10.1016/0034-6667(94)90128-7.

Grimaldi, D., Shedrinsky, A., and Wampler, T.P., 2000, A remarkable deposit of fossiliferous amber from the Upper Cretaceous (Turonian) of New jersey, in Grimaldi, D., ed., Studies in amber with particular reference to the Cretaceous of New Jersey: Leiden, Backhuys, p. 1–76.

Grimaldi, D.A., Engel, M.S., and Nascimbene, P.C., 2002, Fossiliferous Cretaceous amber from Myanmar (Burma): Its rediscovery, biotic diversity, and paleontological significance: American Museum Novitates, no. 3361, p. 1–71.

Hahn, G., Hahn, R., and Brauckmann, C., 1986, Zur Kenntnis von *Arthropleura* (Myriapoda; Ober-Karbon): Geologica et Palaeontologica, v. 20, p. 125–137.

Hall, J.W., 1975, *Ariadnaesporites* and *Glomerisporites* in the Late Cretaceous ancestral Salviniaceae: American Journal of Botany, v. 62, p. 359–369.

Hamer, J.H., and Rothwell, G.W., 1988, The vegetative structure of *Medullosa endocentrica* (Pteridospermopsida): Canadian Journal of Botany, v. 66, p. 375–387.

Hansen, M.C., 1996, Phylum Chordata-Vertebrate fossils, in Feldman, R.M., and Hackathorn, M., eds., Fossils of Ohio: Ohio Department of Natural Resources, Division of the Geological Survey, Bulletin 70, p. 288–369.

Havlena, V., 1961, Die flöznahe und flözfremde Flora des oberschlesichen Namur A und B: Palaeontographica, Abt. B, v. 105, p. 1–2, 22–38.

Hernández-Castillo, G.R., and Cevallos-Ferriz, S.R.S., 1999, Reproductive and vegetative organs with affinities to Haloragaceae from the Upper Cretaceous Huepec Chert locality of Sonora, Mexico: American Journal of Botany, v. 86, p. 1717–1734.

Hickey, L.J., and Doyle, J.A., 1997, Early Cretaceous fossil evidence for angiosperm species: Botanical Review, v. 43, p. 3–104.

Hill, R.S., and Dettman, M.E., 1996, Origin and diversification of the genus *Nothofagus*, in Veblen, T.T., Hill, R.S., and Read, J., eds., The ecology and biogeography of *Nothofagus* forests: New Haven, Yale University Press, p. 11–24.

Hogarth, P.J., 1999, The biology of mangroves: Oxford, Oxford University Press, 228 p.

Hook, R.W., and Baird, D., 1986, The Diamond Coal Mine of Linton, Ohio, and its Pennsylvanian-age vertebrates: Journal of Vertebrate Paleontology, v. 6, p. 174–190.

Hotton, C.L., Hueber, F.M., Griffing, D.H., and Bridge, J.S., 2001, Early terrestrial plant environments: An example from the Emsian of Gaspé, Canada, in Gensel, P.G., and Edwards, E., eds., Plants invade the land: New York, Columbia University Press, p. 179–121.

Hou, L., and Ericson, P.G.P., 2002, A Middle Eocene shorebird from China: Condor, p. 896–899.

Iturralde-Vinent, M., and Hartstein, E., 1998, Miocene amber and lignitic deposits in Puerto Rico: Caribbean Journal of Science, v. 34, p. 308–312.

Janis, C.M., Colbert, M.W., Coombs, M.C., Lambert, W.D., and MacFadden, B.J., Mader, B.J., Prothero, D.R., Schoch, R.M., Shoshani, J., and Wall, W.P., 1998, Perissodactyla and Proboscidea, in Janis, C.M., Scott, K.M., and Jacobs, L.L., eds., Evolution of Tertiary mammals of North America: Cambridge, UK, Cambridge University Press, p. 511–524.

Jeram, A.J., Selden, P.A., and Edwards, D., 1990, Land animals in the Silurian: arachnids and myriapods from Shropshire, England: Science, v. 250, p. 658–661.

Jillson, W.R., 1968, The extinct vertebrata of the Pleistocene in Kentucky: Frankfort, Kentucky, Roberts Printing Company, 122 p.

Johnson, G.A.L., 1999, Cordaites tree trunks in the British coal measures: Geology Today, v. 15, p. 106–109.

Johnson, K.R., Nichols, D.J., Attrep, M., Jr., and Orth, C.J., 1989, High-resolution leaf-fossil record spanning the Cretaceous-Tertiary boundary: Nature, v. 340, p. 708–711, doi: 10.1038/340708a0.

Jones, C.G., Lawton, J.H., and Shachak, M., 1994, Organisms as ecosystem engineers: Oikos, v. 69, p. 373–386.

Jones, K.B., 1986, Amphibians and reptiles, in Cooperider, A.Y., Boyd, R.J., and Stuart, H.R., eds., Inventory and monitoring of wildlife habitat: Denver, U.S. Bureau of Land Management, p. 267–290.

Jones, M.B., 1988, Photosynthetic responses of C_3 and C_4 wetland species in a tropical swamp: Journal of Ecology, v. 76, p. 253–262.

Julien, M.H., Center, T.D., and Tipping, P.W., 2002, Floating fern (*Salvinia*) In Van Driesche, R., Lyon, S., Blossey, B., Hoddle, M., and Reardon, R., cords., Biological control of invasive plants in the eastern United States, USDA Forest Service Publication FHTET-2002–04, 413 p. (http://www.invasive.org/eastern/biocontrol/2FloatingFern.html).

Juniper, B.E., Robins, R.J., and Joel, D.M., 1989, The Carnivorous Plants: San Diego, Academic Press, 353 p.

Junk, W.J., 1983, Ecology of swamps on the Middle Amazon, in Gore, A.J.P., ed., Mires: Swamp, bog, fen and moor: New York, Elsevier, Ecosystems of the World, v. 4A, p. 269–292.

Kasper, A.E., and Andrews, H.N., 1972, *Pertica*, a new genus of Devonian plants from northern Maine: American Journal of Botany, v. 59, p. 897–911.

Keddy, P.A., 2000, Wetland ecology—Principles and conservation: Cambridge, United Kingdom, Cambridge University Press, Studies in Ecology, 614 p.

Kellogg, E.A., 2001, Evolutionary history of the grasses: Plant Physiology, v. 125, p. 1198–1205, doi: 10.1104/pp.125.3.1198.

Kenrick, P., and Crane, P.R., 1997, The origin and early evolution of plants on land: Nature, v. 389, p. 33–39, doi: 10.1038/37918.

Kershaw, A.P., Bolger, P.F., Sluiter, I.R.K., Baird, J.G., and Whitelow, M., 1991, The nature and evolution of lithotypes in the Tertiary brown coals of the Latrobe Valley, southeastern Australia: International Journal of Coal Geology, v. 18, p. 233, doi: 10.1016/0166-5162(91)90052-K.

Kerp, H., Abu Hamad, A.M.B., Bandel, K., Niemann, B., and Eshet, Y., 2004, A Late Permian flora with *Dicroidium* from the Dead Sea region, Jordon: VII International Organization of Paleobotany Conference, Bariloche, Argentina, Abstracts, p. 64–65.

Kerp, J.H.F., 1988, Aspects of Permian palaeobotany and palynology. X. The West- and Central European species of the genus *Autunia* Krasser emend. Kerp (Peltaspermaceae) and the form-genus *Rhachiphyllum* Kerp (callipterid foliage): Review of Palaeobotany and Palynology, v. 54, p. 249–360, doi: 10.1016/0034-6667(88)90017-6.

Klein, R., 1987, Problems and prospects in understanding how early people exploited animals, in Nitecki, M.H., and Nitecki, D.V., eds., The evolution of human hunting: New York, Plenum Press, p. 11–45.

Knobloch, E., 1970, The Tertiary floras of Moravia (Czechoslovakia): Paläontologische Abhandlungen, Abt. B, Paleobotanik, v. 3, p. 381–390.

Knobloch, E., and Mai, D.H., 1984, Neue Gattungen nach Früchten und Samen aus dem Cenoman bis Maastricht (Kreide) von Mitteleuropa: Berlin, Feddes Repert, v. 95, p. 3–41.

Knoll, A.H., 1985, Exceptional preservation of photosynthetic organisms in silicified carbonates and silicified peats: Royal Society of London Philosophical Transactions, Bulletin, v. 311, p. 111–122.

Kordos, L., and Begun, D.R., 2002, Rudabánya: A Late Miocene subtropical swamp deposit with evidence of the origin of the African apes and humans: Evolutionary Anthropology, v. 11, p. 45–57, doi: 10.1002/evan.10010.

Kovar-Eder, J., Kvacek, Z., and Meller, B., 2001, Comparing Early to Middle Miocene floras and probable vegetation types of Oberdorf N Voitsberg (Austria), Bohemia (Czech Republic), and Wackersdorf (Germany): Review of Palaeobotany and Palynology, v. 114, p. 83–125, doi: 10.1016/S0034-6667(00)00070-1.

Kozlowski, T.T., 1997, Responses of woody plants to flooding and salinity: Tree Physiology Monograph no. 1: British Columbia, Victoria, Heron Publishing, 29 p.

Krasilov, V.A., 1975, Paleoecology of terrestrial plants: Basic principles and techniques: New York, John Wiley and Sons, 283 p.

Kron, D.G., and Manning, E., 1998, *Anthracotheriidae, in* Janis, C.M., Scott, K.M., and Jacobs, L.L., eds., Evolution of Tertiary Mammals of North America: Cambridge, UK, Cambridge University Press, p. 381–388.

Kroodsma, D.E., 1979, Habitat values for nongame wetland birds, *in* Greeson, P.E., Clark, J.R., and Clark, J.E. eds., Wetland functions and values–The state of our understanding: Minneapolis, American Water Resources Association, p. 320–343.

Krull, E.S., 1999, Permian palsa mires as paleoenvironmental proxies: Palaios, v. 14, p. 520–544.

Kukalova-Peck, J., 1978, Origin and evolution of insect wings, and their relation to metamorphosis, as documented by the fossil record: Journal of Morphology, v. 156, p. 53–125, doi: 10.1002/jmor.1051560104.

Kukalova-Peck, J., 1983, Origin of the insect wing and wing articulation from the arthropodan leg: Canadian Journal of Zoology, v. 61, p. 933–955.

Kuo, J., and den Hartog, C., 2000, Seagrasses: A profile of an ecological group: Biologica Marina Mediterranea, Genova, v. 7, p. 3–17.

Kurten, B., and Anderson, E., 1980, Pleistocene mammals of North America: New York, Columbia University Press, 442 p.

Kvaček, Z., 1995, *Limnobiophyllum* Krassilov—A fossil link between the Araceae and the Lemnaceae: Aquatic Botany, v. 50, p. 49–61, doi: 10.1016/0304-3770(94)00442-O.

Kvaček, Z., 1998, Bílina: A window on Early Miocene marshland environments: Reviews of Palaeobotany and Palynology, v. 101, p. 111–123.

Labandeira, C.C., and Eble, G., 2006, The fossil record of insect diversity and disparity, *in* Anderson, J., Thackeray, F., Van Wyke, B., and DeWit M., eds., Gondwana Alive–Biodiversity and the Evolving Biosphere: Johannesburg, Witwatersrand University Press, 54 p. (in press).

Labandeira, C.C., and Phillips, T.L., 1996, Insect fluid-feeding on Upper Pennsylvanian tree ferns (Palaeodictyoptera and Marattiales) and the early history of the piercing and sucking functional feeding group: Annals of the Entomological Society of America, v. 89, p. 157–183.

Labandeira, C.C., Beall, B.S., and Hueber, F.M., 1988, Early insect diversification: Evidence from a Lower Devonian bristletail from Québec: Science, v. 242, p. 913–916.

Lacovera, K.J., Smith, J.R., Smith, J.B., and Lamanna, M.C., 2002, Evidence of semi-diurnal tides along the African coast of the Cretaceous Tethys seaway, Bahariya Oasis, Egypt: Geological Society of America Abstracts with Programs, v. 34, p. 32.

Lang, W.H., and Cookson, I.C., 1935, On a flora, including vascular plants associated with *Monograptus,* in rocks of Silurian age, from Victoria, Australia: Royal Society of London Philosophical Transactions, B., v. 224, p. 421–449.

Lancucka-Srodoniowa, M., 1966, Tortonian flora from the "Gdów Bay" in the south of Poland: Acta Palaeobotanica, v. 7, 135 p.

LaPasha, C.A., and Miller, C.N., 1985, Flora of the Early Cretaceous Kootenai Formation in Montana, bryophytes and tracheophytes excluding conifers: Palaeontographica, Abt. B, Paläophytologie, v. 196, p. 111–145.

Larsson, S.G., 1978, Baltic amber: A palaeobiological study: Klampenborg, Denmark, Scandinavian Science Press, 192 p.

Laurin, M., and Reisz, R.R., 1999, A new study of *Solenodonsaurus janenschi,* and a reconsideration of amniote origins and stegocephalian evolution: Canadian Journal of Earth Sciences, v. 36, p. 1239–1255, doi: 10.1139/cjes-36-8-1239.

Lenz, O.K., and Riegel, W., 2001, Isopollen maps as a tool for the reconstruction of a coastal swamp from the middle Eocene at Helmstedt (northern Germany): Facies: International Journal of Paleontology, Sedimentology and Geology, v. 45, p. 177–194.

Les, D.H., Cleland, M.A., and Waycott, M., 1997, Phylogenetic studies in Alismatidae, II: Evolution of marine angiosperms (seagrasses) and hydrophily: Systematic Biology, v. 22, p. 443–463.

Lidgard, S., and Crane, P.R., 1988, Quantitative analyses of the early angiosperm radiation: Nature, v. 331, p. 344–346, doi: 10.1038/331344a0.

Lister, A., and Bahn, P., 1994, Mammoths: New York, Macmillan, 168 p.

Liu, Y., and Gastaldo, R.A., 1992, Characteristics of a Pennsylvanian ravinement surface: Sedimentary Geology, v. 77, p. 197–214, doi: 10.1016/0037-0738(92)90126-C.

Long, R.A., and Padian, K., 1986, Vertebrate biostratigraphy of the Late Triassic Chinle Formation, Petrified Forest National Park, Arizona—Preliminary results, *in* Padian, K., ed., The beginning of the Age of Dinosaurs: Faunal change across the Triassic-Jurassic boundary: Cambridge, UK, Cambridge University Press, p. 161–169.

Looy, C.V., Brugman, W.A., Dilcher, D.L., and Visscher, H., 1999, The delayed resurgence of equatorial forests after the Permian-Triassic ecologic crisis: Proceedings of the National Academy of Sciences of the United States of America, v. 96, p. 13857–13862, doi: 10.1073/pnas.96.24.13857.

Looy, C.V., Twitchett, R.J., Dilcher, D.L., Van Konijnenburg-Van Cittert, J.H.A., and Visscher, H., 2001, Life in the end-Permian dead zone: Proceedings of the National Academy of Sciences of the United States of America, v. 98, no. 14, p. 7879–7883, doi: 10.1073/pnas.131218098.

Lumbert, S.H., den Hartog, C., Phillips, R.C., and Olsen, S.F., 1984, The occurrence of fossil seagrasses in the Avon Park Formation (late Middle Eocene), Levy County, Florida (U.S.A.): Aquatic Botany, v. 20, p. 121–129.

Lupia, R., Schneider, H., Moeser, G.M., Pryer, K.M., and Crane, P.R., 2000, Marsileaceae sporocarps and spores from the Late Cretaceous: International Journal of Plant Sciences, v. 161, p. 975–988, doi: 10.1086/317567.

MacGinitie, H.D., 1969, The Eocene Green River flora of northwestern Colorado and northeastern Utah: University of California Publications in Geological Sciences, v. 83, p. 1–203.

Mai, D.H., 1985, Entwicklung der Wasser-und Sumpfpflanzen-Gesellschafeten Europas von der Kreide bis ins Quarter: Flora, v. 176, p. 449–511.

Mapes, G., and Gastaldo, R.A., 1986, Late Paleozoic non-peat accumulating floras, *in* Broadhead, T.W., ed., Land plants—Notes for a short course: University of Tennessee, Department of Geological Sciences, Studies in Geology, v. 15, p. 115–127.

Maraven, R.V., 1998, Biogeographical and evolutionary considerations of Mauritia (Arecaceae), based on palynological evidence: Review of Palaeobotany and Palynology, v. 100, p. 109–122, doi: 10.1016/S0034-6667(97)00060-2.

Marden, J.H., and Kramer, M.G., 1994, Surface-skimming stoneflies: A possible intermediate stage in insect flight evolution: Science, v. 266, p. 427–430.

Martill, D.M., and Naish, D., 2001, Dinosaurs of the Isle of Wight: The Palaeontological Association Field Guide to Fossils, no. 10. 433 pp.

Martin, T., 2000, Overview of the Guimarota ecosystem, *in* Martin, T., and Krebs, B., eds., Guimarota—A Jurassic ecosystem: Munich, Verlag Dr. Friedrich Pfeil, p. 143–146.

Martins-Neto, R.G., 2003, The fossil tabanids (Diptera Tabanidae): When they began to appreciate warm blood and when they began transmit diseases?: Rio de Janeiro, Memoirs of the Institute of Oswaldo Cruz, v. 98, supplement 1, p. 29–34.

McCoy, E.D., and Heck, K.L., Jr., 1976, Biogeography of corals, seagrasses, and mangroves: An alternative to the center of origin concept: Systematic Zoology, v. 25, p. 201–210.

MacLeod, K.G., Smith, R.M.H., Koch, P.L., and Ward, P.D., 2000, Timing of mammal-like reptile extinctions across the Permian-Triassic boundary in South Africa: Geology, v. 28, p. 227–230, doi: 10.1130/0091-7613(2000)028<0227:TOMLRE>2.3.CO;2.

McElwain, J.C., Beerling, D.J., and Woodward, F.I., 1999, Fossil plants and global warming at the Triassic-Jurassic boundary: Science, v. 285, p. 1386–1390, doi: 10.1126/science.285.5432.1386.

Menon, S., 1997, The people of the bog: Discover, v. 18, p. 60–68.

Meyen, S.V., 1982, The Carboniferous and Permian floras of Angaraland (a synthesis): Lucknow, India, International Publishers, 109 p.

Meyer-Berthaud, B., Scheckler, S.E., and Wendt, J., 1999, *Archaeopteris* is the earliest known modern tree: Nature, v. 398, p. 700–701, doi: 10.1038/19516.

Miao, F., Qian, L., and Zhang, X., 1989, Peat-forming processes and evolution of swamp sequences–Case analysis of a Jurassic inland coal basin in China: International Journal of Coal Geology, v. 12, p. 733–765, doi: 10.1016/0166-5162(89)90070-0.

Miller, C.N., Jr., 1987, Land plants of the northern Rocky Mountains before the appearance of flowering plants: Annals of the Missouri Botanical Garden, v. 74, p. 692–706.

Miller, E.R., Rasmussen, D.T., and Simons, E.L., 1998, Fossil storks (Ciconiidae) from the Late Eocene and Early Miocene of Egypt: Ostrich, v. 68, p. 23–26.

Milner, A.R., and Sequeira, S.E.K., 1994. The temnospondyl amphibians from the Visean of East Kirkton, West Lothian, Scotland, *in* Rolfe, W.D.I., Clarkson, E.N.K., and Panchen, A.L., eds., Volcanism and early terrestrial biotas: Transactions of the Royal Society of Edinburgh: Earth Sciences, v. 84, p. 331–361.

Milner, A.R., Smithson, T.R., Milner, A.C., Coates, M.I., and Rolfe, W.D.I., 1986, The search for early tetrapods: Modern Geology, v. 10, p. 1–28.

Mitsch, W.J., and Gosselink, J.G., 2000, Wetlands (2nd edition): New York, Van Nostrand Reinhold, 539 p.

Moore, P.D., 1989, The ecology of peat-forming processes—A review: International Journal of Coal Geology, v. 12, p. 89–103, doi: 10.1016/0166-5162(89)90048-7.

Moore, P.D., 1995, Biological processes controlling the development of modern peat-forming ecosystems: International Journal of Coal Geology, v. 28, p. 99–110, doi: 10.1016/0166-5162(95)00015-1.

Mosbrugger, V., Gee, C.T., Belz, G., and Ashraf, A.R., 1994, Three-dimensional reconstruction of an in-situ Miocene peat forest from the Lower Rhine Embayment, northwestern Germany: New methods in palaeovegetation analysis: Palaeogeography, Palaeoclimatology, Palaeoecology, v. 110, p. 295–317, doi: 10.1016/0031-0182(94)90089-2.

Muller, J., 1980, Fossil pollen records of extant angiosperms: Botanical Review, v. 47, p. 1–142.

Muller, J., 1984, Significance of fossil pollen for angiosperm history: Annals of the Missouri Botanical Garden, v. 71, p. 419–443.

Myers, R.L., 1990, Palm swamps, in Lugo, A.E., Brinson, M., and Brown, S., eds., Forested wetlands: New York, Elsevier, Ecosystems of the World, v. 15, p. 267–286.

Naiman, R.J., and Rogers, K.H., 1997, Large animals and system-level characteristics in river corridors: Implications for river management: BioSciences, v. 47, p. 521–529.

National Research Council, 1995, Wetlands: Characteristics and boundaries: Washington, D.C.: National Academies Press, 328 pp.

Neiburg, M.F., 1958, Permian true mosses of Angaraland: Journal of the Paleontological Society of India, v. 3, p. 22–29.

Nguyen Tu, T., Kvaček, J., Uličny, D., Bocherens, H., Mariotti, A., and Broutin, J., 2002, Isotope reconstruction of plant paleoecology case study of Cenomanian floras from Bohemia: Palaeogeography, Palaeoclimatology, Palaeoecology, v. 183, p. 43–70, doi: 10.1016/S0031-0182(01)00447-3.

Nicholas, G.P., 1998, Wetlands and hunter-gatherers; a global perspective: Current Anthropology, v. 39, p. 720–731, doi: 10.1086/204795.

Nichols, D.J., 1995, The role of palynology in paleoecological analyses of Tertiary coals: International Journal of Coal Geology, v. 28, p. 139–159, doi: 10.1016/0166-5162(95)00017-8.

Nichols, D.J., and Fleming, R.F., 1990, Plant microfossil record of the terminal Cretaceous event in the western United States and Canada, in Sharpton, V.L., and Ward, P.D., eds., Global catastrophes in earth history: An interdisciplinary conference on impacts, volcanism, and mass mortality: Geological Society of America Special Paper 247, p. 445–455.

Nichols, D.J., and Pillmore, C.L., 2000, Palynology of the K-T boundary in the Raton Basin, Colorado and New Mexico—New data and interpretations from the birthplace of K-T plant microfossil studies in nonmarine rocks, in Catastrophic events and mass extinctions: Impacts and beyond, Catastrophic Events Conference, July 9–12, 2000, Vienna, Austria, Conference Guide to Technical Sessions and Activities: Lunar and Planetary Institute Report 1053, p. 150–151.

Nitecki, M.H., ed., 1979, Mazon Creek Fossils: New York, Academic Press, 581p.

Norman, D.B., 1980, On the Ornithischian dinosaur *Iguanodon bernissartensis* from the Lower Cretaceous of Bernissart (Belgium): Brussels, Institut Royal des Sciences Naturelles de Belgique Memoir 178, 103 p.

Neuburg, M.F., 1958, Permian true mosses of Angaraland: Journal of the Paleontological Society of India, v. 3, p. 22–29.

Olson, S.L., and Feduccia, A., 1980, *Presbyornis* and the origin of the Anseriformes (Aves: Charadriomorphae): Smithsonian Contributions to Zoology, v. 323, p. 1–24.

Olson, S.L., and Rasmussen, T., 1986, Paleoenvironment of the earliest hominoids: New evidence from the Oligocene avifauna of Egypt: Science, v. 233, p. 1202–1204.

Oshurkova, M.V., 1996, Paleoecological parallelism between the Angaran and Euramerican phytogeographic provinces: Review of Palaeobotany and Palynology, v. 90, p. 99–111, doi: 10.1016/0034-6667(95)00026-7.

Packnall, D.T., 1989, Late Eocene to Early Miocene vegetation and climate history of New Zealand: Journal of the Royal Society of New Zealand, v. 19, p. 1–18.

Parker, L.R., 1975, Paleoecology of the fluvial coal-forming swamps and associated flood-plain environments in the Blackhawk Formation of central Utah: Geological Society of America Abstracts with Programs, v. 7, p. 1225.

Parker, L.R., and Rowley, R.L., Jr., 1989, Dinosaur footprints from a coal mine in east-central Utah, in Gillette, D.D., and Lockley, M.G., eds., Dinosaur tracks and traces: Cambridge, United Kingdom, Cambridge University Press, p. 354–359.

Patil, G.V. and Singh, R.B., 1978, Fossil *Eichhornia* from the Eocene Deccan intertrappen beds, India: Paleontographica, Abt. B, v. 167, p. 1–7.

Pelzer, G., Riegel, W., and Volker, W., 1992, Depositional controls on the Lower Cretaceous Wealdon coals of northwest Germany, in McCabe, P.J., and Parrish, J.T., eds., Controls on the distribution and quality of Cretaceous coals: Geological Society of America Special Publication, v. 267, p.227–244.

Person, C.P., and Delevoryas, T., 1982, The Middle Jurassic flora of Oaxaca, Mexico: Palaeontographica, Abt. B, v. 180, p. 82–119.

Peterson, W., 1924, Dinosaur tracks in the roofs of coal mines: Natural History, v. 24, p. 388–391.

Pfefferkorn, H.W., 1972, Distribution of *Stigmaria wedingtonensis* (Lycopsida) in the Chesterian (Upper Mississippian) of North America: American Midland Naturalist, v. 88, p. 225–231.

Pfefferkorn, H.W., 1995, We are temperate climate chauvinists: Palaios, v. 10, p. 389–391.

Pfefferkorn, H.W., and Thomson, M.C., 1982, Changes in dominance patterns in Upper Carboniferous plant-fossil assemblages: Geology, v. 10, p. 641–644, doi: 10.1130/0091-7613(1982)10<641:CIDPIU>2.0.CO;2.

Pfefferkorn, H.W., Archer, A.W., and Zodrow, E.L., 2001, Modern tropical analogs for Carboniferous standing forests: Comparison of extinct *Mesocalamites* with extant *Montrichardia*: Historical Biology, v. 15, p. 235–250.

Phillips, T.L., and Cross, A.T., 1991, Paleobotany and paleoecology of coal, in Gluskoter, H.J., Rice, D.D., and Taylor, R.B., eds., Economic geology, U.S.: Boulder, Colorado, Geological Society of America, Geology of North America, v. P-2, p. 483–502.

Phillips, T.L., and DiMichele, W.A., 1992, Comparative ecology and life-history biology of arborescent lycopods in Late Carboniferous swamps of Euramerica: Annals of the Missouri Botanical Garden, v. 79, p. 560–588.

Phillips, T.L., Peppers, R.A., Avcin, M.J., and Laughnan, P.F., 1974, Fossil plants and coal: Patterns of change in Pennsylvanian coal swamps of the Illinois Basin: Science, v. 184, p. 18–49.

Phillips, T.L., Peppers, R.A., and DiMichele, W.A., 1985, Stratigraphic and interregional changes in Pennsylvanian coal-swamp vegetation: Environmental inferences: International Journal of Coal Geology, v. 5, p. 43–109, doi: 10.1016/0166-5162(85)90010-2.

Pigg, K.B., 1992, Evolution of Isoetalean lycopsids: Annals of the Missouri Botanical Garden, v. 79, p. 589–612.

Pigg, K.B., Davis, W.C., and Ash, S., 1993, A new permineralized Upper Triassic flora from the Petrified Forest National Park, Arizona: A preliminary report, in Lucas, S.G., and Morales, M., eds., The non-marine Triassic: New Mexico Museum of Natural History and Science Bulletin 3, p. 411–413.

Plaziat, J.-C., 1995, Modern and fossil mangroves and mangals: Their climatic and biogeographic variability, in Bosence, D.W. J. and Allison, P.A., eds., Marine palaeoenvironmental analysis from fossils: Geological Society [London], Special Publication 83, p. 73–96.

Plunkett, C.M., Soltis, D.E., Soltis, P.S., and Brooks, R.E., 1995, Phylogenetic relationships between Juncaceae and Cyperaceae: Insights from rbcL sequence data: American Journal of Botany, v. 82, p. 520–525.

Pocknall, D.T., 1985, Palynology of Waikato Coal Measures (Late Eocene-Late Oligocene) from the Raglan area, North Island, New Zealand: New Zealand Journal of Geology and Geophysics, v. 28, p. 329–349.

Poinar, G.O., Jr., 1992, Life in amber: Stanford, California, Stanford University Press, 350 p.

Pole, M.S., 1996, Eocene Nypa from Regatta Point, Tasmania: Review of Palaeobotany and Palynology, v. 92, p. 55–67, doi: 10.1016/0034-6667(95)00099-2.

Potts, R., and Behrensmeyer, A.K., 1992, Late Cenozoic Terrestrial Systems, in Behrensmeyer, A.K., Damuth, J.D., DiMichele, W.A., Potts, R., Sues, H.-D., and Wing, S.L., eds., Terrestrial ecosystems through time: Chicago, University of Chicago Press, p. 418–541.

Purkynová, E., 1977, Namurian flora of the Moravian part of the Upper Silesian coal basin, in Holub, V.M., and Wagner, R.H., eds., Symposium on Carboniferous Stratigraphy: Prague, Geological Survey, p. 289–303.

Pryer, K.M., 1999, Phylogeny of marsileaceous ferns and relationships of the fossil *Hydropteris pinnata* reconsidered: International Journal of Plant Sciences, v. 160, p. 931–954, doi: 10.1086/314177.

Pryor, J.S., 1993, Patterns of ecological succession within the Upper Pennsylvanian Duquesne Coal of Ohio: Evolutionary Trends in Plants, v. 7, p. 57–66.

Randall, D.J., Burggren, W.W., Farrell, A.P., and Haswell, M.S., 1981, The evolution of air-breathing vertebrates: Cambridge, United Kingdom, Cambridge University Press, 133 p.

Raymond, A., 1988, The paleoecology of a coal ball deposit from the Middle Pennsylvanian of Iowa dominated by cordaitalean gymnosperms: Review of Palaeobotany and Palynology, v. 53, p. 233–250.

Raymond, A., and Phillips, T.L., 1983, Evidence for an Upper Carboniferous mangrove community, in Teas, H.J., ed., Tasks for vegetation science: The Hague, Dr. W. Junk Publishers, p. 19–30.

Raymond, A., Phillips, M.K., Gennett, J.A., and Comet, P.A., 1997, Palynology and paleoecology of lignites from the Manning Formation (Jackson Group) outcrop in the Lake Somerville spillway of east-central Texas: International Journal of Coal Geology, v. 34, p. 195–223, doi: 10.1016/S0166-5162(97)00023-2.

Raymond, A., Costanza, S.H., and Slone, E.D.J., 2001, Was Cordaites a Late Carboniferous mangrove?: Geological Society of America Abstracts with Programs, v. 33, p. 172–176.

Raymond, A., Phillips, M.K., Gennett, J.A., and Comet, P.A., 1997, Palynology and paleoecology of lignites from the Manning Formation (Jackson Group) outcrop in the Lake Somerville spillway of east-central Texas: International Journal of Coal Geology, v. 34, p. 195–223.

Rayner, R.J., 1992, *Phyllotheca*: the pastures of the Late Permian: Palaeogeography, Palaeoclimatology, Palaeoecology, v. 92, p. 31–40, doi: 10.1016/0031-0182(92)90133-P.

Rees, P.M., Ziegler, A.M., and Valdes, P.J., 2000, Jurassic phytogeography and climates: New data and model comparisons, *in* Huber, B.T., McLeod, K.G., and Wing, S.L., eds., Warm climates in earth history: Cambridge, UK, Cambridge University Press, p. 297–318.

Rees, P.M., Ziegler, A.M., Gibbs, M.T., Kutzbach, J.E., Behling, P.J., and Rowley, D.B., 2002, Permian phytogeographic patterns and climate data/model comparisons: Journal of Geology, v. 110, p. 1–31, doi: 10.1086/324203.

Remy, W., 1982, Lower Devonian gametophytes—Relation to the phylogeny of land plants: Science, v. 215, p. 1625–1627.

Remy, W., and Remy, R., 1980, *Lyonophyton rhyniensis* nov. gen. et nov. sp., ein Gametophyt aus dem Chert von Rhynie (Unterdevon, Schottland): Argumenta Palaeobotanica, v. 8, p. 69–117.

Retallack, G.J., 1980, Late Carboniferous to Middle Triassic megafossil floras from the Sydney Basin, *in* Herbert, C., and Helby, R.J., eds., A guide to the Sydney Basin: Geological Survey of New South Whales, Bulletin, v. 26, p. 384–430.

Retallack, G.J., 1985, Fossil soils as grounds for interpreting the advent of large plants and animals on land: Royal Society of London Philosophical Transactions, ser. B, v. 309, p. 108–142.

Retallack, G.J., 1992, What to call early plant formations on land: Palaios, v. 7, p. 508–520.

Retallack, G.J., 1995, Permian-Triassic life crisis on land: Science, v. 267, no. 5194, p. 77–80.

Retallack, G.J., 1997, Earliest Triassic origin of *Isoetes* and quillwort evolutionary radiation: Journal of Paleontology, v. 71, p. 500–521.

Retallack, G.J., 2000, Ordovician life on land and early Paleozoic global change, *in* Gastaldo, R.A. and DiMichele, W.A., eds., Phanerozoic terrestrial ecosystems: Paleontological Society Papers, v. 6, p. 21–45.

Retallack, G.J., and Dilcher, D.L., 1981, Arguments for a glossopterid ancestry of angiosperms: Paleobiology, v. 7, p. 54–67.

Retallack, G.J., and Feakes, C.R., 1987, Trace fossil evidence for Late Ordovician animals on land: Science, v. 235, p. 61–63.

Retallack, G.J., Veevers, J.J., and Morante, R., 1996, Global coal gap between Permian-Triassic extinction and Middle Triassic recovery of peat-forming plants: Geology, v. 108, p. 195–207.

Rice, C.M., Ashcroft, W.A., Batten, D.J., Boyce, A.J., Caulfield, J.B.D., Fallick, A.E., Hole, M.J., Jones, E., Pearson, M.J., Rogers, G., Saxton, J.M., Stuart, F.M., Trewin, N.H., and Turner, G., 1995, A Devonian auriferous hot spring system, Rhynie, Scotland: London: Journal of the Geological Society, v. 152, p. 229–250.

Rice, C.M., Trewin, N.H., and Anderson, L.I., 2002, Geological setting of the Early Devonian Rhynie cherts, Aberdeenshire, Scotland: an early terrestrial hot spring system: London: Journal of the Geological Society of London, v. 159, p. 203–214.

Rich, F.L., Pirkle, F.L., and Arenberg, E., 2002, Palynology and paleoecology of strata associated with the Ohoopee River dune field, Emanuel County, Georgia: Palynology, v. 26, p. 239–256, doi: 10.2113/0260239.

Rolfe, W.D.I., 1980, Early invertebrate terrestrial faunas, *in* Panchen, A.L., ed., The terrestrial environment and the origin of land vertebrates: London, Academic Press, p. 117–157.

Rolfe, W.D.I., 1985, Early terrestrial arthropods: A fragmentary record: Royal Society of London Philosophical Transactions, ser., B., v. 309, p. 207–218.

Rolfe, W.D.I., 1990, Seeking the arthropods of Eden: Nature, v. 348, p. 112–113.

Rolfe, W.D.I., Durant, G.P., Fallick, A.E., Hall, A.J., Large, D.J., Scott, A.C., Smithson, T.R., and Walkden, G., 1990, An early terrestrial biota preserved by Viséan vulcanicity in Scotland, *in* Lockley, M.G., and Rice, A., eds., Volcanism and fossil biotas: Geological Society of America Special Paper 244, p. 1 3–24.

Rothwell, G.W., 1984, The apex of *Stigmaria* (Lycopsida), rooting organ of *Lepidodendrales*: American Journal of Botany, v. 71, p. 1031–1034.

Rubidge, B.S., ed., 1995, Biostratigraphy of the Beaufort Group (Karoo Basin): South African Committee for Stratigraphy, Biostratigraphic Series, no. 1, 46 p.

Rudolph, D.C., and Dickson, J.G., 1990, Streamside zone width and amphibian and reptile abundance: The Southwestern Naturalist, v. 35, p. 472–476.

Rudwick, M.J.S., 1997, Georges Cuvier, fossil bones, and geological catastrophes: Chicago, University of Chicago Press, 301 p.

Rull, V., 1998, Middle Eocene mangroves and vegetation changes in the Maracaibo Basin, Venezuela: Palaios, v. 13, p. 287–296.

Rull, V., 1999, Palaeofloristic and palaeovegetational changes across the Paleocene/Eocene boundary in northern South America: Review of Palaeobotany and Palynology, v. 107, p. 83–95, doi: 10.1016/S0034-6667(99)00014-7.

Sage, R.F., 2001, Environmental and evolutionary preconditions for the origin and diversification of the C_4 photosynthetic syndrome: Plant Biology, v. 3, p. 202–228, doi: 10.1055/s-2001-15206.

Sánchez-Villagra, M.R., Aguilera, O., and Horovitz, I., 2003, The anatomy of the world's largest extinct rodent: Science, v. 301, p. 1708–1710.

Sanders, R.B., 1968, Devonian spores of the Cedar Valley coal of Iowa, U.S.A: Journal of Palynology, v. 2–3, p. 17–32.

Savage, R.J.G., Domning, D.P., and Thewissen, J.G.M., 1994, Fossil Sirenia of the west Atlantic and Caribbean region. V. The most primitive known sirenian, *Prorastomus sirenoides* Owen, 1855: Journal of Vertebrate Paleontology, v. 14, p. 427–449.

Saward, S.A., 1992, A global view of Cretaceous vegetational patterns *in* McCabe, P.J., and Parrish, J.T., eds., Controls on the distribution and quality of Cretaceous coals: Geological Society of America Special Publication, v. 267, p. 17–35.

Schaal, S., and Ziegler, W., 1992, Messel—An insight into the history of life and the Earth: Oxford, Clarendon Press, 322 p.

Scheckler, S.E., 1986a, Floras of the Devonian-Mississippian transition, *in* Broadhead, T.W., ed., Land plants—Notes for a short course: University of Tennessee, Department of Geological Sciences, Studies in Geology, v. 15, p. 81–96.

Scheckler, S.E., 1986b, Geology, floristics, and palaeoecology of Late Devonian coal swamps from Laurentia (USA): Annales de la Société Géologique de Belgique, v. 109, p. 209–222.

Schneider, W., 1992, Floral successions in Miocene swamps and bogs in central Europe: Zeitschrift für Geologie Wissenschaften, v. 20, p. 55 5–570.

Schneider, W., 1995, Palaeohistological studies on Miocene brown coals of central Europe: International Journal of Coal Geology, v. 28, p. 229–248, doi: 10.1016/0166-5162(95)00019-4.

Schumm, S.A., 1968, Speculations concerning paleohydrologic controls of terrestrial sedimentation: Geological Society of America Bulletin, v. 79, 1573–1588.

Schwimmer, D.R., 2002, King of the crocodylians: The paleobiology of *Deinosuchus*: Bloomington, Indiana University Press, 220 p.

Scotese, C.R., 2001, Atlas of Earth History, Volume 1, Paleogeography: PALEOMAP Project, Arlington, Texas, 52 p. (www.scotese.com/earth.htm).

Scott, A.C., 1978, Sedimentological and ecological control of Westphalian B plant assemblages from west Yorkshire: Proceedings of the Yorkshire Geological Society, v. 41, p. 461–508.

Scott, A.C., 1980, The ecology of some Upper Paleozoic floras, *in* Panchen, A.L., ed., The terrestrial environment and the origin of land vertebrates: London, Academic Press, p. 87–115.

Scott, A.C., 1989, Observations on the nature and origin of fusain: International Journal of Coal Geology, v. 12, p. 443–476.

Scott, A.C., 2001, Roasted alive in the Carboniferous: Geoscientist, v. 11, p. 4–7.

Scott, A.C., and Rex, G.M., 1987, The accumulation and preservation of Dinantian plants from Scotland and its borders, *in* Miller, J., Adams, A.E., and Wright, V.P., eds., European Dinantian environments: Geological Journal Special Issue 12, p. 329–344.

Scott, A.C., Brown, R., Galtier, J., and Meyer-Berthaud, B., 1994, Fossil plants from the Viséan of East Kirkton, West Lothian, Scotland: Transactions of the Royal Society of Edinburgh: Earth Sciences, v. 84, p. 249–260.

Scott, A.C., Stephenson, J., and Chaloner, W.G., 1992, Interaction and coevolution of plants and arthropods during the Paleozoic and Mesozoic: Royal Society of London Philosophical Transactions, ser. B, v. 335, no. 1274, p. 129–165.

Scott, R.A., Barghoorn, E.S., and Leopold, E.B., 1960, How old are the angiosperms?: American Journal of Science, v. 258, p. 284–299.

Selden, P.A., Shear, W.A., and Bonamo, P.M., 1991, A spider and other arachnids from the Devonian of New York, and reinterpretations of Devonian Araneae: Palaeontology, v. 34, p. 241–281.

Semonin, P., 2000, American monster: How the nation's first prehistoric creature became a symbol of national identity: New York, NYU Press, 483 p.

Shear, W.A., Bonamo, P.M., Grierson, J.D., Rolfe, W.D.I., Smith, E.I., and Norton, R., 1984, Early land animals in North America—Evidence

from Devonian age arthropods from Gilboa, New York: Science, v. 224, p. 492–494.
Shear, W.A., Palmer, J.M., Coddington, J.A., and Bonama, P.M., 1989, A Devonian spineret—Early evidence of spiders and silk use: Science, v. 246, p. 479–481.
Shear, W.A., Gensel, P.G., and Jeram, A.J., 1996, Fossils of large terrestrial arthropods from the Lower Devonian of Canada: Nature, v. 384, p. 555–557, doi: 10.1038/384555a0.
Shear, W.A., and Selden, P.A., 2001, Rustling in the undergrowth: animals in early terrestrial ecosystems, in Gensel, P.G., and Edwards, D., eds., Plants invade the land: New York, Columbia University Press, p. 29–51.
Shearer, J.C., Staub, J.R., and Moore, T.A., 1994, The conundrum of coal bed thickness—A theory for stacked mire sequences: Journal of Geology, v. 102, p. 611–617.
Shearer, J.C., Moore, T.A., and Demchuk, T.D., 1995, Delineation of the distinctive nature of Tertiary coal beds: International Journal of Coal Geology, v. 28, p. 71–98, doi: 10.1016/0166-5162(95)00014-3.
Shinaq, R., and Bandel, K., 1998, The flora of an estuarine channel margin in the Early Cretaceous of Jordan: Freiberger, Forschungshefte, v. C474, p. 39–57.
Silverman, A.J., and Harris, W.L., 1967, Stratigraphy and economic geology of the Great Falls-Lewistown coal field, central Montana: Montana Bureau of Mines and Geology Bulletin, v. 56, p. 1–20.
Singh, R.A., 1999, Diversity of Nypa in the Indian subcontinent: Late Cretaceous to Recent: Palaeobotanist, v. 48, p. 147–154.
Skelly, D.K., Werner, E.E., and Cortwright, S.A., 1999, Long-term distributional dynamics of a Michigan amphibian assemblage: Ecology, v. 80, p. 2326-2337.
Skelly, D.K., Freidenburg, L.K., and Kiesecker, J.M., 2002, Forest canopy and the performance of larval amphibians: Ecology, v. 83, p. 983–992.
Skog, J.E., and Dilcher, D.L., 1992, A new species of Marsilea from the Dakota Formation in central Kansas: American Journal of Botany, v. 79, p. 982–988.
Sluiter, I.R.K., Kershaw, A.P., Holdgate, G.R., and Bulman, D., 1995, Biogeographic, ecological and stratigraphic relationships of the Miocene brown coal floras, Latrobe Valley, Victoria, Australia, in Demchuck, T.D., Shearer, J.C., and Moore, T.A., eds., International Journal of Coal Geology, v. 28, p. 277–302.
Smith, A.H.V., 1957, The sequence of microspore assemblages associated with the occurrence of crassidurite in coal seams of Yorkshire: Geological Magazine, v. 94, p. 345–363.
Smith, A.H.V., 1962, The palaeoecology of Carboniferous peats based on miospores and petrography of bituminous coals: Proceedings of the Yorkshire Geological Society, v. 33, p. 423–463.
Smith, J.B., Lamanna, M.C., Lacovera, K.J., Dodson, P., Smith, J.R., Poole, J.C., Giegengack, R., and Attia, Y., 2001, A giant sauropod dinosaur from an Upper Cretaceous mangrove deposit in Egypt: Science, v. 292, 5522, p. 1704–1706. 2001.
Smith, R.M.H., 1995, Changing fluvial environments across the Permian-Triassic boundary in the Karoo Basin, South Africa and possible causes of tetrapod extinction: Palaeogeography, Palaeoclimatology, Palaeoecology, v. 117, p. 81–104, doi: 10.1016/0031-0182(94)00119-S.
Smithson, T.R., 1989, The earliest known reptile: Nature, v. 342, p. 676–678, doi: 10.1038/342676a0.
Smithson, T.R., Carroll, R.L., Panchen, A.L., and Andrews, S.M., 1994, Westlothiana lizziae from the Visean of East Kirkton, West Lothian, Scotland, and the amniote stem: Transactions of the Royal Society of Edinburgh, v. 84, p. 383–412.
Speight, M.C.D., and Blackith, R.E., 1983, The animals, in Gore, A.J.P., ed., Mires: Swamp, bog, fen and moor: New York, Elsevier, Ecosystems of the World, v. 4A, p. 349–382.
Spicer, B., 2002, Changing climate and biota, in Skelton, P.W., Spicer, R.A., Kelley, S.P., and Gilmour, I., eds., The Cretaceous world: Cambridge, United Kingdom, Cambridge University Press, 360 p.
Spicer, R.A., and Parrish, J.T., 1986, Paleobotanical evidence for cool north polar climates in Middle Cretaceous (Albian Cenomanian): Geology, v. 14, p. 703–706, doi: 10.1130/0091-7613(1986)14<703:PEFCNP>2.0.CO;2.
Spicer, R.A., Parrish, J.T., and Grant, P.R., 1992, Evolution of vegetation and coal-forming environments in the Late Cretaceous of the north slope of Alaska, in McCabe, P.J., and Parrish, J.T., eds., Controls on the distribution and quality of Cretaceous coals: Geological Society of America Special Publication, v. 267, p. 177–192.
Statz, G., 1994, Neue dipteran (Nematocera) aus dem Oberoligozän von Rot: V. Familie Culiciden (steckmueken): Palaeontographica, v. 95, p. 108–121.
Steere, W.C., 1946, Cenozoic and Mesozoic bryophytes of North America: American Midland Naturalist, v. 36, p. 298–324.

Stein, W.E., Wight, C., and Beck, C.B., 1984, Possible alternatives for the origin of Sphenopsida: Systematic Botany, v. 9, p. 102–118.
Stevenson, J.C., 1988, Comparative ecology of submersed grass beds in freshwater, estuarine, and marine environments: Limnology and Oceanography, v. 33, p. 867–893.
Stewart, J.H., Poole, F.G., and Wilson, R.F., 1972, Stratigraphy and origin of the Chinle Formation and related Upper Triassic strata in the Colorado Plateau region: U.S. Geological Survey Professional Paper 690, 335 pp.
Stewart, R.E., Jr., 1996, Technical aspects of wetlands–Wetlands as bird habitats, in National water summary on wetland resources: U.S. Geological Survey Water Supply Paper 2425, 51 p.
Stewart, W.N., and Rothwell, G.R., 1993, Paleobotany and the evolution of plants: Cambridge, United Kingdom, Cambridge University Press, 521 p.
Stebbins, G.L., 1974. Flowering Plants: Evolution above the species level: Cambridge, Massachusetts, Harvard University Press, 399 p.
Stebbins, G.L., 1976. Seeds, seedlings, and the origin of angiosperms, in Beck, C.B., ed., Origin and early evolution of angiosperms: New York, Columbia University Press, p. 300–311.
Stockey, R.A., 1982, The Araucariaceae; an evolutionary perspective, in Taylor, T.N., and Delevoryas, T., eds., Gymnosperms; Paleozoic and Mesozoic: Review of Paleobotany and Palynology, v. 37, p. 133–154.
Stockey, J.R.A., Hoffman, G.L., and Rothwell, G.W., 1997, The fossil monocot Limnobiophyllum scutatum: Resolving the phylogeny of Lemnaceae: American Journal of Botany, v. 84, p. 355–368.
Størmer, L., 1976, Arthropods from the Lower Devonian (Lower Emsian) of Alken an der Mosel, Germany. Part 5: Myriapoda and additional forms, with general remarks on fauna and problems regarding invasion of land by arthropods: Senckenbergiana Lethaea, v. 57, p. 87–183.
Sun, G., and Dilcher, D.L., 2002, Early angiosperms from the Lower Cretaceous of Jixi, eastern Heilongjiang, China: Review of Palaeobotany and Palynology, v. 121, p. 91–112, doi: 10.1016/S0034-6667(02)00083-0.
Sun, G., Dilcher, D.L., Zheng, S., and Zhou, Z., 1998, In search of the first flower: A Jurassic angiosperm, Archaefructus, from northeast China: Science, v. 282, p. 1692–1695, doi: 10.1126/science.282.5394.1692.
Sun, G., Ji, Q., Dilcher, D.L., Zheng, S.L., Nixon, K., and Wang, X.F., 2002, Archaeofruntaceae, a new basal angiosperm family: Science, v. 296, p. 899–904, doi: 10.1126/science.1069439.
Taylor, W.C., and Hickey, R.J., 1992, Habitat, evolution, and speciation in Isoetes: Annals of the Missouri Botanical Garden, v. 79, p. 613–622.
Taylor, E.L., and Taylor, T.N., 1990, Antarctic paleobiology: Its role in the reconstruction of Gondwanaland: New York, Springer-Verlag, 261 p.
Taylor, T.N., 1988, The origin of land plants: Some answers, more questions: Taxon, v. 37, p. 805–833.
Tedford, R.H., and Harington, C.R., 2003, An arctic mammal fauna from the early Pliocene of North America: Nature, v. 425, p. 388–390, doi: 10.1038/nature01892.
Teichmüller, M., 1958, Rekonstruktion verscheidener Moortypen des Hauptflözen der Niederrheinischen Braunkohle: Fortschritte in der Geologie von Rheinland und Westfalen, v. 2, p. 599–612.
Teichmüller, M., 1962, Die Genese der Kohle: Compte Rendu du quatrième Congrès pour l'avancement des etudes de Géologie du Carbonifère, Heerlen, 1958, vol. 3, Ernst von Aelst, Maestricht, p. 699–722.
Teichmüller, M., 1982, Origin of the petrographic constituents of coal, in Stach, D.E., Mackowsky, M.-Th, Teichmüller, M., Taylor, G.H., Chandra, D., and Teichmüller, R., eds., Stach's textbook of coal petrology: Gebrüder Borntraeger, Berlin, p. 219–294.
Teichmüller, M., 1990, Genesis of coal from the viewpoint of coal geology: International Journal of Coal Geology, v. 16, p. 121–124, doi: 10.1016/0166-5162(90)90016-R.
Therrien, F., and Fastovsky, D.E., 2000, Paleoenvironments of early theropods, Chinle Formation (Late Triassic), Petrified Forest National Park, Arizona: Palaios, v. 15, p. 194–211.
Thewissen, J.G.M., Williams, E.M., Roe, L.J., and Hussain, S.T., 2001, Skeletons of terrestrial cetaceans and the relationship of whales to artiodactyls: Nature, v. 413, p. 277–281, doi: 10.1038/35095005.
Thieme, H., 1997, Lower Paleolithic hunting spears from Germany: Nature, v. 385, p. 807, doi: 10.1038/385807a0.
Thiessen, R., 1925, Origin of boghead coals: U.S. Geological Survey Professional Paper 1321, p. 121–138.
Thomas, L., 2002, Coal geology: New York, John Wiley and Sons, 384 p.
Thomasson, J.R., 1986, Fossil grasses: 1820–1986 and beyond, in Soderstrom, T.R., Khidir, W.H., Campbell, C.S., and Barkworth, M.E., eds., Grass systematics and evolution: Washington, D.C., Smithsonian Institution Press, p. 159–169.

Tidwell, W.D., 1975, Common fossil plants of western North America: Provo, Utah, Brigham Young University Press, 198 p.

Trewin, N.H., 1996, The Rhynie Cherts—An early Devonian ecosystem preserved by hydrothermal activity, in Bock, G.R., and Goode, J.A., eds., Evolution of hydrothermal ecosystems on Earth (and Mars?): Chichester, UK, John Wiley, Ciba Foundation Symposium 202, pp. 131–149.

Trewin, N.H., 1994, Depositional environment and preservation of biota in the Lower Devonian hot springs of Rhynie, Aberdeenshire, Scotland: Transactions of the Royal Society of Edinburgh: Earth Sciences, v. 84, p. 433–442.

Trewin, N.H., and Rice, C.M., 1992, Stratigraphy and sedimentology of the Devonian Rhynie Chert locality: Scottish Journal of Geology, v. 28, p. 37–47.

Truman, J.W., and Riddiford, L.M., 1999, The origins of insect metamorphosis: Nature, v. 401, p. 447–452, doi: 10.1038/46737.

Tschudy, R.H., Pillmore, C.L., Orth, C.J., Gilmore, J.S., and Knight, J.D., 1984, Disruption of the terrestrial plant ecosystem at the Cretaceous-Tertiary boundary, Western Interior: Science, v. 225, p. 1030–1032.

Tuross, N., and Dillehay, T.D., 1995, The mechanism of organic preservation at Monte Verde, Chile, and one use of biomolecules in archeological interpretation: Journal of Field Archeology, p. 97–110.

Udomkan, B., Ratanasthien, B., Takayasu, K., Fyfe, W.S., Sato, S., Kandharosa, W., Wongpornchai, P., and Kusakabe, M., 2003, Fluctuation of depositional environment in the Bang Mark Coal deposit, Krabi mine, southern Thailand: Stable isotope implication: ScienceAsia, v. 29, p. 307–317.

Uhl, N.W., and Dransfield, J., 1987, Genera Palmarum-a classification of palms based on the work of Harold E. Moore, Jr.: Lawrence, Kansas, Allen Press, 610 p.

Upchurch, G.R., and Dolye, J.A., 1981, Paleoecology of the conifers *Frenelopsis* and *Pseudofrenelopsis* (Cheirolepidiaceae) from the Cretaceous Potomac Group of Maryland and Virginia, in Romans, R.C., ed., Geobotany II: New York, Plenum Press, p. 167–202.

Vajda, V., Raine, J.I., and Hollis, C.J., 2001, Indication of global deforestation at the Cretaceous-Tertiary boundary by New Zealand fern spike: Science, v. 294, p. 1700–1702, doi: 10.1126/science.1064706.

Van der Burgh, J., and Zetter, R., 1998, Plant mega- and microfossil assemblages from the Brunssumian of "Hambach" near Dueren, B.R.D., in Ferguson, D.K., ed., Case Studies in the Cenophytic Paleobotany of Central Europe: Review of Palaeobotany and Palynology, v. 101, p. 209–256.

Van-Erve, A.W., and Mohr, B., 1988, Palynological investigations of the Late Jurassic microflora from the vertebrate locality Guimarota coal mine (Leiria, central Portugal): Neues Jahrbuch für Geologie und Paläontologie, Monatshefte, v. 4, p. 246–262.

Visscher, H., Brinkhuis, H., Dilcher, D.L., Elsik, W.C., Eshet, Y., Looy, C.V., Rampino, M.R., and Traverse, A., 1996, The terminal Paleozoic fungal event: Evidence of terrestrial ecosystem destabilization and collapse: Proceedings of the National Academy of Sciences of the United States of America (Ecology), v. 93, p. 2155–2158.

Wagner, R.H., Winkler Prins, C.F., and Granados, L.F., 1983, The Carboniferous of the world. I. China, Korea, Japan, and S.E. Asia: Madrid, Instituto Geológico y Minero de España, 243 p.

Walker, S., 2000, Major coalfields of the World: London, International Energy Agency, Coal Research, 130 p.

Wall, W.P., 1998, *Amynodontidae*, in Janis, C.M., Scott, K.M., and Jacobs, L.L., eds., Evolution of Tertiary mammals of North America: Cambridge, UK, Cambridge University Press, p. 583–588.

Wartmann, R., 1969, Studie uber die papillen-formingen Verdickungen auf der Kutikule bei *Cordaites* an material aus dem Westphal C des Saar-Karbons: Argumenta Palaeobotanica, v. 3, p. 199–207.

Weller, M.W., 1994, Freshwater marshes: Minneapolis, University of Minnesota Press, 192 p.

Westgate, J.W., and Gee, C., 1990, Paleoecology of a middle Eocene mangrove biota (vertebrates, plants, and invertebrates) from Southwest Texas: Palaeogeography, Palaeoclimatology, Palaeoecology, v. 78, p. 163–177, doi: 10.1016/0031-0182(90)90210-X.

White, M.E., 1990, The flowering of Gondwana: Princeton, New Jersey, Princeton University Press, 256 p.

Williams, C.J., Johnson, A.H., LePage, B.A., Vann, D.R., and Taylor, K.D., 2003a, Reconstruction of Tertiary *Metasequoia* forests. I. Test of a method for biomass determination based on stem dimensions: Paleobiology, v. 29, p. 256–270.

Williams, C.J., Johnson, A.H., LePage, B.A., Vann, D.R., and Sweda, T., 2003b, Reconstruction of Tertiary *Metasequoia* forests. II. Structure, biomass, and productivity of Eocene floodplain forests in the Canadian Arctic: Paleobiology, v. 29, p. 271–292.

Wilson, H.M., and Anderson, L.I., 2004, Morphology and taxonomy of Paleozoic millipedes (Diplopoda, Chilognatha, Archipolypoda) from Scotland: Journal of Paleontology, v. 78, p. 169–184.

Wing, S.L., 1987, Eocene and Oligocene floras and vegetation of the Rocky Mountains: Annals of the Missouri Botanical Garden, v. 74, p. 748–784.

Wing, S.L., and Boucher, L.D., 1998, Ecological aspects of the Cretaceous flowering plant radiation: Annual Review of Earth and Planetary Sciences, v. 26, p. 379–421, doi: 10.1146/annurev.earth.26.1.379.

Wing, S.L., and Sues, H.-D., 1992, Mesozoic and early Cenozoic terrestrial ecosystems in Behrensmeyer, A.K., Damuth, J.D., DiMichele, W.A., Potts, R., Sues, H.-D., and Wing, S.L., eds., Terrestrial ecosystems through time: Chicago, University of Chicago Press, p. 326–416.

Wnuk, C., and Pfefferkorn, H.W., 1984, Life habits and paleoecology of Middle Pennsylvanian medullosan pteridosperms based on an in situ assemblage from the Bernice Basin (Sullivan County, Pennsylvania, U.S.A.): Review of Palaeobotany and Palynology, v. 41, p. 329–351, doi: 10.1016/0034-6667(84)90053-8.

Wolbach, W.S., Gilmour, I., and Anderson, E., 1990, Major wildfires at the Cretaceous/Tertiary boundary in Sharpton, V.L., and Ward, P.D., eds., Global catastrophes in earth history: An interdisciplinary conference on impacts, volcanism, and mass mortality: Geological Society of America Special Paper 247, p. 391–400.

Wright, V.P., Taylor, K.G., and Beck, V.H., 2000, The palaeohydrology of Lower Cretaceous seasonal wetlands, Isle of Wight, Southern England: Journal of Sedimentary Research, v. 70, p. 619–663.

Xingxue, L., and Xiuyhan, W., 1996, Late Paleozoic phytogeographic provinces in China and its adjacent region: Review of Palaeobotany and Palynology, v. 90, p. 41–62, doi: 10.1016/0034-6667(95)00023-2.

Yamada, T., and Kato, M., 2002, *Regnellites nagashimae* gen. et sp. nov., The oldest macrofossil of Marsileaceae, from the Upper Jurassic to Lower Cretaceous of western Japan: International Journal of Plant Sciences, v. 163, p. 715–723, doi: 10.1086/342036.

Yang, S.-Y., Lockley, M.G., Greben, R., Erikson, B.R., and Lim, S.-Y., 1994, Flamingo and duck-like bird tracks from the Late Cretaceous and Early Tertiary: Evidence and implications: Ichnos, v. 4, p. 21–34.

Ziegler, A.M., 1990, Phytogeographic patterns and continental configurations during the Permian period, in McKerrow, W.S., and Scotese, C.R., eds., Palaeozoic Palaeogeography and biogeography: Geological Society [London], Memoir 12, p. 363–377.

Zoltai, S.C., and Pollett, F.C., 1983, Wetlands in Canada: Their classification, distribution, and use, in Gore, A.J.P., ed., Mires: Swamp, bog, fen and moor: New York, Elsevier, Ecosystems of the World, v. 4A, p. 245–266.

MANUSCRIPT ACCEPTED BY THE SOCIETY 28 JUNE 2005

Wetlands before tracheophytes: Thalloid terrestrial communities of the Early Silurian Passage Creek biota (Virginia)

Alexandru Mihail Florian Tomescu*
Gar W. Rothwell
Department of Environmental and Plant Biology, Ohio University, Athens, Ohio 45701-2979, USA

ABSTRACT

Early Silurian (Llandoverian) macrofossils from the lower Massanutten Sandstone at Passage Creek in Virginia represent the oldest known terrestrial wetland communities. Fossils are preserved as compressions in overbank deposits of a braided fluvial system. Specimens with entire margins and specimens forming extensive crusts provide evidence for in situ preservation, whereas pre-burial cracks in the fossils demonstrate subaerial exposure. Developed in river flood plains that provided the wettest available environments on land at the time, these communities occupied settings similar to present-day riverine wetlands. Compared with the latter, which are continuously wet by virtue of the moisture retention capabilities of soils and vegetation, Early Silurian flood-plain wetlands were principally abiotically wet, depending on climate and fluctuations of the rivers for moisture supply. Varying in size from <1 cm to >10 cm, fossils exhibit predominantly thalloid morphologies but some are strap-shaped or form crusts. Their abundance indicates that a well-developed terrestrial groundcover was present by the Early Silurian. Morphological and anatomical diversity of specimens suggests that this groundcover consisted of several types of organisms and organismal associations, some characterized by complex internal organization. Earlier microfossil finds at Passage Creek corroborate an image of systematically diverse but structurally simple communities, consisting only of primary producers and decomposers. Ten to fifteen million years older than the oldest previously known complex terrestrial organisms (e.g., *Cooksonia*), they provide a new perspective on the early stages of land colonization by complex organisms, whereby the earliest terrestrial communities were built by a guild of thalloid organisms and associations of organisms comparable to extant biological soil crusts.

Keywords: macrofossils, fluvial, Llandovery, complex, diversity, soil crusts, braided.

INTRODUCTION

In the modern world, the term "wetlands" brings to mind lush vegetation and water-rich soils. We tend to think that wetland ecosystems result primarily from high volumes of relatively constant moisture input. In reality, the moisture retention capabilities of the system play an equally important role as moisture input, and it is largely because of the abundant vegetation, with its soil-forming and water retention capacities, that wetlands are the continuously water-soaked environments that we know today.

---

*Present address: Department of Biological Sciences, Humboldt State University, Arcata, California 95521, USA

Tomescu, A.M.F., and Rothwell, G.W., 2006, Wetlands before tracheophytes: Thalloid terrestrial communities of the Early Silurian Passage Creek biota (Virginia), in Greb, S.F., and DiMichele, W.A., Wetlands through time: Geological Society of America Special Paper 399, p. 41–56, doi: 10.1130/2006.2399(02). For permission to copy, contact editing@geosociety.org. ©2006 Geological Society of America. All rights reserved.

This has been the situation for a good part of the interval since life colonized land surfaces, especially since the advent of lignophytes (Algeo et al., 2001). By contrast, during the earliest stages of land colonization, before the formation of organic soils and the development of a thick, multistoried groundcover, wetland landscapes must have been of a distinctly different nature. Until recently, a notably scarce fossil record for the earliest phases of terrestrial plant community development has left a distinct gap in our understanding of how wetlands developed through time. However, a growing body of new micro- and macrofossil evidence from basal Phanerozoic deposits now provides an opportunity to fill this hiatus. In this paper we review the oldest known wetland biotas, introduce recently discovered macrofossils from the Llandoverian of Virginia, and explore their implications for the evolution of wetland ecosystems and terrestrial life.

THE OLDEST WETLAND BIOTAS

Biotas of the Rhynie Chert and Battery Point Formation

The oldest currently known wetland biota preserved in situ is that of the Rhynie Chert (Fig. 1). World famous for exquisite cellular preservation of the fossils, this Early Devonian (Pragian) biota (Rice et al., 1995) includes terrestrial and fresh-water communities. Trewin et al. (2003) provide a comprehensive compilation of information on the genesis and environments of the Rhynie Chert. Communities of the Rhynie Chert occupied several ecological niches within a river system (flood plain, ponds, and lakes). Plants, animals, algae, cyanobacteria, and fungi are preserved as autochthonous or allochthonous assemblages by hydrothermal activity. The hot springs provided a source of silica-rich solutions to permineralize the organisms. Extensive studies of the different organisms in the Rhynie Chert have accumulated evidence that this biota includes most of the elements of the modern trophic chain (primary producers, decomposers, detritivores, and carnivores) and features several types of mutualistic as well as antagonistic associations (Taylor and Taylor, 2000; Shear and Selden, 2001, and references therein).

Only slightly younger, the Emsian (Early Devonian) Battery Point Formation of Gaspé (Quebec, Canada) preserves another well-documented wetland biota (Fig. 1). Hotton et al. (2001) produced detailed reconstructions of terrestrial communities inhabiting different environments of a fluvial-deltaic landscape on a tidally influenced coastal plain. Based on the distribution of compressed remains, those authors defined three types of fossil associations correlated with different environments inferred from sedimentary facies. These correlations suggested clade-based niche partitioning among dysaerobic wetland sites within interdistributary basins (dominated by zosterophylls and *Renalia*), more ephemeral near-channel environments (preferentially occupied by trimerophytes), and fully terrestrial riparian environments (with *Spongiophyton* and *Prototaxites*), which the authors tentatively related to contrasting life-history strategies.

The biotas of the Rhynie Chert and Battery Point Formation provide solid evidence that by the end of the Early Devonian diverse and complex wetland communities were present in fluvial systems. Although ancient, these communities reveal that Early Devonian ecosystems already had differentiated into

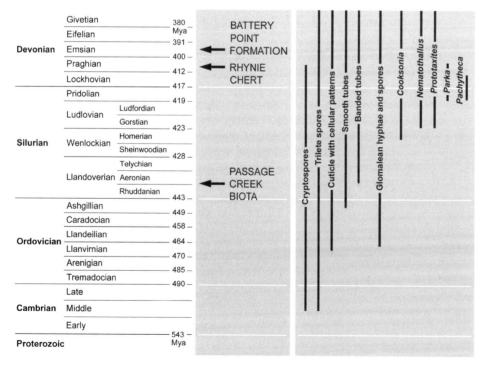

Figure 1. Stratigraphic ranges of microfossil types and macrofossil taxa considered in discussions of the colonization of land, and positions of localities significant for the fossil record of early wetland communities. Absolute ages from Palmer and Geissman (1999).

multiple trophic levels and that they were partitioned into several ecological niches occupied by organisms within the same trophic level (at least for the primary producers). In fact, by the Pragian riverine wetlands harbored communities that lacked only few of the physiognomic and trophic attributes of their modern counterparts (e.g., multistoried vegetation and, arguably, herbivory).

The presence of herbivory in the Rhynie Chert and Battery Point Formation biotas is a debated issue. Using a very restrictive definition of herbivory—as being exerted only by animals that routinely feed on leaves, shoots, and roots of living plants—Shear and Selden (2001, p. 49) have questioned the presence of this trophic level in the two biotas. On the other hand, use of a more inclusive definition allowed Labandeira (2002) to infer herbivory represented by boring, piercing and sucking, and spore consumption, on the basis of plant damage and dispersed coprolites containing plant fragments that are reported in the Rhynie Chert and Battery Point Formation biotas.

Not only are these well-characterized Early Devonian wetland biotas quite complex ecologically, but they also comprise phylogenetically derived primary producers: the bulk of the preserved biomass is contributed by a combination of non-tracheophytic and tracheophytic polysporangiophytes. Among embryophytes, a clade most commonly equated to Kingdom Plantae in modern systematics, polysporangiophytes form the crown group characterized by branching sporophytes bearing more than one sporangium. Defined as a clade by the presence of a multicellular sporophyte in the life cycle, embryophytes include three bryophytic clades (liverworts, hornworts, and mosses) in positions basal to and/or sister to the polysporangiophytes. Such phylogenetic relationships suggest that the fossil record should preserve wetland communities older than the Rhynie Chert and Battery Point Formation, and which would include earlier, bryophytic stages of embryophytic evolution. However, in the pre-Devonian fossil record microfossil evidence for embryophytes is not paralleled by abundant terrestrial and wetland macrofossils or by in situ preservation.

The Pre-Devonian Fossil Record

Microfossil evidence in the form of putative trilete spores from the Middle Cambrian Rogersville Shale in Tennessee (Strother, 2000) suggests that embryophytes were indeed present much earlier than previously documented by macrofossil evidence. This sets a minimum age for the earliest embryophytes (Fig. 1), but the marine depositional environment of the rocks (Rankey et al., 1994) leaves open all three alternatives for the origin of the spores, from an ecological standpoint: marine, freshwater, or terrestrial environments. The embryophyte macrofossil record begins much later, in the late Wenlockian (Fig. 1), and consists of polysporangiophytes (e.g., *Cooksonia*; Edwards and Feehan, 1980), rather than the bryophytic grade forms predicted by modern systematic studies (Mishler and Churchill, 1985). Attempts to document bryophytes in the earliest macro- and mesofossil embryophyte assemblages (Edwards, 2000) have been frustrated until now by the fragmentary state of fossils. Claims of bryophytic-grade plants based on a few problematic Ordovician fossils (reviewed by Retallack, 2000) have not been substantiated by further results. Current perceptions also are hampered by relatively narrow definitions of bryophyte-grade plants that are based on extant taxa, and that do not accommodate the inclusion of organisms exhibiting novel combinations of characters.

The Late Silurian witnessed rapid radiation and geographic expansion of embryophytes, as documented by the list of Silurian and Early Devonian fossil localities compiled by Edwards and Wellman (2001). At least some of these fossils represent terrestrial plants very likely growing in wetland communities, and the considerable taxonomic diversity displayed by embryophytes as early as the Late Silurian suggests that such communities were quite complex. However, all of these pre–Rhynie Chert embryophyte assemblages are allochthonous, and although they provide a broad idea of the types of plants, the transported state of the fossils restricts our ability to characterize the communities in which they grew.

Embryophytes are not the only macroscopic organisms found in the early terrestrial fossil record. Nematophytes and other problematic macrofossils sometimes considered in discussions of land plant origins (e.g., *Pachytheca*, *Parka*, briefly reviewed by Taylor, 1988), are also known as early as the Silurian. Of these, nematophytes appear earliest in the fossil record, and include several taxa characterized principally by a tubular internal organization. The biological relationships of nematophytes have been tentatively placed with different extant groups of algae and fungi, or with groups that have no living descendants, but a lack of preserved reproductive structures has frustrated attempts to conclusively resolve their systematic affinities. The oldest macrofossil nematophyte occurrence *(Nematothallus, Prototaxites)* is reported from late Wenlockian to Ludlovian strata in the Bloomsburg Formation of Pennsylvania (Strother, 1988) and is roughly contemporaneous with the oldest *Cooksonia* (Fig. 1). The Bloomsburg Formation represents fluvial (Epstein and Epstein, 1969; Fail and Wells, 1974; Dennison and Wheeler, 1975; Retallack, 1985; Driese et al., 1992) and tidally influenced (Epstein, 1993) settings at different locations. Detailed sedimentological studies will be needed to characterize the depositional environment at the fossil locality with confidence, but current interpretations and the age of the rocks place these fossils among the oldest macrofossil occurrences of both potentially terrestrial and wetland organisms. Problematic carbonaceous tubes lacking internal anatomy *(Eohostimella)* were described by Schopf et al. (1966) from nearshore marine deposits of late Llandoverian age in Maine, and geochemical studies by Niklas (1976) have suggested a vascular plantlike chemistry for these fossils. However, the marine depositional environment and the inconclusive morphology and anatomy have precluded unequivocal taxonomic assignment of *Eohostimella*. Although animals are not the focus of this paper, it is worth mentioning that occurrences of probable fresh-water arthropods have been reported from Ordovician nonmarine deposits (review in Retallack, 2000), and that the earliest unequivocal terrestrial animals (arachnids, centipedes, and scorpions) are known in the Pridolian (Shear and Selden, 2001).

A wide diversity of cuticle-like fragments and tubes that apparently lack extant counterparts also occur as dispersed microfossils as early as the Middle Ordovician (Fig. 1), and they have been compared with a wide range of fossil and living groups by different authors (e.g., Edwards, 1982, 1986; Gray and Boucot, 1977; Banks, 1975a, 1975b; Kroken et al., 1996; Kodner and Graham, 2001; Graham and Gray, 2001). However, their biological relationships remain largely unresolved because of a lack of diagnostic systematic characters and the dispersed nature of these fossils. Together with cryptospores and spores, cuticle-like fragments and tubes are traditionally associated with land organisms on the basis of (1) their resistance to degradation, and (2) their occasional recovery from continental (fluvial) deposits.

Resistance to degradation is interpreted as reflecting origin in desiccation-resistant organisms, or parts of organisms (Graham and Gray, 2001) that otherwise may have been submerged. Nevertheless, it is noteworthy that resistant microfossils (i.e., not degradable by palynological extraction techniques) are not produced exclusively by terrestrial organisms. They also occur as acritarchs, dinoflagellates, prasinophyceans and other algal cysts, chitinozoans, fungal spores, and eurypterid cuticle, among others. This indicates that resistant, fossilizable substances are not limited to organisms living in desiccation-prone environments, and therefore their presence alone cannot be used as evidence for life in terrestrial environments. More biochemical and geochemical studies on the different types of such resilient substances and their distribution among extant and fossil taxa are needed before this statement can be made with any degree of certainty. The fact that cryptospores, tubular fossils, and cuticle-like fragments have sometimes been recovered from fluvial deposits can be, and has been, used as evidence for their origin in nonmarine organisms. However, their dispersed nature precludes unequivocal attribution to organisms occupying specific continental environments (i.e., fresh-water versus terrestrial).

The Immediate Challenge

Prior to the late Wenlockian the fossil record of terrestrial organisms is so incomplete that the search for wetland biotas becomes synonymous with the search for terrestrial life. Communities of terrestrial microorganisms in the form of cyanobacteria and bacteria were probably established relatively early in the Precambrian, as has been proposed by numerous authors (e.g., Barghoorn, 1977; Siegel, 1977; Campbell, 1979; Golubic and Campbell, 1979; Gay and Grandstaff, 1980), and Retallack (2001) provides a comprehensive review of the types of evidence accumulated to date. Such communities may have produced the microfossils of uncertain affinities reported by Horodyski and Knauth (1994) from paleokarst cavities in 1.2 billion and 800 million year old rocks in California. It is natural to think that early terrestrial microbial communities would have thrived in the wettest available environments, in the proximity of marine and fresh-water bodies. Therefore, they probably represent the first terrestrial wetland communities.

The challenge before us today is to discover the origins of complex terrestrial life, and to decipher the succession of wetland communities that has culminated in the establishment of modern wetlands. Colonization of land by complex organisms appears to be a Phanerozoic phenomenon that may be intimately associated with the evolution of new clades and of new mutualistic associations (Pirozynski and Malloch, 1975; Pirozynski, 1981; Selosse and Le Tacon, 1998; Knoll and Bambach, 2000). In the interval between the establishment of terrestrial microbial communities and the first macrofossils of terrestrial organisms (e.g., *Cooksonia*), the fossil record previously has yielded only dispersed microfossils traditionally but not conclusively associated with land organisms (as discussed above). The newly recognized macrofossil evidence from this interval that is introduced below provides a first opportunity to begin to fill this gap by providing an organismal context for understanding the microfossil evidence, and by establishing a paleoecological framework for the most ancient complex terrestrial communities.

THE PASSAGE CREEK BIOTA

One notable exception to the general paucity of terrestrial macrofossils prior to the Wenlockian is the Passage Creek biota (Fig. 1). The Early Silurian (Llandoverian) lower Massanutten Sandstone preserves at Passage Creek, in Virginia, the oldest macroscopic evidence for complex terrestrial groundcover. Embedded in fine-grained partings representing overbank deposition in a braided river system, the Passage Creek fossils are the remains of wetland communities occupying riverine flood plains (see sections on the age and depositional environments of the fossil assemblages below).

Previous Studies

This locality, first reported by Pratt et al. (1975a, 1975b), is well known for microfossils, but it also yields the oldest macrofossil assemblages known from nonmarine deposits. Fossils occur in the lower member of the Massanutten Sandstone, the age of which is early–middle Llandoverian (Early Silurian). Pratt et al. (1978) were the first to study the fossils of the Massanutten Sandstone carefully. They mention and briefly describe macroscopic compressions up to several centimeters in length, exhibiting foliose habit or axial construction, and that tend to be fragmented, with irregular outlines. However, after clearing and examination of coalified matter with scanning electron microscopy (SEM), the compressions showed no internal cellular preservation or external details. The study by Pratt et al. (1978) focuses on organic residues obtained by bulk maceration of rock fragments. These authors describe several types of dispersed microfossils: smooth and banded tubular elements, cuticular and cellular sheet fragments, trilete spores, spore tetrads, cryptospores, and septate hyphal filaments, sometimes apparently associated in different combinations. Because of the fluvial origin of the fossiliferous rocks, Pratt et al. (1978) consider this fossil assemblage to represent the oldest evidence of nonvascular, thalloid land plants, which they compare to nematophytes.

In a subsequent study, Niklas and Pratt (1980) subjected carbonaceous compressions from the Massanutten Sandstone to biogeochemical analyses and found small amounts of chemical constituents that they interpreted as potential degradative by-products of a lignin-like moiety. Niklas and Smocovitis (1983) studied shale partings from Passage Creek containing small (0.26–2.75 mm) compressions. Macerating individual fossils, they described cuticle-like fragments and three types of tubular elements, including an interesting fragment of tissue that incorporates longitudinally aligned smooth and banded tubes. On the basis of the disposition of the tubes (several smooth walled tubes surrounding two to three larger banded tubes), and by analogy with present-day embryophytes, they tentatively interpreted the strand of tubular cells as conducting tissue. The authors nevertheless emphasize that until more is known about anatomy and morphology, the systematic affinities of organisms represented by the fossils will remain conjectural.

Material and Methods

Fossils occur in the lower Massanutten Sandstone, along Route 678 at the gap of Passage Creek in Shenandoah County, 6 km southeast of Strasburg, Virginia, USA (38°56'N; 78°18'W) (see Figure 1 in Pratt et al., 1978). The Llandoverian sedimentary succession is well exposed in road cuts and on the slopes of Green Mountain. Lithology of the formation consists of thick and very hard sandstone beds and discontinuous, finely to thickly laminated fossiliferous siltstone, and very fine sandstone with minor shale, forming discontinuous partings (Fig. 2). Thickness of the partings ranges from a few centimeters to over 10 cm. The somewhat softer shale tends to weather back from the surface, and the hardness of the adjacent sandstone beds renders sampling difficult, especially along road cuts. Sampling the more extensive outcrops on the slopes of Green Mountain allowed us to maximize the size of shale samples and consequently the surface of bedding planes exposed in them. Two distinct siltstone and shale partings 15–25 cm thick were sampled on their whole thickness, 0.8–1 m of length, and depths ranging from 0.3 to 0.8 m. Samples thus obtained totaled hundreds of fossils visible on freshly split surfaces, and considerably more on the closely spaced (1–5 mm) unexposed bedding planes. This was important for observations on the size and morphology of the fossils, as well as on their patterns of spatial distribution.

The abundant macrofossils are preserved as compressions that form black carbonaceous films of varying thicknesses. Fresh breaks in the rock usually reveal most of each compression on the bedding plane, where the outlines of the fossils seem angular and fragmentary at first glance. However, this appearance is often misleading. Due to the low fissility of the rock and to the position of the fossils, which are rarely perfectly flat, the original margins of the fossils most often remain unexposed. The cause for this is very likely the sedimentary microstructure of the fossiliferous siltstones that reflects irregular sedimentation at the millimeter and centimeter scale. Careful degagement of both part and counterpart is therefore needed to uncover the original outline of each fossil. Prior to degagement rock samples were placed in 20% hydrofluoric acid for 1–2 min. This enhanced the contrast between the fossils and the surrounding mineral matrix, and softened the surface layer of the sample, allowing for easier preparation. Degagement was then performed under a dissecting microscope, with the samples immersed in water, using dissecting needles and a scalpel. Most of the images that document external morphology of the fossils were taken with the samples immersed in water.

The carbonaceous material of the compressions is very brittle. Because of profuse fissuring it tends to disaggregate into minute pieces upon maceration of the rocks, prohibiting separation of whole fossils from the sedimentary matrix. Therefore, we documented internal organization of the fossils from individual unitary carbonaceous fragments removed by hand from the compressions, or obtained by bulk maceration. When accompanied by mineral matrix, the fragments were passed through 40% hydrofluoric acid to remove siliceous sediment, and then rinsed in distilled water. Other carbonaceous fragments were recovered from bulk maceration of small pieces of fossiliferous rock.

The fragments were subjected to different treatments depending on the investigation techniques that were to be utilized. Some fragments were bleached using household-grade sodium hypochlorite for eight to nine days, followed by progressive rinsing in distilled water, dehydration in a graded ethanol series and xylene, and mounting on microscope slides with Eukitt

Figure 2. Lower Massanutten Sandstone, Passage Creek, Shenandoah County, Virginia, Llandoverian. Fossiliferous siltstone and very fine sandstone (between arrows) forms relatively thin, finely laminated, and discontinuous partings separating the sandstone beds at Green Mountain. Hammer for scale, 27.5 cm.

(Calibrated Instruments, Hawthorne, New York). For SEM the carbonaceous fragments were dehydrated in ethanol, mounted on aluminum stubs, and gold-coated. Thin sectioning for light microscopy was performed in parallel with ultra-thin sectioning for transmission electron microscopy (TEM). Carbonaceous fragments dehydrated in an ethanol series were immersed in propylene oxide and subsequently embedded in Epon-type resin (Electron Microscopy Sciences, Fort Washington, Pennsylvania). Thin sections (0.5–0.7 μm) and ultra-thin sections (60–80 nm) were cut using glass knives on a Reichert Ultracut microtome. Thin sections were mounted on glass slides for light microscopy, using Eukitt, and ultra-thin sections were picked on copper grids. Experiments showed that fixation prior to resin embedding, and staining of ultra-thin sections on the copper grids have no influence on the quality of the sections or differential contrast of the images. Therefore these procedures were omitted.

Imaging of the specimens was realized with Leaf Lumina (Leaf Systems Inc., Southboro, MA), and PhotoPhase (Phase One A/S, Frederiksberg, Denmark) digital scanning cameras, using a macro lens mounted on a Leitz Aristophot bellows camera, or using the Aristophot in conjunction with a Zeiss WL compound microscope for light microscopy. Scanning electron microscopy was performed on a Zeiss DSM 962 digital scanning microscope, and ultra-thin sections were observed and imaged at 80 kV using a Zeiss transmission electron microscope. Specimens are reposited in the Ohio University Paleobotanical Herbarium as numbers 15980–16001.

Extensive systematic studies of the Passage Creek fossils are in course of completion. Because of the specifics of both the methods used and the fossil preservation, such studies are time-consuming and will be the focus of future publications. The results presented here are based on qualitative surveys of the samples and on the most significant finds of the fossil characterization work completed to date. Visual survey of all available specimens allowed us to broadly constrain the size range of fossils, as well as to assess morphological diversity within the assemblages. Internal organization of the fossils was observed on thin sections from 35 distinct unitary fragments out of which 16 came from five individual fossils, and 19 were recovered by bulk maceration of small samples. Even at this stage, the finds nevertheless allow us to build an unforeseen picture of some of the earliest wetland biotas that we considered of high interest for the broad paleoecological and evolutionary foci of the present volume.

Diversity and Structure of Early Silurian Wetland Communities

The external morphology and internal anatomy of the macrofossils in the Passage Creek assemblages, as well as the wide spectrum of dispersed microfossils described by Pratt et al. (1978) in association with the macrofossils, all contribute to an image of considerable diversity of the Llandoverian wetland communities. The organic material in the fossils exhibits a variety of textures, and the fossils occur as crusts or discrete specimens displaying ellipsoidal, lobed, elongated, and irregular morphologies, as well as a wide range of sizes. Examination of the internal organization of the fossils reveals several distinct types of construction, and indicates the presence of complex organisms.

Morphological Diversity of Macrofossils

Macrofossils range in size from smaller than 1 mm to over 11 cm in greatest dimension (Figs. 3A–D, 4A–F). At least two different general types of textures have been observed in the compressions (Fig. 3C at arrows). One type is represented by more-or-less smooth, relatively thick, black and continuous organic films, while the other is characteristic of much thinner films with profuse small-scale discontinuities that form a checkered, salt-and-pepper pattern and give the appearance of gray color. Most of the fossils for which we documented external morphology exhibit rounded outlines suggestive of very little or no fragmentation. Sometimes the edges of the fossils are recurved. For example, the edges of the fossil in Figure 4B go down into the matrix (arrows). Some of the fossils display numerous cracks in the carbonaceous material that are identical to those that form in modern biological soil crusts (Fig. 3C at arrowhead). Careful observations showed that the cracks in the fossils are not related to discontinuities of the surrounding rock. This demonstrates that the cracks formed prior to burial in sediment and fossilization of the organisms, and therefore most likely resulted from desiccation due to subaerial exposure.

Morphologically, a first distinction can be made between largely continuous crusts and discrete fossils. The largest dimension of a continuous crust observed to date is 11 cm (Fig. 3D) and is limited by the size of the rock sample. The crust displays highly irregular voids with sizes ranging from less than 1 mm to several cm in size. Both types of texture outlined above occur in this crust, but the limits between them are difficult to determine.

Discrete fossils fall into two main morphological classes: (1) fossils with roughly isodiametric outlines that we term thalloid (Fig. 4A–D), and (2) elongated, strap-shaped fossils (Fig. 4E, 4F). Thalloid fossils largely dominate the assemblages at Passage Creek. They range from less than 1 cm up to several centimeters across, and often display rounded margins and more-or-less pronounced lobes (lobe sinuses at arrowheads on Figures 4A–D). We have documented specimens with deltoid (Fig. 4A, 4D) and oval (Fig. 4C) outlines, as well as compressions with more irregular outlines (Fig. 4B, 4D). Some of the fossils exhibit short protrusions (Fig. 4A, 4C, 4D, at arrows) that may have been involved in attachment of the organisms to the substrate. However, additional specimens that display this type of feature need to be studied, and anatomical evidence for differentiation of an attachment structure needs to be substantiated. The surface of the compressions sometimes reveals differences in the thickness of organic matter, as seen in Figure 4A between the thicker (darker) central part of the specimen and the thinner (lighter) two margins that diverge from the base. The thickness of the fossils rarely exceeds 50 μm. Other morphological features, such as potential reproductive structures, are absent from the studied specimens.

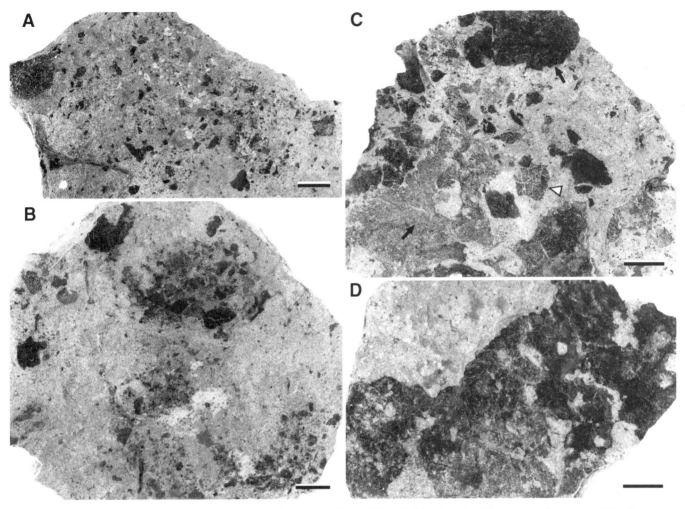

Figure 3. Lower Massanutten Sandstone, Passage Creek, Shenandoah County, Virginia, Llandoverian. Decimeter-scale patterns of fossil arrangement on bedding planes. (A) Predominance of small-sized fossils, evenly distributed. OUPH 15980. (B) Predominance of small-sized fossils, forming agglomerations. OUPH 15981. (C) Predominance of larger fossils. Two different types of texture can be observed: one is represented by thicker, black, and continuous compressions with definite margins (upper arrow), while the other is characteristic of thinner compressions with profuse small-scale discontinuities (lower arrow), sometimes lacking definite margins. Some of the fossil crusts display cracks (arrowhead) indicative of desiccation due to subaerial exposure. OUPH 15982. (D) Extensive carbonaceous crust. The limit at the upper left is not the edge of the fossil, but rather mineral matrix overlying the fossil. OUPH 15983. All scale bars 1 cm.

Elongated, strap-shaped fossils are quite rare at Passage Creek. Their morphology bears resemblance to axial forms usually associated to embryophytes. However, axial external morphology implies axial arrangement of anatomical features both at the outside and inside the body part, and is generally associated with radial symmetry. The ends of the two elongated fossils found so far at Passage Creek (Fig. 4E, 4F) exhibit no conclusive evidence for breaking from longer organisms, and until we find evidence for radial symmetry and axial anatomy they cannot be considered axial. A similar approach was taken by Strother (1988) in describing ribbon-shaped *Nematothallus* specimens from the Bloomsburg Formation, which he termed taenioid. At Passage Creek strap-shaped fossils are considerably thicker than thalloid ones and can reach 160 µm. The two fossils are 5.6 × 0.6 cm and 8 × 1.2 cm in size, and exhibit somewhat wavy margins. While the smaller fossil (Fig. 4E) is slightly curved, with apparently rounded terminations, the larger one (Fig. 4F) is straight and displays a two-pronged termination (the other end is truncated by the edge of the rock sample).

Internal Organization

Thin sectioning of the carbonaceous material has shown that sections 0.5–0.7 µm thick are translucent and appear colored in hues of brown, whereas sections thicker than 0.9 µm are opaque and black. Sectioning, together with scanning electron microscopy, also revealed dramatic effects of diagenesis that obscured the cellular structure of the original organisms. The internal anatomy of macrofossils as observed in light microscopy of

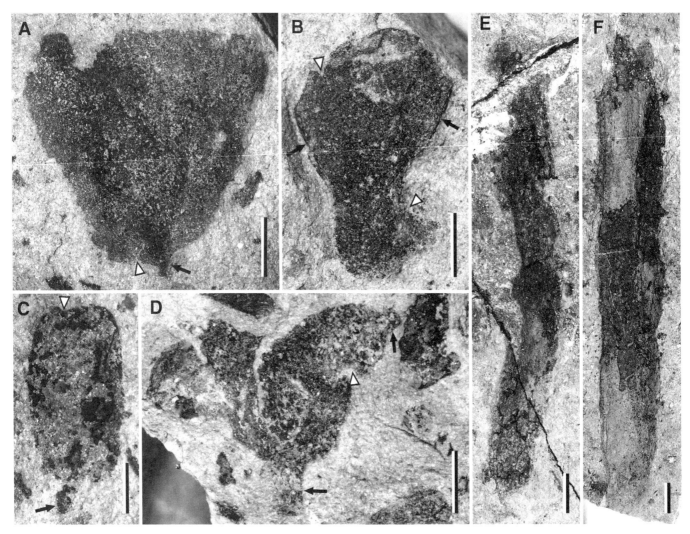

Figure 4. Lower Massanutten Sandstone, Passage Creek, Shenandoah County, Virginia, Llandoverian. Thalloid (A–D), and strap-shaped (E, F) macrofossils. (A) Thalloid specimen with a deltoid shape, undulating margins, a possible attachment protrusion (arrow), and a strongly incised lobe sinus (arrowhead). It displays a thicker and darker central area and thinner areas along the two margins that diverge from the base; the upper left side feature is an artifact of rock splitting, and not an actual lobe, as shown by degagement of the counterpart of the fossil. OUPH 15984. Scale bar 5 mm. (B) Thalloid specimen with irregular shape, lobes (lobe sinuses at arrowheads), and edges recurved down into the mineral matrix (at arrows). OUPH 15985. Scale bar 5 mm. (C) Oval thalloid specimen with possible attachment protrusion (arrow) and very shallow lobation (lobe sinus at arrowhead). OUPH 15986. Scale bar 2 mm. (D) The central part of the image shows two partly overlapping thalloid specimens, one with a deltoid shape and a possible attachment protrusion (lower arrow), and the other with a possible attachment protrusion (upper arrow) and lobed (lobe sinus at arrowhead). OUPH 15987. Scale bar 2 mm. (E) Strap-shaped specimen with rounded terminations. OUPH 15988. Scale bar 5 mm. (F) Strap-shaped specimen with a two-pronged termination. OUPH 15989. Scale bar 5 mm.

cross sections can be nevertheless classified into five main types (Fig. 5A–J) based on layering and on the types and organization of organic matter within the different layers. In some instances, anatomical features observed under light microscopy in cross sections could be related to features revealed by SEM and TEM.

Type 1 (Fig. 5A). Type 1 fossils exhibit an anastomosing pattern that consists of micron-thin vertically undulating features. Representing either filaments or highly discontinuous sheets of organic matter, they form a continuous intertwining network 25–90 μm thick in the plan of the compression. The network is denser toward the two faces of the carbonaceous film. Voids within the fossil (arrow) have consistent outlines that can be followed in serial sections.

Type 2 (Figs. 5B–D, G, 6A, 6C, 6D). This type consists of two layers of dense organic matter sometimes separated by a median region containing sporadic organic material. The two outer layers display irregular surfaces, occasional discontinuities and punctures, variable thickness (2–12 μm), tangential fissures, and may be in turn split into thinner sub-units (Figs. 5B, 6A, 6C). Figures 5C and 5D represent the part and counterpart of a

Figure 5. Lower Massanutten Sandstone, Passage Creek, Shenandoah County, Virginia, Llandoverian. Transmitted light micrographs of cross sections illustrating the main types of internal anatomy displayed by the fossils (see text for detailed description of types). (A) Type 1. The arrow points to a void with consistent outlines that can be followed in serial cross sections. OUPH 15990. (B) Type 2. The black arrowhead indicates organic material of a different nature reflected by lighter color and different texture. The white arrowhead points to agglomeration of minute dark dots nested between the two outer layers (similar with the feature shown in SEM, Fig.5C, 5D). OUPH 15991. (C, D) Separated part and counterpart of a specimen with Type 2 anatomy. OUPH 15992 and 15993, respectively. (E) Type 3. OUPH 15994. (F) Thicker section of the same specimen as in Figure 5E, showing diaphanous organic matter in the median region (arrowhead). OUPH 15994. (G) Part of a specimen with Type 2 anatomy displaying diaphanous organic matter (light gray, at center). OUPH 15995. (H) Type 4. OUPH 15996. (I) Detail of the same specimen as in Figure 5H featuring molds of large euhedral crystals (arrowhead), and of agglomerations of very small crystals (arrow). OUPH 15996. (J) Type 5. Arrowhead indicates mold of euhedral crystal. OUPH 15997. All scale bars 20 μm.

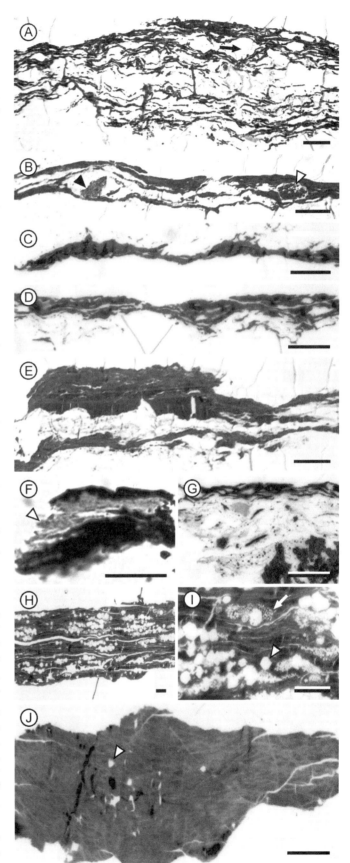

specimen with Type 2 internal anatomy. The outer layers exhibit a laminar structure consisting of largely parallel sheets (around 1 μm thick) of variable densities (lighter and darker), not very well individualized. Where the separating median region is absent the two layers appear continuous (right side of Fig. 5B). The median region has variable thickness and contains fragments similar to the outer layers, as well as organic material of an apparently different nature reflected by lighter color and different texture (Fig. 5B, black arrowhead; 5G).

Regions where minute dark dots can be seen in transverse sections are sometimes nested between the two outer layers (Fig. 5B, white arrowhead). In SEM they appear as features consisting of more-or-less regularly constricted filaments (1 μm or smaller across) with a beadlike appearance, tightly packed in a tangle with no consistent pattern (Figs. 6C at arrow, and 5D). These features are reminiscent of the packing of vesicular arbuscular mycorrhizal hyphae inside plant root cells.

Type 3 (Figs. 5E, 5F, 6B). Type 3 internal organization is similar to Type 2 in that it involves two outer layers separated by a median region (Fig. 5E). While one of the outer layers is similar to those of Type 2 fossils, the other layer (shown as the upper layer; orientation of the fossil is arbitrary) consists of two sub-layers and reaches 27 μm in thickness. The outermost sub-layer exhibits a laminar structure comparable to the one described in Type 2. The inner sub-layer is denser (darker) and massive, and has variable thickness and vertical cracks. A three-dimensional SEM image of the two sub-layers is shown in Figure 6B. The two-parted structure of the outer layer can be replaced laterally by a structure similar to that encountered in Type 2 (right side of Fig. 5E), suggesting that Type 2 and Type 3 may represent variations of the same type of internal organization.

The median region contains organic material that is less dense and very light in color, referred to here as diaphanous organic matter. Almost transparent in 0.5-μm-thick sections, it becomes more obvious only in thicker sections (Fig. 5F at arrowhead). The presence of diaphanous organic matter (Fig. 5G) associated with the specimen shown in Figure 5D suggests that it is probably characteristic of the Type 2 internal organization as well.

Difficult to observe in light microscopy, the structure of diaphanous organic matter is revealed by TEM (Fig. 7) and demonstrates preservation of unexpected levels of detail in the organic material. The diaphanous material forms multiple discrete, mostly thin layers (100–150 nm) of homogeneous density, but also thicker layers (460 nm, at arrow) displaying variations of density. In TEM the denser organic matter of the outer layers displays a stratified structure of tightly packed microlaminae 80–230 nm thick, with gradational boundaries between them (Fig. 7, lower right).

Type 4 (Fig. 5H, 5I). This type consists of material with laminar structure that is disrupted by tangential fissures and crystal growth. The laminar structure is similar to that described in Type 2, with undulating darker and lighter laminae, but the thickness

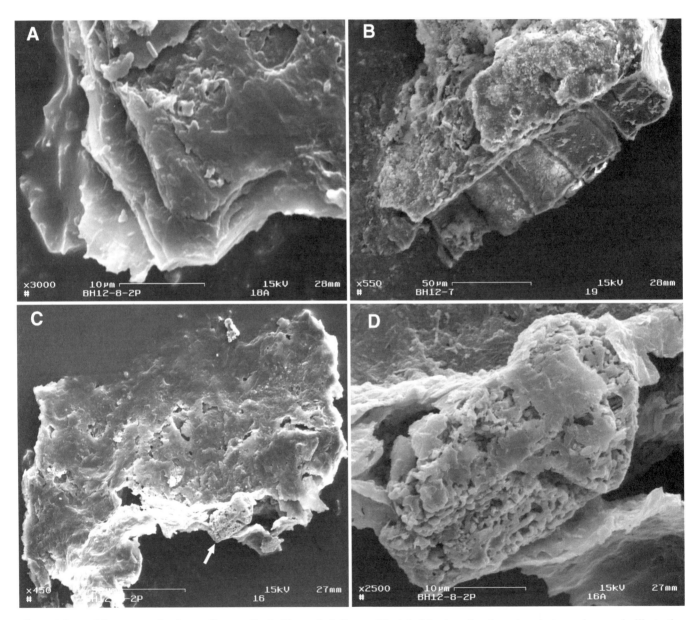

Figure 6. Lower Massanutten Sandstone, Passage Creek, Shenandoah County, Virginia, Llandoverian. Scanning electron micrographs illustrating the three-dimensional geometry of specimens with Type 2 (A, C, and D) and Type 3 (B) internal anatomy. (A) Type 2 anatomy. Surface features (punctures, discontinuities) and irregular layering. OUPH 15998. (B) Type 3 anatomy. Thin outermost laminar sub-layer and subjacent massive layer with cracks perpendicular to layering. OUPH 15999. (C) Type 2 anatomy. Specimen with irregular, punctured surface and feature (arrow) similar to the agglomeration in Figure 5B. OUPH 15998. (D) Detail of feature in Figure 6C. OUPH 15998.

of the fossil reaches 160 μm. There are lenticular areas 10–25 μm long and 5–15 μm thick that represent molds of crystals or clusters of crystals removed during hydrofluoric acid dissolution for fossil extraction. Crystals fall into two size classes: the larger tend to be isodiametric, 4–12 μm across, whereas the smaller are 0.5 μm or smaller and are grouped in areas roughly the size of the larger crystals (Fig. 5I at arrow). The crystal molds exhibit mainly hexagonal outlines in section (Fig. 5I at arrowhead), and their euhedral habit with sharp angles indicates that they developed within the fossils by solution precipitation. Examination of relationships between voids left by the crystals and the laminar structure of the fossil demonstrates that growth of crystals was mainly disruptive. This suggests that crystal growth occurred after burial of the fossils, when it could not be accommodated through expansion of the organic material, tightly embedded in the sedimentary matrix.

Type 5 (Fig. 5J). Type 5 fossils exhibit a massive structure traversed by what appears to be a very dense network of fine fissures (1 μm or thinner) that are filled with light organic matter. The network is superimposed on a pattern of diffuse zones of darker and lighter material. Presence of a few euhedral crystal molds or voids (arrowhead) that disrupt the fissures indicates that if the fissures are the result of postdepositional deformation of the fossils, then the growth of crystals occurred after fossilization of the organic material.

Apart from the above methods of investigation, sodium hypochlorite bleaching of the carbonaceous material in the fossils can reveal recognizable structures. Such was the case with the strap-shaped specimen shown in Figure 4E, where prolonged bleaching brought the jet-black coaly material to hues of light brown, orange, and yellow, and revealed the presence of filamentous structures embedded in an apparently amorphous matrix (Fig. 8). Although the filaments appear lighter in color in reflected light (Fig. 8A), viewed in transmitted light they consist of denser, opaque matter (Fig. 8B). The filaments run roughly parallel to each other and to the length of the fossil, and are 30–40 μm wide. Some of them can be followed over lengths of around 1 mm, but show no evidence for branching.

Systematic Affinities of the Passage Creek Fossils

All of these observations reflect the presence of communities comprising primarily thalloid organisms and associations of organisms that can be aggregated into more extensive mats. Morphological and anatomical data obtained so far reveal not only diversity, but also complexity of the fossils, indicating that several types of complex organisms were present in the groundcover of Llandoverian wetlands.

Thalloid growth is not restricted to a single group of organisms but characterizes a very diverse assortment of extant and fossil groups and mutualistic symbiotic associations, all of which are potential producers of the Massanutten Sandstone fossils. These groups include cyanobacterial colonies, algae (charophyceans and others), fungi, lichens, and bryophytes (liverworts and hornworts), all of which are also encountered as constituents of the

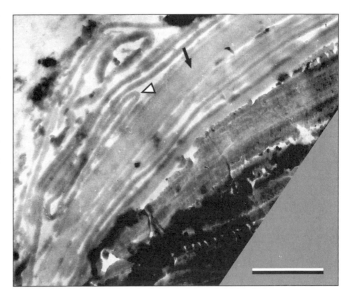

Figure 7. Lower Massanutten Sandstone, Passage Creek, Shenandoah County, Virginia, Llandoverian. Transmission electron micrograph illustrating diaphanous organic matter (light gray) that forms multiple discrete and mostly thin layers of homogeneous density, sometimes folded (arrowhead), and thicker layers (arrow) with variations of density. Denser organic matter of the outer layers (darker, at lower right) displays a stratified structure of tightly packed microlaminae with gradational boundaries between them. OUPH 16000. Scale bar 2 μm.

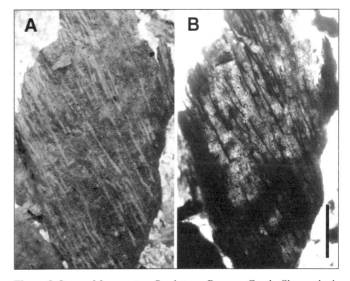

Figure 8. Lower Massanutten Sandstone, Passage Creek, Shenandoah County, Virginia, Llandoverian. A fragment of the strap-shaped specimen in Figure 4E bleached in sodium hypochlorite reveals filamentous structures more or less parallel to each other and to the length of the fossil, and incorporated in an apparently amorphous matrix. The filaments appear lighter in color in reflected light (A), but viewed in transmitted light they consist of denser, opaque matter (B). OUPH 16001. Scale bar 0.5 mm.

more loosely defined associations known as biological soil crusts, as well as the fossils *Nematothallus, Parka, Spongiophyton,* and *Protosalvinia.*

An animal origin for these fossils is unlikely as none of the numerous compressions yielded by our extensive sampling exhibit consistent and regular shapes characteristic of fragments from animals with a unitary body. Cuticle-like fragments for which a possible animal origin has been discussed (Gensel et al., 1990) have been found at this locality (Pratt et al., 1978, and our observations). However, no fragments of appendages or other unequivocal animal fossils like those described by Gray and Boucot (1994) from the Tuscarora Formation are present in the microscopic fraction of maceration preparations.

Resolution of the systematic affinities of the Passage Creek thalloid fossils necessarily involves comparisons with the morphology and anatomy of all of the above organisms and associations of organisms of the thalloid guild. But before such comparisons are possible, we need to document and clarify three main types of correlations that are crucial in understanding the organisms that produced the fossils: (1) Correlation between anatomy and external morphology. This involves correlations among anatomical features revealed by different investigation techniques (cross sections, bleached specimens, SEM), and correlations between these and the external morphology of the fossils. (2) Correlation between anatomy and ultrastructure. This involves understanding the intimate structure of different types of anatomical elements, and of the relationships between them. (3) Correlation between features of the macrofossils and the components of microfossil assemblages described in the same rocks. Thorough documentation of all of these relationships is very important for understanding and reconstructing the original organisms. Such documentation will allow us to compare the fossils with different living groups of similar morphology, in order to reveal the systematic affinities of the fossils.

Even at this stage of research, we can know something about potential systematic affinities of the Passage Creek organisms from the dispersed microfossils described from the same sediments by Pratt et al. (1978), and these complement the image of diversity projected by the macrofossils. Trilete spores and tetrahedral tetrads are hallmarks of the embryophyte reproduction, and their presence in the Massanutten Sandstone microfossil assemblages suggests that embryophytes were present among the thalloid communities in the Llandoverian wetlands. This corroborates the morphology predicted for the hypothetical embryophyte archetype by Mishler and Churchill (1985), which features a thalloid gametophyte with single sessile sporangium.

The dispersed fungal hyphae are compared by Pratt et al. (1978) to dematiaceous hyphomycetes of the Fungi Imperfecti, on the basis of morphological features such as size, branching, and septation. As such they are interpreted as evidence (the earliest) for higher fungi. Presently some in the scientific community are reluctant to accept reports of earliest occurrences of dispersed fungal microfossils in the absence of conclusive evidence that sample contamination with recent material during sampling, handling, and processing can be ruled out completely. Before this issue is addressed with scientific arguments dealing with the fossil material itself, the Passage Creek fungi need to be treated with caution. However, more recently Redecker et al. (2000) have reported dispersed glomalean hyphae and spores from the mid-Ordovician Guttenberg Formation of Wisconsin (Fig. 1), in a context that excluded the possibility of contamination with recent material (Redecker et al., 2000, p. 1921, note 16). This finding indicates that fungal groups that are presently encountered only in mycorrhizal associations with embryophytes were present long before the Silurian, and suggests that some of the thalloid bionts at Passage Creek may indeed represent or include a fungal component.

The systematic affinities of membranous cellular sheets, sometimes featuring cuticular coverings, and tubular elements with or without wall thickenings, occurring as dispersed microfossils are commonly placed with the nematophytes; as emphasized earlier, however, the systematics of such fossils remain largely unresolved. Nevertheless, Graham and Gray (2001) made an important observation by pointing out disparities between the stratigraphic record of cuticle-like fragments and tubes, thus suggesting that the two types of fossils probably originated from different types of organisms.

Structure of the Passage Creek Communities

Decimeter-scale arrangement of compression fossils on bedding planes shows several patterns. Some of the samples show a predominance of small size fossils (<5 mm) and only rare larger (>1 cm) fossils (Fig. 3A, 3B), whereas others exhibit an abundance of larger fossils around 1 cm or larger, and a few small fossils (Fig. 3C). The end member of this continuum is represented by surfaces where the compressions form extensive crusts of highly irregular outline (Fig. 3D). Fossils can be evenly distributed on the bedding planes (Fig. 3A), or they can form agglomerations a few centimeters across (Fig. 3B). The area covered by fossils is small on bedding planes where small fossils predominate (Fig. 3A, 3B), and considerably bigger where larger fossils are dominant (Fig. 3C) and in the case of crusts (Fig. 3D). Sedimentologic and taphonomic evidence indicates that at least some of the fossils at Passage Creek are preserved in situ (see section on depositional environments of the fossil assemblages). However, until in-depth treatments of the sedimentology and taphonomy of the fossiliferous layers are completed, it is difficult to estimate to what extent fossil arrangement on bedding planes can be interpreted in terms of community physiognomy.

From a trophic standpoint the Early Silurian communities at Passage Creek were relatively simple, including primary producers and probably decomposers. Primary producers at the base of the trophic structure included embryophytes, probably at a thalloid bryophytic grade of organization, as suggested by trilete spores and tetrads in the microfossil assemblages. Other groups of organisms were very likely also contributing to the biomass of this trophic level. Cyanobacteria and green algae have long been hypothesized as early terrestrial invaders (e.g., Stebbins and

Hill, 1980; Wright, 1985), and the finds of Horodyski and Knauth (1994) in the Precambrian may represent such organisms. Fungi were almost certainly present at Passage Creek as well, probably as decomposers or in associations with photoautotrophs, or both. Evidence for an animal component in the Massanutten Sandstone biota is absent from the micro- and macrofossil assemblages to date. However, Gray and Boucot (1994) recovered dispersed microfossils of animal origin along with plant and fungal spores from stratigraphically equivalent fluvial strata in the Tuscarora Formation of Pennsylvania. Shear and Selden (2001) interpret these fragments as representing fresh-water or terrestrial animals, and their presence in stratigraphically and depositionally equivalent strata leaves open the possibility that the Passage Creek biota may include an as yet unidentified animal component. Continuing study and resolution of the systematic affinities of the macrofossils is needed to clarify and substantiate these preliminary interpretations, and to shed more light on the structure of the Passage Creek communities.

Age of the Fossils

In their description of the stratigraphy of the Massanutten Sandstone along Passage Creek and in the surrounding region, Rader and Biggs (1976) recognize two distinct units. Based on stratigraphic relationships and lithology, they consider the lower unit roughly equivalent to the Tuscarora Formation, and possibly older. Previously, Yeakel (1962) had included the Massanutten Sandstone of Massanutten Mountain in his study of the Tuscarora Formation in the Central Appalachians, and Dennison and Wheeler (1975) had suggested a lower Massanutten-Tuscarora equivalence. This equivalence was inferred again by Whisonant (1977) in Virginia, and was reiterated by Pratt et al. (1978). The Passage Creek fossil assemblage occurs in the lower unit of Rader and Biggs (1976), which Pratt et al. (1978) informally named the Tuscarora member to reflect the stratigraphic equivalence.

The Tuscarora member and the Tuscarora Formation generally lack body fossils except for those reported by Pratt et al. (1978) and in this study, the biostratigraphic significance of which has yet to be assessed. The age of the Tuscarora member is consequently constrained biostratigraphically on the basis of marine assemblages in the underlying and overlying strata. These limit the age to an interval between the Ashgillian (underlying Martinsburg Formation) and the Ludlovian (top of the upper unit of the Massanutten Sandstone) (Pratt et al., 1978). Taking into account the thickness of the upper unit (Clinton Member) of the Massanutten, the latter authors consider the age of the Tuscarora member most likely early to middle Llandoverian (stages A–B). This age is in accord with that proposed for the Tuscarora Formation in Pennsylvania by Cotter (1983), i.e., early to lower-late Llandoverian (stages A–C_{2-3} of Berry and Boucot, 1970). The age of the Passage Creek fossils can therefore be placed with confidence in the Llandoverian, somewhere between the beginning of the Llandoverian and the basal late Llandoverian.

Depositional Environments of the Fossil Assemblages

Pratt et al. (1978) interpret the lower Massanutten Sandstone at Passage Creek as nonmarine, on the basis of the absence of marine fossils and nearshore indicator trace fossils (*Skolithus*, *Arthrophycus*). Although no detailed sedimentological study of the Passage Creek locality has been published, the lower Massanutten Sandstone is included in studies of the depositional environments of the Tuscarora Formation.

In an early study, Folk (1960) considered the Tuscarora in West Virginia a transitional nearshore marine to beach deposit, mainly on the basis of textural features. However, several other studies (e.g., Yeakel, 1962; Smith, 1970) strongly argue for a fluvial origin of at least most of the Tuscarora and its equivalents in Pennsylvania, New Jersey, New York, Maryland, Virginia, and West Virginia. Evidence used to support this interpretation is summarized by Fail and Wells (1974) and includes among others tabular sets of cross-beds, consistent dip directions of the cross-beds, thin, lenticular siltstones and shales, irregular bedding surfaces, abundant cut-and-fill structures, the presence of shale pebbles, and the systematic decrease in maximum pebble size in the direction of cross-bedding dip vectors. Subsequent work by Whisonant (1977), Cotter (1978), and Cotter (1983) is in agreement with the fluvial interpretation of the depositional environment. Two independent and extensive regional studies by Yeakel (1962) and Whisonant (1977) include the lower Massanutten Sandstone in their data sets. On the basis of paleocurrent directions and regional patterns of grain size distribution, both authors agree upon its location closest to the source of sediment, on a coastal plain that sloped toward the northwest.

Smith (1970) is the first to have suggested a braided style for the fluvial systems that deposited the Tuscarora, using comparisons of various sedimentary features with those of the modern South Platte and Platte Rivers of Colorado and Nebraska. His interpretation is supported by the studies of Pratt (1978) and Cotter (1978, 1983). In central Pennsylvania, Cotter (1983) interprets the basal part of the formation as beach deposits, the main body as braided-fluvial in southeastern, proximal facies, and the topmost part of the formation as coastal, sand or mud flat deposits. His paleogeographic reconstruction shows a southeast to northwest facies transition down the Llandoverian paleoslope from most proximal, alluvial fan complexes, through coastal alluvial plain facies of braided river systems, beach-strandplain, lagoon and estuary settings, to distal, shelf sand wave complexes. Cotter's (1983) reconstruction supports the paleogeographic interpretations of Yeakel (1962) and Whisonant (1977), and all provide good evidence to consider the Tuscarora as deposited by braided river systems carrying terrigenous material northwestward to the coast, from the Taconic Highlands that formed a linear source area in the southeast. The position of the lower Massanutten Sandstone in a setting proximal to the source area within this large-scale sedimentary system justifies its interpretation as deposits of braided rivers.

The macrofossils at Passage Creek occur in fine-grained partings that form thin, discontinuous layers between thicker, coarser beds of sandstone and fine conglomerate (Fig. 2). In fluvial sequences, as a result of the characteristic partitioning of sedimentation by grain size, fine-grained facies represent sedimentation outside of active channels: overbank, waning flood, or backswamp deposits (Miall, 1978, 1996). Such partitioning is very marked in braided streams and separates coarse channel deposits from finer material deposited outside active braid channels. Although "flood plain" is a term often avoided in describing the geomorphology of braided river systems, such systems include river flats – elevated surfaces within the channel tract and adjacent to active braid channels. As pointed out by Nanson and Croke (1992), once removed vertically or laterally from the proximity of active braid channels, these surfaces accumulate overbank fines in the same way as other flood plains. From the point of view the sediment partitioning by grain size such settings are therefore equivalent to the classic flood plains of meandering streams. The finer grain size of the sediments that preserve the fossils at Passage Creek (mainly siltstone to fine sandstone, with minor shale) indicates that they were deposited outside of the channels, broadly speaking in the flood plain of the river system that deposited the Massanutten Sandstone.

The preservation of fossils in flood-plain deposits does not automatically eliminate the possibility that they represent remains of fresh-water organisms transported and buried in overbank settings by flood events. The question then becomes, are the fossils autochthonous, preserved in situ and hence representing the flood-plain ecosystem, or are they allochthonous, transported material? In the latter case it would be difficult to determine whether they were transported from other locations on the flood plain, or from fresh-water ecosystems of the braid channels. Conclusive evidence in support of autochthony awaits careful sedimentological and taphonomic study of the fossiliferous layers, but several observations indicate that at least some of the Passage Creek fossils were buried in situ or underwent minimal transport. Pratt (1978) cites carbonaceous streaks perpendicular to the sedimentary lamination as evidence for in situ preservation of the fossils. Braided rivers are notorious for transporting coarse-grained bedload that acts as an extremely effective "grinding mill" on rock fragments, even over distances on the order of meters (G.C. Nadon, personal commun., 2003). Especially during floods, when it is greatly enhanced, this action would have reduced to minute fragments any organic material transported. However, several large specimens uncovered at Passage Creek by careful degagement preserve entire margins, and others form extensive organic crusts, indicating minimal, or more likely no, transport. In this context crusts with cracked surfaces were almost certainly preserved in situ.

Fossils preserved in situ on the flood plain can represent terrestrial organisms living in the aerial realm, but they also could represent fresh-water organisms of backswamp ponds or lakes. The relatively coarse texture of the sediment that incorporates the fossils (mainly siltstone and fine sand) disproves the latter alternative: silt-grade material is not well suited for retaining the water needed to form ponds. Isotopic $\delta^{13}C$ values obtained by Niklas and Smocovitis (1983) on Passage Creek fossils range from −25.6 to −26.4, favoring a terrestrial origin for the fossils. Additional evidence is provided by desiccation cracks in the fossiliferous siltstones, and by occurrences of fossils displaying desiccation cracks (e.g., Fig. 3C, at arrowhead). All of this evidence indicates that the Passage Creek fossils represent terrestrial organisms occupying wetland settings that were at least periodically emergent, and probably submerged primarily during the floods that buried them in silty sediment.

Importance of the Passage Creek Biota

The present level of understanding of the Passage Creek biota allows for several inferences of considerable importance for the colonization of land and the role played by wetland environments. A first important conclusion is that the Passage Creek fossils represent terrestrial organisms. This interpretation was also suggested by previous workers (Pratt et al., 1978; Niklas and Pratt, 1980; Niklas and Smocovitis, 1983) and is supported by data on the depositional environments. Given their early- to mid-Llandoverian age, these fossils represent the oldest direct, macroscopic evidence for terrestrial life. Ten to fifteen million years older than the oldest previously known terrestrial organisms (polysporangiophytic embryophytes and nematophytes), the Passage Creek biota provides an unprecedented perspective for understanding the early phases of the colonization of land by macroscopic organisms.

The Passage Creek biota reveals an abundance of fossil preservation, indicating that a well-developed terrestrial groundcover was present by the Early Silurian. Developed in the flood plains of river systems, this groundcover represents communities occupying settings that are geomorphologically homologous to present-day riverine wetlands. However, compared with modern wetlands that are continuously wet by virtue of the moisture retention capabilities of soils and vegetation, the Early Silurian flood-plain wetlands were only abiotically wet, and were dependent on climate and the fluctuations of the river system for their moisture supply. Moisture retention capabilities of the groundcover were probably very limited, and the flood-plain settings were prone to desiccation between floods, as suggested by the cracks on the surface of some of the fossils. Even so, these settings provided the wettest available environments on land at the time. It is no coincidence, therefore, that these wettest environments were home to the earliest well-developed terrestrial communities of complex organisms.

Fossil assemblages preserved at Passage Creek encompass a considerable level of structural diversity, in terms of both external morphology and internal anatomy. This diversity represents evidence for the presence of systematically diverse terrestrial communities in the groundcover of Early Silurian wetlands. The different types of internal organization documented at Passage Creek indicate that these communities were built by several types of complex organisms and/or associations of organisms. Dispersed microfossils from the same rocks provide evidence for embryophyte and probably fungal components in the biota. The observation that these organisms are thalloid provides the first direct evidence for developing an interesting new perspective on the earliest stages of the colonization of land by complex organisms.

The traditional embryophyte-focused view of the colonization of land has forged a search image for early land colonists based on one of the embryophytic synapomorphies, the axial sporophyte. However, the macrofossil record of axial sporophytes begins only in the late Wenlockian, a limit below which only tetrahedral spore tetrads and trilete spores, the microscopic embryophyte hallmarks, previously have been known. In the absence of macrofossils, spores alone do not reveal the habitat of early embryophyte growth, and the problem of terrestriality remains in the realm of hypothesis and speculation. The absence of axial sporophyte fossils and a predominance of thalloid forms in this earliest known terrestrial biota suggest that an alternative search image may provide greater success in the search for early terrestrial colonists. Rather than searching for axial sporophyte fossils in older and older deposits, a broadened focus that includes thalloid gametophytes, lichens, and other mutualistic associations such as the biological soil crusts may prove to be a more fruitful endeavor.

CONCLUSIONS

The Llandoverian Passage Creek biota of Virginia comprises the oldest macrofossil evidence for complex terrestrial life, preserved in fluvial wetland deposits. Although only preliminarily characterized, these fossils demonstrate that a well-developed, though discontinuous, groundcover was present in wetlands by the Early Silurian. This groundcover consisted of communities formed by a fairly diverse guild of thalloid organisms or associations of organisms, and of organic crusts comparable to extant biological soil crusts.

The Passage Creek biota emphasizes the importance of wetlands for the colonization of land by complex forms of life, and for the study of this process. Even though mainly abiotically wet in the absence of the considerable moisture retention capabilities conferred by soils and tracheophytic vegetation, Early Silurian river flood plains represented some of the least water-stressed environments on land at the time. As such, these wetlands offered the most favorable conditions both for the development and for the preservation of communities consisting of complex terrestrial organisms and associations of organisms. Likewise, the fossils they preserve have great potential for revealing the earliest stages of the colonization of land.

ACKNOWLEDGMENTS

We thank G.C. Nadon for helpful discussions on fluvial sedimentology, and R. Hikida for permission and help in using TEM facilities. We also thank W.A. DiMichele and G. Retallack for thoughtful and constructive comments. This research is funded by U.S. National Science Foundation grants EAR-0308931 (GWR) and DEB-0308806 (GWR and AMFT), as well as grants from the Geological Society of America, the Systematics Association, the Paleontological Society, the Botanical Society of America, Sigma Xi, and the Ohio University John Houk Memorial Research Fund (AMFT).

REFERENCES CITED

Algeo, T.J., Scheckler, S.E., and Maynard, J.B., 2001, Effects of the Middle to Late Devonian spread of vascular land plants on weathering regimes, marine biotas, and global climate, *in* Gensel, P.G., and Edwards, D., eds., Plants invade the land: Evolutionary and environmental perspectives: New York, Columbia University Press, p. 213–236.

Banks, H.P., 1975a, The oldest vascular land plants: A note of caution: Review of Palaeobotany and Palynology, v. 20, p. 13–25, doi: 10.1016/0034-6667(75)90004-4.

Banks, H.P., 1975b, Early vascular land plants: Proof and conjecture: Bioscience, v. 25, p. 730–737.

Barghoorn, E.S., 1977, *Eoastrion* and *Metallogenium, in* Ponnamperuma, C., ed., Chemical evolution of the early Precambrian: New York, Academic Press, p. 185–187.

Berry, W.B.N., and Boucot, A.J., 1970, Correlation of the North American Silurian rocks: Geological Society of America Special Paper 102, 289 p.

Campbell, S.E., 1979, Soil stabilization by a prokaryotic desert crust: Implications for a Precambrian land biota: Origins of Life, v. 9, p. 335–348, doi: 10.1007/BF00926826.

Cotter, E., 1978, The evolution of fluvial style, with special reference to the Central Appalachian Paleozoic, *in* Miall, A.D., ed., Fluvial sedimentology: Calgary, Alberta, Canadian Society of Petroleum Geologists, p. 361–383.

Cotter, E., 1983, Shelf, paralic, and fluvial environments and eustatic sea-level fluctuations in the origin of the Tuscarora Formation (Lower Silurian) of central Pennsylvania: Journal of Sedimentary Petrology, v. 53, no. 1, p. 25–49.

Dennison, J.M., and Wheeler, W.H., 1975, Stratigraphy of Precambrian through Cretaceous strata of probable fluvial origin in southeastern United States and their potential as uranium host rocks: Southeastern Geology Special Publication 5, 210 p.

Driese, S.G., Mora, C.I., Cotter, E., and Foreman, J.L., 1992, Paleopedology and stable isotope chemistry of Late Silurian vertic paleosols, Bloomsburg Formation, Central Pennsylvania: Journal of Sedimentary Petrology, v. 62, no. 5, p. 825–841.

Edwards, D., 1982, Fragmentary non-vascular plant microfossils from the late Silurian of Wales: Botanical Journal of the Linnean Society, v. 84, p. 223–256.

Edwards, D., 1986, Dispersed cuticles of putative non-vascular plants from the Lower Devonian of Britain: Botanical Journal of the Linnean Society, v. 93, p. 259–275.

Edwards, D., 2000, The role of Mid-Palaeozoic mesofossils in the detection of early bryophytes: Royal Society of London Philosophical Transactions, ser. B, v. 355, p. 733–755.

Edwards, D., and Feehan, J., 1980, Records of *Cooksonia*-type sporangia from late Wenlock strata in Ireland: Nature, v. 287, p. 41–42, doi: 10.1038/287041a0.

Edwards, D., and Wellman, C., 2001, Embryophytes on land: the Ordovician to Lochkovian (Lower Devonian) record, *in* Gensel, P.G., and Edwards, D., eds., Plants invade the land: Evolutionary and environmental perspectives: New York, Columbia University Press, p. 3–28.

Epstein, J.B., 1993, Stratigraphy of Silurian rocks in Shawangunk Mountain, southeastern New York, including a historical review of nomenclature: U.S. Geological Survey Bulletin 1839-L, 40 p.

Epstein, J.B., and Epstein, A.G., 1969, Geology of the Valley and Ridge Province between Delaware Water Gap and Lehigh Gap, Pennsylvania, *in* Subitzky, S., ed., Geology of selected areas in New Jersey and eastern Pennsylvania and guidebook of excursions: New Brunswick, New Jersey, Rutgers University Press, p. 132–205.

Fail, R.T., and Wells, R.B., 1974, Geology and mineral resources of the Millerstown Quadrangle, Perry, Juniata, and Snyder counties, Pennsylvania: Pennsylvania Geological Survey Atlas 136, 276 p.

Folk, R.L., 1960, Petrography and origin of the Tuscarora, Rose Hill, and Keefer Formations, Lower and Middle Silurian of eastern West Virginia: Journal of Sedimentary Petrology, v. 30, no. 1, p. 1–58.

Gay, A.L., and Grandstaff, D.E., 1980, Chemistry and mineralogy of Precambrian paleosols at Elliot Lake, Ontario, Canada: Precambrian Research, v. 12, p. 349–373, doi: 10.1016/0301-9268(80)90035-2.

Gensel, P.G., Johnson, N.G., and Strother, P.K., 1990, Early land plant debris (Hooker's "waifs and strays"?): Palaios, v. 5, p. 520–547.

Gray, J., and Boucot, A.J., 1977, Early vascular plants: Proof and conjecture: Lethaia, v. 10, p. 145–174.

Gray, J., and Boucot, A.J., 1994, Early Silurian nonmarine animal remains and the nature of the early continental ecosystem: Acta Palaeontologica Polonica, v. 38, no. 3–4, p. 303–328.

Golubic, S., and Campbell, S.E., 1979, Analogous microbial forms in recent subaerial habitats and in Precambrian cherts, *Gloeothece caerulea* Geitler and *Eosynechococcus moorei* Hofmann: Precambrian Research, v. 8, p. 201–217, doi: 10.1016/0301-9268(79)90029-9.

Graham, L.E., and Gray, J., 2001, The origin, morphology and ecophysiology of early embryophytes: neontological and paleontological perspectives, *in* Gensel, P.G., and Edwards, D., eds., Plants invade the land: Evolutionary and environmental perspectives: New York, Columbia University Press, p. 140–158.

Horodyski, R.J., and Knauth, P.L., 1994, Life on land in the Precambrian: Science, v. 263, p. 494–498.

Hotton, C.L., Hueber, F.M., Griffing, D.H., and Bridge, J.S., 2001, Early terrestrial plant environments: An example from the Emsian of Gaspé, Canada, *in* Gensel, P.G., and Edwards, D., eds., Plants invade the land: Evolutionary and environmental perspectives: New York, Columbia University Press, p. 179–212.

Knoll, A.H., and Bambach, R.K., 2000, Directionality in the history of life: diffusion from the left wall or repeated scaling of the right? *in* Erwin, D.H., and Wing, S.L., eds., Deep time: *Paleobiology*'s perspective: Paleobiology, supplement to v. 26, no. 4, p. 1–14.

Kodner, R.B., and Graham, L.E., 2001, High-temperature, acid-hydrolyzed remains of Polytrichum (Musci, Polytrichaceae) resemble enigmatic Silurian-Devonian tubular microfossils: American Journal of Botany, v. 88, no. 3, p. 462–466.

Kroken, S.B., Graham, L.E., and Cook, M.E., 1996, Occurrence and evolutionary significance of resistant cell walls in charophytes and bryophytes: American Journal of Botany, v. 83, no. 10, p. 1241–1254.

Labandeira, C.C., 2002, The history of associations between plants and animals, *in* Herrera, C.M., and Pellmyr, O., eds., Plant-animal interactions: An evolutionary approach: London, Blackwell Science, p. 26–74, and 248–261.

Miall, A.D., 1978, Lithofacies types and vertical profile models in braided river deposits: A summary, *in* Miall, A.D., ed., Fluvial sedimentology: Calgary, Alberta, Canadian Society of Petroleum Geologists, p. 597–604.

Miall, A.D., 1996, The geology of fluvial deposits: Sedimentary facies, basin analysis, and petroleum geology: Berlin, Springer.

Mishler, B.D., and Churchill, S.P., 1985, Transition to a land flora: Phylogenetic relationships of the green algae and bryophytes: Cladistics, v. 1, no. 4, p. 305–328.

Nanson, G.C., and Croke, J.C., 1992, A genetic classification of floodplains: Geomorphology, v. 4, p. 459–486, doi: 10.1016/0169-555X(92)90039-Q.

Niklas, K.J., 1976, Chemical examinations of some non-vascular Paleozoic plants: Brittonia, v. 28, p. 113–137.

Niklas, K.J., and Pratt, L.M., 1980, Evidence for lignin-like constituents in early Silurian (Llandoverian) plant fossils: Science, v. 209, p. 396–397.

Niklas, K.J., and Smocovitis, V., 1983, Evidence for a conductive strand in early Silurian (Llandoverian) plants: implications for the evolution of the land plants: Paleobiology, v. 9, no. 2, p. 126–137.

Palmer, A.R., and Geissman, J., compilers, 1999, Geologic time scale: Boulder, Colorado, Geological Society of America.

Pirozynski, K.A., 1981, Interactions between fungi and plants through the ages: Canadian Journal of Botany, v. 59, p. 1824–1827.

Pirozynski, K.A., and Malloch, D.W., 1975, The origin of land plants: a matter of mycotrophism: Bio Systems, v. 6, p. 153–164.

Pratt, L.M., 1978, Interpretation of the depositional environment and habitat of a widespread Early Silurian land flora from the lower Massanutten Sandstone in Virginia: Geological Society of America Abstracts with Programs, v. 10, p. 473–474.

Pratt, L.M., Phillips, T.L., and Dennison, J.M., 1975a, Evidence of nematophytes in the early Silurian (Llandoverian) of Virginia, U.S.A.: Botanical Society of America Annual Meeting Abstracts, p. 23–24.

Pratt, L.M., Phillips, T.L., and Dennison, J.M., 1975b, Nematophytes from early Silurian (Llandoverian) of Virginia provide oldest record of probable land plants in Americas: Geological Society of America Abstracts with Programs, v. 7, p. 1233–1234.

Pratt, L.M., Phillips, T.L., and Dennison, J.M., 1978, Evidence of non-vascular land plants from the Early Silurian (Llandoverian) of Virginia, U.S.A: Review of Palaeobotany and Palynology, v. 25, p. 121–149, doi: 10.1016/0034-6667(78)90034-9.

Rader, E.K., and Biggs, T.H., 1976, Geology of the Strasburg and Toms Brook quadrangles, Virginia: Virginia Division of Mineral Resources Report of Investigations, no. 45, 104 p.

Rankey, E.C., Walker, K.R., and Srinivasan, K., 1994, Gradual establishment of iapetan passive margin sedimentation—Stratigraphic consequences of Cambrian episodic tectonism and eustasy, Southern Appalachians: Journal of Sedimentary Research, Section B: Stratigraphy and Global Studies, v. 64, no. 3, p. 298–310.

Redecker, D., Kodner, R., and Graham, L.E., 2000, Glomalean fungi from the Ordovician: Science, v. 289, p. 1920–1921, doi: 10.1126/science.289.5486.1920.

Retallack, G.J., 1985, Fossil soils as grounds for interpreting the advent of large plants and animals on land: Philosophical Transactions of the Royal Society of London, ser. B, v. 309, p. 105–142.

Retallack, G.J., 2000, Ordovician life on land and Early Paleozoic global change: Paleontological Society Papers, v. 6, p. 21–45.

Retallack, G.J., 2001, Soils of the past. An introduction to paleopedology (second edition): Oxford, Blackwell Science.

Rice, C.M., Ashcroft, W.A., Batten, D.J., Boyce, A.J., Caulfield, J.B.D., Fallick, A.E., Hole, M.J., Jones, E., Pearson, M.J., Rogers, G., Saxton, J.M., Stuart, F.M., Trewin, N.H., and Turner, G., 1995, A Devonian auriferous hot spring system, Rhynie, Scotland: Journal of the Geological Society, London, v. 152, p. 229–250.

Schopf, J.M., Mencher, E., Boucot, A.J., and Andrews, H.N., 1966, Erect plants in the Early Silurian of Maine: U.S. Geological Survey Professional Paper 550-D, p. D69–D75.

Selosse, M.-A., and Le Tacon, F., 1998, The land flora: A phototroph-fungus partnership?: Trends in Ecology and Evolution, v. 13, no. 1, p. 15–20, doi: 10.1016/S0169-5347(97)01230-5.

Shear, W.A., and Selden, P.A., 2001, Rustling in the undergrowth: Animals in early terrestrial ecosystems, *in* Gensel, P.G., and Edwards, D., eds., Plants invade the land: Evolutionary and environmental perspectives: New York, Columbia University Press, p. 29–51.

Siegel, B.Z., 1977, *Kakabekia*, a review of its physiological and environmental features and their relations to its possible ancient affinities, *in* Ponnamperuma, C., ed., Chemical evolution of the early Precambrian: New York, Academic Press, p. 143–154.

Smith, N.D., 1970, The braided stream depositional environment: Comparison of the Platte River with some Silurian clastic rocks, North-Central Appalachians: Geological Society of America Bulletin, v. 81, p. 2993–3014.

Stebbins, G.L., and Hill, G.J.C., 1980, Did multicellular plants invade the land?: American Naturalist, v. 115, p. 342–353, doi: 10.1086/283565.

Strother, P.K., 1988, New species of *Nematothallus* from the Silurian Bloomsburg Formation of Pennsylvania: Journal of Paleontology, v. 62, no. 6, p. 967–982.

Strother, P.K., 2000, Cryptospores: The origin and early evolution of the terrestrial flora: Paleontological Society Papers, v. 6, p. 3–20.

Taylor, T.N., 1988, The origin of land plants: Some answers, more questions: Taxon, v. 37, no. 4, p. 805–833.

Taylor, T.N., and Taylor, E.L., 2000, The Rhynie Chert ecosystem: A model for understanding fungal interactions, *in* Bacon, C.W., and White, J.F., eds., Microbial endophytes: New York, Marcel Dekker, p. 31–47.

Trewin, N.H., Fayers, S., and Anderson, L.I., 2003, The biota of early terrestrial ecosystems: The Rhynie Chert (a teaching and learning resource). http://www.abdn.ac.uk/rhynie/

Whisonant, R.C., 1977, Lower Silurian Tuscarora (Clinch) dispersal patterns in western Virginia: Geological Society of America Bulletin, v. 88, p. 215–220, doi: 10.1130/0016-7606(1977)88<215:LSTCDP>2.0.CO;2.

Wright, P.V., 1985, The precursor environment for vascular plant colonization: Royal Society of London Philosophical Transactions, ser. B, v. 309, p. 143–145.

Yeakel, L.S., Jr., 1962, Tuscarora, Juniata, and Bald Eagle paleocurrents and paleogeography in the Central Appalachians: Geological Society of America Bulletin, v. 73, p. 1515–1540.

MANUSCRIPT ACCEPTED BY THE SOCIETY 28 JUNE 2005

Sedimentology and taphonomy of the Early to Middle Devonian plant-bearing beds of the Trout Valley Formation, Maine

Jonathan P. Allen*
Department of Geosciences, 214 Bessey Hall, University of Nebraska, Lincoln, Nebraska 68588-0340, USA

Robert A. Gastaldo*
Department of Geology, 5800 Mayflower Hill, Colby College, Waterville, Maine 04901-8858, USA

ABSTRACT

The Trout Valley Formation of Emsian–Eifelian age in Baxter State Park, Maine, consists of fluvial and coastal deposits that preserve early land plants (embryophytes). Seven facies are recognized and represent deposits of main river channels (Facies 1, 2), flood basin (Facies 4), storm-influenced nearshore shelf bars (Facies 3), a paleosol (Facies 5), and tidal flats and channels (Facies 6, 7). The majority of plant assemblages are preserved in siltstones and are allochthonous and parautochthonous, with only one autochthonous assemblage identified in the sequence above an apparent paleosol horizon. Taphonomic analysis reveals that plant material within allochthonous assemblages is highly fragmented, poorly preserved, and decayed. Plant material within parautochthonous assemblages shows evidence of minimal transport, is well preserved, and shows signs of biologic response after burial. The one autochthonous assemblage contains small root traces. Trimerophytes *(Psilophyton* and *Pertica quadrifaria)*, rhyniophytes (cf. *Taeniocrada)*, and lycopods *(Drepanophycus* and *Kaulangiophyton)* are the most common taxa in estuarine environments. *Psilophyton* taxa, *Pertica*, cf. *Taeniocrada*, and *Drepanophycus* are found also in fluvial settings. The presence of tidal influence in deposits where parautochthonous and autochthonous assemblages occur shows that these plants occupied coastal-estuarine areas. However, the effects on the growth and colonization of plants of the physical conditions (e.g., salinity) that exist in these settings in the Early to Middle Devonian are unknown.

Keywords: sedimentology, plant taphonomy, paleobiology, paleobotany, Emsian, Eifelian.

INTRODUCTION

The Devonian marked a time of rapid diversification of vascular land plants. According to Beerbower (1985) and DiMichele and Hook (1992), increased root and rhizoid activity that accompanied vascular land-plant diversification stabilized Devonian substrates, increased both physical and chemical weathering, and increased nutrient availability. The spread of vascular land plants also has been proposed to account for the decreased atmospheric concentration of CO_2 at the end of the Paleozoic (Berner, 1997; 1998;

---

*E-mails: jallen19@bigred.unl.edu; ragastal@colby.edu. Institution where work was carried out: Department of Geology, 5800 Mayflower Hill, Colby College, Waterville, Maine 04901-8858, USA.

Allen, J.P., and Gastaldo, R.A., 2006, Sedimentology and taphonomy of the Early to Middle Devonian plant-bearing beds of the Trout Valley Formation, Maine, *in* Greb, S.F., and DiMichele, W.A., Wetlands through time: Geological Society of America Special Paper 399, p. 57–78, doi: 10.1130/2006.2399(03). For permission to copy, contact editing@geosociety.org. ©2006 Geological Society of America. All rights reserved.

Elick et al., 1998) and the initiation of Carboniferous glaciation. Major Middle to Late Devonian marine-bottom anoxic events also are considered to be, in part, the result of Devonian land-plant radiation (Algeo and Scheckler, 1998; Algeo et al., 2001).

Despite the importance of early land plants, there have been few studies that detail their ecologic setting (e.g., Andrews et al., 1977; Gensel and Andrews, 1984; Edwards and Fanning, 1985; Griffing et al., 2000; Hotton et al., 2001). The wetland environments in which these Early Devonian land plants grew have been interpreted to range from coastal lowland marshes to terrestrial freshwater settings such as stream banks, exposed bar forms, and backwater swales (Edwards, 1980; Gensel and Andrews, 1984; Beerbower, 1985; DiMichele and Hook, 1992). On the basis of these studies, Early Devonian land-plant communities have been interpreted as consisting of an array of vegetational patches, where each was dominated by a single taxon (DiMichele and Hook, 1992). Plants in this low-diversity, patchy landscape developed a space-occupation pattern referred to as "turfing in," allowing them to control access to limited nutrients and water (DiMichele and Hook, 1992).

The Trout Valley Formation has been the focus of numerous paleobotanical studies since Dorf and Rankin (1962) first described its fossiliferous character and geologic setting. They described the Trout Valley Formation as a heterogeneous mix of light-blue, gray-to-black shale, siltstone, sandstone, and conglomerate, with minor sideritic sandstone and ironstone. Fossilized plant remains occurred in thin zones of limited lateral extent. Dorf and Rankin (1962) interpreted the depositional environment as a shallow, brackish-water setting on the slope of a volcanic island.

Previous to the present work, studies focused on the morphology and anatomy of the fossil plants recovered from these rocks (Andrews et al., 1968; Gensel et al., 1969; Kasper and Andrews, 1972; Kasper et al., 1974; Andrews et al., 1977; Kasper and Forbes, 1979; and Kasper et al., 1988). Two new genera and six new species of early vascular and nonvascular land plants were identified. Locality data were given as collections along Trout Brook and, in most accounts, the actual lithology in which the specimen was preserved was not identified. The depositional context of the plant fossils was based upon Dorf and Rankin's (1962) interpretation.

The details surrounding the depositional context of the Trout Valley plant fossils must be understood because these assemblages have been used, in part, to reconstruct the stereotypical Early to Middle Devonian plant community. Without an integrated sedimentologic and taphonomic analysis, paleoecologic interpretations cannot be substantiated. Plant fossil assemblages may be preserved in both their growth position and habitat (autochthonous), in their growth environment but not in situ (parautochthonous), or as transported material out of their growth environment (allochthonous) (Gastaldo, 2001). This information is pivotal for interpreting the paleocommunities in which these fossil plants lived. There is, however, no indication in the published literature whether a certain assemblage horizon represents an allochthonous accumulation of detritus that was transported or an autochthonous assemblage that was buried in situ. Plant taphonomic studies were not envisioned when the collections were made, and the foci of the published studies were biological, not geological. Fully integrated studies of pre-Carboniferous plant assemblages have only recently been conducted (e.g., Scheckler, 1985; Powell et al., 2000; Jarvis, 2000; Griffing et al., 2000; Hotton et al., 2001).

The project goal was to conduct an integrated examination of the sedimentologic, stratigraphic, and plant taphonomic character of the Trout Valley Formation to test the hypotheses that (1) the depositional setting of the formation was a terrestrial brackish marsh and (2) the fossil plant assemblages are autochthonous as suggested by Andrews et al. (1977).

REGIONAL GEOLOGIC SETTING AND AGE

The strata of the Trout Valley Formation are part of a thick succession of clastic rocks deposited in a foreland basin northwest of the Acadian orogen (Bradley et al., 2000). The paleolatitude was ~20°–30° S in the Emsian (Scotese and McKerrow, 1990); hence, climate was presumably subtropical with pronounced wet/dry seasonality. The aerial distribution of the formation truncates major tectonic structures in the Traveler Rhyolite, and Rankin (1968) placed an unconformity between these units. The Trout Valley Formation is of latest Emsian to earliest Eifelian age based on plant fossils (Kasper et al., 1988) and palynomorphs corresponding to the *douglastownese-eurypterota* spore assemblage zone (McGregor, 1992).

STUDY AREA

The Trout Valley Formation is located in the northwest section of Baxter State Park in northern Maine, located in T6 R9, T5 R9, and T5 R10 (Frost Pond and Wassataquoik, Maine, USGS 7.5′ quadrangles; Fig. 1). The present study was conducted in agreement with park officials; any and all material collected from the Trout Valley Formation is the sole property of the Baxter State Park Authority. Unauthorized collecting in the park is strictly prohibited.

The outcrop localities of Dorf and Rankin (1962) and Andrews et al. (1977) were used in field reconnaissance, and several additional outcrops not reported previously were found and described (Fig. 1). Strata are exposed along Trout Brook and South Branch Ponds Brook, and dip to the NW at 15° (Bradley et al., 2000). The longest outcrop exposure is 100 m in length. The maximum vertical extent of any particular outcrop is 7.1 m in height. Three normal faults, with a maximum displacement of 1.3 m, have been observed along Trout Brook.

METHODS

Detailed measured sections were logged and photomosaics were taken of all exposed outcrops to describe bed geometries and interpret depositional environments. Hand samples of all sedimentological and plant-bearing facies were collected for laboratory analysis. Sedimentologic analysis included standard lithologic

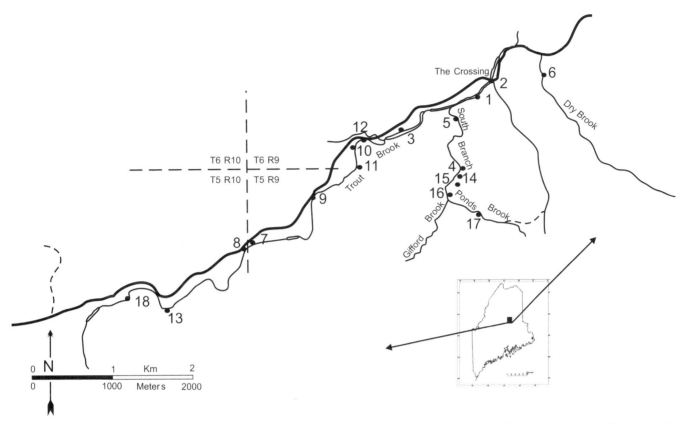

Figure 1. Map of the Trout Valley Formation and all sections in the present study (T6 R9, T5 R9, and T5 R10, Frost Pond and Wassataquoik, Maine, USGS 7.5′ quadrangles). Note that minor faults have not been mapped.

identification and description of primary sedimentary structures, presence or absence of bioturbation, and thin section evaluation.

Fossil-bearing samples were split along bedding planes for taphonomic analyses using modified methods of Krassilov (1975). These include the relationships between plant material and sediments, arrangement of axes in sediment (prostrate or erect, flat-lying or dispersed three-dimensionally), concentrated or dispersed assemblages, isomeric (plant parts of one type) versus heteromeric (an array of different plant parts) part composition, ordered (plants have a "dynamic" orientation; e.g., parallel arrangement) versus disordered (plants are oriented randomly on bedding planes), examination of crosscutting relationships (between the plant parts and the matrix in outcrop and thin section), and sediment fining/coarsening sequences relative to plant axes. Thin section analysis also was used to evaluate the microstratigraphic relationships between organic debris and the entombing sediments, using both petrographic and binocular stereo microscopes.

SEDIMENTOLOGY

Seven facies are recognized in the Trout Valley Formation. Details of these facies and their depositional interpretations are given in Table 1, and their relative vertical and lateral variations are provided in Figure 2.

Facies 1—Conglomerate

Description

Facies 1 consists of an extraformational clast-supported conglomerate of ~155 m in thickness. Clasts range in size from 3 to 15 cm, are poorly sorted, subrounded to rounded, ovate to platy in shape, and entirely rhyolite in composition. The conglomerate unconformably overlies the Traveler Rhyolite (Rankin and Hon, 1987). The base of the conglomerate appears massive; however, crude meter-scale trough cross-bedding is observed in the upper part of the facies (Fig. 3). The upper portion also exhibits northwardly oriented clast imbrication.

Interbedded, lenticular dark-gray sandy siltstone occurs in the upper portion of this facies. These beds are restricted laterally and are in sharp erosional contact with overlying conglomerate beds. Siltstone lenses are 15–80 cm thick with centimeter scale undulose/wavy bedding in which plant fossils assigned to cf. *Taeniocrada* (ribbon-like axes with wavy margins) are preserved. The contact with Facies 2 is abrupt and erosional.

Interpretation

This facies is either alluvial fan or fluvial in origin. In the context of the original Dorf and Rankin (1962) interpretations, the conglomerate could represent braided channel systems

TABLE 1. SUMMARY OF LITHOLOGIC, SEDIMENTOLOGIC, AND BIOGENIC FEATURES, AND INTERPRETED DEPOSITIONAL ENVIRONMENTS OF THE TROUT VALLEY FORMATION

Facies	Lithology	Locality	Sedimentary characteristics	Fossils	Trace fossils	Depositional environment
1	Clast-supported, poorly sorted, pebble-cobble extraformational rhyolitc conglomerate, upward fining with interbedded siltstone lenses	14, 15, 16, 17	Massive or crudely developed cross-bedding or horizontal stratification, imbricated clasts, siltstone lenses truncated by successive conglomerate beds.	Plant debris	N/A	Braided fluvial or alluvial fan
2	Poor-moderate sorted, sub-angular-sub-rounded, coarse-fine lithic arenites and wackes, fining-upward units <2m thick	4, 5, 6, 7, 8, 13	Channel geometries, trough cross-bedding, planar bedding, ripple cross-lamination, loading, reactiviation structures	Plant axes and debris	N/A	Migrating braided channels
3	Very coarse-fine siltstone	4, 5, 6, 7, 10, 13	Laterally continuous beds, massive bedding, laminated bedding	Plant axes	N/A	Flood basin/overbank
4	Fine-medium grained quartz arenites and wackes	2, 11, 12	Lenticular geometries, ripple laminated, current modified wave ripples, hummocky cross-stratification	N/A	(?) *Skolithos*, (?) *Helminthopsis*, meniscate burrows	Storm-influenced nearshore shelf sand bars
5	Pedogenically altered medium siltstone	10	Planar bedding, slickensides, sideritc glaebules	Plant axes, apparent root traces	N/A	Protosol (incipient wetland soil horizon)
6	Interbedded fine sandstone and siltstone	1, 2, 3, 9, 10	Sheet and channel geometries; trough cross-bedding; planar bedding; climbing, wave, and starved ripples; ripple cross-lamination; low-angle cross-stratification; herringbone cross-stratification; mud cracks	Plant axes	(?) *Skolithos*, (?) *Helminthopsis*, (?) *Diplocraterion*, meniscate burrows, fecal pellet aggradations	Estuarine/tidal flats and channels
7	Coarse siltstone	18	Lenticular bedforms	Bivalves, gastropods, ostracods, a single eurypterid	N/A	(?) Tidal channels

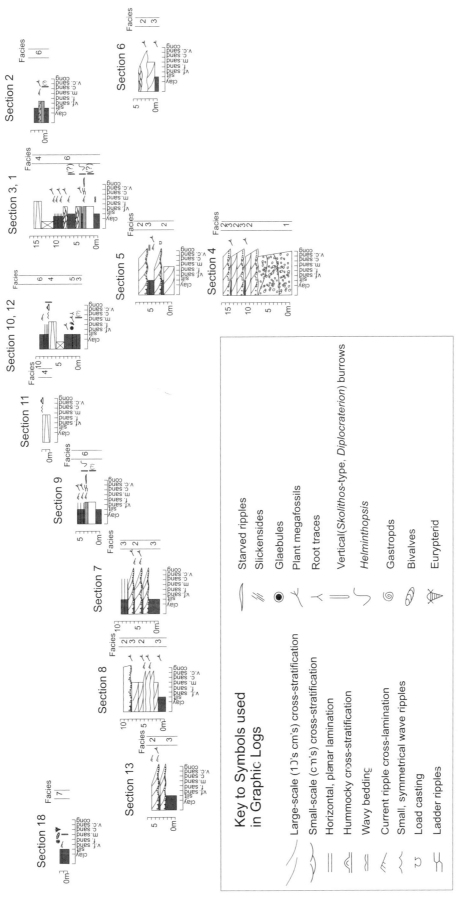

Figure 2. Graphic logs of measured sections for the Trout Valley Formation based on outcrop exposures along Trout Brook, South Branch Ponds Brook, and Dry Brook, Baxter State Park, Maine. Graphic logs detail the relative stratigraphic positioning and lateral variation of outcrops. Due to the limited exposure, physical correlation is not possible except at certain locations along Trout Brook.

Figure 3. Upper portion of Facies 1—conglomerate. Arrow points to crudely developed trough cross-bedding. The best examples of this feature occur in the upper part of this facies.

entrenched in an alluvial fan complex. Modern alluvial fan channels are characterized by thick deposits in which imbricated clasts define crude bedding (Steel and Thompson, 1983; Nemec and Steel, 1984). Vos and Tankard (1981), López-Gómez and Arche (1997), and Yagishita (1997) have described similar conglomerates as proximal fan deposits. The poor sorting and well-rounded, oblate to prolate clast shapes are features of shallow, gravel braided channels (Pettijohn et al., 1987; Miall, 1996), which are indicative of a braided fluvial influence.

These deposits could have been deposited in braided stream channels, alluvial fan channels, or mass flows; however, in the overall context that these were deposits from eroding highlands, a braided stream channel environment seems likely.

Facies 2—Trough Cross-Bedded Lithic Sandstone

Description

Facies 2 is in sharp, erosional contact and consists of beds of very coarse sandstone with granule clasts at the base fining upward to medium- to fine-grained sandstones. Sand clasts are sub-angular to sub-rounded and are poorly to moderately sorted. Compositionally, these sandstones are lithic arenites or wackes (depending on section), with abundant quartz and rhyolite. Interbedded siltstones are common within fining-upward cycles.

Sandstone beds are typically 1–2 m in thickness and sheetlike in overall geometry. Beds have sharp erosional bases that display loading, and are channel-form in some cases (Fig. 4). Upper contacts, where preserved, may be either a gradational or sharp contact with siltstones of Facies 3 (see description below). The fine-grained sandstone and siltstone of the overlying facies commonly are truncated by a successive fining-upward sandstone body. Locally, flame structures penetrate material overlying the upper contact.

Trough cross-bedding is the dominant internal sedimentary structure within sandstone bodies, with sets typically 0.1–0.5 m in thickness. Fossiliferous coarse siltstone displaying lenticular bedding, similar to that of Facies 1, commonly occurs in the troughs. Small-scale sedimentary structures, including trough cross-bedding (centimeter-scale), planar bedding, and ripple cross-lamination, are common at the tops of beds. Reactivation surfaces at the upper contact with Facies 3 are observed locally. Paleocurrent orientations measured from cross-beds are to the northwest (Fig. 5). Plant remains are preserved within the medium- and fine-grained sediments of Facies 2.

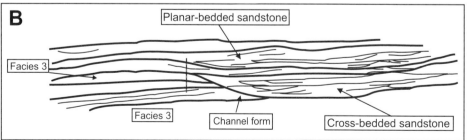

Figure 4. Typical unit of Facies 2—trough cross-bedded lithic sandstone. (A) Photomosaic showing one channel form that truncates a fining-upward sequence of cross-bedded and planar sandstone and siltstone (Facies 3—massive siltstone) along Trout Brook (section 8). (B) Line drawing illustrating bounding surfaces that are difficult to discern because of similar weathering patterns between lithotypes. Scale = 2 m.

Interpretation

The trough cross-bedded lithic sandstone represents migrating braided fluvial channels. The dominance of granule clasts within channel-form geometries indicates a high proportion of fluvial channel-lag and bar deposits similar to those described in time-equivalent settings from Schoharie Valley, New York (Bridge and Jarvis, 1998), the Battery Point Formation, Quebec (Griffing et al., 2000), and other localities (Miall, 1977, 1996).

The coarse-grained nature and low paleocurrent variance are similar to modern shallow braided channel systems (Nyambe, 1999). According to Rust (1978), the dominance of framework-supported grains is a key diagnostic feature for distinguishing braided from meandering systems. In addition, the lithic components of Facies 2 are of the same mineralogy as the clasts of the conglomerate facies, which suggests continued mechanical weathering and accumulation within a more mature braided environment than the underlying conglomerate (Facies 1). Nyambe (1999) reported a similar association in which the dominant clasts within a coarse sandstone were derived from a micro-conglomerate, indicative of a braided fluvial influence.

The relatively low variance in paleocurrent direction also is more indicative of a braidplain channel as opposed to a meandering system (Miall, 1978). This is because channel migration in the latter results in highly variable paleocurrent orientations reflecting the degree of channel sinuosity. Paleocurrent direction has been inferred to reflect mean channel direction (Rust, 1972; Chakraborty, 1999) and, as such, inferred channel axes, based on trough cross-bed measurements in this facies, were to the northwest. This orientation is toward the inferred paleoshoreline (Bradley et al., 2000). Ripple cross-lamination directed to the northwest also supports this interpretation.

Reactivation surfaces occur in the upper part of Facies 2 and may represent evidence for change in direction of discharge or environments influenced by other processes (McCabe and Jones, 1977; Weimer et al., 1982). Also, the juxtaposition of this facies fining upward into bioturbated siltstone suggests that these channels were in a coastal plain, and possibly within an estuarine-fluvial setting. Hence, this part of the interval may be transitional between fluvial and tidal environments.

Facies 3—Massive Siltstone

Description

Facies 3 consists of massive medium- to dark-gray siltstone/sandy siltstone and locally thinly interbedded very fine to fine-grained sandstone. In some instances, it is intercalated with medium-grained sandstone. Its massive appearance is attributed to the nature in which it weathers, because locally it is thinly laminated and millimeter-scale, fining-upward sequences are observed in thin section. This facies can be at least 2 m in thickness but, in many instances, it is truncated erosionally by channel-fills of Facies 2 (Fig. 4). Plant fragments are preserved throughout this lithofacies.

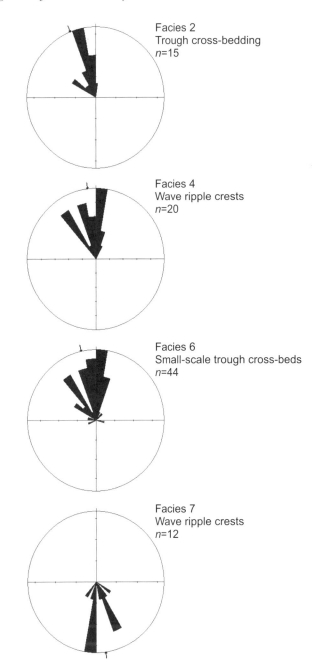

Figure 5. Rose diagrams summarizing paleocurrent data for several facies of the Trout Valley Formation. Facies 2 orientations measured from trough cross-beds; Facies 4 orientations based on measurements of wave ripples; Facies 6 orientations based on measurements from small-scale trough cross-beds; and Facies 7 orientations measured from ripple crests.

Facies 3 has either gradational or sharply bounded bases where it overlies Facies 2 (Fig. 2), as noted above, and where it is overlain by Facies 5 (section 10; see below). This facies occurs as discrete lenses in cross-bed troughs of Facies 1 and 2, but it is predominantly laterally continuous and displays a sheetlike geometry in the study area.

Interpretation

The massive siltstone is interpreted as an overbank (flood basin) environment. On the basis of its association with Facies 2 and a sheetlike geometry, these sediments were deposited in a non-channel setting. The presence of intercalated sand indicates that these clasts were deposited proximal to channel margins. The occurrence of Facies 3 as lenses in which plant fragments are preserved may represent either overbank deposits along the margin of the channel or abandoned scours of channel-fill sequences.

Facies 4—Lenticular Quartz-Rich Sandstone

Description

Facies 4 consists of subrounded to rounded, well-sorted, medium- to fine-grained quartz arenites (section 1 and 12) and wackes (section 11) that weather to a pale brown. Sandstone beds are sharply bounded, lenticular in geometry, range in thickness from 0.8 to 2.5 m, and thin laterally (Fig. 6). Beds exhibit low-angle trough forms, with ~16 m between troughs. Several troughs are filled with dark-gray siltstone preserving fragmentary remains of *Psilophyton*.

Internal structures are absent; however, the tops of these sandstone beds are ripple laminated and locally contain current-modified wave ripples and *Skolithos*-like trace fossils. Hummocky cross-stratification also is observed locally with wavelengths on the order of 10 cm between swales. Paleocurrents, based on measurements taken from wave ripples, are to the north, but vary from the northwest to the northeast (Fig. 5).

Interpretation

These quartz-rich sandstones are interpreted as nearshore storm-influenced shelf sand bars. Their thin lenticular nature, lithologic composition, bioturbated bed tops, minimal amount of terrestrial detritus, and *en echelon* arrangement are consistent with reported nearshore deposits (Reineck and Singh, 1980; McCubbin, 1982; and others). Measured paleocurrents toward the north and northeast differ from the underlying lithic sandstone facies and indicate sediment transport parallel to the inferred shoreline (Bradley et al., 2000). Northeast paleocurrents suggest that these sand bars migrated laterally across the shelf and possibly shoreward. However, the absence of shoreface deposits precludes an interpretation of onshore sand-wave migration.

These sandstones are compositionally more mature and better sorted than others in the formation, which suggests that the depositional environment was exposed to wave reworking. The absence of internal bedding structures, the presence of modified ripples at bed contacts, and hummocky cross-stratification at the upper bed contacts are consistent with storm deposition above wave base (Dott and Bourgeois, 1982; Aigner, 1985; Nottvedt and Kreisa, 1987; Collinson and Thompson, 1989; Duke et al., 1991).

Facies 5—Pedogenically Altered Siltstone

Description

Facies 5, present only at section 10 (Fig. 2), consists of 10 cm of dark-gray siltstone. This lithology is similar to Facies 3; however, there are several important differences. Pedogenic features, including slickensides, and thin, organic structures oriented vertically downward, are present. Aerial debris of *Psilophyton* is concentrated within the upper few centimeters of the bed. Petrographic analysis shows the presence of geopedally oriented organic structures and sideritic glaebules, ranging from 0.5 to 1.5 mm in diameter (Fig. 7). These features have been observed only within this facies.

Interpretation

The slickensides, sideritic glaebules, concentrated plant debris, and vertically oriented axes, possibly roots, are indicative of a paleosol. Sideritic nodules are characteristic of permanently waterlogged soils (Altschuler et al., 1983; Moore et al., 1992). The presence of iron carbonates is a potential product of original soil formation (Ludvigson et al., 1998) and, as such, this paleosol is best described as a protosol (Mack et al., 1993) based on the characteristic features and poorly developed horizonation (see Plant Taphonomy).

Facies 6—Bioturbated Interbedded Sandstone and Siltstone

Description

This facies consists of coarse siltstone and interbedded fine- to very fine grained sandstone. Siltstones are heavily bioturbated, with *Helminthopsis*-like traces preserved within the upper portions as well as burrows and fecal pellets observed in thin section. Locally, the upper 18–30 cm are massive siltstone alternating

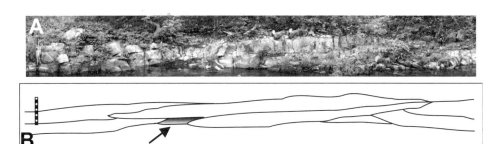

Figure 6. Photomosaic of section 11 where Facies 4 (lenticular quartz-rich sandstone) is best exposed. This lithofacies consists of *en echelon* stacked lenticular bodies of quartz arenites (section 11) or quartz wackes (section 10). Arrow points to a siltstone lens. Scale at left = 1 m.

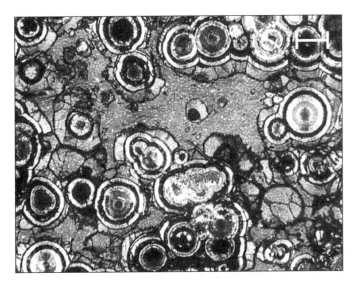

Figure 7. Sideritic glaebules, 0.5–1.5 mm in diameter, occur in Facies 5 (pedogenically altered siltstone; section 11). These structures are indicative of original reducing conditions. Scale = 1 mm.

with ripple cross-stratified siltstone (1–6 cm thickness). Mudcracks also are preserved at one locality.

The contacts with sandstone beds are sharp and bed thickness ranges from 0.2 to 1 m. Sandstones are heavily bioturbated and structureless, planar bedded, or trough cross-bedded. Their upper surfaces display ripple cross-lamination, symmetrical ladder ripples, starved ripples, and climbing ripple stratification (Fig. 8). Trace fossils are also common at the tops of planar-bedded sandstone. Traces (Fig. 9) include vertically compressed burrows averaging 5 mm in diameter (?*Skolithos*) and, in some instances, appear paired (cf. *Diplocraterion*). Burrow densities are high, with an average of 120 burrows per 10 cm^2. Horizontal traces, ~2 mm in diameter (?*Helminthopsis*), are less common in mud drapes.

These rocks typically display a sheetlike geometry that is laterally continuous over at least 100 m. At several sections (1, 3), this facies occurs in channel-forms (Fig. 10). Channel geometries measure 6 m in width and 0.5 m in depth, and are stacked *en echelon*. Here, bioturbated siltstone is overlain by low-angle cross-stratified (~10°) sandstone and interbedded ripple cross-laminated siltstone. A fossiliferous rippled siltstone of variable thickness occurs at the top of these channel fills (Fig. 11). Fossils are restricted to the rippled siltstone and include cf. *Taeniocrada*, *Psilophyton forbesii*, *P. princeps*, *P.* sp., and *Kaulangiophyton akantha*. Possible herringbone cross-stratification is present at one locality; however, it is truncated by an overlying sandstone bed, making identification of this structure equivocal. Due to reworking of ripples and obfuscation of ripple crests, only one paleocurrent measurement to the southeast was taken. Overall paleocurrent direction for this facies, based upon small-scale trough cross-beds, varies from northwest to northeast (Fig. 5).

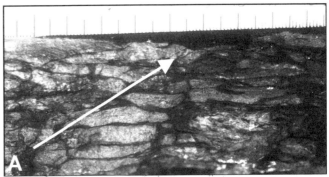
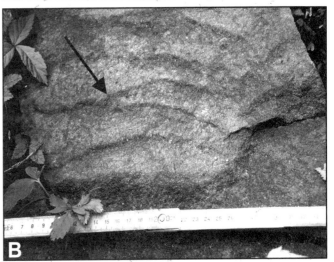
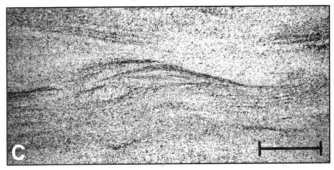

Figure 8. Sedimentological features of Facies 6—bioturbated interbedded sandstone and siltstone. (A) Cross section of siltstone in which climbing ripples can be seen. Arrow points to the direction of climb. Scale in cm. (B) Bedding surface exposure of very fine sandstone/coarse siltstone showing ladder ripples. Arrow shows "Y" split. (section 9); scale in cm. (C) Thin section in which micro cross-stratification is preserved in very fine sandstone to coarse siltstone. Scale = 1 mm.

Interpretation

This bioturbated and interbedded facies is interpreted as a coastal setting, probably within an estuarine intertidal flat. The sediments are fine grained, including very fine sand and coarse silt. Primary and biogenic structures and the lateral extent of beds are similar to intertidal deposits reported from the North Sea (Weimer et al., 1982) and other areas (van Straaten, 1954;

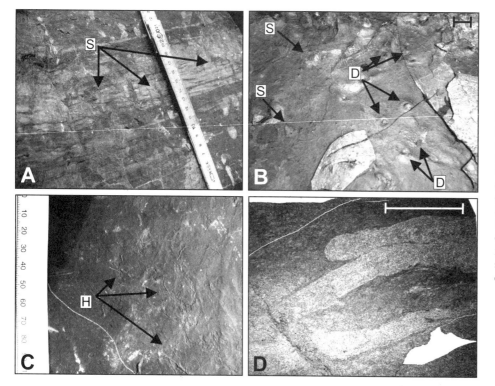

Figure 9. Trace fossils within Facies 6—Bioturbated interbedded sandstone and siltstone. (A) *Skolithos* (S) on vertical surface exposure at section 1. Scale in cm. (B) *Skolithos* (S) and ?*Diplocraterion* sp. (D) on bedding surface of a homogenous siltstone, section 9. Scale = 1 cm. (C) Horizontal traces, ?*Helminthopsis* (H), on bedding surface of homogenous siltstone, section 9. Scale in mm. (D) Thin section of homogenous siltstone in which infilled burrow system can be seen. Scale = 1 cm.

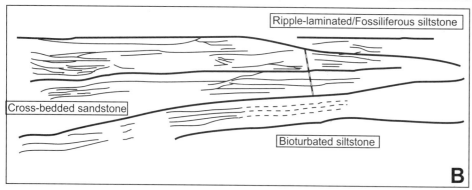

Figure 10. Facies 6—bioturbated interbedded sandstone and siltstone. (A) Photomosaic of this facies occurring as channel geometries (section 1). (B) Line drawing interpretation detailing the characteristic broad and shallow channel geometries. Scale = 1 m.

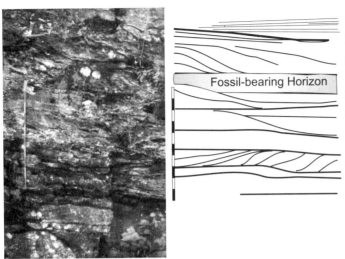

Figure 11. Alternating beds of cross-stratified sandstone and ripple-laminated siltstone in Facies 6 channel fills (section 1). Plant-bearing horizons occur within ripple-stratified beds found at the top of channel-fill sequences. Scale = 1 m.

Baldwin and Johnson, 1977; Reineck and Singh, 1980; Clifton, 1982; Terwindt, 1988). The absence of a macrofauna, the high degree of bioturbation, and dominance of *Skolithos*-type ichnofacies in the coarse siltstone/fine sandstone are typical of deposits in coastal areas such as tidal flats (Miller 1984; Miller and Woodrow, 1991; Bridge and Jarvis, 1998; Griffing et al., 2000).

Parallel and tabular cross-stratified beds adjacent to channel forms are interpreted as channel-bank deposits associated with shallow migrating tidal channels. In turn, these structures are overlain by tidal-flat deposits. Other evidence supporting a tidal-flat setting includes the presence of starved ripples (Singh and Singh, 1995) as well as ladder and wave-modified ripples, indicative of sedimentation during tidal slack water.

A tidal channel environment is interpreted for Facies 6 where channel-form features are found. Channel-form geometries characterized as broad (up to 6 m) and shallow (0.5 m) and *en echelon* stacked, as well as possible herringbone cross-stratification indicating bi-directional flow, is evidence for this type of setting (Boothroyd, 1985; Singh and Singh, 1995). The alternating nature of parallel-bedded siltstone, grading into alternations of cross-stratified sandstone and rippled siltstone from channel to channel, represents the changes in hydrodynamic conditions from higher flow velocities and open channels to lower flow velocities following channel filling. The presence of well-preserved plant axes, ranging from small (1–2 mm) fragments to large (19 cm) entire plants in the ripple-laminated, final stages of channel fill suggests detritus originated via erosion of the tidal channel margin during bank undercutting (see discussion).

Facies 7—Bioturbated Siltstone

Description

Facies 7 is restricted to one isolated locality exposed only at low water (Figs. 1, 2). It consists of dark-gray bioturbated coarse siltstone/very fine sandstone. Beds are lenticular in geometry with wavelengths of 6 m and a maximum thickness of 17 cm. Beds thin laterally and wave ripples are preserved at the contact between beds. Numerous dispersed macroinvertebrates, including bivalves, gastropods, ostracods, and a eurypterid, are preserved as impressions beneath bed-form crests. Paleocurrents measured from the wave-ripple crests are oriented to the south (Fig. 5).

Interpretation

The bioturbated siltstone facies records the migration of megaripples within a tidal channel. Megaripples are a common structure within tidal or tidally influenced channels (Boersma et al., 1968; Reineck and Singh, 1980; Terwindt, 1981). The presence of marine and brackish macroinvertebrates, including *Phthonia sectifrons* and the eurypterid cf. *Erieopterus* sp., in the crests of ripples indicates that these animals were transported into these channels and concentrated at megaripple crests (Selover et al., 2005). The concentration of shelly detritus within megaripple crests, rather than in the troughs, supports an interpretation emplacement during a high-energy event. Southward-directed paleocurrents are opposite to those measured in the underlying facies, and suggest a flood-dominated or possibly storm-influenced depositional event(s). The transition from this fossiliferous, megarippled siltstone into overlying bioturbated siltstone records the transition from high- to relatively low energy deposits (Miller and Woodrow, 1991). The bioturbated siltstone may have been reworked at the margins of these channels.

PLANT TAPHONOMY

Twelve plant taxa are reported from the Trout Valley Formation (Table 2). Megafossils are preserved in all facies, with concentrated assemblages occurring within Facies 2 (trough cross-bedded lithic sandstone), 4 (lenticular quartz-rich sandstone), and 6 (bioturbated interbedded sandstone and siltstone). Trimerophytes (*Pertica* and *Psilophyton*) are the most common components of all

TABLE 2. ASSOCIATION OF PLANT REMAINS AND FACIES IN THE TROUT VALLEY FORMATION BASED ON RESULTS OF ANDREWS ET AL. (1977), KASPER ET AL. (1988), AND THIS STUDY

Plant Megafossils	Facies						
	1	2	3	4	5	6	7
	Fluvial				Estuarine		
Thallophytes							
Prototaxites sp.*	X*						
Embryophytes							
Sciadophtyon sp.*						X*	
Bryophytes							
Sporongonites sp.*						X*	
Rhyniophytes							
Taeniocrada	X	X				X	
Trimerophytes							
Pertica quadrifaria		X	X			X	
Psilophyton dapsile						X	
P. forbesii		X	X		X	X	
P. microspinosum*						X*	
P. princeps		X					
Psilophyton sp.		X	X		X	X	
Lycophytes							
Drepanophycus gaspianus		X				X	
Drepanophycus sp.		X					
Kaulangiophyton akantha						X	
Kaulangiophyton sp.						X	
Leclercqia complexa*		X*				X*	
Leclercqia sp.			X			X	
						X	
Unidentified		X	X				X

*Reported by other authors, but not found in present study.

plant assemblages. Associated floristic elements, such as rhyniophytes and lycopods, either constitute a minor component (a few fragmented individuals) of any one assemblage or are absent. As a result, assemblage characteristics and the following discussion are focused mainly on trimerophytes.

Facies 1

Plant-bearing intervals occur in the upper part of the conglomerate facies where fine-grained deposits are found near the contact with the overlying trough cross-bedded lithic sandstone (Facies 2; Fig. 2). Plant remains occur in the troughs of cross-beds preserved in lenticular bodies of dark-gray, coarse siltstone that pinch out laterally across a distance of a few meters (Fig. 12A). Compressions of cf. *Taeniocrada* sp. dominate at section 4 (Fig. 12B) with *Psilophyton* sp. and cf. *Taeniocrada* on some bedding planes. Both homogeneous and heterogeneous assemblages occur in these channel fills. Plants are moderately well preserved fragments, ranging in size from 5 to 20 mm on the surface of bedding planes. Axes are oriented randomly and concentrated at the base of siltstone lenses (~15 cm in larger lenses) and become increasingly dispersed upward.

Facies 2

Plant assemblages are preserved in coarse-to-medium sandstone that fines into siltstone and can be traced laterally for meters across outcrop at sections 5, 7, 8, and 13. Different taxa are preserved at different stratigraphic horizons. *Pertica quadrifaria* is common at the base of plant-bearing intervals. Fragments average 13 cm in length, but axes as much as 50 cm in length have been collected (Fig. 13A). Plants consist of both main and lateral axes (some specimens display second-order dichotomizing branching) that occur parallel to bedding. Primary axes of *Pertica* from a single locality are preserved in random orientations (Fig. 13B). The concentration and diversity of axes increases upward as the sediment fines to medium-fine sandstone. *Psilophyton* sp., *P. princeps*, and, to a lesser degree, *P. forbesii* dominate the upper

Figure 12. Plant fossil assemblages of Facies 1—conglomerate. (A) Fossiliferous siltstone lens (15 cm in thickness) intercalated within the conglomerate. Picture occurs in the upper part of this facies. Note overlying conglomerate truncates siltstone lens at arrow. (B) Typical feature of cf. *Taeniocrada*-dominated assemblages recovered from siltstone lens. Scale in mm.

Figure 13. Plant macrofossils from fluvial facies (Facies 2, 3). (A) Large main axis with lateral axes assigned to *Pertica quadrifaria* preserved in fine sandstone (locality 12). Scale in mm. (B) Fragmented main axes and laterals of *P. quadrifaria* in very fine sandstone/coarse siltstone intercalated with medium sandstone. These features are typical of assemblages found at the base of plant-bearing intervals. Scale = 2 cm. (C) Fragmentary nature of most plant axes recovered from fluvial facies. Scale = 2 cm. (D) Axes of *Psilophyton* sp. that crosscut bedding in upper part of plant-bearing fluvial intervals. Inclined axes are not indicative of in situ burial (see text for explanation). Scale = 2 cm.

portions of plant assemblages within this facies (*Drepanophycus* sp. at section 5), and are best preserved in medium-fine sandstone. Axial fragments, 1–13 cm long, occur as disordered assemblages and are restricted to bedding planes (Fig. 13C). Sandstones fine upward into siltstones of Facies 3 where the number of *Pertica quadrifaria* and *Psilophyton* sp. axial remains decreases.

Facies 3

Plant assemblages in the massive siltstone facies typically are disordered and consist of fragmented *Psilophyton* sp. axes. However, dense, matlike concentrations of *Pertica* and *Psilophyton* axes occur in coarse siltstone intercalated with medium-fine sandstone ~40 cm above a non-fossiliferous interval (section 5). Here, axes are concentrated, disordered, heteromeric, and typically flat-lying. However, some axes crosscut bedding at angles ranging from 10° to 35° (Fig. 13D). This interval is truncated by a coarse channel sandstone of Facies 2. *Pertica quadrifaria* is preserved parallel to bedding above these matlike concentrations, similar to assemblages previously described from the trough cross-bedded lithic sandstone (Facies 2).

Millimeter-scale plant fragments are dispersed throughout the matrix, with orientations ranging from horizontal to sub-vertical in thin section. Plant material is typically concentrated in medium-to-fine sandstone and coarse siltstone, but in thin section occurs in both the coarse and fine fractions of micro-fining-upward sequences. Plant detritus overlies the contact between fine and coarse sediments (Fig. 14A, 14B). In several thin sections, sediment grains are observed to have migrated within bedload over flat-lying plant material, creating a scour surface on the down-current side of plant detritus that was filled subsequently with finer sediment (Fig. 14C). Plant material also is observed as casts, with axes filled with fine mud and silt (Fig. 14D). Several axes are contorted and overlie ripple crests (Fig. 14E).

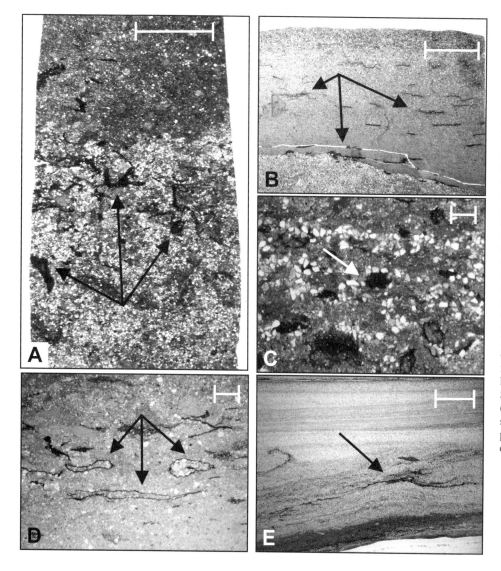

Figure 14. Photomicrographs of plant-bearing intervals in fluvial assemblages. (A) Dispersal and arrangement of plant matter in fining-upward sequence typical of Facies 3. Plant material (arrows) are oriented variously within both the coarse and fine sediment intervals. Scale = 1 cm. (B) Plant debris (arrows) oriented parallel to bedding and concordant with the contact between the coarse and fine sediments. Plant axes also are oriented parallel within the fine sediment interval. Scale = 1 cm. (C) Axis at arrow shows individual silt clasts that were transported over the plant material, scouring the downstream side. Scale = 1 mm. (D) Plant axes of *Psilophyton* sp. (arrows) that are infilled with coarse silt, indicating that axes decayed and were hollow prior to deposition and burial. Scale = 1 mm. (E) Contorted plant axis (arrow) that overlies a primary ripple structure, indicating settling from suspension load following a decrease in discharge. Scale = 1 cm.

Facies 4

Plant remains occur as sparse, fragmentary, disordered axes of *Psilophyton* sp. preserved within one lens of siltstone of the lenticular, quartz-rich sandstone. This assemblage is not mentioned in the later generalized taphonomic discussion.

Facies 5

Plant-bearing horizons at section 11 are similar to those at section 2 (Facies 6; Fig. 1). Plant remains are very concentrated, disordered, and isomeric within a 25 cm interval. The assemblage is dominated by *Psilophyton* sp. (?*P. princeps*), with minor contributions from *P. forbesii*. Aerial axes are inclined upward and crosscut bedding, whereas positively geopedal, vertically oriented structures, smaller in size than associated axes, crosscut the siltstone. These are interpreted as rooting structures (see Discussion), and this is the only location where possible original (primary) rooting is observed (recent roadwork near Trout Brook has exposed another paleosol horizon which has not been evaluated, to date. R.A. Gastaldo, September 2005, personal observation). Aerial detritus overlies this possible rooting horizon. The megaflora is associated with sideritic glaebules (Fig. 7) in the upper part of this plant-bearing interval. Several positive geopedal structures, also interpreted as roots, occur in oriented thin section and crosscut bedding into the subjacent coarser sediment (Fig. 15).

Figure 15. Geopedally oriented rootlike axes in estuarine Facies 5. Apparent roots on right display well-defined bifurcation and are similar to those illustrated in Figure 16. Scale = 1 mm.

Facies 6

Here, plant remains occur in coarse siltstone that underlies planar-bedded, very coarse siltstone or very fine sandstone. Plant intervals vary in thickness, ranging from 10 to 40 cm. *Psilophyton forbesii* and cf. *Taeniocrada* axes are preserved at the base of these beds and are dispersed within the matrix. Axes are primarily in random orientations and several *Taeniocrada*-like axes, restricted to the basal intervals of the siltstone, show three-dimensional curvature. *Psilophyton microspinosum*, although not abundant, occurs in association with cf. *Taeniocrada* in the coarse siltstone. Axes are dispersed at the base of the beds and become more concentrated upsection, with fragments parallel aligned and restricted to bedding planes.

Plants occur in planar siltstone below tabular cross-stratified beds at section 2 (Figs. 1, 2). Above the contact, *Psilophyton* sp. occurs as relatively sparse, flat-lying axes in an interval ~14 cm in thickness. A maximum of four, dense matlike intervals overlie this dispersed assemblage and consist of axial concentrations of *P. forbesii*, *P. dapsile*, *P. princeps*, and *Pertica quadrifaria*. Each interval varies from 1.5 to 3 cm in thickness and preserves a disordered plant assemblage. Each layer, however, is separated from the overlying assemblage by 2–7 cm of non-fossiliferous siltstone. Inclined axes (10° to 45°) originate from flat-lying axes and can be traced across bedding for several centimeters (Fig. 16A). Axes also vertically crosscut bedding and, in some instances, small axes (0.2–1.5 mm in width) crosscut bedding in

Figure 16. Plant macrofossils from estuarine Facies 6—bioturbated interbedded sandstone and siltstone. (A) Oblique view of inclined *Psilophyton* axes penetrating bedding. (B) Naked, sigmoidal small axis extending down from bedding surface. Scales = 2 cm.

a sigmoidal orientation (Fig. 16B). *Pertica quadrifaria* is concentrated at the base of the matlike accumulations without preferential orientation; this taxon becomes rare in stratigraphically higher beds. Orientations of *Psilophyton forbesii* and *P. dapsile* become more ordered in the uppermost matlike accumulations and approach parallel orientation. Axial orientations range from 50° to 266°, with predominant north and northwest directions. These vectors are similar to paleocurrent measurements based on trough cross-beds within this facies (Fig. 5).

In thin section, plant material is scattered within the coarser silt with up to several millimeters of matrix separating stratigraphically successive axes. Plant material is relatively concentrated in the fine silt where less than 15 mm of sediment is found between overlying axes. Plant material is parallel to bedding, and generally parallels the bedding contact between the coarse and fine silt in millimeter-scale fining-upward sequences (Fig. 17A). Several axes appear to be sub-horizontal relative to bedding, although they overlie small-scale, poorly defined cross-lamination (Fig. 17B). Contorted axes occur in both coarse and fine intervals.

Remains of *Psilophyton forbesii* are well preserved in concentrated, monotypic, isomeric assemblages in channel-form geometries, where observations were limited to bedding surface exposures on 0.5 m² float blocks. Axes are concentrated throughout the rippled siltstone intervals of channel fills and occur parallel to each other on every bedding surface (Fig. 18), although successive bedding surfaces display different axial orientations. Each plant-bearing interval, generally 2–3 cm in thickness, is overlain by barren ripple-laminated siltstone that is 3–5 cm in thickness. Primary structures in these unfossiliferous intervals include bi-directional, ladder, and current-modified ripples. Plant-bearing beds are restricted to the troughs of meso-scale bedforms on the order of 30 cm in wavelength with no indication that plants extend into waveform crests. While *P. forbesii* is the dominant, and at times the only, taxon present, *Psilophyton* sp., *P. princeps*, cf. *Taeniocrada sp.*, and *Kaulangiphyton akantha* also occur (Table 2).

Thin section analysis of samples from section 1 shows plant axes in various orientations. Plant material may be distorted and recurved, and associated with soft-sediment deformation. Detritus typically occurs at the top of fining-upward sequences. There is evidence of coarse silt migration over several plant axes (? lateral axes), creating a scour surface on the down current side of the organic remains (similar to Facies 2). Apparent rooting structures, originating from flat-lying axes, vertically crosscut bedding, and these organically stained structures disrupt bedding (Fig. 17C). Three different horizons of possible rooting structures occur in a series of fining-upward sequences, with each originating from the top of the sequence (Fig. 17D).

Facies 7

Plant remains in the bioturbated siltstone facies are small, <1 cm in greatest dimension, and occur only as isolated fragments. Because of their fragmentary nature and poor preservation, systematic identification is not possible. Axes show no preferential

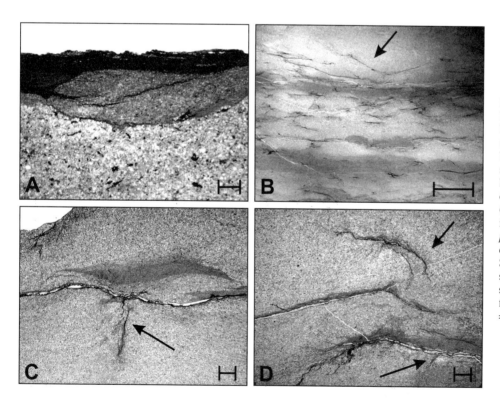

Figure 17. Photomicrographs of plant-bearing intervals in estuarine Facies 6—bioturbated interbedded sandstone and siltstone. (A) Plant material paralleling bedding contact between coarse and fine silt fraction. Scale = 1 mm. (B) Sub-horizontal axes (arrow) overlying poorly defined ripple-lamination. Scale = 1 cm. (C) Possible rooting structures originating from flat-lying aerial axes of *Psilophyton*. Rooting structures (at arrow) disrupt bedding below a starved ripple. Scale = 1 mm. (D) Three fining-upward sequences in which possible rooting structures can be seen originating from flat-lying aerial axes (at arrows). Scale = 1 mm.

orientation, are parallel to bedding, and are found only beneath the crests of bedforms that constitute the major geometry of this unit.

DISCUSSION

From the extensive investigation of plant morphology and anatomy (Andrews et al., 1968; Gensel et al., 1969; Kasper and Andrews, 1972; Kasper et al., 1974; Kasper and Forbes, 1979; and Kasper et al., 1988), Andrews et al. (1977) interpreted the Trout Valley plant assemblages as having been preserved in a terrestrial brackish or freshwater marsh setting surrounded by modest elevation. Using the dense, parallel-aligned assemblages of *Psilophyton* in the bioturbated interbedded sandstone and siltstone (Facies 6, this study), the dominance of a single taxon in many collection sites with relatively few accessory taxa, isolated pockets of single taxa, the high preservational quality of plant remains, and modern (marsh) analogues, Andrews et al. (1977) concluded that these Early to Middle Devonian assemblages represented monotypic, low-diversity wetland communities. The presence of parallel-aligned axes was used to interpret dense stands of plants that periodically were flattened and preserved in situ by floodwaters (Andrews et al., 1977). This interpretation has been propagated as an archetypical model for plant burial in the Devonian, and has been cited by various authors when interpreting the paleoecological significance of early land plants (Edwards, 1980; Gensel, 1982; Gensel and Andrews, 1984; Edwards and Fanning, 1985; Gensel and Andrews, 1987; Kasper et al., 1988; DiMichele and Hook, 1992; Hotton et al., 2001).

The present study has identified a range of continental to nearshore marine environments wherein well-preserved plant debris was restricted primarily to two environments: fluvial (Facies 2, 3) and estuarine/tidal (Facies 6) settings. Within these settings, only one interpreted autochthonous assemblage is identified (Facies 5). Nevertheless, many in situ plant communities must have been present to have supplied abundant axes preserved in these rocks.

Plant Taphonomy in Fluvial Assemblages

The plant taphonomic data indicate that all plant assemblages within fluvial environments are allochthonous. The quality of preservation, degree of fragmentation, and arrangement of axes are all characteristic of transported plant assemblages (Bateman, 1991; Behrensmeyer and Hook, 1992; Gastaldo et al., 1995, 2005).

The plant-bearing intervals within fluvial facies are thin (a few centimeters in thickness), laterally continuous over several meters, and stratified with respect to plant size. Larger plant fragments occur at the base of the assemblages and decrease in size upward through the plant-bearing interval. Plants are found primarily flat-lying, with few axes oriented at some angle to bedding. This stratification is a reflection of settling characteristics of the various-sized plant detritus from suspension during waning flow. As overbank velocity decreased, the larger and denser detritus, such as *Pertica quadrifaria* axes, settled first; these were followed

Figure 18. Parallel-oriented axes of *Psilophyton forbesii* in estuarine Facies 6—bioturbated interbedded sandstone and siltstone. These plants are typical of previously reported collections (e.g., Andrews et al., 1977). Plant fossil assemblages are restricted to the ripple-laminated intervals in the uppermost parts of interpreted tidal channel fills (see Fig. 10). Scale in cm.

by the smaller *P. quadrifaria* and *Psilophyton* sp. axial fragments. Plant parts are parallel to bedding unless the axes retained their structural fidelity and three-dimensional architecture. In the latter case, axes then may be oriented subhorizontal to bedding.

The presence of interbedded medium sandstone throughout these plant intervals, occurring with both large and small axes, indicates that flood-stage velocities fluctuated. Plant assemblages are dominated by axial components deposited within both channels and overbank settings during high-flow events.

The orientation of plant material that crosscut bedding is not believed to be a response to burial of autochthonous communities. No evidence has been observed for subjacent rooting structures or pedogenic alteration. The plants that crosscut bedding are generally fragments, ~5–6 cm in length, and originate from flat-lying axes; these fragments crosscut bedding at low angles. Hence, these assemblages are interpreted as transported (allochthonous). Inclined axes in this setting represent plants that maintained their structural integrity following deposition and burial, rather than exhibiting a biological response to burial.

An interpretation of allochthony also is supported by the microstratigraphic relationships between plant material and their entombing sediments. Plant fragments are scattered in random orientations within the matrix, suggesting that individuals were disarticulated before transport and deposition. If plants were buried in place, plant material would be expected to be concentrated at the contacts of a rich organic horizon with downward-projecting roots anchoring them in place. Additionally, the plants are preserved as casts rather than adpressions. Hence, axes underwent decay and hollowing, and subsequently were infilled with coarse silt and mud (e.g., Gastaldo et al., 1989). Decay is consistent with deposition in channels and overbank deposits (Kosters, 1989;

Alexander et al., 1999). Plant debris introduced from the floodplain would be entrained directly into the floodwaters and incorporated into the suspended load of the channel. Following decay, axes settled to the bedload and the hollow void was infilled by migrating bedload sediment (Degges and Gastaldo, 1989). Under high-discharge conditions, these small sediment-filled axes could be re-entrained and transported into overbank settings. Infilled axes, as well as other flat-lying detritus, are contorted and overlie primary ripple structures, which indicates settling from suspension load transport to bedload where hydrodynamic processes were in flux. Scouring on one side of plant material has been observed, implying a baffling effect that deflected sediments over the top of the plant fragment (see Rygel et al., 2004). Scouring on the down-flow side resulted in subsequent infill by finer-grained sediment and burial. This evidence points toward allochthonous accumulations in this setting (Gastaldo, 2004).

Plant Taphonomy in Estuarine/Tidal Assemblages

Fluvial and marine processes may affect estuarine environments, and plant assemblages preserved in these Trout Valley Formation regimes span the spectrum of possible preservational modes. Most assemblages are interpreted as parautochthonous, with only one allochthonous assemblage identified (Facies 7—bioturbated siltstone facies) and only one autochthonous assemblage encountered (Facies 5—pedogenically altered siltstone).

Autochthonous Facies 5

The presence of geopedally oriented structures that crosscut bedding within an inferred protosol suggests that the plants in the pedogenically altered siltstone were buried in situ. It may be argued that the geopedal structures are laterally forked branches of *Psilophyton* that were oriented downward during burial. However, if these were lateral axes emplaced when the plant fell over, more of these structures would be expected in this horizon owing to the organization and architecture of these taxa (e.g., Andrews et al., 1968; Kasper et al., 1974). A forked, lateral, aerial axis would have been deformed when emplaced into soft mud (Gastaldo, 1984), yet no such deformation is observed; the structures in the facies are straight and unkinked (Fig. 15). These structures are smaller in diameter than the other axial *Psilophyton* fragments observed within this facies (and the formation as a whole), as well as having features similar to other structures identified as roots in coeval floral assemblages (Griffing et al., 2000; Hotton et al., 2001). These lines of evidence lead to the conclusion that these structures are roots. Roots originate from an organic-rich horizon (poorly developed O-horizon) consisting of concentrated axial fragments that crosscut into the subjacent coarser sediment. Aerial plant remains preserved above the organic-rich horizon are similar in arrangement and preservation to axes in Facies 6 (bioturbated interbedded sandstone and siltstone). These are oriented subvertical to vertical, crosscut bedding, and are adpressions. Axes were neither hollowed nor infilled, as in the fluvial assemblages, indicating that individuals were not decayed prior to burial. This is the only locality where rooting structures are present and, thus, the only locality where an autochthonous assemblage is preserved.

Parautochthonous Facies 6

Here, plant assemblages are characterized by concentrated, well-preserved axial fragments, in both random and preferred orientations, and may be flat-lying, inclined, or vertically crosscut bedding. These accumulations are interpreted as parautochthonous assemblages. The relative size sorting seen in the fluvial assemblages with the larger plants (e.g., *Pertica quadrifaria*) concentrated at the bottom and smaller plants (e.g., *Psilophyton*) concentrated at the top of the plant-bearing intervals also is seen in the bioturbated interbedded sandstone and siltstone. This arrangement indicates settling in decreasing flow velocities. The small angle between dichotomizing branches noted in *Psilophyton microspinosum*, which is not typical of the plant (Andrews et al., 1977), is also attributed to transport and grounding processes.

Axes are found in three-dimensional arrangement throughout Facies 6 assemblages, where they crosscut bedding. However, bedding is neither interrupted nor disrupted by any inclined axis. This condition indicates that axial orientation occurred during burial, and was not a subsequent biological response by the plant to burial (regeneration and/or positive phototropic response). Plant response to burial was reported in several assemblages from the Battery Point Formation (Hotton et al., 2001) and observed in other Trout Valley Formation localities. Hotton et al. (2001) used re-anchorage, preservation of complete axes in fine detail (although an equivocal criterion as demonstrated by the present study), and axes oblique or perpendicular to the bedding plane as evidence for plant response to burial.

The assemblages preserved in interpreted tidal channels are characterized by concentrated, well-preserved axial segments with lateral axes in sub-parallel to parallel-aligned orientations. Several lines of evidence suggest that the alignment is a result of transport and not in situ burial as previously interpreted (Andrews et al., 1977). The presence of barren siliciclastic intervals between each plant-bearing horizon is evidence that these are not in situ assemblages. A basal layer of organic detritus (incipient or actual O-horizon) from which positive geopedal structures originate generally characterizes in situ assemblages. Although the assemblage mat could be such an organic layer, there is no macroscopic or microscopic evidence for primary rooting originating from these horizons. Each bedding surface on which large fragments are preserved is underlain by coarser clastics in which neither rhizomes nor rhizoids exist. Aerial axes are found in three-dimensional arrangement throughout the plant-bearing interval, where inclined axes are both sub-horizontal and vertical. These plant parts are in their original burial orientation rather than indicative of a later biological response to burial.

The presence of barren intervals above and below each assemblage, in which wave and tidally modified primary structures are preserved, also indicates transport before emplacement. Aerial plant detritus occurs only in the troughs of large-scale ripples, indicating a relationship between transport and settlement as the current velocities decreased (Gastaldo, 2004). Although these assemblages occur within interpreted tidal channel deposits and may have represented colonization of adjacent incipient wetlands, the absence of rooting horizons, the multiple stratified assemblages intercalated with barren intervals, and restriction of plant axes to ripple troughs are indicative of some degree of transport.

However, these plants probably were not transported out of their habitat to the burial sites. The plants in these assemblages apparently were neither dead nor decayed at the time of transport and burial. Nearly all the plants retain their three-dimensional architecture along with the presence of epidermal features (e.g., spines), intact terminal and lateral sporangia, and complex lateral branching. The primary sedimentary structures indicate a transported assemblage, but microstratigraphic analysis suggests that these plants were still alive when buried. Infrequent, forked, possibly root-like structures that appear to originate from flat-lying aerial axes suggest regeneration. Although regeneration has not been observed in other Devonian plant assemblages, it has been documented in Carboniferous *Calamites* (Gastaldo, 1992a).

There is a possibility that the plant material was transferred from inland sites into the nearshore setting and reworked into tidal channels. In general, such freshwater-discharged plant material gets entrained within tidal cycles and degraded, chemically and physically, to fragmentary materials before burial in tidal flats and channels (Gastaldo et al., 1987; Gastaldo, 1994; Gastaldo et al., 1993). Rarely do aerial architectures remain intact following tidal and wave activity. Rather, well-preserved and entire aerial plant debris is found where tidal channels have been occluded or blocked (e.g., Gastaldo and Huc, 1992). The presence of nearly complete, well-preserved plant axes up to 30 cm in length in these channel forms negates the possibility of long-distance fluvial transport and reworking into a coastal, estuarine channel system. Thus, these plants are interpreted as having lived close to the depositional setting, possibly on the margins of tidal channels and/or intertidal flats. Channel erosion during high-velocity spring flood or tidal cycles enhanced by storm processes would have undercut channel bank margins, allowing sediments and plant material to be incorporated into the adjacent channel.

Allochthonous Facies 7

The poorly preserved nature of the plant remains, their random orientations, and the concentration of material only in bedform crests in the bioturbated siltstone facies, along with their association with marine and brackish-water macroinvertebrates, indicate that this is an allochthonous assemblage (Selover et al., 2005).

Taphonomic and Paleoecologic Implications

The occurrence of trimerophytes in both fluvial assemblages and estuarine/tidal assemblages suggests that these plants occupied a range of habitats. Larger trimerophytes, such as *Pertica quadrifaria*, are found in the coarser sediments, indicating that they grew near the margins of active channels and were transported during flood events. Smaller trimerophytes (e.g., *Psilophyton*) also occur in channel margin settings, especially those of Facies 6 (bioturbated interbedded sandstone and siltstone; see below). However, their occurrence within finer sediments above an interpreted rooting horizon suggests that they may have also colonized coastal flood basin and wetland (marsh) settings.

There is no evidence in any of the Trout Valley Formation plant assemblages for rhizomes attached or associated with any aerial axes in the present or in previously reported studies (Andrews et al., 1968; Gensel et al., 1969; Kasper and Andrews, 1972; Kasper et al., 1974; Kasper and Forbes, 1979). Therefore, there are two ways to account for this observation. The first is that rhizomatous specimens have not been collected, because of limited exposure and a Park rule that prohibits extensive excavation of outcrops. The other explanation to account for the presence of large, well-preserved, monopodial sporophyte axes with sterile and/or fertile laterals, but lacking rooting parts, in transported assemblages is that individuals were uprooted by some physical mechanism associated with a hydrological event (wind-derived traumatic events would rip and fragment individual plants; Gastaldo, 1992b) prior to burial. This is particularly true of assemblages in the bioturbated interbedded sandstone and siltstone (Facies 6). Here, plants are preserved within the final stages of tidal channel infill (the origin of many previously reported taxa; Table 2), and this is the interpretation in the present study. Hence, these plants represent initial, r-strategist juveniles/adults established on tidal channel margins and intertidal flats that were subjected to channel-margin erosion as each tidal channel migrated across the transitional zone. Bank undercutting would have been effortless because of the shallow rooting horizon that developed following colonization, introducing aerial biomass into the channels. Plants are restricted to the uppermost part of each channel fill within troughs of meso- to macro-scale ripples (developed during bedload clastic transport) following suspension-load transfer. Individual plants may have originated from other areas within the tidal flat and not immediately adjacent to their preservational site.

Trout Valley Formation plant assemblages are very low in diversity, with one or two taxa dominating any assemblage (Table 2). This condition reaffirms earlier conclusions that early land plants "turfed in" as monodominant stands (DiMichele and Hook, 1992; Hotton et al., 2001). Where plants occur within estuarine/tidal settings, no clear marine indicators are present, suggesting that these plants lived in primarily freshwater habitats. However, the associations with nearshore sand deposits (Facies 4) both above and below estuarine/tidal deposits suggest that part of this section was within the marine-influenced zone where the plants may have been exposed to saline (fully marine or brackish water) conditions. However, due to a lack of data, such as halite pseudomorphs in the siltstones, there is no direct evidence to suggest that these plants either were tolerant of or were killed by brackish-water conditions.

Comparison with Other Devonian Plant Localities

Several time-equivalent Devonian plant localities have been reported from northern Maine, New Brunswick, and Quebec. However, only the Cap-aux-Os Member floral assemblage of the Battery Point Formation from Gaspé Bay, Quebec, has been studied in detail (Griffing et al., 2000; Hotton et al., 2001). Both the Trout Valley and Cap-aux-Os localities are interpreted as fluvial channels that migrated across coastal plains. The Trout Valley Formation and Cap-aux-Os Member floras are very similar floristically, but there are significantly more autochthonous assemblages preserved in the Cap-aux-Os Member. There, plant megafossils occupy specific areas of the landscape, which are inferred to represent clade-related niche partitioning (Hotton et al., 2001). Trimerophytes and rhyniopsids occupied fully fluvial ephemeral, near-channel environments, whereas monotypic zosterophyll assemblages occur in mud-dominated wetlands such as backswamps and marshes (Griffing et al., 2000; Hotton et al., 2001). In the Trout Valley Formation trimerophytes also are interpreted as having occupied fully fluvial environments, similar to the Cap-aux-Os assemblages, as well as coastal flood basin and wetland tidal-flat settings. *Prototaxites*, an enigmatic fungus (Hueber, 2001), is reported to occur exclusively in terrestrial fluvial environments within both the Trout Valley Formation and the Cap-aux-Os.

SUMMARY AND CONCLUSIONS

The Trout Valley Formation of north-central Maine was deposited initially in a relatively steep alluvial fan complex with high-velocity channels flowing to a coastal plain setting in which estuarine environments dominated. The presence of nearshore shelf sands, in addition to the increasing proportion of fine clastics upsection, is indicative of increasing marine influence in the area.

Trimerophytes are preserved in both fluvial and estuarine settings, suggesting that these plants probably occupied a wider range of habitats than previously interpreted. The plant assemblages consist of monodominant stands that occupied fluvial and estuarine/tidal channel margins, as well as coastal flood basin and wetland settings, that were primarily freshwater hydrological regimes. These conclusions reaffirm similar observations on early land plant habitats (DiMichele and Hook, 1992; Griffing et al., 2000; Hotton et al., 2001). The question of whether these plants were tolerant of brackish water conditions common in coastal areas (e.g., sea spray, storm-generated marine incursions) remains elusive due to a lack of sedimentologic and taphonomic evidence. The effect(s) of such physical conditions on early land plants only can be answered by further paleoecological studies based on other Early-Middle Devonian land-plant localities.

ACKNOWLEDGMENTS

This research was supported by National Science Foundation grant EAR 0087433 to R.A. Gastaldo and R.E. Nelson, Colby College. The authors would like to acknowledge the Baxter State Park Research Commission with special thanks to Jean Hoekwater, naturalist, and Buzz Caverly, Park Director, for permission to work in the park and their support throughout this project; Patricia Gensel (University of North Carolina–Chapel Hill) for valuable field and laboratory assistance; Cynthia Jones (University of Connecticut–Storrs) for access to the Andrews Collection; and Robert Selover, Michael Terkla, and Ivan Mihajlov for all their help in the field. The constructive reviews of S. Scheckler and J. Bridge and the advice of S. Greb are greatly appreciated.

REFERENCES CITED

Aigner, T., 1985, Storm depositional systems, dynamic stratigraphy in modern and ancient shallow-marine sequences: New York, Springer, 174 p.

Alexander, J., Fielding, C.R., and Jenkins, G., 1999, Plant-material deposition in the tropical Burdekin River, Australia: Implications for ancient fluvial sediments: Palaeogeography, Palaeoclimatology, Palaeoecology, v. 153, p. 105–125, doi: 10.1016/S0031-0182(99)00073-5.

Algeo, T.J., and Scheckler, S.E., 1998, Terrestrial-marine teleconnections in the Devonian: Links between the evolution of land plants, weathering processes, and marine anoxic events: Royal Society of London Philosophical Transactions, ser. B, v. 353, p. 113–130, doi: 10.1098/rstb.1998.0195.

Algeo, T.J., Scheckler, S.E., and Maynard, J.B., 2001, Effects of early vascular land plants on weathering processes and global chemical fluxes during the Middle and Late Devonian, *in* Gensel, P. G., and Edwards, D., eds., Plants invade the land: Evolutionary and environmental perspectives: New York, Columbia University Press, p. 83–102.

Altschuler, Z.S., Schnepfe, M.M., Silber, C.C., and Simon, F.O., 1983, Sulfur diagenesis in Everglades peat and the origin of pyrite in coal: Science, v. 281, p. 1659–1662.

Andrews, H.N., Kasper, A.E., and Mencher, E., 1968, *Psilophyton forbesii*, a new Devonian plant from northern Maine: Bulletin of the Torrey Botanical Club, v. 95, p. 1–11.

Andrews, H.N., Kasper, A.E., Forbes, W.H., Gensel, P.G., and Chaloner, W.G., 1977, Early Devonian flora of the Trout Valley Formation of northern Maine: Review of Palaeobotany and Palynology, v. 23, p. 255–285, doi: 10.1016/0034-6667(77)90052-5.

Baldwin, C.T., and Johnson, H.D., 1977, Sandstone mounds and associated facies sequences in some late Precambrian and Cambro-Ordovician inshore tidal flat/lagoonal deposits: Sedimentology, v. 24, p. 801–818.

Bateman, R.M., 1991, Paleoecology, *in* Cleal, C.J., ed., Plant fossils in geological investigation: The Paleozoic: New York, Ellis Hardwood, p. 34–116.

Beerbower, J., 1985, Early development of continental ecosystems, *in* Tiffney, B.H., ed., Geological factors and the evolution of plants: New Haven, Connecticut, Yale University Press, p. 47–92.

Behrensmeyer, A.K., and Hook, R.W., rapporteurs, 1992, Paleoenvironmental contexts and taphonomic models, *in* Behrensmeyer, A.K., Damuth, J.D., DiMichele, W.A., Potts, R., Sues, H.D., and Wing, S.L., eds., Terrestrial ecosystems through time: Chicago, University of Chicago Press, p. 15–136.

Berner, R.A., 1997, The rise of plants and their effect on weathering and atmospheric CO_2: Science, v. 276, p. 544–546, doi: 10.1126/science.276.5312.544.

Berner, R.A., 1998, The carbon cycle and CO_2 over Phanerozoic time: the role of land plants: Royal Society of London Philosophical Transactions, ser. B, v. 353, p. 75–82, doi: 10.1098/rstb.1998.0192.

Boothroyd, J.C., 1985, Tidal flats and tidal deltas, *in* Davis, R.A., ed., Coastal sedimentary environments: New York, Springer, p. 445–525.

Boersma, J.R., van de Meene, E.A., and Tjalsma, R.C., 1968, Intricated cross-stratification due to interaction of a mega ripple with its lee-side system of backflow ripple (upper-pointbar deposits, Lower Rhine): Sedimentology, v. 11, p. 147–162.

Bradley, D.C., Tucker, R.D., Lux, D.R., Harris, A.G., and McGregor, D.C., 2000, Migration of the Acadian orogen and foreland basin across the northern Appalachians of Maine and adjacent areas: U.S. Geological Survey Professional Paper 1624, p. 1–48.

Bridge, J., and Jarvis, S., 1998, Devonian fluvial to shallow marine strata, Schoharie Valley, New York, in Nashlund, H.R., 70th annual meeting of the New York State Geological Association, field trip guide: New York, New York State Geological Association, p. 43–69.

Clifton, H.E., 1982, Estuarine Deposits, in Scholle, P.A., and Spearing, D., eds., Sandstone depositional systems: Tulsa, American Association of Petroleum Geologists, p. 179–189.

Chakraborty, T., 1999, Reconstruction of fluvial bars from the Proterozoic Mancheral Quartzite, Pranhita-Godavari Valley, India, in Smith, N.D., and Rogers, J., eds., Fluvial sedimentology VI: Oxford, International Association of Sedimentologists, p. 451–466.

Collinson, J.D., and Thompson, D.B., 1989, Sedimentary structures: London, Unwin Hyman, 201 p.

Degges, C.W., and Gastaldo, R.A., 1989, Mechanisms of Carboniferous pith cast infilling: Geological Society of America Abstracts with Programs, v. 21, no. 3, p. 12.

DiMichele, W.A., and Hook, R.W., rapporteurs, 1992, Paleozoic terrestrial ecosystems, in Behrensmeyer, A.K., Damuth, J.D., DiMichele, W.A., Potts, R., Sues, H.D., and Wing, S.L., eds., Terrestrial ecosystems through time: Chicago, University of Chicago Press, p. 205–326.

Dott, R.H., and Bourgeois, J., 1982, Hummocky stratification: Significance of its variable bedding sequences: Geological Society of America Bulletin, v. 93, p. 663–680, doi: 10.1130/0016-7606(1982)93<663: HSSOIV>2.0.CO;2.

Dorf, E., and Rankin, D.W., 1962, Early Devonian plants from the Traveler Mountain area, Maine: Journal of Paleontology, v. 36, p. 999–1004.

Duke, W.L., Arnott, R.W.C., and Cheel, R.J., 1991, Shelf sandstones and hummocky cross-stratification: New insights on a stormy debate: Geology, v. 19, p. 625–628, doi: 10.1130/0091-7613(1991)019<0625:SSAHCS>2.3.CO;2.

Edwards, D.H., 1980, Early land floras, in Panchen, A.L., ed., The terrestrial environment and the origin of land vertebrates: New York, Academic Press, p. 55–85.

Edwards, D.H., and Fanning, U., 1985, Evolution and environment in the late Silurian-early Devonian: The rise of the pteridophytes: Royal Society of London Philosophical Transactions, ser. B, v. 309, p. 147–165.

Elick, J.M., Driese, S.G., and Mora, C.L., 1998, Very large plant and root traces from the Early to Middle Devonian: Implications for early terrestrial ecosystems and atmospheric p(CO_2) estimations: Geology, v. 26, p. 143–146, doi: 10.1130/0091-7613(1998)026<0143:VLPART>2.3.CO;2.

Gastaldo, R.A., 1984, A case against pelagochthony: The untenability of Carboniferous arborescent lycopod-dominated floating peat mats, in Walker, K.R., ed., The evolution-creation controversy: Perspectives on religion, philosophy, science and education—A handbook: Knoxville, Tennessee, Paleontological Society Special Publication 1, p. 97–116.

Gastaldo, R.A., 1992a, Regenerative growth in fossil horsetails (Calamites) following burial by alluvium: Historical Biology, v. 6, p. 203–220.

Gastaldo, R.A., 1992b, Taphonomic considerations for plant evolutionary investigations: The Paleobotanist, v. 41, p. 211–223.

Gastaldo, R.A., 1994, The genesis and sedimentation of phytoclasts with examples from coastal environments, in Traverse, A., ed., Sedimentation of organic particles: Cambridge, UK, Cambridge University Press, p. 103–127.

Gastaldo, R.A., 2001, Plant taphonomy, in Briggs, D.E.G., and Crowther, P.R., eds., Palaeobiology II: Oxford, Blackwell Scientific, p. 314–317.

Gastaldo, R.A., 2004, The relationship between bedform and log orientation in a Paleogene fluvial channel, Weißelster basin, Germany: Implications for the use of coarse woody debris for paleocurrent analysis: Palaios, v. 19, p. 587–597.

Gastaldo, R.A., and Huc, A.Y., 1992, Sediment facies, depositional environments, and distribution of phytoclasts in the Recent Mahakam River delta, Kalimantan, Indonesia: Palaios, v. 7, p. 574–591.

Gastaldo, R.A., Allen, G.P., and Huc, A.Y., 1993, Detrital peat formation in the tropical Mahakam River delta, Kalimantan, eastern Borneo: Formation, plant composition, and geochemistry, in Cobb, J.C., and Cecil, C.B., eds., Modern and ancient coal-forming environments: Geological Society of America Special Paper 286, p. 107–118.

Gastaldo, R.A., Douglass, D.P., and McCarroll, S.M., 1987, Origin, characteristics and provenance of plant macrodetritus in a Holocene crevasse splay, Mobile delta, Alabama: Palaios, v. 2, p. 229–240.

Gastaldo, R.A., Pfefferkorn, H.W., and DiMichele, W.A., 1995, Taphonomic and sedimentologic characterization of "roof-shale" floras, in Lyons, P., Wagner, R.H., and Morey, E., ed., Historical perspective of early twentieth century Carboniferous paleobotany in North America: Geological Society of America Memoir 185, p. 341–352.

Gastaldo, R.A., Demko, T.M., Liu, Y., Keefer, W.D., and Abston, S.L., 1989, Biostratinomic processes for the development of mud-cast logs in Carboniferous and Holocene swamps: Palaios, v. 4, p. 356–365.

Gastaldo, R.A., Adendorff, R., Bamford, M.K., Labandeira, C.K., Neveling, J., and Sims, H.J., 2005, Taphonomic trends of macrofloral assemblages across the Permian-Triassic boundary, Karoo Basin, South Africa: Palaios, v. 20, p. 478–497.

Gensel, P.G., 1982, Orcilla, a new genus referable to the zosterophyllophytes from the late Early Devonian of northern New Brunswick: Review of Palaeobotany and Palynology, v. 37, p. 345–359, doi: 10.1016/0034-6667(82)90007-0.

Gensel, P.G., and Andrews, H.N., 1984, Plant life in the Devonian: New York, Praeger, 380 p.

Gensel, P.G., and Andrews, H.N., 1987, The evolution of early land plants: American Scientist, v. 75, p. 478–489.

Gensel, P.G., Kasper, A.E., and Andrews, H.N., 1969, Kaulangiophyton, a new genus of plants from the Devonian of Maine: Bulletin of the Torrey Botanical Club, v. 96, p. 265–276.

Griffing, D.H., Bridge, J.S., and Hotton, C.L., 2000, Coastal-fluvial palaeoenvironments and plant palaeoecology of the Lower Devonian (Emsian) Gaspé Bay, Quebec, Canada, in Friend, P.F., and Williams, B.P.J., eds., New perspectives on the Old Red Sandstone, Geological Society [London] Special Publication, 180, p. 61–84.

Hotton, C.L., Hueber, F.M., Griffing, D.H., and Bridge, J.S., 2001, Early terrestrial plant environments: An example from the Emsian of Gaspé, Canada, in Gensel, P.G. and Edwards, D.H., eds., Plants invade the land: Evolutionary and environmental perspectives: New York, Columbia University Press, p. 179–212.

Hueber, F.M., 2001, Rotted wood–alga–fungus: The history and life of Prototaxites Dawson 1859: Review of Palaeobotany and Palynology, v. 116, p. 123–158, doi: 10.1016/S0034-6667(01)00058-6.

Jarvis, D.E., 2000, Palaeoenvironment of the plant bearing horizons of the Devonian-Carboniferous Kiltorcan Formation, Kiltorcan Hill, Co. Kilkenny, Ireland, in Friend, P.F., and Williams, B.P.J., eds., New perspectives on the Old Red Sandstone, Geological Society [London] Special Publication 180, p. 333–341.

Kasper, A.E., and Andrews, H.N., 1972, Pertica, a new genus of Devonian plants from northern Maine: American Journal of Botany, v. 59, p. 892–911.

Kasper, A.E., and Forbes, W.H., 1979, The Devonian lycopod Leclercqia from the Trout Valley Formation of Maine: Geological Society of Maine Bulletin, v. 1, p. 49–59.

Kasper, A.E., Andrews, H.N., and Forbes, W.H., 1974, New fertile species of Psilophyton from the Devonian of Maine: American Journal of Botany, v. 61, p. 339–359.

Kasper, A.E., Gensel, P.G., Forbes, W.H., and Andrews, H.N., 1988, Plant paleontology in the state of Maine: A review: Maine Geological Survey, Studies in Geology, v. 1, p. 109–128.

Kosters, E.C., 1989, Organic-clastic relationships and chronostratigraphy of the Barataria Interlobe Basin, Mississippi Delta Plain: Journal of Sedimentary Petrology, v. 59, p. 98–113.

Krassilov, V.A., 1975, Paleoecology of terrestrial plants: Basic principles and techniques: New York, Wiley, 283 p.

López-Gómez, J., and Arche, A., 1997, The Upper Permian Boniches Conglomerates Formation: Evolution from alluvial fan to fluvial system environments and accompanying tectonic and climatic controls in the southeast Iberian Ranges, central Spain: Sedimentary Geology, v. 114, p. 267–294, doi: 10.1016/S0037-0738(97)00062-6.

Ludvigson, G.A., Gonzalez, L.A., and Metzger, R.A., 1998, Meteoric sphaerosiderite lines and their use in paleohydrology and paleoclimatology: Geology, v. 26, p. 1039–1042, doi: 10.1130/0091-7613(1998)026<1039: MSLATU>2.3.CO;2.

Mack, G.H., Calvin, J.W., and Curtis, M.H., 1993, Classification of paleosols: Geological Society of America Bulletin, v. 105, p. 129–136, doi: 10.1130/0016-7606(1993)105<0129:COP>2.3.CO;2.

McCabe, P.J., and Jones, C.M., 1977, Formation of reactivation surfaces within superimposed deltas and bedforms: Journal of Sedimentary Petrology, v. 47, p. 707–715.

McCubbin, D.G., 1982, Barrier-island and strand plain facies, in Scholle, P.A., and Spearing, D., eds., Sandstone depositional systems: Tulsa, American Association of Petroleum Geologists, p. 247–281.

McGregor, D.C., 1992, Palynomorph evidence for the age of the Trout Valley Formation of northern Maine: Geological Survey of Canada Report F1-9-1992-DCM, 4 p.

Miall, A.D., 1977, A review of the braided-river depositional environment: Earth Science Reviews, v. 13, p. 1–62, doi: 10.1016/0012-8252(77)90055-1.

Miall, A.D., 1978, Lithofacies types and vertical profile models in braided river deposits, in Miall, A.D., ed., Fluvial sedimentology: Canadian Society of Petroleum Geologists Memoir 5, p. 597–604.

Miall, A.D., 1996, The geology of fluvial deposits, sedimentary facies, basin analysis, and petroleum geology: New York, Springer, 582 p.

Miller, M.F., 1984, Distribution of biogenic structures in Paleozoic nonmarine and marine-margin sequences: An actualistic model: Journal of Paleontology, v. 58, p. 550–570.

Miller, M.F., and Woodrow, D.L., 1991, Shoreline deposits of the Catskill Deltaic Complex, Schoharie Valley, New York, in Landing, E., and Brett, C.E., Dynamic stratigraphy and depositional environments of the Hamilton Group (Middle Devonian) in New York State, Part II: Albany, New York State Geological Survey, p. 153–177.

Moore, S.E., Ferrell, R.E., and Aharon, P., 1992, Diagenetic siderite and other ferroan carbonates in a subsiding marsh sequence: Journal of Sedimentary Petrology, v. 62, p. 357–366.

Nemec, W., and Steel, R., 1984, Alluvial and coastal conglomerates: Their significant features and some comments on gravelly mass-flow deposits, in Koster, E., and Steel, R, eds., Sedimentology of gravels and conglomerates: Canadian Society of Petroleum Geologists Memoir 10, p. 1–31.

Nottvedt, A., and Kreisa, R.D., 1987, Model for the combined flow origin of hummocky cross-stratification: Geology, v. 15, p. 357–361, doi: 10.1130/0091-7613(1987)15<357:MFTCOO>2.0.CO;2.

Nyambe, I.A., 1999, Sedimentology of the Gwembe Coal Formation (Permian), Lower Karoo Group, mid-Zambezi Valley, southern Zambia, in Smith, N.D., and Rogers, J., eds., Fluvial sedimentology VI: Oxford, International Association of Sedimentologists, p. 409–434.

Pettijohn, F.J., Potter, P.E., and Siever, R., 1987, Sand and sandstone: New York, Springer, 553 p.

Powell, C.L., Trewin, N.H., and Edwards, D., 2000, Palaeoecology and plant succession in a borehole through the Rhynie cherts, Lower Old Red Sandstone, Scotland, in Friend, P.F., and Williams, B.P.J., eds., New perspectives on the Old Red Sandstone, Geological Society [London] Special Publication 180, p. 439–457.

Rankin, D.W., 1968, Volcanism related to tectonism in the Piscataquis volcanic belt: An island arc of early Devonian age in north-central Maine, in Zen, E-an, White, W.S., Hadley, J.B., and Thompson, J.B., Jr., eds., Studies of Appalachian geology, northern and maritime: New York, Wiley Interscience, p. 355–369.

Rankin, D.W., and Hon, R., 1987, Traveler Rhyolite and overlying Trout Valley Formation and the Katahdin Pluton: A record of basin sedimentation and Acadian magmatism, northcentral Maine: Geological Society of America Centennial Field Guide, Northeastern Section, p. 293–301.

Reineck, H.-E., and Singh, I.B., 1980, Depositional sedimentary environments, with reference to terrigenous clastics: New York, Springer, 549 p.

Rust, B.R., 1972, Structure and processes in a braided river: Sedimentology, v. 18, p. 221–245.

Rust, B.R., 1978, Depositional models for braided alluvium, in Miall, A.D., ed., Fluvial sedimentology: Calgary, Canadian Society of Petroleum Geologists, p. 605–625.

Rygel, M.C., Gibling, M.R., and Calder, J.H., 2004, Vegetation-induced sedimentary structures from fossil forests in the Pennsylvanian Joggins Formation, Nova Scotia: Sedimentology, v. 51, p. 531–552, doi: 10.1111/j.1365-3091.2004.00635.x.

Scheckler, S.E., 1985, Seed plant diversity in the Late Devonian (Famennian): American Journal of Botany, v. 72.

Scotese, C.R., and McKerrow, W.S., 1990, Revised world maps and introduction, in McKerrow, W.S., and Scotese, C.R., eds., Palaeozoic palaeogeography and biogeography, Geological Society [London] Memoir 12, p. 1–21.

Selover, R.W., Nelson, R.E., and Gastaldo, R.A., 2005, An estuarine assemblage from the Middle Devonian Trout Valley Formation of northern Maine: Palaios, v. 20, p. 192–197, doi: 10.2110/palo.2004.p04-16.

Singh, B.P., and Singh, H., 1995, Tidal influence in the Murree Group, India, in Flemming, B.W., and Bartholomä, A., eds., Tidal signatures in modern and ancient sediments: Oxford, International Association of Sedimentologists Special Publication 24, p. 343–351.

Steel, R., and Thompson, D., 1983, Structures and textures in Triassic braided conglomerates ('Bunter' Pebble Beds) in the Sherwood Sandstone Group, North Staffordshire, England: Sedimentology, v. 30, p. 341–367.

Terwindt, J.H.J., 1981, Origin and sequences of sedimentary structures in inshore mesotidal deposits of the North Sea, in Nio, S.D., Shüttenhelm, R.T.E., and van Weering, Tj.C.E., eds., Holocene marine sedimentation in the North Sea Basin: Oxford, International Association of Sedimentologists Special Publication 5, p. 3–37.

Terwindt, J.H.J., 1988, Paleo-tidal reconstructions of inshore tidal depositional environments, in de Boer, P.L., van Gelder, A., and Nio, S.D., eds., Tide-influenced sedimentary environments and facies: Dordrecht, D. Reidel, p. 233–263.

van Straaten, L.M.J.U., 1954, Composition and structure of recent marine sediments in the Netherlands: Leidse Geologische Mededelingen, v. 19, p. 1–110.

Vos, R.G., and Tankard, A.J., 1981, Braided fluvial sedimentation in the Lower Paleozoic Cape Basin, South Africa: Sedimentary Geology, v. 29, p. 171–193, doi: 10.1016/0037-0738(81)90006-3.

Weimer, R.J., Howard, J.D., and Lindsay, D.R., 1982, Tidal flats, in Scholle, P.A., and Spearing, D., eds., Sandstone depositional systems: Tulsa, American Association of Petroleum Geologists, p. 191–245.

Yagishita, K., 1997, Paleocurrent and fabric analysis of fluvial conglomerates of the Paleogene Noda Group, northeast Japan: Sedimentary Geology, v. 109, p. 53–71, doi: 10.1016/S0037-0738(96)00058-9.

MANUSCRIPT ACCEPTED BY THE SOCIETY 28 JUNE 2005

Plant paleoecology of the Late Devonian Red Hill locality, north-central Pennsylvania, an Archaeopteris-dominated wetland plant community and early tetrapod site

Walter L. Cressler III
Francis Harvey Green Library, 29 West Rosedale Avenue, West Chester University, West Chester, Pennsylvania 19383, USA

ABSTRACT

The Late Devonian Red Hill locality in north-central Pennsylvania contains an *Archaeopteris*-dominated plant fossil assemblage, a diverse fossil fauna, and an extensive sedimentary sequence ideal for investigating the landscapes and biotic associations of the earliest forest ecosystems. Sedimentological analysis of the main plant-fossil bearing layer at Red Hill indicates that it was a flood-plain pond. A seasonal wet-and-dry climate is indicated by well-developed paleovertisols. The presence of charcoal interspersed with plant fossils indicates that fires occurred in this landscape. Fires appear to have primarily affected the fern *Rhacophyton*. The specificity of the fires, the distribution profile of the plant remains deposited in the pond, and additional taphonomic evidence all support a model of niche partitioning of the Late Devonian landscape by plants at a high taxonomic level. At Red Hill, *Archaeopteris* was growing on the well-drained areas; *Rhacophyton* was growing in widespread monotypic stands; cormose lycopsids grew along the pond edge; and gymnosperms and *Gillespiea* were possibly opportunists following disturbances. Tetrapod fossils have been described from Red Hill—therefore, this paleoecological analysis is the first systematic interpretation of a specific site that reflects the type of wetland environment within which the earliest tetrapods evolved.

Keywords: *Archaeopteris*; charcoal; cormose lycopsid; *Gillespiea*; gymnosperm; Late Devonian; paleoecology; *Rhacophyton*; tetrapod; wetlands.

INTRODUCTION

By the Late Devonian, plant communities had developed many characteristics that can be observed in modern terrestrial ecosystems (DiMichele et al., 1992). Lowland vegetation had differentiated into forest trees, shrubs, ground cover, vines, and specialized swamp plants. The morphological and ecological complexity of land plants had increased steadily throughout the Devonian. At the beginning of the period, vegetation consisted of small plants with simple architecture occupying low-diversity wetland patches (Edwards and Fanning, 1985). By the end of the Devonian, most of the major vegetative and reproductive features of vascular plants had evolved, and they appear in complex ecosystems (Chaloner and Sheerin, 1979; Algeo and Scheckler, 1998).

The development of plant-animal interactions lagged behind the development of plant community structure (Beerbower, 1985). Terrestrial arthropods throughout the Devonian are all evidently predators and detritivores (Beerbower, 1985; Shear, 1991). Evidence of plant damage repair due to herbivory is negligible and equivocal at best for the period (DiMichele et al., 1992). The earliest tetrapods appeared during the Late Devonian but were all predators that were ecologically linked to aquatic ecosystems (Beerbower, 1985). Late Devonian terrestrial ecosystems continued to be detritus based, but the plant biomass from which the detritus was derived increased dramatically during the interval due to the development of forests and the colonization of extrabasinal landscapes by seed plants (Algeo and Scheckler, 1998). Fully integrated terrestrial ecosystems with a complete range of arthropod herbivore guilds and a wide spectrum of low- to high-fiber vertebrate herbivory did not develop until the Late Carboniferous (Labandeira and Sepkoski, 1993; Hotton et al., 1997; Labandeira, 1998).

Although the earliest widespread forests were Late Devonian in age (Meyer-Berthaud, Scheckler, and Wendt, 1999), the first plant lineages to have members that developed secondary growth and the robust architecture of trees appeared during the Middle Devonian. These were aneurophytalean progymnosperms, cladoxylaleans, and lepidosigillarioid lycopsids (DiMichele et al., 1992; Algeo and Scheckler, 1998). Lycopsids formed wetland groves, and aneurophytes and cladoxylaleans grew in drier scrub thickets. The areal extent of these ecosystems was on a much smaller scale than the forests of the Late Devonian (Scheckler, 2001). By the beginning of the Late Devonian, archaeopteridalean progymnosperms became the dominant trees of widespread forests that ranged from tropical to boreal regions of the globe (Beck, 1964; Algeo and Scheckler, 1998). They were moderately sized trees that could reach an estimated height of 20–30 m. They bore the earliest known modern wood but had a free-sporing reproductive biology (Phillips et al., 1972; Meyer-Berthaud, Scheckler, and Wendt, 1999; Meyer-Berthaud, Scheckler, and Bousquet, 2000).

Nearly every Late Devonian plant-bearing terrestrial deposit has yielded fossil material of the progymnosperm tree *Archaeopteris* Dawson (Fairon-Demaret, 1986). The foliage genus *Archaeopteris* Dawson has been synonymous with the wood genus *Callixylon* Zalessky since the two were found in attachment (Beck, 1960). *Archaeopteris/Callixylon* is often found in association with the zygopterid fern *Rhacophyton* Crépin, another Late Devonian biostratigraphic indicator (Banks, 1980). Additional plant taxa that are found in Late Devonian plant fossil assemblages are rhizomorphic lycopsids, the earliest gymnosperms, barinophytaleans, cladoxylaleans, stauropteridalean ferns, and sphenopsids (Scheckler, 1986a). Red Hill, the fossil locality investigated in this study, has most of these floral components (Table 1). Red Hill is unusual for the association of its plant fossils with vertebrates, which include fish as well as early tetrapods (Daeschler et al., 1994; Daeschler, 1998, 2000a, 2000b); for its association with terrestrial invertebrate remains (Shear, 2000); and for the extent of its sedimentological profile (Woodrow et al., 1995). For these reasons, the Red Hill locality is ideal for conducting detailed paleoecological studies and for understanding one set of wetland environments in which the earliest tetrapods evolved.

The paleoecological analysis described here was undertaken at Red Hill with the objective to test the hypothesis of phylogenetic partitioning of the environment by plants in a Late Devonian lowland ecosystem. Observations of plant fossil assemblages in the Pennsylvanian Period have shown that plants partitioned lowland ecosystems at a high taxonomic level during that period (Peppers and Pfefferkorn, 1970; DiMichele and Bateman, 1996). Rhizomorphic lycopsids dominated the wetlands; ferns initially dominated in disturbed environments; sphenopsids dominated aggradational environments; and gymnosperms dominated on well- to poorly drained clastic substrates. This pattern persisted until climatic drying during the Middle-to-Late Pennsylvanian transition caused extinctions that disrupted the trend of within-clade replacement in each environment. Opportunistic ferns then dominated in many lowland environments during the Late Pennsylvanian. They were succeeded by seed plants in the Permian, which have been dominant in nearly every vegetated environment since that time (DiMichele et al., 1992; DiMichele and Bateman, 1996).

The class-level taxa that dominated the various Pennsylvanian lowland environments were well established by the end of the Late Devonian–Early Mississippian transition (Scheckler, 1986a). Apparently the ecological patterns of these plant groups were established as part of their evolutionary origin and radiation as basic plant body plans (DiMichele and Phillips, 1996). Observations of Late Devonian plant fossil assemblages have suggested habitat partitioning by plants during that interval also (Scheckler, 1986a, 1986c; Rothwell and Scheckler, 1988; Scheckler et al., 1999). The suggested pattern for the Late Devonian is that cormose lycopsids occupied permanent wetlands, zygopterid ferns occupied ephemeral wetlands, the early gymnosperms were pioneer plants on disturbed habitats, and *Archaeopteris* occupied the better-drained part of the overbanks (Scheckler, 1986a, 1986c; Rothwell and Scheckler, 1988; Scheckler et al., 1999). Niche partitioning among plants in the Shermans Creek Member of the Late Devonian Catskill Formation of Pennsylvania has also been indicated by a study of root traces that vary in different depositional environments (Harvey, 1998). The extinction of *Archaeopteris* after the end of the Devonian and the decline in importance of the zygopteridalean ferns facilitated a transition to the pattern of phylogenetic partitioning seen in the Pennsylvanian. The Late Devonian to Middle Pennsylvanian partitioning of lowland environments by plants at a high taxonomic level is an ecological pattern that is not repeated as clearly in subsequent geological periods.

The abundant and relatively diverse plant fossil material at Red Hill and the site's clear sedimentological relationships have provided an opportunity to test the model of habitat partitioning among the components of Late Devonian lowland vegetation. In addition, the flora of the Catskill Formation and related rocks of Pennsylvania has historically been poorly sampled, except by Arnold (1939). He described several species of *Archaeopteris*, "seed-like objects" referred to as *Calathiops* Arnold that were later recognized as seed-bearing cupules called *Archaeosperma arnoldii* by Pettitt and Beck (1968), the lycopsids *Prolepidodendron breviinternodium* Arnold and *Lepidostrobus gallowayi* Arnold, and an enigmatic plant of unknown affinities, *Hostimella crispa* Arnold. In contrast to this situation, the Hampshire Formation flora of equivalent age in Virginia and West Virginia is well known (Kräusel and Weyland, 1941; Andrews and Phillips,

1968; Phillips, Andrews, and Gensel, 1972; Gillespie, Rothwell, and Scheckler, 1981; Scheckler, 1986b; Rothwell, Scheckler, and Gillespie, 1989). The assemblage described here increases our knowledge of the Catskill flora and provides a better basis for comparison with the Hampshire flora and other Late Devonian plant assemblages.

MATERIALS AND METHODS

Geologic Setting—The Catskill Formation

The Red Hill outcrop where this paleoecological study was undertaken is a roadcut exposure of the Duncannon Member of the Catskill Formation (Woodrow et al., 1995), which represents the upper alluvial plain facies of the Late Devonian Catskill Delta Complex (Diemer, 1992). The Catskill Delta Complex was an enormous wedge of sediment that was shed from mountains formed during the collision between ancestral North America, Europe, the Avalon Terrane, and other microcontinents during the Acadian orogeny, as part of the early stages in the closure of the Iapetus Ocean and the formation of Pangaea (Ettensohn, 1985). Sediments were shed from the mountains and prograded toward the west into the foreland basin adjacent to the Acadian fold-thrust belt (Gordon and Bridge, 1987). During the Late Devonian, the subaerial portions of the clastic wedge formed a wide, low-gradient plain traversed by meandering streams that flowed into a shallow epicontinental sea. This fluvial sequence in Pennsylvania constitutes the Catskill Formation (Sevon, 1985).

The Catskill Formation overlies dark marine shales of the Trimmers Rock and Lock Haven Formations, and where they come in contact marks the inception of delta progradation (Glaeser, 1974). In most of north-central Pennsylvania, three members of the Catskill Formation are recognized according to their depositional characteristics with respect to their location on the coastal plain. The Irish Valley Member represents sedimentation along a muddy, low-energy tidal flat (Walker, 1971; Walker and Harms, 1971). The Sherman Creek Member represents deposition on a broad inactive coastal plain, and the Duncannon Member represents deposition higher on the coastal plain where the meandering river facies is most fully expressed (Rahmanian, 1979).

Geologic Setting—The Red Hill Outcrop

Red Hill is the informal local name for a large roadcut in the Duncannon Member of the Catskill Formation on PA highway 120 between the villages of Hyner and North Bend in Chapman Township, Clinton County, Pennsylvania (41°20′30″ N latitude, 77°40′30″ W longitude; Fig. 1). A clean vertical exposure extends for 1 km along an east-west stretch of the highway, and an additional 170 m of rock to the west is banked and overgrown. The outcrop consists of nearly horizontal layers of fluvial sandstones, siltstones, mudstones, and paleosols (Fig. 2). These lithologies are typical of the Duncannon Member of the Catskill Formation. The layers dip slightly to the west. The Duncannon Member

TABLE 1. RED HILL FLORAL LIST

Plantae
  Tracheophyta
    Zosterophyllopsida
      Barinophytales
        Barinophyton obscurum (Dun) White
        Barinophyton sibericum Petrosian
    Lycopsida
      Isoetales
        Otzinachsonia beerboweri Cressler and Pfefferkorn
        cf. ***Lepidodendropsis*** Lutz
    Filicopsida
      Zygopteridales
        Rhacophyton ceratangium Andrews and Phillips
      Stauropteridales
        Gillespiea randolphensis Erwin and Rothwell
    Progymnospermopsida
      Archaeopteridales
        Archaeopteris macilenta (Lesq.) Carluccio et al.
        Archaeopteris hibernica (Forbes) Dawson
        Archaeopteris obtusa Lesquereaux
        Archaeopteris halliana (Göppert) Dawson
    Gymnospermopsida
      Pteridospermales
        cf. ***Aglosperma quadripartita*** Hilton and Edwards
        cupulate gymnosperms, indet.

(Classification scheme based on Stewart and Rothwell, 1993)

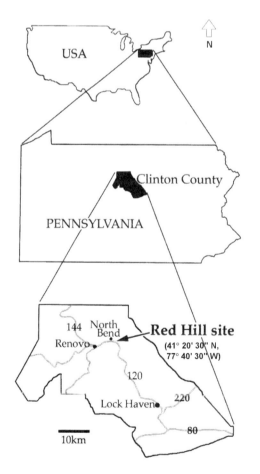

Figure 1. Location of the Red Hill site, Clinton County, Pennsylvania, U.S.A. (after Daeschler, Frumes, and Mullison, 2003).

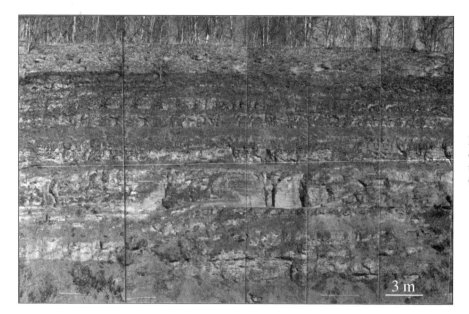

Figure 2. Uncorrected photomosaic of portion of Red Hill outcrop featuring the eastern portion of the plant layer (after Cressler, 2001). © SEPM (Society for Sedimentary Geology).

represents an upper alluvial plain facies that was characteristically far inland from marine deposition, and there is no evidence of marine influence at Red Hill.

A stratigraphic column of a sequence that includes the plant-fossil-bearing layer sampled for this paleoecological study is shown in Figure 3. The sampling of plant fossils took place within a 1 m thick predominantly dark-gray siltstone layer shown in Figure 3 as the "plant layer." The stratigraphic column was measured down through the plant layer at a point 15 m to the west of where it pinches out. The upper and lower margins of the plant layer consist of ~15 cm of unfossiliferous greenish-gray siltstone, not shown in the stratigraphic column. They form uneven zones that grade into the overlying and underlying red beds where the contact area between them is mottled with reduction spots. Above and below the plant layer are alternating sequences of red laminated siltstones and red paleosols. Toward the top of the stratigraphic sequence in the measured section is a gray sandstone lens partially covered by soil and vegetation. The dark-gray and greenish-gray siltstones constitute Lithofacies 3 of Woodrow et al.'s (1995), "Greenish-Gray Mudstone and Very Fine Grained Sandstone." The red laminated siltstones correspond to Lithofacies 1 of Woodrow et al.'s (1995), "Red Hackly-Weathering Mudstone." The red paleosols correspond to Lithofacies 2 of Woodrow et al.'s (1995) "Red Pedogenic Mudstone." Woodrow et al. (1995) refer to the sandstone bodies at Red Hill as Lithofacies 4, "Flat-Laminated Gray Sandstone." Not all of the sandstone bodies are flat-laminated, however.

Geologic Setting—The Dark-Gray Plant-Fossil-Bearing Siltstone Layer (Plant Layer)

As implied by its name, most of Red Hill consists of highly oxidized rocks with hematite coating the sediment grains. Plant fossils are poorly preserved or nonexistent in these red rocks. The reduced horizons are where the well-preserved plant fossils have been found. The only large and easily accessible reduced horizon at Red Hill is the one that was sampled for the paleoecological analysis described in this paper, the "plant layer."

Figures 4 and 5 show the sedimentological relationships of the sampled plant layer at the two ends of where it is exposed in the outcrop. Where it pinches out at its eastern end (Fig. 4) it truncates two layers below it, a red paleosol and a red laminated siltstone. The pinched-out lenticular portion of the plant layer is end-to-end with the apex of a thin sandstone wedge that thickens to the east. The sandstone wedge exhibits small-scale cross-bedded laminations accreting toward the apex of the wedge. Another sandstone body 74 m to the west of the pinched-out eastern end of the plant layer is laterally continuous with the plant layer and marks the western end of the sampled horizon (Fig. 5). The flat lower surface of the western sandstone body is on the same horizon with the lower boundary of the plant layer. The western sandstone body is convex upward for the same thickness as the plant layer. It also exhibits small-scale cross-bedded laminations. Where the western sandstone body crosscuts the plant layer, the reduced horizon is mottled with red at a transition zone and changes to completely red in proximity to the sandstone. On the western side of this sandstone body, the dark-gray siltstone layer continues to the west for another 93 m until it plunges below talus piles and the road, owing to the slight western dip of all the layers at Red Hill.

The total reduced horizon, including the 74-m-long portion of the plant layer where sampling occurred, averages 1 m in thickness over its exposed length of 167 m. The portion of it west of the convex-upward sandstone body has sandstone stringers 1–2 cm thick interspersed regularly throughout the depth of the dark-gray siltstone. The western portion of the plant layer also differs from the eastern portion in that the 15 cm upper margin of greenish-gray siltstone is absent.

Stratigraphic Analyses

To produce the photomosaic of the Red Hill outcrop (Fig. 2), color photographs were taken with a tripod-mounted Pentax K1000 at successive intervals along the face of the outcrop so that there was at least 30% overlap between successive photographs. Photographs were taken from the shoulder of the highway on the opposite side from the exposure. The height of the outcrop necessitated two photographs, one above the other, at each horizontal position. The foreshortening of the images was left uncorrected in the photographs. The outcrop profile of Red Hill was prepared by drawing a diagram of the exposure based on this photomosaic (Figs. 4 and 5).

A stratigraphic section was measured down through the outcrop at a place that included the plant layer at a point 15 m from the eastern end of the layer (Fig. 3). A rope with 1 m, 5 m, and 10 m intervals marked in different colored tape was lowered from the top of the exposure. The widths of the stratigraphic layers were read off the rope by using a pair of binoculars, and the measurements were recorded in a field notebook. Correcting the measurements to account for the angle of the slope was considered unnecessary and impractical due to the relatively short height and uneven nature of the exposure.

Age of Beds

Palynological analyses on several layers including the plant layer at Red Hill place it within the palynomorph biozone Famennian 2c (Fa2c; Traverse, 2003). It is equivalent to the VCo miospore zone, which is distinguished by the palynomorph index species *Grandispora cornuta* Higgs, *Rugispora flexuosa* (Juschko) Streel, and others (Richardson and McGregor, 1986; Streel and Scheckler, 1990).

Paleoecological Sampling

Plant fossils and fossil charcoal were systematically sampled by excavating 12 small quarries (between 0.25 m^3 and 0.5 m^3) along a 64 m transect of the plant layer. The transect followed the exposed face of the layer between the two previously described sandstone bodies, which are 74 m apart.

Within each quarry, bedding surfaces were exposed at successive centimeter-scale intervals throughout the vertical dimension of the plant layer. Plant fossil diversity and abundance and the number of charcoal fragments were counted on these successive bedding surfaces using a line-intercept method. For this method, a grid with recording dimensions of 24 cm x 24 cm was used, consisting of a wooden frame with strings arrayed at right angles to each other. The intersections of the strings were 2 cm apart. The grid was laid over the bedding surfaces and the fossils

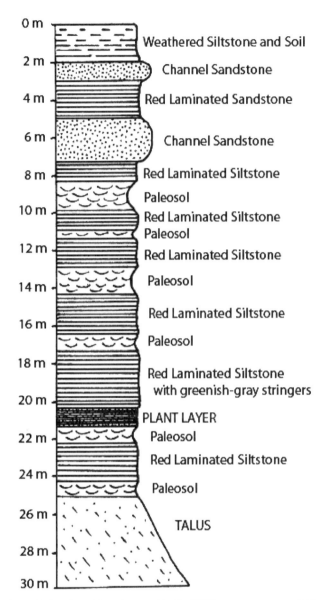

Figure 3. Stratigraphic column of the Red Hill outcrop measured 15 m to the west of the eastern end of the plant layer (after Cressler, 2001). © SEPM (Society for Sedimentary Geology).

at the intersections of the strings were identified, counted, and recorded.

The 12 quarries were arranged in pairs in order to determine variation of plant fossil distribution at a smaller spatial scale if that was deemed important upon analysis of the data. Red Hill is a roadcut with a south-facing exposure. The quarries were numbered along the plant layer from west to east. The quarries within each pair were adjacent to each other, with the exception of quarries 2a and 2b, which were 2 m apart. These two quarries could not be made adjacent to each other because of the excessive amount of overburden above the plant layer at this location in the outcrop. The following chart delineates the distances between quarry pairs:

> **Western Sandstone Body** – 10 m – **1ab** – 9 m – **2a** – 2 m – **2b** – 15 m – **3ab** – 19 m – **4ab** – 11 m – **5ab** – 8 m – **6ab**

Quarry 4b was excavated farther into the cut face at the same position as quarry 4a after the rock from quarry 4a had been removed. Excavation lateral to quarry 4a was impractical due to a large amount of overburden. Quarries 6a and 6b were excavated at the extreme eastern end of the plant layer where it pinches out and makes lateral contact with the apex of the wedge-shaped sandstone body. Overall, the unequal distances between pairs of quarries result from the uneven accessibility of the plant layer due to the varying amounts of overburden along its length.

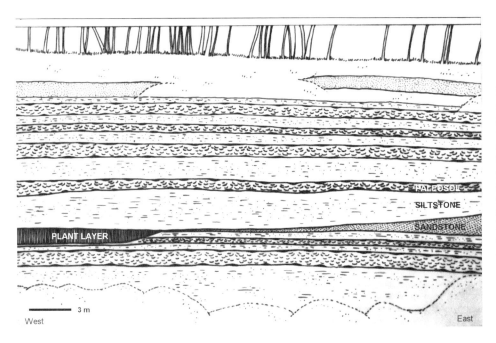

Figure 4. Diagram of the Red Hill outcrop profile that includes 11 m of the eastern portion of the 64 m plant layer where the sampling took place (after Cressler, 2001). © SEPM (Society for Sedimentary Geology). This diagram roughly corresponds to the photomosaic in Figure 2. Dots and dashes = Siltstones/Woodrow's Lithofacies 1. Arcuate stippling = Paleosols/Woodrow's Lithofacies 2. Dark shading = Plant layer/Woodrow's Lithofacies 3. Stippling = Sandstones/Woodrow's Lithofacies 4.

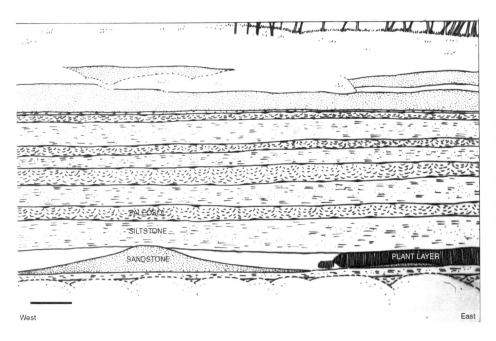

Figure 5. Diagram of the Red Hill outcrop profile that includes 10 m of the western portion of the 64 m plant layer where the sampling took place. Note that ~10 m of highly oxidized siltstone that is laterally equivalent to the sampled plant layer onlaps the sandstone body to the west of it.

Quarries were excavated from between the contact of the gray plant layer with the overlying red laminated siltstone down to the contact between the plant layer with either an underlying red paleosol or an underlying sandstone. The depth of the plant layer varied from 33 cm to 106 cm among quarries.

Quarries	Depth of Plant Layer
1a/1b	33 cm
2a/2b	106 cm
3a/3b	83 cm
4a/4b	93 cm
6a/6b	102 cm

The small depth of the sampled portion of the plant layer at quarries 1a and 1b is due to the more highly oxidized nature of the layer where it is in proximity to the western sandstone body. At that location, the presence of the reduced plant-bearing gray siltstone gradually is replaced laterally by red siltstone. Otherwise, the range of thickness throughout the reduced portion of the plant layer only varies between 83 cm and 106 cm. The variation in depth is mostly due to the uneven amount of mottling at the contact between the plant layer and the overlying red laminated siltstone.

The entire depth of the plant layer was sampled in each of the quarries, with the exception of quarries 2b and 6b. The upper 41 cm was missing from quarry 2b and the upper 33 cm was missing from quarry 6b. This material had been removed during previous excavations at these locations in the outcrop. While this activity provided access to the plant layer for the purposes of this study, it also removed some of the data.

An attempt was made to clear bedding surfaces at each quarry that were between 0.25 and 0.5 sq m in area. The exact size and dimension of each set of bedding surfaces were dictated by the accessibility of the surfaces within each quarry. The plant fossil census counts were normalized to 0.5 m² for each bedding surface by multiplying each original count times the quantity of 0.5 m² divided by the original area where that count was made. The bedding surface areas that were measured for each quarry are as follows:

Quarry	Bedding Surface	Quarry	Bedding Surface
1a	0.28 m²	4a	0.22 m²
1b	0.27 m²	4b	0.13 m²
2a	0.51 m²	5a	0.18 m²
2b	0.26 s m²	5b	0.20 m²
3a	0.35 m²	6a	0.34 m²
3b	0.29 m²	6b	0.34 m²

Quarries were excavated using a 3 lb crack hammer and the chisel edge of a sedimentary rock pick. Sometimes a pry bar was necessary. To remove large amounts of red overburden in order to get down to the plant layer, a gas-powered hammer was used on a number of occasions.

The intervals between counted bedding surfaces varied between 3 cm and 8 cm. The size and sequence of the intervals was established initially by finding the sedimentation breaks in the first quarry excavated, and then varied slightly from quarry to quarry depending on the individual characteristics of the quarry. The sampling intervals were at a smaller scale than the thickness of separate depositional events, so many of the surfaces counted were not smooth bedding planes.

The rock of each successive interval was removed in fragments. The fragments were turned over and laid next to each other on a flat area next to the quarry. The line-intercept counts were made on the undersides of the fragments. These combined surfaces were a close counterpart of the cleared quarry surface, and fossils on them were easier to see because the quarry surface was covered with the dust from the excavation process. For the purposes of comparison, line-intercept counts were initially made on both the cleared in situ surface and on the reassembled counterpart fragments. The counts were virtually identical every time, so for the remainder of the sampling process the line-intercept counts were made only on the more easily discernible reassembled counterpart fragments.

Plant Identification

Most of the plant fossils that were counted during the paleoecological sampling in this study were axis fragments of various sizes. Many of the axis fragments were unidentifiable (52%). They had no diagnostic characteristics that made them assignable to any known taxon. The unidentifiable plant remains were carbonized compressions of plant axes ranging in length from a few millimeters to several centimeters and were smooth and featureless.

Identifiable plant fragments were assigned to five taxa. These five taxa were (1) lycopsid, (2) gymnosperm, (3) *Rhacophyton* Crépin, (4) *Gillespiea* Erwin and Rothwell, and (5) *Archaeopteris* Dawson. The nature of this material is further described in the Results. Figured specimens are on repository at the Academy of Natural Sciences in Philadelphia (ANSP).

In addition, many of the axis fragments found during the paleoecological sampling had the appearance of charred and fragmented pieces of wood. Macroscopic and microscopic analysis of this material, including SEM imaging, has led to its identification as charcoal (Cressler, 2001). The charcoal fragments were counted as a separate category during the plant fossil census.

During removal of the red laminated siltstone to reach the plant layer, some poorly preserved plant material was occasionally encountered. This material consisted mainly of unidentifiable carbonized stems within reduction spots, and limonitized *Archaeopteris* axes. They were not counted during the sampling process because of their poor preservation and lack of stratigraphic control outside the plant layer.

RESULTS

The plant layer and its associated sediments and paleosols at Red Hill are consistent with an interpretation of avulsion processes at work on an aggrading flood plain (Slingerland and Smith, 2004). The lateral migration of a channel belt can leave in its path an anastomosing network of sand deposits and ponded flood-plain scours. The plant layer at Red Hill may represent such ponded water between two sand bodies.

Quarries 1a and 1b have their lower contact with the tapering right extension of the western sandstone body. The sandstone is 25 cm thick beneath quarry 1a and 5 cm thick beneath quarry 1b. The plant layer is highly red-mottled in this region. Another small sandstone body several meters long extends beneath quarry 3a at its eastern end where it is 12 cm thick. The lower contacts of all the other quarries are with the underlying red paleosol. Quarries 6a and 6b are at the pinched-out eastern end of the plant layer, end to end with the wedge-shaped eastern sand body. This location is interpretable as a clear shoreline profile.

The sampling transect goes from the shoreline of the pond to a sand body within the pond some distance from the shore (in reverse order of the numbered quarries). The western end of the 64 m transect may be much less than 64 m away from the nearest shoreline. This cannot be determined with certainty because of the two-dimensional nature of the exposure, which does not permit a map view of the pond. The face of the roadcut may have cut the pond at a tangent, in such a way that none of the plant material found would have been very far away from the shoreline.

The reduction of the plant layer was apparently primary, or syndepositional, and resulted in the preservation of large numbers of plant fossils. Because of their mottled character, the unfossiliferous greenish-gray margins above and below the plant layer are interpreted as possibly being secondarily reduced. The contact of the upper greenish-gray siltstone with the overlying red laminated siltstone is not even throughout its length. There are some dips in its generally horizontal configuration, but the relative thickness of the upper greenish-gray margin stays essentially the same throughout. This irregularity appears to result from uneven geochemical changes, and not disturbances in the flat lamination of sediments.

The thick (3 m) red laminated siltstone layer (Figs. 2–5) above the plant layer and its laterally associated sandstones apparently results from a series of overbank floods probably due to a change in proximity of the main active channel. This thick red siltstone layer has abundant articulated and disarticulated fish fossils found in strand-line lenses as well as a series of greenish-gray reduced stringers. Above this layer is a well-developed paleosol, which records an interval during which pedogenic processes exceeded the rate of deposition on the flood plain. Three additional cycles of alternating red laminated siltstones and paleosols continue up the section until they are interrupted by a sandstone layer (Figs. 2–5). Flood-plain strata bounded by paleosols in this fashion are characteristic of avulsion-derived sediment that is subsequently weathered during periods of isolation from flooding (Slingerland and Smith, 2004). The sandstone marks a return to active channel scouring and deposition in this part of the landscape.

Most of the deposition in the pond was through low-energy sedimentation of a suspended load of silt. Individual beds cannot be followed for more than a meter within the plant layer. The plant layer in quarries 5a and 5b was examined closely for subtle changes in lithology throughout its thickness. Slight changes in color and grain size suggest that no more than ten depositional events occurred over the 1 m thickness of the entire reduced horizon.

Occasional higher-energy movement of sediment into the pond is also indicated by observations of sharply curved bedding surfaces such as those that were encountered in quarry 4b and by the presence of some mud pebble conglomerate layers. Thin (1 cm) mud pebble conglomerate layers were found in the lower levels of quarries 4a and 4b and quarries 5a and 5b as well as at

Figure 6. *Otzinachsonia beerboweri* Cressler and Pfefferkorn rooting organ and stem (ANSP 4512). Scale in cm.

a spot that was excavated a few meters west of quarries 4a and 4b during preliminary excavations. The rounded mud pebbles are 2–8 mm in diameter and are densely packed on the bedding planes where they occur. Slightly larger pieces of rounded bone and fish scales are scattered among the mud pebbles in some places. The mud pebble conglomerate layers are immediately above the ~15 cm of the lower greenish-gray margin, except in quarry 5a where there is 5 cm of the dark-gray organic-rich siltstone in between. Two layers of mud pebble conglomerate were found 1 cm apart in quarry 5b with unfossiliferous greenish-gray siltstone between them. The mud pebble conglomerate layers represent a higher-energy influx of water into the pond containing a bedload of organic matter and pellets of ripped-up, previously deposited mud. These influxes took place when the pond was occasionally accessible to higher energy flow on the highly dynamic flood plain.

Simultaneous with the plant census, a record was kept of the vertebrate remains found during the sampling effort. Numerous isolated vertebrate fragments were found scattered relatively evenly throughout the lateral and vertical extent of the plant layer. Remains included *Hyneria* teeth and scales, megalichthyid scales, groenlandaspid plates, acanthodian spines, and articulated palaeoniscoid fish.

Determination of Plant Fossil Taxa

Lycopsids

Lycopsid stem fragments constitute 4.5% of the identifiable plant remains in the plant layer at Red Hill. The lycopsid stem fragments vary in size from 1 cm to 10 cm in width and 2 cm to 20 cm in length. They represent various levels of lycopsid stem decortication but are all characterized by the distinctive spiral pattern of lycopsid leaf traces (Stewart and Rothwell, 1993). Two specimens exhibit leaf traces arranged in rows of pseudowhorls and on that basis can be tentatively assigned to the genus *Lepidodendropsis* Lutz pending study of leaf cushion details (Jennings, 1975; Jennings, Karrfalt, and Rothwell, 1983). Most of the other decorticated lycopsid stem specimens resemble *Cyclostigma* in gross appearance or are assignable to form genera that are now only considered to reflect levels of stem decortication. Therefore, all lycopsid fragments have simply been given the designation "lycopsid" for the purpose of the paleoecological analysis.

In addition to stem fragments, several lycopsid rooting organs have been found at Red Hill (Fig. 6). These were found in the plant layer, but not during the paleoecological sampling. They vary in diameter from 2.5 to 10 cm. These extremes in size are represented on one bedding surface (Fig. 6). The rooting organs consist of four-lobed bases of stems with masses of attached rootlets representing the new taxon *Otzinachsonia beerboweri* (Cressler and Pfefferkorn, 2005).

Ferns

Remains of *Rhacophyton ceratangium* Andrews and Phillips are abundant at Red Hill (Fig. 7). They constitute 38% of the identifiable plant fossils and consist primarily of foliar axes with a uniform width of ~3 mm that have a distinctive groove down the middle due to the bilobed morphology of the stele (Andrews and Phillips, 1968). Some *Rhacophyton* foliage was found during the paleoecological sampling. The foliage is highly divided and entangled with the distinctive stems in dense mats. There were also several *Rhacophyton* foliar axes found together that were 1 m long and 1 cm in width. These were probably main foliar axes. *Rhacophyton* was a zygopterid fern that has been reconstructed as a 1–2 m high plant and has some wood in its larger foliar axes (Kräusel and Weyland, 1941; Andrews and Phillips, 1968; Dittrich, Matten, and Phillips, 1983).

Gillespiea randolphensis Erwin and Rothwell was discovered at Red Hill during the paleoecological sampling (Fig. 8).

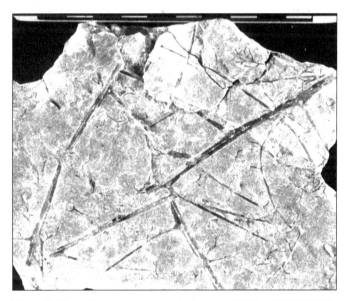

Figure 7. *Rhacophyton ceratangium* Andrews and Philips (ANSP 4500). Scale in cm.

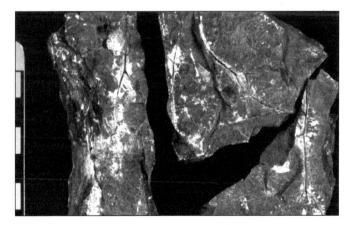

Figure 8. *Gillespiea randolphensis* Erwin and Rothwell (left to right: ANSP 4501, ANSP 4502, ANSP 4503). Scale in cm.

Figure 9. *Archaeopteris hibernica* (Forbes) Dawson (ANSP 4506). Scale in cm.

Figure 10. *Archaeopteris macilenta* (Lesquereux) Carluccio et al. (ANSP 4507). Scale in cm.

Figure 11. *Archaeopteris obtusa* Lesquereux (ANSP 4508). Incisions in leaves are taphonomic features. Scale in cm.

Gillespiea constitutes 2% of the identifiable plant remains at Red Hill. Axes of *Gillespiea* are 1 mm wide and less, often tapering to a very thin curving distal portion. As is characteristic of *Gillespiea*, small fertile structures emerge from between the dichotomies in some of the axes found at Red Hill. Some specimens of *Gillespiea* have been described elsewhere as having a sclerenchymatous cortex (Erwin and Rothwell, 1989). The stauropteridalean fern *Gillespiea* was a heterosporous plant of slender proportions and wiry, flexible construction (Erwin and Rothwell, 1989).

Progymnosperms

The remains of *Archaeopteris* Dawson (Figs. 9–12) dominate the plant-fossil assemblage at Red Hill, where they constitute 55% of the identifiable plant biomass. *Archaeopteris* remains at Red Hill consist almost entirely of 1-cm-wide penultimate branches, usually with attached biseriate ultimate branches. The penultimate branches have both distinctive longitudinal lineations from the underlying vascular strands and a transverse rugose pattern from clusters of cortical sclereids that help to identify *Archaeopteris* axes even when there is no foliage. The foliage of two *Archaeopteris* species, *A. hibernica* (Forbes) Dawson and *A. macilenta* (Lesquereux) Carluccio et al., and fertile branches with sporangia, were identified during the census study, but by far most of the *Archaeopteris* remains were axes unidentifiable as to species. The designation of *Archaeopteris* species is based on leaf morphologies. Arnold (1939) and Kräusel and Weyland (1941) provided useful keys and synonymy lists that are widely followed. The leaves of *Archaeopteris hibernica* often exceed 2 cm in length and are rounded or obovate (Fig. 9). The leaves of *Archaeopteris macilenta* are less than 1.5 cm in length and are rounded or broadly obovate with deeply cut margins (Fig. 10).

In addition to *A. hibernica* and *A. macilenta*, two other less abundant species of *Archaeopteris* have been collected from Red Hill that are assignable to *A. obtusa* Lesquereux (Fig. 11) and *A. halliana* (Göppert) Dawson (Fig. 12).

Gymnosperms

Gymnosperms at Red Hill consist of both cupulate and acupulate forms of the earliest grade of gymnosperm evolution, constituting 0.5% of the plant fossil assemblage. Their vegetative structures are morphologically similar to those of other Late Devonian gymnosperms such as *Elkinsia polymorpha* Rothwell et al. from West Virginia (Gillespie et al., 1981; Rothwell, 1989; Serbet and Rothwell, 1992) and *Moresnetia* Stockmans (Stockmans, 1948; Fairon-Demaret and Scheckler, 1987) from Belgium. Both are also from the same Famennian 2c time horizon as Red Hill. Their reproductive axes are dichotomously branching and get progressively thinner with each distal dichotomy. They range in width from 1 cm to 1 mm. Cupules, ovules, or both are usually found at the terminations of the dichotomous branches. Because the dichotomous branches of the various gymnosperms are indistinguishable from one another and do not always have fructifications, they have all been designated "gymnosperm" for the purpose of this paleoecological analysis.

Figure 12. *Archaeopteris halliana* (Göppert) Dawson (ANSP 4509). Scale in cm.

Figure 13. Gymnosperm cupule from plant layer showing three ovules (ANSP 4531).

The cupulate gymnosperms found in the plant layer have associated ovules and look different from *Elkinsia* and *Moresnetia* but may not be preserved well enough to be adequately described as a new taxon (Fig. 13). The acupulate gymnosperms are similar to *Aglosperma quadripartita*, previously described from younger Late Devonian rocks in Wales (Hilton and Edwards, 1996). Cupules found outside the sampling area in a loose block that fell from the far western end of the outcrop in October 1997 are different in morphology from those found in the plant layer. They are more fused than any other Late Devonian gymnosperm cupule besides *Dorinnotheca streelii* Fairon-Demaret (Fairon-Demaret, 1996), but unlike *Dorinnotheca* they are not inverted, and thus far have not been found associated with any ovules.

Barinophytes

Barinophyton White is rare, but isolated specimens of two species have been recovered from Red Hill. They were not recorded during the paleoecological sampling. One specimen of *Barinophyton obscurum* (Dun) White (Fig. 14) was found in a loose block of gray siltstone during a visit to the outcrop in October 1997. It had fallen from a vertical portion of the exposure toward the western end of the outcrop, several hundred meters from the plant layer that was sampled for this study. The loose block was collected because it contained abundant gymnosperm cupules. During further preparation in the lab, a distal branch of *Barinophyton obscurum* was uncovered, which includes five strobili. The sessile morphology of the strobili, with no pedicels at their attachment to the main stem, is diagnostic of *Barinophyton obscurum* (Brauer, 1980). This is in contrast to *Barinophyton citrulliforme* Arnold,

Figure 14. *Barinophyton obscurum* (Dun) White (ANSP 4504). Scale in cm.

first described by Arnold (1939) from New York specimens, but further elaborated by Brauer (1980) on the basis of a large collection from a Catskill Formation site in Potter County, Pennsylvania. *Barinophyton citrulliforme* has pedicels on the strobili where they attach to the main stem. Brauer also found poorly preserved specimens of *Barinophyton* cf. *obscurum* and what he called *Protobarinophyton pennsylvanicum* (Brauer, 1981).

Two fragmentary specimens of the long, slender strobili of cf. *Barinophyton sibericum* Petrosian (Fig. 15) were found

Figure 15. *Barinophyton sibericum* Petrosyan (ANSP 4505). Scale in cm.

by Norm Delaney some years before the systematic paleoecological sampling was conducted. He had found the specimens in the plant layer, but further specimens were not discovered during the extensive sampling there. *Barinophyton sibericum* was also reported by Gillespie et al. (1981) and Scheckler (1986c) for exposures of the Hampshire Formation. This taxon has long pedicels that often bend sharply at the attachments to strobili.

Distribution of Plant Fossils in the Deposit

The distribution of the plant fossils and charcoal in the plant layer at Red Hill as determined by the sampling procedure outlined above is shown in Figures 16–21. The plant fossil designations have been restricted to the following five taxa for the purpose of this analysis: *Archaeopteris*, *Rhacophyton*, *Gillespiea*, lycopsids, and gymnosperms.

Along the vertical axis of each figure are centimeter intervals measured up from the boundary between the plant layer and the underlying red paleosol or sandstone. The average thickness of the reduced horizon is ~1 m, but the plant fossils were found only between 15 cm and 78 cm from the bottom, with an isolated occurrence of gymnosperms at 93 cm from the bottom of quarry 4.

The horizontal axis represents the length in meters of the plant layer between the flat-bottom, convex-upward sandstone lens to the west and the wedge of sandstone to the east. Although the plant layer is shown laterally compressed and vertically exaggerated, the distances between the quarries are shown proportionally in these figures. The pairs of quarries are combined in the figures. The density distributions of the plant fossil taxa are shown as vertical bars beneath the number of the quarries in which they were found. The density distributions, with taxon counts normalized to 0.5 m^2 of bedding surface at each level sampled, are simplified to reflect differences of order of magnitude only. The width of the vertical bars reflects the maximum density of a particular plant fossil taxon within that portion of a particular quarry. Obviously, the distribution densities traverse the vertical dimension of time in the depositional profile. This reflects a particular taxon as having a persistent point source in the landscape through at least several depositional events into the pond.

The right edge of Figures 16–21 corresponds to the shoreline of pond. The left edge of the figures corresponds to the sandstone body at the other end of the transect. As can be seen in Figure 16, the densest concentration of *Archaeopteris* remains is close to the shoreline in the middle levels of quarry 5 and quarry 6. The line-intercept counts of *Archaeopteris* on these bedding planes was consistently in the hundreds. *Archaeopteris* is also best preserved on these bedding planes and has recognizable foliage. Lower concentrations of less well preserved *Archaeopteris* branches are distributed out to quarry 2, farther into the pond.

Rhacophyton was found in its densest and best-preserved concentrations in quarry 4 (Fig. 17), where the line-intercept counts numbered in the thousands on bedding planes between 30 and 40 cm from the bottom of the reduced horizon. These quarries are 19 m along the transect from the shoreline. *Rhacophyton* axes by the hundreds were scattered on bedding planes in the lower portions of quarries 4, 5, and 6. Lighter concentrations of *Rhacophyton* axis fragments were counted in the upper portions of those quarries and in quarry 3.

The distribution of lycopsid stems (Fig. 18) lies in between the densest concentration of *Archaeopteris* and the densest concentration of *Rhacophyton* in the distribution profile. They are concentrated in a diagonal distribution from the bottom of quarry 6 to the top of quarry 4. Some sorting of lycopsid stems by size appears to be in evidence, with the larger stems found toward the shoreline of the pond.

The most significant concentrations of *Gillespiea* were sampled from quarry 2 (Fig. 19) with minor amounts found in quarry 3. The main concentration of *Gillespiea* was centered 53 m along the transect from the shoreline. *Gillespiea* was found in quarry 2 several centimeters above the densest concentrations of charcoal found during the entire sampling process (Fig. 20). Few other recognizable plant remains were found in this quarry.

The gymnosperms that were found during the sampling process were concentrated in two small areas toward the top of quarries 4 and 5 (Fig. 21).

Previous excavations in the plant layer had also yielded significant plant remains. A quarry a few meters west of quarry 4 that E.B. Daeschler designated "9702" yielded numerous *Archaeopteris* branches, cormose lycopsids, and sphenopteroid foliage with associated axes and roots. Seven meters to the east of quarry 4 a trial quarry was excavated just prior to embarking on the systematic sampling. From this quarry were removed cupulate and acupulate gymnosperm material with charcoal on the same bedding planes. The subsequent discovery of gymnosperms in

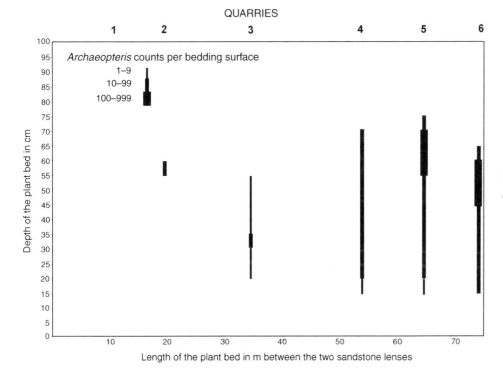

Figure 16. Distribution profile of *Archaeopteris* fossils in plant layer.

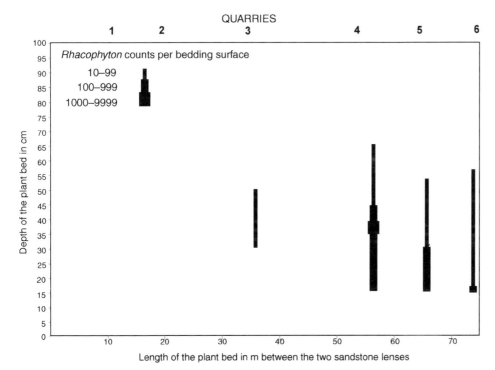

Figure 17. Distribution profile of *Rhacophyton* fossils in plant layer.

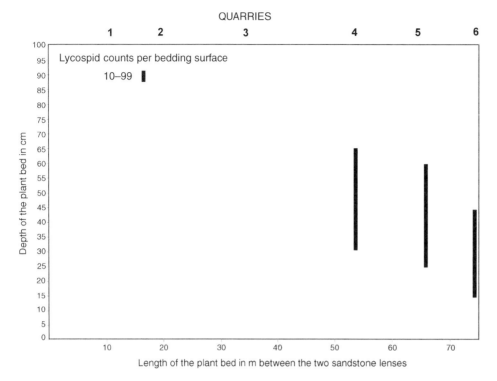

Figure 18. Distribution profile of lycopsid fossils in plant layer.

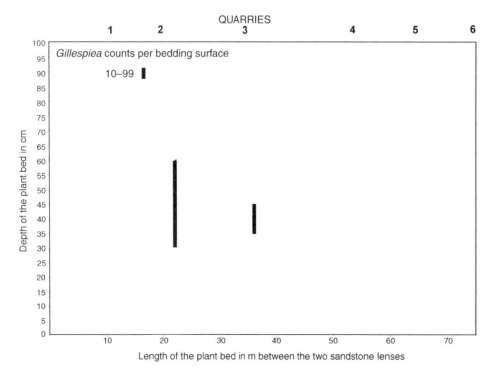

Figure 19. Distribution profile of *Gillespiea* fossils in plant layer.

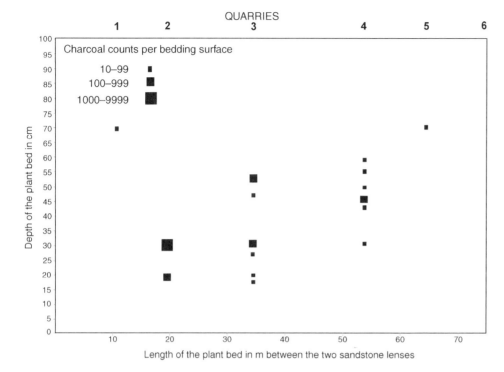

Figure 20. Distribution profile of fossil charcoal fragments in plant layer.

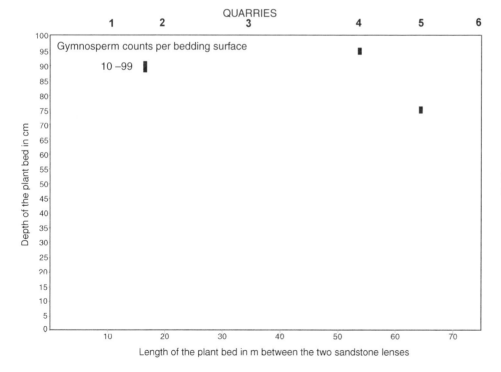

Figure 21. Distribution profile of gymnosperm fossils in plant layer.

quarry 4 and 5 to either side of the initial discovery reinforces the impression that their distribution is concentrated in this portion of the plant layer.

DISCUSSION

The following factors working in combination explain the distribution profile of plant fossils in the deposit:

(1) the limitations of the sampling process;

(2) the differential preservation of the plant fossils for both biological and geochemical reasons;

(3) the different hydrodynamic and aerodynamic properties of the plant parts as they entered the deposit under various flow regimes;

(4) the timing of cyclic and stochastic events before and during deposition, such as wind, fire, floods, and seasonal adaptations of plants, such as the seasonal branch shedding of *Archaeopteris*;

(5) the original distribution of the plants growing in the landscape.

Limitations of the Sampling Process

Several phenomena no doubt contributed to some error in the process of counting the plant fossils. Many of the surfaces on which plant fossils were counted did not cleanly break along bedding surfaces. Frequently, the fragments broke out of a quarry into small uneven pieces. Several of these roughly breaking intervals would regularly occur between two cleanly breaking bedding surfaces. The sampling was simply taking place at a finer scale than the thickness of the depositional packages. Counting the plant fossils on the rough fragmentary surfaces would probably not have contributed any significant error to the plant census. Because the flow regime appears to have remained the same throughout each depositional event, these counts merely add finer resolution to the overall plant count. Even though it is also possible that not every clean sedimentation break was found in every quarry, the sampling was fine enough to have included part of every depositional event.

Another difficulty was that the bedding surface area of the quarries changed slightly as the excavations proceeded. Rock was increasingly difficult to remove from areas immediately adjacent to the back wall of the quarries. Some weathering was required to express bedding planes. This potential decrease in the size of the measured surface area for each interval was always compensated by an increase in surface area along the proximal margin of the quarry due to the sloping hillside. Measuring precisely how well this process maintained uniform area throughout the quarry was impractical, though.

The measured sets of centimeter-scale intervals from quarry to quarry probably do not correspond precisely to each other because the measurements were made for each increment after the previous surface was cleared in each quarry. Distance from the lower contact of the reduced horizon was used as a rough proxy for correlating beds from quarry to quarry. Any imprecision would not have confounded the interpretations made in this paper, since only the coarsest trends in the vertical and lateral distribution of the plant fossils and charcoal are considered in the analysis.

In contrast to the fine-scale vertical sampling within the quarries, the lateral spacing of the quarries left large portions of the plant layer unsampled. Patchiness in the distribution of the plant fossil taxa may have gone undetected as a result. The number and location of the quarries were dictated by practical considerations of the time and effort it would have taken to remove overburden. Differences in spatial scale variation in the distribution of the plant fossils was taken into account by making some quarries adjacent to each other and others some distance apart. A pattern of concentric density gradients of the plant fossils in the distribution profile suggests that significant patchiness does not occur at a smaller scale for the taxa that occur at high densities. At the scale of sampling for this study, the rare taxa do not occur at high enough densities to produce a distribution pattern discernible from statistical noise.

Differential Preservation

Studies of organic detritus in Holocene oxbow lakes and other alluvial settings have determined that the most decay-resistant elements of the local vegetation get preserved in compression-impression assemblages (Scheihing and Pfefferkorn, 1984; Gastaldo et al., 1989). Decay-resistant elements of a Late Devonian tropical lowland flora are all that are represented at Red Hill. The reducing conditions of the pond sediments into which the plant remains were deposited promoted plant preservation. Reduction of sediments was largely maintained by the decay of the less resistant organic detritus. The plant layer underwent this geochemical process. Its reduction state was produced simultaneously with deposition. The mostly unfossiliferous greenish-gray siltstone margins above and below the dark-gray layer were probably secondarily reduced, perhaps by anoxic groundwater migration. The upper greenish-gray siltstone, for example, does not always conform to bedding. There are occasional poorly preserved plant fossils in the immediately overlying red laminated siltstone that are surrounded by greenish-gray reduction spots. These conditions place an upper and lower boundary on the horizon of well-preserved plant fossils.

Laterally, the reducing conditions in the plant layer appear to have been fairly uniform except in quarries 1a and 1b. Excavations in these quarries did not yield any recognizable plant fossils. Conditions there were more highly oxygenated as indicated by extensive red mottling. These conditions are perhaps due to the close proximity of the sandstone body. The high porosity of the sand perhaps promoted groundwater flow at a higher oxygenation state within the accumulating silt in the immediate vicinity.

The waters of the pond itself were probably well oxygenated most of the time as indicated by the numerous articulated specimens of actinopterygian fishes found among the plant remains in the plant layer (Daeschler, 2000a). An occasional anoxic episode in the pond may have killed off schools of these fish, ultimately leading to their incorporation into the fossil record. Supporting evidence for general oxygenating conditions in the pond water itself is the discovery of the green alga *Courvoisiella ctenomorpha* Niklas in a hydrofluoric acid maceration of the plant layer dark-gray siltstone (Niklas, 1976; Pat Gensel, 1998, personal commun.).

Transport into the Deposit

All of the plants fossilized in the plant layer at Red Hill appear to have been transported into the deposit from a range of relatively short distances. Variation in preservational states suggests that some were originally growing along the edge of the pond, and some were growing at somewhat farther distances away on the surrounding flood plain. The pond that the plant layer represents was silting up on the flood plain under predominately low-energy conditions. Silt moved over the landscape suspended in flood waters, which also carried with it plant matter, animal remains, and charcoal. The arrangement of the plant remains and charcoal on all the bedding planes that were examined does not indicate any clear flow direction. Silt-laden waters entered the pond either from a general rise of water level on the flood plain resulting from a nearby active channel overtopping its banks, or from a breach in the pond shoreline that gave access to waters from a more active channel through sporadic crevasse splays. The condition of the plant fossils and the lack of flow-direction indicators imply that a general rise of water level was the more prevalent process. The presence of at least two mud pebble conglomerate layers, however, suggests that shoreline breaches also occasionally occurred during times of higher-energy deposition.

Archaeopteris remains were found scattered throughout the sampling transect, but the densest and best-preserved concentration of *Archaeopteris* remains occurred along the shoreline of the pond. All *Archaeopteris* found consisted entirely of distal branch portions, some of which have attached leaves and sporangia. Their source was probably from the canopies of nearby small to medium-sized trees. The intact condition of many of the branches with attached leaves and sporangia suggests that they were wind carried directly into the deposit. The less well preserved condition of many of the other *Archaeopteris* branches implies that some of the material was dried on the ground before being carried into the deposit by silt-laden floodwaters. Either the windblown branches may have sunk where they landed on the edge of the pond, or they may have accumulated by wind-drift along the shoreline.

Like *Archaeopteris*, the *Rhacophyton* remains were found distributed widely in the plant layer. The condition of these fossils varied widely as well. The densest and best preserved concentration of *Rhacophyton* was found in quarries 4a and 4b, a location more distal from the lakeshore, as indicated by the transect line, than the main concentrations of *Archaeopteris* and lycopsid stems. The condition of these *Rhacophyton* remains indicates that they were transported from close by into the deposit, and in fact were probably growing along the edge of the pond. If the face of the roadcut slices the pond nearly parallel to one of its shorelines, the source of the well-preserved *Rhacophyton* remains may have been from an area of the landscape where they were growing in concentration. If the transect is more perpendicular to the shoreline, *Rhacophyton* may have been more easily transportable into the distal portions of the deposit. Almost all of the *Rhacophyton* material in the rest of the plant layer consists of small, highly transported broken axes.

The decorticated lycopsid stems show evidence for transport not only by their condition, but also by sorting of stem sizes. Smaller stems were found farther out into the deposit away from the shoreline and larger stems were found closer to the shoreline. They may have come in with flood waters from elsewhere in the landscape, but the cormose lycopsid rooting organs with attached rootlets were probably transported from a negligible distance.

The main concentration of *Gillespiea* remains was found farthest from shoreline. They were also found closer to shoreline, but in extremely small concentrations in proportion to the other plant remains. *Gillespiea* was found entirely as small, thin axes. Transport may be the best explanation for its occurrence away from the shoreline compared with the other plant remains, but their low numbers may also indicate that no pattern of distribution is discernible. The highest concentrations of charcoal were also found in the same distal quarries as the highest concentration of *Gillespiea*. Charcoal is light and easily transported, and perhaps *Gillespiea* axes have similar characteristics. On the other hand, *Gillespiea* may have been transported a relatively shorter distance such as from the top of the western sandstone body or a shoreline of the pond that is not discernible from the transect configuration.

Most of the gymnosperms found in the plant layer were associated with fragments of charcoal. They were transported into the deposit together from nearby on the surrounding flood plain.

Aside from the articulated actinopterygian fish, which can be inferred to have died in their original habitat, the isolated vertebrate elements found in the plant layer were most likely transported into the pond from a main river channel. The fish died and became disarticulated at some point between a main channel and the pond, and their isolated elements were part of the bedload of the slowly moving flood waters.

Cyclic and Stochastic Events Influencing Plant Fossil Distribution

Timing of floods and fires due to seasonal influences, as well as any seasonal adaptations of the plants, such as deciduousness, would influence the pattern of plant distribution in the Red Hill pond deposit. For example, *Archaeopteris* penultimate branches are interpreted to be deciduous (Scheckler, 1978), and possibly shed as a dry-season adaptation (DiMichele et al., 1992). When complete they have a swollen abscission region at their base and often occur as dense mats of foliage and penultimate branches only, such as at Red Hill. Within the seasonal parameters, stochastic phenomena such as variation in storm intensities and wind gusts, and the magnitude and directionality of floods also have an influence on plant fossil deposition patterns.

The most direct evidence for seasonality at Red Hill is the abundant paleovertisols. The Red Hill outcrop includes a series of stacked paleovertisols above and below the plant layer where the sampling took place. There are at least two thin paleovertisols below the plant layer, and at least four well-developed paleovertisols above the plant layer (Figs. 3–5).

The shrink-and-swell clays in the paleovertisols of the Catskill Formation have been diagenetically altered to illite (Harvey, 1998). The first paleosol above the plant layer and the intervening red laminated siltstone is 3 m thick. This is a minimum thickness, because the soil horizon could have been partly truncated by erosion or scour without leaving any evidence for it, prior to being buried by the next series of red laminated siltstones. The paleosols have vertical cracks and pedogenic slickensided surfaces arcing away from the cracks in the manner distinctive of mukkara structures (Dudal and Eswaran, 1988). The cracks have been filled, and are distinguished by being reduced to a gray color. The mukkara have angles that range between 20° and 50° with respect to horizontal. They are concentrated between 80 and 180 cm below the apparent surface of the paleovertisol. The paleovertisols have lost all indication of stratification through pedoturbation. They have angular peds ranging in size from 1.0 - 1.5 cm in their upper portions, and from 2 - 4 cm in their lower portions. They also contain numerous carbonate nodules.

The paleovertisols at Red Hill are indicative of a seasonally wet and dry climate when the fossil flora was growing on the landscape (Driese and Mora, 1993). Even though there are no paleovertisols that are laterally continuous with the plant layer, they are found both above and below it. To assume that the plant layer was a free-standing body of water on the flood plain while elsewhere vertisols were being formed in the landscape is not unreasonable. Free-standing bodies of water are found on landscapes where vertisols are being formed today (Dudal and Eswaran, 1988). In this model, deposition of silt into the pond would have also been subject to seasonal changes.

Paleogeographic reconstructions showing the central Appalachian Basin at 20° south of the equator (Scotese and McKerrow, 1990) imply a subtropical region in a latitudinal zone of a single annual wet season and dry season, which possibly resulted from monsoonal circulation patterns in the "Proto-Tethys" sea (Bambach et al., 1999). A seasonally wet and dry climate for the central Appalachian Basin in the Late Devonian is further supported by lithological and paleontological studies in addition to those conducted on the paleosols. Evaporites, calcretes, the structures of stream and flood-plain deposits, and traces of lungfish burrows have led to an estimate of an annual precipitation of 75 cm arriving in the wet season during this Late Devonian climate regime (Woodrow et al., 1973).

The subtle changes in grain size and color observed throughout the entire 1 m thickness of the reduced horizon indicates that only ten depositional events are represented. If deposition in the pond occurred primarily during annual wet-season floods, the pond could have been entirely silted up in as little as ten years.

One outcome of the wet-and-dry seasonality at Red Hill during the Late Devonian was the occurrence of wildfires during the dry season (Cressler, 2001). Small black organic fragments that were subsequently identified as charcoal were discovered in the plant layer when the paleoecological sampling was conducted. Scanning electron microscopy revealed that the anatomically preserved charcoal was *Rhacophyton*.

Distribution of the charcoal in the plant layer is shown in Figure 20. Some bedding planes are densely covered with charcoal fragments along distinct horizons that might connect across sampling quarries. Charcoal is also sparsely scattered throughout many other parts of the deposit. There is no preferred orientation of the charcoal fragments on any of the bedding surfaces. Some bedding planes have varying amounts of charcoal in association with unburned plant material. Dense accumulations of charcoal co-occur with recognizable but poorly preserved penultimate branches of *Archaeopteris*. Dense accumulations of charcoal also occur on the same bedding planes with well-preserved gymnosperms. Most of the gymnosperm material found in the plant layer was found in association with charcoal. Even on the hand specimens where charcoal and unburned plant material occur together, all of the identifiable charcoal fragments are *Rhacophyton*. Well-preserved *Rhacophyton* has not been found on the same bedding planes with charcoal, although occasionally scattered charcoal fragments and small *Rhacophyton* fragments have been found on the same bedding plane. The quarries with the most abundant charcoal (2a and 2b) also had the most abundant *Gillespiea* remains. Abundant *Gillespiea* occurs within 15 cm above a large pulse of charcoal in quarry 2a, and within 30 cm above a large pulse of charcoal in quarry 2b, but *Gillespiea* and charcoal have not been found together on the same bedding plane. Whether the charcoal in quarries 2a and 2b is on the same horizon is ambiguous. Charcoal in quarry 2a is 19 cm above the contact between the greenish-gray siltstone and the underlying red siltstone, while that in quarry 2b is 30 cm above the contact. The two quarries are 2 m apart and beds cannot be followed from one to the other.

Original Growth Position of the Vegetation in the Landscape

Because of the entirely parautochthonous nature of the plant fossils that were sampled, these data provide a less direct test of hypotheses concerning the original growth position of the plants than if they were found in situ. In combination with the available taphonomic information, however, the sampling profile of this parautochthonous assemblage does place many more constraints on plant community models than would an allochthonous assemblage. The model being tested here is whether the Late Devonian tropical lowland plants at Red Hill had phylogenetically partitioned the landscape by environment sensu DiMichele and Phillips (1996).

Plant fossils sampled during this paleoecological analysis all came from the same sedimentological layer. They may have come from different habitats in the landscape, but they ended up buried in the same depositional environment, the silt of a floodplain pond. The evidence indicates that the plant fragments had been transported from various distances, but the condition of the identifiable plant remains indicates that they were deposited no more than several meters from their growth position. Any discernible habitat partitioning by the plants at Red Hill would therefore have been over a spatial scale on the order of tens or hundreds of square meters rather than square kilometers.

The *Archaeopteris* trees that shed their branches into the deposit were probably growing the farthest away from the shores of the pond of all the plant taxa sampled. There are no remains of *Archaeopteris* trees in the plant layer that are larger than 1-cm-thick penultimate branches. All *Archaeopteris* remains must have arrived in the deposit as wind- or water-borne fragments. The transport distance of the fragments would have been enhanced by their having dropped from the canopies of sizable trees at the edge of a nearby forest. The nearest *Archaeopteris* forest to the pond apparently did not grow up to the edge of the pond, otherwise the trees would have dropped larger branches into the deposit. They probably were growing higher on the flood plain or on a nearby levee where the water table was lower. The *Archaeopteris* forest that supplied the well-preserved, apparently windblown, branches along the pond shoreline may have been growing some distance upslope from the shoreline where their branches accumulated. Alternatively, the concentration of *Archaeopteris* remains along the edge of the pond is due to accumulation by wind transport along the surface of the water. In either case, the *Archaeopteris* forest was not growing very near the pond, but where the water table was consistently lower.

Archaeopteris is believed to have been capable of forming closed canopy forests (Retallack, 1985; DiMichele et al., 1992). The flood-plain paleosols at Red Hill have not yielded any clear evidence of *Archaeopteris* root casts or tree stumps. At the extreme western end of the outcrop is a suggestive reduction halo in a paleosol constituting two 85 cm parallel vertical lines ~50 cm apart. The greenish-gray lines flare out from each other at the base, suggestive of a stump with roots. The structure is weathering out from the outcrop in a manner also suggestive of a three-dimensional stump cast. This is the only putative evidence of an in situ plant of *Archaeopteris* proportions in the entire 1-km-long exposure of Red Hill. The scarcity of in situ tree stumps and root casts in the flood-plain paleosols of the Catskill Delta Complex has been addressed by Driese et al. (1997). They list three possible reasons: (1) flood plains may have been unfavorable environments for large trees, (2) preservation is poor due to pedoturbation in fine-grained flood-plain paleosols, or (3) the rates of sediment accumulation were insufficient to bury stumps and root casts. Of these, the second reason seems the most plausible.

The presence of locally concentrated and well-preserved remains of *Rhacophyton* as well as scattered fragmentary remains throughout the deposit indicates that *Rhacophyton* was widespread on the surrounding landscape and grew up to the edge of the pond. *Rhacophyton* probably grew in monotypic clonal patches (Scheckler, 1986a, 1986c) as suggested by the specificity of the wildfires and the evidence for distinct deep-rooted and shallow-rooted areas in the Late Devonian landscape (Harvey, 1998). *Rhacophyton* was shallow rooted (Scheckler, 1986c) and would have dried up even with small reductions of wetland water tables. It was subject to desiccation and ignition during the dry season. In contrast, *Archaeopteris* was a deep-rooted tree that would have been little affected by a temporary, drought-induced drop in the water table (Algeo and Scheckler, 1998; Algeo et al., 2001). Burned-over areas formerly occupied by *Rhacophyton* perhaps provided an opportunity for other plants to colonize, such as the gymnosperms and perhaps *Gillespiea*. At Elkins, West Virginia, study of contemporaneous sediments shows that gymnosperms were pioneers that were quickly overgrown and replaced by *Rhacophyton* (Scheckler, 1986c; Rothwell and Scheckler, 1988), which suggests a complex ecological interaction.

The diagonal profile of lycopsid distribution in the deposit suggests a progradation of lycopsids along the shore of a shallowing pond. This evidence is based on the distribution of decorticated stems, which may have actually come from plants growing in the surrounding flood plain. The discovery of several lycopsid rooting organs with attached rootlets is firmer evidence that cormose lycopsids were growing along the edge of the pond, however. They would have been growing in soft sediment along the edge of the pond, which slumped into the water during floods. For part of the year at least, the cormose lycopsids had their lower portions submerged in the water. Red Hill lacks the coal horizons, similar to those described by Scheckler (1986c) and by Goodarzi et al. (1989, 1994) for other Devonian sites. This suggests that the Red Hill wetlands either periodically dried out or that their waters were too shallow and too well oxygenated to sustain massive carbon burial. The cormose lycopsids show evidence for constituting a multi-aged community, as indicated by specimens of different sizes on one bedding plane (Fig. 6). Other lycopsids, such as *Lepidodendropsis*, may have been growing elsewhere on the flood plain and their decorticated stems washed into the pond during flood events. In the Early Carboniferous, for example, *Lepidodendropsis* inhabited both peat- and non-peat-accumulating wetlands (Scheckler, 1986a).

Gillespiea remains are concentrated at a point along the transect farthest away from the identifiable pond shoreline. This bias could be due to their having been preferentially transported farther because of their lightness and buoyancy. The *Gillespiea* remains are highly fragmented. They could have transported into the deposit from a growth position anywhere in the surrounding landscape, including as close by as the top of the western sandstone body. *Gillespiea* seems to be loosely associated with the presence of charcoal, but that may be due to similar transport characteristics. Nevertheless, the small construction of *Gillespiea* suggests that it might be a quick-growing opportunistic plant of disturbed areas, a lifestyle perhaps aided by its heterosporous reproduction.

Gymnosperms were found on only a few bedding planes during the entire sampling effort. Any pattern of their distribution may be insignificant due to their relative rarity and disarticulation. All gymnosperm remains were found toward the top of the plant layer. Their upper placement in the deposit implies an appearance during the later stages of the pond's existence. Frequent association of gymnosperms on the same bedding planes as charcoal may indicate that they were pioneer plants in the areas of the landscape affected by fires. Gymnosperms at Red Hill show a range of morphologies, from acupulate forms to cupulate forms with differing degrees of fusion of the cupules. This range of reproductive morphologies likely reflects a variety of ecological niches already occupied by gymnosperms as early as their first fossil appearance in the Famennian 2c.

The sum total of taphonomic evidence, the distinct distributions of the plant taxa in the depositional profile, and the specificity of the charcoal modestly support a model of small-scale heterogeneity of the Red Hill landscape with respect to its vegetation. The ecological partitioning of the landscape by the plants at Red Hill does appear to correspond to higher-order taxonomic groups, especially in light of low apparent diversity at the species level. However, since low species diversity necessitated the pre-selection of higher-order categories for the analysis at the outset, some of the higher-order patterning attributed to these results may reflect this conceptual bias. Phylogenetic partitioning of the landscape by major plant groups close to the time of their origin has important implications for understanding macroevolution. The evolutionary appearance of clade-specific, distinct plant architectures appears to be associated with specific ecological modes, which may reflect the divergent biology that underlies these morphotypes (DiMichele and Bateman, 1996; DiMichele and Phillips, 1996).

Comparison of Red Hill with Other Late Devonian Floras

The Late Devonian (Famennian) Hampshire Formation in West Virginia and Virginia includes plant fossil assemblages that represent some of the earliest known coal swamps (Scheckler, 1986c). Older coal beds are now known from the Givetian and Frasnian of Melville Island, Arctic Canada (Goodarzi et al., 1989; Goodarzi and Goodbody, 1990; Goodarzi et al., 1994). Subsequent examination of these deposits reveals that the coal-forming plants were arborescent lycopsids (Stephen Scheckler, 2002, personal commun.). In the Famennian, *Rhacophyton* dominated both deltaic marshes and flood-plain backswamps (Scheckler, 1986c). The Elkins, West Virginia locality in the Hampshire Formation is a deltaic shoreline deposit of Famennian 2c age that has the most diverse Famennian plant assemblage in North America. Five species of *Archaeopteris* are included: *A. macilenta*, *A. halliana*, *A. hibernica*, *A. obtusa*, *A. sphenophyllifolia*, as well as the affiliated progymnosperm wood taxon *Callixylon erianum*. *Rhacophyton ceratangium*, *Gillespiea randolphensis*, and *Barinophyton sibericum* are also present. Seed plant material at Elkins consists of *Sphenopteris* foliage, abundant cupules with ovules, cupuliferous branches, and foliage of *Elkinsia polymorpha*, and cupule fragments of *Condrusia*. Petrified axes of the cladoxylalean *Hierogramma* were also recovered. The Elkins assemblage includes the sphenopsids *Sphenophyllum subtenerrimmum* and cf. *Eviostachya* sp., as well as an arborescent lycopsid (Scheckler, 1986b). Despite the difference in environmental setting, the Red Hill flora is closely comparable to the Elkins flora, except that Red Hill does not have any gymnospermous synangia or any sphenopsids. *Barinophyton* has been found only in the shoreline setting of the Elkins locality in the Hampshire Formation, and has therefore been interpreted as a plant of marine deltaic environments. Two species of *Barinophyton* have now been found at Red Hill, a freshwater fluvial setting high on the alluvial plain. Red Hill is similar in environment to the Rawley Springs, Virginia, locality in the Hampshire Formation, which is at the top of a fluvial fining-upward cycle. The coal-forming *Rhacophyton* swamp at Rawley Springs is interpreted as forming in an oxbow or a backswamp continuous with the adjacent soil. The only other plants found at Rawley Springs are *Callixylon* sp. and an arborescent lycopsid (Scheckler, 1986c).

The flora of the Evieux Formation in Belgium (Stockmans, 1948) is of the same Famennian 2c age as the flora of Red Hill and Elkins, West Virginia. The Evieux flora is dominated by *Rhacophyton* and *Archaeopteris* (Kenrick and Fairon-Demaret, 1991) but also includes the most diverse assemblage of Late Devonian gymnosperm taxa. At least three cupulate taxa (Fairon-Demaret and Scheckler, 1987; Fairon-Demaret, 1996) and two acupulate taxa (Hilton, 1999) are known from the Evieux Formation. Stockmans (1948) described many forms of possible gymnosperm foliage (Rothwell and Scheckler, 1988). The Evieux flora also includes *Eviostachya*, *Barinophyton*, *Condrusia* (Stockmans, 1948), and *Barsostrobus*, a large heterosporous lycopsid cone (Fairon-Demaret, 1977; Fairon-Demaret, 1991). Aside from the possible sphenopsid and several other minor elements, the Red Hill flora is similar to the Belgian Evieux flora.

The Oswayo Formation in northern Pennsylvania has yielded plant fossils of a slightly younger age (Fa2d to Tn1a). *Archaeopteris* has been recovered from the Oswayo Formation (Arnold, 1939), in addition to cupules with ovules of the gymnosperm *Archaeosperma arnoldii* (Pettitt and Beck, 1968).

In Great Britain, the Taffs Well assemblage of South Wales (Tn1a to lower Tn1b; Hilton and Edwards, 1996) includes the acupulate gymnosperm *Aglosperma quadripartita*, similar forms of which are now also known from Red Hill, and the organ genera *Telangiopsis* and *Platyphyllum*. Cupulate gymnosperms are also present at Taffs Well. The highly disarticulated nature of the plant remains there indicates that it is an allochthonous assemblage (Hilton and Edwards, 1996). The Avon Gorge assemblage from near Bristol, England (lower Tn1b) is dominated by *Chlidanophyton dublinensis*, a plant originally identified as *Rhacophyton* (Utting and Neves, 1969) but of unknown affinities (Hilton, 1999). This assemblage also includes the acupulate gymnosperm *Aglosperma*, *Platyphyllum*, the gymnosperm synangium *Telangiopsis*, *Alcicornepteris* sp., and other plant organs of unknown affinities (Hilton, 1999). The environment of deposition for this flora has been interpreted as fluvial. The plant fossils occur either on green, fine-grained sheet-flood-derived sandstones, or on micaceous mudstones within a sedimentological context that suggests flood-plain deposits in a deltaic or estuarine system. Occasional marine incursions are indicated by acritarchs recovered in macerated samples. Elsewhere in Great Britain, the Baggy Beds of North Devon contain the gymnosperm *Xenotheca devonica* (Arber and Goode) emend. Hilton and Edwards (see Figs. 2.3.13 and 2.3.14 *in* Rothwell and Scheckler, 1988), along with gymnospermous foliage *Sphenopteris* sp. and *Sphenopteridium* sp., and gymnospermous synangia *Telangiopsis* sp. (Hilton and Edwards, 1999).

In southwest Ireland, the Kiltorcan flora of County Kilkenny is a diverse assemblage of latest Devonian and earliest Carboniferous age (Tn1a to Tn1b; Fairon-Demaret, 1986; Jarvis, 1990). *Archaeopteris hibernica* is dominant and the lycopsid *Cyclostigma kiltorkense* is common. Other members of the flora are *Ginkgophyllum kiltorkense*, *Sphenopteris hookeri*, *Lepidodendropsis* sp., *Rhacophyton* sp., and *Spermolithus devonicus* (Chaloner et al., 1977). The Kiltorcan flora is similar to the Red Hill flora, but it lacks the diversity of early gymnosperms that Red Hill has. The Hook Head flora from County Wexford (Tn1a to lower Tn1b) is similar to the Kiltorcan flora in having *Archaeopteris hibernica* and cf. *Cyclostigma*, as well as cf. *Pitus*, cf. *Barinophyton*, and the lycopsid *Wexfordia* (Matten, 1995). The Ballyheigue locality in County Kerry (upper Tn1a to lower Tn1b) is less comparable to the Red Hill flora (Klavins and Matten, 1996). This predominantly gymnosperm flora is known from siliceous permineralizations in a deposit interpreted as a crevasse splay (Klavins and Matten, 1996). The gymnosperm domination of the Ballyheigue flora is similar to the younger Cementstone flora of Scotland (Scott et al., 1984; Bateman and Rothwell, 1990) and may mark the transition from progymnosperm dominance to gymnosperm dominance, at least in certain environments (Matten, 1995).

The flora of Bear Island, Norway (Fa2d to Tn1b; Nathorst, 1900; Nathorst, 1902; Scheckler, 1986a; Fairon-Demaret, 1986) is similar to that of the Kiltorcan flora. *Archaeopteris* and *Cyclostigma* are dominant elements, along with the sphenopsid *Pseudobornia ursina* (Kaiser, 1970) and *Cephalopteris mirabilis*, which is similar to *Rhacophyton*. Other plants, including many tree lycopsids, are less common.

Archaeopteris remains are also known from southeastern New York (Carluccio et al., 1966), from the early Frasnian Yahatinda Formation of Alberta (Scheckler, 1978) and many localities of the Frasnian and Famennian of Ellesmere and Melville Islands in Arctic Canada (Nathorst, 1904; Andrews, Phillips, and Radforth, 1965; Hill et al., 1997), South America (Berry and Edwards, 1996), Eastern Europe and Russia (Snigirevskaya, 1982, 1988, 1995a, 1995b), Siberia (Petrosyan, 1968), China (Cai, 1981, 1989; Cai et al., 1987; Cai and Wang, 1995), North Africa (Galtier et al., 1996; Meyer-Berthaud et al., 1997, 1999, 2000), South Africa (Anderson et al., 1995), Australia (White, 1986; Scheckler, 1998, personal commun.), and possibly Antarctica (Retallack, 1997). Investigations at Red Hill add another increment to our understanding of this worldwide flora during the time of the earliest forests.

An Emerging Picture of a Late Devonian Continental Ecosystem

In addition to plants, Red Hill is the source of abundant Late Devonian animal fossils. Red Hill is emerging as the richest known Late Devonian continental fossil site. More components of a Late Devonian terrestrial ecosystem are found at Red Hill than at any other locality. Animal fossils at Red Hill include vertebrates and terrestrial arthropods such as scorpions, myriapods, and a trigonotarbid arachnid (Shear, 2000). Rivers were inhabited by large predatory fish such as tristichopterid, megalichthyid, and rhizodontid lobefins, as well as groenlandaspidid and phyllolepidid placoderms, acanthodians, and ageleodid and ctenacanthid sharks (Daeschler, 1998). The pond deposit yielded numerous palaeoniscoid ray-finned fish. They were found in close association with penultimate branches of *Archaeopteris*, among which they may have been seeking refuge from the larger predators. Red Hill is also the locality for some of the earliest known tetrapods. *Hynerpeton bassetti* (Daeschler et al., 1994) and *Densignathus rowei* (Daeschler, 2000b) were discovered here.

No evidence for herbivory (leaves with chew marks or feeding trails) has been found among the fossils at Red Hill. There is no sign of herbivore damage repair among the plant fossils, and no herbivores have been discovered in the faunal assemblage. It appears that this freshwater-terrestrial ecosystem was largely supported by detritus from the abundant plant biomass growing on the land. The frequent floods across the flood plain transported this organic matter into the aquatic ecosystem. The seasonally rising and falling waters, the migrating channels and ever-changing flood-plain environment, the lush seasonal vegetation growing along the water's edge, and the abundant submerged plant material created a blurred habitat boundary between the aquatic and terrestrial realms. It was in this environment that lobe-finned fish thrived. Members of one lineage of lobe-finned fish, the earliest tetrapods, emerged in this type of environment (Westenberg, 1999).

ACKNOWLEDGMENTS

The author thanks E. B. Daeschler, D. Rowe, N. Delaney, F. Mullison, and R. Shapley for their collaboration on fieldwork at Red Hill; G. I. Omar, J. Smith, C. Williams, G. Harrison, and W. Romanow for technical assistance; S. E. Scheckler, W. A. DiMichele, H. W. Pfefferkorn, B. LePage, and P. Wilf for valuable discussions and reviews of drafts; W. E. Stein for his thoughtful review; and Claire Brill for her support throughout. This work was supported by student grants from the Paleontological Society, the Geological Society of America, Sigma Xi, the Paleontological Research Institution, and a summer stipend from the Penn Summer Research Stipends in Paleontology instituted by an anonymous donor.

REFERENCES CITED

Algeo, T.J., and Scheckler, S.E., 1998, Terrestrial-marine teleconnections in the Devonian: Links between the evolution of land plants, weathering processes, and marine anoxic events: Royal Society of London Philosophical Transactions, ser. B, v. 353, p.113–130.

Algeo, T.J., Scheckler, S.E., and Maynard, J.B., 2001, Effects of Middle to Late Devonian spread of vascular land plants on weathering regimes, marine biotas, and global climate, in Gensel, P.G., and Edwards, D., eds. Plants invade the land: Evolutionary and environmental perspectives: New York, Columbia University Press, p. 213–236.

Anderson, H.M., Hiller, N., and Gess, R.W., 1995, *Archaeopteris* (Progymnospermopsida) from the Devonian of southern Africa: Botanical Journal of the Linnean Society, v. 117, p. 305–320, doi: 10.1006/bojl.1995.0021.

Andrews, H.N., and Phillips, T.L., 1968, *Rhacophyton* from the Upper Devonian of West Virginia: Journal of the Linnean Society, v. 61, p. 37–64.

Andrews, H.N., Phillips, T.L., and Radforth, N.W., 1965, Paleobotanical studies in Arctic Canada. I. *Archaeopteris* from Ellesmere Island: Canadian Journal of Botany, v. 43, p. 545–556.

Arnold, C.A., 1939, Observations on fossil plants from the Devonian of eastern North America, IV. Plant remains from the Catskill delta deposits of northern Pennsylvania and southern New York: Contributions from the Museum of Paleontology, University of Michigan, v. 5, p. 271–314.

Bambach, R.K., Algeo, T.J., Thorez, J., and Witzke, B.G., 1999, Paleogeography and paleoclimatic implications for expanding *Archaeopteris* forests: International Botanical Congress, 16th, St. Louis, Missouri, Abstracts, p.13.

Banks, H.P., 1980, Floral assemblages in the Siluro-Devonian, in Dilcher, D.L., and Taylor, T.N., Biostratigraphy of fossil plants: Successional and paleoecological analyses: Stroudsburg, Pennsylvania, Dowden, Hutchinson, and Ross, p. 1–24.

Bateman, R.M., and Rothwell, G.W., 1990, A reappraisal of the Dinantian floras at Oxroad Bay, East Lothian, Scotland. 1. Floristics and development of whole-plant concepts: Transactions of the Royal Society of Edinburgh, v. 81, p. 127–159.

Beck, C.B., 1960, Connection between *Archaeopteris* and *Callixylon*: Science, v. 131, p. 1524–1525.

Beck, C.B., 1964, Predominance of *Archaeopteris* in Upper Devonian flora of western Catskills and adjacent Pennsylvania: Botanical Gazette, v. 125, p. 126–128, doi: 10.1086/336257.

Beerbower, R., 1985, Early development of continental ecosystems, in Tiffney, B.H., Geological factors and the evolution of plants: New Haven, Connecticut, Yale University Press, p. 47–91.

Berry, C.M., and Edwards, D., 1996, The herbaceous lycophyte *Haskinsia* Grierson and Banks from the Devonian of western Venezuela, with observations on leaf morphology and fertile specimens: Botanical Society of the Linnean Society, v. 122, p. 103–122, doi: 10.1006/bojl.1996.0053.

Brauer, D.F., 1980, *Barinophyton citrulliforme* (Barinophytales incertae sedis, Barinophytaceae) from the Upper Devonian of Pennsylvania: American Journal of Botany, v. 67, p. 1186–1206.

Brauer, D.F., 1981, Heterosporous, barinophytacean plants from the Upper Devonian of North America and a discussion of the possible affinities of the Barinophytaceae: Review of Palaeobotany and Palynology, v. 33, p. 347–362, doi: 10.1016/0034-6667(81)90092-0.

Cai Chongyang, 1981, On the occurrence of *Archaeopteris* in China: Acta Palaeontologica Sinica, v. 20, p. 75–80.

Cai Chongyang, 1989, Two *Callixylon* species from the Upper Devonian of Junggar Basin, Xinjiang: Acta Palaeontologica Sinica, v. 28, p. 571–578.

Cai Chongyang, and Wang Yi, 1995, Devonian floras, in Li Xingxue, Zhou Zhiyan, Cai Chongyang, Sun Ge, Ouyang Shu, and Deng Longhua, eds., Fossil floras of China through the geological ages: Guangzhou, China, Guangdong Science and Technology Press, p. 28–77.

Cai Chongyang, Wen Yaoguang, and Chen Peiquan, 1987, *Archaeopteris* florula from Upper Devonian of Xinhui County, central Guangdong and its stratigraphical significance: Acta Palaeontologica Sinica, v. 26, p. 55–64. [In Chinese with English summary.]

Carluccio, L.M., Hueber, F.M., and Banks, H.P., 1966, *Archaeopteris macilenta*, anatomy and morphology of its frond: American Journal of Botany, v. 53, p. 719–730.

Chaloner, W., and Sheerin, A., 1979, Devonian macrofloras, in House, M.R., Scrutton, C.T., and Bassett, M.G., The Devonian system: London, Palaeontological Association Special Papers in Palaeontology, v. 23, p. 145–161.

Chaloner, W., Hill, A.J., and Lacey, W.S., 1977, First Devonian platyspermic seed and its implications for gymnosperm evolution: Nature, v. 265, p. 233–235, doi: 10.1038/265233a0.

Cressler, W.L., 2001, Evidence of earliest known wildfires: Palaios, v. 16, p. 171–174.

Cressler, W.L., and Pfefferkorn, H.W., 2005, A Late Devonian isoetalean lycopsid, *Otzinachsonia beerboweri*, gen et sp. nov., from north-central Pennsylvania, USA: American Journal of Botany, v. 92, p. 1131–1140.

Daeschler, E.B., Shubin, N.H., Thomson, K.S., and Amaral, W.W., 1994, A Devonian tetrapod from North America: Science, v. 265, p. 639–642.

Daeschler, E.B., 1998, Vertebrate fauna from the non-marine facies of the Catskill Formation (Late Devonian) in Pennsylvania [Ph.D. thesis]: Philadelphia, University of Pennsylvania, 166 p.

Daeschler, E.B., 2000a, An early actinopterygian fish from the Catskill Formation (Late Devonian, Famennian) in Pennsylvania, U.S.A: Proceedings of the Academy of Natural Sciences of Philadelphia, v. 150, p. 181–192.

Daeschler, E.B., 2000b, Early tetrapod jaws from the Late Devonian of Pennsylvania, USA: Journal of Paleontology, v. 74, p. 301–308.

Daeschler, E.B., Frumes, A.C., and Mullison, C.F., 2003, Groenlandaspidid placoderm fishes from the Late Devonian of North America: Records of the Australian Museum, v. 55, p. 45–60.

Diemer, J.A., 1992, Sedimentology and alluvial stratigraphy of the upper Catskill Formation, south-central Pennsylvania: Northeastern Geology, v. 14, p. 121–136.

DiMichele, W.A., and Bateman, R.M., 1996, Plant paleoecology and evolutionary inference: Two examples from the Paleozoic: Review of Palaeobotany and Palynology, v. 90, p. 223–247, doi: 10.1016/0034-6667(95)00085-2.

DiMichele, W.A., and Hook, R.W., rapporteurs; Beerbower, R., Boy, J.A., Gastaldo, R.A., Hotton III, N., Phillips, T.L., Scheckler, S.E., Shear W.A., and Sues, H.-D., contributors. 1992, Paleozoic terrestrial ecosystems, in Behrensmeyer, A.K., Damuth, J.D., DiMichele, W.A., Potts, R., Sues, H.-D., and Wing, S.L., Terrestrial ecosystems through time: Chicago, University of Chicago Press, p. 205–325.

DiMichele, W.A., and Phillips, T.L., 1996, Clades, ecological amplitudes, and ecomorphs: Phylogenetic effects and persistence of primitive plant communities in the Pennsylvanian wetland tropics: Palaeogeography, Palaeoclimatology, Palaeoecology, v. 127, p. 83–105, doi: 10.1016/S0031-0182(96)00089-2.

Dittrich, H.S., Matten, L.C., and Phillips, T.L., 1983, Anatomy of *Rhacophyton ceratangium* from the Upper Devonian (Famennian) of West Virginia: Review of Palaeobotany and Palynology, v. 40, p. 127–147, doi: 10.1016/0034-6667(83)90007-6.

Driese, S.G., and Mora, C.I., 1993, Physico-chemical environment of pedogenic carbonate formation in Devonian vertic paleosols, central Appalachians, U.S.A.: Sedimentology, v. 40, p. 199–216.

Driese, S.G., Mora, C.I., and Elick, J.M., 1997, Morphology and taphonomy of root and stump casts of the earliest trees (Middle to Late Devonian), Pennsylvania and New York, U.S.A.: Palaios, v. 12, p. 524–537.

Dudal, R., and Eswaran, H., 1988, Distribution, properties, and classification of vertisols, in Wilding, L.P., and Puentes, R., eds., Vertisols: Their distribution, properties, classification and management, College Station, Texas, Texas A&M University, p. 1–22.

Edwards, D., and Fanning, U., 1985, Evolution and environment in the late Silurian-early Devonian: rise of the pteridophytes: Royal Society of London Philosophical Transactions, ser. B, v. 309, p. 147–165.

Erwin, D.M., and Rothwell, G.W., 1989, *Gillespeia randolphensis* gen. et sp. nov. (Stauropteridales), from the Upper Devonian of West Virginia: Canadian Journal of Botany, v. 67, p. 3063–3077.

Ettensohn, F.R., 1985, The Catskill Delta complex and the Acadian orogeny: A model, in Woodrow, D.L., and Sevon, W.D., eds., The Catskill Delta: Geological Society of America Special Paper 201, p. 39–49.

Fairon-Demaret, M., 1977, A new lycophyte cone from the Upper Devonian of Belgium: Palaeontographica B, v. 162, p. 51–63.

Fairon-Demaret, M., 1986, Some uppermost Devonian megafloras: A stratigraphical review: Annales de la Société Géologique de Belgique, v. 109, p. 43–48.

Fairon-Demaret, M., 1991, The Upper Famennian lycopods from the Dinant Sinclinorium (Belgium): Neues Jahrbuch für Geologischen und Paläontologischen, Abhandlungen, v. 183, p. 87–101.

Fairon-Demaret, M., 1996, *Dorinnotheca streelii* Fairon-Demaret, gen. et sp. nov., a new early seed plant from the upper Famennian of Belgium: Review of Palaeobotany and Palynology, v. 93, p. 217–233, doi: 10.1016/0034-6667(95)00127-1.

Fairon-Demaret, M., and Scheckler, S.E., 1987, Typification and redescription of *Moresnetia zalesskyi* Stockmans, 1948, an early seed plant from the Upper Famennian of Belgium: Bulletin de l'Institut Royal des Sciences Naturelles de Belgique, Sciences de la Terre, v. 57, p. 183–199.

Galtier, J., Paris, F., and Aouad-Debbaj, Z.E., 1996, La présence de *Callixylon* dans le Dévonien supérieur du Maroc et sa signification paléogéographique: Comptes Rendus de l'Académie des Sciences de Paris, v. 322, p. 893–900.

Gastaldo, R.A., Bearce, S.C., Degges, C.W., Hunt, R.J., Peebles, M.W., and Violette, D.L., 1989, Biostratinomy of a Holocene oxbow lake: A backswamp to mid-channel transect: Review of Palaeobotany and Palynology, v. 58, p. 47–59, doi: 10.1016/0034-6667(89)90056-0.

Gillespie, W.H., Rothwell, G.W., and Scheckler, S.E., 1981, The earliest seeds: Nature, v. 293, p. 462–464, doi: 10.1038/293462a0.

Glaeser, J.D., 1974, Upper Devonian stratigraphy and sedimentary environments in northeastern Pennsylvania: Pennsylvania Geological Survey General Geology Report, v. 63, 89 p.

Goodarzi, F., Gentzis, T., and Embry, A.F., 1989, Organic petrology of two coal-bearing sequences from the Middle to Upper Devonian of Melville Island, Arctic Canada: Calgary, Geological Survey of Canada, Contributions to Canadian Coal Geoscience, no. 89-8, p. 120–130.

Goodarzi, F., Gentzis, T., and Harrison, J.C., 1994, Petrology and depositional environment of Upper Devonian coals from eastern Melville Island, Arctic Canada, *in* Christie, R.L., and McMillan, N.J., The geology of Melville Island, Arctic Canada: Geological Survey of Canada Bulletin, v. 450, p. 203–213.

Goodarzi, F., and Goodbody, Q., 1990, Nature and depositional environment of Devonian coals from western Melville Island, Arctic Canada: International Journal of Coal Geology, v. 14, p. 175–196, doi: 10.1016/0166-5162(90)90002-G.

Gordon, E.A., and Bridge, J.S., 1987, Evolution of Catskill (Upper Devonian) river systems: Intra- and extrabasinal controls: Journal of Sedimentary Petrology, v. 57, p. 234–249.

Harvey, A., 1998. A paleoenvironmental reconstruction in the Devonian Sherman Creek Member of the Catskill Formation in central Pennsylvania [Master's thesis]: Philadelphia, Temple University, 90 p.

Hill, S.A., Scheckler, S.E., and Basinger, J.F., 1997, Ellesmeris sphenopteroides, gen. et sp. nov., a new zygopterid fern from the Upper Devonian (Frasnian) of Ellesmere, N.W.T., Arctic Canada: American Journal of Botany, v. 84, p. 85–103.

Hilton, J., 1999, A Late Devonian plant assemblage from the Avon Gorge, west England: taxonomic, phylogenetic and stratigraphic implications: Botanical Journal of the Linnean Society, v. 129, p. 1–54, doi: 10.1006/bojl.1998.0209.

Hilton, J., and Edwards, D., 1996, A new Late Devonian acupulate preovule from the Taff Gorge, South Wales: Review of Palaeobotany and Palynology, v. 93, p. 235–252, doi: 10.1016/0034-6667(95)00128-X.

Hilton, J., and Edwards, D., 1999, New data on *Xenotheca devonica* Arber and Goode, an enigmatic seed plant cupule with preovules, *in* Kurmann, M., and Hemsley, A.R., eds., The evolution of plant architecture: London, Kew Botanic Gardens Press, p. 75–90.

Hotton, N., Olson, E.C., and Beerbower, R., 1997, Amniote origins and the discovery of herbivory, *in* Sumida, S.S., and Martin, K.L.M., eds., Amniote origins: San Diego, Academic Press, p. 207–264.

Jarvis, E., 1990, New palynological data on the age of the Kiltorkan flora of Co. Kilkenny, Ireland: Journal of Micropalaeontology, v. 9, p. 87–94.

Jennings, J.R., 1975, *Protostigmaria*, a new plant organ from the Lower Mississippian of Virginia: Palaeontology, v. 18, p. 19–24.

Jennings, J.R., Karrfalt, E.E., and Rothwell, G.W., 1983, Structure and affinities of *Protostigmaria eggertiana*: American Journal of Botany, v. 70, p. 963–974.

Kaiser, H., 1970, Die Oberdevon-Flora der Bäreninsel. 3. Mikroflora des höheren Oberdevons und des Unterkarbons: Palaeontographica Abteilung B, v. 129, p. 71–124.

Kenrick, P., and Fairon-Demaret, M., 1991, *Archaeopteris roemeriana* (Göppert) *sensu* Stockmans, 1948 from the Upper Famennian of Belgium: Anatomy and leaf polymorphism: Bulletin de l'Institut Royal des Sciences Naturelles de Belgique, Sciences de la Terre, v. 61, p. 179–195.

Klavins, S.D., and Matten, L.C., 1996, Reconstruction of the frond of Laceya hibernica, a lyginopterid pteridosperm from the uppermost Devonian of Ireland: Review of Palaeobotany and Palynology, v. 93, p. 253–268

Kräusel, R., and Weyland, H., 1941, Pflanzenreste aus dem Devon von Nord-Amerika. II. Die Oberdevonischen Floren von Elkins, West-Virginien, und Perry, Maine, mit Berücksichtigung einiger Stücke von der Chaleur-Bai, Canada: Palaeontographica Abteilung B, v. 86, p. 3–78.

Labandeira, C.C., 1998, Plant-insect associations from the fossil record: Geotimes, v. 43, p. 18–24.

Labandeira, C.C., and Sepkoski, J.J., 1993, Insect diversity in the fossil record: Science, v. 261, p. 310–315.

Matten, L.C., 1995, The megafossil flora from the uppermost Devonian of Hook Head, County Wexford, Ireland: Birbal Sahni Centenary Volume, p. 167–173.

Meyer-Berthaud, B., Scheckler, S.E., and Bousquet, J.-L., 2000, The development of *Archaeopteris*: New evolutionary characters from the structural analysis of an Early Famennian trunk from southeast Morocco: American Journal of Botany, v. 87, p. 456–468.

Meyer-Berthaud, B., Scheckler, S.E., and Wendt, J., 1999, *Archaeopteris* is the earliest known modern tree: Nature, v. 398, p. 700–701, doi: 10.1038/19516.

Meyer-Berthaud, B., Wendt, J., and Galtier, J., 1997, First record of a large *Callixylon* trunk from the Late Devonian of Gondwana: Geological Magazine, v. 134, p. 847–853, doi: 10.1017/S0016756897007814.

Nathorst, A.G., 1900, Die oberdevonische Flora (die "Ursaflora") der Bären Insel: Bulletin of the Geological Institution of the University of Uppsala, no. 8, v. IV, part 2, p. 1-5.

Nathorst, A.G., 1902, Zur oberdevonische Flora der Bären-Insel: Kongliga Svenska Vetenskaps-Akademiens Handlingar, v. 36, p. 1–60.

Nathorst, A.G., 1904, Die oberdevonische Flora des Ellesmere-landes: Report of the Second Norwegian Arctic Expedition in the *Fram*, 1898–1902, no. 1, p. 1–22.

Niklas, K.J., 1976, Morphological and chemical examination of *Courvoisiella ctenomorpha* gen. and sp. nov., a siphonous alga from the Upper Devonian, West Virginia, U.S.A: Review of Palaeobotany and Palynology, v. 21, p. 187–203, doi: 10.1016/0034-6667(76)90048-8.

Peppers, R.A., and Pfefferkorn, H.W., 1970, A comparison of the floras of the Colchester (No. 2) Coal and the Francis Creek Shale, *in* Smith, W.H., and others, eds., Depositional environments in parts of the Carbondale Formation—Western and Northern Illinois: Springfield, Illinois, Illinois State Geological Survey Field Guidebook Series, no. 8, p. 61–74.

Petrosyan, N.M., 1968, Stratigraphic importance of the Devonian flora of the USSR *in* Oswald, D. H., ed., International Symposium on the Devonian System, Volume 1: Calgary, Alberta, Canada, Alberta Society of Petroleum Geology, p. 579–586.

Pettitt, J.M., and Beck, C.B., 1968, *Archaeosperma arnoldii*—A cupulate seed from the Upper Devonian of North America: Contributions of the Museum of Paleontology, University of Michigan, v. 22, p. 139–154.

Phillips, T.L., Andrews, H.N., and Gensel, P.G., 1972, Two heterosporous species of *Archaeopteris* from the Upper Devonian of West Virginia: Palaeontographica Abteilung B, v. 139, p. 47–71.

Rahmanian, V., 1979, Stratigraphy and sedimentology of the Upper Devonian Catskill and uppermost Trimmers Rock Formation in central Pennsylvania [Ph.D. thesis]: State College, Pennsylvania, Pennsylvania State University, 340 p.

Retallack, G.J., 1985, Fossil soils as grounds for interpreting the advent of large plants and animals on land: Royal Society of London Philosophical Transactions, ser. B, v. 309, p. 105–142.

Retallack, G.J., 1997, Early forest soils and their role in Devonian global change: Science, v. 276, p. 583–585, doi: 10.1126/science.276.5312.583.

Richardson, J.B., and McGregor, D.C., 1986, Silurian and Devonian spore zones of the Old Red Sandstone continent and adjacent regions: Geological Survey of Canada Bulletin, v. 364, p. 1–79.

Rothwell, G.W., and Scheckler, S.E., 1988, Biology of ancestral gymnosperms, *in* Beck, C.B., Origins and evolution of gymnosperms: New York, Columbia University Press, p. 85–134.

Rothwell, G.W., Scheckler, S.E., and Gillespie, W.H., 1989, *Elkinsia* gen. nov., a Late Devonian gymnosperm with cupulate ovules: Botanical Gazette, v. 150, p. 170–189, doi: 10.1086/337763.

Scheckler, S.E., 1978, Ontogeny of progymnosperms. II. Shoots of Upper Devonian Archaeopteridales: Canadian Journal of Botany, v. 56, p. 3136–3170.

Scheckler, S.E., 1986a, Floras of the Devonian-Mississippian transition, *in* Gastaldo, R.A., and Broadhead, T.W., Land plants: Notes for a short course: University of Tennessee Department of Geological Sciences Studies in Geology, v. 15, p. 81–96.

Scheckler, S.E., 1986b, Old Red Continent facies in the Late Devonian and Early Carboniferous of Appalachian North America: Annales de la Société Géologique de Belgique, v. 109, p. 223–236.

Scheckler, S.E., 1986c, Geology, floristics and paleoecology of Late Devonian coal swamps from Appalachian Laurentia (U.S.A.): Annales de la Société Géologique de Belgique, v. 109, p. 209–222.

Scheckler, S.E., 2001. Afforestation—The first forests, in Briggs, D.E.G., and Crowther, P., Palaeobiology II: Oxford, Blackwell Science, p. 67–71.

Scheckler, S.E., Cressler, W.L., Connery, T., Klavins, S., and Postnikoff, D., 1999, Devonian shrub and tree dominated landscapes: International Botanical Congress, 16th, St. Louis, Missouri, Abstracts, v. 16, p. 13.

Scheihing, S.E., and Pfefferkorn, H.W., 1984, The taphonomy of land plants in the Orinoco Delta: A model for the incorporation of plant parts in clastic sediments of Late Carboniferous age of Euramerica: Review of Palaeobotany and Palynology, v. 41, p. 205–240, doi: 10.1016/0034-6667(84)90047-2.

Scotese, C.R., and McKerrow, W.S., 1990, Revised world maps and introduction, in McKerrow, W.S., and Scotese, C.R., Palaeozoic palaeogeography and biogeography: Geological Society [London] Memoir 12, p.1–21.

Scott, A.C., Galtier, J., and Clayton, G., 1984, Distribution of anatomically preserved floras in the Lower Carboniferous in Western Europe: Transactions of the Royal Society of Edinburgh, v. 75, p. 311–340.

Serbet, R., and Rothwell, G.W., 1992, Characterizing the most primitive seed ferns. I. A reconstruction of *Elkinsia polymorpha*: International Journal of Plant Sciences, v. 153, p. 602–621, doi: 10.1086/297083.

Sevon, W.D., 1985, Nonmarine facies of the Middle and Late Devonian Catskill coastal alluvial plain, in Woodrow, D.L., and Sevon, W.D., eds., The Catskill Delta: Geological Society of America Special Paper 201, p. 79–90.

Shear, W.A., 1991, The early development of terrestrial ecosystems: Nature, v. 351, p. 283–289, doi: 10.1038/351283a0.

Shear, W.A., 2000, *Gigantocharinus szatmaryi*, a new trigonotarbid arachnid from the Late Devonian of North America (Chelicerata, Arachnida, Trigonotarbida): Journal of Paleontology, v. 74, p. 25–31.

Slingerland, R., and Smith, N.D., 2004, River avulsions and their deposits: Annual Review of Earth and Planetary Sciences, v. 32, p. 257–285, doi: 10.1146/annurev.earth.32.101802.120201.

Snigirevskaya, N.S., 1982, The shoot of *Archaeopteris archetypus* with preserved anatomical structure: Botanicheskii Zhurnal (Academy of Sciences, SSSR), v. 67, p. 1237–1243, (in Russian with English summary).

Snigirevskaya, N.S., 1988, The Late Devonian—The time of the appearance of forests as a natural phenomenon, in Contributed papers: The formation and evolution of the continental biotas, L.: 31st Session of the All-Union Palaeontological Society, p. 115–124 (in Russian).

Snigirevskaya, N.S., 1995a, Archaeopterids and their role in the land plant cover evolution: Botanicheskii Zhurnal (Academy of Sciences, SSSR), v. 80, p. 70–75.

Snigirevskaya, N.S., 1995b, Early evolution of forest ecosystems (the Late Devonian): International Symposium on Ecosystem Evolution, Moscow, Palaeontological Institute, Russian Academy of Sciences, 1995, Abstracts, p. 87–88.

Stewart, W.N., and Rothwell, G.W., 1993, Paleobotany and the evolution of plants: Cambridge, UK, Cambridge University Press, 521 p.

Stockmans, F., 1948, Végétaux du Dévonien Supérieur de la Belgique: Mémoires du Musée Royal d'Histoire Naturelle de Belgique, v. 110, p. 1–85.

Streel, M., and Scheckler, S.E., 1990, Miospore lateral distribution in upper Famennian alluvial lagoonal to tidal facies from eastern United States and Belgium: Review of Palaeobotany and Palynology, v. 64, p. 315–324, doi: 10.1016/0034-6667(90)90147-B.

Traverse, A., 2003, Dating the earliest tetrapods: A Catskill palynological problem in Pennsylvania: Courier Forschungs-Institut Senckenberg, v. 241, p. 19–29.

Utting, J., and Neves, R., 1969, Palynology of the Lower Limestone Shale Group (Basal Carboniferous Limestone series) and Portishead Beds, (Upper Old Red Sandstone) of the Avon Gorge, Bristol, England: Colloque sur la stratigraphie du Carbonifère Congrès et colloques, Université de Liège, p. 411–422.

Walker, R.G., 1971, Nondeltaic depositional environments in the Catskill clastic wedge (Upper Devonian) of central Pennsylvania: Geological Society of America Bulletin, v. 82, p. 1305–1326.

Walker, R.G., and Harms, J.C., 1971, The "Catskill Delta": A prograding muddy shoreline in central Pennsylvania: Journal of Geology, v. 79, p. 381–399.

Westenberg, K., 1999, The rise of life on Earth: From fins to feet: National Geographic, v. 195, p. 114–127.

White, M.E., 1986, The greening of Gondwana: Frenchs Forest, New South Wales, Reed Books, p. 84.

Woodrow, D.L., Fletcher, F.W., and Ahrnsbrak, W.F., 1973, Paleogeography and paleoclimate at the deposition sites of the Devonian Catskill and Old Red facies: Geological Society of America Bulletin, v. 84, p. 3051–3064, doi: 10.1130/0016-7606(1973)84<3051:PAPATD>2.0.CO;2.

Woodrow, D.L., Robinson, R.A.J., Prave, A.R., Traverse, A., Daeschler, E.B., Rowe, N.D., and DeLaney, N.A., 1995, Stratigraphic, sedimentologic, and temporal framework of Red Hill (Upper Devonian Catskill Formation) near Hyner, Clinton County, Pennsylvania: Site of the oldest amphibian known from North America, in Guidebook, 60th Annual Field Conference of Pennsylvania Geologists, 16 p.

MANUSCRIPT ACCEPTED BY THE SOCIETY 28 JUNE 2005

Tournaisian forested wetlands in the Horton Group of Atlantic Canada

Michael C. Rygel*
Department of Earth Sciences, Dalhousie University, Halifax, Nova Scotia, Canada, B3H 3J5

John H. Calder
Nova Scotia Department of Natural Resources, P.O. Box 698, Halifax, Nova Scotia, Canada, B3J 2T9

Martin R. Gibling
Department of Earth Sciences, Dalhousie University, Halifax, Nova Scotia, Canada, B3H 3J5

Murray K. Gingras
Department of Earth and Atmospheric Sciences, University of Alberta, Edmonton, Alberta, Canada, T6G 2E3

Camilla S.A. Melrose
Department of Earth Sciences, Dalhousie University, Halifax, Nova Scotia, Canada, B3H 3J5

ABSTRACT

The Horton Group (late Famennian to Tournaisian) of Atlantic Canada provides an unusually complete record of Early Mississippian wetland biota. Best known for tetrapod fossils from "Romer's Gap," this unit also contains numerous horizons with standing vegetation. The taphonomy and taxonomy of Horton Group fossil forests have remained enigmatic because of poor preservation, curious stump cast morphology, and failure to recognize the unusual sedimentary structures formed around standing plants.

Four forested horizons within the Horton Group are preserved as cryptic casts and vegetation-induced sedimentary structures formed by the interaction of detrital sediment with in situ plants. *Protostigmaria*, the lobed base of the arborescent lycopsid *Lepidodendropsis*, occur as sandstone-filled casts attached to dense root masses. Mudstone-filled hollows formed when a partially entombed plant decayed, leaving a void that was later infilled by muddy sediment. A scratch semi-circle formed where a current bent a small plant, causing it to inscribe concentric grooves into the adjacent muddy substrate. Obstacle marks developed where flood waters excavated erosional scours into sandy sediment surrounding juvenile *Lepidodendropsis*. These cryptic lycopsid forests had considerably higher densities than their Pennsylvanian counterparts.

Vegetation-induced sedimentary structures are abundant in Horton Group strata and could easily be misidentified as purely hydrodynamic or soft sediment

---

*Present address: Department of Geosciences, 224 Bessey Hall, University of Nebraska, Lincoln, Nebraska 68588-0340; mrygel@unlnotes.unl.edu.

Rygel, M.C., Calder, J.H., Gibling, M.R., Gingras, M.K., and Melrose, C.S.A., 2006, Tournaisian forested wetlands in the Horton Group of Atlantic Canada, *in* Greb, S.F., and DiMichele, W.A., Wetlands through time: Geological Society of America Special Paper 399, p. 103–126, doi: 10.1130/2006.2399(05). For permission to copy, contact editing@geosociety.org. ©2006 Geological Society of America. All rights reserved.

deformation structures without careful analysis. Recognition of these structures in early Paleozoic strata has great potential to expand our knowledge about the distribution of early land plants.

Keywords: Horton Bluff Formation, *Lepidodendropsis*, *Protostigmaria*, vegetation-induced sedimentary structures, scratch circle, Mississippian.

INTRODUCTION

Home to dense lycopsid forests and diverse tetrapod communities, Tournaisian wetlands were the predecessors of the "coal swamp" ecosystem that dominated tropical Euramerica throughout the Pennsylvanian (Scheckler, 1985; Ahlberg and Milner, 1994). Following the demise of *Leptophloeum* and *Archaeopteris* forests in the latest Devonian, forested tracts continued to expand into wetland and dryland areas during the early Mississippian (Scott et al., 1984; Scheckler, 1986a, 1986b; Falcon-Lang, 2000). During this time, animals were rapidly diversifying to fill the multitude of available niches in the terrestrial realm. The paucity of Early Mississippian wetland deposits has limited our knowledge of terrestrial ecosystems during this formative stage (Raymond et al., 1985) and is largely responsible for "Romer's Gap," a 20-million-year hiatus in the fossil record of early tetrapods (Coates and Clack, 1995; Carroll, 2002; Clack, 2002).

Evolution of the arborescent habit in the Middle Devonian resulted in the preservation of numerous Late Paleozoic "fossil forest" horizons (Scott and Calder, 1994). Although Tournaisian plant-bearing localities occur worldwide (Scott et al., 1984; Scott and Rex, 1987; DiMichele and Hook, 1992), this paper provides one of the first detailed descriptions of in situ Tournaisian wetlands. Tetrapod footprints and vertebrate skeletons from the Horton Group of Atlantic Canada (late Famennian to Tournaisian) are well known (Carroll et al., 1972; Sarjeant and Mossman, 1978a), but in situ vegetation remains poorly described despite several decades of study (Bell, 1960; Carter and Pickerill, 1985a; Martel and Gibling, 1991).

In the present paper we provide a review of the Horton Group biota and place this information in a modern paleoecological context. Our description of cryptic stump casts and sedimentary structures formed around standing plants provides valuable information about the paleoecology and dynamics of these wetland ecosystems.

GEOLOGICAL SETTING

Horton Group

The Horton Group comprises 600–3200 m of Late Devonian–Mississippian strata that were deposited across much of the Maritimes Basin of Atlantic Canada (van de Poll et al., 1995). Late Paleozoic sedimentation in this area began when post-Acadian extension and granite intrusion initiated a period of half-graben development from New Brunswick to southern Newfoundland (Marillier et al., 1989; Piper, 1994). The resulting depocenters are considered "basins" in their own right and together form the regional Maritimes Basin. The Devonian history of these basins is recorded in the organic-rich deposits of the Murphy Brook and McAdams Lake Formations and the continental rift facies of the Fountain Lake Group (Calder, 1998). Overlying clastics of the Horton Group (Late Devonian to Tournaisian) generally exhibit a tripartite stratigraphy of fine-grained "lacustrine" deposits over- and underlain by coarse-grained alluvial deposits (Hamblin and Rust, 1989).

The Tournaisian forests described in this paper lie within the Horton Bluff Formation of Nova Scotia and the Albert Formation of New Brunswick (Fig. 1). These formations represent the fine-grained portion of the Horton Group in the Windsor (Nova Scotia) and Moncton (New Brunswick) Basins, respectively. During deposition of the Horton Group, the Maritimes Basin was at ~12° S, in the heart of southern Euramerica's tropical dry belt (Van der Zwan, 1981; Scotese, 2001).

Horton Bluff Formation

The late Famennian to late Tournaisian Horton Bluff Formation is subdivided into the Harding Brook, Curry Brook, Blue Beach, and Hurd Creek Members (Fig. 2). Three forested intervals within the Hurd Creek Member were studied at the Blue Beach North locality, near Avonport, Nova Scotia (Calder et al., 1998). This youngest member of the Horton Bluff Formation contains spores of late Tournaisian (Ivorian) age (Utting et al., 1989; Martel and Gibling, 1996).

Strata of the Hurd Creek and underlying Blue Beach Members are organized into well-developed, shallowing-upward cycles (Martel and Gibling, 1991). Cycle bases consist of laminated gray shale (low-energy, offshore deposits), that passes up into shale with interbedded hummocky and wave-rippled sandstone (wave-influenced nearshore deposits). The cycles are capped by green rooted mudstone with nodular and stratified dolomite ("marsh" deposits).

Cycle architecture, sedimentology, and fossil assemblages are similar in the two members, but the relative proportion of terrestrial ("marsh"- and sandstone-dominated) facies increases in Hurd Creek cycles (Martel and Gibling, 1991). The gradational contact between these members is marked by an increase in sandstone proportion and decrease in ostracod abundance (Martel and Gibling, 1996).

Strata of the Horton Bluff Formation were interpreted as strictly freshwater (Martel and Gibling, 1991) until the discovery of brackish-marine ostracods and agglutinated foraminifera by

Tibert and Scott (1999). Tibert and Scott (1999) concluded their study at cycle 20 of the Hurd Creek Member, which constitutes the top of the section in their study area (Blue Beach South). A detailed micropaleontological study would be necessary to verify the presence of marine indicators in every cycle at every location but, in general, terrestrial strata represent "wetlands" (see Application of Wetland Terminology) bordering a body of water that was, at times, marine influenced.

Albert Formation

This study also considers a forested horizon in the Albert Formation in central New Brunswick (Fig. 1). This and other outcrops of the Albert Formation in the Sussex area are best viewed along Provincial Highway 1 between kilometer markers 168 and 180.

The Albert Formation comprises the medial shale-dominated interval in the Horton Group of the Moncton Basin (Fig. 2 in St. Peter, 1993; Keighley, 2000). It has been divided into three members in the type area near Albert Mines (Greiner, 1962; Carter and Pickerill, 1985b), but this scheme is of limited utility because it oversimplifies the varied and intertonguing lithofacies that constitute the formation (St. Peter, 1993). Outcrops north of Albert Mines bear spores of late Hastarian to early Ivorian age (Utting, 1987), but samples from other locations in the basin suggest an early Hastarian to Ivorian age (St. Peter, 1993). Outcrops of the Albert Formation near Sussex exhibit paleontological and sedimentological similarities to those of the type area and are generally regarded as equivalent (Pickerill, 1992; Miller and McGovern, 1996).

Outcrop and core data from the Albert Formation typically indicate nonmarine deposition (Pickerill, 1992). Strata of the Albert Formation were deposited within a fault-bounded valley and represent a collage of sedimentary environments including lakes, fan deltas, alluvial fans, and fluvial systems (van de Poll, 1978; Keighley, 2000; Gingras, 2002).

PREVIOUS STUDIES OF BIOTA

Floral Record

The pioneering study of W.A. Bell (1960) remains the only comprehensive paleobotanical study of the Horton Group. Following Bell (1960), flora from all subdivisions of the Horton Group are discussed to give a complete overview (Table 1) of the plants that may have inhabited the Tournaisian forests described herein.

Compressions of the arborescent lycopsid *Lepidodendropsis* are by far the most abundant plant fossil in the Horton Group (Fig. 3A). These compressions exhibit false leaf scars arranged in pseudowhorls and may be found with the presumed cones (*Lepidostrobophyllum* sp.) and megaspores (*Triletes* sp.) of the plant (Bell, 1960). Bell (1960) noted the presence of a poorly preserved stem that he tentatively classified as *Lepidodendron*, but this is the only specimen (out of thousands) reported from the Horton Group and we therefore question its identification. The absence of lepidodendrids (lycopsids with stigmarian rhizomorphs) is

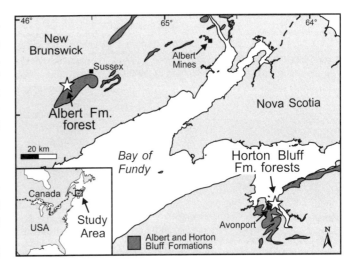

Figure 1. Location map showing the distribution of the Horton Bluff and Albert Formations (after Macauley and Ball, 1982; St. Peter et al., 1997; Keppie, 2000). Stars denote study areas near Avonport, Nova Scotia, and Sussex, New Brunswick.

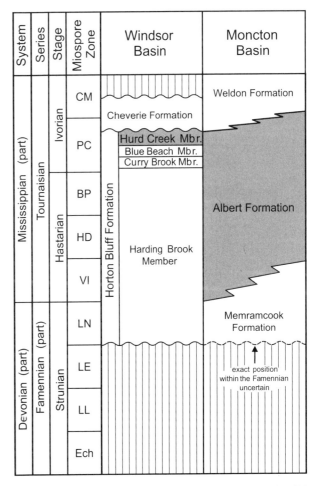

Figure 2. Stratigraphy of the Horton Group in the Windsor Basin of Nova Scotia and the Moncton Basin of New Brunswick (after Carr, 1968; Utting, 1987; Utting et al., 1989; Martel et al., 1993; St. Peter, 1993).

TABLE 1. FOSSIL FLORA OF THE HORTON GROUP, ATLANTIC CANADA

Class	Order	Genus and Species	Type of Fossil	Relevant Occurrences
Lycopsida	Isoetales (lycopsids with corm-like bases)	*Lepidodendropsis corrugatum*	bark	Albert and Horton Bluff Formations
		Lepidodendropsis sp.	bark	Horton Bluff Formation
		Protostigmaria sp.	trunk base	Horton Bluff Formation
		Lepidophyllum (*Lepidostrobophyllum*) *fibriatum*	cone	Horton Bluff Formation
		Lepidostrobophyllum sp.	cone	Albert Formation
		Triletes glaber	large megaspore	Horton Bluff Formation
		Triletes cheveriensis	small megaspore	Albert and Cheverie Formations
	Lepidodendrales (lycopsids with stigmarian rootstocks)	?*Lepidodendron* sp.	bark	Horton Bluff Formation
Sphenopsida	Equisetales (horsetails and their relatives)	*Asterocalamites* (=*Archaeocalamites*) *scrobiculatus*	stem	Horton Bluff Formation
		Nematophyllum sp. (*incertae sedis*)	stem	Kennebecasis Formation (Albert Formation equivalent)
Filicopsida	Filicales ('true ferns')	*Adiantites tenuifolius*	foliage	Cheverie Formation (overlies the Horton Bluff Formation)
Gymnospermopsida	Pteridospermales (seed ferns)	*Aneimites acadica*	foliage	Albert and Horton Bluff Formations
		Sphenopteridium macconochiei?	foliage	Albert Formation
		Sphenopteridium sp.	foliage	Cheverie Formation
		Sphenopteris strigosa	foliage	Cheverie Formation
		Triphyllopteris minor	foliage	Cheverie Formation
		Triphyllopteris virginiana	foliage	Horton Group (undifferentiated)
		Diplotmema patentissimum	foliage	Albert and Horton Bluff Formations
		indeterminate	stem	Horton Bluff Formation
	Incertae sedis	*Carpolithus tenellus*	seed	Horton Bluff and Albert Formations

Note: Modified from Calder (1998).

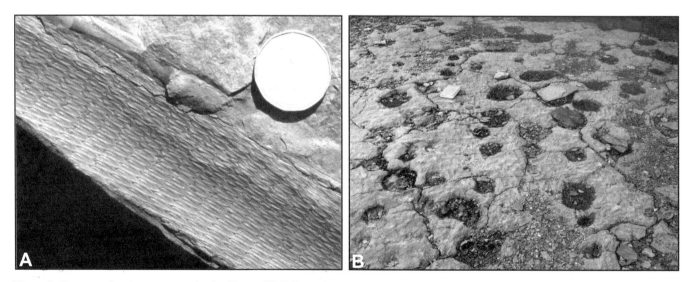

Figure 3. Common floral occurrences in the Horton Bluff Formation. (A) Well-preserved compression of *Lepidodendropsis*. Coin for scale, 26 mm. (B) Poorly preserved molds of in situ trees. Although no identifiable plant material is preserved in direct association with the molds, their size and nearby adpressions suggests that they are after *Lepidodendropsis*. Similar molds observed by Bell (1960) and Martel and Gibling (1996, Fig. 9) were attributed to *Lepidodendropsis* and possibly *Archaeocalamites*. Book for scale, 19 cm long.

supported by Bell's (1929) observation that, although slender roots resembling those connected to stigmarian rhizomorphs are abundant, the rootstocks themselves are completely absent—an observation repeatedly confirmed over the last 70 years.

Although seemingly paradoxical at the time, Bell's (1929) observations foreshadowed the discovery that *Lepidodendropsis* was a non-lepidodendrid lycopsid whose roots attached directly to *Protostigmaria*, the corm-like base of the trunk (Jennings et al., 1983; Pigg, 2001; Gensel and Pigg, 2002). Following Stewart (1947) and Stewart and Rothwell (1993), we use the term "root" in reference to the hairlike, water-absorbing "true" roots of lycopsids and the term "rhizomorph" for the lobed or dichotomously branching organ to which they attach. Additionally, the absence of *Stigmaria* supports Read and Mamay's (1964) and Scheckler's (1986a) observation that although lepidodendrids appear in the middle Tournaisian in Europe, they may not have reached Appalachia until the Visean.

In situ vegetation has been reported at many locations in the Horton Group, but poor preservation has limited our knowledge about the taxonomic affinity and paleoecology of these intervals (Fig. 3B). Bell (1929) noted that paleosols could be identified by their "not uncommon association with upright tree stems" and that at one horizon "120 feet by 15 feet, ninety-six such stems were counted in the basal 1 foot of a 6-foot soil bed." These stems, which reached a diameter of 30 cm, lacked stigmarian rhizomorphs and were tentatively classified as pteridosperms (Bell, 1929, p. 34). In a later memoir, Bell (1960) provided a more detailed description, changed his opinion about the affinity of the trunks, and provided insightful commentary on the probable root structure of *Lepidodendropsis*:

Several siltstone soil zones with erect stems were noted by the writer towards the top of the middle [Blue Beach] member, but in the overlying upper [Hurd Creek] member these become numerous and highly characteristic. Commonly they stand out prominently to the eye on account of their "hackly" weathering and peculiar yellowish green coloration....[E]rect stems occurring in the rootlet-bearing zones rarely exceed a foot in diameter, and 4 to 8 inches is most common. Generally, too, they are not more than a foot or so high. A few are marked by an external thin layer of coal, but most are arenaceous casts with no coal or surface markings. Stems were seen in some instances to have traces of rootlets directly attached to them and organisms resembling *Stigmaria* were extremely rare among the drifted plant remains. If the small stems belong to *Lepidodendropsis*, as seems probable, rather than to a pteridosperm, that genus seemingly lacked stigmarian rhizomes. (Bell, 1960)

Bell (1960) observed that one 216 ft^2 bedding plane had 58 stems (2.9 stems/m^2) and a second 45 ft^2 surface had 19 stems (4.5 stems/m^2). Standing vegetation figures prominently in the sedimentary model of the Horton Bluff Formation depicted by Martel and Gibling (1991, their Fig. 10), but they mention only that beds of green mudstone may "show moulds of roots and tree trunks with diameters up to 40 cm," and attribute the molds to *Lepidodendropsis* and/or *Archaeocalamites*. Preliminary reports of the forests documented below were presented by Gingras (2002) and Melrose and Gibling (2003). While this paper was in review, Falcon-Lang (2004) carried out and published a study of fossil forests in the Albert Formation near Sussex, New Brunswick.

Given the seasonally arid Tournaisian climate of the Euramerican paleotropics (Van der Zwan, 1981; Falcon-Lang, 1999), including the Maritimes Basin (Calder, 1998), low-lying, groundwater-fed (coastal?) areas may have been one of the few environments capable of supporting dense vegetation. Martel and Gibling (1991) attributed calm conditions along the shore to wave attenuation by dense stands of *Archaeocalamites* growing in the shallow water of the nearshore (their facies 2c), which were in turn surrounded by *Lepidodendropsis* "marshes." Evidence for these wetland communities comes from the abundance of their detritus and brief descriptions of in situ vegetation. Accessory gymnosperm material was probably transported from somewhat drier areas where these floras thrived (Scheckler, 1985; Falcon-Lang, 2000).

Faunal Record

Invertebrates

From a paleoecological perspective, one of the most important discoveries within the Horton Group was the documentation of brackish/marine ostracods (*Copelandella*, *Shemonaella*, *Camishaella*, *Carbonita*, *Bairdia*, *Geisina*, and *Youngiella*) and agglutinated foraminifera (*Trochammina*) within the Blue Beach and Hurd Creek Members of the Horton Bluff Formation (Tibert and Scott, 1999). Tibert and Scott (1999) interpreted strata with paraparchitacean ostracods, glaucony grains, palaeoniscid fish, and serpulid worms as distal lagoon, low-energy bay, and restricted nearshore deposits. Assemblages with carbonitacean ostracods, palaeoniscid fish, and serpulid worms were attributed to coastal ponds, and lycopsid-bearing strata with agglutinated foraminifera to "salt marshes."

The discovery of a trilobite pygidium (Order Proetida) in talus from Horton Bluff Formation provides convincing evidence of a marine connection (Weir, 2002). Although the exact stratigraphic position of the block is uncertain, it yielded Tournaisian spores (Dolby, 2003) and appears to be from a nearshore sandstone in the upper Horton Bluff Formation.

Invertebrate activity in the Horton Group is commonly recorded by the *Isopodichnus*, *Palaeophycus*, *Planolites*, *Rusophycus*, *Cruziana*, and *Lockeia* ichnogenera (Martel and Gibling, 1991; Pickerill, 1992; Wood, 1999; Weir, 2002). The wide range of depositional environments preserved within the group is mirrored by the diversity of trace fossil assemblages; previous paleoenvironmental interpretations range from fluvio-lacustrine (Pickerill, 1992) to open marine (Wood, 1999).

The most common ichnogenera of the Horton Bluff Formation (*Isopodichnus*, *Palaeophycus*, and *Planolites*; Martel and Gibling, 1991) represent various trophic-generalist behaviors. Such general and mixed ethological strategies are strongly linked to the exploitation of adverse depositional environments. This, and low trace fossil diversities coupled with high population densities, has been consistently associated with brackish-water deposits (Pemberton et al., 1982). This assemblage is also similar to those described by Buatois et al. (2002) from Pennsylvanian strata in Kansas where a low-diversity, opportunistic, impoverished ichnofaunal assemblage constituted a mixed, depauperate *Cruziana* and *Skolithos* ichnofacies.

Fish

Scales, bones, and teeth of bony fish are ubiquitous in the Horton Bluff and Albert Formations. Scales from the palaeoniscid (ray-finned) genera *Rhadinichthys*, *Elonichthys*, and *Canobius* are particularly abundant in the Albert Formation (Lambe, 1910; Miller and McGovern, 1996). Spines from acanthodians (*Gyracanthus* and *Acanthodes*) are also present (Carroll et al., 1972; Cameron et al., 1992). Lobe-finned fish are represented by *Glyptolepis* scales (Martel and Gibling, 1991) as well as bones of rhizodontid crossopterygians (*Rhizodus hardingi*) and lungfish (*?Ctenodus*) (Carroll et al., 1972). Cartilaginous fish are represented by the elasmobranch shark *Stethacanthus* (Bell, 1929). If the preliminary report of a single dorsal plate holds to be that of an antiarch fish (Cameron et al., 1992), it would represent the only known Mississippian survivor of the heavily armored placoderms.

In addition to a wide variety of sole structures (Martel and Gibling, 1994) and clastic dykes (Martel and Gibling, 1993), fish trace fossils are commonly preserved as casts on the base of nearshore sandstones. Common forms include sinusoidal trails left by fins and tails of swimming fish (ichnogenus *Undichna*) and isolated marks of fins and spines (Wood and Cameron, 1998).

Tetrapods

Until recently, amphibian bones and footprints from the Horton Bluff Formation were the only known tetrapod fossils from the Tournaisian (Milner et al., 1986; Clack and Carroll, 2000; Clack, 2002). Osteological remains of Devonian tetrapods are now known from the United States, Greenland, Scotland, Latvia, Russia, and Australia (Daeschler et al., 1994; Warren and Turner, 2004), but these early amphibians were predominantly aquatic forms with many fish-like characteristics (Milner et al., 1986; DiMichele and Hook, 1992; Coates and Clack, 1995). Although the Horton Bluff Formation contains a diverse assemblage of skeletal remains (Anderson et al., 2005) from what are assumed to be principally aquatic forms, it also contains the remains of seymouriamorph anthracosaurs—close relatives of early reptiles whose terrestrial lifestyle was underpinned by a femur structure fully adapted to dryland locomotion (Carroll et al., 1972; Cameron et al., 1992).

The Horton Bluff trackway record is exceeded in age by Devonian sites in Australia (Warren and Wakefield, 1972) and Brazil (Leonardi, 1983), but the Horton Bluff type area is singular in its abundance of Tournaisian tetrapod footprints (Carroll et al., 1972; Sarjeant and Mossman, 1978a, 1978b; Hunt et al., 2004; Lucas et al., 2004). At least six footprint morphotypes are present in the Horton Bluff Formation, most of which were made on subaerially-exposed substrates (Lucas et al., 2004). Included in the record are the first known Paleozoic footprints, collected in 1841 by Sir William Logan and assigned to the ichnogenus *Hylopus* (Sarjeant and Mossman, 1978a).

APPLICATION OF WETLAND TERMINOLOGY

Confusion and contradiction can arise in the use of wetland terms because similar terms are used in classification schemes based on either hydrologic or vegetative parameters (see Gore, 1983 for discussion). Although the term "marsh" as used by Martel and Gibling (1991) is consistent with a non-peat-forming wetland in the sense of Moore (1987), when used prevalently to reflect herbaceous vegetation the term marsh has the potential to be misconstrued. The term "swamp" offers less clarity still. The Horton Group forests are consistent with the vernacular usage of the term swamp in the United States, where it is commonly used to describe a forested wetland irrespective of peat formation (e.g., the cypress swamps of the Mississippi). Gastaldo (1987) recognized this fact by differentiating "clastic swamps" from those that are peat forming. In the hydrologically based nomenclature advocated by Moore (1987), however, swamp refers to a rheotrophic wetland that is peat forming (a mire). Whereas the Horton Group forests described herein are not rooted in a coaly (peat) substrate, they are not mires per se and are best described as "non-peat-forming, forested wetlands."

EVIDENCE OF FORESTED WETLANDS IN THE HORTON GROUP

The forested levels described here comprise cryptic stump casts less than 47 cm high that yield limited taxonomic information, giving an initial impression that vegetation provided only a modest contribution to the landscape. However, a fuller analysis showed that vegetation-induced sedimentary structures are widespread in the strata, both in association with stump casts and where little original vegetation is preserved. These primary sedimentary structures formed by the interaction of in situ plants with detrital sediment (Rygel et al., 2004). Analysis of features in the Pennsylvanian Joggins Formation suggests that vegetation-induced sedimentary structures reflect three main types of process: (1) hydrodynamic processes around a plant (centroclinal cross-strata, vegetation shadows, coalesced scour fills, scour-and-mound beds, and obstacle marks); (2) passive contact between the plant and the substrate (scratch circles and semicircles); and (3) sedimentation associated with a decaying plant (mudstone-filled hollows and downturned beds). In situ plants need not be preserved to identify vegetation-induced sedimentary structures positively, but a firm link to vegetation must be established to distinguish them from similar structures developed around mobile or nonvegetative obstacles. Identification of all three process types of vegetation-induced sedimentary structures in the Horton Group has greatly expanded our understanding of the nature and extent of these cryptic forested horizons.

We examined four horizons with suspected in situ vegetation in the Horton Group and described them using detailed sedimentological analysis (sensu Driese et al., 1997). Particular attention was dedicated to description of vegetation-induced sedimentary structures (Rygel et al., 2004) and rooting systems (using the classification scheme of Pfefferkorn and Fuchs, 1991). Cliff exposures of the Horton Bluff Formation were described at the centimeter scale and the locations of standing vegetation were recorded on a photomosaic calibrated to a measured traverse. The Horton Bluff Formation forests described in this study are representative of the numerous other levels with poorly preserved stump casts. The forested bedding plane in the Albert Formation was divided into 1 m grids, photographed, and the exact position of each stem and sedimentary structure was noted on a photomosaic calibrated to a measured traverse.

Stump/Root Casts

Description

Twenty-one stump casts were described along a 110 m traverse in cycle 36 of the Horton Bluff Formation (Hurd Creek Member) at Blue Beach North. The in situ stump casts are entombed within the uppermost of three coarsening-upward packages in a 40-cm-thick mudstone (Figs. 4 and 5). This unit marks the initial phase of terrestrial deposition and overlies a *Lepidodendropsis*-bearing, nearshore sandstone (Fig. 6; Bed 8).

Preserved cast height ranges from 6 to 28 cm depending on the preservation of casting material and the relative abundance of sandstone and mudstone within (Fig. 7). Casts extend no higher than the overlying sandstone bed 16–28 cm above (Bed 10; bed thicknesses vary along the outcrop). Partial removal of the entombing mudstone reveals that casts have a columnar shape and a flared, convex-down base (Fig. 7A). The maximum (basal) cast diameter of the 20 measurable casts ranges from 16 to 109 cm, with a mean value of 50 cm. The bottom of the casts (<5 cm) commonly flare at ~60° creating small (<7 cm wide) "toes" in the adjacent sediment (Fig. 8). In all cases, these structures terminate within a few centimeters of the trunk and do not continue into stigmarian rhizomorphs. Many casts are too short (<28 cm) to assess trunk diameter above the basal flare accurately, but the tops of particularly well preserved specimens taper to between 53% and 89% (74% mean) of the basal width.

A partially calcified organic film coats newly exposed specimens (Fig. 7A). Subtle vertical ridges and furrows with less than 1 mm relief ornament the film and the adjacent sandstone (Fig. 7B). The mudstone encasing one specimen exhibits similar vertical features and had small pits ~20 cm above the base of the trunk suggestive of leaf scars. Shortly after

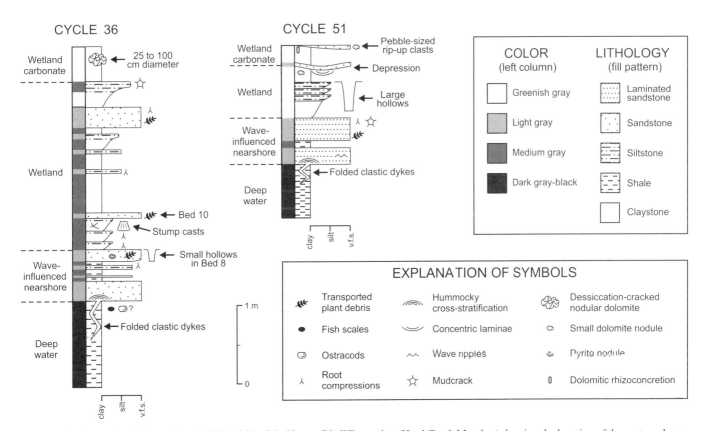

Figure 4. Sedimentological logs of cycles 36 and 51 of the Horton Bluff Formation (Hurd Creek Member) showing the location of the casts and vegetation-induced sedimentary structures. A full account of the sedimentology of the Horton Bluff Formation is given by Martel and Gibling (1996).

Figure 5. Photograph of nearshore and basal wetland strata in cycle 36 of the Horton Bluff Formation (Hurd Creek Member), Blue Beach North. Photograph accompanies the log in Figure 4; HCS = hummocky cross-strata. Hammer for scale, 32 cm.

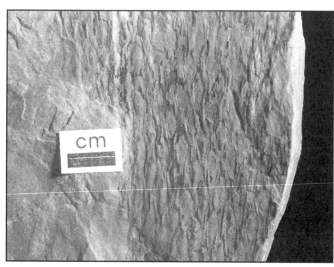

Figure 6. Poorly preserved compression of *Lepidodendropsis*? from Bed 8 in cycle 36.

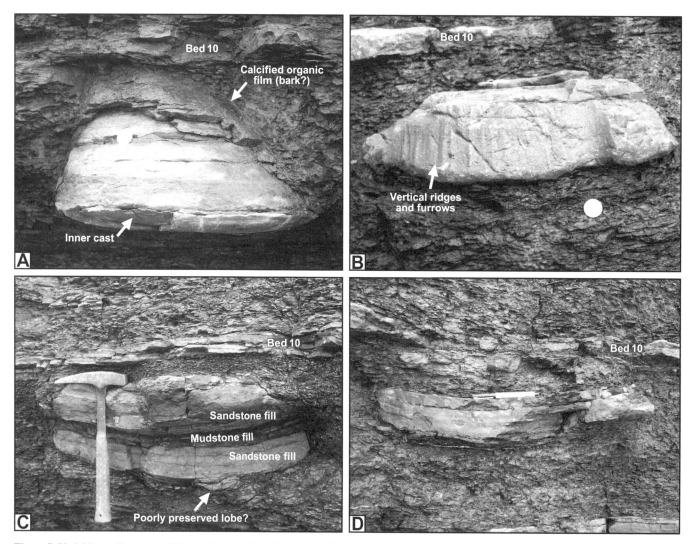

Figure 7. Variable-quality casts of *Protostigmaria*, the lobed base of *Lepidodendropsis*. All casts are in situ and seen in cross section. (A) Well-preserved cast showing how the outer surface weathers away to expose the laminated inner cast. Coin for scale, 18 mm. (B) Poorly preserved cast with vertical ridges and furrows in sandstone fill. Coin for scale, 18 mm. (C, D) Partially eroded casts, which crop out as cryptic, 2D forms. Hammer for scale in C, 32 cm; pen for scale in D, 14 cm.

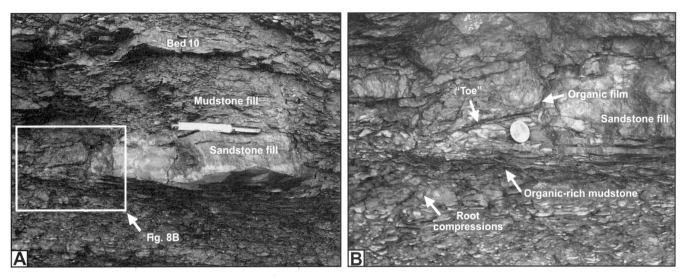

Figure 8. (A) Cross sectional view of *Protostigmaria* with "toe" extending into adjacent sediment. Pen for scale, 14 cm. (B) Detail of "toe" showing the surrounding organic film, underlying organic-rich mudstone, and attached roots extending into the underlying sediment. Coin for scale, 19 mm.

exposure, the organic film weathers away to expose the cast fill. Internally, casts generally comprise a basal layer of very fine-grained sandstone (>5 cm thick) overlain by varying proportions of interbedded very fine-grained sandstone and mudstone (Fig. 7C). Sandy casting material is thinly laminated, and gently onlaps or is truncated against the margin. Unless laminae in the casting mudstone are well preserved, the fill is indistinguishable from the entombing mudstone, which is invariably structureless. Laminae within the casting material are horizontal or convex down and form-concordant with the base. Casts quickly weather to form two-dimensional bodies that would be difficult to interpret without comparison with fresh examples (Fig. 7C, 7D).

Casts and "toes" are underlain by 1–4 cm of organic rich mudstone, which is not present in the adjacent strata (Fig. 9). Dense concentrations of slender, fibrous roots extend away from the carbonaceous shale beneath the "toes" (Fig. 9B, 9C). These roots exhibit a radial arrangement, penetrate downward 20–40 cm into the subjacent strata, and are oriented 5° to 20° from horizontal (Type E-I/II-b rooting system). Scattered vertical roots are present to a depth of 20 cm directly below the casts. Individual roots are preserved as unbranched carbonized compressions 3–8 mm wide (Fig. 9C) that occur almost exclusively in association with the casts.

Two particularly well preserved stump casts exposed in cross section have <5 cm deep furrows that divide the basal surface of the sandstone fill and associated carbonaceous film into discrete lobes (Fig. 9B, 9D–F). The smaller of the two (45 cm diameter) has at least five lobes and the larger, partially eroded specimen (75 cm diameter) has at least three lobes. Poor preservation and the fissile nature of the casts/organic layer prevented fuller description of the lobed base of other specimens.

Interpretation

Although several types of plants inhabited this area during Horton time, these casts are interpreted as the lobed protostigmarian bases of *Lepidodendropsis* because of their size (larger than *Archaeocalamites*), root and rhizomorph morphology, the ubiquity of *Lepidodendropsis* compressions throughout the over- and underlying beds, and because pteridosperms and filicopsids are infrequently preserved in this fashion (Gastaldo et al., 1989). The well-known reconstruction of the original, three- to four-lobed specimens of *Protostigmaria* depict a plant with a distinctive corm-like base (Jennings, 1975), but larger specimens are poorly preserved and difficult to interpret (Fig. 10; G. Rothwell, 2003, personal commun.). Excluding differences in size, the casts described in this study are nearly identical to the multilobed *Protostigmaria eggertiana* described by Jennings et al. (1983) from the Price Formation of Virginia. Their largest (ten-lobed) specimen had a basal diameter of 32 cm, whereas the Nova Scotian examples can be two to three times larger.

The 26% (mean) decrease in trunk width within the basal 22 cm of the Horton Bluff *Protostigmaria* is comparable to the flaring seen in the Price specimens. A particularly well preserved specimen from Pulaski, Virginia, has a 28 cm base that tapers to 17.5 cm (37.5% decrease) in the first 39 cm; above this level the taper is very gradual (Jennings et al., 1983).

Small Mudstone-Filled Hollows

Description

Fifteen small mudstone-filled hollows were described in cycle 36 of the Horton Bluff Formation (Hurd Creek Member) at Blue Beach North. The hollow fills occur within a 5–13 cm thick, planar-bedded sandstone (Bed 8) along the 110 m traverse

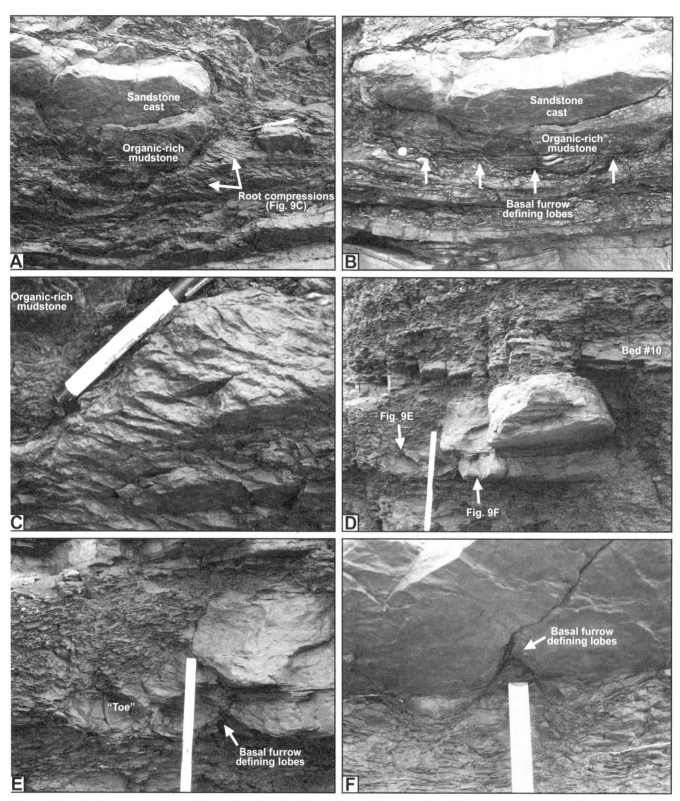

Figure 9. Lobed rhizomes and roots of *Protostigmaria*. (A) Oblique view of the underside of *Protostigmaria* showing the relationship of the sandstone fill to the organic-rich mudstone and roots. Pen for scale in A and C, 14 cm. (B) Side view of the cast in A showing probable furrows and lobes in the organic-rich mudstone. Coin for scale, 19 mm. (C) Detail of root compressions shown in A. (D) Large, partially eroded *Protostigmaria* with well-developed toe and basal furrows. The furrow shown in F is not visible from this angle, but its position is shown. Ruler for scale in D–F, 15 mm wide. (E) Detail of toe and adjacent basal furrow. (F) Detail of a second well-developed furrow; view is nearly perpendicular to that in D.

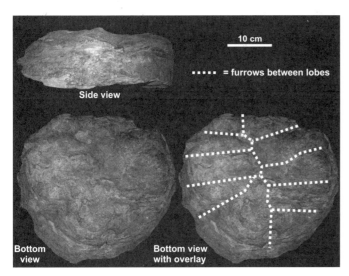

Figure 10. Photos of a ten-lobed *Protostigmaria* from the Price Formation showing cryptic furrows and lobes in a large specimen. Images courtesy G. Rothwell; see Jennings et al. (1983, Figs. 9 and 10) for original description. Ohio University Paleobotanical Herbarium specimen 8113.

(Figs. 4 and 5). This *Lepidodendropsis*-bearing unit (Fig. 6) caps the heterolithic nearshore deposits and is overlain by the cast-bearing unit described above.

Hollow fills are discrete, mudstone bodies within an otherwise continuous, very fine-grained sandstone (Fig. 11). Mean hollow width is 21.5 cm and ranges from 7 to 43 cm, and fill thickness (5–13 cm) equals that of the surrounding sandstone. The mudstone fill is generally structureless, but some specimens contain 2–4 cm pyrite nodules. The surrounding sandstone layer has a sharp contact with the hollow fill and locally thins toward or gently onlaps the hollow fill, although layer thickness is typically unaffected by the presence of the fill. Partial erosion of one example revealed that the original hollow had a columnar form. Four specimens tapered strongly downward with a basal diameter that is only 20–80% of that at the top (Fig. 11B, 11D). The 15 hollows occurred only in a single sandstone bed, and one extended untapered through 16 cm of underlying sediment (Fig. 11C).

In 12 examples, vertical carbonaceous streaks are present in strata below the fills, where bedding was commonly disturbed or destroyed. These features occur only below hollow fills and are inferred to represent poorly preserved roots (Type D-II-b).

In five examples, a thin organic film lined the contact with the adjacent sandstone (Fig. 11D). Removal of the surrounding sandstone in one example revealed a poorly preserved, carbonized bark impression with vertical ridges and furrows with possible rounded leaf or root scars (~3 mm diameter). The underlying sediment was disturbed and had poorly preserved organic streaks that may represent poorly preserved roots (Type D-II-b). Adjacent, weathered mudstone within the hollow was massive and lacked a carbonaceous film.

Interpretation

Although a few hollow fills may be poorly preserved stump casts (including the example with a bordering bark impression), most hollows have no features diagnostic of plants (cf. Gastaldo et al., 1989) and can be identified only by their apparently columnar shape and the contrast between mudstone fills and adjacent sandstone. However, roots and disturbed sediment below commonly indicate the former presence of a plant, and we infer that much of the hollow may have filled after the standing plant decayed.

The scarcity and poor preservation of plant material associated with the hollows cloud the affinity of the plant(s) responsible for their formation. A single, poorly preserved *Lepidodendropsis* compression (Fig. 6) and several poorly preserved/decorticated specimens of similar affinity were recovered from this bed. Hollow size is similar to the *Lepidodendropsis* casts described by Bell (1960), but the mean value of 21.5 cm is less than half that of the *Protostigmaria* casts described above. We suggest that these structures formed in association with juvenile *Lepidodendropsis* or around the smaller *Archaeocalamites*. In view of the position of hollow fills below large stump casts in shoaling-upward cycles, the latter scenario would support Martel and Gibling's (1991) suggestion that *Archaeocalamites* dominated the nearshore and *Lepidodendropsis* favored more landward swamps.

The tapering of mudstone-filled hollows could be an apparent phenomenon caused by a former plant inclined at an angle to the present outcrop face or by slumping of sandstone into a hollow left by a decayed plant. Alternately, the hollow may be genuinely tapered, a morphology that could result from infilling of a tuberoid stem base (Bateman, 1991) or a stem that tapered near its point of attachment to a rhizome. Either of the last two options would be consistent with a plant of (archaeo)calamitean affinity.

Large Mudstone-Filled Hollows

Description

Twelve large mudstone-filled hollows are present along a 50 m traverse in cycle 51 of the Horton Bluff Formation (Hurd Creek Member) at Blue Beach North (Figs. 4 and 12), where they lie within the basal 47 cm of terrestrial strata in the cycle. These hollows (Fig. 13) have a maximum (basal) diameter of 37–106 cm (64.5 cm mean); one abnormally large hollow with an apparent diameter of 195 cm may have been a composite of smaller structures and was excluded from calculation of the mean. Hollows are most noticeable where they pass through a 25-cm-thick heterolithic (silt/claystone) unit, but closer inspection shows that they continue downward through 22 cm of underlying sediment to terminate at organic-rich zones 1–2 cm above the uppermost nearshore sandstone. The upper boundary of the hollows occurs at the top of the heterolithic unit and all specimens are 47 cm deep. Strata above the heterolithic unit are intensely pedoturbated, and hollow fills, if formerly present, are indistinguishable from the associated sediment. Although generally exposed in two dimensions along the cliff face, one hollow fill was partially exposed in the third dimension and had a columnar form (Fig. 14A). Eleven

Figure 11. Small mudstone-filled hollows in Bed 8, cycle 36. (A) Hollow partially exposed in 3D; further excavation revealed that this feature has a columnar shape. Hammer for scale, 32 cm. (B) Strongly tapered mudstone-filled hollow. Pen for scale in B–D, 14 cm. (C) Hollow extending through 16 cm of underlying strata. (D) Mudstone-filled hollow with root compressions and bark. Continued weathering would destroy the organic material, leaving only the mudstone fill.

hollows had nearly vertical margins; one tapered from 63 cm width at the base to 38 cm at top, a 40% reduction in the basal 47 cm. The base of one fill flared 6 cm into the adjacent strata. Most fills comprise laminated, concave-up claystone that onlaps the margins (Fig. 14A); two were filled with structureless claystone (Fig. 13A).

The hollow fill with a columnar form also exhibited a glassy, bark-like impression (cf. DiMichele et al., 1996) with vertical ridges and furrows less than 0.5 mm wide and deep (Fig. 14B). Although glassy surfaces associated with pedogenic slickensides occur in Hurd Creek strata, these ped surfaces are smooth and lack the ridged texture. Additionally, the impression wraps around half the circumference of the hollow and is much larger than the observed ped surfaces. We interpret the feature as a bark impression overprinted by a pedogenic slickenside.

Ten large hollows are underlain by slender, unbranched carbonized roots up to 8 mm wide, 94 cm long (Fig. 14), with a medial ridge along their length (Fig. 14C, 14F) (Type E-I-b). They are particularly abundant near hollow margins, where they radiate into the underlying strata (Fig. 14E) up to a depth of 93 cm. Although scattered roots occur at this level, the majority occur in association with mudstone-filled hollows.

Concentrically filled depressions 16–35 cm deep and 37–210 cm wide are present above six of the 12 large hollow fills (Fig. 15). These depressions are centered on, and are up to 74 cm wider than, the underlying hollow fills. Depressions extend down from a level ~10 cm above the top of the hollow fills and contain concentrically laminated sandstone and mudstone. It is unknown whether they are hemispherical or elongate in three-dimensional form.

Figure 12. Photograph of nearshore and wetland strata in cycle 51 of the Horton Bluff Formation (Hurd Creek Member), Blue Beach North; photograph accompanies the log in Figure 4. Arrows point to the location of large mudstone-filled hollows; dotted line marks contact with "deep-water" strata of the overlying cycle. Book for scale, 19 cm tall.

(64.5 cm mean) is slightly larger than the mean basal diameter of the stump casts (50 cm), and both occur in association with poorly preserved, ridged bark impressions and roots of lycopsid affinity. As with the casts, roots are most abundant near the bottom edges of the hollow fills, where one flared "toe" was also preserved.

The depressions above the large hollows probably filled after complete decay of the lycopsid periderm. Such pre-existing lows within the wetland would have localized flow, resulting in broader depressions and deposition of concentric fills, possibly with some slumping. Depressions probably did not develop atop stumps that had filled to the level of the adjacent marsh when the periderm decayed. Centroclinal cross-strata, which reflect hydrodynamic scour around an upright tree (Underwood and Lambert, 1974), differ from concentrically filled depressions because no tree is present in the center of the feature and because the hollows probably formed after complete decay of the tree. Without the presence of a mudstone-filled hollow or roots below, the concentrically filled depressions seen in cross section would be nearly impossible to distinguish from small floodplain scours or channels.

Scratch Semicircle

Description

Scratch circles are concentric grooves formed where a fixed obstacle (in this case a plant) rotates around its axis and inscribes grooves in the adjacent substrate (Richter, 1926; Allen, 1982). Such grooves rarely form a complete circle, and those that comprise arcs are more accurately termed "scratch semicircles" (Kukal and Al Naqash, 1970). One of these structures (Fig. 16) was found in talus of the Hurd Creek Member at Blue Beach North. The exact stratigraphic position could not be determined, but Martel and Gibling (1991) described these structures on the

Interpretation

As with the small hollow fills, we infer that these large fills record the former presence of standing vegetation, much of which decayed before infilling of the hollow. Although only one hollow indicates the former presence of periderm, roots are particularly abundant beneath these structures.

Similarities between the large hollows in cycle 51 and the casts in cycle 36 suggest that the former also developed in association with *Lepidodendropsis*. The diameter of the hollow fills

Figure 13. Large mudstone-filled hollows in cycle 51. Hammer for scale, 32 cm. (A) Large hollow with massive fill, slightly tapered form, and no overlying concentrically filled depression. (B) Large hollow overlain by well-developed concentrically filled depression.

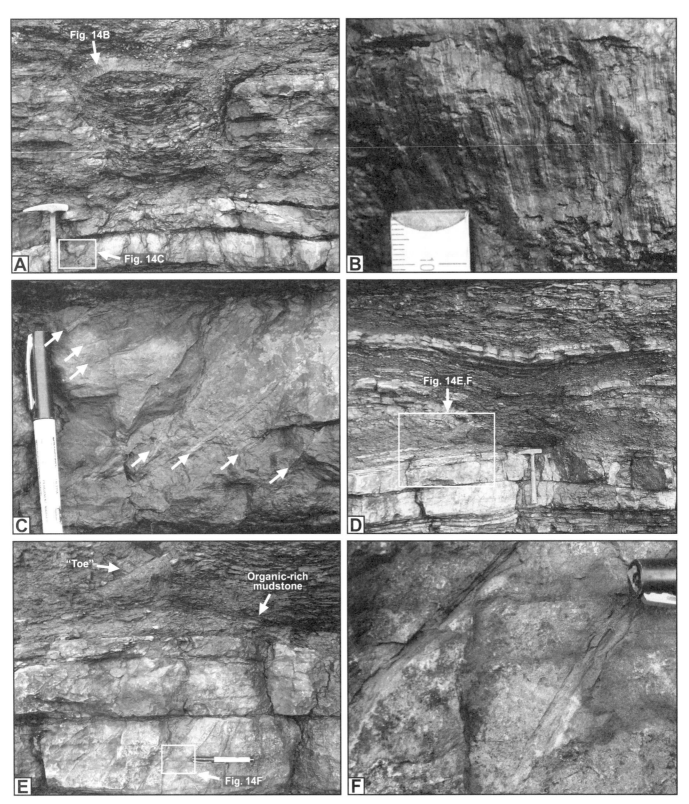

Figure 14. Evidence linking large mudstone-filled hollows to *Lepidodendropsis*. Scale: hammer in A and D, 32 cm; gradations on ruler in B, 1 mm; pen in C, E and F, 14 cm. (A) Large, columnar hollow surrounded by bark impression (B) and underlying slender, lycopsid-like roots (C). (D) Large hollow with toe (E) and well-preserved roots (F).

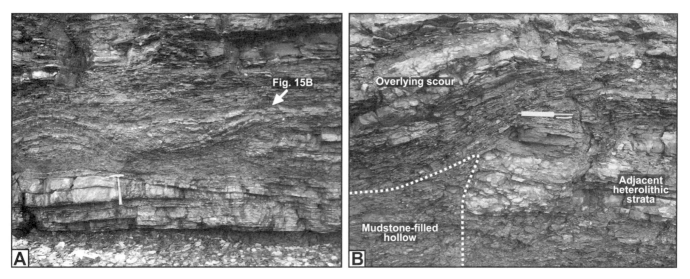

Figure 15. (A) Concentrically filled depression overlying an exceptionally large mudstone-filled hollow. The distribution of roots below this structure suggests that its large size may reflect the decay of two closely spaced trees. Hammer for scale, 32 cm. (B) Detail of the contact between the large mudstone-filled hollow, the concentrically filled depression, and the adjacent heterolithic strata. Pen for scale, 14 cm.

basal surfaces of nearshore sandstones—an observation consistent with the lithology of this specimen.

The scratch semicircle occurs as an incomplete mold on the base of a block of very fine grained sandstone. It comprises a series of ridges 1–5 mm high that form arcs of at least 87° (incomplete exposure). No fewer than 16 distinct grooves occur between 1.5 and 12.2 cm from the center of the circle. The attachment point of the plant was just off the slab. Some grooves exhibit fretting where delicate parts of the plant brushed the sediment surface, ornamenting the deeper grooves (see Allen, 1982).

Interpretation

The scratch semicircle formed where a small plant, bent by a current in wind or water, inscribed grooves into the adjacent sediment. The plant was likely in situ, because plant debris of this size would be easily transported unless firmly affixed by roots. Candidates for the scratchmaker (see Table 1) include ferns or small pteridosperms or sphenopsid branches. Circular tool marks have been attributed to vegetation transported by swirling flood waters (Rigby, 1959), but such structures lack the concentric form of scratch (semi-)circles. The structure provides little information about the generating current because circle form is largely controlled by the torsional strength of the stem and plant architecture (Allen, 1982) and because the circle is incompletely exposed. From morphology alone it is difficult to determine whether this structure formed subaqueously or subaerially, but its nearshore setting suggests a shallow subaqueous origin. The Horton Bluff specimen is the oldest known scratch circle formed by plants; previously described examples come from Pennsylvanian to Cretaceous strata in the United Kingdom, United States, and Canada (Prentice, 1962; Metz, 1991, 1999; Rygel et al., 2004).

Obstacle Marks

Description

Obstacle marks are formed where a current excavates sediment from the front and sides of an obstacle (Allen, 1982), in this case a plant. Obstacle marks around in situ vegetation are common in modern environments (see reviews *in* Nakayama et al., 2002 and Rygel et al., 2004), but this is the first report of an ancient example.

Figure 16. Mold of a scratch semicircle on the base of a nearshore sandstone from the Hurd Creek Member of the Horton Bluff Formation. Nova Scotia Museum specimen NSM004GF016.002.

Twenty obstacle marks were examined on a 37 m² bedding plane exposure of the Albert Formation located at kilometer marker 171 on the east side (northbound lane) of Highway 1. Although this exposure contains 136 in situ *Lepidodendropsis* stems (Fig. 17), only 20 had clearly distinguishable obstacle marks developed around them. The surface lies within a fine-grained sandstone in the upper part of a fluvio-deltaic coarsening-upward cycle (Fig. 18). Only the bedding surface was only exposed across much of the outcrop and it was impossible to determine if all the stems were rooted in exactly the same horizon. Much of the surface was covered with an "egg crate" texture characteristic of closely spaced obstacle marks (Fig. 19), but the reported number is a minimum because only those with a plant preserved in the center (Fig. 20) were counted and studied in detail. There is no obvious pattern as to which plants have obstacle marks and which do not. Mean stem diameter for plants with obstacle marks (4.2 cm) is similar to those without (3.8 cm), and there are no systematic trends in distribution of obstacle marks across the outcrop face. Obstacle marks occur both around individual plants that are isolated and around those that are closely spaced (Fig. 20A, 20C).

The obstacle marks are U-shaped scours that open in a downstream direction and have a stem at their center (Fig. 20A). Maximum scour depth of 1.5–11 cm (4.1 cm mean) occurs just upstream of the obstacle-forming plant. Despite the similarity between mean values for scour depth and stem diameter, scour depth ranges from 30% to 170% of stem width. The downstream-trailing segments of most scours become indistinguishable within ~10 cm of the stem, but six examples extend 8–110 cm behind the plant. The trend of these elongated obstacle marks fall within a 25° range. It is difficult to distinguish uneroded sediment from a vegetation shadow accumulated behind the plant, but two stems appear to have elongate sediment accumulations 1–1.5 cm high in their wake (Fig. 20C). Closely spaced stems may act as a single obstruction or each may have a discrete obstacle mark (Fig. 20D). Although there is no clear correlation between individual stem diameter and scour depth, the two deepest scours (9.5

Figure 17. *Lepidodendropsis* stem with attached microphyllous leaves collected from the Albert Formation at kilometer marker 171 (northbound lane) of Highway 1. Specimen was in situ and is identical to the stems with obstacle marks developed around them. New Brunswick Museum specimen NBMG 12157.

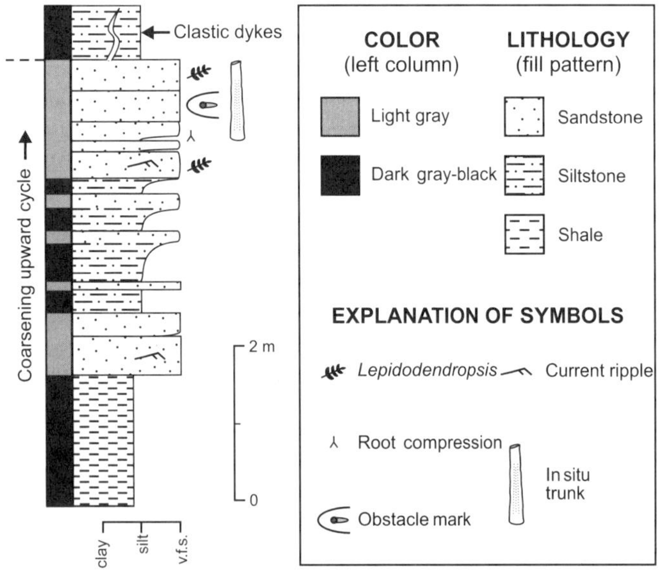

Figure 18. Sedimentological log of a portion of the Albert Formation exposed at kilometer marker 171 (northbound lane) of Highway 1 showing the location of in situ vegetation and obstacle marks.

Figure 19. Bedding plane exposure of the Albert Formation with numerous obstacle marks that give the outcrop its characteristic "egg crate" texture. Strata are vertical and stratigraphic up is to the right. Arrow points to hammer for scale, 32 cm.

and 11.0 cm) occur around stems separated by a distance less than two times their diameter.

Interpretation

The Sussex obstacle marks represent an episode of erosional unidirectional flow through a juvenile *Lepidodendropsis* forest growing on a fluvio-lacustrine delta plain. Obstacle marks form where erosive downflow degrades the sediment bed in front of an obstacle (plant stem) while horseshoe vortices remove sediment from the sides and wake (Allen, 1982). Recent studies of bridge scour have established that maximum scour depth is primarily a function of obstacle width and flow depth, and that flow velocity determines only how quickly the scour reaches maximum depth (Jain, 1981; Melville and Coleman, 2000). These studies provide equations relating flow depth to obstacle width, but they disregard the influence of Froude number and cannot be applied to obstructions less than 10 cm in diameter (Melville and Coleman, 2000). Unraveling the conditions responsible for scour formation is further complicated by interference between plants that develop when the stem centers are closer than 12 times their diameter (Karcz, 1968)—a condition that is true for the majority of stems in this outcrop. Insufficient experimental data are available to attempt paleohydraulic reconstructions for these obstacle marks, but the occurrence of the deepest scours with the widest (composite) obstructions suggests that scour depth is related to obstacle width even at this scale.

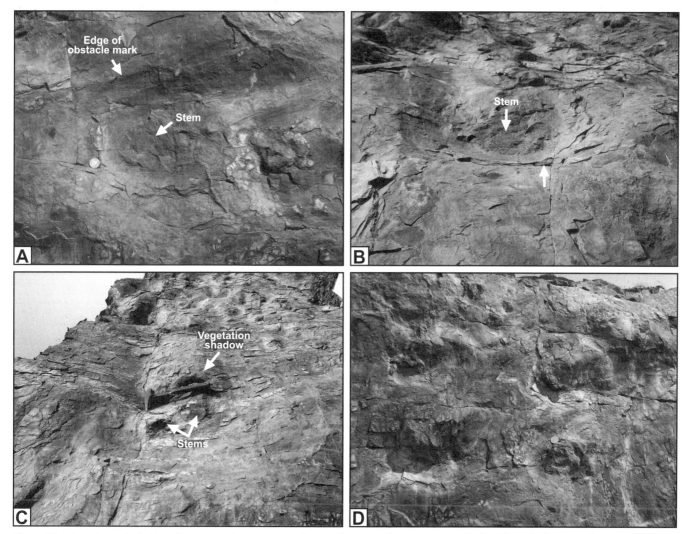

Figure 20. Obstacle marks in the Albert Formation. Photographs are taken from the perspective of being above the bedding plane. (A) Well-preserved three-dimensional obstacle mark developed around in situ stem. Coin for scale, 19 mm. (B) Downstream view of obstacle mark shown in A; this view emphasizes the three-dimensional form of the structure. Arrow points to coin for scale, 19 mm. (C) Downstream view of a large, three-dimensional obstacle mark developed around two closely spaced stems. Note large body of sediment, possibly a vegetation shadow, in the lee of the obstruction. Hammer for scale, 32 cm. (D) Four closely spaced obstacle marks developed around in situ stems. Coin for scale above bottom right obstacle mark, 24 mm.

DENSITY OF FORESTED HORIZONS

The superb two-dimensional cliff exposures of the Horton Bluff Formation and the bedding plane exposure of the Albert Formation provide insight into the organization and density of these ancient forests. Spatial analyses of lepidodendrid lycopsid forests have been reported for the Pennsylvanian (Beckett, 1845; Gastaldo, 1986; DiMichele and DeMaris, 1987; DiMichele et al., 1996) and juvenile *Lepidodendropsis* from the Tournaisian (Falcon-Lang, 2004), but this is the first such study to document mature stands of *Lepidodendropsis*. At Horton Bluff, forest density was estimated by modifying the strip-cruise method of timber inventory (Chapman and Meyer, 1949; Avery and Burkhart, 1983). Specifically, the outcrop face was treated as a swath through the forest whose width is equal to twice the mean basal diameter of the casts. Forest density of the Albert Formation horizon was calculated by counting all the in situ stems in the measured area and extrapolating these values to a hectare; a similar approach was used by Gastaldo (1986). Application of forestry techniques to fossil forests can be problematic because the basic unit of mensuration is trunk diameter at breast height (1.3 m)—a unit that is challenging to measure given the limitations of outcrop exposure and incomplete preservation. Fortunately, lycopsid trunks taper very gradually and a reasonable estimate of diameter at breast height can be achieved if the trunk is measured above the basal flange (Calder et al., 1996). Using the trunk diameter, we calculated biomass (basal stem area [m^2] per hectare) and compared it with that of previously described lycopsid forests (Table 2). We did not perform statistical analysis of plant distribution (Hayek and Buzas, 1997) because of the narrow width of the exposures.

Given the 26% mean reduction in trunk width for the *Protostigmaria/Lepidodendropsis* in cycle 36 of the Horton Bluff Formation, the casts probably attached to trunks that were no more that 12–82 cm wide (36.9 cm mean) at breast height (Fig. 21A). The distribution of trunk diameters is unimodal and left skewed, suggesting that many trees were mature at the time of death (DiMichele and DeMaris, 1987). All but one of the 47-cm-tall mudstone-filled hollows in cycle 51 had vertical margins, indicating that the original tree did not have a significantly flared base and that the measured widths are representative of trunk diameter. *Lepidodendropsis* from cycle 51 ranged from 37 to 106 cm in diameter, with a mean of 62.2 cm (Fig. 21B). These trunk diameters are unimodal and right skewed, a pattern suggestive of a juvenile forest or one that experienced multiple phases of colonization (Calder et al., 1996). We did not plot the size distribution of stems in the Albert Formation because the stems may not have been rooted at the same level.

TABLE 2. COMPARISON OF PENNSYLVANIAN LEPIDODENDRID FORESTS AND TOURNAISIAN *LEPIDODENDROPSIS* FORESTS

Author	Outcrop area (m^2)	Mean trunk diameter (cm)	Density (trees/hectare)	Biomass) (basal stem area (m^2)/hectare)
Pennsylvanian lepidodendrid forests				
Beckett (1845)	1,005	30*	763	215*
Gastaldo (1986)	4,250	20.8	87	38
DiMichele and DeMaris (1987)	130	95	1,769	992
DiMichele et al., (1996)	14,200	60.2	573	164
Tournaisian *Lepidodendropsis* forests (previous studies)				
Bell (1929)	167	all <30 cm	17,609	N.D.
Bell (1960)	20	N.D.	28,892	N.D.
	4	N.D.	45,192	N.D.
Falcon-Lang (2004)	72	4.3†	18,472	27
	35	3.8†	19,428	22
	10.5	12.4†	10,476	127
	7	8.2†	11,428	61
	149	4.4†	29,060	16
Tournaisian *Lepidodendropsis* forests (this study)				
cycle 36, Horton Bluff Fm.	110	36.9§	1,909	204
cycle 51, Horton Bluff Fm.	65	62.2#	1,846	561
Albert Formation	37	3.8#	37,000	50

Note: Lepidodendrid data compiled by DiMichele et al., (1996), N.D. = no data.
* Value estimated by DiMichele et al., (1996).
† value reduced by 40% to correct for basal flare.
§ value reduced by 26% to correct for basal flare.
# little/no flare, no correction applied.

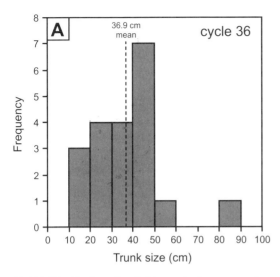
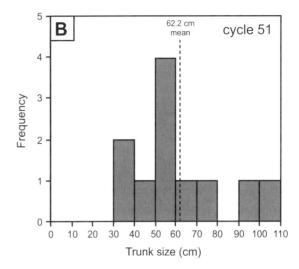

Figure 21. (A) Distribution of probable trunk diameters at breast height for the 20 measurable *Protostigmaria/Lepidodendropsis* casts in cycle 36 of the Horton Bluff Formation. (B) Distribution of probable trunk diameters at breast height for 11 large mudstone-filled hollows observed in cycle 51 of the Horton Bluff Formation.

In cycle 36 of the Horton Bluff Formation, 21 *Protostigmaria* casts (rooted in the same horizon) were present in a 110 m² traverse. At this density, one hectare in this forest contained 1909 trees (0.19 trunks/m²) and had a biomass of 204 m² per hectare (36.9 cm mean diameter). A density estimate for the underlying mudstone-filled hollows was not calculated because they may have been rooted in multiple horizons and were not unequivocally lycopsid. The forested horizon in cycle 51 had a density of 1846 trees per hectare (0.18 trunks/m²), and their considerably larger trunks (62.2 cm mean diameter) yielded a biomass of 561 m² per hectare.

It is unclear whether all the stems in the Albert Formation are rooted at the same level, but density estimates are useful for comparison with other horizons. Regardless of where they were rooted, lycopsids could not propagate adventitiously and the sediment-laden flood waters that inundated the forest encountered a density of standing trees identical to the one now exposed. Average stem density in the Albert Formation is 3.7 stems/m², with a range of 1–12 stems/m². At this density, a hectare would contain 37,000 stems with a basal trunk area of only 50 m² (0.5% coverage). Extrapolating the observed value to a hectare plot makes the tenuous assumption that the 37 m² bedding plane is representative of the remaining 99.6% of the plot. This horizon occurs at approximately the same level as two of the forests studied by Falcon-Lang (2004; his sites 2.105 and 2.112), intervals for which he reported densities of 18,500 and 19,500 stems per hectare, respectively. Our larger estimate could reflect differences in sampling strategy, or it could indicate that the bedding surface described in this study contained stems rooted in two stacked horizons. Calculated densities for the *Lepidodendropsis* forests described in this study are within or below the 29,000–45,000 trees/hectare values reported by Bell (1960).

Our data show that *Lepidodendropsis* forests had more trees per hectare than their Pennsylvanian counterparts (Fig. 22, Table 2). Horton Bluff forests were as much as three times more dense than lepidodendrid forests of similar tree size and biomass. This difference could reflect local variations in environment or substrate, but it could also be the product of diminished competition/interaction between protostigmarian rooting systems (compared with the large and extensive *Stigmaria* of lepidodendrids).

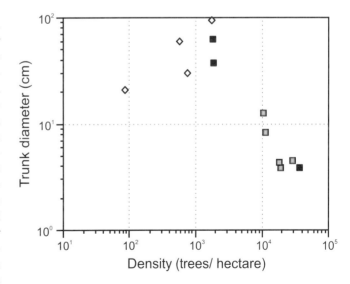

◇ Pennsylvanian lepidodendrids (see Table 2)
▫ Tournaisian *Lepidodendropsis* (Falcon-Lang, 2004)
■ Tournaisian *Lepidodendropsis* (this study)

Figure 22. Relationship between trunk diameter and tree density for Pennsylvanian lepidodendrid and Tournaisian *Lepidodendropsis* forests (data from Table 2).

The small size of the Albert Formation trees makes them more difficult to compare with the lepidodendrids, but our size and density values compare well with the range reported by Falcon-Lang (2004). Comparing density values among mature and immature *Lepidodendropsis* forests (Fig. 22), the relatively low density of the mature Horton Bluff examples supports Falcon-Lang's (2004) interpretation that these forests experienced competition-induced mortality (self-thinning) as they matured.

DISCUSSION

Protostigmaria Variations

The Horton Group of Atlantic Canada is one of the few *Protostigmaria*-bearing units described to date. The rarity of these specimens is surprising given the nearly global distribution of the Early Mississippian "*Lepidodendropsis* flora" (Chaloner and Lacey, 1973; Wagner, 1979; Raymond et al., 1985). Although poorly preserved, the 21 specimens described from Blue Beach North during 2002–2003 demonstrate a surprising level of diversity within the genus. These specimens are commonly twice, and possibly as much as three times, as wide as the largest specimen from Virginia (Jennings, 1975; Jennings et al., 1983). In addition to variations in size, the Horton Bluff specimens confirm Jennings et al.'s (1983) observation that lobes in large *Protostigmaria* become difficult to identify.

Horton Bluff *Protostigmaria* provide information about how the arborescent *Lepidodendropsis* anchored itself in the substrate and maintained its upright form. Bell (1960) and Jennings (1975) noted that *Lepidodendropsis* had roots similar to those extending away from *Stigmaria*, but lacked the rhizomes themselves. Without this wide support base, Jennings et al. (1983) speculated that the lobed base must have anchored itself in the substrate in a *Isoëtes*-like fashion. As the plant grew new lobes, it was able to support its increasing height as the base became buried in the substrate and the roots anchored themselves (Karrfalt, 1977). Although the sediment adjacent to the trunks and hollows is thoroughly pedoturbated, the "toes" at the base of the plants seem particularly well suited for soil penetration and may have provided a means of entrenchment. If so, their point of attachment with the trunk may represent the depth to which *Protostigmaria* was buried in the soil (generally ~5 cm plus the depth to which the concave-down base extended below the margins). The radiation of roots down and away from the base of the trunk may have provided pre-stigmarian support against wind shear (Jennings et al., 1983; Mosbrugger, 1990). The 93 cm penetration depths attained by these rootlets are much greater than previously reported for early arborescent lycopsids and rival those of Devonian progymnosperms (Algeo and Scheckler, 1998).

Tournaisian Wetland Ecosystems

The recent reinterpretation of these strata as marine-influenced through the identification of key ostracod genera, agglutinated foraminifera, and a trilobite specimen at Horton Bluff profoundly affects our understanding of this Tournaisian wetland ecosystem. Tibert and Scott (1999) documented agglutinated foraminifera in green rooted mudstones that bracket intervals with standing lycopsids in equivalent strata elsewhere at Blue Beach. Their model agrees well with the setting envisioned for the Price Formation of Virginia, where back barrier *Lepidodendropsis* "swamps" lay just behind sandy beaches bordering the sea (Scheckler and Beeler, 1984; Scheckler, 1985, 1986a, 1986b). There is equivocal evidence that Pennsylvanian lycopsids may have tolerated some salinity (Gastaldo, 1986), but it is unclear whether Horton Bluff *Lepidodendropsis* was salt tolerant or if the forests were later inundated (killed?) by marine waters. Regardless of their salinity tolerance, it is clear that the *Lepidodendropsis* forests and their inhabitants thrived in close proximity to a body of water that was, at times, marine influenced.

We confirm Martel and Gibling's (1991) suggestion that the Horton shoreline was surrounded by *Lepidodendropsis* wetlands, but the most basinward vegetation (those elements lowest in the cycles) has proved more difficult to identify. The size and morphology of the small mudstone-filled hollows in nearshore sandstones is consistent with both *Archaeocalamites* and juvenile *Lepidodendropsis*, although direct evidence for *Archaeocalamites* is lacking. Detailed paleobotanical studies of plant macrodetritus or additional study of nearshore vegetation-induced sedimentary structures may help to confirm the identity of nearshore floral communities.

The localities provide little information about inland or upland vegetation during this period. Transported *Lepidodendropsis* stems are ubiquitous throughout the wetland and nearshore facies of the Horton Bluff Formation, but the uncertain affinity of the stump casts previously made the source of the plant macrodetritus difficult to pinpoint. The identification of *Lepidodendropsis* forests suggests that much of the plant macrodetritus accumulated close to its source. In a study of the Mobile Delta, Gastaldo et al., (1987) observed that floral elements from distant locations (>13 km) were overrepresented within the site of accumulation. A few pteridosperm axes, probably derived from well-drained floodplain niches (Scheckler, 1985), are the only evidence of plant material originating from outside the *Lepidodendropsis* forests. Given the ubiquity of *Lepidodendropsis* compressions in the Horton Bluff Formation and the presence of in situ lycopsid trunks, it appears that much of the plant material accumulated essentially in situ. The abundance of locally derived plant remains may reflect the paucity of fluvial channels in the Hurd Creek and Blue Beach Members, even though terrestrial strata are important components of the cycles. Without major channels to entrain plant macrodetritus, plant material probably spread by floating in standing bodies of water—a potentially difficult dispersal mechanism in densely vegetated areas. Alternatively, the low-diversity, *Lepidodendropsis* forests could have been so widespread that other flora were too distant to contribute significant amounts of plant material.

A growing body of evidence suggests that, unlike most extant species, early amphibians inhabited brackish or even fully marine environments (see reviews in Coates et al., 2000; Laurin et al., 2000). Similarly, several types of fish (acanthodians, lungfish, and elasmobranch sharks) are known to have been tolerant of salt water (Carroll, 1988). Because Horton Bluff tetrapod bones are frequently found in talus or by casual collectors, their precise location relative to known brackish-marine zones is rarely recorded. Although tetrapod footprints have been found within known marine intervals (Bell, 1960; Tibert and Scott, 1999), it is possible that tetrapods occupied discrete freshwater areas within this environment. Without coordinated micro- and vertebrate paleontological study, the salinity tolerance of the Horton Bluff tetrapod community will remain enigmatic.

Plant-Sediment Interaction: Low-energy Environments and Insight into Early Land Plants

The vegetation-induced sedimentary structures described from the Horton Group, particularly those from Horton Bluff, were deposited in a low-energy depositional environment. These structures provide an important counterpoint to the meter-deep erosional structures that feature prominently in the original description of vegetation-induced sedimentary structures from the Pennsylvanian Joggins Formation (Rygel et al., 2004). Although formed within fine-grained sediment deposited under relatively calm conditions, the scale of Horton Bluff vegetation-induced sedimentary structures approaches that of many features from Joggins.

Obstacle marks around plants in the Albert Formation are the oldest known examples from the ancient record. These structures represent the midpoint between the low-energy environment at Horton Bluff and the intense floods of the Joggins Formation (Calder et al., this volume), and further illustrate that vegetation-induced sedimentary structures can form under a wide range of hydrodynamic conditions.

Unlike traditional mud-cast logs and stumps, mudstone-filled hollows generally bear no direct evidence of having formed in association with a plant (cf. Gastaldo et al., 1989). These structures can only be attributed to formation in association with plants by their sedimentary context and by examining underlying strata for roots or phytoturbation. Without careful study, these structures could easily be mistaken for purely hydrodynamic or soft-sediment deformation features.

Land plants evolved in the Ordovician (Wellman et al., 2003), but in situ specimens were rarely preserved until the evolution of arborescent progymnosperms in the Middle Devonian (Driese et al., 1997; Elick et al., 1998). Because vegetation-induced sedimentary structures persist despite the destruction of organic material, identification of these structures has the potential to expand our knowledge about early plant colonization of the terrestrial landscape. The vegetation-induced sedimentary structures described in this study formed in calm, swampy environments similar to those inhabited by early bryophytes and vascular plants. Preservation of delicate scratch circles shows that even small individual stems can interact with the substrate to form sedimentary structures with preservation potential much higher than that of the in situ stem itself. Because vegetation-induced sedimentary structures also occur in dryland settings (Rygel et al., 2004) where organic material is unlikely to be preserved, their identification in red bed sequences could be vital in constraining Paleozoic phytogeography.

CONCLUSIONS

This study records the former presence of *Lepidodendropsis* forests in Tournaisian strata of the Horton Group, Atlantic Canada. We describe new variations in *Protostigmaria* morphology and resolve the taxonomic affinity of cryptic stump casts. The *Protostigmaria* described here expand upon the morphological differences described by Jennings et al. (1983) and provide a contrast to the small individuals commonly depicted in textbooks. Additionally, this study provides compelling evidence that *Lepidodendropsis* inhabited coastal wetlands similar to the ones envisioned for the Price Formation of Virginia (Scheckler and Beeler, 1984; Scheckler, 1985, 1986a).

Vegetation-induced sedimentary structures that developed in the low-energy swamps of the Horton Group provide an interesting contrast to the erosion-dominated forms described from the Pennsylvanian Joggins Formation (Rygel et al., 2004). Without careful sedimentological analysis, the mud-dominated vegetation-induced sedimentary structures of the Horton Bluff Formation could easily be mistaken for channels, hydrodynamic scours, or soft-sediment deformation structures. The expanded descriptions of vegetation-induced sedimentary structures will contribute to our ability to recognize standing vegetation in facies likely to have been colonized by early land plants. Vegetation-induced sedimentary structures provide direct evidence of plant-sediment interaction, an activity that has profoundly influenced clastic depositional systems for over 425 million years.

ACKNOWLEDGMENTS

Insightful reviews by Brigitte Meyer-Berthaud and Hermann Pfefferkorn greatly improved this manuscript. Discussions with Howard Falcon-Lang, Gar Rothwell, Patricia Gensel, John Utting, Robert Carroll, and Clint St. Peter also contributed to the development of this manuscript. Chris Mansky and Sonja Wood are acknowledged for providing us with a tour of the exquisite fossils housed at the Blue Beach Fossil Museum, 127 Blue Beach Road, Hantsport, Nova Scotia, Canada, B0P 1P0. Adrienne Rygel provided valuable assistance in the field. This study was made possible by a Nova Scotia Museum Research Grant and a Killam Predoctoral Scholarship awarded to M. Rygel and a Natural Sciences and Engineering Research Council of Canada Grant to M. Gibling.

REFERENCES CITED

Ahlberg, P.E., and Milner, A.R., 1994, The origin and diversification of early tetrapods: Nature, v. 368, p. 507–514, doi: 10.1038/368507a0.

Algeo, T.J., and Scheckler, S.E., 1998, Terrestrial-marine teleconnections in the Devonian: Links between the evolution of land plants, weathering processes, and marine anoxic events: Royal Society of London Philosophical Transactions, ser. B, v. 353, p. 113–130.

Allen, J.R.L., 1982, Sedimentary structures: Their character and physical basis, Volume 2: Amsterdam, Elsevier, 663 p.

Anderson, J.S., Mansky, C., Wood, S., Godfrey, R., and Carroll, R.L., 2005, New tetrapod fossils from the Lower Carboniferous of Blue Beach (Horton Bluff Formation), Nova Scotia, in North American Paleontology Convention, Halifax, Nova Scotia, June 19–25: Paleobios, v. 25, supplement to no. 2, p. 13–14.

Avery, T.E., and Burkhart, H.E., 1983, Forest measurements: New York, McGraw-Hill, 331 p.

Bateman, R.M., 1991, Palaeobiological and phylogenetic implications of anatomically-preserved archaeocalamites from the Dinantian of Oxroad Bay and Loch Humphrey Burn, southern Scotland: Palaeontographica Abteilung B, v. 223, p. 1–59.

Beckett, H., 1845, On a fossil forest in the Parkfield Colliery near Wolverhampton: Quarterly Journal of the Geological Society [London], v. 1, p. 41–43.

Bell, W.A., 1929, Horton-Windsor District, Nova Scotia: Geological Survey of Canada Memoir 155, 268 p.

Bell, W.A., 1960, Mississippian Horton Group of Type Windsor-Horton District, Nova Scotia: Geological Survey of Canada Memoir 314, 112 p.

Buatois, L.A., Mangano, M.G., Alissa-Abdulrahman, A., and Carr, T.R., 2002, Sequence stratigraphic and sedimentologic significance of biogenic structures from a late Paleozoic marginal-to open-marine reservoir, Morrow Sandstone, subsurface of Southwest Kansas, USA: Sedimentary Geology, v. 152, p. 99–132, doi: 10.1016/S0037-0738(01)00287-1.

Calder, J.H., 1998, The Carboniferous evolution of Nova Scotia, in Blundell, D. J., and Scott, A. C., eds., Lyell, the past is the key to the present: Geological Society [London] Special Publication 143, p. 261–302.

Calder, J.H., Gibling, M.R., Eble, C.F., Scott, A.C., and MacNeil, D.J., 1996, The Westphalian D fossil lepidodendrid forest at Table Head, Sydney Basin, Nova Scotia: Sedimentary, paleoecology and floral response to changing edaphic conditions: International Journal of Coal Geology, v. 31, p. 277–313, doi: 10.1016/S0166-5162(96)00020-1.

Calder, J.H., Boehner, R.C., Brown, D.E., Gibling, M.R., Mukhopadhyay, P.K., Ryan, R.J., and Skilliter, D.M., 1998, Horton Bluffs coastal section (Stop 2), in Classic Carboniferous sections of the Minas and Cumberland Basins in Nova Scotia: Nova Scotia Department of Natural Resources Open File Report ME 1998-5, p. 25–35.

Cameron, B., Van Dommelen, R., White, P.D., Domanski, D., Rogers, D., and Jones, J.R., 1992, Vertebrate fossils from the middle member of the Early Carboniferous Horton Bluff Formation, in MacDonald, D.R., and Mills, K.A., eds., Sixteenth annual open house and review of activities—program and summaries: Nova Scotia Department of Mines and Energy Report 92-4, p. 48.

Carr, P.A., 1968, Stratigraphy and spore assemblages, Moncton map-area, New Brunswick: Geologic Survey of Canada Paper 67-29, 47 p.

Carroll, R.L., 1988, Vertebrate paleontology and evolution: New York, W.H. Freeman, 689 p.

Carroll, R.L., 2002, Early land vertebrates: Nature, v. 418, p. 35–36, doi: 10.1038/418035a.

Carroll, R.L., Belt, E.S., Dineley, D.L., Baird, D., and McGregor, D.C., 1972, Horton Bluff, Lower Mississippian, in Glass, D.J., ed. Vertebrate palaeontology of eastern Canada: Montreal, International Geological Congress, 24th, Excursion A59, Guidebook, p. 17–21.

Carter, D.C., and Pickerill, R.K., 1985a, Algal swamp, marginal and shallow evaporitic lacustrine lithofacies from the late Devonian-early Carboniferous Albert Formation, southeastern New Brunswick, Canada: Maritime Sediments and Atlantic Geology, v. 21, p. 69–86.

Carter, D.C., and Pickerill, R.K., 1985b, Lithostratigraphy of the Late Devonian-Early Carboniferous Horton Group of the Moncton Subbasin, southern New Brunswick: Maritime Sediments and Atlantic Geology, v. 21, p. 11–24.

Chaloner, W.G., and Lacey, W.S., 1973, The distribution of Late Paleozoic floras, in Hughes, N. F., ed., Organisms and continents through time: London, Palaeontological Association Special Papers in Palaeontology 12, p. 271–289.

Chapman, H.H., and Meyer, W.H., 1949, Forest mensuration: New York, McGraw-Hill, 522 p.

Clack, J.A., 2002, An early tetrapod from Romer's Gap: Nature, v. 418, p. 72–76, doi: 10.1038/nature00824.

Clack, J.A., and Carroll, R.L., 2000, Early Carboniferous tetrapods, in Heatwole, H., and Carroll, R.L., eds., Amphibian Biology: Chipping Norton, Surrey Beatty and Sons, v. 4, p. 1030–1043.

Coates, M.I., and Clack, J.A., 1995, Romer's gap: Tetrapod original and terrestriality: Bulletin du Muséum National d'Histoire Naturelle, v. 17, p. 373–388.

Coates, M.I., Ruta, M., and Milner, A.R., 2000, Early tetrapod evolution: Trends in Ecology and Evolution, v. 15, p. 327–328, doi: 10.1016/S0169-5347(00)01927-3.

Daeschler, E.B., Shubin, N.H., Thomson, K.S., and Amaral, W.H., 1994, A Devonian tetrapod from North America: Science, v. 265, p. 639–642.

DiMichele, W.A., and DeMaris, P.J., 1987, Structure and dynamics of a Pennsylvanian-age *Lepidodendron* forest: Colonizers of a disturbed swamp habitat in the Herrin (No. 6) coal of Illinois: Palaios, v. 2, p. 146–157.

DiMichele, W.A., and Hook, R.H., 1992, Paleozoic terrestrial ecosystems, in Behrensmeyer, A.K., Damuth, W.A., DiMichele, W.A., Potts, R., Sues, H.-D., and Wing, S.L., eds., Terrestrial ecosystems through time: Chicago, University of Chicago Press, p. 205–325.

DiMichele, W.A., Eble, C.F., and Chaney, D.S., 1996, A drowned lycopsid forest above the Mahoning coal (Conemaugh Group, Upper Pennsylvanian) in eastern Ohio, U.S.A: International Journal of Coal Geology, v. 31, p. 249–276, doi: 10.1016/S0166-5162(96)00019-5.

Dolby, G., 2003, Palynological analysis of ten outcrop and core hole samples from Nova Scotia: Nova Scotia Department of Natural Resources Open-File Report ME 2003-5, 8 p.

Driese, S.G., Mora, C.I., and Elick, J.M., 1997, Morphology and taphonomy of root and stump casts of the earliest trees (Middle to Late Devonian), Pennsylvania and New York, U.S.A.: Palaios, v. 12, p. 534–537.

Elick, J.M., Driese, S.G., and Mora, C.I., 1998, Very large plant and root traces from the Early to Middle Devonian: Implications for early terrestrial ecosystems and atmospheric $p(CO_2)$: Geology, v. 26, p. 143–146, doi: 10.1130/0091-7613(1998)026<0143:VLPART>2.3.CO;2.

Falcon-Lang, H., 1999, The Early Carboniferous (Asbian–Brigantian) seasonal tropical climate of northern Britain: Palaios, v. 14, p. 116–126.

Falcon-Lang, H., 2000, Fire ecology of the Carboniferous tropical zone: Palaeogeography, Palaeoclimatology, Palaeoecology, v. 164, p. 339–355, doi: 10.1016/S0031-0182(00)00193-0.

Falcon-Lang, H.J., 2004, Early Mississippian lycopsid forests in a delta-plain setting at Norton, near Sussex, New Brunswick, Canada: Journal of the Geological Society [London], v. 161, p. 969–981.

Gastaldo, R.A., 1986, Implications on the paleoecology of autochthonous lycopsids in clastic sedimentary environments of the Early Pennsylvanian of Alabama: Palaeogeography, Palaeoclimatology, Palaeoecology, v. 53, p. 191–212, doi: 10.1016/0031-0182(86)90044-1.

Gastaldo, R.A., 1987, Confirmation of Carboniferous clastic swamp communities: Nature, v. 326, p. 869–871, doi: 10.1038/326869a0.

Gastaldo, R.A., Douglass, D.P., and McCarroll, S.M., 1987, Origin, characteristics, and provenance of plant macrodetritus in a Holocene crevasse splay, Mobile Delta, Alabama: Palaios, v. 2, p. 229–240.

Gastaldo, R.A., Demko, T.M., Liu, Y., Keefer, W.D., and Abston, S.L., 1989, Biostratinomic processes for the development of mud-cast logs in Carboniferous and Holocene swamps: Palaios, v. 4, p. 356–365.

Gensel, P.G., and Pigg, K.B., 2002, Reconstruction of the *Lepidodendropsis/Protostigmaria* plant from the Mississippian Price Formation of Virginia, USA: Botany 2002, Madison, Wisconsin, Abstracts, p. 27.

Gingras, M.K., 2002, Stop 1 (km 168), in Armitage, I., Gingras, M.K., and Keighley, D.G., eds., Field trip number four: Lacustrine sedimentology, ichnology, and sequence stratigraphy of Early Carboniferous rift basins (Albert Fm.), Sussex, New Brunswick: Frederick, New Brunswick, Atlantic Universities Geologic Conference, Guidebook, p. 4–6.

Gore, A.J.P., ed., 1983, Mires: Swamp, bog, fen and moor: Amsterdam, Elsevier, 440 p.

Greiner, H.R., 1962, Facies and sedimentary environments of the Albert shale, New Brunswick: American Association of Petroleum Geologists Bulletin, v. 46, p. 219–234.

Hamblin, A.P., and Rust, B.R., 1989, Tectono-sedimentary analysis of alternate-polarity half-graben basin-fill succession: Late Devonian-Early Carboniferous Horton Group, Cape Breton Island, Nova Scotia: Basin Research, v. 2, p. 239–255.

Hayek, L.C., and Buzas, M., 1997, Surveying natural populations: New York, Columbia University Press, 563 p.
Hunt, A.P., Lucas, S.G., Calder, J.H., Van Allen, H.E.K., George, E., Gibling, M.R., Hebert, B.L., Mansky, C., and Reid, D.R., 2004, Tetrapod footprints from Nova Scotia: The Rosetta Stone for Carboniferous tetrapod ichnology: Geological Society of America Abstracts with Programs, v. 36, no. 6, p. 66.
Jain, S.C., 1981, Maximum clear-water scour around cylindrical piers: Journal of Hydraulic Engineering, v. 107, p. 611–625.
Jennings, J.R., 1975, *Protostigmaria*, a new plant organ from the Lower Mississippian of Virginia: Palaeontology, v. 18, p. 19–24.
Jennings, J.R., Karrfalt, E.C., and Rothwell, G.W., 1983, Structure and affinities of *Protostigmaria eggertiana*: American Journal of Botany, v. 70, p. 963–974.
Karcz, I., 1968, Fluviatile obstacle marks from the wadis of the Negev (southern Israel): Journal of Sedimentary Petrology, v. 38, p. 1000–1012.
Karrfalt, E.E., 1977, Substrate penetration by the corm of *Isoëtes*: American Fern Journal, v. 67, p. 1–4.
Keighley, D., 2000, A preliminary sequence stratigraphy of the Horton Group, southeast Moncton Basin, southeast New Brunswick: Interpretation of the Dawson Settlement Member of the Carboniferous Albert Formation, Shell Albert Mines #4 well: New Brunswick Department of Natural Resources and Energy, Minerals and Energy Division Mineral Resource Report, p. 17–29.
Keppie, J.D., 2000, Geological map of the province of Nova Scotia: Nova Scotia Department of Natural Resources, Minerals and Energy Branch, scale 1:500 000.
Kukal, Z., and Al Naqash, A.B., 1970, Scratch circles on the Pliocene sandstones in central Iraq: Zeitschrift für Geomorphologie, v. 14, p. 329–334.
Lambe, L.M., 1910, Palaeoniscid fishes from the Albert Shales of New Brunswick: Contributions to Canadian Paleontology, Geological Survey of Canada Memoir 3, p. 1–69.
Laurin, M., Girondot, M., and Ricqlès, A. de, 2000, Early tetrapod evolution: Trends in Ecology and Evolution, v. 15, p. 118–123, doi: 10.1016/S0169-5347(99)01780-2.
Leonardi, G., 1983, *Notopus petri*, n. gen., sp.; an amphibian imprint in the Devonian of Paraná, Brazil: Geobios, v. 16, p. 233–239.
Lucas, S.G., Hunt, A.P., Mansky, C., and Calder, J.H., 2004, The oldest tetrapod footprint ichnofauna, from the lower Mississippian Horton Bluff Formation, Nova Scotia, Canada: Geological Society of America Abstracts with Programs, v. 36, no. 5, p. 66.
Macauley, G., and Ball, F.D., 1982, Oil Shales of the Albert Formation, New Brunswick: New Brunswick Department of Natural Resources and the Geologic Survey of Canada, Institute of Sedimentary and Petroleum Geology Open-File Report 82-12, 173 p.
Marillier, F., Keen, C.E., Stockmal, G.S., Quinlan, G., Williams, H., Colman-Sadd, S.P., and O'Brien, S.J., 1989, Crustal structure and surface zonation of the Canadian Appalachians: Implications of deep seismic reflection data: Canadian Journal of Earth Sciences, v. 26, p. 305–321.
Martel, A.T., and Gibling, M.R., 1991, Wave-dominated lacustrine facies and tectonically controlled cyclicity in the Lower Carboniferous Horton Bluff Formation, Nova Scotia: Canada International Association of Sedimentologists Special Publication 13, p. 223–243.
Martel, A.T., and Gibling, M.R., 1993, Clastic dykes of the Devono-Carboniferous Horton Bluff Formation, Nova Scotia: Storm related structures in shallow lakes: Sedimentary Geology, v. 87, p. 103–119, doi: 10.1016/0037-0738(93)90038-7.
Martel, A.T., and Gibling, M.R., 1994, Combined-flow generation of sole structures, including recurved groove casts, associated with Lower Carboniferous lacustrine storm deposits in Nova Scotia, Canada: Journal of Sedimentary Research, v. A64, p. 508–517.
Martel, A.T., and Gibling, M.R., 1996, Stratigraphy and tectonic history of the Upper Devonian to Lower Carboniferous Horton Bluff Formation, Nova Scotia: Atlantic Geology, v. 32, p. 13–38.
Martel, A.T., McGregor, D.C., and Utting, J., 1993, Stratigraphic significance of Upper Devonian and Lower Carboniferous miospores from the type area of the Horton Group, Nova Scotia: Canadian Journal of Earth Sciences, v. 30, p. 1091–1098.
Melrose, C.S.A., and Gibling, M.R., 2003, Fossilized forests of the Lower Carboniferous Horton Bluff Formation, Nova Scotia: Geological Society of America Abstracts with Programs, v. 35, p. 92.
Melville, B.W., and Coleman, S.E., 2000, Bridge scour: Highlands Ranch, Colorado, Water Resources Publications, 550 p.

Metz, R., 1991, Scratch circles from the Towaco Formation (Lower Jurassic), Riker Hill, Roseland, New Jersey: Ichnos, v. 1, p. 233–235.
Metz, R., 1999, Scratch circles; a new specimen from a lake-margin deposit of the Passaic Formation (Upper Triassic), Douglassville, Pennsylvania: Northeastern Geology and Environmental Sciences, v. 21, p. 179–180.
Miller, R.F., and McGovern, J.H., 1996, Preliminary report of fossil fish (Actinopterygii; Palaeonisciformes) from the Lower Carboniferous Albert Formation at Norton, New Brunswick (NTS 21 H/12), *in* Carroll, B.M.W., ed., Current research 1996: New Brunswick Department of Natural Resources and Energy; Minerals, Policy and Planning Division, Mineral Resource Report MRR 97-4, p. 191–200.
Milner, A.R., Smithson, T.R., Milner, A.C., Coates, M.I., and Rolfe, W.D.I., 1986, The search for early tetrapods: Modern Geology, v. 10, p. 1–28.
Moore, P.D., 1987, Ecological and hydrological aspects of peat formation, *in* Scott, A.C., ed., Coal and coal-bearing strata: Recent advances: Geological Society [London] Special Publication 32, p. 7–15.
Mosbrugger, V., 1990, The tree habit in land plants: Berlin, Springer, Lecture Notes in Earth Sciences 28, 161 p.
Nakayama, K., Fielding, C.R., and Alexander, J., 2002, Variations in character and preservation potential of obstacle marks in the variable discharge Burdekin River north of Queensland, Australia: Sedimentary Geology, v. 149, p. 199–218, doi: 10.1016/S0037-0738(01)00173-7.
Pemberton, S.G., Flach, P.D., and Mossop, G.D., 1982, Trace fossils from the Athabasca Oil Sands, Alberta, Canada: Science, v. 217, p. 825–827.
Pfefferkorn, H.W., and Fuchs, K., 1991, A field classification of fossil plant substrate interactions: Neues Jahrbuch für Geologie und Paläontologie, Abhandlungen, v. 183, p. 17–36.
Pickerill, R.K., 1992, Carboniferous nonmarine invertebrate ichnocoenoses from southern New Brunswick, Eastern Canada: Ichnos, v. 2, p. 21–35.
Pigg, K.B., 2001, Isoetalean lycopsid evolution: From the Devonian to the present: American Fern Journal, v. 91, p. 99–114.
Piper, D.J., 1994, Late Devonian–earliest Carboniferous basin formation and relationship to plutonism, Cobequid Highlands, Nova Scotia: Ottawa, Geological Survey of Canada, Current Research 1994-D, p. 109–112.
Prentice, J.E., 1962, Some sedimentary structures from a Weald Clay sandstone at Warnham brickworks, Horsham, Sussex: Proceedings of the Geologists' Association, v. 73, part 2, p. 171–185.
Raymond, A., Parker, W.C., and Parrish, J.T., 1985, Phytogeography and paleoclimate of the Early Carboniferous, *in* Tiffney, B. H., ed., Geological factors and the evolution of plants: New Haven, Connecticut, Yale University Press, p. 169–222.
Read, C.B., and Mamay, S.H., 1964, Upper Paleozoic floral zones and floral provinces of the United States: U.S. Geological Survey Professional Paper 454-K, p. K1–K35.
Richter, R., 1926, Eine geologische Exkursion in das Wattenmeer: Natur und Museum, v. 56, p. 389–407.
Rigby, J.K., 1959, Possible eddy markings in the Shinarump Conglomerate of north-eastern Utah: Journal of Sedimentary Petrology, v. 29, p. 283–284.
Rygel, M.C., Gibling, M.R., and Calder, J.H., 2004, Vegetation-induced sedimentary structures from fossil forests in the Pennsylvanian Joggins Formation, Nova Scotia: Sedimentology, v. 51, p. 531–552, doi: 10.1111/j.1365-3091.2004.00635.x.
Sarjeant, W.A.S., and Mossman, D.J., 1978a, Vertebrate footprints from the Carboniferous sediments of Nova Scotia: A historical review and description of newly discovered forms: Palaeogeography, Palaeoclimatology, Palaeoecology, v. 23, p. 279–306, doi: 10.1016/0031-0182(78)90097-4.
Sarjeant, W.A.S., and Mossman, D.J., 1978b, *Peratodactylopus*, new name for the vertebrate footprint ichnogenus *Anticheiropus* Sarjeant and Mossman, 1978, non Hitchcock, 1865: Journal of Paleontology, v. 52, p. 1102.
Scheckler, S.E., 1985, Origins of the coal swamp biome: Evidence from the southern Appalachians: Geological Society of America Abstracts with Programs, v. 17, p. 134.
Scheckler, S.E., 1986a, Floras of the Devonian-Mississippian transition, *in* Broadhead, T. W., ed., Land plants: Notes for a short course: University of Tennessee Department of Geological Sciences Studies in Geology, v. 15, p. 81–96.
Scheckler, S.E., 1986b, Old Red Continent facies in the Late Devonian and Early Carboniferous of Appalachian North America: Annales de la Société Géologique de Belgique, v. 109, p. 223–236.
Scheckler, S.E., and Beeler, H.E., 1984, Early Carboniferous coal swamp floras from eastern USA (Virginia): International Organization of Palaeobotany Conference, 2nd, Edmonton, Alberta, Abstracts, p. 36.

Scotese, C.R., 2001, Atlas of earth history, Volume 1, Paleogeography: Arlington, Texas, Paleomap Project, 52 p.

Scott, A.C., and Calder, J.H., 1994, Carboniferous fossil forests: Geology Today, v. 10, p. 213–217.

Scott, A.C., and Rex, G.M., 1987, The accumulation and preservation of Dinantian plants from Scotland and its borders, in Miller, J., Adams, A.E., and Wright, V.P., eds., European Dinantian environments: Chichester, UK, John Wiley and Sons, p. 181–200.

Scott, A.C., Galtier, J., and Clayton, G., 1984, Distribution of anatomically preserved floras in the Lower Carboniferous of Western Europe: Transactions of the Royal Society of Edinburgh, Earth Sciences, v. 75, p. 311–340.

St. Peter, C., 1993, Maritimes Basin evolution; key geologic and seismic evidence from the Moncton Subbasin of New Brunswick: Atlantic Geology, v. 29, p. 233–270.

St. Peter, C., Johnson, S.C., Deblonde, C., and Lynch, G., 1997, Compilation map, geology, west central Nova Scotia, southeastern New Brunswick and western Prince Edward Island: Geological Survey of Canada Open-File Report 3521, scale 1:250 000, 1 sheet.

Stewart, W.N., 1947, A comparative study of stigmarian appendages and *Isoëtes* roots: American Journal of Botany, v. 34, p. 315–324.

Stewart, W.N., and Rothwell, G.W., 1993, Paleobotany and the evolution of plants: Cambridge, UK, Cambridge University Press, 521 p.

Tibert, N.E., and Scott, D.B., 1999, Ostracodes and agglutinated foraminifera as indicators of paleoenvironmental change in an Early Carboniferous brackish bay, Atlantic Canada: Palaios, v. 14, p. 246–260.

Underwood, J.R., and Lambert, W., 1974, Centroclinal cross strata, a distinctive sedimentary structure: Journal of Sedimentary Petrology, v. 44, p. 1111–1113.

Utting, J., 1987, Palynostratigraphic investigation of the Albert Formation (Lower Carboniferous) of New Brunswick, Canada: Palynology, v. 11, p. 73–96.

Utting, J., Keppie, J.D., and Giles, P.S., 1989, Palynology and age of the Lower Carboniferous Horton Group, Nova Scotia: Contributions to Canadian Palaeontology, Geological Survey of Canada Bulletin, v. 396, p. 117–143.

van de Poll, H.W., 1978, Paleoclimatic control and stratigraphic limits of synsedimentary mineral occurrences in Mississippian–Early Pennsylvanian strata of eastern Canada: Economic Geology and the Bulletin of the Society of Economic Geologists, v. 73, p. 1069–1081.

van de Poll, H.W., Gibling, M.R., and Hyde, R.S., 1995, Upper Paleozoic rocks, in Williams, H., ed., Geology of the Appalachian-Caledonian orogen in Canada and Greenland: Geological Society of America, Geology of North America, v. F-1, p. 449–566.

Van der Zwan, C.J., 1981, Palynology, phytogeography and climate of the Lower Carboniferous: Palaeogeography, Palaeoclimatology, Palaeoecology, v. 33, p. 279–310, doi: 10.1016/0031-0182(81)90023-7.

Wagner, R.H., 1979, Megafloral zones of the Carboniferous, in Sutherland, P. K., and Manger, W. L., eds., Compte Rendu, International Congress on Carboniferous Stratigraphy and Geology, 9th, Washington, D.C., and University of Illinois at Urbana-Champaign: Carbondale and Edwardsville, Southern Illinois University Press, p. 109–134.

Warren, A., and Turner, S., 2004, The first stem tetrapod from the Lower Carboniferous of Gondwana: Palaeontology, v. 47, p. 151–184.

Warren, J.W., and Wakefield, N.A., 1972, Trackways of tetrapod vertebrates from the Upper Devonian of Victoria, Australia: Nature, v. 238, p. 469–470, doi: 10.1038/238469a0.

Weir, S.L., 2002, Invertebrate ichnofossils of the Horton Bluff Formation in the collections of the Nova Scotia Museum of Natural History [bachelor's thesis]: Halifax, Nova Scotia, St. Mary's University, 89 p.

Wellman, C.H., Osterloff, P.L., and Mohiuddin, U., 2003, Fragments of the earliest land plants: Nature, v. 425, p. 282–285, doi: 10.1038/nature01884.

Wood, D., and Cameron, B., 1998, Fish trace fossils from the Horton Bluff Formation (Lower Carboniferous) of Nova Scotia: Atlantic Geology, v. 34, p. 81.

Wood, D.A., 1999, Vertebrate and invertebrate surficial trace fossils from the Horton Bluff Formation (Lower Carboniferous) near Avonport, Nova Scotia [bachelor's thesis]: Wolfville, Nova Scotia, Acadia University, 114 p.

MANUSCRIPT ACCEPTED BY THE SOCIETY 28 JUNE 2005

The Fayetteville Flora of Arkansas (USA): A snapshot of terrestrial vegetation patterns within a clastic swamp at Late Mississippian time

Michael T. Dunn
Department of Biological Sciences, Cameron University, Lawton, Oklahoma 73505, USA

Gar W. Rothwell
Gene Mapes
Department of Environmental and Plant Biology, Ohio University, Athens, Ohio 45701, USA

ABSTRACT

The Fayetteville Formation of northwestern Arkansas (upper Mississippian/middle Chesterian) contains two compression plant fossil assemblages (one in situ) that represent plant communities, and an allochthonous permineralized assemblage recovered from marine strata that represents the landscape. This preservation of spatial ecological subunits (communities) nested within a larger subunit (landscape) provides a snapshot of vegetation patterns within a Late Mississippian clastic swamp. Fifteen whole plants are recognized. Seed ferns are the most speciose group and lycopsids account for most biomass. Seed fern taxa known only as permineralized specimens include one canopy tree (*Megaloxylon*), two understory trees, and five herbaceous layer plants. Two herbaceous layer seed ferns are observed only as compressions. Lycopsids are represented as two canopy trees that are known from both permineralizations and compressions. *Archaeocalamites* is also known from both permineralizations and compressions but was an understory tree. Ferns are rare and are preserved only as fragments of permineralized rachises from two species. As revealed by the in situ compression assemblage, the two species of lycopsid canopy trees co-occur and they formed communities that occupied ever-wet bottomlands, with *Archaeocalamites* occupying the understory, and a single species of seed fern comprising the herbaceous layer. Lycopsids do not co-occur with *Megaloxylon*. *Megaloxylon* probably formed a second community type in somewhat water-stressed areas of the swamp with an understory of small arborescent seed ferns, some *Archaeocalamites*, and an herbaceous-layer seed fern. Ferns probably formed a third type of community in disturbed sites.

Keywords: clastic swamp, community, compression, landscape, Mississippian, permineralization.

INTRODUCTION

The Fayetteville Formation of northwestern Arkansas is primarily a marine black shale unit from which a number of allochthonous, anatomically preserved plant morphotaxa have been described (e.g., Mapes, 1966; Taylor and Eggert, 1967a, 1968; Mapes and Rothwell, 1980; Mapes, 1985; Tomescu et al., 2001; Dunn et al., 2002, 2003a, 2003b). Until the Pennsylvanian, plants were restricted to ever-wet to occasionally dry lowlands (Scott, 1980), so the plants that produced these organs probably grew in several lowland communities and plant remains were transported into the marine basin and ultimately preserved offshore. Cumulatively, these communities represent the terrestrial landscape. Within the marine shales is a terrestrial/deltaic lens that contains at least two compression plant assemblages—the Wedington Sandstone assemblage (White, 1937) and the Wedington Shale assemblage (Eggert and Taylor, 1971; Dunn, 2004)—that represent two distinct plant communities. The occurrence of plant communities, nested within a larger landscape, has not been previously recognized in North American Mississippian plant fossil assemblages and provides a unique snapshot of terrestrial vegetation structure in the Late Mississippian, a very important (Scott et al., 1984) but poorly known episode in the evolution of life on land.

Land plants experienced the first of several major radiations from the Late Silurian to the Early Devonian (Chaloner and Sheerin, 1979; Gensel and Andrews, 1987; DiMichele and Hook, 1992); however, plant community structure was relatively simple. Communities were restricted to persistent wetlands and watercourses, were unstratified, and consisted of small, monospecific patches of plants, or low-diversity patches with a single dominant species (Andrews et al., 1977; Edwards, 1980; Edwards and Fanning, 1985). From the Early to the Late Devonian, the number of plant genera remained relatively constant, with new taxa replacing those going extinct, and it is among these new taxa that modern plant organography evolved (Gensel and Andrews, 1987).

The second major plant radiation spanned the Late Devonian and Early to Middle Mississippian and established the major lineages of vascular plants, i.e., Lycopsida, Sphenopsida, Pteropsida, Progymnospermopsida, and Spermatopsida (DiMichele and Bateman, 1996). This radiation led to the evolution of the structurally modern stratified forest, consisting of arborescent canopy and understory layers and an herbaceous layer of scrambling vinelike plants and/or lianas. However, diversity within each community remained low (Bateman 1991; DiMichele and Bateman, 1996), with usually just one species occupying each tier in each stratified community (Scheckler, 1986a; DiMichele and Hook, 1992). By the Visean, community heterogeneity included peat-forming communities, clastic lycopsid communities, fern communities, and several different gymnosperm communities (Rex, 1986; Rex and Scott, 1987). Like diversity patterns in these ancient plant communities, dominance patterns in the early forests were distinctly different from those of modern forests. Most modern plant communities are dominated by a single group, the flowering plants, whereas in pre-Late Carboniferous communities, community structure (broadly defined as canopy, understory, and herbaceous layer) was divided among the Lycopsida, Sphenopsida, Pteropsida, and Spermatopsida (DiMichele and Bateman, 1996; DiMichele and Phillips, 1996; DiMichele et al., 2001). This class-level partitioning of the ecosystem persisted until the end of the Westphalian (middle Pennsylvanian) in much of the world (DiMichele et al., 2001), when a series of climate-driven extinctions led to the rise of seed plant, and ultimately angiosperm dominance (DiMichele and Bateman, 1996).

Patterns of evolution within the older, "primeval" communities (after DiMichele et al., 2001) are relatively well documented for the Early and Middle Mississippian (e.g., Scott, et al., 1984), and are particularly well known for the Early Pennsylvanian (DiMichele and Hook, 1992), but little is known about the Late Mississippian terrestrial ecosystem (Taylor and Eggert, 1967b; Mapes and Rothwell, 1980; Tomescu et al., 2001). Thus, hypotheses as to the mode and tempo of late Paleozoic ecosystem evolution are hindered by a lack of pertinent data. The purpose of this paper is to use the unique preservation of the vegetation of the Fayetteville Formation of Arkansas to illustrate Late Mississippian plant ecosystem structure and diversity in a clastic swamp, at the community and landscape levels.

GEOLOGICAL SETTING

The Fayetteville Formation extends for ~46,800 km^2 (Cate, 1962) on the south flank of the Ozark Dome (Meeks, 1997) in eastern Oklahoma and northwestern Arkansas (Fig. 1). The strata are divided into upper and lower Fayetteville Shale units (Fig. 2) by the terrestrial/deltaic Wedington Sandstone Member (Saunders, Manger, and Gordon, 1977). Both units of the Fayetteville Shale are black, pyritic, and organic-rich, and are interpreted as a deep muddy shelf deposit (Handford and Manger, 1993). The upper unit contains a benthic faunal assemblage and the lower unit contains a nektonic assemblage, primarily cephalopods (Meeks et al., 1997) and the permineralized plant remains of this report. On the basis of cephalopods, conodonts, and foraminifers (Meeks et al., 1997) and miospores (Owens et al., 1979; Rezaie, 1980) the strata are dated as Chesterian (upper Mississippian), equivalent to the Pendelian (E_1) Stage of the Namurian A, or lower Serpukhovian.

The Wedington Member contains two plant compression assemblages (Fig. 3) and may represent a delta system fed by a low-sinuosity braided stream (Price, 1981); however, conclusive data on the sedimentology of this unit are not available. Plant remains from the Wedington Member were first reported by White (1937) from what were interpreted as nearshore marine strata (the Wedington Sandstone assemblage of this report): the matrix is a coarse to conglomeratic, ripple-marked and cross-bedded sandstone. Specimens are fragmentary and show evidence of "water wear" (White, 1937) and, as discussed below, may represent more than one plant community.

A second compression assemblage (the Wedington Shale assemblage of this report) is interpreted as an (at least in part) in situ plant community, on the basis of the presence of lycopsid rooting organs with attached rootlets (Fig. 4A). The matrix of the Wedington Shale is a fissile, gray to tan siltstone to claystone lens.

METHODOLOGY

Permineralized Fossils

Permineralized plant fossils were recovered from marine black shales of the lower Fayetteville shale unit where these rocks

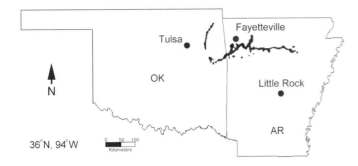

Figure 1. Location of the Fayetteville Formation. OK—Oklahoma; AR—Arkansas.

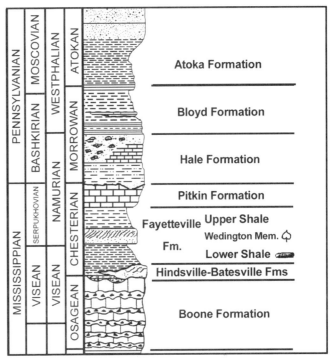

Figure 2. Stratigraphic section and correlation of strata in northwestern Arkansas. Modified from Meeks (1997).

are exposed by construction, or more commonly from stream channels that erode through the shales. Plant remains are relatively rare and are recovered as part of an extensive invertebrate fossil collecting effort, primarily by R.H. and G. Mapes and colleagues. Fossilization is by calcareous cellular permineralization (Schopf, 1975) with abundant pyrite. One hundred forty-nine specimens were analyzed for this report, with numerous other specimens lost to pyrite oxidation. Detailed analysis of morphotaxa, analytical techniques, and locality data are reported in Dunn (2004), and Dunn et al., (2002, 2003a, 2003b).

Compression Fossils

The Wedington Shale assemblage is ~25 km southeast of Fayetteville, Arkansas, and was exposed during construction of the Lincoln Dam. The site was originally collected in 1969 by a team from the University of Illinois at Chicago Circle, led by T.N. Taylor and D.A. Eggert. Plant fossils were "grab sampled" from the site that in 1969 measured ~100 m^2; however, individual bedding planes could not be quantified (R.H. Mapes, 2002, personal commun.). The assemblage was revisited by M.T.D., R.H. Mapes, D. Kidder, and K. Tanabe in the summer of 2001; however, the site is revegetated and fossils are now only exposed in eroded gullies. Seven hundred twenty hand samples were analyzed for this project and abundance data are calculated after Pfefferkorn et al. (1975). This method treats each hand sample as a quadrat and scores the presence of each taxon in the sample equally, regardless of the number of each taxon in the sample. Biomass estimates are based on frequency analysis of abundance by calculating the percent of the total number of quadrats in which a species occurs. Because specimens were not collected under rigorous ecological sampling techniques, only a general impression can be reported here, rather than the results of a more in-depth statistical analysis.

The Wedington Sandstone assemblage is reported in this paper as described by White (1937) and subsequent revisions to his taxonomy (e.g., Lacey and Eggert, 1964; Eggert and Taylor, 1971; Dunn, 2004). Those sections are no longer exposed and the type specimens were not reanalyzed. Both the total number of

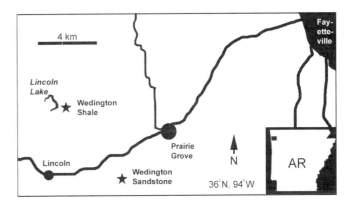

Figure 3. Location of the Wedington Shale and Wedington Sandstone compression assemblages. AR—Arkansas.

specimens analyzed by White and quantitative data for each morphotaxon are unknown; therefore abundance data for compression specimens reflect only the Wedington Shale assemblage.

Whole plant species are identified by a modified "minimum species concept" (Knoll et al., 1979; Knoll 1986) based on stem morphotaxa because of the reliability of "real taxon" identification based on stems in this assemblage (Dunn, 2004). Organs that cannot be assigned to a particular stem species are divided among stems of the same higher taxon, based on the proportion of said stems in the assemblage. For example, the 19 specimens of

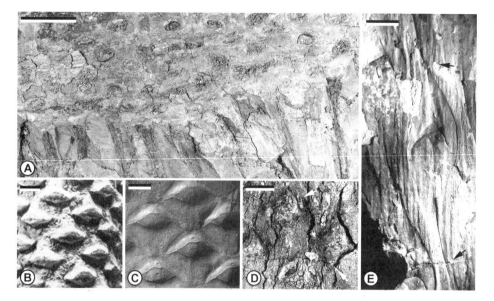

Figure 4. Arborescent lycopsids. (A) *Stigmaria ficoides* with attached rootlets from the Wedington Shale; M3350, OUPH15084; scale bar = 2 cm. (B) Lycopsid sp. A, permineralized leaf cushions.; M218; scale bar = 6 mm. From Mapes (1966), reprinted with permission. (C) *Lepidodendron volkmannianum*, compression leaf cushions. Note imbrication, rhombic shape, and distal leaf traces on 4B and 4C; M3822, OUPH15077; scale bar = 3 mm. (D) Lycopsid sp. B, permineralized leaf cushions; M3159, OUPH15076; scale bar = 6 mm. (E) *Lepidodendron veltheimii*, compression leaf cushions. Note widely spaced cushions with sinuous acute apices (at arrows), and central leaf traces on 4D and 4E; M3792, OUPH15079; scale bar = 4 mm.

permineralized lycopsid cones (Dunn, 2004) are divided among the two lycopsid stems according to the proportion of occurrence of lycopsid sp. A. (46) to lycopsid sp. B (3).

RESULTS AND DISCUSSION

Whole Plants

Forty-one morphospecies were identified within the Fayetteville assemblages: 19 permineralized species and 22 compression species that represent at least 15 whole plants (Dunn, 2004). These 15 whole plants include three canopy trees, three understory trees, seven herbaceous layer seed ferns, and two ferns (Table 1). Four classes of vascular plants are represented (Table 2); Spermatopsida (all seed ferns) are the most species rich (ten species), followed by Lycopsida and Pteropsida (two species each), and Sphenopsida (one species).

The dominant canopy tree in the permineralized (landscape) assemblage as well as both compression (community) assemblages is represented by the permineralized stem Lycopsid sp. A and the compression stem *Lepidodendron volkmannianum* Sternberg. These two organ species are interpreted as representing different modes of preservation of the same whole plant on the basis of shared leaf cushion morphology and their dominance in both assemblages. Leaf cushions of both morphospecies are in vertical and horizontal alignment, imbricated, and rhombic in shape, they vary in size on the same specimen, and they have leaf trace scars at the upper part of the cushion (Fig. 4B, 4C). This species constitutes 43% of the permineralized assemblage and 52% of the Wedington Shale compression assemblage (Table 1).

A second arborescent lycopsid co-occurs with lycopsid sp. A / *Lepidodendron volkmannianum* but is relatively rare. The permineralized form, Lycopsid sp. B, and the compression form, *Lepidodendron veltheimii* Sternberg, both produced widely spaced leaf cushions with sinuous, acute proximal apices (Fig. 4D, 4E at arrows). In the permineralized assemblage lycopsid sp. B constitutes 3% of the specimens, and in the compression assemblage *L. veltheimii* constitutes 8% of the specimens (Table 1).

Megaloxylon wheelerae Mapes has not been reconstructed as a whole plant (Mapes, 1980), but it is interpreted as a canopy tree on the basis of the production of abundant biomass, the dense wood, and the large size of the recovered stems. This species constitutes 21% of the permineralized assemblage and has not been identified in the compression assemblages (Table 1).

Carboniferous sphenopsids are commonly considered small understory trees (Darrah, 1960; Taylor and Taylor, 1993). In North America, all specimens of *Archaeocalamites* have been placed in the permineralized species *Archaeocalamites esnostensis* or the compression species *Archaeocalamites radiatus* (Mamay and Bateman, 1991). Permineralized and compression specimens from the Fayetteville Formation fall within the ranges of variation of these two species respectively (Dunn, 2004) and are interpreted as representing a single whole-plant species. *Archaeocalamites radiatus* is present in 10% of the compression samples and *A. esnostensis* constitutes 3% of the permineralized specimens (Table 1).

Additional understory trees include *Quaestora amplecta* Mapes and Rothwell and a new unnamed lyginopterid; these plants represent 10% and 3% of the permineralized assemblage, respectively, and have not been identified in the compression assemblage. These two plants are known only from stem specimens, and the production of dense wood suggests upright growth architecture, but the relatively small size of the stems along with low biomass suggests relatively small trees.

TABLE 1. DISTRIBUTION OF WHOLE PLANTS

Taxon	Abundance: permineralized	Frequency: permineralized	Abundance: compression	Frequency: compression	Functional Group
Lycopsid sp. A-/*L. volkmannianum*	64	43	376	52	Canopy tree
Lycopsid sp. B/*L. veltheimii*	4	3	55	8	Canopy tree
Archaeocalamites	4	3	69	10	Understory tree
Etapteris	2	1	NA	NA	Herbaceous fern
Ankyropteris brongniartii	1	<1	NA	NA	Herbaceous fern
Lyginopteris royalii	1	<1	NA	NA	Herbaceous seed fern
Rhetinangium arberi	4	3	NA	NA	Herbaceous seed fern
Trivena arkansana	7	5	NA	NA	Herbaceous seed fern
Megaloxylon wheelerae	32	21	NA	NA	Canopy tree
Arborescent lyginopterid	5	3	NA	NA	Understory tree
Vinelike lyginopterid	1	<1	NA	NA	Herbaceous seed fern
Quaestora amplecta	15	10	NA	NA	Understory tree
Medullosa steinii	9	6	NA	NA	Herbaceous seed fern
Sphenopteris sp.	NA	NA	450	63	Herbaceous seed fern
Sphenopteris mississippiana	NA	NA	NA	NA	Herbaceous seed fern

Notes: Abundance permineralized is the number of whole plants of each species in the assemblage as calculated by a modified minimum species concept (see text for details). Frequency permineralized is the percentage of each species in the assemblage (*n* = 149). Abundance compression is the total number of hand samples (quadrats) that contain at least one specimen of that species. Frequency compression is the percentage of hand samples that contain at least one specimen of that species (*n* = 720). Compression data are based only on the Wedington Shale assemblage because abundance data are not reported by White (1937) for the Wedington Sandstone assemblage.

Seven herbaceous layer seed ferns have been identified, including two compression species—*Sphenopteris* sp. and *S. mississippiana* White—and five permineralized stem morphotaxa—*Lyginopteris royalii* Tomescu et al., *Rhetinangium arberi* Gordon, *Trivena arkansana* Dunn et al., *Medullosa steinii* Dunn et al., and an as yet unnamed lyginopterid (Dunn, 2004). The growth architecture of these plants has been interpreted to be clinging, climbing, or vinelike on the basis of a variety of characters, including the production of manoxylic wood with numerous tall multiseriate rays, long internodes, and narrow diameter stems with large rachis bases (Tomescu et al., 2001; Gordon, 1912; Dunn et al., 2003a, 2003b; Dunn, 2004). None of the specimens from the Fayetteville assemblage have been observed in connection to a substrate, thereby making it is impossible to determine with certainty if these plants were ground cover or obligate vines, so they are placed in the broadly defined herbaceous layer. All of these permineralized stem taxa are relatively rare: *Lyginopteris royalii*, <1%; *Rhetinangium arberi*, 3%; *Trivena arkansana*, 5%; *Medullosa steinii*, 6%; and the unnamed lyginopterid, <1%. The compression species *Sphenopteris* sp. is abundant, occurring in 63% of the Wedington Shale assemblage samples, but this morphospecies cannot be correlated with any of the permineralized taxa. Abundance data are unknown for *S. mississippiana* from the Wedington Sandstone (Table 1).

TABLE 2. LANDSCAPE SPECIES DIVERSITY BY CLASS

Class	No. of Species
Spermatopsida (all seed ferns)	10
Lycopsida	2
Pteropsida	2
Sphenopsida	1

Ferns are the rarest taxa in the assemblage (Table 1) and are known only from permineralized rachis fragments. One specimen of the rachis of *Ankyropteris brongniartii*, and two specimens of *Etapteris* sp., the rachis of *Zygopteris* (Sahni, 1932; Baxter, 1952; Dennis, 1974), have been recovered. No compression fern remains have been identified.

Community Assemblages

Two or possibly three types of community species assemblages have been identified within the clastic swamp biome of the Fayetteville assemblage. These are lycopsid canopy communities in ever-wet bottomlands, *Megaloxylon* canopy communities in intermittently water-stressed areas of the swamp, and possibly fern communities colonizing disturbed sites.

Lycopsid Canopy Communities

Wedington Shale. The Wedington Shale compression assemblage represents a lycopsid canopy community, and these trees are interpreted as being preserved in situ on the basis of the presence of the lycopsid rooting organ *Stigmaria* with attached rootlets (Fig. 4A). The Wedington Shale has a species richness of 4. By functional group, this breaks down as follows: canopy (2), arborescent understory (1), and herbaceous layer (1). Species richness by class is as follows: Lycopsida (2), Sphenopsida (1), Spermatopsida (1), Pteropsida (0).

Arborescent lycopsids are commonly reported to have inhabited the wettest parts (bottomlands) of the coastal swamps of the Namurian A (Scheckler, 1986a, 1986b; DiMichele and Phillips, 1996). In the Wedington Shale community, Lycopsid sp. A/*Lepidodendron volkmannianum* and Lycopsid sp. B/*Lepidodendron veltheimii* share the canopy tier, with the former dominant and the latter relatively rare (Table 1). Gastaldo (1987) reports that in a lycopsid canopy community in a Westphalian A clastic swamp assemblage from Georgia, USA, this tier is generically partitioned with *Lepidophloios* found in the wettest part of the swamp, *Lepidodendron* capable of intermediate water tolerance, and *Sigillaria* found in the drier areas. But as mentioned above, sampling of the Wedington Shale was not spatially controlled, so it is not known how or whether the canopy was partitioned in this Namurian A swamp.

The only observed component of the arborescent understory in the Wedington Shale assemblage is *Archaeocalamites radiatus*. This plant was a much branched small tree or shrub (Fig. 5) and is present in 10% of the samples. Very little is known about the biology and phylogeny of Mississippian sphenopsids (Smoot et al., 1982; Stein et al., 1984), but Pennsylvanian sphenopsids probably grew in dense stands (Scott, 1979) of clonal shoots produced by a common rhizome (Tiffney and Niklas, 1985; Bateman, 1988). Dispersal by floods was common (Scheckler, 1986a), and these plants regenerated readily after such events (Gastaldo, 1992). It is possible that Mississippian sphenopsids exhibited patterns of growth and reproduction analogous to those of Pennsylvanian sphenopsids, as has been hypothesized for roots (Rex, 1986), and habitat partitioning (Scott, 1980). Sphenopsids were common throughout the Mississippian wetlands but dominated only in areas of high disturbance such as riparian and lacustrine zones and in flood basins (Scheckler, 1986a; DiMichele and Phillips, 1996; DiMichele et al., 2001; but see Rex, 1986).

The herbaceous layer was most likely occupied by a lyginopterid seed fern that produced foliage assigned to *Sphenopteris* sp. (Dunn, 2004). These specimens are present in 63% of the samples and constitute over 95% of the non-lycopsid, non-sphenopsid stems over 5 mm in diameter. Axes of *Sphenopteris* sp. are up to 25 mm in diameter and are flexuous (Fig. 6A) with up to five orders of pinnae. Young pinnae unfurl by circinate vernation and are covered by numerous trichomes (Fig. 6B). Trichomes are slender, acute, and up to 6 mm long, and they often abscise, leaving a distinct scar (Fig. 6C); scar density may be as high as 300/cm^2. Decorticated specimens exhibit parallel vertical bands of cortical sclerenchyma that do not anastomose (Fig. 6D). Pinnules are constricted at the base with three to (usually) five rounded lobes (Fig. 6E), but are often overlapping and may appear to be fused (Fig. 6F at arrow). The abundance of this foliage and the lack of evidence of transport (Dunn, 2004) suggest that this plant grew in or very near the lycopsid-dominated bottomland community.

Figure 5. *Archaeocalamites radiatus*; note abundant branching; M3228, OUPH15086; scale bar = 2 cm.

Cardiopteridium sp. (Fig. 6G) is present in 13% of the samples and represents sterile foliage with pinnae that branch at least once and pedicellate pinnules. *Cardiopteridium* is unlike the foliage of any known Mississippian filicalean ferns (Galtier, 1981) and may have been produced by either pteridosperms or progymnosperms (Gensel, 1988). Scheckler (1986a) suggests that *Cardiopteridium* preserved in the coastal coal swamps of the mid-Namurian A Bluefield and Hinton Formations of West Virginia was produced on plants that grew upstream and was transported into the lycopsid canopy community by storms. Wedington Shale specimens are most commonly found abscised from the pedicel and often occur in dense mats (Fig. 6H), suggesting pre-depositional transport, perhaps from an upstream community as suggested by Scheckler. For this reason they are not interpreted here as part of the lycopsid canopy community and are only included for completeness.

Wedington Sandstone. The Wedington Sandstone assemblage, at least in part, represents a lycopsid canopy community that was deposited in a near shore marine environment and may therefore contain parts of additional communities. The Wedington Sandstone has a species richness of 8 (including four fragmentary foliage species); broken down by functional group, species richness is as follows: canopy (2), arborescent understory

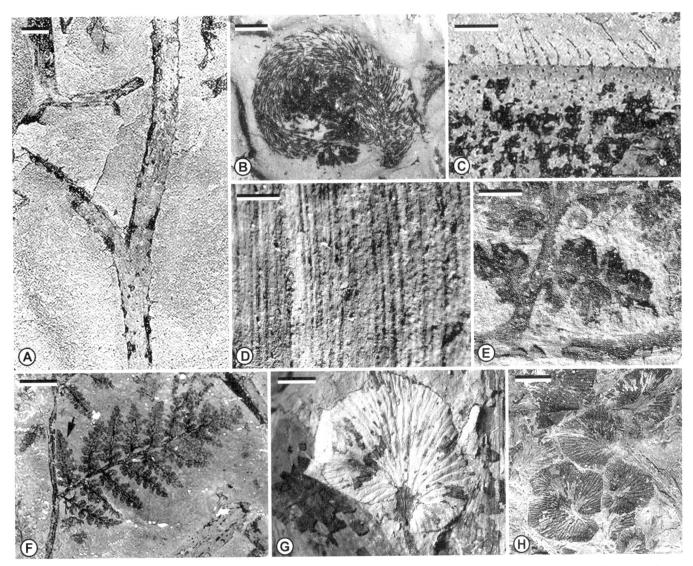

Figure 6. Compression foliage. (A) *Sphenopteris* sp., flexuous, branching axis; M3051, OUPH16000; scale bar = 1 cm. (B) *Sphenopteris* sp., trichome-covered crozier; M3964, OUPH15098; scale bar = 2 mm. (C) *Sphenopteris* sp., axis with trichomes and trichome scars; M3240, OUPH15096; scale bar = 3 mm. (D) *Sphenopteris* sp., axis with *Sparganum*-type outer cortex, note parallel vertical bands of sclerenchyma that do not anastomose; M3310, OUPH15097; scale bar = 1 mm. (E) *Sphenopteris* sp., pinnule; note constricted base and rounded lobes; M3776, OUPH15100; scale bar = 1 mm. (F) *Sphenopteris* sp., fourth- and fifth-order pinnae. Note overlapping pinnules that falsely appear to be fused on fifth-order pinnule (at arrow); M3632, OUPH15101; scale bar = 5 mm. (G) *Cardiopteridium* sp. with attached pedicel; scale bar = 2 mm. (H) *Cardiopteridium* sp., mat of abscised pinnules; M3292, OUPH15214; scale bar = 4 mm.

(1), herbaceous layer (1), unknown growth architecture (4); no quantitative data are available.

The most important locality of the Wedington Sandstone assemblage is ~6 km southeast of the Wedington Shale locality (Fig. 3). The canopy is composed of *Lepidodendron volkmannianum* and *L. veltheimii* and the understory is occupied by *Archaeocalamites radiatus* (Lacey and Eggert, 1964), as in the Wedington Shale assemblage. However, the herbaceous layer is occupied by a different species of *Sphenopteris* in each of the two compression assemblages. White (1937) described *Sphenopteris mississippiana* from the Wedington Sandstone for foliage with several orders of pinnae and pinnules that conform to *Sphenopteris* (White, 1937, Plate 4, Figs. 11, 13–15, 17, 18, 21–23). However, White's species is found attached to axes with anastomosing (*Dictyoxylon*-type) sclerenchyma bands (White, 1937, Plate 4, Figs. 4, 5) and stout "spines" (White, 1937, Plate 4, Figs. 36, 39) and, as noted above, *Sphenopteris*-bearing axes from the Wedington Shale produce parallel vertical bands of cortical sclerenchyma (Fig. 6D) that do not anastomose (*Sparganum*-type), and they bear numerous slender trichomes or trichome dehiscence scars (Fig. 6C). Additional foliage White (1937) identified (i.e., *Neuropteris* sp., *Rhodea* cf. *subpetioleata*, *Sphenopteris* cf.

schimperiana, *Cardiopteridium hirta*), from an unknown number of fragmentary specimens, suggests additional whole-plant species, but because this is an allochthonous shallow marine assemblage it is not known how many communities are represented.

Megaloxylon Communities

Although seed plants most likely arose in wetland settings (Gillespie et al., 1981; Scheckler, 1986a, 1986b; but see DiMichele et al., 1989), it was in areas of periodic moisture limitation that these plants radiated and dominated during the Mississippian (Matten et al., 1984; Retallack and Dilcher, 1988; DiMichele and Phillips, 1996) and in the Fayetteville flora *Megaloxylon wheelerae* probably occupied the canopy niche in such drier sites. The abundance (21%) of *M. wheelerae* in the permineralized assemblage, but its absence in the compression (lycopsid community) assemblages, suggests that it was a widely distributed species but did not co-exist with the arborescent lycopsids in the ever-wet bottomlands.

As noted above *Quaestora amplecta* and an unnamed lyginopterid are interpreted as understory trees, and because no evidence of these plants was found in the lycopsid-dominated compression assemblages, they may have inhabited drier sites where *Megaloxylon* was the dominant canopy tree. However, it is also possible that these plants grew in as yet unknown communities. As in the lycopsid dominated bottomlands, *Archaeocalamites* was probably common along streams, shores, and at the edges of these drier communities within the clastic swamp, but may also have been a minor component within the canopy communities. However, because no in situ evidence of the *Megaloxylon* community type has been observed, habitat partitioning and plant distribution within these communities remain unknown.

Fern Communities

As mentioned above, ferns are absent from the compression assemblages and are quite rare in the permineralized assemblage, consisting of only fragments of rachises of *Zygopteris* and *Ankyropteris*. Like other Carboniferous species of *Zygopteris* (see Dunn, 2004), those from the Fayetteville assemblage probably had a prostrate growth architecture (Dennis, 1974). The growth architecture of *Ankyropteris brongniartii* (Renault) Bertrand may have been that of a liana (Scott and Galtier, 1985; Rothwell, 1996; Rößler, 2000) or a "facultative" climber, growing prostrate across the ground until a climbing substrate was encountered (DiMichele and Phillips, 1996).

In general, the Namurian represents a gap in our knowledge of fern biology, evolution, and ecology (Scott and Galtier, 1985). However most upper Paleozoic ferns are interpreted to be ecological opportunists (Rothwell, 1987). Many are reported to have been colonizers of disturbed sites (DiMichele et al., 2001 and references therein), and particularly the zygopterid ferns may have been opportunists able to tolerate a range of conditions from peat swamps to fire- and ash-stressed volcanic settings (Rex and Scott, 1987). Therefore, if previous interpretations of upper Paleozoic fern biology and ecology are correct, the fern remains in the Fayetteville assemblage may represent one or more disturbance communities within the larger landscape.

Landscape Patterns

After DiMichele's (1994) definition, a community is an assemblage of species, and species assemblages form multi-community landscapes. Therefore, the landscape of the Fayetteville Formation consists of a time-averaged assemblage of plants from the two compression communities discussed above, plus an unknown number of additional communities represented by allochthonous permineralized plant fossils. The establishment of landscapes composed of ecologically sorted communities (i.e., lycopsid canopy communities in the ever-wet bottomlands, seed-fern canopy communities in drier parts of the swamp, and fern communities colonizing disturbed sites) is well documented for Pennsylvanian floras and was apparently established by the late Tournaisian (DiMichele and Hook, 1992 and references therein). This pattern of ecosystem partitioning is suggested in the Fayetteville assemblage, and judging from biomass, lycopsid canopy communities were dominant, with seed-fern canopy communities subordinate but not rare, and fern communities quite rare. Overall landscape species richness in the Fayetteville Formation is 15; species richness broken down by functional group is as follows: canopy (3), arborescent understory (3), herbaceous layer [including ferns] (9); landscape species richness by class: Spermatopsida [all seed ferns] (10), Lycopsida (2), Pteropsida (2), Sphenopsida (1).

Lycopsid sp. A/*L. volkmannianum* is the most abundant species in all of the Fayetteville assemblages where quantity data are known (Table 1). Organs produced by this plant constitute 43% of the permineralized specimens, and are present in 52% of the Wedington Shale samples. In addition, this plant is present in the Wedington Sandstone community, although quantitative data are unavailable. Lycopsid sp. B/*L. veltheimii* was also part of the canopy niche of lycopsid canopy communities but in much lower abundance. This species constitutes 3% of the permineralized specimens, is present in 8% of the Wedington Shale samples, and is also present in the Wedington Sandstone community (Table 1).

It is interesting that Lycopsid sp. A/*L. volkmannianum* and Lycopsid sp. B/*L. veltheimii* apparently share the canopy niche in the bottomlands of the ecosystem because (1) previous reports have suggested that Namurian A swamp floras were dominated by a single species of arborescent lycopsid, and (2) the dominant species is thought not to have arisen from a species within the flora, but rather to have migrated from an adjacent assemblage and outcompeted the established dominant (Scheckler, 1986a). If that hypothesis is correct, the presence of two arborescent lycopsids, distributed across the landscape but in distinctly different abundance, may indicate either the beginning or the end of a cycle of canopy dominance turnover. However, it is also possible that this pattern documents the origin of habitat partitioning that has been reported in lycopsid dominated communities of the Pennsylvanian (Phillips and Peppers, 1984; DiMichele and Phillips, 1985, 1994; Gastaldo, 1987).

Archaeocalamites is present in both the permineralized and compression assemblages, and was apparently widely distributed across the landscape. The difference in abundance between the permineralized assemblage (3%) and the Wedington Shale compression assemblage (10%) may suggest heterogeneous distribution; however, this discrepancy may be an artifact of taphonomy. This is because, unlike extant *Equisetum*, which disarticulates at nodes where a membrane forms a buoyant capsule (personal observation), *Archaeocalamites* does not disarticulate at nodes. Instead, *Archaeocalamites* often has nodes that are differentially preserved and breakage occurs internodally, leaving "frayed" ribs (Fig. 7), that may not transport readily.

As noted above, the arborescent seed fern *Megaloxylon wheelerae* most likely occupied the canopy niche in communities in drier areas of the swamp, possibly with *Quaestora amplecta* and the unnamed lyginopterid occupying the understory. However, because these plants are known only from allochthonous permineralized remains, the distribution of these seed-fern taxa is speculative.

At the landscape level, the herbaceous layer is shared by at least nine plants including two ferns and seven seed ferns. As noted above, ferns are the rarest group in the assemblage and may have formed a third community type by colonizing disturbed sites, but because of the rarity of fern specimens in the assemblage, this interpretation is speculative.

Seed ferns with vinelike, or clinging-climbing, growth architecture constitute the most speciose group in the assemblage, with seven whole plants identified at the landscape level. However, in the lycopsid canopy communities the herbaceous layer is occupied by a single, but different, species, in each community: *Sphenopteris* sp. in the Wedington Shale and *Sphenopteris mississippiana* in the Wedington Sandstone. Therefore, despite relatively low diversity of canopy and understory trees across the landscape, herbaceous-layer plants are diverse and heterogeneously distributed at this spatial ecological level. This suggests that herbaceous-layer seed ferns are the source of much diversity and innovation in upper Mississippian clastic swamps. In addition, the obvious disparity in diversity between spatial subunits of the ecosystem serves as a reminder that when measurements of ancient diversity are attempted, the ecological subunit must be clearly defined biologically to avoid distorting true diversity patterns.

CONCLUSIONS

The stratigraphic position of the Fayetteville Formation, which was deposited during the end of the Devonian-Mississippian vascular plant radiation but before the establishment of Pennsylvanian coal swamps, along with the presence of permineralized and compression plant remains representing distinct spatial subunits of the ecosystem, reveals important details about upper Mississippian terrestrial ecosystem evolution:

1. Within-community diversity in lycopsid canopy communities is low at the herbaceous and understory layers (one species per tier), but the canopy consists of two co-occurring species. The

Figure 7. *Archaeocalamites* at node; note branch scar (br) and frayed ribs (at arrows). M3276, OUPH15090; scale bar = 4 mm

herbaceous layer is occupied by a single scrambling, or vinelike, seed fern; the understory is occupied by *Archaeocalamites*; and the canopy is shared by two arborescent lycopsids—one species is common and one is rare.

2. Community heterogeneity includes the dominant lycopsid canopy communities in the ever-wet bottomlands, *Megaloxylon* canopy communities in drier regions of the swamp, and fern communities in disturbed sites.

3. Species richness across the landscape includes at least 15 whole plants. Seed ferns are the most diverse group, consisting of the canopy tree *Megaloxylon wheelerae*; two understory trees, *Quaestora amplecta* and an unnamed lyginopterid; and seven herbaceous layer plants, *Lyginopteris royalii*, *Rhetinangium arberi*, *Medullosa steinii*, *Trivena arkansana*, an unnamed lyginopterid, *Sphenopteris* sp., and *S. mississippiana*. Lycopsids are represented by two canopy trees that dominate the ecosystem in terms of biomass: Lycopsid sp. A/*L. volkmannianum*, which is common, and Lycopsid sp. B/*L. veltheimii*, which is relatively rare. Two ferns, *Ankyropteris brongniartii* and *Zygopteris*, are present but are rare and known only from fragments of rachises. *Archaeocalamites* is the only representative of the Sphenopsida.

4. The high diversity of herbaceous layer seed ferns across the landscape, but low diversity of this functional group at the community level, and the abundance of canopy lycopsids across the landscape, but low diversity suggests different modes and tempos of evolution between classes and/or across ecospace.

5. Continued investigation, particularly involving re-collection of the Wedington Shale under rigorous ecological protocol, is needed to determine habitat partitioning and plant distribution within the lycopsid canopy community.

ACKNOWLEDGMENTS

This research was supported in part by the Geological Society of America, Graduate Student Research Grant 6876-01; Botanical Society of America, Karling Graduate Student Research Award; Paleontological Society of America, Student Grant-in Aid; and Ohio University, Student Enhancement Award SEA02-08 to M.T.D.

We thank two anonymous reviewers, and the editors, W.A. DiMichele and S.F. Greb, for comments that greatly improved this manuscript.

REFERENCES CITED

Andrews, H.N., Casper, A.E., Forbes, W.H., Gensel, P.G., and Chaloner, W.G., 1977, Early Devonian flora of the Trout Valley formation of Northern Maine: Review of Palaeobotany and Palynology, v. 23, p. 255–285, doi: 10.1016/0034-6667(77)90052-5.

Bateman, R.M., 1991, Paleoecology, in Cleal, C.J., ed., Plant fossils in geological investigation: The Paleozoic. Chichester, UK, Ellis Horwood, p. 34–116.

Bateman, R.M., 1988, Palaeobotany and Palaeoenvironments of Lower Carboniferous floras from two volcanigenic terrains in the Scottish Midland Valley [Ph. D. thesis]: London, University College, 384 p.

Baxter, R.W., 1952, The coal-age flora of Kansas. II. On the relationships among the genera *Etapteris, Scleropteris,* and *Botrychioxylon*: American Journal of Botany, v. 39, no. 4, p. 263–274.

Cate, P.D., 1962, The Geology of the Fayetteville Quadrangle, Washington County, Arkansas [master's thesis]: Fayetteville, Arkansas, University of Arkansas, 112 p.

Chaloner, W. G., and Sheerin, A., 1979, Devonian macrofloras, in House, M.R., Scrutton, C.T., and Bassett, M.G., eds., The Devonian system: London, Palaeontological Association, Special Papers in Palaeontology 23, p. 145–161.

Darrah, W.C., 1960, Principles of paleobotany: New York, Ronald Press, 295 p.

Dennis, R.L., 1974, Studies of Paleozoic ferns: *Zygopteris* from the Middle and Late Pennsylvanian of the United States: Palaeontographica, Abteilung B, v. 148, p. 95–136.

DiMichele, W.A., 1994, Ecological patterns in time and space: Paleobiology, v. 20, no. 2, p. 89–92.

DiMichele, W.A., and Bateman, R.M., 1996, Plant paleoecology and evolutionary inference: Two example from the Paleozoic: Review of Palaeobotany and Palynology, v. 90, p. 223–247, doi: 10.1016/0034-6667(95)00085-2.

DiMichele, W.A., and Hook, R.W., rapporteurs, 1992, Paleozoic terrestrial ecosystems, in Behrensmeyer, A.K., Damuth, J.D., DiMichele, W.A., Potts, R., Sues, H-D., and Wing, S.L., eds., Terrestrial ecosystems through time: Chicago, University of Chicago Press, p. 205–325.

DiMichele, W.A., and Phillips, T.L., 1985, Arborescent lycopod reproduction and paleoecology in a coal-swamp environment of late Middle Pennsylvanian age (Herrin Coal, Illinois, USA): Review of Palaeobotany and Palynology, v. 44, p. 1–26, doi: 10.1016/0034-6667(85)90026-0.

DiMichele, W.A., and Phillips, T.L., 1994, Paleobotanical and paleoecological constraints on models of peat formation in the Late Carboniferous: Palaeogeography, Palaeoclimatology, Palaeoecology, v. 106, p. 39–90, doi: 10.1016/0031-0182(94)90004-3.

DiMichele, W.A., and Phillips, T.L., 1996, Clades, ecological amplitudes, and ecomorphs: Phylogenetic effects and persistence of primitive plant communities in the Pennsylvanian-age tropical wetlands: Palaeogeography, Palaeoclimatology, Palaeoecology, v. 127, p. 83–105, doi: 10.1016/S0031-0182(96)00089-2.

DiMichele, W.A., Davis, J.I., and Olmstead, R.G., 1989, Origins of heterospory and the seed habit: The role of heterochrony: Taxon, v. 38, no. 1, p. 1–11.

DiMichele, W.A., Stein, W.E., and Bateman, R.M., 2001, Ecological sorting of vascular plants classes during the Paleozoic evolutionary radiation, in Allmon, W.D., and Bottjer, D.J., eds., Evolutionary paleoecology: The ecological context of macroevolutionary change: Columbia University Press, New York: 285–335.

Dunn, M.T., 2004, The Fayetteville Flora I: Upper Mississippian (middle Chesterian/lower Namurian A) plant assemblage of permineralized and compression remains from Arkansas, USA: Review of Palaeobotany and Palynology, v. 132, p. 79–102.

Dunn, M.T., Rothwell, G.W., and Mapes, G., 2002, Additional observations on *Rhynchosperma quinnii* (Medullosaceae): A permineralized ovule from the Chesterian (Upper Mississippian) Fayetteville Formation of Arkansas: American Journal of Botany, v. 89, no. 11, p. 1799–1808.

Dunn, M.T., Rothwell, G.W., and Mapes, G., 2003a, On Paleozoic plants from marine strata: *Trivena arkansana* gen. et sp. nov., a lyginopterid from the Fayetteville Formation (middle Chesterian/Upper Mississippian) of Arkansas, USA: American Journal of Botany, v. 90, p. 1239–1252.

Dunn, M.T., Krings, M., Mapes, G., Rothwell, G.W., Mapes, R.H., and Keqin, S., 2003b, *Medullosa steinii* sp. nov., a seed fern vine from the Late Mississippian: Review of Palaeobotany and Palynology, v. 124, p. 307–324, doi: 10.1016/S0034-6667(02)00254-3.

Edwards, D., 1980, Early land floras, in Panchen, A.L., ed., The terrestrial environment and the origin of land vertebrates: London, Academic Press, Systematics Association Special Volume 15, p. 55–85.

Edwards, D., and Fanning, U., 1985, Evolution and environment in the Late Silurian-Early Devonian: The rise of the pteridophytes: Royal Society of London Philosophical Transactions, ser. B, v. 309, p. 147–165.

Eggert, D.A., and Taylor, T.N., 1971, *Telangiopsis* gen. nov., and upper Mississippian pollen organ from Arkansas: Botanical Gazette, v. 132, no. 1, p. 30–37, doi: 10.1086/336559.

Galtier, J., 1981, Structures foliares de fougères et ptéridospermales du Carbonifère inférieur et leur signification évolutive: Palaeontographica, Abteilung B, v. 142, p. 1–38.

Gastaldo, R.A., 1987, Confirmation of Carboniferous clastic swamp communities: Nature, v. 326, no. 6116, p. 869–871, doi: 10.1038/326869a0.

Gastaldo, R.A., 1992, Regenerative growth in fossil horsetails following burial by alluvium: Historical Biology, v. 6, p. 203–219.

Gensel, P.G., 1988, On *Neuropteris* Brongniart and *Cardiopteridium* Nathorst from the early Carboniferous Price Formation, southwestern Virginia, U.S.A: Review of Palaeobotany and Palynology, v. 54, p. 105–119, doi: 10.1016/0034-6667(88)90007-3.

Gensel, P.G., and Andrews, H.N., 1987, The evolution of early land plants: American Scientist, v. 75, p. 478–489.

Gillespie, W.H., Rothwell, G.W., and Scheckler, S.E., 1981, The earliest seeds: Nature, v. 293, p. 462–464, doi: 10.1038/293462a0.

Gordon, W.T., 1912, On *Rhetinangium arberi*, a new genus of Cycadofilices from the Calciferous Sandstone Series: Transactions of the Royal Society of Edinburgh, v. 48, p. 813–825.

Handford, C.R., and Manger, W.L., 1993, Sequence stratigraphy of a Mississippian carbonate ramp, north Arkansas and southwestern Missouri: New Orleans Geological Society, AAPG Annual Convention Field Guide.

Knoll, A.H., 1986, Patterns of change in plant communities through geologic time, in Diamond, J., and Case, T.K., eds., Community ecology: New York, Harper and Row, p. 126–141.

Knoll, A.H., Niklas, K.J., and Tiffney, B.H., 1979, Phanerozoic land-plant diversity in North America: Science, v. 206, p. 1400–1402.

Lacey, W.S., and Eggert, D.A., 1964, A flora from the Chester Series (Upper Mississippian) of southern Illinois: American Journal of Botany, v. 51, no. 9, p. 976–985.

Mamay, S. H., and Bateman R, M., 1991, *Archaeocalamites lazarii* sp. nov.: the range of Archaeocalamitaceae extended from the lowermost Pennsylvanian to the mid-lower Permian: American Journal of Botany, v. 78(4), p. 489–496.

Mapes, G., 1985, *Megaloxylon* in Mid-continent North America: Botanical Gazette, v. 146, no. 1, p. 157–167, doi: 10.1086/337511.

Mapes, G., and Rothwell, G.W., 1980, *Quaestora, amplecta* gen et sp. n., a structurally simple medullosan stem from the Upper Mississippian of Arkansas: American Journal of Botany, v. 67, no. 5, p. 636–647.

Mapes, R., 1966, Late Mississippian lycopod branch from Arkansas: Oklahoma Geology Notes, v. 26, no. 4, p. 117–120.

Matten, L.C., Tanner, W.R., and Lacey, W.S., 1984, Additions to the silicified Upper Devonian/Lower Carboniferous flora from Ballyheigue, Ireland: Review of Palaeobotany and Palynology, v. 43, p. 303–320, doi: 10.1016/0034-6667(84)90002-2.

Meeks, L.K., 1997, Ammonoid taphonomy, biostratigraphy, and lithostratigraphy of the Fayetteville Formation (Mississippian-Chesterian), in its type area, northwest Arkansas [Ph.D. thesis]: Iowa City, Iowa, University of Iowa: 154p.

Meeks, L.K., Titus, A.L., and Manger, W.L., 1997, Taphonomy and biostratigraphy of ammonoid cephalopods, Fayetteville Shale (middle Chesterian, Mississippian), Northern Arkansas, United States: Proceedings, International Congress on the Carboniferous and Permian, 13th, 1995, Krakow, p. 311–317.

Owens, B., Loboziak, S., and Coquel, R., 1979, Late Mississippian-Early Pennsylvanian miospore assemblages from northern Arkansas: in Sutherland, P.K., and Manger, W.L., eds., Compte Rendu, International Congress on Carboniferous Stratigraphy and Geology, 9th, Washington, D.C., and University of Illinois at Urbana-Champaign: Carbondale and Edwardsville, Southern Illinois University Press, v. 2, 377–384.

Pfefferkorn, H.W., Mustafa, H., and Haas, H., 1975, Quantitative charakterisierung ober-karboner Abdruckfloren: Neues Jahrbuch für Geologie und Paläontologie, Abdruckfloren, v. 150, p. 253–269.

Phillips, T.L., and Peppers, R.A., 1984, Changing patterns of Pennsylvanian coal-swamp vegetation and implications of climatic control on coal occurrence: International Journal of Coal Geology, v. 3, p. 205–255, doi: 10.1016/0166-5162(84)90019-3.

Price, C.R., 1981, Transportational and depositional history of the Wedington Sandstone (Mississippian), northwest Arkansas [master's thesis]: Fayetteville, Arkansas, University of Arkansas, 146 p.

Retallack, G.J., and Dilcher, D.L., 1988, Reconstructions of selected seed ferns: Annals of the Missouri Botanical Garden, v. 75, no. 3, p. 1010–1057.

Rex, G., 1986, The preservation and palaeoecology of the Lower Carboniferous silicified plant deposits at Esnost, near Autun, France: Geobios, v. 19, p. 733–800.

Rex, G., and Scott, A.C., 1987, The sedimentology, palaeoecology and preservation of the Lower Carboniferous plant deposits at Pettycur, Fife, Scotland: Geological Magazine, v. 124, no. 1, p. 43–66.

Rezaie, P.A.J., 1980, Palynological correlation of upper Chesterian and Morrowan strata of northwest Arkansas with their Western European equivalents [M.S. thesis]: Fayetteville, Arkansas, University of Arkansas: 187 p.

Rößler, R., 2000, The late Paleozoic tree fern *Psaronius*—An ecosystem unto itself: Review of Palaeobotany and Palynology, v. 108, p. 55–74, doi: 10.1016/S0034-6667(99)00033-0.

Rothwell, G.W., 1987, Complex Paleozoic Filicales in the evolutionary radiation of ferns: American Journal of Botany, v. 74, p. 458–461.

Rothwell, G.W., 1996, Pteridophytic evolution: An often underappreciated phytological success story: Review of Palaeobotany and Palynology, v. 90, p. 209–222, doi: 10.1016/0034-6667(95)00084-4.

Sahni, B., 1932, On the structure of *Zygopteris primaria* (Cotta) and on the relations between the genera *Zygopteris*, *Etapteris*, and *Botrychioxylon*: Royal Society of London Philosophical Transactions, ser. B, v. 222, p. 29–45.

Saunders, W.B., Manger, W.L., and Gordon, M.G., 1977, Upper Mississippian and Lower and Middle Pennsylvanian biostratigraphy of northern Arkansas, *in* Sutherland, P.K., and Manger, W.L., eds., Mississippian and Pennsylvanian boundary in northwestern Oklahoma and northeastern Arkansas: Oklahoma Geological Survey Guidebook 18, p. 117–137.

Scheckler, S.E., 1986a, Floras of the Devonian-Mississippian transition: University of Tennessee, Department of Geological Sciences, Studies in Geology, v. 15, p. 81–96.

Scheckler, S.E., 1986b, Geology, floristics and paleoecology of Late Devonian coal swamps from Appalachian Laurentia (USA): Annales de la Société Géologique de Belgique, T. 109: 209–222.

Schopf, J.M., 1975, Modes of fossil preservation: Review of Palaeobotany and Palynology, v. 20, p. 27–53, doi: 10.1016/0034-6667(75)90005-6.

Scott, A.C., 1979, The ecology of Coal Measure floras from northern Britain: Proceedings of the Geologists' Association, v. 90, p. 97–116.

Scott, A.C., 1980, The ecology of some Upper Paleozoic floras, *in* Panchen, A.L., ed., The terrestrial environment and the origin of land vertebrates: London, Academic Press, Systematics Association Special Volume 15, p. 87–115.

Scott, A.C., and Galtier, J., 1985, Distribution and ecology of early ferns: Proceedings of the Royal Society of Edinburgh, Biological Sciences, v. 86, p. 141–149.

Scott, A.C., Galtier, J., and Clayton, G., 1984, Distribution of anatomically-preserved floras in the Lower Carboniferous in Western Europe: Transactions of the Royal Society of Edinburgh, Earth Sciences, v. 75, p. 311–340.

Smoot, E.L., Taylor, T.N., and Serlin, B.S., 1982, *Archaeocalamites* from the Upper Mississippian of Arkansas: Review of Palaeobotany and Palynology, v. 36, p. 325–334, doi: 10.1016/0034-6667(82)90027-6.

Stein, W.E., Jr., Wight, D.C., and Beck, C.B., 1984, Possible alternatives for the origin of Sphenopsida: Systematic Botany, v. 9, no. 1, p. 102–118.

Taylor, T.N., and Eggert, D.A., 1967a, Petrified plants from the Upper Mississippian of North America. I: The seed *Rhynchosperma* gen. nov.: American Journal of Botany, v. 54, no. 8, p. 984–992.

Taylor, T.N., and Eggert, D.A., 1967b, Petrified plants from the Upper Mississippian (Chester Series) of Arkansas: Transactions of the American Microscopic Society, v. 86, no. 4, p. 412–416.

Taylor, T.N., and Eggert, D.A., 1968, Petrified plants from the Upper Mississippian of North America. II: *Lepidostrobus fayettevillense* sp. nov.: American Journal of Botany, v. 55, no. 3, p. 306–313.

Taylor, T.N., and Taylor, E.L., 1993, The biology and evolution of fossil plants: Englewood-Cliffs, New Jersey, Prentice-Hall, 982 p.

Tiffney, B.H., and Niklas, K.J., 1985, Clonal growth in land plants: A paleobotanical perspective, *in* Jackson, J.B.C., Buss, L.W., and Cook, R.E., eds., Population biology and evolution of clonal organisms: New Haven, Connecticut, Yale University Press, p. 35–66.

Tomescu, A.M.F., Rothwell, G.W., and Mapes, G., 2001, *Lyginopteris royalii* sp. nov. from the Upper Mississippian of North America: Review of Palaeobotany and Palynology, v. 116, p. 159–173, doi: 10.1016/S0034-6667(01)00055-0.

White, D.W., 1937. Fossil flora of the Wedington Sandstone Member of the Fayetteville Shale: U.S. Geological Survey Professional Paper 186-B.

MANUSCRIPT ACCEPTED BY THE SOCIETY 28 JUNE 2005

A Late Mississippian back-barrier marsh ecosystem in the Black Warrior and Appalachian Basins

Robert A. Gastaldo
Department of Geology, Colby College, Waterville, Maine 04901, USA

Michael A. Gibson
Departments of Geology, Geography, and Physics, University of Tennessee, Martin, Tennessee 38238-5039, USA

Allyn Blanton-Hooks
Department of Geology & Geography, Auburn University, Alabama 36849-5301, USA

ABSTRACT

An outcrop of the Mississippian Hartselle Sandstone in north-central Alabama preserves in situ, erect cormose lycopsids, assigned to *Hartsellea dowensis* gen. and sp. nov., in association with a low diversity bivalve assemblage dominated by *Edmondia*. The isoetalean lycopsids are rooted in a silty claystone in which the bivalve assemblage occurs, representing the transition from tidal flat and tidal channel regime into a poorly developed inceptisol. Two paleosols are preserved in the sequence and each is overlain by a fine-grained quartz arenite, responsible for casting aerial stems and cormose bases of the entombed plants. The massive quartz arenites are in sharp contact with interpreted O-horizons of the paleosol, and the lower sandstone displays a lobate geometry. The plant assemblages are interpreted as back-barrier marshes, the first unequivocal marshlands in the stratigraphic record, preserved by overwash processes associated with intense storm surges in a Transgressive Systems Tract. A sample suite curated in the National Museum of Natural History, collected by David White at the turn of the last century in the Greenbrier Limestone of West Virginia, preserves rooting structures, leaves and sporophylls, and sporangia and megaspores of *H. dowensis* in a mixed carbonate mud (micrite). The presence of isoetalean lycopsids in both siliciclastic and carbonate peritidal environments within nearshore shelf settings of the Early Carboniferous indicates that adaptation to periodic brackish water, if not tolerance to infrequent fully marine-water inundation during storm surges, had evolved in these marsh plants by the late Paleozoic.

Keywords: Hartselle Sandstone, lycopsid, paleobotany, paleoecology, Alabama.

INTRODUCTION

Marshes are a type of wetland frequently or continuously inundated with water wherein emergent nonwoody plants are adapted to saturated soils generally of an immature nature. There are several different classification schemes for marshlands (Gore, 1983; Keddy, 2000), although a broad dichotomy can be made into non-tidal (freshwater or brackish) or tidally influenced (freshwater, brackish, or fully marine) settings. In coastal environments, the main criterion that separates a marshland from a tidal flat is the presence of rooted herbaceous plants. Presently, such marshes are restricted primarily to temperate zones because woody mangrove taxa predominate in frost-free latitudes. This probably was not the case prior to the evolution of woody higher plants, with marshes dominating the latitudinal spectrum once plants colonized peritidal environments. Although inference has been made that marshlands are a common component of transitional coastal environments since the evolution of higher plants (e.g., Bateman et al., 1998; Retallack, 1992, 2000; Strother, 2000), few substantive geologic, paleontologic, or taphonomic data exist to support this claim (see Allen and Gastaldo, this volume).

There may be several reasons to account for an absence of autochthonous marsh assemblages in the stratigraphic record. Marshes are restricted geographically along a narrow band of land fringing coastal zones. Sediment accretion in these areas results in better-drained soils and ecosystem replacement by other communities; marsh-to-swamp transitions may be the result of less than a 30 cm change in elevation in modern coastal regimes (Gastaldo et al., 1987). Following sediment compaction, this transition may be recorded in only a few centimeters of stratigraphic section. Sediment accretion is most common in prograding (offlap) sequences that build out onto the shelf, with the last formed marsh at the distal reaches of the wedge. Hence, unless sufficient accommodation during progradation is generated quickly to allow for catastrophic burial (Gastaldo et al., 2004), it is unlikely that recognizable marshlands can be preserved along the way. Coastal-zone transgression (onlap) often results in erosion of coastal plain deposits as the coastline is moved inland by sea-level rise (ravinement; e.g., Liu and Gastaldo, 1992). During ravinement, marsh deposits are reworked and removed from the sediment record. Therefore, it is unusual to encounter equivocal marshes and even rarer to find unequivocal examples of this wetland in the stratigraphic record.

The upper Mississippian (Chesterian [Hombergian]) Hartselle Sandstone crops out in a NW-SE trend along the western margin of the Cumberland Plateau from the Kentucky/Tennessee border south to northeast Alabama, with isolated exposures along the flanks of the Birmingham anticlinorium (Thomas, 1972; Stapor and Cleaves, 1992). It conformably overlies the Pride Mountain Formation, interpreted as part of a lowstand wedge, and represents mainly beach (Stapor et al., 1992), offshore barrier bar and sand-rich shelf deposits (Thomas and Mack, 1982), and tidal facies of an onlap sequence (Stapor and Cleaves, 1992). A small outcrop of the Hartselle Sandstone in Jefferson County, Alabama, exposes in situ erect and rooted cormose lycopsids in an inceptisol that also preserves a macroinvertebrate assemblage indicative of tidal estuarine conditions. A coeval plant assemblage, collected in the early part of the twentieth century by David White in the calcareous Greenbrier Formation, West Virginia, has been identified in the collections of the National Museum of Natural History (USNM), providing for insight into the range of depositional settings for this Late Mississippian plant. Hence, this report provides the first unequivocal evidence for in situ marshlands in the Paleozoic.

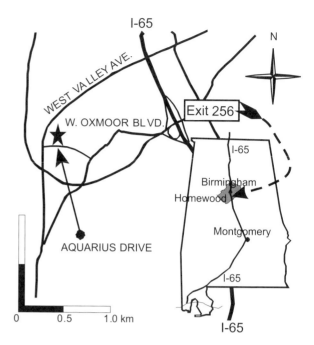

Figure 1. Locality of Hartselle Sandstone outcrop in Homewood, Walker County, Alabama (shaded county within insert map). The collection site is behind a commercial building on Aquarius Drive (NW ¼, NE ¼, Sec. 22, T16S, R3W, Birmingham South 7.5′ Quadrangle, Jefferson County).

STUDY SITE

Collections were made in Homewood, Alabama (NW ¼, NE ¼, Sec. 22, T16S, R3W, Birmingham South 7.5′ Quadrangle, Jefferson County [GPS]), approximately three miles west of I-65 (Homewood-Oxmoor Road exit) behind an industrial building on Aquarius Drive (Fig. 1). Obscured by a stand of pine trees and underbrush, the section crops out in an arcuate trend with a total vertical thickness of 3.5 m. The section consists of two sequences, each beginning with a basal silty claystone overlain by a quartz arenite assignable to the Hartselle Sandstone (Thomas, 1972; Rindsberg, 1994; Fig. 2); a basal quartz arenite underlies the outcrop, which strikes N55E and dips 12° to the southeast.

The silty claystones are massive and primarily white (10 YR 8/1) with pale red (10R 6/3) and reddish-yellow (7.5YR 7/8) mottling. Mineralogically, the claystone consists of quartz, muscovite, and illite; plant fossils are preserved in both intervals. The basal silty claystone also preserves an erect assemblage, with vertically oriented sandstone casts of cormose bases within the claystone and aerial axes extending more than 35 cm into the lowermost arenite (Figs. 2B, 3). A low-diversity macroinvertebrate assemblage is preserved within the basal rooting interval and a lenticular accumulation lateral to the plant assemblage, as well as the claystone that overlies the lowermost arenite. This upper claystone also preserves a concentrated plant-fossil assemblage consisting of the same systematic diversity found in the lower bed.

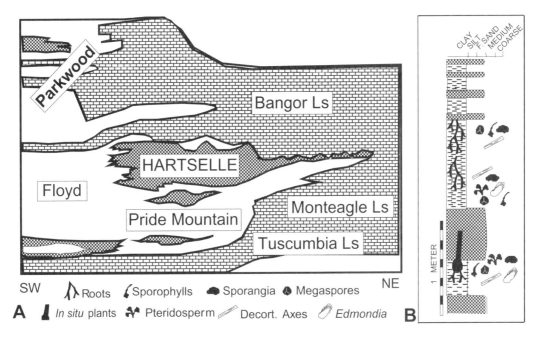

Figure 2. Stratigraphic context of the Hartselle Sandstone and the outcrop from which the present assemblage is described. (A) Southwest-northeast transect of the Mississippian stratigraphy in the Black Warrior Basin showing the intercalated position of the Hartselle Sandstone between the Pride Mountain Formation (lowstand wedge of Stapor and Cleaves, 1992) and Bangor Limestone (highstand deposits) (after Thomas, 1972). The Hartselle is interpreted as offshore beach and barrier deposits in a transgressive systems tract (Stapor and Cleaves, 1992). (B) Stratigraphic column of the Homewood outcrop of the Hartselle Sandstone indicating the position of erect *Hartsellea*, litter horizons of two inceptisols, and macroinvertebrate assemblages. Scale = 1 m.

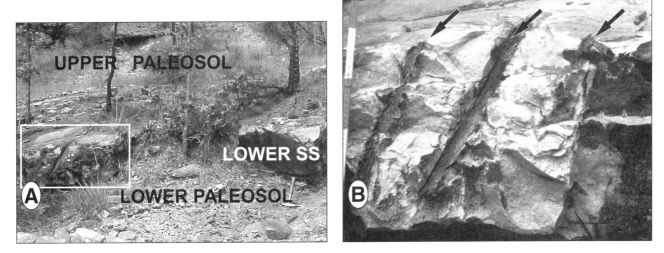

Figure 3. Outcrop photographs of the Homewood locality. (A) The lower paleosol lies above a massive quartz arenite of undetermined thickness, which, in turn, is in sharp contact with the lower sandstone unit in which erect *Hartsellea* are preserved (lower left in box). The lower sandstone is overlain by another interval of rooted silty claystone (inceptisol), which is overlain by thinner beds of resistant quartz arenite (upper center). (B) Enlargement of lower quartz arenite (box in A) in which three erect *Hartsellea* stems were preserved (arrows), indicating the density of the marsh vegetation. Specimens were removed prior to the photograph, with molds of specimens remaining.

Each claystone is overlain by a well-sorted, very fine to fine-grained ($\bar{x} = 3\phi$), white (Gley 1 N8) to light-gray (Gley 1 N7), moderately well sorted quartz arenite. The arenites are massive and in sharp contact with the underlying claystone; tool and sole marks are common at the bed contact. The basalmost arenite is exposed in three dimensions along the outcrop by the weathering and erosion of the overlying silty claystone. That exposure displays a massive sandstone that has a lobate surficial geometry consisting of two lobes, each of which is ~5 m across and a few dm in height. Mineralogically, the arenite consists of monocrystalline, non-undulatory quartz, with minor components including chert, lithics (sedimentary and metamorphic), and mica.

METHODS

Casts of cormose bases were evaluated to identify sites of root initiation, which were mapped to document the range of first-order root diameters and to detect rhizotaxy. Features along aerial stems were analyzed using acetate overlays to determine phyllotaxy. Estimates for growth height were made using the allometric equations of Niklas (1994) based on the six longest stems. Changes in stem diameter were recorded as a function of distance from the stem base. For each specimen, the distances from the stem base were regressed against the stem diameters to determine the allometry of stem taper. Stem diameters taken from levels 5, 10, and 15 cm above each stem base were $\log_{10}$-transformed and inserted into both the "nonwoody and woody" and "nonwoody" species equations (Niklas, 1992) to determine estimated plant height. The choice to use both algorithms was based on previously reported observations by Pigg (1992), Pigg and Rothwell (1983a), and Rothwell and Erwin (1985) on the internal anatomy of isoetalean lycopsids.

Adpressions and casts of aerial axes, subterranean axes, and megaspores were studied using scanning electron microscopy (SEM) in addition to standard microscopic techniques. SEM resulted in characterization of megaspores and cellular details of roots, but no cellular or stomatal patterns of leaves or sporophylls were preserved. Megaspore measurements were made along the polar and equatorial axes using a digital micrometer to acquire length and width data. Two 5–10 g samples from each silty claystone interval preserving megaspores were macerated to isolate mega- and/or microspores. Preparation techniques followed the standard palynological practices as described by Traverse (1988).

Invertebrate-bearing horizons were mined when encountered during section measuring, rather than in a systematic stratigraphic interval, primarily due to the low numbers of preserved specimens in localized occurrences and their poor preservational state. Friable claystone blocks were broken in the field to expose invertebrates, and additional blocks were returned to the laboratory to systematically split under more controlled conditions. The two quartz arenite beds also were examined for the presence of trace fossils.

TAPHONOMY

Plant Macrofossils

Sandstone- (Fig. 4A–C) and mudstone-cast (Fig. 4D) bulbous plant bases are preserved in the assemblage. Radiating bifurcate rooting structures are found originating from these bases in addition to adpression bifurcating rooting structures in the lower silty claystone. These bifurcate roots occur horizontally and at angles up to 40° penetrating the claystone, with several nodes traceable throughout the fine-grained matrix up to 0.5 m depth (i.e., upper claystone paleosol). Roots are more densely concentrated in the uppermost 15 cm and, although no root sheaths have been recovered because of degradation associated with groundwater leaching, cellular epidermal patterns are preserved. Other rooting structures also occur in the matrix as adpressions, along with aerial plant parts of the herbaceous lycopsids and pteridosperms. Aerial plant parts are a heteromeric assemblage and show no signs of sorting, with parts ranging from centimeter-diameter axes and entire leaves to millimeter-sized megaspores with exine ornamentation. All plant parts are concentrated within the uppermost 20 cm of the bed, with a thin condensed, poorly preserved assemblage directly at the contact with each overlying sandstone. Sporophylls and ?leaves are found typically in isolated zones within 15 cm of the top of the bed, whereas sporangia and megaspores occur throughout the bed.

Erect plants are found originating from within the basal claystone and extend upward into the overlying arenite (Fig. 3). These plants are cast by the same lithology that buried them, but they display no primary structures that can be discerned on the exterior of the sandstone fill. Cast specimens may be oriented either vertically (80°–90°) or in a semi-prostrate position. Aerial axes preserve leaf traces in a helical pattern or as longitudinal striae when axes are decorticated. Often the stele can be observed erect and offset within the cast, and neither leaves nor sporophylls are preserved attached to these stems. Although subterranean corms closely resemble their original shape, aerial axes are elliptical in cross section and somewhat compressed vertically due to compaction and dewatering of the entombing sandstone. Hence, both assemblages at this locality represent autochthonous communities buried in situ by an event that emplaced the overlying, lobate, thick quartz arenite. Hence, the claystone interval in which plant parts are concentrated in the uppermost part of the section is interpreted as an original O-horizon.

Macroinvertebrates

Extensive groundwater leaching has resulted in a poorly preserved macrofauna. Nearly all calcitic shell material is dissolved completely, but some vestiges of highly degraded shell appear to adhere to a few specimens. The invertebrates consists almost entirely of internal, external, and composite claystone molds of bivalve molluscs, with a single partial external mold of a gastropod spire. Compaction has distorted some specimens, producing

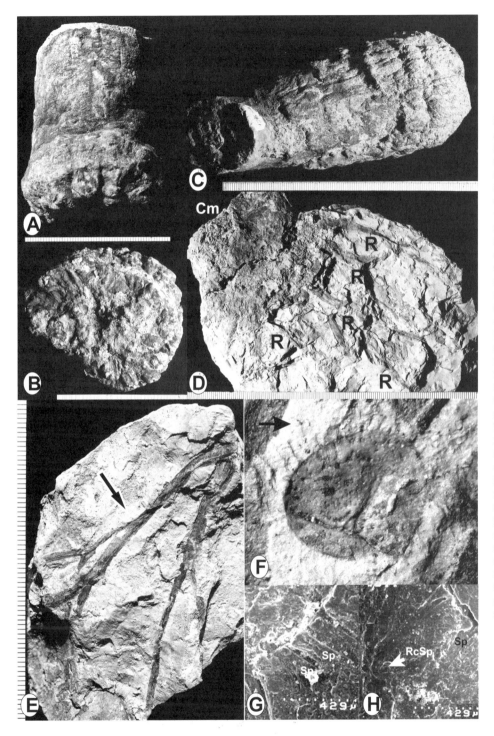

Figure 4. Corms of *Hartsellea dowensis* sp. nov. Scales in A–E in mm. (A) Sandstone-cast basal corm and lower stem transitional area showing the absence of leaf traces. Root bases occur as rounded projections from the bottom of the cast. USNM 527757. (B) Bottom view of specimen shown in A showing root initiation points and central furrow. (C) Top view of a cormose base that was dislodged from vertical during the life of the plant and continued to grow in a subhorizontal orientation. Sandstone-cast roots can be seen to envelope the corm, and the erect stem is shown in cross section at the lower left. Viewed from the side, this specimen is golf-club shaped. USNM 527759. (D) Bottom view of a claystone-cast corm (Cm; upper left) from which bifurcate roots (R) emerge and permeate the inceptisol. USNM 527760. (E) Adpression specimen showing three consecutive bifurcations of the root system. USNM 527761. Arrow points to bifurcation of root. (F) Stereomicrograph of isolated, dispersed megaspore with distorted trilete mark and punctae on the proximal side, as well as radiating apicular and bifurcate (at arrow) spines. USNM 527762. (G) SEM micrograph of proximal megaspore surface showing punctae (Pt) and radiating spines (Sp) along the spore margin. Scale = 429 µm. (H) SEM micrograph showing apicular to retuse (RcSp) nature of megaspore spines (Sp) depending upon the degree of matrix cover. Scale = 429 µm.

undulating surfaces on the shell molds, whereas concentric surface ornamentation generally is preserved, if only slightly distorted. Compaction also has produced composite shells in which impressions of bent and folded sporophyll/leaves of *Hartsellea* are impressed into the mold. These sometimes resemble horizontal sinuous shell borings but are distinguished easily by tracing the lamina off of the shell into the adjacent claystone matrix (Fig. 5D). Rooting structures also are found to cross cut the invertebrate-rich horizons, and are in association with disarticulated valves. The highly degraded state of the shells hinders taxonomic identification beyond the level of genus. External ornamentation, where preserved, is usually restricted to small areas and not the entire specimen (Fig. 5C). In spite of the difficulty of species-level identification, the shell outlines and orientations are

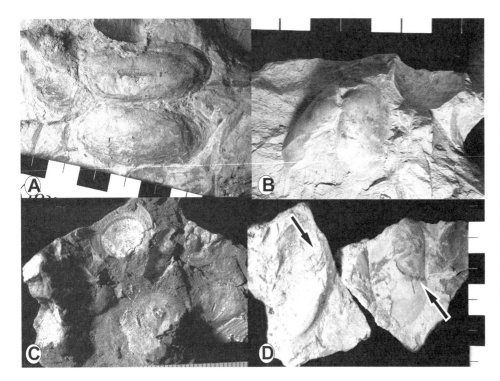

Figure 5. Macroinvertebrates collected from the lower inceptisol and a lenticular assemblage laterally adjacent to the inceptisol. (A) Butterflied *Edmondia*? with portions of other *Edmondia* shells. Note lack of abrasion, breakage, or shell material. Scale in cm. USNM 527753. (B) Butterflied *Edmondia* cutting across claystone laminae. Scale in mm. USNM 527754. (C) Claystone block with several partial *Streblochondria* clustered together. Scale in mm. USNM 527755. (D) Two claystone blocks showing *Hartsellea* gen. nov. roots (arrows) adpressed onto *Edmondia* shells, resembling burrowing structures. Scale in cm. USNM 527756.

preserved sufficiently well to allow for taphonomic observations concerning the assemblage.

Bivalve shells consist of disarticulated or butterflied valves oriented horizontal to bedding (Fig. 5A, 5B). Occasionally valves are found inclined across claystone laminae (Fig. 5B); however, these appear to be slightly agape, butterflied, or isolated valves indicating that specimens are not in life position. There is no difference between the claystone infilling the concave surfaces and that of the exterior convex interior surfaces of the valves. The bivalve shells are not broken, nor do they show obvious signs of abrasion, although the latter would be difficult to demonstrate conclusively considering the lack of preserved shell material. Size range of the bivalves, as measured by variation in maximum length within taxa, indicates some degree of probable sorting; no very small or juvenile shells were encountered. But, there is no evidence for size sorting or preferred orientation of valves on bedding surfaces. Taken together, these characteristics suggest that the invertebrates represent localized sparse shell accumulations preserved within their habitat rather than in situ assemblages.

HARTSELLE SANDSTONE ASSEMBLAGE

Cormose Lycopsid Base

The vertically oriented bases (rhizomorphs) are roughly club-shaped with an upright, unbranched, cylindrical aerial stem

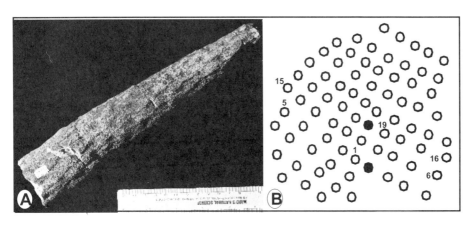

Figure 6. *Hartsellea* stem characters. (A) Erect, sandstone-cast, partially decorticated stem on which leaf traces can be identified as approximately equidimensional ellipsoid or rounded scars arranged in a helical pattern. Scale in cm. USNM 527763. (B) Leaf-trace pattern traced onto a cellulose-acetate overlay from which a 2/19 phyllotaxis can be seen. The base of the axis shows some distortion by compression, as is evident from the displacement of the leaf scars.

(Fig. 4); semi-prostrate bases resemble a golf club covered in roots (Fig. 4C). Basal diameters range from 4.1 to 10.5 cm ($\bar{x}$ = 6.64 cm), and numerous, spirally arranged roots appear to diverge from a centrally located circular furrow (Fig. 4B). There is no evidence that the bases are quadripartite or quadrilobate (K. Pigg, 2004, personal commun.). Rootlet initiation points on casts are preserved as elongate bulges, whereas the first-order roots are cylindrical, imparting an overall irregular surface to the rhizomorph. The number and diameter of roots originating from the furrow are related to the diameter of the corm. First-order roots range from 2.5 to 7.8 mm in diameter ($\bar{x}$ = 5.0 mm ± 1.2 mm; n = 90), and smaller rhizomorphs produce smaller first-order rootlets. Although cast specimens preserve root initiation points and primary roots, it has not been possible to determine the rhizotaxy. Above the basal corm is a transitional zone where neither leaf nor root traces are preserved (Fig. 4A).

Roots

Preserved roots are geopedally oriented, bifurcate, and cross-cut the silty claystone at low to moderate angles, with complete rooting systems generally occupying several bedding planes (Fig. 4D). The depth of preserved root penetration is between 5 and 7 cm, based upon measurements from the more densely rooted horizons (decompaction of the claystone at 15:1 would result in original root penetration of at least a half meter). The longest root segment collected is 4.8 cm, with an average root segment being less than 2 cm. A centrally located vascular trace, identified as a darkened carbonaceous line, runs the length of the roots. All roots bifurcate, and there are at least five orders of axes in the rooting zone with specimens typically displaying three orders of branching (Fig. 4E). Roots bifurcate at angles ranging from 32° to 52° ($\bar{x}$ = 43°; n = 50). Diameters of second- and higher-order roots range from 3.8 to 0.6 mm with an average length between each successive bifurcation of 5.75 mm. Root lengths between successive bifurcations are variable and range from 0.30 to 20.85 mm. Root epidermis is composed of rectangular epidermal cells that average 76 × 43 µm (n = 30).

Stems

Aerial, decorticated axes are preserved primarily as in situ erect sandstone casts, although vestiges of decayed axes occur prostrate in the assemblages as adpressions and claystone-infilled stems. Erect axes range in height from 13.0 to 43.0 cm with basal axial diameters equal to that of each plant's transitional zone (Fig. 6). Aerial axes taper upward, show no signs of branching, and exhibit helically arranged ellipsoid or rounded leaf scars that are approximately equi-dimensional (6.2 × 5.6 mm; n = 20). Decortication of most axes prevented estimation of phyllotaxy; however, one specimen preserved faint but definite scars from which a 2/19 phyllotaxis was determined (Fig. 6B). Owing to the nature of the casting lithology, none of the leaf scars provide evidence of a vascular trace or parichnos tissue, and no leaves have

been found attached to any stem axis. Remnants of a central stele are preserved in several cast specimens, and the "star-shaped" configuration is suggestive of exarch maturation. Using the allometric equations developed by Niklas (1994), predicted growth heights are estimated to be 3.35 m ("nonwoody" species equation) and 3.22 m ("woody and nonwoody" species equation).

Leaves (Sporophylls) and Sporangia

Leaves and/or sporophylls are preserved isolated and mostly incomplete because of fracturing across bedding; complete specimens are observed rarely (Fig. 7). The best, most

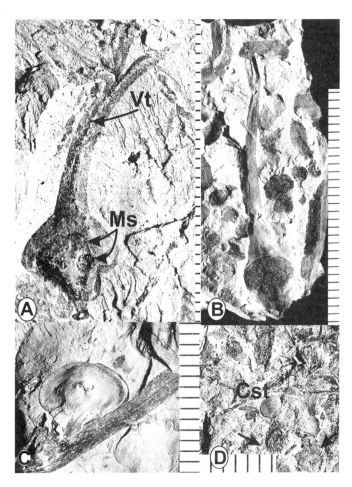

Figure 7. *Hartsellea* sporophylls and leaves. (A) Adaxial side of megasporophyll on which 5 megaspores (Ms) can be seen. Sporophyll pedicels are more inflated than those of vegetative leaves, providing surface area for the development of the sporangium. A linear-lanceolate lamina with central vascular trace (Vt) developed distal to the pedicel. UNSM 527764. (B) ?Abaxial side of a leaf showing non-inflated pedicel congruent with the linear-lanceolate lamina with central vascular trace. Note compressed megaspores in the matrix. USNM 527765. (C) Isolated, reniform sporangium. USNM 627766. (D) Dispersed, isolated megaspores assigned to *Triletes* preserved as impressions with apicular ornamentation (arrows) and claystone casts (Cst) on which no ornamentation can be seen. Scales in mm. USNM 627767.

complete leaves/sporophylls are lanceolate and taper logarithmically in an apical direction. A central vascular strand constitutes approximately one-fifth of the leaf width. Overall, leaves range between 1.3 and 4.1 cm in length (× = 22 mm; $n = 11$) and possess an inflated, diamond-shaped base that is typically 1.7–5.5 mm high and 3.1–12.2 mm wide (Fig. 7B). The lanceolate lamina expands where it is contiguous with the inflated base. Because of specimen orientation within the matrix and the presence often of only the distal lanceolate lamina, it is not possible to distinguish between vegetative leaves and sporophylls in many instances. Sporophylls in the Hartselle suite can be distinguished from vegetative leaves only when the adaxial surface of the diamond-shaped base is visible and a sporangium is present (compare Fig. 7A and 7B). No evidence of strobilar organization has been found.

Compressed, isolated sporangia and sporangia attached to sporophylls are kidney shaped (reniform), averaging 7.5 mm high and 8.3 mm long (Fig. 7A, 7C). The surfaces of sporangia are smooth. No microsporangia have been identified; however, several of the sporangia preserve densely packed megaspores (Fig. 7D). The maximum number of megaspores per megasporangium is unknown, but sporangia containing four and eight megaspores have been observed (Fig. 7A).

Palynomorphs

Isolated trilete megaspores are found in abundant numbers throughout the rooted interval. These may be preserved as either compressions or claystone casts retaining three-dimensional features; both preservational modes exhibit ornamentation originating from the outer wall and extending into the matrix (Figs. 4F, 7D). The megaspores are circular to triangular in outline with a mean diameter of 1.9 mm (standard deviation = 0.3 mm; $n = 200$). The distal side of each is covered with punctae, small bumps, or both, whereas the proximal side is relatively smooth, with a trilete mark that extends nearly to the margin of the megaspore. Punctae are moderately dense, averaging 57 per 0.5 mm^2. There is no pronounced cingulum. Most megaspores in the matrix exhibit long spines that protrude from the margins. The most complete spines exhibit a taper and appear apicular. In rare instances a bifurcation, or "hook," can be seen terminating the spine (Fig. 4H). Spines are up to 600 μm in length, and under high magnification it is possible to see a linear thickening within the spine axis, with ridges running parallel to the middle of the spine. Spine density based on three well-preserved specimens is 15 spines per mm^2. Evidence of spore-wall structure also is visible, but megaspore ultrastructure is unknown. No microspores have been recovered from processed claystone samples.

ASSOCIATED HARTSELLE MEGAFLORA

Preserved in association with the cormose lycopsid, within both rooted intervals and at the contact with the overlying quartz arenite, are a variety of rooting, axial, vegetative leaf, and reproductive structures (Fig. 8). Non-bifurcate rooting structures include elongate and contorted axes (due to compression) that are less than 5 mm in width and several centimeters in preserved length. Small, millimeter-diameter, lateral roots emerge at enlarged nodal areas and penetrate the matrix (Fig. 8A). Aerial axial fragments often are decorticated and occur in two morphologies: bifurcate axes reminiscent of pteridosperm rachides (Fig. 8B, 8C) and linear axes of a smooth and striated nature. Axes bearing helically arranged scars also are encountered. Bifurcate axes may be up to 0.75 m in length and 7.24 cm in width, with most axes on the order of a centimeter in width. Pinnatifid leaves, which probably are of pteridosperm affinity, are found isolated from any axial fragment. They possess a petiole and central axis from which sessile, broad, rhombic and trilobed pinnules occur (Fig. 8D), and conform to *Genselia* (R. Ianuzzi, 2004, personal commun.). Fertile structures conforming to the pteridosperm reproductive structure *Telangiopsis* consist of free sporangia, fused at the base into a synangial organ, that are borne in clusters terminally on branching axes that lack foliar units (Fig. 8F). Rare, isolated seedlike structures also are preserved (Fig. 8E).

GREENBRIER FOSSILS

The small specimen suite curated in the paleobotany collections of the USNM (SM7727) is from the Greenbrier Limestone, of middle Chesterian (Hombergian) age (Yeilding, 1984). The exact collection site is not recorded in the museum's records. The fossil assemblage consists only of plant material—roots, leaves and sporophylls, sporangia, and megaspores—preserved in a light-gray (5Y 7/1) mixed carbonate-siliciclastic lithology (Fig. 9). The rock is predominantly micrite with scattered polycrystalline, medium-sand grains within this matrix.

Greenbrier roots are bifurcate with diameters ranging between 3.2 and 0.7 mm, and angles of bifurcation averaging 42°. Only three orders of bifurcation are displayed in the sample suite, representing distal parts of the root system (Fig. 9B). A single, medial vascular strand runs the length of each root and divides within the axial bifurcation.

The Greenbrier suite appears to preserve both vegetative leaves and sporophylls. Vegetative leaves are distinguished by a greater overall laminar length and less well developed inflated base (Fig. 9A). These are linear-lanceolate in shape and up to 7 cm in length. Although expanded, the leaf base is less inflated and presents a more gentle spindle-shaped taper. Sporophyll morphology is similar to that of the Hartselle material, with leaf bases characterized by flared, diamond-shaped inflations above which short, linear-lanceolate laminae extend for a maximum of 3 cm (Fig. 9D). Sporophyll bases appear more bulbous, not broad and flattened as in the Hartselle specimens, and this may be a taphonomic bias because the latter have been compressed significantly.

Sporangia are isolated and reniform, measuring 4.9 mm in height and 6.8 mm length (Fig. 9D). The surfaces of sporangia

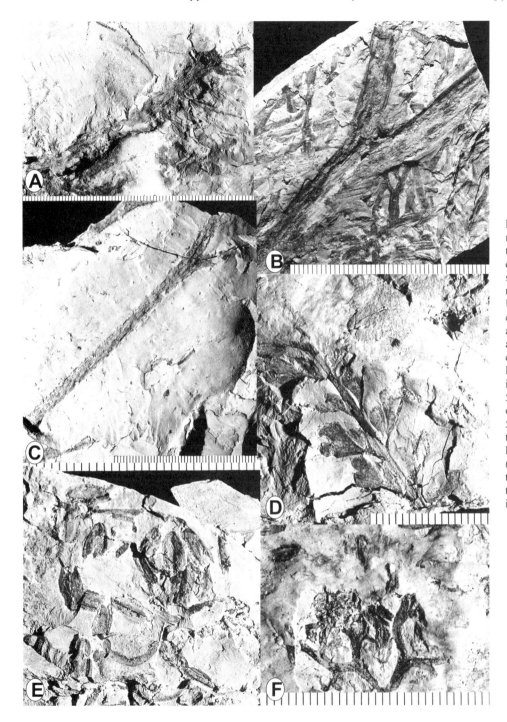

Figure 8. Non-lycopsid macrofloral elements preserved within Hartselle inceptisols. (A) Non-bifurcate roots that are elongate and contorted from compression with small, mm-diameter, lateral roots emergent at enlarged nodal areas. USNM 527768. (B) Large, bifurcate, decorticated axis of undetermined systematic affinity. USNM 527769. (C) Small axis reminiscent of a pteridosperm rachis consisting of elongate petiole and basal leaf bifurcation. The surface of the axis is covered with small punctae. USNM 527770. (D) Isolated pinnatifid leaves of an unidentified pteridosperm. USNM 527771. (E) Isolated small seedlike structures found in association with pinnatifid leaves and *Telangiopsis*. USNM 527772. (F) Elongate sporangia developed from the apices of bifurcate branching system assigned to *Telangiopsis*. All scales in mm. USNM 527773.

are smooth, although carbonaceous residue imparts a punctate appearance. Isolated megaspores average 2 mm in diameter, are circular to triangular, and exhibit the same surficial distribution of long, apicular spines along the megaspore (Fig. 9C, 9D). The number of megaspores per sporangium could not be determined from the available specimens, and no microspores have been recovered from the macerations.

SYSTEMATIC PALEONTOLOGY

The combined Hartselle/Greenbrier specimen suite consists of material that, in part, conforms with *Cormophyton mazonensis* (Pigg and Taylor, 1985), an authigenically preserved cormose lycopsid base discovered in nodules of the Francis Creek Shale, Carbondale Formation, Middle Pennsylvanian of Illinois

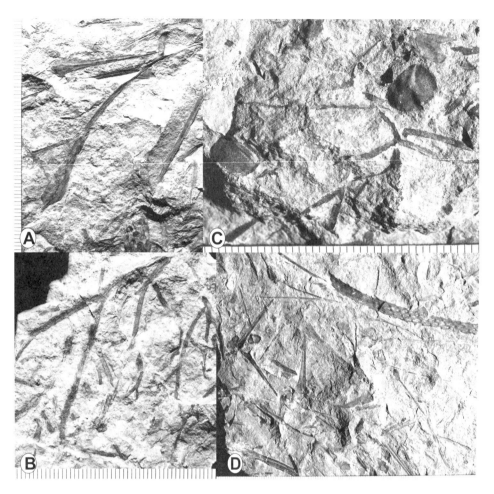

Figure 9. Representative specimens of *H. dowensis* from the Greenbrier Limestone, West Virginia (USNM SM7727). (A) Isolated, vegetative leaves preserved within a micrite inceptisol. (B) Inceptisol roots showing successive bifurcations and spinose megaspore (lower center). (C) *Triletes* micrite-cast megaspores with proximal surface exposed within rooted micritic limestone. (D) Dispersed megasporophylls, sporangia, and punctate axis (?pteridosperm rachis).

(Westphalian D). *Hartsellea* is interpreted as a plant similar to *Chaloneria cormosa*; however, preservational criteria and the absence of anatomical detail preclude its incorporation into this latter genus. Bases from Mazon Creek and those organs described herein provide a wealth of information concerning external features of *Chaloneria*-type plants.

Class **Lycopsida**
Order **Isoetales**
Family **Chaloneriaceae** Pigg and Rothwell, 1983a
Genus *Hartsellea* gen. nov. Gastaldo, Gibson, and Blanton-Hooks

Diagnosis. Herbaceous plant with rounded, club-shaped base and conical, unbranched stem; circular root scars arranged helically; bifurcate roots; leaves and sporophylls arranged helically; rounded leaf scars, wider than high. *Triletes*-type dispersed megaspores.

Hartsellea dowensis **sp. nov.** Gastaldo, Gibson, and Blanton-Hooks

Type species. *Hartsellea dowensis*

Diagnosis. Cormose base; bifurcate roots from central basal furrow, up to 5× bifurcate; lanceolate, elongate leaf lamina with single, central vascular strand and dilated base; megasporophyll pedicels diamond-shaped and inflated with short, lanceolate lamina; adaxial reniform sporangium; trilete megaspores, four to eight per megasporangium, rounded to triangular, spines on distal surface appear apicular, hooked, or bifurcate.

Type locality. Jefferson County, Alabama (NW ¼, NE ¼, Sec. 22, T16S, R3W, Birmingham South USGS 7.5' Quadrangle, Jefferson County, Alabama [33° 27.675' N, 86° 50.340' W])

Stratigraphic Position. Upper Mississippian Hartselle Sandstone

Age. Late Mississippian, Chesterian

Etymology. The generic designation *Hartsellea* refers to the Mississippian Formation in which the plant is preserved. The specific epithet is named after Doug Owens who discovered the fossil assemblage and collected the erect plants for the present study.

Type specimen. Holotype—USNM 527757: Figure 4A. Paratypes—USNM 527759: Figure 4C; USNM 527760: Figure 4D; USNM 5277561: Figure 4E; USNM 5277562: Figure 4F; USNM 527763: Figure 6; USNM 527764: Figure 7A; USNM 5277645: Figure 7B; USNM 527766: Figure 7C; USNM 527767: Figure 7D.

Repository. National Museum of Natural History, Department of Paleobiology, USNM Loc. 42130.

Description

Late Mississippian *Hartsellea* attained a calculated height of ~3.5 m and consisted of an unbranched, monopodial aerial stem that developed from a cormose root base. Aerial axes attained basal diameters of at least 10 cm, with subterranean corms expanded beneath the stem-corm transition. A small, central "star-shaped" stele encountered in several specimens is used to infer that these plants were nonwoody. Aerial stems tapered apically, but the apical terminus is unknown. It is also unknown whether the plant underwent a determinate or indeterminate growth strategy.

Aerial axes were enveloped with helically arranged vegetative leaves up to 7 cm in length and smaller sporophylls, both of which typically produced near-equidimensional leaf scars when the axis was decorticated. Linear lanceolate leaves consist of a distal lamina and a distinct pedicel or keel. Leaf bases are flared, or diamond shaped, with vegetative leaf bases less expanded than those of sporophylls. Leaves narrow distally from the widest portion of the leaf base and sporophylls taper logarithmically. Reniform sporangia occur on the adaxial surface of sporophylls, but it is not known if the plant produced fertile regions interspersed with vegetative zones (as in *Chaloneria*; Pigg and Rothwell, 1983a) or terminated in a strobilar structure (as in *Clevelandodendron*; Chitaley and Pigg, 1996). Only megasporangia have been identified, each of which contains four to eight megaspores. Trilete megaspores, assignable to the dispersed taxon *Triletes* (Bennie and Kidston) ex Zerndt 1930, average 2 mm in diameter and possess numerous, bifurcate spines that may be up to 600 μm long. Because retuse or bifurcate spines (depending upon preservational state) are not always evident owing to sediment cover, megaspores appear apiculate. In the dispersed condition, this ornamentation would have been lost during sample preparation. An unsuccessful search was made for reported dispersed Mississippian megaspores with these morphological features, although those of *Bothrodendrostrobus–Setosisporites* are similar (Stubblefield and Rothwell, 1981). Microspores are unknown.

Cormose bases grew up to 10.5 cm or more in diameter, and bore numerous root-initiation points that average 5 mm in diameter on the largest specimens. Roots originate from a central basal furrow on the corm, and it is not possible to determine whether the corm is more than bilobed. Primary roots can be preserved as elongate cylinders or bulges that give the base an overall irregular appearance. In compressions, these can be seen as departing from the base and then bifurcating within 5 cm. Five orders of roots bifurcate at angles of ~42° and are characterized by a single, central vascular strand. Roots penetrate a weakly developed soil (inceptisol) and extend geopedally for up to 15 cm in compacted sediment. The total length of the rooting system may have been up to 0.5 m. Evidence indicates that once a plant was displaced from a vertical position, growth continued with elongation of the cormose base in a prostrate orientation (Fig. 4C).

COMPARISON WITH OTHER CORMOSE LYCOPSIDS

Cormose lycopsids are assigned to the Isoetales, an extant clade whose origin was thought to be somewhere between the Late Devonian and Early Mississippian (Pigg, 1992; Retallack, 1997), but now is known since the Late Devonian (Cressler and Pfefferkorn, 2005). The primary character distinguishing this clade is an unusual rhizomorph generally described as a swollen, nonbranching, or lobed base, although Chitaley and Pigg (1996) used a monopodial growth habit and a terminal bisporangiate strobilus as criteria for systematic assignment of a Late Devonian lycopsid to the order. Debate continues as to the phylogenetic origin of the cormose structure and derived lineages (Jennings, 1975; Jennings et al., 1983; Rothwell and Erwin, 1985; Pigg, 1992; Bateman et al., 1992; Bateman, 1992). Some authors favor the evolution of *Stigmaria*-type rooting, found in the Lepidodendrales, as an evolutionary derivation from a cormose base (Bateman et al., 1992), while earlier workers proposed that cormose bases are the result of reductionism (e.g., Mägdefrau, 1956) or that structures in lepidodendraleans and isoetaleans had separate origins (Jennings, 1975; Stubblefield and Rothwell, 1981). There is general agreement that rhizomorphic lycopsids belong to one monophyletic "plexus" (Pigg, 1992; DiMichele and Bateman, 1996). Regardless, early representatives of the clade were identified solely on the basis of aerial (Chitaley and Pigg, 1996) or subterranean parts (Cressler and Pfefferkorn, 2005; Bateman, 1992).

Of all the non-arborescent taxa described from the Late Devonian, Mississippian, or Pennsylvanian, morphological features of the present material conform most closely with younger described isoetaleans. Two Late Devonian forms—*Cyclostigma* (Chaloner, 1984) and *Otzinachsonia* (Cressler and Pfefferkorn, 2005)—have similar basal configurations, but Pigg (1992) considered the bilobed nature of *Cyclostigma* to be ambiguous. And, although *Otzinachsonia* possesses a lobe-and-furrow architecture, the cormose base is definitely four-lobed. The compact anchoring structures in the Early Mississippian form, *Oxroadia* (Bateman and Rothwell, 1990; Bateman, 1992) are unlike those in *Hartsellea dowensis*.

Two taxa of the Pennsylvanian-aged Chaloneriaceae—*C. periodica* and *C. cormosa*—are most similar architecturally to the plant described herein. The ligulate taxon *Chaloneria periodica* is known from permineralized peat of the Middle Pennsylvanian, and is characterized by an aerial axis that possessed alternating vegetative and fertile zones (DiMichele et al., 1979; Pigg and Rothwell, 1983a, 1873b, 1985; Pigg, 1992). Neither vegetative leaves nor sporophylls show evidence of dehiscence, and these structures differ considerably in size. Sporophylls are smaller and thinner than the vegetative leaves, with adaxial sporangia preserved in the axils of some sporophylls (DiMichele et al., 1979). No basal organs are known for this plant. *Chaloneria cormosa*, a permineralized Pennsylvanian form, is an unbranched, upright plant with a cormose base (Pigg and Rothwell, 1979; 1983a, 1983b; Pigg, 1992). The plant is estimated to have been ~2 m in height, with a maximum stem diameter of ~10 cm.

Helically arranged leaves were dehiscent and produced irregular axial leaf scars on the axis. The plant was heterosporous with fertile regions located toward the stem apex (Pigg and Rothwell, 1983a; Pigg, 1992). Dispersed megaspores and microspores are assigned to *Valvisisporites auritus* and *Endosporites globosus*, respectively (Gastaldo, 1981; Pigg and Rothwell, 1983a; Pigg, 1992). Hence, although the morphological features and overall bauplan of *Hartsellea downesi* are comparable to *Chaloneria*, the megaspores of these plants are very different.

Megaspores are more similar to those in *Bothrodendrostrobus* (Stubblefield and Rothwell, 1981). *Setosisporites praetextus* is similar in overall shape, possessing a robust trilete mark. The spore is smaller in maximum dimension (1.5 mm) with a smooth proximal surface adjacent to the trilete mark, whereas the distal surface is covered with branched, elongate spines. Spines may be up to 250 µm (reportedly broken at the spine tips) and when completely broken off leave small, rounded punctae on the distal surface. Stubblefield and Rothwell (1981) documented the embryogenesis of this plant and compared it with that of extant Isoetales. They concluded that *Bothrodendrostrobus* was probably a small plant with a cormose rooting system and, hence, similar in bauplan to *Hartsellea*.

HARTSELLE INVERTEBRATES

The invertebrate assemblage is restricted to discrete horizons within the claystones. These are not laterally continuous; rather, they occur in lenticular pockets where the claystone thickens above thinner parts of the quartz arenites. The macrofauna is nearly a monospecific accumulation of bivalves, with only three tentatively identified taxa ($n = 34$ total specimens from two excavated pockets). *Edmondia* (Fig. 5A, 5B; $n = 29$) is found primarily as disarticulated valves occurring on 29 claystone blocks, and three specimens articulated and butterflied open. *Streblopteria* ($n = 2$) is uncommon and encountered as disarticulated valves, whereas *Streblochondria* ($n = 3$) is found only as disarticulated valves, although two valves may belong to the same individual. *Edmondia* shows the greatest size range, with length ranging from 1 to 4.5 cm; unfortunately other taxonomic features, such as hingeline characteristics and musculature, are not preserved. Poorly preserved *Streblopteria* and *Streblochondria* specimens all occur closely clustered on two claystone blocks (Fig. 5C).

Some very small cf. *Lockeia* (Rindsberg, 1994) traces were encountered on the surface of quartz arenite blocks, but their morphology varied enough to preclude positive identification. No shelly fauna corresponding to the size of the potential *Lockeia* were found. A few sinuous structures resembling horizontal traces, similar to *Planolites*, were found to be distinct enough to differentiate them from tool marks.

DISCUSSION

The Chesterian Series in the Black Warrior (BWB) and Appalachian Basins consists of mixed carbonate-siliciclastic deposits that reflect major changes in the tectonic and paleogeographic setting of the region. Most of the Mississippian in the BWB is characterized by limestones that accumulated within a variety of carbonate-ramp environments (Pashin and Gastaldo, in press) including shoals and tidal flats. Subsequently, the area was influenced by siliciclastic input from the developing Alleghenian orogenic province to the east-southeast (Mack et al., 1983). Carbonates accumulated when the area was under highstand conditions followed by cyclically interbedded shale, sandstone, and limestone of the Pride Mountain Formation (BWB), interpreted as a lowstand wedge within a lowstand systems tract (Stapor et al., 1992). During sea-level rise, the Hartselle Sandstone, consisting of beach, barrier island, and tidal facies (Thomas, 1982; Smith, 1983), was deposited within a nearshore regime of the overlying Transgressive Systems Tract. Hartselle sandstones are barrier strandplains, each of which was deposited during a stillstand, with barriers and back-barrier lagoonal and tidal-flat facies sequentially stacked in the sequence (J. Pashin, 2004, personal commun.). Hence, nearshore subaqueous and subaerial, tidally influenced environments were available for colonization by plants capable of inhabiting poorly developed soils (inceptisol) that were subjected to tidally driven brackish and fully marine waters.

Sedimentological evidence indicates that the outcrop of claystone and quartz arenites at Homewood represents a siliciclastic-dominated nearshore marine setting characterized as a barrier island with back-barrier deposits (e.g., Walker, 1984; Boggs, 1995). A basal, moderately well-sorted quartz arenite is overlain by a poorly developed silty, clay-rich, deeply rooted paleosol in which the herbaceous lycopsid *Hartsellea* and other plants grew. The erect isoetaleans and their soil-litter (O) horizon were covered by a well-sorted, massive quartz arenite exhibiting a lobate surface geometry deposited in a single event. Death and subsequent decay of the erect axes resulted in an arenite infill by the overlying entombing sediment, indicating shifting sands across the top of the deposit following void development. This sequence was repeated at the site with the overlying thicker inceptisol in which the same plant assemblage is preserved as the result of the emplacement of another fine-grained, well-sorted, quartz-rich sand body. Hence, a repetitive process was responsible for organic matter preservation in this regime, overtopping a herbaceous wetland community that is, by definition, a marsh.

Laterally adjacent lenticular assemblages within the silty claystone preserve macroinvertebrates that were co-opted as part of the inceptisol, the result of channel accretion and fill to where the sediment surface was raised sufficiently to allow for plant colonization. The Homewood assemblage is very low diversity when compared with other Late Mississippian coastal, deltaic, and shallow-marine faunas, such as the bivalve-rich Redoak Hollow (e.g., Elias, 1957), the Greenbrier Limestone, and Mauch Chunk Formations (Kammer and Lake, 2000). The near predominance of *Edmondia*, the butterflied and disarticulated nature of the shell occurrences, and the localized distribution restricted to pockets of claystone within low swale areas of the quartz arenite are consistent with a transported thanatocenosis rather than an in situ biocenosis.

A fully marine habitat for the Homewood invertebrates is well established in the literature by authors using a combination of shell morphology, faunal associations, and lithologic characteristics. The absence of brachiopods restricts the assemblage to shallower-water settings (e.g., Bretsky, 1969), and although Kammer and Lake (2000) concluded that *Edmondia* and *Streblochondria* were euryhaline taxa, *Edmondia* generally is interpreted as a shallow-burrowing marine form (e.g., Craig, 1956; Calver, 1968; Runnegar and Newell, 1974; Hoare, Sturgeon, and Kindt, 1979; Gibson and Gastaldo, 1987). This interpretation is based upon association with other known marine invertebrates and such morphological characteristics as the elongate, oval shell form, nearly equilateral valve shapes, position and nature of external ornamentation, and internal anatomy (e.g., Stanley, 1970, 1972). Stanley (1970) notes that many of the analodesmatids, to which *Edmondia* belongs, were likely endobyssate, a life-habit adaptation for substrate stabilization. Such adaptations in bivalves are indicative of life that is just below the sediment-water interface or, perhaps, semi-infaunal (Stanley, 1972). *Streblopteria* and *Streblochondria* usually are interpreted as epibyssate shallow-nearshore to deeper-water offshore epifauna (e.g., Watkins, 1975; Hoare et al., 1979; Kammer and Lake, 2000), with *Strebolopteria* perhaps restricted to lower oxygen conditions as indicated by its occurrence in black shale facies. Considering the depositional setting and the relationship of in situ plants to the mollusc-dominated macroinvertebrates, the invertebrate assemblage represents a localized, reworked marine fauna trapped within peritidal clastics (tidal channels, tidal flats, or the marsh, proper).

The absence of invertebrate trace fossils within both the quartz arenites and claystones, as compared with other Hartselle occurrences (Rindsberg, 1994) or other Mississippian units such as the Pennington (e.g., Sheehan, 1988) or Borden (e.g., Chaplin, 1982) Formations, indicates that either burrowing organisms were not common in the Homewood substrates or evidence for burrowing was not preserved. In the case of the sandstone, this is due probably to mobility of the substrate and/or subaerial exposure.

The presence of rooted *H. dowensis* in micritic limestones of the Greenbrier Group expands the habitat range of this plant from strictly siliciclastic muddy, back-barrier environments to a carbonate mud substrate. Muddy carbonate depositional environments are indicative of settings surrounded by fully marine conditions, away from significant siliciclastic influence (e.g., Walker, 1984). The Greenbrier is considered to be the equivalent of the Newman Limestone (Big Lime) in Kentucky (Harris and Sparks, 2000), a shallow-marine carbonate sequence that overlies the deeper marine carbonates of the Fort Payne Formation. The oolites of the Big Lime were deposited in shoals or tidal channels (MacQuown and Pear, 1983) in a more offshore position than time-equivalent lithologies in West Virginia where peritidal lime mudstone, peloidal, ooid and skeletal grainstones formed (Smosna, 1996; Wynn and Read, 2002). Hence, the preservation of *Hartsellea* in these peritidal environments indicates that it was a colonizer of wetlands throughout the transitional to marine regime.

Herbaceous wetland plants possess shallow rooting structures within the upper few decimeters of poorly developed soils (inceptisols). In modern marshes, shallow rooting is due to several factors. These include periodically saturated near-surface or surficial pore waters following rainfall or incursion of daily tides; the availability of only a shallow freshwater lens, derived from meteoric waters, overlying saline pore water at some depth; and changing redox potentials at depth in response to organic accumulation, methanogenesis, and the activity of sulfur-reducing bacteria. Autochthonous *Harstellea* in both siliciclastic and carbonate peritidal settings indicates that this plant may have been tolerant of brackish, if not fully marine waters during times when storm washover events and/or high (King) tides brought saline waters into this wetland setting. The presence of fully marine macroinvertebrates in the paleosol is evidence that these bivalves were transported into this setting before plant colonization. These animals represent both dead and living bivalves, as indicated by disarticulated valves and articulated specimens, respectively. This indicates that the bivalves lived in close proximity to the marshland, were transported into a channel feature (the lenticular geometry of the assemblage in an otherwise nondescript claystone), and subsequently buried. The energy required to move these benthic, infaunal bivalves is associated with hurricane-style storm events during which salt spray and inundation of fully ocean waters are common in back-barrier settings. Hence, it is parsimonious to propose that *Hartsellea* was tolerant of some saline influence in its habitat.

Hartsellea probably was not the only Carboniferous lycopsid tolerant of at least brackish waters. Gastaldo (1986) noted that the presence of in situ, autochthonous *Stigmaria* preserved in Mississippian bioclastic carbonates of the Battleship Wash Formation, Arizona, (Pfefferkorn, 1972), was evidence for brackish and/or salt-water tolerance of arborescent lycopsids in the Early Carboniferous. Hence, it appears that the lycopsid clade had evolved this physiological tolerance by at least the late Paleozoic.

The modern Gulf Coast barrier island system off Alabama, as well as the coastal barrier-island chain along the eastern seaboard, paralleling the coasts of Georgia to New Jersey, may serve as a modern environmental and process analogue for the wetland vegetation and associated macroinvertebrate assemblages (Dardeau et al., 1992; Kopaska-Merkel et al., 2000; see discussion of modern nearshore and primary-to-tertiary bay faunas and their ecologies in Britton and Morton, 1989; Fig. 10). Wetland marshes are composed primarily of herbaceous plants growing in back-barrier settings on the more protected, restricted marine side of these islands. These wetlands are the result of colonization of peritidal deposits that have accreted to near the air-water interface. A low-diversity bivalve community colonizes the sediment-water interface within tidal channels that cross the back-barrier marshes, although individuals may be found within the marshes byssally attached to the roots of grasses (e.g., *Spartina*; Prezant et al., 2002). Within the tidal channels of St. Catherines Island,

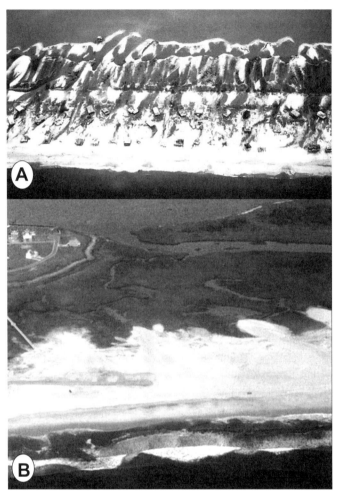

Figure 10. Oblique aerial photographs of recent overwash-fan deposits on the Atlantic and Gulf Coasts. (A) Overwash fan on Dauphin Island photographed the day after Hurricane Frederick, 1979, with wind speeds of 143 mph recorded on the island (photograph by J. Dindo, Dauphin Island Sea Lab, Alabama). (B) Overwash fan deposits on Carolina Beach, south of Wilmington, North Carolina, deposited during Hurricane Bonnie in 1998. Note how the overwash sand was deposited over the back-barrier marsh vegetation (photograph by J. Buie)

Georgia, larger bivalves tend to burrow in muddy point bars within tidal creeks, and small to large oyster bars occasionally are found to occupy creek banks (Prezant et al., 2002). Reworking of such assemblages during intense storm activity (e.g., hurricane-force winds) results in shell concentrations within the tidal channels or within the marshes (Donnelly et al., 2001); vertical accretion of tidally transported mud within the channels results in their infill and subsequent colonization by marsh plants. Storm surges associated with hurricanes also are responsible for the redistribution of foreshore and upper shoreface sands up and over dunes into the back-barrier wetlands (Fig. 10B; Donnelly et al., 2001). Such overwash sands interbedded with back-barrier marsh-and-lake deposits are considered to be formed in response to the most intense hurricane strikes (Liu and Fearn, 1993, 2000; Donnelly et al., 2001). The contact between the underlying marsh mud or peat and the overlying well-sorted quartz sand is abrupt and often displays soft-sediment deformation. The overwash sand may be up to several decimeters closest to the barrier, thinning to less than 1 cm in distal sites. Depending upon the stem diameter and structural rigidity of the marsh plants in the path of the overwash sediments, these may be entombed either in an erect, standing orientation or in a prostrate position at the upper soil horizon in contact with the overlying sand.

CONCLUSIONS

Cormose lycopods assigned to *Hartsellea dowensis* gen. and sp. nov. occur in siliciclastic- and carbonate-dominated marine settings of the Mississippian within the eastern United States. These isoetaleans are preserved in peritidal environments in back-barrier wetlands that provide evidence for the first unequivocal marshes in the stratigraphic record. This interpretation is based on the herbaceous vegetational character of the assemblages, dominated by cormose lycopsids assigned to the Chaloneriaceae (Isoetales), and on leaves, seeds, and pollen organs of an unidentified pteridosperm within sedimentological context. The Homewood, Alabama, locality preserves an autochthonous assemblage of in situ plants with cormose bases rooted in a silty claystone inceptisol, and erect aerial stems entombed and cast by overlying fine-grained quartz arenite. Two stratigraphically successional marshes are preserved in the sequence, with the standing vegetation restricted to the basal paleosol. This exposure of the Hartselle Sandstone records barrier-island overwash deposits into back-barrier marshes during an interval of sea-level rise (TST). Here, a depauperate, low-diversity bivalve assemblage is preserved within the paleosol, as well as in lenticular beds lateral to the soil horizon that represent tidal channel fills. Invertebrates preserved in these beds—*Edmondia*, *Streblopteria*, and *Streblochondria*—are considered to be fully marine, and their presence in this depositional setting indicates that these taxa probably were transported into the back-barrier environment and buried during high-energy storm conditions. The rooted presence of *H. dowensis* in the Greenbrier Limestone of West Virginia, preserved in a carbonate mudstone, indicates that these isoetaleans were capable of colonizing both siliciclastic and carbonate tidal-flat deposits. Hence, the herbaceous lycopsids of the Chaloneriaceae, normally found within peat-accumulating wetlands, were well adapted to other wetland settings, and were the principal component of Carboniferous marshes in North America.

ACKNOWLEDGMENTS

The authors are indebted to Doug Owen of Birmingham, Alabama, of Milo's Famous Sweet Tea, for finding and collecting the cormose lycopsids from behind the factory site in Homewood, Alabama, and contacting the University of Alabama and Auburn University in the early 1990s. We would like to thank the following individuals who helped in the collection and analysis of the samples over the subsequent decade: Todd Grant, Nikolai Pedentchouk, Alex Webster, and Jonathan Collier of Auburn University; Matthew Charles of Colby College, and Kathleen Pigg, Arizona State University. Reviews by Kathleen Pigg, Cort Eble, and Steve Greb are greatly appreciated and were used as the basis for manuscript revision.

REFERENCES CITED

Bateman, R.M., 1992, Morphometric reconstruction, paleobiology and phylogeny of *Oxroadia gracililis* Alvin emend. and *O. conferta* sp. nov., anatomically-preserved rhyzomorphic lycopsids from the Dinantian of Oxroad Bay, SE Scotland: Palaeontographica, Abteilung B, v. 228, p. 29–103.

Bateman, R.M., and Rothwell, G.W., 1990, A reappraisal of the Dinantian floras at Oxroad Bay, East Lothian, Scotland. 1. Floristics and the development of whole-plant concepts: Transactions of the Royal Society of Edinburgh, Biological Sciences, v. 8, p. 127–159.

Bateman, R.M., DiMichele, W.A., and Willard, D.A., 1992, Experimental cladistic analysis of anatomically preserved arborescent lycopsids from the Carboniferous of Euramerica: An essay on paleobotanical phylogenetics: Annals of the Missouri Botanical Garden, v. 79, p. 500–559.

Bateman, R.M., Crane, P.R., DiMichele, W.A., Kenrick, P.R., Rowe, N.P., Speck, T., and Stein, W.E., 1998, Early evolution of land plants: Phylogeny, physiology, and ecology of the primary terrestrial radiation: Annual Review of Ecology and Systematics, v. 29, p. 263–292, doi: 10.1146/annurev.ecolsys.29.1.263.

Boggs, S., 1995, Principles of sedimentology and stratigraphy (2nd edition): Englewood Cliffs, New Jersey, Prentice-Hall, 774 p.

Bretsky, P.W., Jr., 1969, Evolution of Paleozoic benthic marine invertebrate communities: Palaeogeography, Paleaeclimatology, Palaeoecology: v. 6, p. 45–59.

Britton, J.C., and Morton, B., 1989, Shore ecology of the Gulf of Mexico: Austin, University of Texas Press, 387 p.

Calver, M.A., 1968, Distribution of Westphalian marine faunas in northern England and adjoining areas: Proceedings of the Yorkshire Geological Society, v. 37, p. 1–72.

Chaloner, W.G., 1984, Evidence of ontogeny of two late Devonian plants from Kiltorcan, Ireland: International Organization of Palaeobotany Conference, 2nd, Edmonton, Alberta, Canada, abstracts of contributions. Paper and poster sessions.

Chaplin, J.R., 1982, Field guidebook to the paleoenvironments and biostratigraphy of the Borden and parts of the Newman and Breathitt Formations (Mississippian–Pennsylvanian) in Northeastern Kentucky: 12th annual field trip guidebook, Great Lakes Section of the SEPM, p. 1–196.

Chitaley, S., and Pigg, K.B., 1996, *Clevelandodendron ohioensis*, gen. et sp. nov., A slender upright lycopsid from the Late Devonian Cleveland Shale of Ohio: American Journal of Botany, v. 83, p. 781–789.

Craig, G.Y., 1956, The mode of life of certain Carboniferous animals from the West Kirkton Quarry, near Bathgate: Transactions of the Edinburgh Geological Society, v. 16, pt. 3, p. 272–279.

Cressler, W.L., and Pfefferkorn, H.W., 2005, A Late Devonian Isoetalean lycopsid, *Otzinachosonia beerboweri*, gen. et sp. nov., from north-central Pennsylvania, USA: American Journal of Botany, v. 92, p. 1131–1140.

Dardeau, M.R., Modlin, R.F., Stout, J.P., and Schroeder, W.W., 1992, Estuaries, *in* Hackney, C., Adams, M., and Martin, B., eds., Biotic diversity of the southeastern U.S.: Aquatic communities: New York, John Wiley and Sons, p. 614–744.

DiMichele, W.A., and Bateman, R.M., 1996, Plant paleoecology and evolutionary inferences: Two examples from the Paleozoic: Review of Palaeobotany and Palynology, v. 90, p. 223–247, doi: 10.1016/0034-6667(95)00085-2.

DiMichele, W.A., Mahaffy, J.R., and Phillips, T.L., 1979, Lycopods of Pennsylvanian age coals: *Polysporia*: Canadian Journal of Botany, v. 57, p. 1740–1753.

Donnelly, J.P., Roll, S., Wengren, M., Butler, J., Lederer, R., and Webb, T., III, 2001, Sedimentary evidence of intense hurricane strikes from New Jersey: Geology, v. 29, p. 615–618, doi: 10.1130/0091-7613(2001)029<0615:SEOIHS>2.0.CO;2.

Elias, M.K., 1957, Late Mississippian fauna from the Redoak Hollow Formation of southern Oklahoma, Part 3: Pelecypoda: Journal of Paleontology, v. 31, p. 737–784.

Gastaldo, R.A., 1981, An ultrastructural and taxonomic study of *Valvisporites auritus* (Zerndt) Bhardwaj: A lycopsid megaspore from the Middle Pennsylvanian of southern Illinois: Micropaleontology, v. 27, p. 84–93.

Gastaldo, R.A., 1986, Implications on the paleoecology of autochthonous Carboniferous lycopods in clastic sedimentary environments: Palaeogeography, Palaeoclimatology, Palaeoecology, v. 53, p. 191–212, doi: 10.1016/0031-0182(86)90044-1.

Gastaldo, R.A., Douglass, D.P., and McCarroll, S.M., 1987, Origin, characteristics and provenance of plant macrodetritus in a Holocene crevasse splay, Mobile delta, Alabama: Palaios, v. 2, p. 229–240.

Gastaldo, R.A., Stevanović-Walls, I.M., and Ware, W.N., 2004, *In situ*, erect forests are evidence for large-magnitude, coseismic base-level changes within Pennsylvanian cyclothems of the Black Warrior Basin, USA, *in* Pashin, J.C., and Gastaldo, R.A., eds., Sequence stratigraphy, paleoclimate, and tectonics of coal-bearing strata: AAPG Studies in Geology, v. 51, p. 219–238.

Gibson, M.A., and Gastaldo, R.A., 1987, Invertebrate paleoecology of the Upper Cliff coal interval (Pennsylvanian), Plateau Coal Field, Northern Alabama: Journal of Paleontology, v. 61, p. 439–450.

Gore, A.J.P., ed., 1983, Ecosystems of the World, Volume 4A, Mires—Swamp, bog, fen, and moor: Amsterdam, Elsevier, 440 p.

Harris, D.C., and Sparks, T.N., 2000, Regional subsurface geologic cross sections of the Mississippian System, Appalachian Basin, eastern Kentucky: Kentucky Geological Survey, Map and Chart Series 14, 14 p.

Hoare, R.D., Sturgeon, M.T., and Kindt, E.A., 1979, Pennsylvanian marine Bivalvia and Rostroconchia of Ohio: Ohio Division of Geological Survey Bulletin, v. 67, p. 1–77.

Jennings, J.R., 1975, *Protostigmaria*, a new plant organ from the Lower Mississippian of Virginia: Palaeontology, v. 18, p. 19–24.

Jennings, J.R., Karrfalt, E.E., and Rothwell, G.W., 1983, Structure and affinities of *Protostigmaria eggertiana*: American Journal of Botany, v. 70, p. 963–974.

Kammer, T.W., and Lake, A.M., 2000, Salinity ranges of Late Mississippian invertebrates of the Central Appalachian Basin: Southeastern Geology, v. 40, p. 99–116.

Keddy, P.A., 2000, Wetland ecology: Principles and conservation: New York, Cambridge University Press, 614 p.

Kopaska-Merkel, D.C., Rindsberg, A.K., and DeJarnette, S.S., 2000, A guidebook to the Mississippian rocks and fossils of north Alabama: Geological Survey of Alabama Educational Series, v. 13, p. 1–57.

Liu, K., and Fearn, M.L., 1993, Lake-sediment record of late Holocene hurricane activities from coastal Alabama: Geology, v. 21, p. 793–796, doi: 10.1130/0091-7613(1993)021<0793:LSROLH>2.3.CO;2.

Liu, K., and Fearn, M.L., 2000, Reconstruction of prehistoric landfall frequencies of catastrophic hurricanes in northwestern Florida from lake sediment records: Quaternary Research, v. 54, p. 238–245, doi: 10.1006/qres.2000.2166.

Liu, Y., and Gastaldo, R.A., 1992, Characteristics of a Pennsylvanian ravinement surface: Sedimentary Geology, v. 77, p. 197–214, doi: 10.1016/0037-0738(92)90126-C.

Mack, G.H., Thomas, W.A., and Horsey, C.A., 1983, Composition of Carboniferous sandstones and tectonic framework of southern Appalachian-Ouchita orogen: Journal of Sedimentary Petrology, v. 53, p. 931–846.

MacQuown, W.C., and Pear, J.L., 1983, Regional and local geologic factors control Big Lime stratigraphy and exploration for petroleum in eastern Kentucky, *in* Luther, M.K., ed., Proceedings of the technical sessions, Kentucky Oil and Gas Association 44th annual meeting: Kentucky Geological Survey, ser. 11, Special Publication 9, p. 1–20.

Mägdefrau, K., 1956, Paläobiologie der Pflanzen: Jena, Germany, G. Fischer, 314 p.

Niklas, K.J., 1992, Plant biomechanics: An engineering approach to plant form and function: Chicago, University of Chicago Press, 607 p.

Niklas, K.J., 1994, Predicting the height of fossil plant remains: An allometric approach to an old problem: American Journal of Botany, v. 81, p. 1235–1242.

Pashin, J.C., and Gastaldo, R.A., in press, Carboniferous of the Black Warrior Basin, *in* Wagner, R.H., Winkler-Prins, C.F., and Granados, L.F., eds., The Carboniferous of the world: Instituto Geológico y Minero de España.

Pfefferkorn, H.W., 1972, Distribution of *Stigmaria wedingtonensis* (Lycopsida) in the Chesterian (Upper Mississippian) of North America: American Midland Naturalist, v. 88, p. 225–231.

Pigg, K.B., 1992, Evolution of isoetalean lycopsids: Annals of the Missouri Botanical Garden, v. 79, p. 589–612.

Pigg, K.B., and Rothwell, G.W., 1979, Stem-root transition of an Upper Pennsylvanian woody lycopsid: American Journal of Botany, v. 66, p. 914–924.

Pigg, K.B., and Rothwell, G.W., 1983a, *Chaloneria* gen. nov.; heterosporous lycophytes from the Pennsylvanian of North America: Botanical Gazette, v. 144, p. 132–147, doi: 10.1086/337354.

Pigg, K.B., and Rothwell, G.W., 1983b, Megagametophyte development in the Chaloneriaceae fam. nov., permineralized Paleozoic Isoetales (Lycopsida): Botanical Gazette, v. 144, p. 295–302, doi: 10.1086/337376.

Pigg, K.B., and Rothwell, G.W., 1985, Cortical development in *Chaloneria cormosa* (Isoetales), and the biological derivation of compressed lycophyte decortication taxa: Paleontology, v. 28, p. 533–545.

Pigg, K.B., and Taylor, T.N., 1985, *Cormophyton* gen. nov., a cormose lycopod from the Middle Pennsylvanian Mazon Creek flora: Review of Palaeobotany and Palynology, v. 44, p. 165–181, doi: 10.1016/0034-6667(85)90014-4.

Prezant, R.S., Toll, R.B., Rollins, H.B., and Chapman, E.C., 2002, Marine macroinvertebrate diversity of St. Catherines Island, Georgia: American Museum Novitates, no. 3367, p. 1–31.

Retallack, G.J., 1992, What to call early plant formations on land: Palaios, v. 7, p. 508–520.

Retallack, G.J., 1997, Earliest Triassic origin of *Isoetes* and quillwort evolutionary radiation: Journal of Paleontology, v. 7, p. 500–521.

Retallack, G.J., 2000, Ordovician life on land and Early Paleozoic global change, *in* Gastaldo, R.A., and DiMichele, W.A., eds., Phanerozoic terrestrial ecosystems: Paleontological Society Papers, v. 6, p. 20–45.

Rindsberg, A.K., 1994, Ichnology of the Upper Mississippian Hartselle Sandstone of Alabama, with notes on other Carboniferous formations: Geological Survey of Alabama Bulletin, v. 158, p. 1–107.

Rothwell, G.W., and Erwin, D.M., 1985, The rhyzomorphic apex of *Paurodendron*: Implications for homologies among the rooting organs of Lycopsida: American Journal of Botany, v. 72, p. 86–98.

Runnegar, B., and Newell, N.D., 1974, *Edmondia* and the Edmondiacea: Shallow-burrowing Paleozoic pelecypods: American Museum Novitates, no. 2533, p. 1–19.

Sheehan, M.A., 1988, Ichnology, depositional environment, and paleoecology of the Upper Pennington Formation (Upper Mississippian), Dougherty Gap, Walker County, Georgia [M.S. thesis]: Athens, University of Georgia, p. 1–211.

Smith, L., 1983, The depositional environment of the Mississippian age Hartselle Sandstone, *in* Tanner, W.F., ed., Near-shore sedimentology: Proceedings of the sixth symposium on coastal sedimentology: Tallahassee, Florida State University, Department of Geology, p. 217–230.

Smosna, R., 1996, Upper Mississippian Greenbrier/Newman Limestones, *in* Roen, J.B., and Walker, B.J., Eds., The atlas of major Appalachian gas plays: West Virginia Geological Survey Publication V-25, p. 37–40.

Stanley, S.M., 1970, Relation of shell form to life habits of the Bivalvia (Mollusca): Geological Society of America Memoir 125, p. 1–296.

Stanley, S.M., 1972, Functional morphology and evolution of byssally attached mollusks: Journal of Paleontology, v. 46, p. 165–212.

Stapor, F.W., and Cleaves, A.W., 1992, Mississippian (Chesterian) sequence stratigraphy in the Black Warrior Basin: Pride Mountain Formation (lowstand wedge) and Hartselle Sandstone (transgressive systems tract): Gulf Coast Association of Geological Societies Transactions, v. 42, p. 683–696.

Stapor, F.W., Driese, S.G., Srinivasan, K., and Cleaves, A.W., 1992, The Hartselle Sandstone and its contact with the underlying Monteagle Limestone: A Lower Chesterian transgressive systems tract and sequence boundary in central Tennessee: Studies in Geology, University of Tennessee, v. 21, p. 79–108.

Strother, P.K., 2000, Cryptospores: The origin and early evolution of the terrestrial flora, *in* Gastaldo, R.A., and DiMichele, W.A., eds., Phanerozoic terrestrial ecosystems: Paleontological Society Papers, v. 6, p. 3–20.

Stubblefield, S.P., and Rothwell, G.W., 1981, Embryogeny and reproductive biology of *Bothrodendrostrobus mundus* (Lycopsida): American Journal of Botany, v. 68, p. 625–634.

Thomas, W.A., 1972, Mississippian stratigraphy of Alabama: Alabama Geological Survey Monograph 12, 121 p.

Thomas, W.A., 1982, Paleogeographic relationship of a Mississippian barrier-island and shelf-bar system (Hartselle Sandstone) in Alabama to the Appalachian-Ouachita orogenic belt: Geological Society of America Bulletin, v. 93, p. 6–19, doi: 10.1130/0016-7606(1982)93<6:PROAMB>2.0.CO;2.

Thomas, W.A., and Mack, G., 1982, Paleogeographic relationship of a Mississippian barrier-island and shelf-bar system (Hartselle Sandstone) in Alabama to the Appalachian-Ouachita orogenic belt: Geological Society of America Bulletin, v. 93, p. 6–19, doi: 10.1130/0016-7606(1982)93<6:PROAMB>2.0.CO;2.

Traverse, A., 1988, Paleopalynology: London, Unwin Hyman, 600 p.

Walker, R.G., ed., 1984, Facies models (2nd edition): Geoscience Canada Reprint Series 1, 317 p.

Watkins, R., 1975, Paleoecology of some Carboniferous Pectinacea: Lethaia, v. 8, p. 125–131.

Wynn, T.C., and Read, J.F., 2002, Eustasy and tectonics of a Mississippian carbonate ramp, West Virginia, USA: Geological Society of America Abstracts with Programs, v. 34, no. 6, p. 3–11.

Yeilding, C.A., 1984, Stratigraphy and sedimentary tectonics of the Upper Mississippian Greenbrier Group in eastern West Virginia [M.S. thesis]: Chapel Hill, University of North Carolina, 117 p.

MANUSCRIPT ACCEPTED BY THE SOCIETY 28 JUNE 2005

Geological Society of America
Special Paper 399
2006

The Hancock County tetrapod locality: A new Mississippian (Chesterian) wetlands fauna from western Kentucky (USA)

William J. Garcia
Glenn W. Storrs
Cincinnati Museum Center, Geier Collections and Research Center, 1301 Western Avenue, Cincinnati, Ohio 45203, USA, and University of Cincinnati, Department of Geology, ML 0013, Cincinnati, Ohio 45221-0013, USA

Stephen F. Greb
Kentucky Geological Survey, 228 MMRB, University of Kentucky, Lexington, Kentucky 40506, USA

ABSTRACT

The earliest tetrapods are known from a handful of Upper Devonian and Lower Carboniferous localities in Europe, North America, and Australia. All Upper Devonian sites and virtually all Early Carboniferous faunas are regarded as predominantly aquatic and most occur within, or are associated with, wetland habitats. A new mid-Carboniferous (Elvirian, Namurian A) fossil locality in Kentucky preserves the first tetrapod fauna from the eastern portion of the Illinois Basin. Four distinct facies at the locality have yielded vertebrate material. Diverse faunas have been found in an abandoned channel/oxbow facies and a floodplain/lake facies.

The abandoned channel/oxbow facies contains Colosteidae, Embolomeri, Rhizodontida, Dipnoi, Xenacanthiformes, Palaeonisciformes, and Gyracanthidae remains. This assemblage is similar to known Mississippian freshwater and brackish-water faunas, providing further evidence of a cosmopolitan tetrapod province during the Mississippian. A different fauna, rich in tetrapods but lacking fish, is associated with granular carbonate masses, rooting structures, and a paleosol in the floodplain/lake facies. Isolated and associated tetrapod elements from this facies exhibit morphological adaptations that may suggest a fauna of more highly terrestrial vertebrates than previously known from the North American Mississippian.

Keywords: tetrapod, Mississippian, wetland, Illinois Basin.

INTRODUCTION

Stem tetrapods are first recorded from the Frasnian (ca. 370 Ma), with their record extending from the Late Devonian into the Early Carboniferous. Fewer than three dozen Upper Devonian (ca. 377–362 Ma) and Lower Carboniferous (ca. 362–323 Ma) tetrapod localities are known worldwide. In the past 20 years, vigorous investigation of these sites and the discovery of new faunas have greatly advanced our knowledge of early tetrapod morphology (Clack, 1988, 1994, 2002; Coates and Clack, 1991; Lombard and Bolt, 1995; Coates, 1996; Milner and Lindsay, 1998; Warren and Turner, 2004) and distribution (Bolt et al., 1988; Ahlberg, 1995; Thulborn et al., 1996). Reassessment of known fossils (Ahlberg, 1991; Daeschler et al., 1994), and new discoveries (Lebedev and Clack, 1993; Lebedev and Coates, 1995; Daeschler, 2000; Zhu et al., 2002; Clement et al., 2004) have revealed a more diverse array of Devonian taxa than previously suspected.

While the last two decades have seen a great improvement in the quality of the Devonian and Mississippian fossil record, the early record of tetrapods in the Devonian and Carboniferous contains significant gaps. The most notable is a 30-million-year period at the base of the Mississippian that is represented by only two localities (Fig. 1). Known Devonian taxa show numerous adaptations that separate them from their piscine ancestors. However, these forms are clearly distinct from Mississippian tetrapods. The morphological disparity and taxonomic diversity of late Mississippian tetrapods implies that their primary diversification occurred early in the Mississippian, as they adapted to a variety of unoccupied niches.

The origin of terrestrial ecosystems bears directly on the early radiation and diversification of tetrapods, and on the aquatic to terrestrial vertebrate transition. This transition occurred in facies associated with lowland wetland habitats that occupy the transition between aquatic and terrestrial ecosystems. Devonian tetrapods for which extensive material is known were clearly aquatic (Coates and Clack, 1991; 1995; Lebedev and Coates, 1995; Jarvik, 1996; Clack et al., 2003a, 2003b). The presence of lateral line canals, paddle-like limbs, and a caudal fin on the tail argues convincingly that *Acanthostega* and *Ichthyostega* were highly aquatic forms (Coates 1996; Jarvik, 1996; Clack et al., 2003a, 2003b). Additionally, *Acanthostega* appears to have retained functional gills (Coates and Clack, 1991, 1995; Coates, 1996). Although it has been argued that certain morphological features of Devonian forms are associated with terrestrial abilities (Carroll and Green, 2003), the overwhelming morphological evidence indicates that known Devonian tetrapods were highly aquatic. Numerous Carboniferous forms retained aquatic adaptations (lateral line systems in the skull table and cheeks, reduction in vertebral ossification, reduced limbs) but also possessed characters associated with increased terrestriality (Holmes, 1984, 1980; Clack, 2002, 2001). All Devonian and most Carboniferous tetrapods were almost certainly non-amniotes, necessitating placement of eggs in a moist environment. While certain extant lissamphibians have evolved strategies for keeping eggs moist in non-wetland environments, the majority of these taxa are restricted to wetlands (Duellman and Trueb, 1986; Stebbins and Cohen, 1995). Thus, early tetrapods, like modern Lissamphibia, almost certainly possessed physiological constraints that forced them to inhabit wet or moist environments. This apparent duality, adaptations for terrestriality but constraints necessitating access to aquatic environments, explains the great importance of wetland ecosystems for early tetrapods. Environments such as ponds, swamps, small streams, and floodplains provided a wide variety of new ecological niches for early tetrapods to exploit during their diversification in the Early Carboniferous. These niches provided access to both aquatic and terrestrial ecosystems, an important feature for organisms with adaptations to both types of environment. At the same time, the expanding diversity of land plants and arthropods in the Devonian and Carboniferous increased the disparity of wetland niches, providing additional habitats and potential prey for early tetrapods (DiMichele and Hook, 1992; Shear and Selden, 2001; Gensel, 1986; Scheckler, 1986). For these reasons wetland ecosystems are of key importance for understanding the evolution of early tetrapods.

The majority of Early Carboniferous tetrapod localities are located in Great Britain and North America, with the notable exception of the Middle Paddock locality of western Australia. These localities are situated along a belt roughly 20° north and south of the paleoequator (Fig. 2). North American localities are concentrated within the Appalachian and Illinois Basins of the United States and Nova Scotia in Canada, while British localities are predominantly from the Midland Valley of Scotland.

All known Early Carboniferous tetrapod faunas derive from a narrow range of environments that are all wetland ecosystems (Milner et al., 1986; DiMichele and Hook, 1992). Mississippian wetland faunas, typically deposited in paludal, small pond or forest-swamp settings, show a high degree of similarity. Most elements of these faunas are decidedly aquatic in their morphology, and composition of these faunas is broadly comparable (Milner, 1993). One major exception to this pattern is known. The unique thermal pond fauna of East Kirkton, Scotland (Lower Carboniferous, Visean) contains unequivocally terrestrial animals (Fig. 1.) (Milner and Sequeria, 1994; Smithson, 1994; Smithson et al., 1994; Clack, 2001).

These ecological/environmental restrictions are not limited to the Mississippian; they are seen also in the early Pennsylvanian (Westphalian). At this time, most faunas are derived from autochthonous, coal-bearing, fluviodeltaic sequences (Hook and Ferm, 1988; DiMichele and Hook, 1992). These deposits represent tropical lowland wetland ecosystems. Although it has been argued that certain components of these faunas were derived from more upland faunas (Milner, 1980; Boyd, 1984), sedimentological evidence suggests that these species coexisted in a single ecosystem (Hook and Hower, 1988). Additional Westphalian tetrapod faunas derive from more terrestrial assemblages, the best known of which is Joggins, Nova Scotia (Upper Carboniferous, Westphalian A). To date, the earliest known fully terrestrial vertebrates in North America are from Joggins (Fig. 1). Here, early amniotes are found inside *Sigillaria* stumps (Dawson, 1868; Carroll, 1967), a lycopod swamp-forest habitat.

A new Lower Carboniferous (Mississippian, Chesterian) site in Hancock County, Kentucky, records the first known tetrapods in the eastern Illinois Basin (Fig. 3). In this paper we present a preliminary description of the vertebrate fauna from two facies at the Hancock County locality. We compare the fauna from these two facies with other Mississippian freshwater faunas. In addition, we discuss differences in faunal composition and morphological features of tetrapod specimens recovered from a flood plain/lake facies at the Hancock County. These morphological features include terrestrial adaptations in early tetrapods at Hancock County when compared with faunas in lacustrine facies from other Mississippian faunas. The morphological adaptations have implications for the timing of terrestriality in tetrapods.

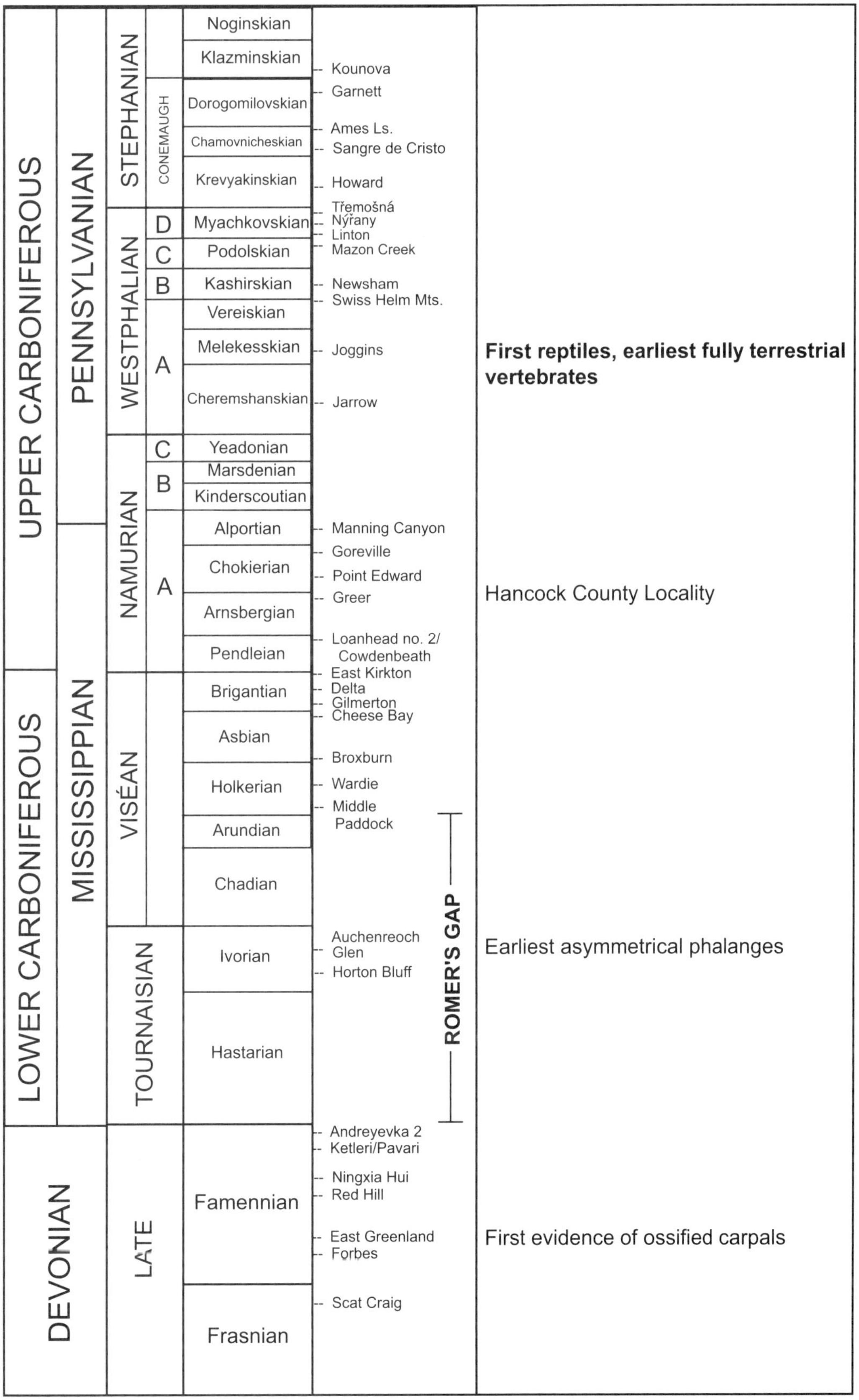

Figure 1. Geological time scale showing occurrences of early tetrapod faunas and timing of major events in the evolution of terrestriality (modified from Clack and Carroll, 2000, and Marshall et al., 1990).

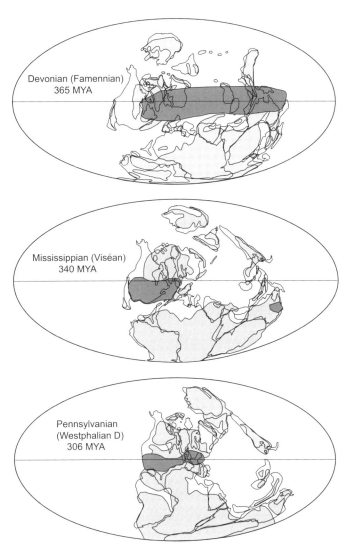

Figure 2. Estimated distribution of tetrapods during the Late Devonian, Early Mississippian, and Pennsylvanian. Gray shading represents terrestrial environments and dark shading represents the distribution of tetrapods. Modified from Scotese, 2000 (Paleomap project) and Milner, 1993.

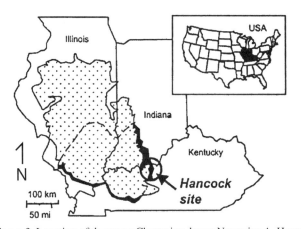

Figure 3. Location of the upper Chesterian, lower Namurian A, Hancock County site in the Illinois (Eastern Interior) Basin of North America.

Stratigraphy and Geologic Setting

The Hancock County site (data on file at Cincinnati Museum Center, CMC) lies on the eastern margin of the Illinois (Eastern Interior) Basin (Fig. 3). Four facies yield well-preserved vertebrate remains. Upper Chesterian rocks in the basin are divided into alternating carbonate and clastic formations, which thin onto the eastern margin of the basin. Along the basin margin in parts of Kentucky, the interval from the base of the Vienna Limestone to the base of Pennsylvanian strata is mapped as the Buffalo Wallow Formation (Weller, 1913; Rice et al., 1979). Some of the carbonate formations that occur basinward are reduced to members or beds within the Buffalo Wallow Formation.

At the Hancock County site, a complex suite of marginal marine and terrestrial rocks above the Menard-equivalent limestone is exposed (Fig. 4). The Clore Limestone is missing here, apparently truncated by the sub-Pennsylvanian unconformity. A heterolithic sandstone above the Menard-equivalent limestones, interpreted as a laterally accreting, fluvial to upper estuarine paleochannel, correlates with the Palestine Sandstone deeper in the basin (personal observation). Carbonaceous shales, cross-cutting paleochannels, and paleosols overlie the Palestine-equivalent sandstone at the Hancock County site (Fig. 4). Some of the facies in this complex are tentatively equated with the upper Palestine because at least one of the cross-cutting channel fills is similar to the underlying heterolithic paleochannel. A pyritic coaly shale drapes a scour and may also be upper Palestine, or mark the beginning of the Clore transgression (personal observation).

At Hancock County, vertebrate material has been found in the Menard Limestone, the heterolithic channel equated to the Palestine Sandstone, a floodplain/lake facies with paleosols above the main heterolithic channel, and the scour-filling carbonaceous shale that may be equivalent to the Clore transgression (Fig. 4, Table 1). Vertebrate material from the lower two facies will be discussed in future papers. Material from the upper two facies is discussed herein.

Rocks from the Hancock site represent distal and nearshore marine, to estuarine and/or lacustrine, to terrestrial/floodplain paleoenvironments. All have produced vertebrate fossils (Table 1). Two of these fossil-bearing facies preserve wetland environments and will be discussed below. Discussion of the faunas from the remaining two facies (distal and nearshore marine rocks) will be dealt with in future papers.

FACIES AND PALEOENVIRONMENTS

Flood Plain/Lake Facies

One of the facies above the main Palestine-equivalent sandstone is a 3-m-thick, tan to buff, unlaminated to poorly laminated shale and claystone (Fig. 5). The contact between this shale and lateral small channel fills is not well exposed, but the shale appears to be truncated eastward. The top of the

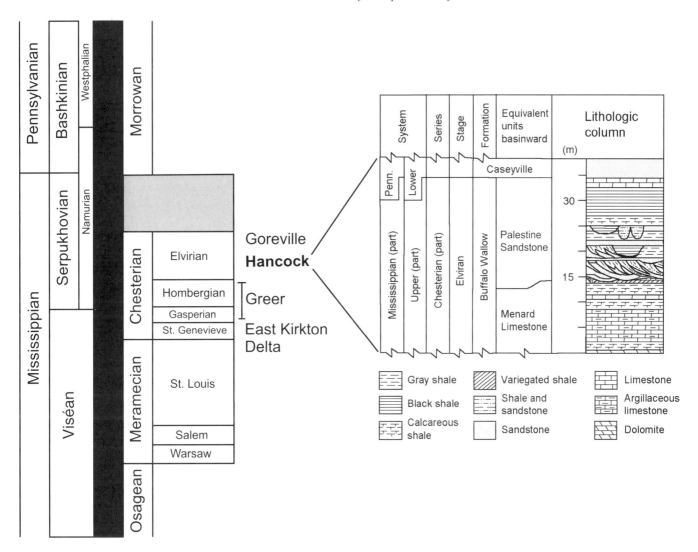

Figure 4. Stratigraphic section at the Hancock County site. Modified from Schultze and Bolt, 1996.

shale is not well exposed, although lateral units are capped by a poorly developed paleosol.

The tan to buff shale contains numerous cross-cutting slickenside surfaces. Siderite nodules and bands (20–35 cm) are concentrated along some slickensides. At least one of the siderite occurrences is bulbous (2–3 cm in diameter) and slightly upward-thickening, as would occur in a root trace. Macroscopic plant remains (carbonaceous fragments on bedding surfaces) occur along semi-continuous, horizontal surfaces in the lower 2 m of the unit but are disrupted by slickensides. The overall concentration of plant remains increases toward the base of the unit.

At least two horizons of well-cemented, granular carbonate masses occur in the lower meter of the unit. The carbonates are laterally discontinuous, up to 14 cm thick, and consist of pisolitic and sub-pisolitic (>2 mm) grains, clay fragments, and dark organic debris. Small hollow tubes (0.5–2 mm) and carbonaceous

TABLE 1. DISTRIBUTION OF MAJOR FAUNAL ELEMENTS

	Menard Limestone (Marine)	Palestine Sandstone (Fluvial)	Palestine Paleosol (Flood-plain)	Palestine Shale (Estuarine)
Holocephali	X			
Rhizodontida		X		X
Gyracanthus				X
Xenacanthiformes				X
Dipnoi				X
Embolomeri		X	X	
Colosteidae				X
Temnospondyli			X	
Whatcheeriid			X	

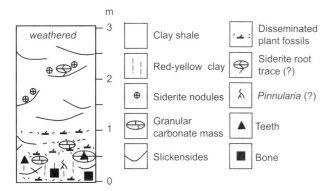

Figure 5. Vertical section through the flood plain/lake facies.

rootlets occur throughout the masses. Pseudolamination is preserved in the carbonate masses but is not preserved in the surrounding shales.

The upper carbonate horizon is overlain by a layer of shale with abundant plant fossil debris and is underlain by a 45–50 cm thick clay-rich horizon. The upper 20 cm of the clay is a mottled tan to yellow in color and contains at least one small, downward-branching feature, similar in appearance to *Pinnularia,* a common Mississippian root structure. The lower 25 cm of the clay-rich interval grades downward from yellow/ochre to red. The lower carbonate horizon is also overlain by thin shale layer with abundant carbonaceous debris.

Disarticulated, associated, and articulated tetrapod bones and teeth are concentrated in a 30 cm interval between the two carbonate horizons. Isolated material, in particular teeth, is found within the upper carbonate horizon.

Interpretation

The shale is situated above a fluvial to fluvio-estuarine channel (personal observation) and is laterally truncated by fluvial facies and an abandoned coaly shale-draped scour. The fine grain size, disseminated plant debris, rooting, and lack of marine trace or body fossils suggest a quiet freshwater environment, such as a distal floodplain or floodplain lake. The possible siderite rootlets and *Pinnularia* indicate intermittent colonization of the flood plain or lake by emergent vegetation. The granular carbonate masses bracket a clay-rich interval. Mottling within the clay indicates mineral leaching and paleosol development. The occurrence of disseminated plant debris and granular carbonates is consistent with an interpretation of a freshwater riparian environment with intermittent or marginal marshes. The unit's fine grain size indicates relative protection from clastic influx. Slickensides suggest seasonal alternations in precipitation (Retallack, 1988), as these features develop from the contraction and expansion of clay-rich substrates in alternating wet and dry seasons. Chesterian paleoclimates in eastern North America are interpreted as alternatively dry and humid (Cecil, 1990; Caudill et al., 1996; Miller and Eriksson, 1999).

Seasonal variation in precipitation may have accompanied this dry to humid alternation.

Abandoned channel/oxbow Facies

One of the scours above the Palestine-equivalent heterolithic channel at the Hancock County site truncates the stratigraphic level of the flood plain/lake facies and lateral channel fills (Fig. 4). The channel is 30–50 m wide. A 5–60 cm layer of black, pyritic, carbonaceous shale drapes the base of the scour. This shale contains vitrainous coal streaks that thicken into the scour (Fig. 6). Palynologic analyses indicate that *Lycospora orbicula, L. micropapillata,* and *Schulzospora,* with subdominant *Calamospora* are the dominant flora of the carbonaceous shale. *Stigmaria* and sideritic nodules (rootlets) underlie the scour and are continuous onto the limbs of the scour, where the coaly shale thins or is absent. On the limbs of the scour, carbonate concretions encase pyritized stigmarian rootlets, forming "coal balls."

Overlying the coaly shale is 30–45 cm of lighter, dark-gray, carbonaceous shale with common pyrite nodules and pyritized fossil wood fragments ranging in length from 5 to 30 cm. The upper contact of this carbonaceous shale is marked by a horizon of small (0.5–2 cm), marcasite nodules (Fig. 6). The marcasite horizon is overlain by 2 m of gray shale lacking pyrite. Vertebrate remains are restricted to this non-pyritic gray shale interval. Palynologic analyses indicate that this interval is dominated by *L. pusilla*, many of which are broken or abraded. The upper part of the non-pyritic gray shale contains two distinct horizons of carbonate concretions. The lowermost of these horizons consists of irregular carbonate concretions. The upper horizon includes elongate, oval-shaped concretions, which contain Dipnoan (lungfish) fossils. The upper concretion horizon is approximately equivalent to the level of coal balls on the limb of the scour (Fig. 6).

Interpretation

Coal-draped scours are common in Carboniferous strata of Kentucky and represent abandoned scours that developed into peat-forming wetlands (Greb and Chesnut, 1992; Eble and Greb, 1997). The presence of pyrite in the lower part of the fill indicates anaerobic substrates typical of wetland soils, and possible marine influences. Rising base level related to the Clore transgression down paleoslope in the basin may have initiated the accumulation of peat to form the coaly shale and the sulfides contained therein (Greb et al., in review).

Lycospora orbicula and *L. micropapillata* are products of *Paralycopodites*, a small, shrubby lycopod (DiMichele and Phillips, 1994). *Lycospora orbicula* and *L. micropapillata* are common in the basal increments of Carboniferous coals and coals that fill scours or abandoned channels (DiMichele and Phillips, 1994; Eble and Greb, 1997). *Schulzospora* is a product of the pteriodosperm (seed fern) *Lyginopteris*, which was common in Lower Carboniferous wetlands. This seed fern had small stems and is interpreted as vinelike with large plannated frond leaves (form genera of the *Sphenopteris* or *Pecopteris* type) (Stewart and Rothwell, 1993). *Lyginopteris* probably relied upon other plants (in this case, *Paralycopodites*) for support. The sphenopsid (rush) *Calamites*, a common inhabitant of Carboniferous wetlands, capable of growing to treelike stature, produced *Calamaspora* (DiMichele and Phillips, 1994). The palynological assemblage in the basal carbonaceous shale suggests a lacustrine scrub-shrub wetland or small forest swamp, which infilled an abandoned scour.

The marcasite nodules mark a distinct change in water chemistry within the fill. Overlying sediments lack pyrite and are fossiliferous, suggesting deposition in more oxygenated waters as the scour filled. The marcasite nodules themselves may represent coprolites that accumulated on top of the buried peaty shale. The overlying non-pyritic shales contain a diverse freshwater fauna typical of a small oxbow lake or pond. The diversity within the scour, in addition to the large size of certain taxa, suggests intermittent connection to a larger lake or river when it was filled. Modern oxbow lakes and ponds are often successively in-filled by flooding from nearby rivers.

The non-pyritic gray shale contains a palynoflora different from the basal carbonaceous shales. Samples from this part of the fill are dominated by *L. pusilla*, the spores of *Lepidodendron*. This arborescent lycopod is a common constituent of Carboniferous swamp forests (DiMichele and Phillips, 1994). Spores are mostly broken, suggesting transport. *Lepidodendron* needed a wet substrate to survive, but its lack of pneumatophores indicates that the lycopod did not occupy a habitat consisting of standing water. Standing water was necessary to support the large rhizodont fish and other fauna found within the shale. The source of these spores may have been *Lepidodendron* swamp forests situated between the oxbow pond and riverine source area, rather than in or adjacent to the pond itself.

The top of the hill is characterized by carbonate concretions containing dipnoans (Fig. 6). This marks a point at which the pond was nearly filled, and perhaps influenced by seasonal drying. The carbonates containing lungfish are at a stratigraphically similar horizon to the rootlet-coal balls beneath the scour on the scour limbs. Carbonate precipitation across the scour boundary suggests a mechanism that was partly independent of depositional facies, such as a change in groundwater level. Water level fluctuations were common in the seasonal paleoclimates inferred for the Chesterian (Cecil, 1990; Caudill et al., 1996; Miller and Eriksson, 1999).

VERTEBRATE FAUNA

Abandoned Channel/Oxbow Facies Fauna

Tetrapods include a nearly complete colosteid (CMC VP7288), with only the anterior third of the skull and the distal half of the caudal vertebral series missing (Fig. 7A, 7B, Table 1). The body is elongate and the limbs small, indicating highly aquatic habits. The disparity in size between the maxillary and larger dentary teeth, tri-radiate parasphenoid, rhachitomous vertebrae, open palate, absence of an otic notch/temporal embayment, and robust single dorsal process of the ilium indicate its colosteid affinity. This specimen is distinguished from known colosteids by the unusual dorsal osteoderms covering it from just anterior of the pectoral girdle to immediately caudad of the pelvis. The thick, rectangular osteoderms are arranged in diagonal rows of 6–10 elements. Each row is oriented 55° posterolaterally to the axis of the vertebral column. Rows from opposite sides of the column alternate position along its axis. The long axes of individual osteoderms parallel the rows. Each is pitted dorsally, while the smooth lateral and posterior edges are beveled for squamous articulation with adjacent scutes. The dorsal scales of the well-known colosteid *Greererpeton* are subcircular and distinctly thinner (Romer, 1972). The scales of CMC VP7288 most closely resemble those of the Pennsylvanian colosteid *Colosteus* (Hook, 1983). Isolated vertebrae, scales, maxilla fragments, and a lower jaw also represent this taxon.

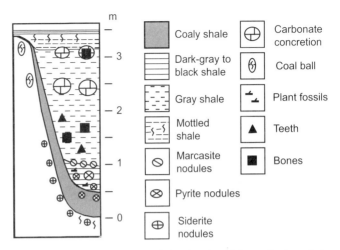

Figure 6. Vertical section through the abandoned channel/oxbow facies.

Dorsal squamation is unusual in early tetrapods but is present in some advanced, presumably terrestrial temnospondyls (e.g., *Peltobatrachus, Cacops, Dissorophus,* and *Broiliellus*). Their large scutes likely served to strengthen the vertebral column (DeMar, 1968). However, the small size and mail-like arrangement of scutes in the Hancock specimen, along with its probable aquatic habit, argue against such a function here. Notably, the occurrence of dermal armor at the Hancock County site coincides with the presence of large, predatory fish in the same deposit (Table 2).

Sarcopterygian rhizodont material is abundant in this facies (Fig. 7D). It includes a partly articulated individual (CMC VP6915) that is the most complete specimen of a giant rhizodont known. It includes skull table and temporal elements, opercular and gular plates, premaxillaries, left maxilla, vomers, complete right mandible, clavicles, cleithra, various fin elements, and numerous scales. Living length of this animal was ~4 m. A second smaller individual is known from an articulated shoulder girdle and partial

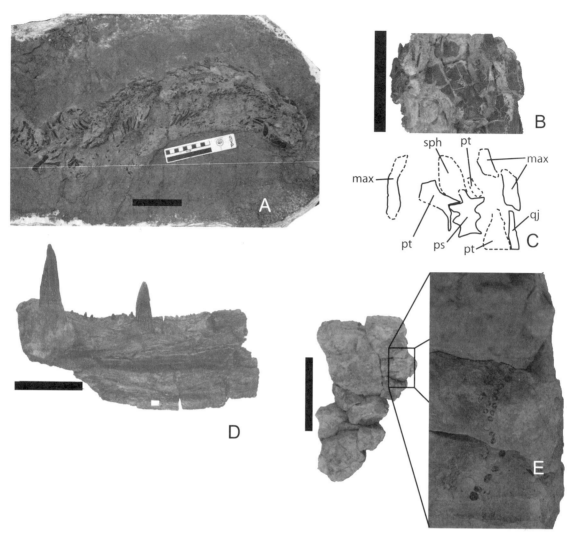

Figure 7. Vertebrate fossils from the abandoned channel/oxbow facies. A. Dorsal view of colosteid. Anterior is to the right. B. Palatal view of colosteid skull. C. Drawing and interpretation of colosteid palatal elements (max = maxillary, ps = parasphenoid, pt = pterygoid, qj = quadratojugal, sph = sphenethmoid) D. Lingual view of rhizodont lower jaw. E. Lungfish burrow with inset of ribs in cross section. All scales are 10 cm.

Table 2. Faunal Comparison of select North American Lower to mid-Mississippian fish and tetrapod localities

Hancock, KY	Goreville, IL	Greer, WV	Delta, IA	Point Edward, NS
Rhizodont		Rhizodont	Rhizodont	?*Strepsodus*
Tranodis	*Tranodis*	*Tranodis*	*Tranodis*	*Sagenodus*
Palaeoniscoids		Palaeoniscoids	Palaeoniscoids	
Xenacanthids				
Gyracanthus		*Gyracanthus*		*Gyracanthus*
Colosteid	?*Greererpeton*	*Greererpeton*	Colosteid	Colosteid
Embolomere	Embolomere	*Proterogyrinus*	Embolomere	Proterogyrinid
		Whatcheeriid		
	Microsaur			
		?*Crassigyrinus*		
				Spathicephalus

cranial material. Although partial remains are distributed worldwide in Upper Devonian and Carboniferous rocks, Rhizodontida is poorly known with few articulated specimens (Andrews, 1985; Jeffery, 2001). The oldest and, to date, most complete rhizodonts are from Antarctica (Young et al., 1992) and Australia (Johanson and Ahlberg, 1998; Johanson et al. 2000; Long, 1989), leading to speculation that this clade, near the base of Tetrapodamorpha, originated in eastern Gondwana (Cloutier and Ahlberg, 1996). North American material is typically incomplete (Daeschler and Shubin, 1998; Davis et al., 2001).

Putative aestivation burrows of lungfish (cf. *Tranodis*) (Fig. 7E) containing entombed skeletons occur within a 30 cm thick horizon near the top of the upper shale at the Hancock County site. These are the oldest known dipnoan burrow structures with skeletons. Carroll (1965) described apparent burrow traces from the Pennsylvanian of the Michigan Basin and *Gnathorhiza* burrows and skeletons are well known from the Lower Permian of New Mexico, Oklahoma, and Texas (Berman, 1979a, 1979b, 1979c; Olson and Daly, 1972; Romer and Olson, 1954; Vaughn, 1964). Smaller burrows, lacking skeletal material, from the Devonian Catskill Group of Pennsylvania are only questionably attributable to dipnoans. Additional fish fossils from the Hancock County are xenacanth coprolites, some of which contain palaeonisciform scales, and numerous, large gyracanthid acanthodian pectoral and pelvic spines.

Floodplain/Lake Facies Tetrapod Fauna

The interval between the two horizons of granular carbonate masses in the lower part of the floodplain/lake facies preserves more than 100 elements, including skull fragments, jaws, vertebrae, ribs, limb girdle bones, propodials, epipodials, phalanges, and teeth. This represents the oldest known paleosol containing tetrapod fossils. Anthracosaur femora of various sizes, intercentra, and pleurocentra are relatively common. Vertebrae of a large temnospondyl have also been recovered. Numerous limb elements of indeterminate taxa are stout with well-developed condyles, suggesting large, robust limbs compatible with a more terrestrial lifestyle than seen in most other Mississippian tetrapods.

An embolomerous vertebra (CMC VP7279, Figure 8A, 8B) with fused pleurocentrum and neural arch represents a previously unknown form. The pleurocentrum is a complete disc, similar to those from known embolomeres such as *Pholiderpeton* (Clack, 1987), *Archeria* (Holmes, 1989a) and *Calligenethlon* (Carroll, 1967), but there is no evidence of a dorsal suture as in *Proterogyrinus* (Holmes, 1984). It is strongly amphicoelous (possessing concave anterior and posterior faces) with a wide notochordal canal rather than the small perforation typical of *Archeria* and *Pteroplax* (Holmes, 1989a; Boyd, 1980). No intercentral facet is apparent. The transverse processes are unusually tall, with long axes directed sub-vertically. Well-developed zygopophyses are set near the midline, but with sub-horizontal articular faces as in seymouriamorphs. This orientation would resist rotational movement of the body and confine the vertebral column to lateral

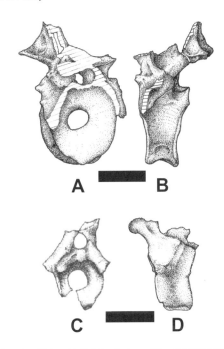

Figure 8. Anterior (A) and left-lateral views (B) of CMC VP 7279 and anterior (C) and left-lateral (D) views of CMC VP 7278. Scale bar = 1 cm

flexion (Panchen, 1977). A supraneural canal is narrow, relative to other anthracosaurs.

A second pleurocentrum and fused neural arch from an unknown taxon (CMC VP7278, Figure 8C, 8D) exhibits wide buttresses that descend from the neural arch and support short, ventrally directed diapophyses. Here, the zygopophyses lie near the midline with medially inclined articular surfaces (~45°). Highly angled zygopophyses in this specimen may equate with complex trunk motion, namely, lateral flexure coupled with rotation around the spinal axis (Holmes, 1989b). Morphological differences between these vertebral types possess potential taxonomic significance, though these differences could represent variation among vertebrae of the same column. Resolution of these issues must await the collection of more material.

Additional elements exhibiting morphological adaptations to terrestrial locomotion include several ilia recovered from the paleosol. Three ilia represent various sizes of the same taxon. The largest (CMC VP7328) (Fig. 9) displays dorsal and posterior processes somewhat resembling those of *Whatcheeria* and *Pederpes*. The dorsal process is fan shaped and is 2.0 cm at its greatest length. Postdepositional compression appears to have flattened portions of the specimen, including the dorsal process; the narrow fan-shaped appearance may be an artifact of preservation. The posterior process is short and relatively stout, broadening slightly over its posterior half. It is distinctly necked dorsoventrally as it emerges from the body of the ilium. The most notable feature of this specimen is the broad dorsal roof of the acetabulum. The acetabular surface is oval in shape, with its long axis oriented primarily dorso-ventrally. The surface is inclined

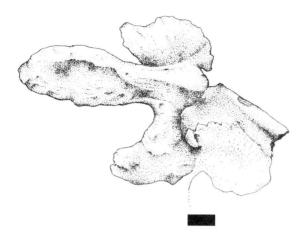

Figure 9. Right lateral view of CMC 7328. Scale bar = 2 cm

such that the articular surface faces ventrolaterally at ~25° to the horizontal plane.

The smaller size class is represented by two left ilia (CMC 7261 and CMC 7664). These specimens also possess a long posterior process and a dorsal process. In neither specimen is the posterior process completely preserved; thus its true length is unknown. A broken rim of bone projecting mediodorsally from the dorsal surface of the ilium indicates the presence of a dorsal process. In both specimens, this projection is broken at its base, suggesting a thin, easily broken process. Like the larger specimen, the acetabulae are oval in shape, but the articular surfaces face more posterior-ventrally in these smaller forms. The ventrally facing dorsal acetabular rim in these specimens provided a surface against which the femoral head would have braced during locomotion. Such a bracing mechanism is unnecessary for aquatic locomotion, as the femur is moved in an anterior/posterior motion when not held against the body. For terrestrial locomotion, this would provide much-needed support.

DISCUSSION

Comparison with Other Mississippian Faunas

Typical Mississippian faunas consist primarily of members of three early tetrapod clades: Embolomeri, Colosteidae, and Baphetidae (Godfrey, 1988). Although other taxa characterize certain Mississippian deposits (*Crassigyrinus*, *Whatcheeria*, *Eoherpeton*), taxa from these three clades are the dominant and most numerous forms. Dominance of Mississippian faunas is not broadly restricted to the clade level, but certain taxa seem to typify these faunas (Table 2). Milner has used this broad faunal similarity to argue that the Euramerican tetrapod fauna represents a single biogeographical province, termed the Mississippian Tetrapod Province (Milner, 1993). Because of ecological bias and the small number of faunas in the early tetrapod fossil record, robust conclusions cannot be drawn from the comparison of these assemblages. The Hancock County assemblage, however, adds evidence for the validity of this province. Preliminary collecting at Hancock has produced a proterogyrinid and an articulated colosteid specimen.

Early Carboniferous tetrapod assemblages are rare, so little is known about their ecosystem structure. However, nearly all known are of fresh- to brackish-water origin, and some comparisons can be made. Tetrapod taxa typical of a Mississippian Tetrapod Province include colosteids, loxommatids, embolomeres, and the aberrant *Crassigyrinus* (Milner, 1993). The abandoned channel/oxbow facies from the Hancock County site contains a typical Mississippian lacustrine/paludal fish fauna with strong similarity to those previously known from North America and Europe. In particular, the Hancock fish fauna is most similar to those of Greer, West Virginia, and Delta, Iowa (Table 2; Godfrey, 1988; Schultze and Bolt, 1996). The depositional setting of the Greer locality has been debated and it has variably been suggested as fluvial point bar to overbank deposits (Busanus, 1974), or marine with periodic encroachment of nearshore mud (Elliott and Taber, 1982).

The other known Illinois Basin Lower Carboniferous tetrapod locality—Goreville, Illinois—has a similar tetrapod fauna (Schultze and Bolt, 1996). Both the Greer and Goreville faunas contain common dipnoans (cf. *Tranodis*), and each preserves a colosteid and an embolomere. The dominant rhizodonts of Hancock are absent from Goreville, although rhizodont material of distinctly smaller taxa is known from Greer (Godfrey, 1989). In addition, the acanthodians, palaeoniscoids, and xenacanthids known from Hancock are absent from Goreville. Differences between the Goreville locality and other Mississippian vertebrate faunas have been ascribed to the relative lack of marine influence at Goreville. Fossil-bearing strata at Goreville represent clastic sedimentation between marine limestones (Schultze and Bolt, 1996).

Implications for Tetrapod Terrestriality

The most significant aspect of the Hancock County site is the preservation of a riparian flood-plain fauna that contains tetrapod elements with morphological characteristics suggesting terrestrial locomotion. Modern riparian areas provide a critical habitat for amphibians and reptiles, due to the abundance of water and accessible transition between aquatic, wetland, and upland habitats.

The 30-million-year "Romer's Gap" of the basal Carboniferous fossil record hinders knowledge of early tetrapod radiations (Coates and Clack, 1995). Recent discovery of the whatcheeriid tetrapod *Pederpes* (whatcheeriids are stem-tetrapods known from the Mississippian of North America and Great Britain) at the base of this gap suggests that even the earliest Mississippian tetrapods were partially adapted to limited terrestrial locomotion, although they were still primarily aquatic (Clack, 2002). *Pederpes* displays asymmetrical phalanges associated with turning of the manus during walking (Clack, 2002). Finds in the Lower Carboniferous at East Kirkton and elsewhere hint at a radiation much broader in scope than previously suspected. The East Kirkton deposit is atypical of Mississippian faunas in that the pond shales are shales deposited within a hot spring setting and interbedded with volcaniclastic sediments (Rolfe et al., 1990). More significantly, the fauna of East Kirkton shows various morphological adaptations for terrestrial locomotion (Clack, 2001; Milner and Sequeira, 1994; Smithson, 1994; Smithson et al., 1994), and many of the specimens were likely washed into the pond from surrounding habitats (Clarkson et al. 1994). The presence of robust limb elements (femora and humeri) and the ossification of carpals and tarsals indicate that the majority of faunal elements from East Kirkton were capable of terrestrial locomotion. Although some of the taxa bearing these features are of uncertain affinity, representatives of at least three early tetrapod clades (Milner and Sequeira, 1994; Smithson, 1994; Smithson et al., 1994) exhibit terrestrial adaptations. Two other taxa of uncertain taxonomic affinities, *Eucritta* (Clack, 2001) and *Eldeceeon* (Smithson, 1994), also exhibit terrestrial morphological adaptations. An exception to the terrestrially dominated nature of the East Kirkton fauna is *Silvanerpeton*, with morphology suggestive of a more aquatic existence (Clack, 1994).

The taxonomic uncertainty of some East Kirkton taxa does not diminish their importance as indicators of great diversity among terrestrially adapted Carboniferous tetrapods. Rather, this evidence indicates that terrestrially adapted tetrapods may have been widespread by the mid-Mississippian. Evidence of highly terrestrial taxa in the Viséan indicates that a potential ecological bias in preservation of Carboniferous assemblages has obscured a portion of the pattern of early tetrapod diversity and disparity. Prior to the discovery of the East Kirkton site, the oldest known Carboniferous locality to yield terrestrial tetrapod remains was Joggins, Nova Scotia—a gap of ~20 million years. Joggins is lower Pennsylvanian (Langsettian, Westphalian A) in age and represents another unique depositional setting. Vertebrates, including the world's oldest reptiles, are preserved inside fossilized lycopod stumps (Dawson, 1868). The reptiles were originally inferred to have been trapped as pitfalls within the trees; more recently, it has been suggested that the animals were trapped within dens during a forest-swamp fire (Calder et al., this volume). Additional support for an early and more widespread terrestrial fauna in the Mississippian comes from Scotland. A single specimen of the terrestrially adapted taxon *Casineria* is known from the Asbian (Lower Carboniferous, Visean) of Cheese Bay (Paton et al., 1999).

Vertebral elaboration was a major adaptation of advanced tetrapods and distinguishes them from their more aquatic precursors. Elaboration of the vertebral column is associated with increased strength and stability of the axial skeleton, and would have allowed greater access to terrestrial settings. Modifications include increased ossification of the central elements and fusion with the neural arch, reduction of the notochord, and development of zygopophyses. However, the earliest tetrapods typically have multipartite centra with separate neural arches. Even probable terrestrial taxa from East Kirkton (with robust limbs, ossified carpals and tarsals) possess multipartite centra (Clack, 1994; Milner and Sequeira, 1994; Smithson, 1994). Apart from lepospondyls, nearly all early tetrapods that possessed single central elements fused to their neural arches are terrestrial. While microsaurs, nectridians, and aïstopods exhibit fusion/suturing, the vast majority of lepospondyls are of small body size, come from aquatic deposits, and/or possess aquatic adaptations (Carroll, 1997; Carroll et al., 1998). Early temnospondyls, as well as colosteids, lack central element/neural arch fusion. In the derived stereospondyls, the intercentra become the dominant elements of the vertebrae, while the pleurocentra are reduced (Schoch and Milner, 2000). Among known embolomeres, fusion does not appear until the Permian. Advanced tetrapods that are thought to have been highly terrestrial, such as seymouriamorphs and diadectomorphs, also exhibit fusion. Their terrestrial adaptations include the reduction or loss of the intercentrum, consequent stiffening of the vertebral column, and a sub-horizontal orientation of the zygopophyses.

Some of the tetrapod remains from the floodplain/lake facies at Hancock County show attributes typically associated with terrestriality among vertebrates. The fauna is thus advanced in comparison to other Lower Carboniferous tetrapods, except those of East Kirkton. At the Hancock County site, specimen CMC VP7279 (Fig. 8A, 8B) exhibits a sub-horizontal orientation of the zygopophyses. As vertebral fusion is typically associated with terrestriality (with the exception of lepospondyls), the morphology of CMC VP7278 and CMC VP7279 is an indication of a terrestrial habit. Ilial morphology strongly indicates adaptations for terrestrial locomotion, namely the robust acetabular roof, which would have braced the femur during locomotion.

The fauna is associated with a paleosol and floodplain or small flood-plain lake setting. The absence of fish in the floodplain/lake facies is different from other Lower Carboniferous tetrapod assemblages (Table 1). The absence of fish also suggests that the vertebrate material was not likely transported from adjacent or nearby fluvial sources. The robust nature of associated limb elements supports the idea that the paleosol taxa were more terrestrially active than most known early tetrapods. The morphology of these elements does not establish terrestrial habits for these taxa, but the absence of fish remains and the presence of paleosol features is strong secondary evidence for some degree of terrestrial activity.

The Hancock County site provides evidence of a potentially more terrestrial fauna in North America during the Chesterian (Namurian A). In particular, morphological adaptations of CMC VP7279 support the notion that anthracosaurs expanded into more terrestrial niches during the Lower Carboniferous and that members of the clade were more terrestrial than other early tetrapod groups (e.g., loxommatids and whatcheeriids) (Smithson, 1994). Typical anthracosaurs have previously been considered highly aquatic (Smithson, 2000). Significantly, the Hancock County site material exhibits new adaptations (ossification and suturing of the vertebrae and pronounced acetabular roof) relative to those seen at East Kirkton. Addition of the Hancock site fauna to our current knowledge of Mississippian tetrapods supports the notion of a widespread distribution of terrestrial tetrapods in the mid-Carboniferous.

ACKNOWLEDGMENTS

This research was supported by grants from the National Science Foundation (EAR-0309747), the University Research Council of the University of Cincinnati, the Geological Society of America, the Theodore Roosevelt Memorial Fund, the Paleobiological Fund, Chiquita Brands International Inc. Foundation, John M. Tate, the Helen B. Vogel Trust, the Grace M. Harvie Foundation, and the Paul Sanders Award. The paper benefited greatly from reviews by Jenny Clack and Ted Daeschler. Katherine Glover provided helpful editorial advice. E. Mohalski Pence aided in the drafting of figures and G. Hardebeck and M. Milam kindly provided the specimen drawings. Field assistance was provided by E. Kvale, D. Chesnut, J. Devera, D. Williams, C. Eble, J. Nelson, D. Phelps, T. Hendricks, A. Horner, J. Sessa, R. Krause, J. Bonelli, T. Bantel, J. Lundquist, A. Watson, J. Bellan, and K. Houck.

REFERENCES CITED

Ahlberg, P.E., 1991, Tetrapod or near-tetrapod fossils from the Upper Devonian of Scotland: Nature, v. 354, p. 298–301, doi: 10.1038/354298a0.

Ahlberg, P.E., 1995, *Elginerpeton pancheni* and the earliest tetrapod clade: Nature, v. 373, p. 420–425, doi: 10.1038/373420a0.

Andrews, S.M., 1985, Rhizodont crossopterygian fish from the Dinantian of Foulden, Berwickshire, Scotland, with a re-evaluation of this group: Transactions of the Royal Society of Edinburgh, Earth Sciences, v. 76, p. 67–95.

Berman, D.S., 1979a, Cranial morphology of the Lower Permian lungfish *Gnathorhiza* (Osteichthyes: Dipnoi): Journal of Paleontology, v. 50, p. 1020–1033.

Berman, D.S., 1979b, Occurrence of *Gnathorhiza* (Osteichthyes: Dipnoi) in aestivation burrows in the Lower Permian of New Mexico with description of a new species: Journal of Paleontology, v. 50, p. 1034–1039.

Berman, D.S., 1979c, *Gnathorhiza bothrotreta* (Osteichthyes: Dipnoi) from the Lower Permian Abo Formation of New Mexico: Annals of the Carnegie Museum, v. 48, p. 211–230.

Bolt, J.R., Mckay, R.M., Witzke, B.J., and McAdams, M.P.A., 1988, A new Lower Carboniferous tetrapod locality in Iowa: Nature, v. 333, p. 768–770, doi: 10.1038/333768a0.

Boyd, M.J., 1980, The axial skeleton of the Carboniferous amphibian *Pteroplax cornutus*: Palaeontology, v. 23, p. 273–285.

Boyd, M.J., 1984, The Upper Carboniferous tetrapod assemblage from Newsham, Northumberland: Palaeontology, v. 27, p. 367–392.

Busanus, J.W., 1974, Paleontology and paleoecology of the Mauch Chunk Group in Northwestern West Virginia [M.S. thesis]: Bowling Green, Ohio, Bowling Green State University, 388 p.

Carroll, R.L., 1965, Lungfish burrows from the Michigan Coal Basin: Science, v. 148, p. 963–964.

Carroll, R.L., 1967, Labyrinthodonts from the Joggins Formation: Journal of Paleontology, v. 41, p. 111–142.

Carroll, R.L., 1997, Limits to knowledge of the fossil record: Zoology (Jena, Germany), v. 100, p. 221–231.

Carroll, R. L., and Green, D., 2003, Origin of terrestrial locomotion in vertebrates: Journal of Vertebrate Paleontology, v. 23, Suppl. to no. 3, p. 39A.

Carroll, R.L., Bossy, K.A., Milner, A.C., Andrews, S.M., and Wellstead, C.F., 1998, Handbuch der Paläoherpetologie: Lepospondyli. Munich, Dr. Friedrich Pfeil, 216 p.

Caudill, M.R., Driese, S.G., and Mora, C.I., 1996, Preservation of paleo-vertisol and an estimate of Late Mississippian precipitation: Journal of Sedimentary Research, v. A66, p. 58–70.

Cecil, C.B., 1990, Paleoclimate controls on stratigraphic repetition of chemical and siliciclastic rocks: Geology, v. 18, p. 533–536, doi: 10.1130/0091-7613(1990)018<0533:PCOSRO>2.3.CO;2.

Clack, J.A., 1987, *Pholiderpeton scutigerum* Huxley, an amphibian from the Yorkshire Coal Measures: Royal Society of London Philosophical Transactions, ser. B, v. 318, p. 1–107.

Clack, J.A., 1988, New material of the early tetrapod *Acanthostega* from the Upper Devonian of East Greenland: Palaeontology, v. 31, p. 699–724.

Clack, J.A., 1994, *Silvanerpeton miripedes*, a new anthracosauroid from the Visean of East Kirkton, West Lothian, Scotland: Transactions of the Royal Society of Edinburgh, Earth Sciences, v. 84, p. 369–376.

Clack, J.A., 2001, *Eucritta melanolimnetes* from the early Carboniferous of Scotland, a stem tetrapod showing a mosaic of characteristics: Transactions of the Royal Society of Edinburgh, Earth Sciences, v. 92, p. 75–95.

Clack, J.A., 2002, An early tetrapod from 'Romer's Gap': Nature, v. 418, p. 72–76, doi: 10.1038/nature00824.

Clack, J.A., and Carroll, R.L., 2000, Early Carboniferous tetrapods, *in* Heatwole, H., and Carroll, R. L., eds., Amphibian biology, Volume 4, Palaeontology: Chipping Norton, Australia, Surrey Beatty and Sons, p. 1030–1043.

Clack, J.A., Ahlberg, P.A., Finney, S.M., Dominguez, A.P., Robinson, J., and Ketcham, R.A., 2003a, A uniquely specialized ear in a very early tetrapod: Nature, v. 425, p. 65–69, doi: 10.1038/nature01904.

Clack, J.A., Blom, H., and Coates, M.I., 2003b, New insights into the postcranial skeleton of *Ichthyostega*: Journal of Vertebrate Paleontology, v. 23 Suppl., p. 41A.

Clarkson, E.N.K., Milner, A.R., and Coates, M.I., 1994, Palaeoecology of the Visean of East Kirkton, West Lothian, Scotland: Transactions of the Royal Society of Edinburgh, Earth Sciences, v. 84, p. 417–425.

Clement, G., Ahlberg, P.E., Blieck, A., Blom, H., Clack, J.A., Poty, E., Thorez, J., and Janvier, P., 2004, Devonian tetrapod from western Europe: Nature, v. 427, p. 412–413, doi: 10.1038/427412a.

Cloutier, R., and Ahlberg, P.E., 1996, Morphology, characters, and the interrelationships of basal sarcopterygians, *in* Stiassny, M. L. J., Parenti, L. R., and Johnson, G. D., eds., Interrelationships of fishes: San Diego, Academic Press, p. 445–479.

Coates, M.I., 1996, The Devonian tetrapod *Acanthostega gunnari* Jarvik: Postcranial anatomy, basal tetrapod interrelationships and patterns of skeletal evolution: Transactions of the Royal Society of Edinburgh, Earth Sciences, v. 87, p. 363–421.

Coates, M.I., and Clack, J.A., 1991, Fish-like gills and breathing in the earliest known tetrapod: Nature, v. 352, p. 234–236, doi: 10.1038/352234a0.

Coates, M.I., and Clack, J.A., 1995, Romer's gap: tetrapod origins and terrestriality: Bulletin du Museum National d'Histoire Naturelle (Paris), ser. 4, 17, p. 373–388.

Daeschler, E.B., 2000, Early tetrapod jaws from the Late Devonian of Pennsylvania, USA: Journal of Paleontology, v. 74, p. 301–308.

Daeschler, E.B., and Shubin, N., 1998, Fish with fingers?: Nature, v. 391, p. 133, doi: 10.1038/34317.

Daeschler, E.B., Shubin, N., Thomson, K.S., and Amaral, W.W., 1994, A Devonian tetrapod from North America: Science, v. 265, p. 639–642.

Davis, M.C., Shubin, N.H., and Daeschler, E.B., 2001, Immature rhizodonts from the Devonian of North America: Bulletin of the Museum of Comparative Zoology, v. 156, p. 171–178.

Dawson, J.W., 1868, Acadian geology: The geological structure, organic remains, and mineral resources of Nova Scotia, New Brunswick, and Prince Edward Island (2nd edition): London, Macmillan and Co., 694 p.

DeMar, R.E., 1968, The Permian labyrinthodont amphibian *Dissorophus multicinctus*, and adaptations and phylogeny of the family Dissorophidae: Journal of Paleontology, v. 42, p. 1210–1242.

DiMichele, W.A., and Hook, R.W., rapporteurs, 1992, Paleozoic terrestrial ecosystems, *in* Behrensmeyer, A. K., Damuth, J. D., DiMichele, W. A., Potts, R., Sues, H.-D., and Wing, S. eds., Terrestrial ecosystems through time: Evolutionary paleoecology of terrestrial plants and animals: Chicago, Chicago University Press, p. 205–325.

DiMichele, W.A., and Phillips, T.L., 1994, Paleobotanical and paleoecological constraints on models of peat formation in the Late Carboniferous of Euramerica: Palaeogeography, Palaeoclimatology, Palaeoecology, v. 106, p. 39–90, doi: 10.1016/0031-0182(94)90004-3.

Duellman, W.E., and Trueb, L., 1986, Biology of amphibians: Baltimore, Johns Hopkins University Press, 670 p.

Eble, C.F., and Greb, S.F., 1997, Channel-fill coals on the western margin of the Eastern Kentucky Coal Field: International Journal of Coal Geology, v. 33, p. 183–207, doi: 10.1016/S0166-5162(96)00048-1.

Elliott, D.K., and Taber, A.C., 1982, Mississippian vertebrates from Greer, West Virginia: Proceedings of the West Virginia Academy of Science, v. 53, p. 73–80.

Gensel, P.G., 1986, Diversification of land plants in the Early and Middle Devonian: University of Tennessee, Department of Geological Sciences, Studies in Geology, v. 15, p. 64–80.

Godfrey, S.J., 1988, Isolated tetrapod remains from the Carboniferous of West Virginia: Kirtlandia, v. 43, p. 27–36.

Godfrey, S.J., 1989, A rhizodontid crossopterygian from the Upper Mississippian at Greer, West Virginia: Acta Musei Reginaehradecensis, ser. A, Scientiae Naturales XXII, p. 89–98.

Greb, S.F., and Chesnut, D.R., Jr., 1992, Transgressive channel filling in the Breathitt Formation (Upper Carboniferous), Eastern Kentucky Coal Field, U.S.A: Sedimentary Geology, v. 75, p. 209–221, doi: 10.1016/0037-0738(92)90093-7.

Holmes, R., 1980, *Proterogyrinus scheelei* and the early evolution of the Labyrinthodont pectoral limb, in Panchen, A. L., ed., The terrestrial environment and the origin of land vertebrates, Systematics Association Special Volume 15, p. 351–376.

Holmes, R., 1984, The Carboniferous amphibian *Proterogyrinus scheelei* Romer, and the early evolution of tetrapods: Royal Society of London Philosophical Transactions, ser. B, v. 306, 431–524.

Holmes, R., 1989a, The skull and axial skeleton of the Lower Permian anthracosauroid amphibian *Archeria crassidisca*: Palaeontographica, Abteilung A., v. 207, p. 161–206.

Holmes, R., 1989b, Functional interpretations of the vertebral structure in Paleozoic labyrinthodont amphibians: Historical Biology, v. 2, p. 111–124.

Hook, R.W., 1983, *Colosteus scutellus* (Newberry), a primitive temnospondyl amphibian from the middle Pennsylvanian of Linton, Ohio: American Museum Novitates, v. 2770, p. 1–41.

Hook, R.W., and Ferm, J.C., 1988, Paleoenvironmental controls on vertebrate abandoned channels in the Upper Carboniferous: Palaeogeography, Palaeoclimatology, Palaeoecology, v. 63, p. 159–181, doi: 10.1016/0031-0182(88)90095-8.

Hook, R.W., and Hower, J.C., 1988, Petrography and taphonomic significance of the vertebrate-bearing cannel coal of Linton, Ohio (Westphalian D, Upper Carboniferous): Journal of Sedimentary Petrology, v. 58, p. 72–80.

Jarvik, E., 1996, The Devonian tetrapod *Ichthyostega*: Fossils and Strata, v. 40, 1–213.

Jeffery, J.E., 2001, Pectoral fins of rhizodontids and the evolution of pectoral appendages in the tetrapod stem-group: Biological Journal of the Linnean Society, v. 74, p. 217–236, doi: 10.1006/bijl.2001.0572.

Johanson, Z., and Ahlberg, P.E., 1998, A complete primitive rhizodont from Australia: Nature, v. 394, p. 569–573, doi: 10.1038/29058.

Johanson, Z., Turner, S., and Warren, A., 2000, First East Gondwanan record of *Strepsodus* (Sarcopterygii, Rhizodontida) from the Lower Carboniferous Ducabrook Formation, central Queensland, Australia: Geodiversitas, v. 22, p. 161–169.

Lebedev, O.A., and Clack, J.A., 1993, Upper Devonian tetrapods from Andreyevka, Tula Region, Russia: Palaeontology, v. 36, p. 721–734.

Lebedev, O.A., and Coates, M.I., 1995, The postcranial skeleton of the Devonian tetrapod *Tulerpeton curtum* Lebedev: Zoological Journal of the Linnean Society, v. 114, p. 307–348, doi: 10.1006/zjls.1995.0027.

Lombard, E., and Bolt, J.R., 1995, A new primitive tetrapod *Whatcheeria deltae* from the Lower Carboniferous of Iowa: Palaeontology, v. 38, p. 471–494.

Long, J.A., 1989, A new rhizodontiform fish from the Early Carboniferous of Victoria, Australia, with remarks on the phylogenetic position of the group: Journal of Vertebrate Paleontology, v. 9, p. 1–17.

Marshall, J.E.A., Astin, T.R., and Clack, J.A., 1999, East Greenland tetrapods are Devonian in age: Geology, v. 27, p. 637–640.

Miller, D.J., and Eriksson, K.A., 1999, Linked sequence development and global climate change: The Upper Mississippian record in the Appalachian Basin: Geology, v. 27, p. 35–38, doi: 10.1130/0091-7613(1999)027<0035:LSDAGC>2.3.CO;2.

Milner, A.C., and Lindsay, W., 1998, Postcranial remains of *Baphetes* and their bearing on the relationships of the Baphetidae (= Loxommatidae): Zoological Journal of the Linnean Society, v. 122, p. 211–235, doi: 10.1006/zjls.1997.0119.

Milner, A.R., 1980, The tetrapod assemblage from Nyrany, Czechoslovakia, in Panchen, A. L., ed., The terrestrial environment and the origin of land vertebrates. Systematics Association Special Volume 15, p. 439–496.

Milner, A.R., 1993, Biogeography of Paleozoic tetrapods, in Long, J. A., ed., Paleozoic vertebrate biostratigraphy and biogeography: London, Belhaven Press, p. 324–353.

Milner, A.R., and Sequeira, S.E.K., 1994, The temnospondyl amphibians from the Viséan of East Kirkton, West Lothian, Scotland: Transactions of the Royal Society of Edinburgh, Earth Sciences, v. 84, p. 331–361.

Milner, A.R., Smithson, T.S., Milner, A.C., Coates, M.I., and Rolfe, W.D.I., 1986, The search for early tetrapods: Modern Geology, v. 10, p. 1–28.

Olson, E.C., and Daly, E., 1972, Notes on *Gnathorhiza* (Osteichthyes, Dipnoi): Journal of Paleontology, v. 46, p. 371–376.

Panchen, A.L., 1977, The origin and early evolution of tetrapod vertebrae, in Andrews, S. M, Miles, R. S., and Walker, A. D., eds., Problems in vertebrate evolution: London, Academic Press, p. 289–318.

Paton, R.L., Smithson, T.R., and Clack, J.A., 1999, An amniote-like skeleton from the Early Carboniferous of Scotland: Nature, v. 398, p. 508–513, doi: 10.1038/19071.

Retallack, G.J., 1988, Field recognition of paleosols, in Reinhardt, J., and Sigleo, W.R., eds., Paleosols and weathering through geologic time: Geological Society of America Special Paper 216, p. 1–20.

Rice, C.L., Sable, E.G., Dever, G.R., Jr., and Kehn, T.M., 1979, The Mississippian and Pennsylvanian (Carboniferous) Systems in the United States—Kentucky: U.S. Geological Survey Professional Paper 1110F, p. F1–F32.

Rolfe, W.D.I., Durant, G.P., Fallick, A.E., Hall, A.J., Large, D.J., Scott, A.C., Smithson, T.R., and Walkden, G.M., 1990, An early terrestrial biota preserved by Visean vulcanicity in Scotland, in Lockley, M.G.R., and Boulder, A., eds., Volcanism and fossil biotas: Geological Society of America Special Paper 244, p. 13–24.

Romer, A.S., 1972, A Carboniferous labyrinthodont with complete dermal armor: Kirtlandia, v. 16, p. 1–8.

Romer, A.S., and Olson, E.C., 1954, Aestivation in a Permian lungfish: Breviora, v. 30, p. 1–8.

Scheckler, S.E., 1986, Floras of the Devonian-Mississippian transition, in Broadhead, T.W., ed., Land plants: University of Tennessee, Department of Geological Sciences, Studies in Geology, v. 15, p. 81–96.

Schoch, R.R., and Milner, A.R., 2000, Handbuch der Paläoherpetologie 3B: Stereospondyli: Munich, Dr. Friedrich Pfeil, 203p.

Schultze, H.-P., and Bolt, J.R., 1996, The lungfish *Tranodis* and the tetrapod fauna from the Upper Mississippian of North America, in Milner, A.R., ed., Studies on Carboniferous and Permian vertebrates: London, Palaeontological Association Special Papers in Palaeontology, v. 52, p. 31–54.

Shear, W.A., and Selden, P.A., 2001, Rustling in the undergrowth: Animals in early terrestrial ecosystems, in Gensel. P. G., and Edwards, D., eds., Plants invade the land: Evolutionary and environmental perspectives: New York, Columbia University Press, p. 29–51.

Smithson, T.R., 1994, *Eldeceeon rolfei*, a new reptilomorph from the Visean of East Kirkton, West Lothian, Scotland: Transactions of the Royal Society of Edinburgh, Earth Sciences, v. 84, p. 377–382.

Smithson, T.R., 2000, Anthracosaurs, in Heatwole, H., and Carroll, R.L., eds., Amphibian biology, Volume 4, Palaeontology: Chipping Norton, Australia, Surrey Beatty and Sons, p. 1053–1063.

Smithson, T.R., Carroll, R.L., Panchen, A.L., and Andrews, S.M., 1994, *Westlothiana lizziae* from the Viséan of East Kirkton, West Lothian, Scotland, and the amniote stem: Transactions of the Royal Society of Edinburgh, Earth Sciences, v. 84, p. 383–412.

Stebbins, R.C., and Cohen, N.W., 1995, A natural history of amphibians: Princeton, New Jersey, Princeton University Press, 316 p.

Stewart, W.N., and Rothwell, G.R., 1993, Paleobotany and the evolution of plants: Cambridge, UK, Cambridge University Press, 521 p.

Thulborn, T., Hamley, T., Turner, S., and Warren, A., 1996, Early Carboniferous tetrapods in Australia: Nature, v. 381, p. 777–780, doi: 10.1038/381777a0.

Vaughn, P.P., 1964, Evidence of aestivating lungfish from the Sangre de Cristo Formation, Lower Permian of northern New Mexico: Contributions to Science, Los Angeles County Museum of Natural History, v. 80, p. 1–8.

Warren, A., and Turner, S., 2004, The first stem tetrapod from the Lower Carboniferous of Gondwana: Palaeontology, v. 47, 151–184.

Weller, S., 1913, Stratigraphy of the Chester Group in southwestern Illinois: Transactions of the Illinois State Academy of Science, v. 118, p. 118–129.

Young, G.C., Long, J.A., and Ritchie, A., 1992, Crossopterygian fishes from the Devonian of Antarctica: Systematics, relationships, and biogeographic significance: Records of the Australian Museum, Suppl. 14, p. 1–77.

Zhu, M., Ahlberg, P.E., and Zhao, W.J., 2002, First Devonian tetrapod from Asia: Nature, v. 420, p. 760–761, doi: 10.1038/420760a.

Manuscript Accepted by the Society 28 June 2005

A fossil lycopsid forest succession in the classic Joggins section of Nova Scotia: Paleoecology of a disturbance-prone Pennsylvanian wetland

John H. Calder
Nova Scotia Department of Natural Resources, P.O. Box 698, Halifax, Nova Scotia B3J 2T9, Canada

Martin R. Gibling
Department of Earth Sciences, Dalhousie University, Halifax, Nova Scotia B3H 3J5, Canada

Andrew C. Scott
Department of Geology, Royal Holloway (University of London), Egham, Surrey TW20 0EX, UK

Sarah J. Davies
Department of Geology, University of Leicester, University Road, Leicester LE1 7RH, UK

Brian L. Hebert
RR 1, Joggins, Nova Scotia B0L 1A0, Canada

ABSTRACT

Standing lycopsid trees occur at 60 or more horizons within the 1425-m-thick coal-bearing interval of the classic Carboniferous section at Joggins, with one of the most consistently productive intervals occurring between Coals 29 (Fundy seam) and 32 of Logan (1845). Erect lepidodendrid trees, invariably rooted within an organic-rich substrate, are best preserved when entombed by heterolithic sandstone/mudstone units on the order of 3–4 m thick, inferred to represent the recurring overtopping of distributary channels of similar thickness. The setting of these forests and associated sediments is interpreted as a disturbance-prone interdistributary wetland system. The heterogeneity and disturbance inherent to this dynamic sedimentary environment are in accord with the floral record of the fossil forests and interpretation of the peat-forming wetlands as topogenous, rheotrophic forest swamps.

Candidates for the erect, *Stigmaria*-bearing trees, which range in diameter (dbh) from 25 to 50 cm, are found in prostrate compressions and represent a broad range of ecological preferences amongst the Lycopsida. This record, which is not significantly time averaged, closely parallels the megaspore record from thin peaty soils in which they are rooted, but differs significantly from the miospore record in studies of other, thicker coals. Dominant megaspores are *Tuberculatisporites mamillarus* and *Cystosporites diabolicus*, derived from *Sigillaria* and *Diaphorodendron/Lepidodendron* respectively. Intervening beds preserve a record of an extramire flora composed in the main of seed-bearing pteridosperms and gymnosperms (and ?progymnosperms). Reproductive adaptation to disturbance appears to have played a key role in ecological partitioning of plant communities within these wetlands. Burial of lycopsid trees by onset of heterolithic deposition resulted in the demise of entire forest stands. Disturbance-tolerant *Calamites* regenerated in the episodically accruing sediment around the dead and dying lycopsid stands, a succession identified here as typical of Euramerican fossil forests. Rapid, ongoing subsidence of the basin accommodated the submergence of the fossil forests, and abiotic disturbance inherent to the seasonal climate facilitated their episodic entombment. Disturbance is inferred to have been mediated by short-term (?seasonal) precipitation flux as suggested by the heterolithic strata and in the record of charred lycopsid trees, recording wildfire most probably ignited by lightning. Within this fossil forest interval is found a glimpse of animal life within the wetland ecosystem beyond the confines of the tree hollows, whence the bulk of the terrestrial faunal record of Joggins historically derives.

Keywords: Joggins, Carboniferous, fossil forests, wetlands, disturbance, lycopsids, paleoecology, paleobotany, coal, vertebrates, invertebrates, Pennsylvanian.

INTRODUCTION

My dear Marianne, —We have just returned from an expedition of three days to the Strait which divides Nova Scotia from New Brunswick, whither I went to see a forest of fossil coal-trees—the most wonderful phenomenon perhaps that I have seen, so upright do the trees stand, or so perpendicular to the strata, in the ever-wasting cliffs, every year a new crop being brought into view, as the violent tides of the Bay of Fundy, and the intense frost of the winters here, combine to destroy, undermine, and sweep away the old one—trees twenty-five feet high, and some have been seen of forty feet, piercing the beds of sandstone and terminating downwards in the same beds, usually coal. This subterranean forest exceeds in extent and quantity of timber all that have been discovered in Europe put together.

—Sir Charles Lyell, in a letter to his sister, July 30, 1842

The standing lycopsid trees of the classic Carboniferous section at Joggins (Gibling, 1987) have been famous since their description by Sir Charles Lyell (1842, 1843, 1845, 1849, 1871) and Sir William Dawson (1855a and others) in the mid nineteenth century, figuring in such seminal works as Darwin's *Origin of Species* (1859, p. 296). Since that time, however, they have received very little systematic study beyond the anecdotal, and the goals espoused by Lyell remain largely valid: to investigate "the peculiar circumstances which favored the preservation of so many fossil trees" (Lyell and Dawson, 1853).

Erect lycopsid trees occur nominally at 60 (Dawson, 1892a) or more (Darwin, 1859) horizons within the 1425-m-thick coal-bearing interval of the Joggins section. One of the most consistently productive intervals over the past 150 yr, constituting true "fossil forests," occurs between Coal 32 and the stratigraphically higher Coal 29a (Fundy seam) of Logan (1845). In this paper, the floral and faunal record of these forests, their sedimentological setting, and their paleoecology are considered, together with the interactive dynamics of the plant communities and sedimentary processes of their depositional setting. The findings that are reported herein were gathered over more than a decade of fieldwork on this particular fossil forest interval. For the first time since the pioneering work of Lyell and Dawson, a contextual, reference description of fossil lycopsid forests will be provided as they occur at Joggins. This will provide a case study that permits assessment of interpretations of the ecological preferences of the paleotropical vegetation of the Pennsylvanian wetlands. In addition, this will better allow comparisons to be drawn with other fossil lycopsid forests across paleoequatorial Euramerica (cf. Scott and Calder, 1994; Kerp, 1996).

PREVIOUS WORK

The earliest known account of Joggins wherein the erect trees were placed in a geological context is that of Richard Brown in 1829. Jackson and Alger (1829) made anecdotal reference to Joggins, and Abraham Gesner (1836) entreated his readers to undertake what was then a rigorous journey to "the place where the delicate herbage of a strange, extinct world is now transmuted in stone." A turning point in the history of research at Joggins, and indeed the world's knowledge of the site, was the first visit of Sir Charles Lyell in 1842, and his subsequent writings (1843, 1845, 1871). In the following year, Sir William Logan undertook the bed-by-bed description of the section as the first field project of the Geological Survey of Canada (GSC), noting in the course of that seminal work the interval of fossil trees described in this paper (Logan, 1845).

The most detailed descriptions of the trees and their occurrence are those of Lyell and Dawson (1853) (Fig. 1) and Dawson (1855a, 1877, 1882). These works in large part were undertaken as a consequence of an unexpected discovery within the cast of a standing tree above Coal 15, fallen from the cliff face at Coal

Fig. 34.—*Section of middle part of Subdivison XV. in which the Dendrerpeton, Land Shells, etc., have been found.*

Figure 1. Woodcut of erect lycopsid and calamites at Joggins, from Lyell and Dawson (1853).

Mine Point. Here, in 1852, Lyell and Dawson were to discover one of the first known terrestrial vertebrate skeletons from the Paleozoic "Coal Age" (Scott, 1998). Subsequent investigations at Coal Mine Point and elsewhere in the section yielded Dawson the skeletal remains of over 100 individual tetrapods (Carroll et al., 1972; Milner, 1987), including the earliest known reptile, *Hylonomus lyelli* (Dawson, 1860). As yet, none have been discovered in the succession of fossil trees described here, although the likelihood of such discovery is high.

Since Dawson's day, there has been a dearth of research on the fossil forests of Joggins. A noteworthy exception is the study of the fire ecology of the Joggins paleoenvironment (Falcon-Lang, 1999, 2000), undertaken concurrently with the research described in this paper.

GEOLOGICAL SETTING

The Joggins section lies within the Athol Syncline, a depocenter of the Cumberland Basin, bordered by the Cobequid Highlands to the south and the Caledonia Highlands to the northwest. The basin records a subsidence history unsurpassed by other Carboniferous coal basins, wherein ~4 km of strata of probable Namurian to Duckmantian age accumulated within ~4 million years (Calder, 1994). Although the timing of fault activity within the basin is poorly constrained, it is clear that many of the major fault systems were active during the Namurian to early Westphalian, with a prominent strike-slip component (Reed et al., 1993; Browne and Plint, 1994) and local thrusting in southeastern New Brunswick at a restraining bend of the Cobequid Fault zone (Plint and van de Poll, 1984; Nance, 1987; Waldron et al., 1989). Intra-Carboniferous halokinesis also may have taken place along basement faults, for example on the Minudie Anticline north of Joggins (Fig. 2; Calder, 1994; Ryan and Boehner, 1994), with evacuation of subsurface salt contributing to basin subsidence (Waldron and Rygel, 2004).

The Cumberland Basin is a local, fault-bounded depocenter of the regional Maritimes Basin, which in turn is a complex of predominantly northeasterly trending intermontane basins, once

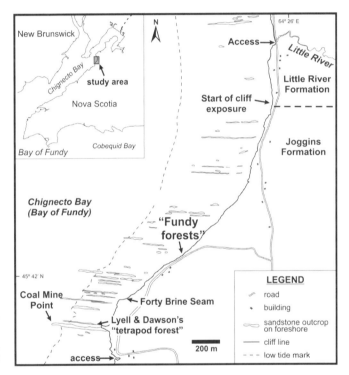

Figure 2. Location map of the "classic" Joggins section and location of Fundy fossil forests sequence. Noted is the location beneath Coal Mine Point of the tetrapod-bearing lycopsid forest of Lyell and Dawson (1853).

variously interconnected and now, as then, defined by intervening massifs of the Avalon, Grenville, and Meguma terranes. Nova Scotia and the Maritimes Basin in the Carboniferous lay within paleoequatorial Euramerica, drifting northward from a paleolatitude of 12° S to cross the equator by the beginning of the Permian (Scotese and McKerrow, 1990). Generally considered a northern part of the Appalachian orogenic belt, the Maritimes Basin lay situated at the paleosoutheastern margin of the Appalachians in a paleogeographic region distinct from the Appalachian Basin to

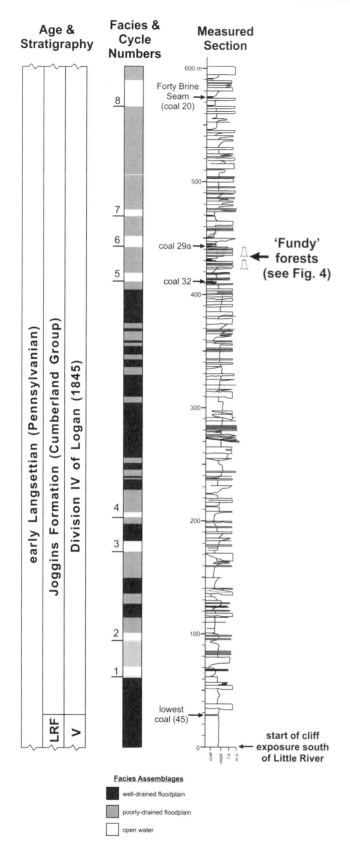

Figure 3. Stratigraphic location of the Fundy fossil forests interval within the basal 600 m of the classic cliff section (after Davies and Gibling, 2003).

the west, called the Maritime-West European Province by Calder (1998, modified after Leeder, 1987).

The general stratigraphy of the Cumberland Basin is shown in Figure 3. The Joggins Formation, described in detail below, rests conformably on red beds of the Little River Formation (Calder et al., 2005) and is overlain conformably by the coal-bearing fluvial Springhill Mines Formation and overlying alluvial units (Rust et al., 1984; Ryan et al., 1991). The Joggins Formation apparently passes southward into a thick conglomerate wedge of the Polly Brook Formation, bordering the Cobequid Highland (Calder, 1994; Ryan et al., 1991). Coal beds such as the heavily mined Forty Brine (Coal 20 of Logan, 1845) and Joggins (Coal 7) seams (Fig. 2) can be traced for tens of kilometers inland (Copeland, 1959) and are commonly associated with bivalve-bearing limestone beds that represent basinwide flooding events (Calder, 1991, 1994; Davies and Gibling, 2003).

JOGGINS FORMATION

The study section lies within the Joggins Formation of the Cumberland Group of Pennsylvanian age. As defined by Ryan et al. (1991), the Joggins Formation includes a 619 m non-coal-bearing red-bed section (Division V of Logan, 1845) and an overlying 1450-m-thick coal-bearing section (Division IV of Logan, 1845). The definition of the Joggins formation as coal bearing with bivalve-bearing limestones and black shales known colloquially as "clam coals," however, led Calder et al., (2005) to assign the basal red beds of Division V to their Little River Formation.

The age of the Joggins Formation is a vexed question, lying near the problematic Namurian-Westphalian boundary. Few floral appearances or extinctions occur across this boundary, which is defined in Europe at the *Gastrioceras subcrenatum* marine band. The absence of goniatites in the Maritimes Basin presents a difficulty exacerbated by the near ecological exclusion from the Maritimes Basin of herbaceous lycopsids that dispersed densospores, a diagnostic Westphalian spore group elsewhere in Euramerica (see Calder, 1998). The section long has been held to be of Duckmantian ("Westphalian B") age (Bell, 1944; Hacquebard and Donaldson, 1964), although later palynological studies (Dolby, 1991) concluded that the Joggins Formation lies wholly within the Langsettian ("Westphalian A"). Recent taxonomic revision of Bell's macrofloral collection similarly favors a Langsettian age, although late Namurian elements are represented, underlining the proximity of the Namurian-Westphalian boundary (R.H. Wagner, 1999, personal commun.).

A marked lithologic change to gray, organic-rich coal-bearing strata is recorded within the Joggins Formation at the horizon of Coal 32, which lies 388.5 m above the base of the Division IV coal measures (Figs. 3, 4). Coal 32 marks the onset of persistent coal formation and is the first coal bed to approach meter-scale thickness, and the lowest mined seam in the formation. Pit props protruding from the cliff face bear witness to early coal mining at Joggins and to the erosive forces of the bay. Four horizons of past mining dating to 1866 (Goudge, 1945) and probably earlier

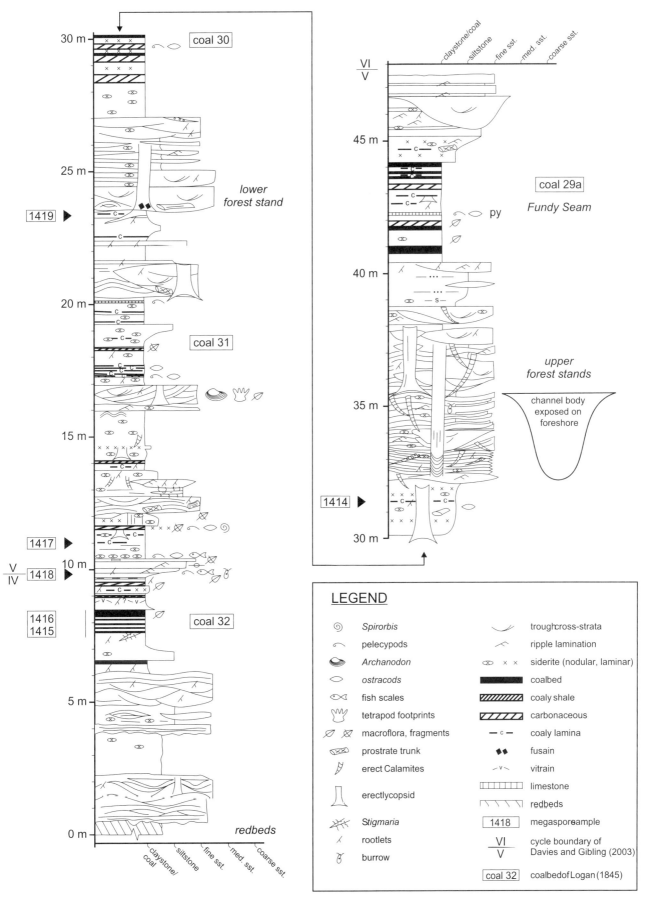

Figure 4. Sedimentological profile of the Fundy fossil forests interval, beginning at 382 m of the Joggins coal measures (Division IV of Logan, 1845). The main forest stands occur entombed within heterolithic mudstone-sandstone units at 23.5–27.0 m (lower stand) and at 32.2–38.8 m (upper stands).

are evident in the 40 m above Coal 32—namely coals 32, 30, 29a, and 29, in ascending order. Iron-rich mine waters issuing from workings on the Fundy Seam (Coal 29a) stain the cliff face a distinctive ochre color, providing a striking visual landmark along the cliffs that identifies the study section. In the 37-m-thick coal-bearing interval from the base of Coal 32 through the roof of Coal 29a—the Fundy Seam—several horizons of standing lycopsid trees occur in the 30-m-high cliffs and on the foreshore.

SEDIMENTOLOGY OF THE FUNDY FORESTS INTERVAL

The section from Coals 32 through 29a (Fig. 4) is rendered lithologically distinct from older strata by the predominance of gray carbonaceous mudrocks and development of thicker coal beds approaching meter-scale. Strata below Coal 32 comprise a thick succession of red, mudstone-dominated flood-plain deposits with channel bodies (Cycle 4 of Davies and Gibling, 2003), capped by ~5 m of gray sandstone and mudstone with thin carbonaceous shales. Coal 32 is ~1 m thick, and comprises numerous thin layers of coal and carbonaceous shale interstratified with gray mudstone. It rests directly on a rooted sandstone and marks a period of sustained wetland conditions that denotes the base of Cycle 5. Within this study interval, three levels of standing lycopsids are consistently exposed, one within the lower heterolithic unit, and one within each body comprising the upper heterolithic unit. Standing lycopsids occur virtually throughout the study interval but are less commonly and obviously exposed within other units; during the decade of study of this interval, however, at least seven other horizons have been observed.

Sheetlike sandstone beds are a striking feature of the Joggins Formation and cliff section. Typically these comprise two types: (1) irregular-based beds, commonly cross-stratified, and (2) sharp-based, planar beds. The prevailing architecture of sandstone bodies remains tabular and sheetlike, but the density of their spacing shows a marked increase, where they form composite heterolithic bodies dominated by sandstone of the first type, but with thinly interstratified, concordant mudrock beds.

The 10 m interval above Coal 32 (Fig. 4) is mudstone dominated, with numerous centimeter-scale beds of coal and carbonaceous shale (including Coal 31), and rare, discrete limestone and shale layers that yield bivalves, ostracods, serpulids, and plant fragments. The organic-rich shales and limestones mark flooding events (Davies and Gibling, 2003) that divide the strata into stacked shoaling-up successions from several decimeters to ~3 m thick. The most prominent successions are capped by discrete, planar-bedded sandstone bodies that are less than 1 m thick (at 10, 12, and 16 m levels). The sandstone bodies have abrupt, erosional bases on which groove marks are locally present. Sedimentary structures include wave-ripples and mounds that superficially resemble hummocky cross-stratification. Mounds are up to 50 cm high with a quasi-regular spacing, form-concordant stratification, and internal truncation surfaces. Sandstone tops are rooted and stigmarian axes common; basal stumps are cast within the sandstones and also within mudstone strata within the shoaling-up successions, suggesting that mounds are related to standing vegetation (vegetation-induced sedimentary structures, or VISS, of Rygel et al., 2004).

The succeeding 23 m includes two heterolithic sandstone/mudstone bodies up to 6m thick, the upper of which is a composite of two stacked units, each succeeded by rooted mudrocks and meter-scale coal beds. Fossil lepidodendrid trees in growth position (Fig. 5A, 5B) are present at several levels, some cast within mudrocks and meter-scale trough cross-stratified channel bodies. The heterolithic beds, however, best preserve the standing lepidodendrid trees and are the primary focus of this study.

At the top of the measured section (approximately 430 m of Division IV), a thin limestone with bivalves and ostracods marks a period of renewed base-level rise associated with the meter-scale Coal 29 and overlying, tabular sandstones and gray siderite-bearing mudstones (the base of Cycle 6 of Davies and Gibling, 2003). Planar-bedded sandstones with hummocky cross-stratification and wave ripples typify this interval, along with platy, gray mudstones with siderite bands. Plant fragments and roots are common, but erect trees in contrast are rare. The base of the succeeding Cycle 7 (~27 m above Coal 29a) is marked by a thick, prominent faunal-rich limestone that overlies Coal 28 and marks the next major rise in base level.

Heterolithic Sandstone/Mudstone Bodies

Two heterolithic units, 3.5 m and 6.6. m thick (23.5–27.0 m and 32.2–38.8 m, Fig. 4), entomb standing lycopsid trees (Fig. 5A, 5B). Trees occur throughout the interval but are best preserved within these units. The heterolithic nature of these units (Facies 3 of Falcon-Lang, 1999) is the result of interbedded mudrock alternating with planar-lenticular sandstone beds that commonly are form-concordant with underlying surfaces. A distinctive feature of the heterolithic units is the abundance of sandstone bodies up to one meter thick infilling apparently symmetrical hollows, in some cases formed around standing trees (see "Scour Hollows," below). Both heterolithic units commence abruptly with erosional bases. Roots and erect calamite stems (Fig. 6) occur throughout.

Sandstone interbeds within the lower heterolithic unit range from 4 to 15 cm thick, are planar-lenticular in form, and exhibit erosional bases in places. Mudrock interbeds are up to 8 cm thick and are moderately well stratified with parallel and lenticular lamination, current-ripple cross-lamination, and possible wave ripples of symmetric form. Lenticular, bedding-parallel siderite concretions are common within the mudrock interbeds.

The upper heterolithic unit comprises the two stacked bodies, each ~3 m in thickness, separated by a reactivation surface below which an upper stand of trees is rooted. The temporal separation of the two stories aside, together they demonstrate an upward thickening of sandstone interbeds, which is apparent within the lower story itself. Rhythmic, coarsening-upward mudrock-sandstone couplets, 7–14 cm in thickness, in the lower

Figure 5. Standing lepidodendrid lycopsid trees of the Fundy forests: (A) from the upper forest stand at 32.2 m; (B) from the lower stand, entombed within heterolithic strata at 23.5 m, with the mold of a tree removed by erosion to the immediate right (hammer 29 cm high).

meter of the heterolithic unit become progressively thicker and less regular upward within the lower story. The 3-m-thick upper story comprises form-concordant mudrock and sandstone interbeds and numerous isolated sandstone bodies up to 0.5 m thick and locally stacked. The uppermost meter of both heterolithic units comprises a paleosol cap in the form of a near homogeneous sandstone with abundant roots, sideritic rhizoconcretions, and local stigmarian axes. Low-angle cross-stratification has been largely destroyed by bioturbation, and the sandstone is overlain by an intensely rooted mudstone below a succeeding coal.

Lying in the foreshore lateral to the lower story (32.2–35.4 m) of the composite heterolithic sandstone unit is an isolated channel sandstone body with incised margin(s) and a low width-to-thickness ratio, measuring 3–4 m thick and 35 m in apparent width. The stratification of this sandstone body is dominated by trough cross beds, with some ripple cross-laminae. The horizon of the channel body can be traced toward the cliff into the basal sandstone layer of the heterolithic unit, and its top appears to correspond approximately to the level of standing lycopsids atop the lower of the two stacked heterolithic bodies (Fig. 4).

Figure 6. Calamite in growth position within heterolithic strata (knife for scale, 9 cm long).

Scour Hollows (Centroclinal Cross-Strata)

The modification of sediment around standing vegetation at Joggins is described in detail in Rygel et al. (2004), and the terms used here are consistent with that study. In both heterolithic bodies, scour hollows up to 1 m deep and 4 m in apparent length are prominent around some erect trees, and are locally connected by sandstone sheets. Most hollows are filled by centroclinally cross-stratified sandstone (Underwood and Lambert, 1974) with sets up to 1 m thick and finely comminuted plant debris on the foreset surfaces. Hollows can be markedly asymmetrical, being better developed (longer) on the western or northwestern side of the tree. In each heterolithic body, an example was observed where the cross-stratification within the hollow-fills has a similar foreset dip direction on either side of an exposed tree ("uniclinal fill" of Rygel et al., 2004). The hollow-fills have slightly mounded tops and are capped with ripple-drift cross-lamination in 1–2 cm sets, with paleoflow direction consistent to that shown by the underlying foresets (see below). At the base of the lower heterolithic body (~24 m level, Fig. 4) a prominent hollow was observed on the upflow side of one erect tree, but a second standing tree 1.14 m downflow was not associated with a noticeable hollow and presumably was shielded from scour effects by the first tree.

A well-developed hollow-fill at the base of the upper heterolithic body is a composite of two distinct sandstone lenses separated by a thin shale that has been removed by local scour. The cross-stratification of the fill has a single foreset dip direction and is capped with ripple cross-laminated sandstone of similar paleoflow direction. The hollow-fill and sandstone lenses are rooted.

Smaller-scale hollow-fills around erect calamites are up to 20 cm thick. Some fills exhibit cross-stratification whereas others show no apparent structure.

Tree Fills

The general infilling pattern described by Lyell (1845, p. 182–183) and documented in detail by Dawson (1882) for tetrapod-bearing trees at Joggins, in general, holds true for the fossil forest succession described here. The tree interiors are void of tissue preservation with the exception in some specimens of the cylindrical stele (Dawson, 1877), which may be broken, bent, or vertical, but invariably off-center. The sediments that infilled the trees represent facies that are seen entombing the trees but in differing proportions and, therefore, at different stratigraphic levels from corresponding strata external to the trees. Most commonly, the bulk of the fill of erect trees entombed within heterolithic units is massive sandstone and this commonly overlies muddy and organic-rich basal fills, which may contain fossil charcoal (fusain). Others exhibit a heterolithic fill of alternating sandstone and mudrock, commonly with meniscus-like internal stratification. In such cases, the thickness of muddy sediments that infill the basal portion of the trees typically is thinner than the corresponding, and presumably coeval, lithology outside and entombing the basal trunks. Sandstone inevitably succeeds the basal mud fill at a lower stratigraphic horizon than the entombing sandstone external to the trunk and usually comprises the bulk of the tree fills. Mud-cast trees occur within mud-rich intervals but have been observed less commonly, defined chiefly by their coaly rind and on occasion by bedding-parallel siderite concentrated within the tree fill.

Paleoflow

Paleoflow patterns are variable through the Joggins section, but with predominantly southeasterly paleoflow in channel bodies of the Boss Point Formation (Browne and Plint, 1994), Little River Formation (Calder et al., 2005), Joggins Formation (M.C. Rygel, 2001, personal commun.), and Springhill Mines Formation (Rust et al., 1984). No large (10 m scale) or multistory channel bodies are present in the studied interval. Paleoflow measurements from cliffs and wave-cut platform for the measured section, including the two heterolithic bodies and related channel body, are shown in Figure 7.

At the base of the lower heterolithic body, planar foresets within scour hollows and connecting sandstone sheets have a predominantly southeasterly paleoflow, and a similar flow direction was recorded within thin, capping sheets of ripple-drift cross-stratification. Sandstone layers at higher levels have a northeasterly paleoflow. The vector mean (VM) for the entire lower heterolithic unit ($n = 8$) is 93° (Fig. 7A). At the base of the upper heterolithic body, planar foresets and capping ripple sets associated with scour fills (*n* = 11) record easterly to southeasterly paleoflow (VM = 105°, Fig. 7B). The channel body on the wave-cut platform that is laterally equivalent to these basal sandstone layers has trough and ripple cross-sets (*n* = 5) with northeasterly paleoflow (VM = 64°, Fig. 7C) that is highly oblique to flow directions recorded from the laterally equivalent scour fills of the basal upper heterolithic body.

Interpretation

... these beds carry our thoughts back to a period when the district was covered by a strange and now extinct vegetation, and when its physical condition resembled that of the Great Dismal Swamp, the Everglades, or the Delta of the Mississippi.

—Sir William Dawson (1855), *Acadian Geology*, p.182

The onset of prolonged periods of peat formation at Joggins, albeit disrupted periodically within virtually every mire, is marked by the development of Coal 32, the first coal bed to approach meter scale and the lowest mined coal seam in the Joggins Formation. The predominantly gray, organic-rich mudrocks and pervasive siderite concretions convey the onset of persistently waterlogged substrate within the section, which can be ascribed to a major base-level rise and a diminished dry season. The 10 m interval above Coal 32 represents periodically higher water levels and perhaps reduced sediment supply relative to the succeeding interval, as shown by flooding surfaces demarcated by aquatic fauna and rare thin limestone beds. Water depths clearly fluctuated, however, and were probably only briefly prohibitively deep for colonization by lepidodendrids, as evident from stigmarian-rooted planar sandstones, lycopsid tree stumps and thin coaly shale beds through much of these nine meters. This interval is interpreted as a shallow interdistributary embayment susceptible to short-term flooding events.

The planar-bedded sandstones in this interval (at 10 m, 12 m, and 15 m) indicate unconfined deposition in open water. The sandstone bed at 10 m was deposited by unidirectional flows in open water too deep for rooted vegetation, and could be interpreted as a distal deltaic accumulation in a large interdistributary bay. Evidence of standing vegetation abounds in sandstone units at 12 m and 16 m, however, and as discussed earlier, these mounds may be ascribed to the vegetation induced sedimentary structures ("VISS") of Rygel et al. (2004).

The rapid infilling of the interdistributary wetland by episodically accruing sediment is marked by the first heterolithic sandstone unit at ~23.5 m. Emplacement of the heterolithic units required accommodation space generated by successive fall in base level. The progressive thickening-upward motif of the upper, composite heterolithic sandstone body implies episodic progradation into an interdistributary embayment, with a hiatus

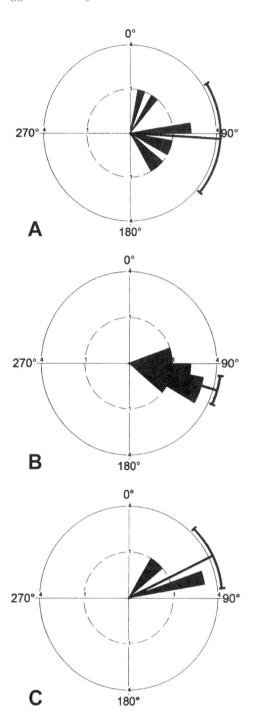

Figure 7. Paleoflow data for the Coal 32–29a interval (Cycle 6 of Davies and Gibling, 2003) in Figure 4. Each rose is plotted with visual correctness; i.e., each additional data point is added with the same area as the first one, so that there is no outward spreading bias with the roses. Vector mean (VM: direction to which flow is directed) shown by heavy radial line, standard deviation by heavy circumferential line. (A) planar foresets (Sp) and ripple sets (Sr), lower heterolithic unit (*n* = 8), VM = 093°. (B) Sp and capping Sr associated with scour fills (*n* = 11), VM = 105°, base of upper heterolithic unit. (C) trough (St) and ripple (Sr) cross-sets (*n* = 5), VM = 64°, for channel body on the wave-cut platform that is laterally equivalent to the lower body of the upper heterolithic unit (see Fig. 4).

in progradation that demarcates the two heterolithic stories. This hiatus was of sufficient duration to allow reestablishment of the lepidodendrid forest, but brief enough that the trunks of dead lepidodendrids buried by the lower story remained intact, protruding among the reestablished forest. The isolated 4-m-thick sandstone body on the foreshore may represent a distributary channel of moderate size (less than 35 m wide) through which sediment bypassed the interdistributary forests but whose bankfull capacity was exceeded regularly during flood events. The paleoflow evidence indicates that sand was transported into the forests by such distributary channels, with sand being swept into the wetlands along trajectories normal to oblique to the channel lines, burying the forests with up to 1 m of sand in single events. Floodstage flow is recorded in centroclinal and unidirectionally infilled scour hollows around standing trees. The regularity of flooding events, to which the rhythmic mudrock-sandstone couplets bear witness, may reflect seasonal or short-term precipitation flux.

Smothering of the shallow stigmarian rootstock of the lycopsid trees, which were unable to regenerate vegetatively, resulted in their rapid demise; decay of their pithy interior ensued as the hollow trunks stood supported by a thick, decay-resistant periderm that was their structural support (DiMichele and Phillips, 1994) both in life and as they stood dead or dying. The nature of the sedimentary tree-fill implies that the trees received only some of the initial, muddy waters that swept through the forest, but were infilled suddenly once waters carrying a sandy bedload overtopped the hollow trunks. The requirement for trees to be overtopped to be infilled may be overstated, however: longitudinal fissures, possibly recording strain from twisting, have been described in lycopsid trees (DiMichele and DeMaris, 1987) and fire scars may have provided basal openings in others.

Entombment and infilling of the standing trees and preservation of form-concordant bedding required accommodation that is inferred to have been provided mainly by the marked regional subsidence of the western Cumberland Basin (Calder, 1994; Davies and Gibling, 2003). Regenerating calamites and pervasive rooting (Fig. 4) indicate that the heterolithic sediments were saturated and by definition resided in the phreatic zone, hence were deposited episodically near the groundwater interface.

Facies within the studied interval at Joggins can be matched closely in modern deltaic and coastal wetlands, for example in the Niger Delta (Allen, 1970) and in the Barataria and Atchafalaya Bay areas of the Mississippi Delta (Coleman and Prior, 1980; van Heerden and Roberts, 1988; Kosters et al., 1987; Tye and Coleman, 1989). Alternating very fine sandstone and mudrocks of similar scale to those that constitute the heterolithic units, planar to ripple laminated and commonly rooted, typify marginal distributary deposits of the Niger delta lower flood plain, which "descend away to swamp-filled interdistributary basins" (Allen, 1970, p. 144). Isolated sandstone lenses with abrupt bases within the heterolithic units may represent dendritic gullies draining into distributary channels. Such drainage conduits occur within mangrove-vegetated tidal flats of the Niger delta (Allen, 1970) although no assertion is made here for a tidal setting. In the Barataria Bay area, sandy barrier systems that border the Gulf of Mexico protect extensive wetlands from wave attack, with distributary channel systems (abandoned bayous), forested regions, and poorly developed peats. In areas adjacent to the present Mississippi River channel, sandy levees and crevasse splays have locally prograded into the wetlands. Seaward parts of these wetlands experience strong tidal influence, but tidal indicators are not evident in the studied Joggins strata. Although components of the sedimentary paleoenvironment can be identified in these modern settings, none serve as a comprehensive analogue for Joggins.

The entombment of organic material beneath sandstones has commonly been attributed to the emplacement of crevasse splays through breaches in the banks of nearby channels. However, Davies-Vollum and Kraus (2001) suggested that burial of carbonaceous shales may represent avulsion of a major channel into a backswamp area, with infilling of the topographically low area with avulsion deposits (Kraus, 1996; Kraus and Wells, 1999) derived from small channels, prior to the consolidation of multiple, short-lived drainage systems into a single, larger channel. For one of the heterolithic bodies at Joggins (32.2–35.4 m, Fig. 4), there is good evidence that the commencement of sand deposition over forested wetlands coincides with the emplacement in the area of a distributary channel of modest width-to-depth ratio (10:1). This channel body probably represents local channel switching within a complex distributary network. Although it could indicate a major avulsion, there is no indication in the section that this modest-sized channel is an early phase in the avulsion of a larger channel system. Rather, the continuity of sandstone sheets from the heterolithic body into the distributary channel fill and the obliquity of paleoflow between the channel fill and the sheets strongly suggest that entombment of organic-rich sediments and erect trees in this instance was due to overbank flooding from the local channel. Channel bodies associated with the heterolithic sandstone units have a width-to-depth aspect ratio similar to that of distributary channel bodies elsewhere in the rock record (e.g., Olsen, 1993; Mjøs and Prestholm, 1993). Both distributary migration and infilling of interdistributary areas may have been linked to local base-level fall.

The occurrence of somewhat thicker, albeit periodically interrupted, coal beds atop each heterolithic sandstone body may indicate that the infilled interdistributary areas afforded relatively stable platforms for mire development that mitigated for a time the effects of regional subsidence (cf. Nemec, 1992; Tibert and Gibling, 1999). Distributary switching away from the site may have contributed to conditions conducive to peat formation by reducing sediment supply locally. The inferred distributary wetland environment that derives from this study accords well with coal thickness trends regionally within the Joggins Formation (Goudge, 1945; Copeland, 1959), which can be described as areally persistent in thin seams, the hallmark of which is significant variation in thickness and seam splitting locally (see Implications for a Coal Depositional Model, below).

In sequence stratigraphic terms, the study interval comprises stacked parasequence sets with both aggradational and progradational aspects (Davies and Gibling, 2003). The base of Cycle 5 above Coal 32 marks a retrogradational phase above red beds and the development of a thin stratal set of shallow open-water facies (8.5–10.5 m, Fig. 4). Above, repeated colonization by lycopsid trees and calamites was permitted by the shoaling effects of episodic progradation of sand bodies, at times to the point of subaerial exposure as witnessed by tetrapod trackways. Subsequently, the shoreline prograded sufficiently to establish the more landward, protected wetland zones with abundant entombed lycopsid trees (23.5–38 m interval). Above the Fundy Seam zone, the thick limestone interval associated with Coal 28 marks a subsequent, retrogradational event with reversion to open-water facies. The dominance of flooding events and absence of sequence boundaries are characteristic of the Joggins Formation as a whole (Davies and Gibling, 2003) and reflect the excessive rate of basin subsidence that is a hallmark of the Cumberland Basin (Calder, 1994).

The erect lepidodendrid trees that constitute the fossil forest stands are not associated with thick coal beds and may be considered by some to represent "clastic swamps"[1] (Gastaldo, 1987) that grew on inorganic mineral substrates, similar to the *Sigillaria* forest at Verdeña (Wagner et al., 2001). Close inspection reveals that in all cases, however, the stigmarian rootstocks coincide with an organic substrate, however thin, that may represent either a surficial A-horizon or onset of histosol development (Smith, 1991). Rather than clastic swamps (sensu Gastaldo, 1987), therefore, we suggest that the ecology of the standing trees is better described as the vegetation, possibly ecotonal, of incipient mires (cf. "ephemeral mires" of Falcon-Lang, 1999). Further discussion of the ecology and environment of the fossil forests follows the presentation of the paleobotanical record.

PALEOBOTANY

Standing Trees

Observed standing lepidodendrids accessible on the cliff face range in diameter from 25 to 50 cm ($n = 12$); the majority of trees higher on the cliffs, which could not be measured, appear to fall within this range. Where possible, measurements were taken at diameter breast height (dbh), a forest mensuration standard set at 1.3 m above ground level (IUFRO, 1959); dbh usefully avoids variable and exaggerated records of diameter arising from basal flaring of tree trunks, which can be problematic in the lepidodendrids (Calder et al., 1996). Trees were spaced as closely as 1.14 m apart along the two-dimensional transect of the forest stands provided in the cliff face. The longevity of exposure for an erect tree in the face of relentless erosion on the shores of the Bay of Fundy, site of the world's highest tides, is ~5 yr. The majority of trees exposed over the course of the study, spanning ten years, are those associated with the heterolithic sandstone/mudstone units (Fig. 5), representing three successive fossil forest stands.

Trunks of standing trees are monopodial and unbranched, and most exhibit little discernible taper over their preserved height, which is dependent on the entombing sediment, accommodation space, and time of exposure. The maximum height observed is 5.5 m, although in this case the tree doubtless extended ~0.5 m lower to the 10-cm-thick coal at 30 m of Figure 4, and an undetermined distance higher into the uppermost body of the upper heterolithic unit. This height (>6 m) is not substantially less than that enthusiastically estimated by Lyell (25 feet, or 7.6 m) in his first written account of his impression of the site (Lyell, 1842).

The erect trees of the study section can be assigned to the lepidodendrid lycopsids on the basis of their stigmarian rootstock. Past literature (e.g., Logan, 1845; Dawson, 1855a and later editions, 1882; Carroll et al., 1972) has assigned all erect trees at Joggins invariably to the genus *Sigillaria*. Although banded periderm of sigillarian character has been obtained from some in situ stump casts (Falcon-Lang, 1999), the diversity of prostrate lycopsid trees and the difficulty inherent in the unequivocal identification of the basal portions of these trees (Calder et al., 1996) suggest the likelihood that other lycopsid genera are represented. Pronounced longitudinal ribbing of some decorticated trunk casts is evocative of the Sigillariae, but one example of a partially decorticated and permineralized *Lepidodendron aculeatum* (see Fig. 10B) revealed similar longitudinal ribbing of the decorticated interior cast. A calcified stele obtained for Dawson (1877) by an "adventurous workman" from a standing tree of the upper forest stand (Fig. 8), rooted at the same horizon as the tallest example described above (30 m, Fig. 4), revealed radially arranged scalariform tissue assigned by Dawson to the form genus *Diploxylon*.

Calcium carbonate–permineralized trunks with a ropy exterior and infilled with charcoal have been encountered rarely, fallen from the cliff face and presumably derived from the same interval as that described by Dawson (1877), above Coal 29a. The charcoal exhibits cellular structure of lycopsid affinity (H. Falcon-Lang, 2002, personal commun.).

Associated with the erect lepidodendrid trunks are calamite stem casts in situ, typically on the order of 5 cm in diameter (Fig. 6). Calamite stems are found throughout the entire thickness of each heterolithic unit. In places, the main apical meristem (growing tip) of a calamite appears to have been severed at the horizon of an erosionally based sandstone bed, whereas a lateral branch passes upward through said horizon, imparting a "snakes and ladders" geometry of vertical persistence through several meters.

Compression Flora

A consistent compression flora has been exposed within the study interval over the past decade of investigation. Prostrate lycopsid logs, most commonly concentrated at the base of heterolithic sandstone bodies, include the genera *Sigillaria*, *Lepidodendron*,

---

[1]Although the term "swamp" as used colloquially in the United States refers to treed wetlands irrespective of peat formation, the term "clastic swamp" is problematic in the peatland terminology advocated by Gore (1983), Moore (1987), and others, wherein "swamp" is defined as a seasonally flooded, rheotrophic peatland or mire.

Fig. 1.—*Surface of the Cliff, showing the position of the Tree.* (From a sketch by Mr. Albert J. Hill.)

a, a. Coal-seams. *b.* Superficial Drift.

Figure 8. Woodcut (from Dawson, 1877) depicting extraction of lycopsid tree from upper forest stand in 1876. This specimen bore a stele with permineralized wood assigned by Dawson to the form genus *Diploxylon*.

and a diverse deciduous branched flora including *Lepidophloios*, *Paralycopodites*, *Bothrodendron* and equivocally, *Diaphorodendron* (Table 1, Figs. 9, 10). Of these, *Sigillaria*, represented by several species *(S. scutellata, S. mamillaris, S.* cf. *rugosa)*, and *Lepidodendron (L. aculeatum, L. lycopodioides)* predominate.

The record of prostrate trees at the base of entombing heterolithic bodies is not significantly time averaged; however, the overall macrofloral record of the measured section includes material fallen from the cliff face directly above, the height of which is ~30 m. This record comprises plants of lycopsid, sphenopsid, pterimdosperm, gymnosperm (both foliar and reproductive organs), fern, and possibly progymnosperm affinity (Table 1, Figs. 11, 12). Of these, the alethopterid foliage of pteridosperms ("seed ferns") is the most commonly encountered, although floral patterns differ between sedimentary lithofacies, reflecting ecological partitioning but also taphonomic bias. Noteworthy within this record are numerous taxa previously unreported from the Joggins Formation (Table 1). Most of these have been identified elsewhere within coeval rocks of the Cumberland and Minas Basins (R.H. Wagner, 2000, personal commun.) and include certain taxa that Wagner (2001) regarded as extra-basinal flora (e.g., *Adiantites adiantoides*, *Pseudadiantites rhomboideus*), and which are described in this paper as extra-mire flora.

TABLE 1. FOSSIL MACROFLORA OF THE FUNDY FORESTS

Class LYCOPSIDA
stem genera
   cf. *Bothrodendron punctatum* Lindley and Hutton*
   ?*Diaphorodendron* sp. DiMichele*
   Lepidodendron aculeatum Sternberg
   L. lycopodoides Sternberg*
   Lepidophloios laricinus Sternberg
   ?*Sigillaria* cf. *laevigata* Brongniart*
   S. mamillaris Brongniart
   S. cf. *rugosa* Brongniart*
   S. scutellata Brongniart
   Ulodendron Lindley and Hutton (= *Paralycopodites* sp.)
foliage
   Cyperites Lindley and Hutton (= *Lepidophylloides*)
cones and bracts
   ?*Lepidostrobophyllum majus* (Brongniart) Hirmer
   ?*Sigillariostrobus* sp. Schimper
rootstock
   Stigmaria ficoides Sternberg

Class SPHENOPSIDA
stem genera
   Calamites carinatus Sternberg*
   C. suckowi Brongniart
   Eucalamites sp.
foliage
   Annularia sp. Sternberg
cones
   Palaeostachya Weiss (? of *C. grandis*)

Class FILICOPSIDA
   Renaultia footneri (Marrat) Kidston*
   R. cf. *schatzlarensis* (Stur) Danzé
   Sphenopteris effusa Kidston*
   indet. circinate tip ("fiddlehead")

Class PROGYMNOSPERMOPSIDA?
   Adiantites adiantoides (Lindley and Hutton) Kidston*
   Pseudadiantites rhomboideus (Ettingshausen) Wagner*

Class PTERIDOSPERMOPSIDA
   Alethopteris decurrens (Artis) Zeiller
   A. discrepans Dawson
   A. cf. *urophylla* Brongniart
   Karinopteris tennesseana (White) Gastaldo & Boersma
   K. cf. *dernonrcourtii* (Zeiller) Boersma*
   Neuropteris cf. *hollandica* Stockmans*
   Neuropteris sp.

Class GYMNOSPERMOPSIDA
   Cordaites palmaeformis (Göppert) Weiss*
   C. principalis (Germar) Geinitz

*Denotes first report from the Joggins Formation.

Megaspore and Palynodebris Analysis

Organic horizons were macerated for megaspore analysis (Table 2, Fig. 4) using the techniques outlined in Calder et al. (1996) and Scott and King (1981). Thin coaly laminae at 398.6 m and 406.4 m coincide with the horizons at which the lepidodendrids constituting two major forest stands, entombed within heterolithic sandstone bodies, are rooted. Other horizons

Figure 9. Lycopsid macroflora of the Fundy fossil forests. (A) *Sigillaria mamillaris* × 1, coll. by BH; (B) permineralized *Sigillariostrobus* sp. × 1, coll. by JC.; (C) *S. mamillaris* × 0.5, coll. by BH; (D) *S. scutellata* × 0.5, coll. by BH; (E) *S.* cf. *scutellata* × 1, coll. by BH.

sampled include Coal 32, and thin organic-rich laminae within the "open water facies" above Coal 32.

Methods

Clastic samples were broken into pieces less than 1 cm across and placed in 40% hydrofluoric acid for two weeks, then washed and sieved using the techniques of Pearson and Scott (1999). Coal samples were crushed to 5–10 mm and macerated using concentrated nitric acid for one week, washed, then placed for one day in dilute ammonia following the procedure of Scott and King (1981). Sieved residues were picked under a binocular microscope. Photographs were taken using a Hitachi 3000N scanning electron microscope and stored digitally.

Results of Megaspore Analyses

The megaspores obtained (Table 2, Fig. 13) are typical of early Westphalian coal-bearing sequences (Scott and Hemsley, 1996). The megaspore record is dominated by *Tuberculatisporites*,

Figure 10. Lycopsid macroflora of the Fundy fossil forests. (A) Lycopsid tree cf. *Bothrodendron punctatum* showing deciduous branch scars, × 0.25; (B) *Lepidodendron aculeatum* × 0.5, coll. by JC: coalified leaf cushions and periderm, illustrating longitudinal ridges where decorticated; (C) *L. aculeatum* × 0.25; (D) non-flattened, prostrate lycopsid trunk cf. *Lepidophloios laricinus*, at base of heterolithic sandstone body at 23.5 m × 0.5.

dispersed by the arborescent lepidodendrid lycopsid *Sigillaria*. The parent plant affinities of the dispersed megaspores are given in Table 3.

Lycopsid rooting horizons are dominated by *Tuberculatisporites*, with subordinate *Lagenoisporites* and *Cystosporites* in the lower forest soil, and *Lagenicula* in the upper forest. Coal 32, the most substantial coal bed in the studied section, was co-dominated by *Cystosporites* and *Tuberculatisporites*, with rare *Lagenicula*. Where fusain was found in abundance, megaspores were comparatively rare. Planar-stratified clastic horizons of shallow open-water intervals were dominated by fusain or degraded plant matter with poor preservation of megaspores.

Comparison of Results

The paleobotanical records that derive from megaspores, miospores, and macrofloral compressions show varying degrees of commonality, but also disparate results that in isolation could

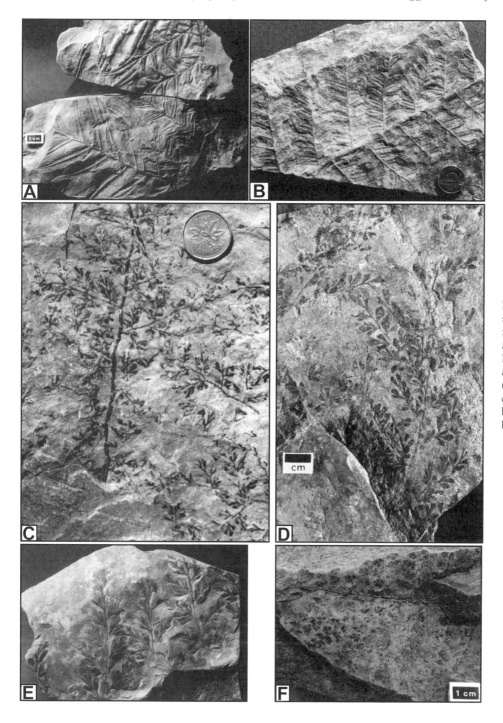

Figure 11. Macroflora of the Fundy fossil forests. (A) *Alethopteris decurrens* × 0.5, coll. by Donald Reid; (B) *Alethopteris discrepans* (cf. *A. lonchitica*) Dawson × 0.5, coll. by BH; (C) cf. *Renaultia schatzlarensis* × 1, coll. by BH; (D) *Sphenopteris effusa* Kidston × 1, coll. by BH; (E) *A. adiantoides* coll. by DR; (F) *Renaultia footneri* × 1, coll. by JC.

lead to significantly different interpretations of floral ecology especially regarding lycopsid communities. As discussed previously, the macrofloral record conveys dominance of *Sigillaria* and *Lepidodendron*. The megaspore record is in close agreement but favors domination of the fossil forests (incipient mires) by polycarpic lycopsids, namely *Sigillaria*, joined by *Diaphorodendron* on thicker peat substrates. The published miospore record of other coal beds at Joggins (Dolby, 1991; Hower et al., 2000), however, indicates that mires were dominated by monocarpic *Lycospora* producers (including *Lepidodendron*). Further research will be required to determine whether this reflects divergence of the miospore record from that provided by macroflora and megaspores or truly different floral composition amongst coal beds of the Joggins Formation.

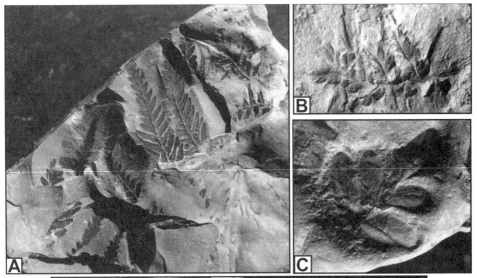

Figure 12. Macroflora of the Fundy fossil forests. (A) *Senftenbergia plumosa* × 1, coll. by DR; (B) cf. *Karinopteris dernoncourtii* × 1, coll. by BH; (C) *Neuropteris* cf. *hollandica* x1, coll. by DR; (D) *Neuropteris* sp. × 1, coll. by BH.

TABLE 2. RESULTS OF MEGASPORE ANALYSIS OF THE FUNDY FORESTS

Sample	Interval	Lithology	Megaspore composition	Associated palynodebris
CP1415	Coal 32 (7.7–8.3 m)	coal	*Cystosporites diabolicus* (rare)	abundant charcoal (fusain); rare pteridosperm cuticle, resin rodlets, and scorpion cuticle
CP1416	top of Coal 32 (8.3–8.5 m)	coal	*Cystosporites diabolicus* (60%) *Tuberculatisporites mammilarius* (40%) *Lagenicula horrida* (rare)	common pteridosperm cuticle; rare resin rodlets and scorpion cuticle
CP1418	1.3m above Coal 32 (9.8 m)	siltstone	?*Lagenicula* sp. (rare)	rare fusain
CP 1417	2.3m above Coal 32 (10.8 m)	organic-rich, platy siltstone	barren	highly degraded plant material
CP1419	rooting horizon of lower Fundy forest (23.4 m)	silty shale with thin coaly laminae	*Tuberculatisporites mammilarius* (80%) *Lagenoisporites rugosus* (10%) *Cystosporites diabolicus* (10%)	rare, poorly preserved pteridosperm cuticle
CP1414	rooting horizon of upper Fundy forest (31.4 m)	silty shale with thin coaly laminae	*Tuberculatisporites mammilarius* (75%) *Lagenicula* sp. (25%)	resin rodlets, scorpion cuticle

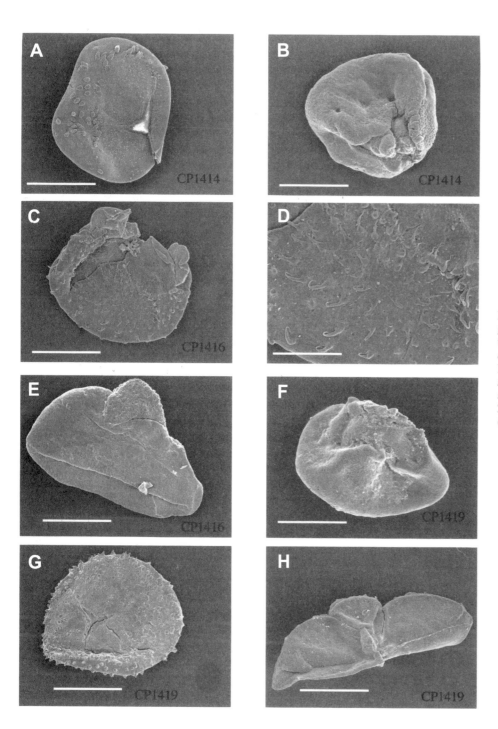

Figure 13. Scanning electron micrographs of representative megaspores of the Fundy fossil forests. (A, G) *Tuberculatisporites mammilarius* (Bartlett) Potonié and Kremp; (B, F, H) *Lagenoisporites rugosus* (Loose) Potonié and Kremp; (C, D) *Lagenicula horrida* Zerndt; (E) *Cystosporites diabolicus* (Scott) Hemsley.

TABLE 3. PARENT PLANT AFFINITIES OF MEGASPORES FROM THE FUNDY FORESTS

Megaspore	Parent plant affinity
Tuberculatisporites mamillaris (Bartlett) Potonié & Kemp	*Sigillaria*
Lagenicula horrida Zerndt	*Lepidodendron* sensu latus
Lagenosoisporites rugosus (Loose) Potonié & Kemp	*Paralycopodites*
Cystosporites diabolicus (Scott) Hemsley	*Diaphorodendron*

FAUNAL ECOLOGY

Although the Joggins section is famed for its faunal record, there have been almost no studies of terrestrial faunal paleoecology in the century that has elapsed since the work of Dawson, save the review in the field guide of Carroll et al. (1972). The published record derives almost exclusively from the "tree stump fauna," which reflects a degree of taphonomic filtering of the fossil record of the wetland ecosystem. Faunal discoveries from the study interval (Table 4, Figs. 14, 15) provide an important introduction to the "extra–tree stump fauna" and to the broader ecology of these lepidodendrid wetlands.

The terrestrial invertebrate record includes a member of the fixed-wing Megasecoptera (Fig. 14A), the whip spider *Graeophonus carbonarius* (Fig. 14B), the millipede *Xyloiulus sigillariae* (Fig. 14 C), and scorpion cuticle recorded in palynodebris from Coal 32 and from the thin coal at the rooting horizon of the upper forest horizon at 31.4 m (see Table 2). The remaining invertebrate specimens were collected at the cliff base and their precise stratigraphic position within the measured section is uncertain.

TABLE 4. FOSSIL TERRESTRIAL AND AQUATIC FAUNA OF THE FUNDY FORESTS INTERVAL

Terrestrial

Phylum ARTHROPODA
    Class ARACHNIDA
        Graeophonus carbonarius
        Scorpion cuticle
    Class INSECTA
        Megasecoptera incertae familiae
    Subphylum MYRIAPODA
    Class DIPLOPODA
        Xyloiulus (Xylobius) sigillariae
Phylum CHORDATA
    Superclass TETRAPODA
        Limnopus vagus*
        Matthewichnus velox*
        Ornithoides trifudus*
        ?*Notalacerta* sp.*
        ?*Pseudobradypus* sp*
    Class AMPHIBIA
        Dendrerpeton acadianum
    Stem amniotes
        Anthracosauria indet.

Aquatic

Phylum MOLLUSCA
    Class BIVALVIA
        Naiadites sp.
        Archanodon westoni
Phyllum ARTHROPODA
    Class OSTRACODA
        indet. ostracodes
Phyllum CHORDATA
    Subphylum VERTEBRATA
    Superclass PISCES
    Class OSTEICHTHYES
        indet. fish scales

*Denotes ichnotaxon, for which taxonomic affinity is inferred.

Two articulated tetrapod specimens of an entire *Dendrerpeton acadianum* and the vertebral column of a reptiliomorph anthracosaur resembling *Callignethelon*, like the *Graeophonus* abdomen, are preserved by authigenic cementation in siderite concretions. This mode of preservation, unlike that of the tree stump fauna from which the bulk of the Joggins tetrapod record derives, favors exceptional preservation of body fossils and articulated skeletons, as in the case of the *Dendrepeton acadianum* specimen, which was described by Holmes et al. (1998) as the best-preserved example of the genus in the world (Fig. 15C). Both tetrapod specimens are believed to derive from the siderite-bearing roof strata of Coal 29a (~45 m) or overlying roof of Coal 29. Complementing the osteological record is the tetrapod trackway record. The fossil forest section has yielded nominally five tetrapod ichnotaxa (Table 4), three of which were unknown previously at Joggins: *Limnopus vagus* (Fig. 15A) of presumed temnospondyl affinity, is a common ichnotaxon of the Appalachian Basin, whereas *Notalacerta* (Fig. 15D), known also from the Appalachian Basin, has been ascribed to early reptiles. The third deserves special mention in that it may represent the top predator of the Joggins wetlands. A trackway of deeply impressed footprints that individually span 10 cm in width (Fig. 15B) is cast by an overlying meter-thick tabular sandstone body at 16 m of the study section. Footprints of this size may record the passage of a large baphetid (loxommatid) stem tetrapod, reptiliomorph, or alternatively, an embolomere amphibian similar to the edyopoid *Edops*.

An aquatic invertebrate fauna of ostracods, bivalve pterioid molluscs, and fish scales occurs within the fossil forest succession. The greatest concentration occurs within the 10-m-thick interval above the roof of Coal 32 (10–20 m, Fig. 4), below the lower forest stand, in association with thin (5–6 cm) limestones, platy, siderite-bearing organic-rich mudrocks, and thin (centimeter-scale) coal beds. This faunal association is characteristic of aquatic faunal horizons generally within the Joggins Formation. Duff and Walton (1973) interpreted such horizons as "aborted marine transgressions" and pointed out their similarities with inland reaches of marine bands within the Western European Carboniferous Basin (Calder, 1998; see also Davies and Gibling, 2003; Skilliter, 2001).

Noteworthy is the youngest known occurrence of the rare, large unionoid bivalve genus *Archanodon (Asthenodonta)*. The *Archanodon* specimens, discovered by GSC paleontologist T.C. Weston in 1892 and described by Whiteaves (1893), apparently derive from the meter-scale, tabular sandstone body at 16 m of Figure 4. Over a century later in 1998, another specimen was discovered by one of the authors (BLH) from this same bed, which also cast the large tetrapod trackway described above. The ecological preferences of this rare bivalve are not well understood; Friedman and Chamberlain (1995) consider *Archanodon catskillensis* from the mid-Devonian of the northeastern United States to be the first freshwater bivalve. It is noteworthy that the most productive Carboniferous locality of *Archanodon* known, from the informally named Hebert sandstone (Calder et al., 1999a) at 267 m of the Division IV coal measures, co-occurs with skeletal

Figure 14. Invertebrate faunal record from the fossil lepidodendrid forest succession. (A) Megasecopterid insect (sans head and prothorax) × 0.5, coll. by DR; (B) abdomen of the whip-spider *Graeophonus carbonarius* × 1, coll. by Jennifer Reid; (C) the millipede *Xyloiulus sigillariae* × 1, coll. by BH.

material consistent with the large trackmaker described above (see Hebert and Calder, 2004; Falcon-Lang et al., 2004). Although the Hebert sandstone has been described by these authors as a dryland waterhole deposit, the occurrence of some of these same taxa within the measured section of this study illustrates that they were not endemic to such settings (Hebert and Calder, 2004).

PALAEOECOLOGY AND DEPOSITIONAL ENVIRONMENT: DISCUSSION

The coincidence of the onset of the coal measures at Joggins and the preservation of prolific stands of lycopsid trees are not unrelated; rather, they bear witness to conditions of their shared paleoenvironment. The paleobotanical reconstruction of the Joggins forests is not considered to be significantly time averaged. Most of the forest soils are thin, centimeter-scale organic horizons representing incipient mires, and the macrofloral record derives largely from prostrate logs at the base of heterolithic units immediately overlying the soils (see Fig. 4, at 23.5 m).

The macrofloral record represents a mixture of ecological preferences (Fig. 16), and includes both mire (lycopsids) and extramire (sphenopsids, pteridosperms, gymnosperms, and ?progymnosperms) plants that can be differentiated by reproductive strategy. The free-sporing lycopsids, discussed further below, represent the framework trees of the mires, most of which were incipient (cf. "clastic swamps" of Gastaldo, 1987). Calamitean sphenopsids represent an intermediate, disturbance-tolerant, and highly adaptive succession to the lycopsids, the advantage accorded by their ability to propagate vegetatively is evident in their vertical persistence. Seed-bearing plants constitute the remaining extra-mire flora.

The lycopsids present within the megaspore and compression records of the study section represent the full spectrum of inferred reproductive strategies and ecological preferences amongst the lepidodendrids (Phillips and DiMichele, 1992; DiMichele and Phillips, 1994). The ongoing, polycarpic reproduction of *Sigillaria* and of deciduous branched trees such as *Bothrodendron*, ?*Diaphorodendron*, and *Paralycopodites* favored their success in the dynamic environment of sediment flux and subsidence at Joggins. Success did not translate into monopoly, however, as ongoing reproduction also meant reduced spore output in comparison to monocarpic lepidodendrids (Phillips and DiMichele, 1992). Rather, the macrofloral record indicates that disturbance translated into diversity for the polycarpic lepidodendrids.

In curious association with the polycarpic lycopsids, apparent in the compression record of prostrate trees and, in part, in the megaspore record of lepidodendrid rooting horizons, are the monocarpic genera *Lepidodendron* and *Lepidophloios*. These genera have been inferred to be the lycopsids least tolerant of

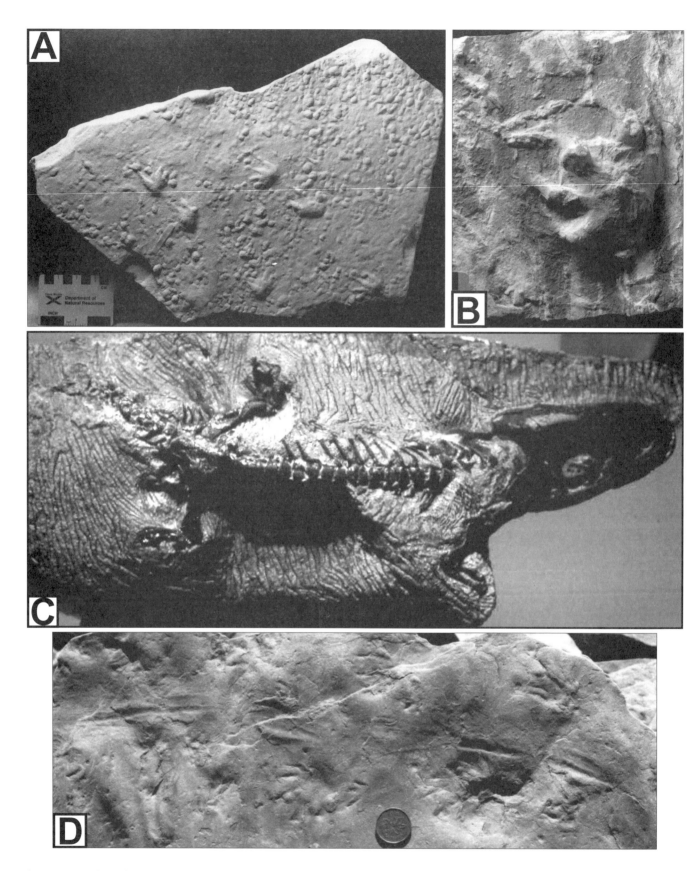

Figure 15. Vertebrate faunal record of the fossil lepidondendrid forest succession: (A) *Limnopus vagus* trackway × 0.25, coll. by DR; (B) single print of the top predator of the Joggins ecosystem × 0.5, coll. by DR; (C) exceptional, articulated specimen of *Dendrerpeton acadianum* × 0.5 (cast of NSM987GF99.1, described by Holmes et al., 1998); (D) ?*Notalacerta* sp. trackway × 0.5, coll. by BH.

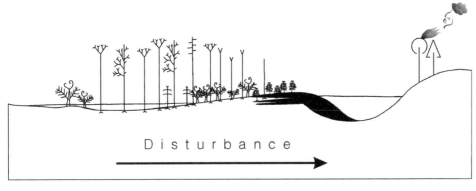

Figure 16. Schematic transect of the interdistributary wetland and surrounding environment at Joggins depicting relative ecological preferences of the flora, derived from combined fossil tree, compression flora, and megaspore data.

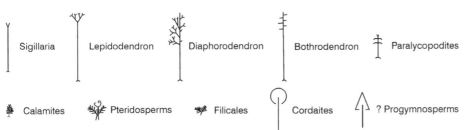

disturbance, their ecological preferences differing from one another only in their nutrient requirements (Phillips and DiMichele, 1992; DiMichele and Phillips, 1994). Their domination of the miospore record of coal beds within the section (Dolby, 1991; see also Hower et al., 2000), representing peaty soils, accords well with such inferences. The diversity of lycopsid genera that co-occur in the compression record, and to a lesser degree in the megaspore record, suggests heterogeneity, temporal or spatial, of the wetland environment (see DiMichele et al., 1985).

The calamites appear to have been almost impervious to the disruptions that spelled an end to entire stands of lepidodendrids. Their vertical persistence, and in particular their apparent ability to reassign the apical meristem functions to a buried branch when the main growing stem had been severed by flood, accorded in them a propagative advantage over other plants during deposition of the heterolithic units. Adventitious rooting in response to burial (Gastaldo, 1992), which required saturated sediment, contributed to this strategy. A modern analogue may be found in *Montrichardia arborescens* from the Orinoco Delta (Pfefferkorn and Zodrow, 1982). Persistence of inundated riparian *Fraxinus* (ash) trees of the Potomac River likewise can be ascribed to transference of apical dominance in response to repeated flood damage and burial (Sigafoos, 1964). The co-occurrence of erect calamites in sequences that entomb erect lepidodendrids is not unique to Joggins, but rather would appear to have been typical of Euramerican fossil forests (Klusemann and Teichmüller, 1954; Teichmüller, 1955; Demko and Gastaldo, 1992; Gradziński and Doktor, 2000; this study).

The association of pteridosperms and gymnosperms within the sequence accords with the inferred tolerance of seed-bearing plants to disturbed habitats and the ability to withstand the stress of moisture-deficit (see DiMichele and Aronson, 1992). The occurrence of elements described by Wagner (2001) from elsewhere in Euramerica (including the Maritimes Basin) as "extra-basinal" should come as no surprise given the disturbed habitat of Joggins. The co-occurrence of *Pseudadiantites rhomboideus* with *Archanodon* shells and ?baphetid tetrapod footprints in the sheet sandstone at 16 m (Fig. 4), for example, is consistent with the interpretation of Wagner (2001) that this represents an extramire floral element (see Hebert and Calder, 2004, and Falcon-Lang et al., 2004, describing a similar fossil assemblage from red beds lower in the Joggins section). The distribution of gymnospermous and pteridospermous floral elements at Joggins has been ascribed to floral turnovers at the landscape scale, with dryland-wetland shifts related to transgressive-regressive cycles (Falcon-Lang, 2003a, 2003b). The preservational record of delicate foliage and reproductive organs within the forest sequence, however, indicates derivation locally and in this instance would seem to argue against wholesale environmental shifts at the landscape or basin scale. It would seem, therefore, that ecological partitioning *within* the wetland landscape may have been equally significant, with such seed plants occupying terra firma areas peripheral to interdistributary areas of higher water level.

The role of wildfire as an agent of disturbance in the Joggins paleoenvironment has been demonstrated by Falcon-Lang (1999) and has been invoked in the hollowing of standing trees and in increased sediment flux. A mat of charcoal (fusain) clasts observed by JHC and ACS within the basal fill of an erect lepidodendrid tree entombed at the base of the lower heterolithic unit (23.5 m, Fig. 4) and in fallen, permineralized trunks bears mute testimony to the passage of such wildfire. Although such charcoal conceivably could have been introduced through flooding from other sources, Falcon-Lang (1999) has demonstrated that the interior of one standing lepidodendrid tree itself was charred and that the majority of fusain within in situ stumps derives from parenchymatous tissue and periderm of lycopsid affinity to the virtual exclusion of other plant taxa present in the landscape. External to the standing lepidodendrids elsewhere in the Joggins Formation, within thin coals representing incipient ("ephemeral") mires, is found charcoal of medullosan and lyginopterid seed ferns, sphenopsid and cordaite affinity (Falcon-Lang, 1999).

Clearly, fire was a significant abiotic influence in the interdistributary wetland ecology of Joggins, contributing to the ecological disturbance of these habitats. Fires during this time may have been exacerbated by extreme atmospheric oxygen levels during the Carboniferous "Icehouse" (Scott and Jones, 1994). The most probable ignition source for wildfire at Joggins is lightning, which is most prevalent in tropical regions of pronounced seasonality (Lottes and Ziegler, 1994). Such seasonality is consistent with the inferred rheotrophic status of paleomires at Joggins (Hower et al., 2000) and elsewhere in the Cumberland Basin (Calder 1994), although permineralized cordaite trunks from the section, albeit presumably from an upland source, have failed to demonstrate seasonal growth rings (Falcon-Lang and Scott, 2000). Although this led these authors to conclude nonseasonality, a similar lack of growth rings within conifer woods in a desert oasis setting from the Lower Jurassic Navajo Sandstone of Utah has been ascribed to tapping of groundwater at depth (Falcon-Lang, 2002, personal commun.), and the existence of deep rooting structures in cordaites has recently been demonstrated (Falcon-Lang and Bashforth, 2004). The inferred plant paleoecology of adaptation to abiotic disturbance accords well with the sedimentary record of episodic but rapid aggradational fills and rapid subsidence of the interdistributary wetland environment in which the lepidodendrid forests repeatedly established themselves and inevitably became buried.

Burial Time

The morphology and internal structure of the lepidodendrid lycopsids suggest rapid, temporary growth. Structural support in the lepidodendrids was achieved through their outer rind of nonwoody periderm, with little investment in the production of wood; the crown is believed to have served primarily as a scaffold for the production and dispersal of cones. Consequently, the life span of these economically constructed trees is thought to have been brief, perhaps as little as 10–15 yr (Phillips and DiMichele, 1992). The time represented by heterolithic sediment accruing around the standing trees is another matter. The length of time that a dying or dead lepidodendrid, with its outer trunk of decay-resistant periderm, could stand probably was considerably longer than its life span, but precisely how long is difficult to measure. The interpretation of the sediments that entombed the trees is equivocal: their episodic depositional history records successive, repetitive flooding events, but at an interval possibly of months, seasons, or longer.

The burial of standing trees can occur by mechanisms whose time frames range from sudden to gradual. These include flood events, both catastrophic and repeated; subsidence induced by compaction or by seismic events; sea-level rise; and volcanism. Flood events can be catastrophic, as at Bijou Creek, Colorado, where one flood event produced scour hollows around trees and deposited 4-m-thick sand sheets (McKee et al., 1967). Catastrophic burial can result from other mechanisms as well, including the sudden subsidence and drowning of wetland forests along the central Mississippi River as a consequence of the 1811–12 New Madrid earthquakes (Lyell, 1849; Fuller, 1912). Catastrophic subsidence induced by seismic activity (Gastaldo et al., 2004) and burial by high-magnitude, low-frequency flood events (Demko and Gastaldo, 1992) have been invoked for lycopsid fossil forests of the Black Warrior Basin. On the slopes of Mount St. Helens, catastrophic burial of trees by mud flows occurred as a consequence of explosive volcanic eruption, and the Carboniferous *Lepidophloios* forests at Arran, Scotland (Scott and Calder, 1994), and *Omphalaphloios* forests of Puertollano, Spain (Wagner et al., 2003), were similarly entombed catastrophically. In contrast, gradual submergence of forests has been documented in settings of active sedimentation and subsidence such as the Niger delta (Allen, 1970) and in coastal settings experiencing sea-level rise, such as the Bay of Fundy shoreline in Nova Scotia (Dawson, 1855b). Episodic burial of flood-plain trees due to repeated flooding has been documented by Sigafoos (1964) along the Potomac River of the eastern United States and by Stone and Vasey (1968) in northern California.

At Joggins, there is no unequivocal evidence that initial submergence of the fossil forests was catastrophic rather than progressive, although coseismic activity clearly would have attended subsidence in the Cumberland Basin given its setting. In the case of lycopsids entombed within heterolithic sandstone bodies, however, it is clear that the subsequent deposition of entombing sediments was episodic rather than catastrophic. (Here a note of caution is warranted in the use of the terms catastrophic and episodic: although the heterolithic strata were emplaced episodically during repetitive events, the onset of inundation was catastrophic to the lycopsid stands.) Clearly, the heterolithic strata were emplaced over a time frame longer than that of a single, catastrophic event, but how much longer? The basal strata of the upper, composite heterolithic unit (32.5–33.4 m, Fig. 4) comprise eight coarsening-up depositional couplets over a 0.90 m interval, each ranging from 7 to 14 cm thick. If these recurrent couplets reflect no more than annual events, then an entombing heterolithic unit 4 m in thickness would have been emplaced in no more than three to six decades and conceivably much less time, given the contribution of thick bedded sandstone lenses. The regeneration of calamite growth through a heterolithic unit and structural integrity of the buried lepidodendrids would seem to support a time frame of this order. In comparison, 15 flood events in northern California resulted in 9 m net aggradation of the alluvial plain over a 1000 year time period (Stone and Vasey, 1968). A burial history at Joggins that was more rapid by as much as an order of magnitude is consistent with the record of profound basinal subsidence (Calder, 1994; Davies and Gibling, 2003) and abiotic disturbance generating sediment flux that facilitated rapid entombment.

IMPLICATIONS FOR A COAL DEPOSITIONAL MODEL

I felt convinced that, if I could verify the accounts of which I had read, of the superposition of so many different tiers of trees, each representing forests which grew in succession on the same area, one above the other; and if I could prove at the same time their connexion with seams of coal, it would go farther than any facts yet recorded to confirm the theory that coal in general is derived from vegetables produced on the spots where the carbonaceous matter is now stored up in the earth.

—Sir Charles Lyell (1845), *Travels in America*, p. 177–178

The promise of insight into conditions of coal formation provided by the fossil forests of Joggins was a promise foretold yet largely unfulfilled in the century that has past since Lyell and Dawson. To the south of Joggins, at the piedmont of the ancestral Cobequid highlands, coals of the Springhill Mines Formation of the Cumberland Group exhibit a distinct and predictable geometric and lithologic zonation that can be linked with confidence to their piedmont-riverine depositional setting (Calder, 1991, 1993, 1994), and to their stratal position in cyclothemic sequences to climate shifts at the scale of the Milankovitch band. For the coals of the underlying Joggins Formation, however, characterization of their depositional setting and a predictive depositional model have proven elusive.

The interdistributary wetland forest setting established herein accords well with spatial and thickness trends of individual coal seams within the Joggins Formation, and with their inferred origins as rheotrophic peats (Hower et al., 2000), the product of regularly inundated topogenous forest swamps. Although correctly interpreting the trophic status of precursor peats is one of the most challenging aspects of coal geology (Teichmüller, 1989), coal beds of the Joggins Formation characteristically convey an archetypal signature of rheotrophic coals (Calder et al., 1996), including high ash yield and poor preservation of tissues when etched. Seams or coal beds of the Joggins Formation typically are thin (centimeter-scale, seldom exceeding 1 m in thickness) and exhibit areal variability in their thickness that was problematic historically in the underground mining of such thin seams (Copeland, 1959; Goudge, 1945). Most represent peat deposits that seldom exceeded one meter in thickness, even using estimates of decompacted plant material (Collinson and Scott, 1987; Winston, 1986) which are probably overestimated (Nadon, 1998) owing to the effects of humification and ongoing compaction that attend peat formation. Ubiquitous clastic partings record their residence as topogenous peats near the groundwater interface. Commonly, these thin coals are associated (overlain or interbedded) with bivalve-bearing limestone beds that record flooding surfaces. Their invariably high sulfur content (4%–7%; Copeland, 1959) is characteristic of coals of the Maritime Basin that are associated with bivalve-bearing strata (Calder and Naylor, 1985), and points suggestively but equivocally to marine influence (see Calder, 1998; Skilliter, 2001).

Coal beds within the Joggins Formation have been interpreted as signatures of flooding surfaces associated with rising base level (Davies and Gibling, 2003); however, not all flooding surfaces within the section are associated with sustained peat accumulation. The complex interaction of allogenic and autogenic processes in peat formation elsewhere in the Cumberland Basin has been described by Calder (1994). At Joggins, climate shifts, accommodation, base level, and distributary processes doubtless figured as well in mire development.

The occurrence of thicker, albeit periodically interrupted, coal beds atop heterolithic sandstone/mudstone body may indicate that these infilled interdistributary areas afforded relatively stable platforms for mire development that mitigated for a time the effects of regional subsidence (cf. Nemec, 1992; Tibert and Gibling, 1999). Distributary switching away from the site may have contributed to conditions conducive to peat formation by reducing sediment supply locally. The inferred distributary wetland environment that derives from this study accords well with coal thickness trends regionally within the Joggins Formation (Copeland, 1959), which can be described as areally persistent in thin seams, the hallmark of which is significant variation in thickness locally.

CONCLUSIONS

The Joggins wetland forests were very much the epitome of the picture painted by Phillips and DiMichele (1992, p. 584) in their commentary on misconceptions and realities of Carboniferous wetland ecology: "The lush reconstructions of Late Carboniferous (Westphalian) lepidodendrid swamps are pictures of environmental uniformity and tranquility, not conveying that such an ecosystem may have been ever on the brink of disaster. These coal swamps were disturbance driven and abiotically controlled." At Joggins, abiotic (allogenic) disturbance was inherent in wetlands that formed part of a shifting distributary system under a climate of seasonal precipitation flux (Calder and Falcon-Lang, 2002). The spatial heterogeneity of this sedimentary environment, and the rapid rate of temporal change that attended its active subsidence and fluctuating precipitation, provided a diverse ecology in which virtually all lepidodendrid lycopsid genera could be represented. Within the interdistributary wetlands, the most resilient, hence successful, of these may have been *Sigillaria*, although *Lepidodendron* and *Lepidophloios* apparently dominated established peat substrates/mires. The ultimate adaptation to this shifting environment, however, is demonstrated not by the Lycopsida, but by the ubiquitous calamitean sphenopsids, whose persistence apparently was aided by vegetative propagation. The occurrence of erect calamites in association with fossil lepidodendrid forests, first described at Joggins by Lyell and Dawson (1853, Fig. 1), was a common successional motif in buried lepidodendrid forests of Euramerica, their co-occurrence having been described from lycopsid forests of the Ruhr Basin of western Europe (Klusemann and Teichmüller, 1954; Teichmüller, 1955), Silesian Basin of eastern Europe (Gradzinski and Doktor, 2000), and Warrior Basin of the southern Appalachians (Demko and Gastaldo, 1992; Gastaldo, 1986). Rather than being co-dominant, however, the regenerating calamites formed a succession to the dead and dying lycopsid stands.

Rapid, ongoing subsidence of the basin clearly facilitated the submergence and entombment of the fossil forests. Recognizing the role played at Joggins by sea-level and climate change or seismic events will continue to be challenging against this backdrop of profound basinal subsidence (Calder, 1994; Davies and Gibling, 2003; Falcon-Lang, 2003a). That said, the historically productive fossil forest stands affiliated with the heterolithic units at Joggins clearly represent repeated flood events of interdistributary wetlands in an area of rapid sediment aggradation and subsidence. Basinal subsidence provided ample accommodation space, while abiotic disturbance inherent to the seasonal climate facilitated its episodic infilling.

In this study, the three sets of paleobotanical data comprising megaspores, miospores, and macroflora were found to be complementary records each of which contributed to reconstruction of lycopsid paleoecology of the fossil forests. The megaspore record is biased toward polycarpic lycopsids, whereas the miospore record is dominated by monocarpic trees. Only the macrofloral record contained both, which points to the importance of this traditional, yet perhaps underutilized, source of paleobotanical data.

The occurrence within the sequence of extra-mire flora (represented by seed-bearing gymnosperms and pteridosperms and possibly by progymnosperms) indicates that edaphic and disturbance variation within the wetland landscape of Joggins was equally important to the ecology of plants as were longer-term climate/transgressive-regressive shifts between drylands and wetlands.

The interdistributary wetland forest setting accords well with spatial and thickness trends of individual coal seams within the Joggins Formation as recorded in a mining history that spanned four centuries. Such a setting is consistent with the interpretation of the coal beds, suggested by their mineral and plant composition, as the products of topogenous, rheotrophic forest swamps and provides the basis for a depositional model for Joggins coals that has long proven elusive.

The temporal and spatial heterogeneity of this interdistributary wetland habitat also has implications for the evolution of its terrestrial fauna, by promoting genetic variation and providing evolutionary opportunity (see DiMichele and Aronson, 1992). This hypothesis has bearing at Joggins, where the terrestrial fauna includes early representatives of invertebrate and vertebrate lineages, including the earliest amniotes.

ACKNOWLEDGMENTS

This study spans more than a decade of fieldwork, during which time many colleagues and field trip participants have provided comment for which we express gratitude. In particular, we thank Donald and Eleanor Reid of the Joggins Fossil Centre for their contributions to the fossil flora and fauna record. Robert Wagner, through his revision of Joggins floral taxonomy for the Geological Survey of Canada, made a major and generous contribution to macrofloral identification. Howard Falcon-Lang, Michael Rygel, and Robert Wagner we thank for their generous advice and sage insight in the field. The manuscript benefited from the reviews of W.A. DiMichele and R.A. Gastaldo. We gratefully acknowledge the support of NATO Collaborative Research Grant CRG 940559 to ACS, JHC, and MRG and a Natural Sciences and Engineering Council of Canada Research Grant to MRG. Janet Webster and Cynthia Phillips of the Graphics Section, Nova Scotia Department of Natural Resources, Andrew Henry, and especially Mike Rygel are thanked for their illustrations.

Note in press: Since the writing of this paper, a formal stratigraphic redefinition of the Joggins section has been completed (Calder et al., 2005). The revised Joggins Formation is now described in detail by Davies et al. (2005), who use different meterage and cycle numbers than those in the diagrams here, which conform to Davies and Gibling (2003).

REFERENCES CITED

Allen, J.R.L., 1970, Sediments of the modern Niger Delta: A summary and review, in Morgan, J.P., ed., Deltaic sedimentation: Society of Economic Paleontologists and Mineralogists Special Publication 15, p. 138–151.

Bell, W.A., 1944, Carboniferous rocks and fossil floras of Nova Scotia: Geological Survey of Canada Memoir 238, 277 p.

Browne, G.H., and Plint, A.G., 1994, Alternating braidplain and lacustrine deposition in a strike-slip setting: The Pennsylvanian Boss Point Formation of the Cumberland Basin, Maritime Canada: Journal of Sedimentary Research, v. B64, p. 40–59.

Brown, R., 1829, Geology and mineralogy [of Nova Scotia], in Haliburton, T.C., An historical and statistical account of Nova Scotia, 2, Section III: Halifax, Joseph Howe, p. 414–453.

Calder, J.H., 1991, Controls on Westphalian peat accumulation: The Springhill Coalfield, Nova Scotia [Ph.D. thesis]: Halifax, Dalhousie University, 310 p.

Calder, J.H., 1993, The evolution of a ground-water influenced (Westphalian B) peat-forming ecosystem in a piedmont setting: The No. 3 seam, Springhill Coalfield, Cumberland Basin, Nova Scotia, in Cobb, J.C., and Cecil, C.B., eds., Modern and ancient coal-forming environments: Geological Society of America Special Paper 286, p. 153–180.

Calder, J.H., 1994, The impact of climate change, tectonism and basin hydrology on the formation of Carboniferous tropical intermontane mires: The Springhill Coalfield, Cumberland Basin, Nova Scotia: Palaeogeography, Palaeoclimatology, Palaeoecology, v. 106, p. 323–351, doi: 10.1016/0031-0182(94)90017-5.

Calder, J.H., 1998, The Carboniferous evolution of Nova Scotia, in Blundell, D.J., and Scott, A.C., eds., Lyell: The past is the key to the present: Geological Society [London] Special Publication 143, p. 261–302.

Calder, J.H., and Falcon-Lang, H.J., 2002, The classic Pennsylvanian locality of Joggins, Nova Scotia: Paleoecology of a disturbance-prone wetland ecosystem: Geological Society of America Abstracts with Programs, v. 34, p. 211.

Calder, J.H., and Naylor, R.D., 1985, Coal deposition in alluvial fan-settings: the Salt Springs and Roslin districts of the Cumberland Basin, in Mills, K.A. and Bates, J.L., eds., Mines and Minerals Branch report of activities, 1984: Nova Scotia Department of Mines and Energy Report 85-1, 5–9.

Calder, J.H., Gibling, M.R., Eble, C.F., Scott, A.C., and MacNeil, D.J., 1996, The Westphalian D fossil lepidodendrid forest at Table Head, Sydney Basin, Nova Scotia: Sedimentology, paleoecology and floral response to changing edaphic conditions: International Journal of Coal Geology, v. 31, p. 277–313, doi: 10.1016/S0166-5162(96)00020-1.

Calder, J.H., Davies, S.J., and Gibling, M.R., 1999a, Classic Upper Carboniferous sections of the Maritimes Basin in Nova Scotia: Field Trip Guide, International Congress of the Carboniferous and Permian, 14th, Calgary, 95 p.

Calder, J.H., Gibling, M.R., Scott, A.C., Davies, S.J., and Hebert, B., 1999b, Palaeoecology and sedimentology of fossil lycopsid forest successions in the classic Upper Carboniferous section at Joggins, Nova Scotia: International Congress on the Carboniferous-Permian, 14th, Calgary, Program with Abstracts, p. 19.

Calder, J.H., Rygel, M.C., Ryan, R.J., Falcon-Lang, H.J., and Hebert, B.L., 2005, Stratigraphy and sedimentology of early Pennsylvanian red beds at Lower Cove, Nova Scotia, Canada: the Little River Formation with redefinition of the Joggins Formation: Atlantic Geology, v. 41, p. 143–167.

Carroll, R.L., Belt, E.S., Dineley, D.L., Baird, D., and McGregor, D.C., 1972, Vertebrate palaeontology of eastern Canada: Guidebook, Excursion A59, International Geological Congress, 24th, Montreal, p. 64–80.

Coleman, J.M., and Prior, D.B., 1980, Deltaic sand bodies: American Association of Petroleum Geologists, Education Course Note Series, no. 15, 171 p.

Collinson, M.E., and Scott, A.C., 1987, Implications of vegetational change through the geological record on models for coal-forming environments, in Scott, A.C., ed., Coal and coal-bearing strata: recent advances: Geological Society (London), Special Publication 32, p. 67–85.

Copeland, M.J., 1959, Coalfields, West Half Cumberland County, N.S.: Geological Survey of Canada Memoir 298, 89 p.

Darwin, C., 1859. The origin of species by means of natural selection (Chapter IX. On the imperfection of the geological record, p. 296.): London, John Murray, 513 p.

Davies, S.J., and Gibling, M.R., 2003, Architecture of coastal and alluvial deposits in an extensional basin: The Carboniferous Joggins Formation of eastern Canada: Sedimentology, v. 50, p. 415–439, doi: 10.1046/j.1365-3091.2003.00553.x.

Davies, S.J., Gibling, M.R., Rygel, M.C., Calder, J.H., and Skilliter, D.M., 2005, The Joggins Formation of Nova Scotia: Sedimentological log and stratigraphic framework of the historic fossil cliffs: Atlantic Geology, v. 41, p. 115–142.

Davies-Vollum, K.S., and Kraus, M.J., 2001, A relationship between alluvial backswamps and avulsion cycles: An example from the Willwood Formation of the Bighorn Basin, Wyoming: Sedimentary Geology, v. 140, p. 235–249, doi: 10.1016/S0037-0738(00)00186-X.

Dawson, J.W., 1855a (and Supplements, 1868, 1878, 1891), The geology of Nova Scotia, New Brunswick and Prince Edward Island or Acadian geology: Edinburgh, Oliver and Boyd, 694 p.

Dawson, J.W., 1855b, On a modern submerged forest at Fort Lawrence, Nova Scotia: Quarterly Journal of the Geological Society [London], v. 11, p. 119–122.

Dawson, J.W., 1860, On a terrestrial mollusk, a millepede, and new reptiles, from the Coal Formation of Nova Scotia: Quarterly Journal of the Geological Society [London], v. 16, p. 268–277.

Dawson, J.W., 1877, Note on a specimen of *Diploxylon* from the Coal-Formation of Nova Scotia: Quarterly Journal of the Geological Society [London], v. 33, p. 836–842.

Dawson, J.W., 1882, On the results of recent explorations of erect trees containing animal remains in the coal formation of Nova Scotia: Royal Society of Canada Philosophical Transactions, Part II, v. 173, p. 621–659.

Dawson, J.W., 1892a, On the mode of occurrence of remains of land animals in erect trees at South Joggins, Nova Scotia: Transactions of the Royal Society of Canada, v. 9, p. 127–128.

Demko, T.M., and Gastaldo, R.A., 1992, Paludal environments of the Mary Lee coal zone, Pottsville Formation, Alabama: Stacked clastic swamps and peat mires: International Journal of Coal Geology, v. 20, p. 23–47, doi: 10.1016/0166-5162(92)90003-F.

DiMichele, W.A., and Aronson, R.B., 1992, The Pennsylvanian-Permian vegetational transition: A terrestrial analogue to the onshore-offshore hypothesis: Evolution (International Journal of Organic Evolution), v. 46, p. 807–824.

DiMichele, W.A., and DeMaris, P.J., 1987, Structure and dynamics of a Pennsylvanian-age *Lepidodendron* forest: Colonizers of a disturbed swamp habitat in the Herrin (No. 6) coal of Illinois: Palaios, v. 2, p. 146–157.

DiMichele, W.A., and Phillips, T.L., 1994, Paleobotanical and paleoecological constraints on models of peat formation in the Late Carboniferous of Euramerica: Palaeogeography, Palaeoclimatology, Palaeoecology, v. 106, p. 39–90, doi: 10.1016/0031-0182(94)90004-3.

DiMichele, W.A., Phillips, T.L., and Peppers, R.A., 1985, The influence of climate and depositional environment on the distribution and evolution of Pennsylvanian coal-swamp plants, in Tiffney, B.H., ed., Geological factors and the evolution of plants: New Haven, Connecticut, Yale University Press, p. 223–256.

Dolby, G., 1991, The palynology of the western Cumberland Basin, Nova Scotia: Nova Scotia Department of Mines and Energy Open-File Report 91-006, 39 p.

Duff, P.McL.D., and Walton, E.K., 1973, Carboniferous sediments at Joggins, Nova Scotia: Septième Congrès International de Stratigraphie et de Géologie du Carbonifère, Krefeld, Compte Rendu, v. 2, p. 365–379.

Falcon-Lang, H., 1999, Fire ecology of a Late Carboniferous flood plain, Joggins, Nova Scotia: Journal of the Geological Society [London], v. 156, p. 137–148.

Falcon-Lang, H., 2000, Fire ecology of the Carboniferous tropical zone: Palaeogeography, Palaeoclimatology, Palaeoecology, v. 164, p. 355–371, doi: 10.1016/S0031-0182(00)00193-0.

Falcon-Lang, H., 2003a, Response of Late Carboniferous tropical vegetation to transgressive-regressive rhythms at Joggins, Nova Scotia: Journal of the Geological Society [London], v. 160, p. 643–647.

Falcon-Lang, H., 2003b, Late Carboniferous dryland tropical vegetation, Joggins, Nova Scotia, Canada: Palaios, v. 18, p. 197–211.

Falcon-Lang, H.J., and Bashforth, A.R., 2005, Morphology, anatomy, and upland ecology of large cordaitalean trees from the Middle Pennsylvanian of Newfoundland: Review of Palaeobotany and Palynology, v. 135, p. 223–243.

Falcon-Lang, H.J., and Scott, A.C., 2000, Upland ecology of some Late Carboniferous cordaitalean trees from Nova Scotia and England: Palaeogeography, Palaeoclimatology, Palaeoecology, v. 156, p. 225–242, doi: 10.1016/S0031-0182(99)00142-X.

Falcon-Lang, H.J., Rygel, M.C., Calder, J.H., and Gibling, M.R., 2004, An early Pennsylvanian waterhole deposit and its fossil biota in a dryland alluvial plain setting, Joggins, Nova Scotia: Journal of the Geological Society [London], v. 161, p. 209–222.

Friedman, G.M., and Chamberlain, J.A., 1995, *Archanodon catskillensis* (Vanuxem): Fresh-water clams from one of the oldest back-swamp fluvial facies (Upper Middle Devonian), Catskill Mountains, New York: Northeastern Geology and Environmental Sciences, v. 17, p. 431–443.

Fuller, M.L., 1912, The New Madrid earthquake: U.S. Geological Survey Bulletin, v. 494.

Gastaldo, R.A., 1986, Implications on the paleoecology of autochthonous lycopods in clastic sedimentary environments of the Early Pennsylvanian of Alabama: Palaeogeography, Palaeoclimatology, Palaeoecology, v. 53, p. 191–212, doi: 10.1016/0031-0182(86)90044-1.

Gastaldo, R.A., 1987, Confirmation of Carboniferous clastic swamp deposits: Nature, v. 326, p. 869–871, doi: 10.1038/326869a0.

Gastaldo, R.A., 1992, Regenerative growth in fossil horsetails following burial by alluvium: Historical Biology, v. 6, p. 203–219.

Gastaldo, R.A., Stevanoviç-Walls, I., and Ware, W.N., 2004, Erect forests are evidence for coseismic base-level changes in Pennsylvanian cyclothems of the Black Warrior Basin, U.S.A., in Pashin, J.C., and Gastaldo, R.A., eds., Sequence stratigraphy, paleoclimate, and tectonics of coal-bearing strata: AAPG Studies in Geology, v. 51, p. 219–238.

Gesner, A., 1836, Remarks on the geology and mineralogy of Nova Scotia: Halifax, Gossip and Coade, 272 p.

Gibling, M.R., 1987, A classic Carboniferous section; Joggins, Nova Scotia, in Geological Society of America Centennial Field Guide, Northeastern Section: p. 409–414.

Gore, A.J.P., ed., 1983, Mires—Swamp, bog, fen and moor: Ecosystems of the world, Volume 4A: Amsterdam, Elsevier.

Goudge, M.G., 1945, Joggins-River Hebert coal district (a review): Nova Scotia Department of Mines, Annual Report, v. 1944, p. 152–182.

Gradzinski, R., and Doktor, M., 2000, Srodowiska sedymentacyjne i systemy depozycyjne weglonosnej sukcesji Zaglebia Gornoslaskiego (Sedimentary environments and depositional systems of coal-bearing strata in the Upper Silesian coal basin), in Lipiarski, I., ed., Geologia formacji weglonosnych Polski (Geology of the coal-bearing strata of Poland): Cracow, Poland, Akademia Gorniczo-Hutnicza, p. 29–33.

Hacquebard, P.A., and Donaldson, J.R., 1964, Stratigraphy and palynology of the Upper Carboniferous coal measures in the Cumberland Basin of Nova Scotia, Canada: International Congress of Carboniferous Stratigraphy and Geology, 5th, v. 9, p. 1157–1169.

Hebert, B.L., and Calder, J.H., 2004, On the discovery of a unique terrestrial faunal assemblage in the classic Pennsylvanian section at Joggins,

Nova Scotia: Canadian Journal of Earth Sciences, v. 41, p. 247–254, doi: 10.1139/e03-096.

Holmes, R.B., Carroll, R.L., and Reisz, R.R., 1998, The first articulated skeleton of *Dendrerpeton acadianum* (Temnospondyli, Dendrerpetonidae) from the Lower Pennsylvanian locality of Joggins, Nova Scotia, and a review of its relationships: Journal of Vertebrate Paleontology, v. 18, p. 64–79.

Hower, J.C., Calder, J.H., Eble, C.F., Scott, A.C., Richardson, J.D., and Blanchard, L.J., 2000, Metalliferous coals of the Westphalian A Joggins Formation, Cumberland Basin, Nova Scotia: Petrology, geochemistry, and palynology: International Journal of Coal Geology, v. 42, p. 185–206, doi: 10.1016/S0166-5162(99)00039-7.

IUFRO, 1959, The standardization of symbols in forest mensuration: International Union of Forest Research Organizations.

Jackson, C.T., and Alger, F., 1829, A description of the mineralogy and geology of a part of Nova Scotia: American Journal of Science and Arts, v. 15, p. 132–160.

Kerp, H., 1996, Der Wandel der Wälder im Laufe des Erdaltertums: Natur und Museum, v. 126, p. 421–430.

Kosters, E.C., Chmura, G.L., and Bailey, A., 1987, Sedimentary and botanical factors affecting peat accumulation in the Mississippi Delta: Journal of the Geological Society [London], v. 144, p. 423–434.

Klusemann, H., and Teichmüller, R., 1954, Begrabene Wälder im Ruhrkohlenbecken: Natur und Volk, v. 84, p. 373–382.

Kraus, M.J., 1996, Avulsion deposits in lower Eocene alluvial rocks, Bighorn Basin, Wyoming: Journal of Sedimentary Research, v. B66, p. 354–363.

Kraus, M.J., and Wells, T.M., 1999, Facies and facies architecture of Paleocene floodplain deposits, Fort Union Formation, Bighorn Basin, Wyoming: The Mountain Geologist, v. 36, p. 57–70.

Leeder, M.R., 1987, Tectonic and palaeogeographic models for Lower Carboniferous Europe, in Miller, J., Adams, A.E., and Wright, V.P., eds., European Dinantian environments: John Wiley and Sons, p. 1–20.

Logan, W.E., 1845, A section of the Nova Scotia coal measures as developed at Joggins on the Bay of Fundy, in descending order, from the neighbourhood of the west Ragged Reef to Minudie, reduced to vertical thickness: Geological Survey of Canada Report of Progress for 1843, Appendix, p. 92–153.

Lottes, A.L., and Ziegler, A.M., 1994, World peat occurrence and the seasonality of climate and vegetation: Palaeogeography, Palaeoclimatology, Palaeoecology, v. 106, p. 23–37, doi: 10.1016/0031-0182(94)90003-5.

Lyell, C., 1842. Letter to Lyell's sister, in Lyell, K.M., 1881, Life, letters and journals of Sir Charles Lyell, 2: London, John Murray, 489 p.

Lyell, C., 1843, On upright fossil trees in the Coal Strata of Cumberland, Nova Scotia: Silliman's Journal, v. 45, p. 353.

Lyell, C., 1845, Travels in North America; with geological observations on the United States, Canada, and Nova Scotia, 2: London, John Murray.

Lyell, C., 1849, A second visit to the United States of North America: London, John Murray.

Lyell, C., 1871, The student's elements of Geology: London, John Murray, 624 p.

Lyell, C., and Dawson, J.W., 1853, On the remains of a reptile (*Dendrerpeton acadianum*) Wyman and Owen), and of a land shell discovered in the interior of an erect fossil tree in the coal measures of Nova Scotia: Quarterly Journal of the Geological Society [London], v. 9, p. 58–63.

McKee, E.D., Crosby, E.J., and Berryhill, H.L., Jr., 1967, Flood deposits, Bijou Creek, Colorado, June, 1965: Journal of Sedimentary Petrology, v. 37, p. 829–851.

Milner, A.C., 1987, The Westphalian tetrapod fauna; some aspects of its geography and ecology: Journal of the Geological Society [London], v. 144, p. 495–506.

Mjøs, R., and Prestholm, E., 1993, The geometry of fluviodeltaic channel sandstones in the Jurassic Saltwick Formation, Yorkshire, England: Sedimentology, v. 40, p. 919–935.

Moore, P.D., 1987, Ecological and hydrological aspects of peat formation, in Scott, A.C. ed., Coal and coal-bearing strata: Recent advances: Geological Society [London] Special Publication 32, p. 7–15.

Nadon, G., 1998, Magnitude and timing of peat-to-coal compaction: Geology, v. 26, p. 727–730, doi: 10.1130/0091-7613(1998)026<0727:MATOPT>2.3.CO;2.

Nance, R.D., 1987, Dextral transpression and Late Carboniferous sedimentation in the Fundy coastal zone of southern New Brunswick, in Beaumont, C., and Tankard, A.J., eds., Sedimentary basins and basin-forming mechanisms: Canadian Society of Petroleum Geologists Memoir 12, p. 363–377.

Nemec, W., 1992, Depositional controls on plant growth and peat accumulation in a braidplain delta environment; Helvetiafjellet Formation (Barremian-Aptian), Svalbard, in McCabe, P.J., and Parrish, J.T., eds., Controls on the distribution and quality of Cretaceous coals: Geological Society of America Special Paper 267, p. 209–226.

Olsen, T., 1993, Large fluvial systems: The Attane Formation, a fluviodeltaic example from the Upper Cretaceous of central West Greenland: Sedimentary Geology, v. 85, p. 457–473, doi: 10.1016/0037-0738(93)90098-P.

Pearson, T., and Scott, A.C., 1999, Large palynomorphs and debris, in Jones, T.P., and Rowe, N.P., eds., Fossil plants and spores: Modern techniques: London, Geological Society, p. 20–25.

Pfefferkorn, H.W., and Zodrow, E.L., 1982, A comparison of standing forests from the Pennsylvanian of Nova Scotia with modern tropical forests: Botanical Society of America, miscellaneous series, Publication 162, p. 62–63.

Phillips, T.L., and DiMichele, W.A., 1992, Comparative ecology and life-history biology of arborescent lycopsids in Late Carboniferous swamps of Euramerica: Annals of the Missouri Botanical Garden, v. 79, p. 560–588.

Plint, A.G., and van de Poll, H.W., 1984, Structural and sedimentary history of the Quaco Head area, southern New Brunswick: Canadian Journal of Earth Sciences, v. 21, p. 753–761.

Reed, B.C., Nance, R.D., Calder, J.H., and Murphy, J.B., 1993, The Athol Syncline: Tectonic evolution of a Westphalian A-B depocenter in the Maritimes Basin, Nova Scotia: Atlantic Geology, v. 29, p. 179–186.

Rust, B.R., Gibling, M.R., and Legun, A.S., 1984, Coal deposition in an anastomosing-fluvial system: The Pennsylvanian Cumberland Group South Joggins, Nova Scotia, Canada: International Association of Sedimentologists Special Publication 7, p. 105–120.

Ryan, R.J., and Boehner, R.C., 1994, Geology of the Cumberland Basin, Cumberland, Colchester and Pictou Counties, Nova Scotia: Nova Scotia Department of Natural Resources, Mines and Energy Branches, Memoir 10, 222 p.

Ryan, R.J., Boehner, R.C., and Calder, J.H., 1991, Lithostratigraphic revision of the Upper Carboniferous to Lower Permian strata in the Cumberland Basin, Nova Scotia and the regional implications for the Maritimes Basin in Atlantic Canada: Canadian Society of Petroleum Geologists Bulletin, v. 39, p. 289–314.

Rygel, M.C., Gibling, M.R., and Calder, J.H., 2004, Vegetation-induced sedimentary structures from fossil forests in the Pennsylvanian Joggins Formation, Nova Scotia: Sedimentology, v. 51, p. 531–552, doi: 10.1111/j.1365-3091.2004.00635.x.

Scotese, C.R., and McKerrow, W.S., 1990, Revised world maps and introduction, in McKerrow, W.S., and Scotese, C.R., eds., Palaeozoic palaeogeography and biogeography: Geological Society [London] Memoir 12, p. 1–21.

Scott, A.C., 1998, The legacy of Charles Lyell: Advances in our knowledge of coal and coal-bearing strata, in Blundell, D., and Scott, A.C., eds., Lyell: The past is the key to the present: Geological Society [London] Special Publication 143, p. 296–331.

Scott, A.C., and Calder, J.H., 1994, Carboniferous fossil forests: Geology Today, v. 10, p. 213–217.

Scott, A.C., and Hemsley, A.R., 1996, Paleozoic megaspores, in Jansonius, J., and McGregor, D.C., eds. Palynology: Principles and applications, Volume 2: Dallas, American Association of Stratigraphic Palynologists Foundation, p. 629–639.

Scott, A.C., and Jones, T.P. 1994, The nature and influence of fire in Carboniferous ecosystems: Palaeogeography, Palaeoclimatology, Palaeoecology, v. 106, p. 91–112.

Scott, A.C., and King, G.R., 1981, Megaspores and coal facies: An example from the Westphalian A of Leicestershire, England: Review of Palaeobotany and Palynology, v. 34, p. 107–113, doi: 10.1016/0034-6667(81)90068-3.

Sigafoos, R.S., 1964, Botanical evidence of floods and flood-plain deposition: U.S. Geological Survey Professional Paper 485-A, 35 p.

Skilliter, D.M., 2001, Distal marine influence in the Forty Brine section, Joggins, Nova Scotia, Canada [M.S. thesis], Chestnut Hill, Massachusetts, Boston College, 96 p.

Smith, M.G., 1991, The flood plain deposits and paleosol profiles of the Late Carboniferous Cumberland coal basin, exposed at Joggins, Nova Scotia, Canada [M.S. thesis]: Guelph, Ontario, University of Guelph, 372 p.

Stone, E.C., and Vasey, R.B., 1968, Preservation of Coast Redwood on alluvial flats: Science, v. 159, p. 157–161.

Teichmüller, M., 1989, The genesis of coal from the viewpoint of coal petrology: International Journal of Coal Geology, v. 12, p. 1–87, doi: 10.1016/0166-5162(89)90047-5.

Teichmüller, R., 1955, Über Küstenmoore der Gegenwart und die Moore des Ruhrkarbons: Eine vergleichende sedimentologische Betrachtung: Geologisches Jahrbüch, v. 71, p. 197–220.

Tibert, N.E., and Gibling, M.R., 1999, Peat accumulation on a drowned coastal braidplain; the Mullins Coal (Upper Carboniferous), Sydney Basin, Nova Scotia: Sedimentary Geology, v. 128, p. 23–38, doi: 10.1016/S0037-0738(99)00059-7.

Tye, R.S., and Coleman, J.M., 1989, Evolution of the Atchafalaya lacustrine deltas, south-central Louisiana: Sedimentary Geology, v. 65, p. 95–112, doi: 10.1016/0037-0738(89)90008-0.

Underwood, J.R., and Lambert, W., 1974, Centroclinal cross strata, a distinctive sedimentary structure: Journal of Sedimentary Petrology, v. 44, p. 1111–1113.

van Heerden, I.L., and Roberts, H.H., 1988, Facies development of Atchafalaya Delta, Louisiana; a modern bayhead delta: American Association of Petroleum Geologists Bulletin, v. 72, p. 439–453.

Wagner, R.H., 2001, The extrabasinal elements in Lower Pennsylvanian floras of the Maritime Provinces, Canada: Description of *Adiantites, Pseudadiantites* and *Rhacopteridium*: Revista Española de Paleontología, v. 16, p. 187–207.

Wagner, R., Delcambre-Broumische, C., and Coquel, R., 2003, Una Pompeya paleobotánica: Historia de una marisma carbonífera sepultada por cenizas volcánicas, *in* Nuche, R., ed., Separata de Patrimonio Geológico de Castilla-La Mancha, Madrid, ENRESA, p. 448–477.

Wagner, R., Diez Ferrer, J.B., and Calvo Murillo, R., 2001, El bosque Carbonífero de Verdeña: Vida y muerte de una comunidad florística, *in* Nuche, R., ed., Patrimonio Geológico de Castilla y León, Madrid, ENRESA, p. 2–15.

Waldron, J.W.F., Piper, D.J.W., and Pe-Piper, G., 1989, Deformation of the Cape Chignecto Pluton, Cobequid Highlands, Nova Scotia: Thrusting at the Meguma-Avalon boundary: Atlantic Geology, v. 25, p. 51–62.

Waldron, J.W.F., and Rygel, M.C., 2004, Evolution of the Cumberland Basin, Nova Scotia: New insights on deposition and salt tectonics from seismic reflection profiles: Atlantic Geoscience Society Colloquium and Annual General Meeting, Program and Abstracts: Fredericton, New Brunswick Department of Natural Resources, p. 30–31.

Whiteaves, J.F., 1893, Note on the recent discovery of large *Unio*-like shells in the Coal Measures at the South Joggins, Nova Scotia: Transactions of the Royal Society of Canada, v. 11, sec. 4, p. 21–25.

Winston, R.B., 1986, Characteristic features and compaction of plant tissues traced from permineralized peat to coal in Pennsylvanian coals (Desmoinsian) from the Illinois Basin: International Journal of Coal Geology, v. 6, p. 21–41.

MANUSCRIPT ACCEPTED BY THE SOCIETY 28 JUNE 2005

Compositional characteristics and inferred origin of three Late Pennsylvanian coal beds from the northern Appalachian Basin

Cortland F. Eble*
Kentucky Geological Survey, University of Kentucky, Lexington, Kentucky, USA

William C. Grady*
West Virginia Geological and Economic Survey, Morgantown, West Virginia, USA

Brenda S. Pierce*
U.S. Geological Survey, Reston, Virginia, USA

ABSTRACT

The Pittsburgh, Redstone, and Sewickley coal beds all occur in the Late Pennsylvanian Pittsburgh Formation of the Monongahela Group in the northern Appalachian Basin. The goal of this study is to compare and contrast the palynology, petrography, and geochemistry of the three coals, specifically with regard to mire formation, and the resulting impacts on coal composition and occurrence. Comparisons between thick (>1.0 m) and thin (<0.3 m) columns of each coal bed are made as well to document any changes that occur between more central and more peripheral areas of the three paleomires.

The Pittsburgh coal bed, which is thick (>1m) and continuous over a very large area (over 17,800 km²), consists of a rider coal zone (several benches of coal intercalated with clastic partings) and a main coal. The main coal contains two widespread bone coal, fusain, and carbonaceous shale partings that divide it into three parts: the breast coal at the top, the brick coal in the middle, and the bottom coal at the base. *Thymospora thiessenii*, a type of tree fern spore, is exceptionally abundant in the Pittsburgh coal and serves to distinguish it palynologically from the Redstone and Sewickley coal beds. Higher percentages of *Crassispora kosankei* (produced by *Sigillaria*, a lycopod tree), gymnosperm pollen, and inertinite are found in association with one of the extensive partings, but not in the other. There is little compositional difference between the thin and thick Pittsburgh columns that were analyzed.

The Redstone coal bed is co-dominated by tree fern and calamite spores and contains no *Thymospora thiessenii*. Rather, *Laevigatosporites minimus*, *Punctatisporites minutus*, and *Punctatisporites parvipunctatus* are the most common tree fern representatives in the Redstone coal. *Endosporites globiformis*, which does not occur in the Pittsburgh coal, is commonly found near the base of the coal bed, and in and around inorganic partings. In this respect, *Endosporites* mimics the distribution of *Crassispora kosankei* in the Pittsburgh coal. Small fern spores are also more abundant in the Redstone coal bed than they are in the Pittsburgh coal. Overall, the Redstone

*E-mails: Eble@uky.edu; Grady@geo.wvu.edu; BPierce@usgs.gov

Eble, C.F., Grady, W.C., and Pierce, B.S., 2006, Compositional characteristics and inferred origin of three Late Pennsylvanian coal beds from the northern Appalachian Basin, *in* Greb, S.F., and DiMichele, W.A., Wetlands through time: Geological Society of America Special Paper 399, p. 197–222, doi: 10.1130/2006.2399(10). For permission to copy, contact editing@geosociety.org. ©2006 Geological Society of America. All rights reserved.

coal bed contains more vitrinite, ash, and sulfur than the Pittsburgh coal. The distribution of the Redstone coal is much more podlike, indicating strong paleotopographic control on its development. Compositionally, there are major differences between the thin and thick Redstone columns, with higher amounts of *Endosporites globiformis*, gymnosperm pollen, inertinite, ash, and sulfur occurring in the thin column.

The Sewickley coal bed is palynologically similar to the Redstone coal in that it is co-dominated by tree fern and calamite spores, with elevated percentages of small fern spores. Tree fern species distribution is different, however, with *Thymospora thiessenii* and *T. pseudothiessenii* being more prevalent in the Sewickley. The distribution of *Crassispora kosankei* in the Sewickley coal bed is similar to that in the Pittsburgh coal, i.e., more abundant at the base of the bed and around inorganic partings. By contrast, *Endosporites* is only rarely seen in the Sewickley coal. The Sewickley is more laterally continuous than the Redstone coal, but not nearly as thick and continuous as the Pittsburgh coal. Overall, the vitrinite content of the Sewickley coal is between that of the Pittsburgh (lowest) and Redstone (highest). Ash yields and sulfur contents are typically higher than in the Pittsburgh or Redstone. The thin and thick Sewickley columns are palynologically and petrographically very similar; ash and sulfur are both higher in the thin column.

Keywords: coal, Appalachian, palynology, petrography, geochemistry.

INTRODUCTION

Geologic Considerations

The purpose of this paper is to use the palynological, petrographic, and geochemical (collectively referred to as compositional) characteristics of the principal coal beds of the Late Pennsylvanian Monongahela Group to interpret the types of mires from which these coal beds formed. The Monongahela Group was named for exposures along the Monongahela River in western Pennsylvania. The group is defined as extending from the base of the Pittsburgh coal bed to the base of the Waynesburg coal bed. The three coal beds discussed in this paper—the Pittsburgh, Redstone, and Sewickley—all occur in the Pittsburgh Formation of the Monongahela Group, and represent major mineable coal beds (Fig. 1). Lithologically, the Monongahela Group consists of ~50% variegated mudrock, 35% sandstone, and 15% shale, nonmarine limestone, and coal (Henry et al., 1979). The Monongahela Group is ~80 m thick along the Ohio River to the west of the study area, and progressively thickens to the east. Nearly 150 m of Monongahela Group sediments have been reported from eastern West Virginia and western Maryland (Cross and Schemel, 1956). Monongahela Group sediments have been interpreted as being deposited in a deltaic setting. The nonmarine limestone, coal, sandstone, and shale formed in lakes, swamps, streams, and interfluvial areas on an aggrading coastal lowland (Donaldson, 1974, 1979; Donaldson and Shumaker, 1979, 1981; Presley, 1979).

Pittsburgh Coal Bed

The Pittsburgh coal bed covers a very large area (Fig. 2), extending over 17,800 km^2 across parts of Maryland, Pennsylvania, West Virginia, and Ohio (Tewalt et al., 2000). This "blanket-like" distribution also includes a large area of thick, mineable coal (>0.7 m). In the present study area (northern West Virginia) much of the coal is greater than 2 m thick (Fig. 3). The Pittsburgh coal consists of two principal parts, a roof coal that is actually multiple coal benches (up to eight) separated by inorganic partings, and the main coal. In surface mine operations, the roof coal is often stripped off and discarded because of high ash yields and sulfur contents. In underground mines, the roof coal usually has to be taken with the main bed because of roof control concerns. The mined coal is then sent to a preparation facility to separate out the higher ash and sulfur coal, and rock material. Whereas the roof coal varies in both thickness and extent, the main coal is much more uniform in occurrence.

The main coal contains two widespread partings that divide it into three parts: the "breast" coal at the top, the "brick" coal, and the "bottom" coal at the base (see Fig. 7). The breast coal often is the thickest part of the main bench. The underlying brick coal is separated from the breast coal by a parting called the bearing-in bench, a thin (usually <15 cm thick), but very laterally persistent, bench of coal bounded by two carbonaceous shale partings. It was named by early pick miners who cut out this bench to help collapse the overlying breast coal. The underlying brick coal derives its name from the tendency to come out in brick-shaped pieces when mined. In many places, the brick and underlying bottom coal are separated by a bone coal layer, though in many other areas this boundary is less clear. Together, the bottom coal and brick coal make up about half of the total bed thickness of the main coal in the central and western areas. In eastern areas, however, the bottom and brick coal benches thin considerably (Thiessen and Staud, 1923; Cross, 1952).

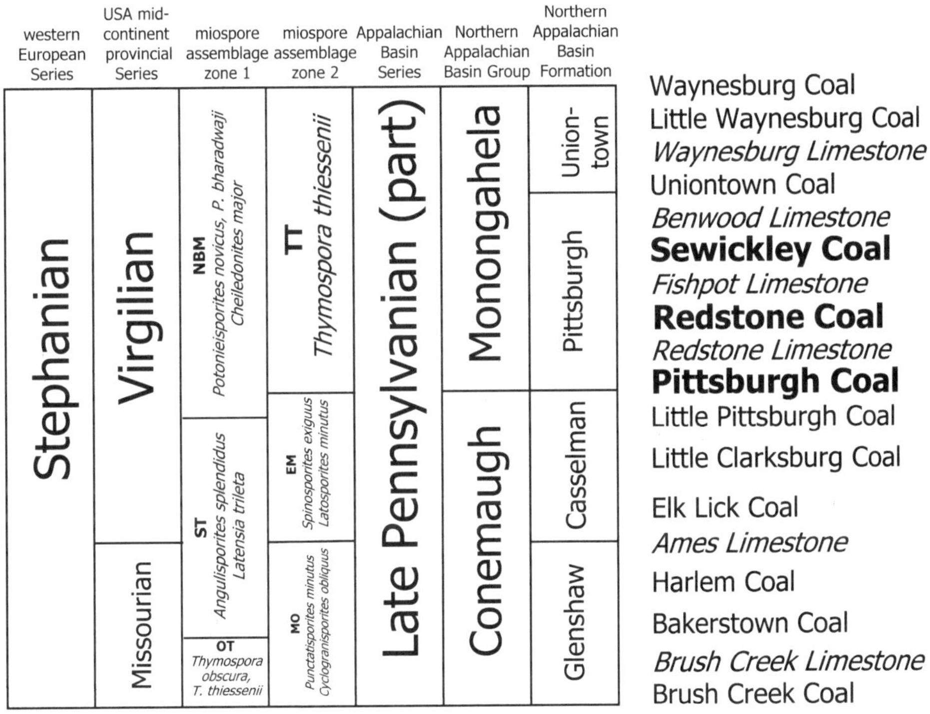

western European Series	USA mid-continent provincial Series	miospore assemblage zone 1	miospore assemblage zone 2	Appalachian Basin Series	Northern Appalachian Basin Group	Northern Appalachian Basin Formation	
Stephanian	Virgilian	NBM *Potonieisporites novicus, P. bharadwaji Cheiledonites major*	TT *Thymospora thiessenii*	Late Pennsylvanian (part)	Monongahela	Uniontown	Waynesburg Coal Little Waynesburg Coal *Waynesburg Limestone* Uniontown Coal *Benwood Limestone*
						Pittsburgh	**Sewickley Coal** *Fishpot Limestone* **Redstone Coal** *Redstone Limestone* **Pittsburgh Coal**
		ST *Angulisporites splendidus Latensia trileta*	EM *Spinosporites exiguus Latosporites minutus*		Conemaugh	Casselman	Little Pittsburgh Coal Little Clarksburg Coal Elk Lick Coal *Ames Limestone*
	Missourian	OT *Thymospora obscura, T. thiessenii*	MO *Punctatisporites minutus Cyclogranisporites obliquus*			Glenshaw	Harlem Coal Bakerstown Coal *Brush Creek Limestone* Brush Creek Coal

Figure 1. Generalized stratigraphic column showing the position of the Pittsburgh, Redstone, and Sewickley coal beds (shown in **bold type**). Miospore assemblage zones are from Clayton et al., 1977 (zone 1), and Peppers, 1996 (zone 2).

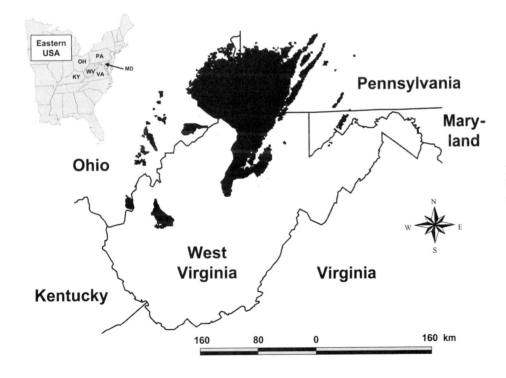

Figure 2. Extent of the Pittsburgh coal bed in the northern Appalachian Basin (from Tewalt et al., 2000).

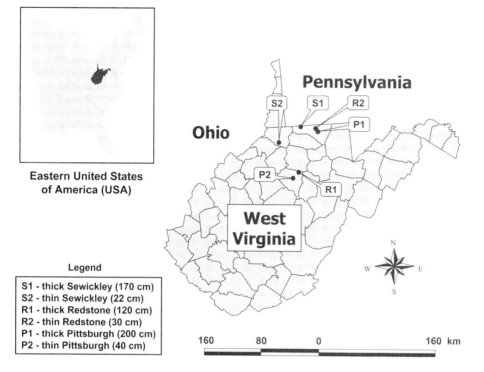

Figure 3. Sample location map.

Redstone Coal Bed

The Redstone coal was named for an exposed coal along the banks of Redstone Creek in southwestern Pennsylvania. The earliest records of the Redstone coal date back to 1759 when the coal was apparently used by Colonel James Burd to fuel his campfire (Thiessen and Staud, 1923). In eastern Ohio, the Redstone coal is called the Pomeroy coal bed, named for its occurrence around the town of Pomeroy, Ohio. In contrast to the Pittsburgh coal, the Redstone coal is thinner, much more discontinuous, and podlike in distribution (Fig. 4). In our study area the Redstone coal is mainly less than 0.6 m thick, with some small, disconnected areas of thicker coal. Most of the thicker pods of Redstone coal here are less than 1 m thick, and in a few areas are up to 2 m thick. The Redstone coal is frequently separated into an upper and lower bench by a claystone or bone parting that is difficult to trace laterally. Other inorganic partings and clay dikes occur regularly as well.

Sewickley Coal Bed

The Sewickley coal bed is named for exposures along Big Sewickley Creek in western Pennsylvania (Thiessen and Staud, 1923). The first official mention of the Sewickley coal bed is in a Report of Progress for Fayette and Westmoreland Counties, located in southwestern Pennsylvania, by the Second Geological Survey of Pennsylvania in 1875 and 1876 (Stevenson, 1877, 1878). The distribution of the Sewickley coal bed is between that of the blanket-like Pittsburgh and podlike Redstone. The Sewickley coal is persistent across our study area, with thicker coal (>1 m) occurring to the north and east, and thinner coal (<1 m) occurring to the south and west (Fig. 5). The Sewickley coal is commonly divided into an upper and lower bench by a thin claystone, or bone parting (similar to the Redstone coal).

The distributions of the Pittsburgh, Redstone, and Sewickley coal beds are highly dissimilar, a disparity that is probably related to paleotopography. The Pittsburgh coal is underlain by fine-grained sediments (claystone, shale, and siltstone) virtually over its entire extent, which provided a very stable platform for widespread peat accumulation. In contrast, the Redstone coal is underlain by the Pittsburgh sandstone, which limited peat development primarily to interfluve areas. Likewise, the distribution of the Sewickley coal appears to have been influenced by the underlying Lower Sewickley sandstone, though not to the degree the Redstone coal was inhibited (Donaldson, 1974, 1979).

PREVIOUS WORK

Thin sections of the Pittsburgh, Redstone, and Sewickley coal were studied by Thiessen and Staud (1923) and Thiessen (1930) in an effort to identify petrographic and palynological characteristics that would serve to distinguish them from one another. Petrographically, the Pittsburgh coal bed is dominated by anthraxylon at the base and attritus at the top. Anthraxylon is essentially the thin-section equivalent of the vitrinite macerals telinite and telocollinite that are recognized in reflected light. The term is derived from "anthra," meaning metamorphosed or coalified, and "xylon," which refers to its origin from woody xylem tissues (at least in

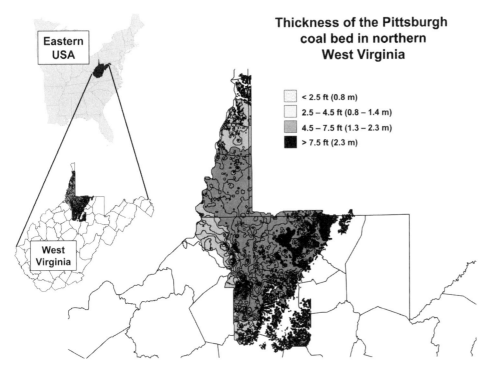

Figure 4. Isopach map of the Pittsburgh coal bed in northern West Virginia. Note the "blanket-like" distribution, thickness, and lateral continuity (adapted from the West Virginia Geological and Economic Survey web-based coal database and electronic map portfolio).

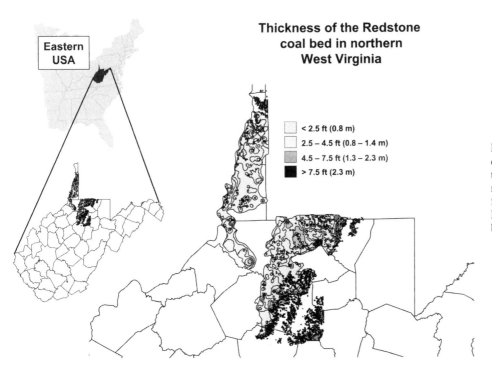

Figure 5. Isopach map of the Redstone coal bed in northern West Virginia. Note the "podlike" distribution, overall thinness, and discontinuous nature (adapted from the West Virginia Geological and Economic Survey web-based coal database and electronic map portfolio).

part). Attritus is the thin-section equivalent of the vitrinite macerals desmocollinite and vitrodetrinite (i.e., disarticulated and/or highly degraded pieces of vitrinite). The term is comparable to the sedimentological term "detritus." The spore content of the Pittsburgh is relatively high, with one type of spore being the dominant contributor. This spore, which wasn't observed in either the Redstone or the Sewickley coal, is described as being small and thick-walled, with numerous closely folded blunt serpentine ridges on the surface (Thiessen and Staud, 1923).

Like the Pittsburgh coal bed the Redstone coal is petrographically more anthraxylous at the base and more attrital at the top. Unlike the Pittsburgh coal, however, the overall spore content of the Redstone coal is relatively low, with the most distinctive spore being a "double-coated, granulate" spore (Thiessen and Staud, 1923).

The Sewickley coal bed is described as being highly laminated with fine layers of anthraxylon and attritus, with the former being more common in the lower part of the coal and the latter near the top of the bed. The overall spore content of the Sewickley is noted as being even lower than that of the Redstone, a feature that helps distinguish it from the Redstone and Pittsburgh coal beds. The most characteristic spore of the Sewickley coal is a "small crinkled one" 20–25 μ in longest dimension, with a surface ornamentation consisting of broad and blunt echinate spines or humps (Thiessen and Staud, 1923).

All three coals have also been studied by using macerated spores and pollen. Kosanke (1943) examined the small spore content of the Pittsburgh and Pomeroy (= Redstone) coals in southeastern Ohio and concluded that the two beds could be separated by palynomorph content. The Pittsburgh coal contains abundant *Thymospora thiessenii*, whereas the overlying Redstone is essentially devoid of this taxon. Cross (1952) performed a detailed study of the Pittsburgh coal throughout its extent in Pennsylvania, Maryland, West Virginia, and Ohio. Seventeen small spore species were identified, and *Laevigatosporites* (= *Thymospora*) *thiessenii* was the most abundant. A subsequent palynological study of the Pittsburgh by Clendening and Gillespie (1963) identified an additional 18 species (total 35) in the Pittsburgh, although the majority of these were found to occur sporadically, and thus be of limited stratigraphic utility.

Palynological investigations of the Redstone include those by Clendening (1965), Habib (1968), Eble (1985), and Grady and Eble (1990). The palynoflora of the Sewickley has been reported on briefly by Cross (1947), and more extensively by Eble et al. (2003), whereas the palynology of the Waynesburg coal was investigated by Gray (1951). Some Monongahela coals and associated strata were also included in an investigation by Kosanke (1988).

MATERIALS AND METHODS

Small-scale (<15 cm) bench samples of the Pittsburgh, Redstone, and Sewickley coal beds were collected from active surface mine and outcrop exposures in northern West Virginia (Fig. 6). We collected two columns of coal for each coal bed—one from an area where the coal is thick (>1 m), and the other from an area

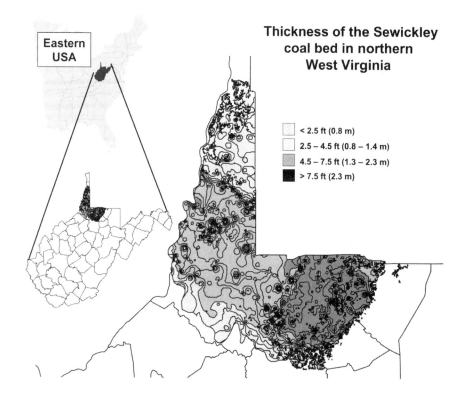

Figure 6. Isopach map of the Sewickley coal bed in northern West Virginia. Note that the Sewickley coal has a distribution pattern similar to that of the Pittsburgh, but that the coal is not as thick (adapted from the West Virginia Geological and Economic Survey web-based coal database and electronic map portfolio).

where the coal is thin (≤0.3 m)—to investigate any compositional differences between areas of thick (mire-central) and thin (mire margin) peat accumulation. Column locations are shown in Figure 6. Coal samples were first crushed to –20 mesh (maximum particle size 850 μ), and then divided to obtain two representative subsplits of ~50 g. One split was used for coal petrographic grain mounts and palynological macerations. The other split was further reduced in size to –60 mesh (maximum particle size 250 μ) for proximate (moisture, volatile matter, fixed carbon and ash yield) and total sulfur analyses.

Coal petrographic pellets were constructed by mixing ~2–3 g of –20 mesh coal with epoxy resin, and then allowing the mixture to cure in 3.2-cm-diameter molds. Once hardened, the pellets were ground and polished in progressive stages to achieve an essentially scratch-free surface with minimum relief among maceral components. Point counting to determine individual maceral percentages followed either ASTM or ISO standards.

Coal macerations followed basic techniques: 3–5 g of –20 mesh or –60 mesh coal were first oxidized in Schulzes solution (HNO$_3$ saturated with KClO$_3$) and then treated with 5% KOH to dissolve the humic fraction of the coal. Macerations of the Pittsburgh and Redstone coal used –20 mesh coal, the residues of which were screened with a –60 mesh screen to separate the coarse and fine fractions. Macerations of the Sewickley coal used –60 mesh coal in an effort to recover and record *Schopfipollenites*, a type of pteridosperm (seed fern) pollen, which because of its large size (>150 μ), is isolated mainly in the coarse residue fraction when screened.

A two-tier count to determine palynomorph abundances was necessitated by the overwhelming abundance of *Punctatisporites minutus* (a.k.a. *Fabasporites*) in all three of the coals. An initial count of 125 spores therefore recorded *Punctatisporites minutus* versus all other taxa. This was followed by another count of 125 palynomorphs, exclusive of *Punctatisporites minutus*, to record the relative percentages of taxa that would have otherwise been masked by the very high percentages (>80% in many cases) of *Punctatisporites minutus*.

Ash yields and total sulfur contents were performed according to ASTM standards using automated equipment (ASTM, 1997a, 1997b). All ash and sulfur values are reported on a dry basis. All acquired data are presented in Appendix 1.

RESULTS

Palynology

With a few exceptions, most of the coal samples were dominated by the tree fern spore *Punctatisporites minutus*, which on average represented >50% of the spore population when included in the statistical counts. Other common tree fern taxa include three species of *Thymospora* (*T. pseudothiessenii*, *T. thiessenii*, and *T. obscura*), three species of *Laevigatosporites* (*L. punctatus*, *L. minimus*, and *L. ovalis*), *Punctasporites minutus*, *Punctatisporites parvipunctatus*, *Spinosporites exiguus*, and *Apiculatasporites saetiger*.

The three coals differ in which species of tree fern spore is dominant. If *Punctatisporites minutus* is excluded from the statistical counts, the Pittsburgh coal contains the most abundant *Thymospora thiessenii* of any Monongahela Group coal bed. This is the "Pittsburgh spore" referred to by Thiessen and Staud (1923). The Redstone coal contains no *Thymospora thiessenii* but contains low percentages of two other species of *Thymospora*, *T. pseudothiessenii* and *T. obscura*.. The Sewickley coal contains all three species of *Thymospora*, with *T. pseudothiessenii* being most abundant.

Small ferns are represented by a variety of miospore taxa, of which *Granulatisporites*, *Leiotriletes*, *Lophotriletes*, and *Punctatisporites* are typically the most common genera. Overall, small fern spores appear to be most abundant in the Redstone coal and least abundant in the Pittsburgh coal (Table 1).

TABLE 1. AVERAGE, MAXIMUM AND MINIMUM VALUES FOR PALYNOMORPHS, GROUPED ACCORDING TO NATURAL AFFINITY

	P. minutus	Lycopod Trees	Small Lycopods	Tree Ferns	Small Ferns	Calamites	Cordaites	Other Gymnosperms
Thick Sewickley	71.4 (100.0, 28.8)	2.8 (16.0, 0.0)	0.0 (0.0, 0.0)	60.9 (100.0, 35.2)	11.1 (27.2, 2.4)	24.7 (37.6, 4.0)	1.7 (4.8, 0.0)	2.3 (8.0, 0.0)
Thin Sewickley.	85.6*	0*	0*	77.6*	3.2*	16.0*	0.0*	3.2*
Thick Redstone	Not recorded	0.0 (0.0, 0.0)	4.1 (36.5, 0.0)	48.7 (68.0, 27.0)	10.5 (25.0, 0.0)	31.3 (49.0, 12.5)	1.3 (3.0, 0.0)	1.4 (3.0, 0.0)
Thin Redstone	Not recorded	0.0 (0.0, 0.0)	7.5 (9.5, 4.0)	47.2 (75.5, 24.5)	26.3 (41.0, 1.0)	5.0 (6.0, 3.5)	0.2 (0.5, 0.0)	13.5 (16.5, 5.5)
Thick Pittsburgh	52.8 (74.4, 30.0)	1.4 (7.6, 0.0)	0.0 (0.0, 0.0)	86.9 (97.2, 50.8)	1.2 (4.4, 0.0)	7.8 (24.4, 1.2)	1.1 (8.4, 0.0)	1.7 (12.0, 0.0)
Thin Pittsburgh	62.4 (88.4, 32.4)	4.0 (5.6, 1.2)	0.1 (0.4, 0.0)	66.4 (78.4, 46.4)	2.3 (6.0, 0.4)	24.9 (36.0, 19.2)	0.8 (2.4, 0.0)	1.2 (2.4, 0.0)

Note: Please refer to Appendix 1 for detailed data.
*The thin Sewickley column only had one 22 cm bench.

Calamite spores are primarily represented by four species of *Laevigatosporites* (*L. medius*, *L. minor*, *L. vulgaris*, and *L. maximus*) and, to a lesser extent, by *Calamospora*. Calamite spores are most abundant in the Redstone coal and least abundant in the Pittsburgh (Table 1). *Florinites*, which is cordaite pollen, and other pollen from other gymnospermous plants (e.g., *Vesicaspora wilsonii*, *Schopfipollenites ellipsoides*, *Pityosporites westphalensis*, *Potonieisporites elegans*) are rare components of all three beds overall but show local abundance.

Lycopods in the Pittsburgh, Redstone, and Sewickley palynofloras are represented by *Crassispora kosankei*, the spore produced by *Sigillaria*. *Lycospora*, a major genus in stratigraphically older lower and middle Pennsylvanian coals, becomes extinct above the Middle-Late Pennsylvanian boundary in the Appalachian Basin (Phillips et al., 1974, 1985), and thus is absent in Late Pennsylvanian coals, including the Sewickley. *Crassispora kosankei* is most abundant in the Sewickley and Pittsburgh coals, especially in, and just above, horizons high in ash and clastic partings.

Seed fern (pteridosperm) pollen was found to be extremely rare in the Sewickley coal palynofloras, even though the preparation techniques used were modified to record its presence. One possibility is that seed ferns were a relatively minor component of the Sewickley flora. Another possibility is that seed ferns produced too little pollen to represent their true biomass.

Petrography and Geochemistry

All of the columns that were studied have overall high vitrinite contents, and correspondingly low liptinite and inertinite contents. Geochemically, all three coals have moderate to high ash yields and total sulfur contents. Tables 2 and 3 present a summary of the average percentages and ranges of maceral, ash yield and total sulfur

TABLE 2. AVERAGE VALUES, AND RANGES OF PETROGRAPHIC AND GEOCHEMICAL PARAMETERS IN FULL-CHANNEL SAMPLES >0.6 M THICK FOR THE PITTSBURGH, REDSTONE AND SEWICKLEY COAL BEDS

Coal Bed	Vitrinite	Liptinite	Inertinite	Ash Yield	Sulfur Content
Sewickley (n = 15, 133)	89.4 (95.0, 82.4)	2.2 (3.7, 0.7)	8.4 (14.7, 3.3)	13.7 (30.4, 6.4)	3.3 (7.0, 0.6)
Redstone (n = 38, 276)	90.1 (95.4, 81.6)	1.8 (3.7, 0.8)	8.0 (16.5, 2.9)	9.3 (30.5, 2.5)	2.9 (6.9, 0.3)
Pittsburgh (n = 36, 864)	82.3 (93.1, 60.9)	3.5 (10.4, 0.0)	14.2 (29.1, 4.6)	8.6 (42.7, 1.4)	2.8 (9.1, 0.5)

Note: All petrographic data are presented on a mineral-matter-free basis. Data are compiled from the West Virginia Geological and Economic Survey coal database. The *n*-values in parentheses represent the number of petrographic and geochemical analyses respectively.

TABLE 3. PETROGRAPHIC AND GEOCHEMICAL DATA FOR SAMPLES USED IN THE PRESENT STUDY

Coal Bed	Vitrinite	Liptinite	Inertinite	Ash Yield	Sulfur Content
Thick Sewickley	82.1 (90.4, 48.0)	5.7 (26.8, 2.4)	12.1 (25.2, 3.2)	12.0 (53.1, 6.3)	3.3 (4.5, 0.7)
Thin Sewickley*	80.0	11.2	8.8	21.9	8.2
Thick Redstone	93.9 (98.4, 85.3)	1.8 (3.4, 0.4)	4.3 (12.5, 0.4)	14.9 (65.2, 5.0)	3.1 (5.4, 1.1)
Thin Redstone	79.9 (83.4, 82.3)	1.2 (1.6, 0.9)	18.7 (16.3, 14.9)	21.6 (45.1, 5.4)	N/D
Thick Pittsburgh	75.5 (89.3, 19.3)	6.1 (22.8, 1.4)	17.7 (58.2, 8.8)	11.5 (40.3, 4.3)	2.4 (5.3, 0.8)
Thin Pittsburgh	86.1 (86.8, 85.4)	1.5 (1.5, 1.4)	9.7 (9.9, 9.4)	5.4 (5.7, 5.1)	2.1 (2.2, 1.9)

Note: All petrographic data is presented on a mineral-matter-free basis. Geochemical data are presented on a moisture-free (dry) basis. Average, maximum and minimum values are shown. Please refer to Appendix 1 for detailed data.
*The thin Sewickley column only had one 22 cm bench.

content for full-channel samples of the Pittsburgh, Redstone, and Sewickley coals, and also for the samples used in this study.

DISCUSSION

Pittsburgh coal bed

The Pittsburgh coal exhibits remarkable thickness and continuity over an extremely large area. Part of the reason for this extensive development is the nature of the underlying strata. The Pittsburgh coal is primarily underlain by fine-grained sediments (claystone, shale, and siltstone), which undoubtedly provided a very stable platform for widespread peat development. However, other factors need to be considered as well. During the time that the Pittsburgh paleomire was developing, basin subsidence rates must have been nearly perfect for accumulating and preserving peat. If a basin subsides too slowly, then peat preservation is negatively affected. However, if subsidence is too rapid then the mire is at risk of drowning or being overwhelmed by sediment influx, both of which would effectively arrest or interrupt peat accumulation.

The climate during Pittsburgh peat formation had to be sufficiently wet to allow for such an extensive accumulation of peat. Climate patterns for the Late Pennsylvanian in the Appalachian Basin have been interpreted to be seasonal on a small scale, and cyclic on a larger scale (Cecil et al., 1985; Cecil, 1990). As can be seen from the stratigraphic column shown in Figure 1, coal beds in the Monongahela Group occur cyclically with nonmarine limestones. Whereas the coal beds are interpreted to represent "wetter" times, the limestones are believed to represent "drier" times in the basin (Cecil, 1990). Cecil further pointed out that the two climate end members actually inhibited clastic sediment release. This may be an important point with regard to the development of the Pittsburgh coal. There are very few small-scale inorganic partings in the Pittsburgh coal that could be attributed to clastic movement into the paleomire (e.g., a crevasse splay). On the contrary, the partings in the Pittsburgh are largely basinal in extent (Cross, 1952), suggesting an allogenic rather than autogenic origin. Therefore, it is possible that the "wet" climate that promoted extensive peat formation also inhibited clastic influx.

Thick Pittsburgh Column

Palynological, petrographic, and geochemical data for a 2-m-thick column of Pittsburgh coal are shown in Figure 7. The ash and sulfur profiles show a high ash (45%), high sulfur (3%) rider coal at the top being separated from the main bed by a claystone layer. The topmost increment of the main bed is also high in ash (40.3%) and sulfur (2%), with the next four increments showing a significant decrease in ash (avg 15.2%); sulfur contents, however, remain high (avg 2.8%). Below the level of the fusain bench marked A on Figure 7, the coal becomes much lower in ash (avg 5.8%) and sulfur (avg 1.8%) down to the basal two increments, which show a return to higher ash (avg 12.3%) and sulfur (avg 4.9%).

Palynologically, the bottom four increments are dominated by tree fern spores, with the basal increment containing slightly higher percentages of *Crassispora kosankei*, calamite spores, and other gymnosperm pollen. *Crassispora kosankei*, which was produced by *Sigillaria*, probably was centered in the clastic lowlands that bounded mires, entering the mire environment during times of edaphic change and/or stress (DiMichele and Phillips, 1994). Likewise, calamites have been ascribed to a riparian flora, occupying clastic levees and bars adjacent to fluvial channels (i.e., non-peat environments). However, it is clear that they also inhabited mire environments.

Coniferous plants (referred to as "other gymnosperms" in this paper to differentiate them from cordaites, which are also gymnosperms) generally are attributed to an upland flora. Seed ferns (also called pteridosperms), on the other hand, were a group of gymnospermous plants that probably dominated the clastic lowland environments adjacent to mires. This type of distribution seems to indicate that seed ferns had a higher nutrient budget than other contemporaneous Carboniferous flora, entering mires only during times of disturbance (e.g., clastic influx, wildfire) (DiMichele and Phillips, 1994). Collectively, the increase of lycopod and calamite spores, and other gymnosperm pollen, is consistent with a transition from a clastic to peat substrate.

A palynological change occurs at a fusain layer about a half a meter above the base of the bed (marked A on Fig. 7). At this level we see the introduction of higher percentages of lycopod and calamite spores, and cordaite and other gymnosperm pollen. Concurrent with this palynological change is a major petrographic change; total vitrinite contents drop from almost 88% (mineral matter free [mmf], basis) to 63%, with a corresponding increase in inertinite from 9.4% to 34.9% (mmf). Ash yields and sulfur contents remain unchanged, however. This palynologic and petrographic change represents the boundary between the bottom coal and brick coal of miner's terminology, a break that can be identified regionally (Cross, 1952).

The cause of this change may have been a drop in water table, possibly caused by decreased rainfall, or a regional decrease in base level; subsequent widespread wildfire probably was also a contributing factor. This would account for the higher inertinite contents, as well as the change in flora. It is likely that the water table continued to fluctuate throughout the deposition of the brick coal as it contains several thin shale partings, lower vitrinite (avg 79.6%, compared with 87.8% for the bottom coal), and a more heterogeneous palynoflora than that of the bottom coal (Fig. 7).

The deposition of a thin, carbonaceous shale, called the "bearing-in bench" (marked B on Fig. 7), indicates another event in the history of the Pittsburgh paleomire. This parting is extremely widespread, and literally can be traced across the entire extent of the Pittsburgh coal bed. At this location, the bearing-in bench contains increased ash (65.2%) and inertinite (58.2%) but has a tree-fern-dominant palynoflora that is indistinguishable from bounding increments.

The cause of this change could be the result of a regional base-level rise, which resulted in a temporary drowning of the mire. Alternatively, the deposition of carbonaceous shale could have been caused by a regional drop in water table, with the increased ash being the result of intense peat decay and the accumulation of authigenic mineral matter (Cecil et al., 1979, 1982). A drop in base level would also help explain the increased amount of inertinite. It is highly unlikely that the bearing-in bench represents a crevasse splay; it simply is too extensive.

The breast coal that overlies the bearing-in bench is strongly dominated by tree fern spores. Petrographically, the breast coal is even lower in overall vitrinite content (avg 76.7%, mmf) than the underlying brick coal; it also contains more ash (avg 15.2%) and sulfur (avg 2.8%). The breast coal is capped by a high ash increment that signals the cessation of peat accumulation in the main Pittsburgh paleomire. A rider coal occurs 0.3 m above the main bed, along with some other carbonaceous shale and claystone layers. Collectively, these layers form the roof coal of mining terminology. The roof coal and shale samples are all dominated by tree fern spores and are very high in ash (Fig. 7).

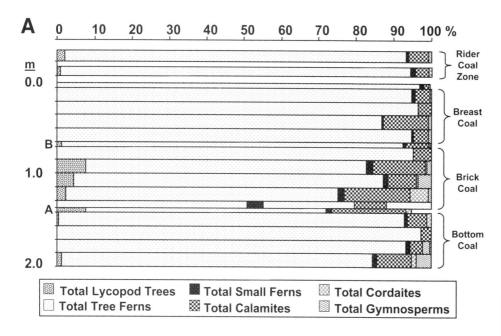

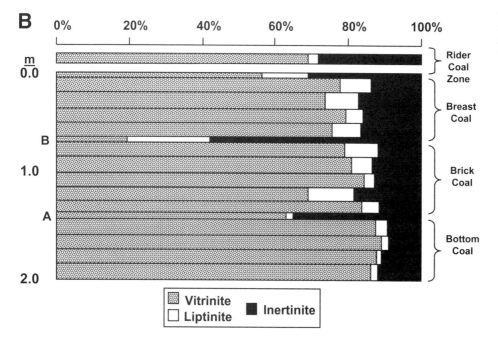

Figure 7 (*continued on following page*). (A) Palynological, (B) petrographic, and (C, D) geochemical data for the thick column of Pittsburgh coal analyzed in this study. (A) represents a widespread parting that separates the Bottom coal from the overlying Brick coal; (B) is another widespread parting called the "bearing-in bench."

Thin Pittsburgh Coal

A thin column (45 cm) of Pittsburgh coal was sampled and analyzed for comparison with the thick Pittsburgh column described above. The two columns are separated geographically by ~70 km (Fig. 6). The basal increment in this column contains a very heterogeneous palynoflora. The top coal increment and overlying carbonaceous shale begin to show more tree fern dominance, but still contain fairly abundant lycopod (*Crassispora kosankei*, avg 5.4%), and calamite spores (avg 27.6%). Both coal increments are high in vitrinite (avg 88.5%, mmf), low in ash yield (avg 5.4%), and moderate in sulfur content (avg 2.1%) (Fig. 8).

These data indicate that the onset of peat accumulation in the central portion of the mire may have preceded the onset of peat accumulation on the mire margin. If correct, peat initially began to accumulate in the more central portions of the paleomire, and then migrated to more distal areas through a

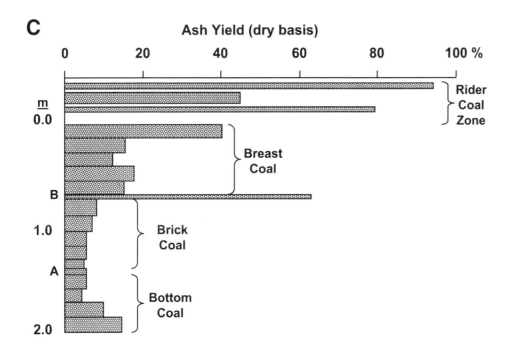

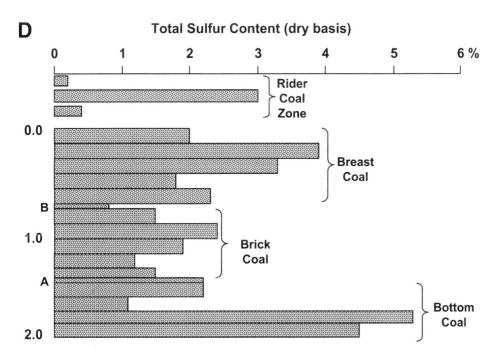

Figure 7 (*continued*).

process called paludification. This process has been observed in modern higher-latitude mires, and also inferred for extensive late Middle Pennsylvanian coal beds in North America. Alternatively, the more heterogeneous palynoflora observed in the thin column may simply reflect its peripheral position, which would have made it more susceptible to influence from surrounding clastic floral elements (in this case *Sigillaria* and gymnosperms). Increased nutrient availability may also have been a factor. If this scenario is correct, then peat accumulation in the central and peripheral portions of the Pittsburgh paleomire probably began at about the same time.

Redstone Coal Bed

The Redstone coal bed is thinner and much more discontinuous than the Pittsburgh coal bed (Fig. 4). As was mentioned previously, this is likely the result of the Redstone paleomire developing on the Pittsburgh sandstone, which essentially limited peat development to interfluve areas. One of the most striking differences between the palynology of the Pittsburgh and Redstone coal beds is the presence of relatively abundant *Endosporites globiformis* and *Punctatisporites parvipunctatus* in the Redstone coal, and their near absence in the Pittsburgh coal. Another major palynological difference between the Pittsburgh and Redstone coals is the relative abundance of *Thymospora thiessenii* in the Pittsburgh coal, and virtual absence of this species in the Redstone coal (Kosanke, 1943). *Endosporites globiformis* and *Punctatisporites parvipunctatus* were found to occur mainly at the base of the Redstone coal, around inorganic partings, and occasionally at the top of the bed. In most cases, these species were associated with higher ash coal layers. They also occurred throughout the vertical extent of the bed in the thin Redstone column. *Florinites* and *Alisporites*, produced by two gymnospermous plants, were also found to occur more frequently in increments with abundant *Endosporites globiformis* and *Punctatisporites parvipunctatus*.

Thick Redstone Column

Figure 9 shows palynological, petrographic, and geochemical data for a 1.2 m thick column of Redstone coal. This column can be subdivided into a lower and upper bench by the presence of a thin (6.1 cm) shale parting near the middle of the bed. The bottom bench is co-dominated by tree fern and calamite spores, with a subdominant small-fern spore component. Overall, spores of small ferns are more abundant in the Redstone coal (avg 10.5%) than in the Pittsburgh coal (avg 1.2%). Calamite spores are also a significant component of the thick Redstone palynoflora (avg 31.3%).

The shale parting is marked by the appearance of higher percentages of *Endosporites globiformis*. *Endosporites globiformis* was produced by *Chaloneria*, a small shrublike lycopod that may have occupied mire margin areas during the Late Pennsylvanian. The increased percentages of *Endosporites* in this parting suggest that *Chaloneria* mainly occupied clastic substrates peripheral to the mire, moving inward during times of environmental change or stress (in this case, a crevasse splay). It is also possible that *Chaloneria* required higher levels of nutrients, and thus is found associated with increased mineral matter. High percentages of *Endosporites* at the base of many other Redstone columns (Habib, 1968; Eble, 1985) further suggest that *Chaloneria* was an important element of the initial flora that developed during the transition from a clastic to peat substrate. Collectively, *Chaloneria* seems to play the same role in the Redstone coal that *Sigillaria* (*Crassispora kosankei*) plays in the Pittsburgh coal. Interestingly, *Crassispora kosankei* is very rarely seen in the Redstone coal.

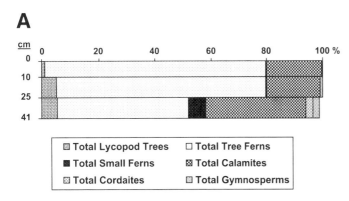

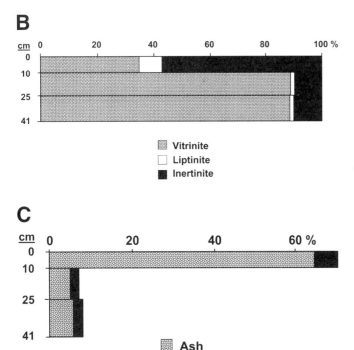

Figure 8. (A) Palynological, (B) petrographic, and (C) geochemical data for the thin column of Pittsburgh coal analyzed in this study.

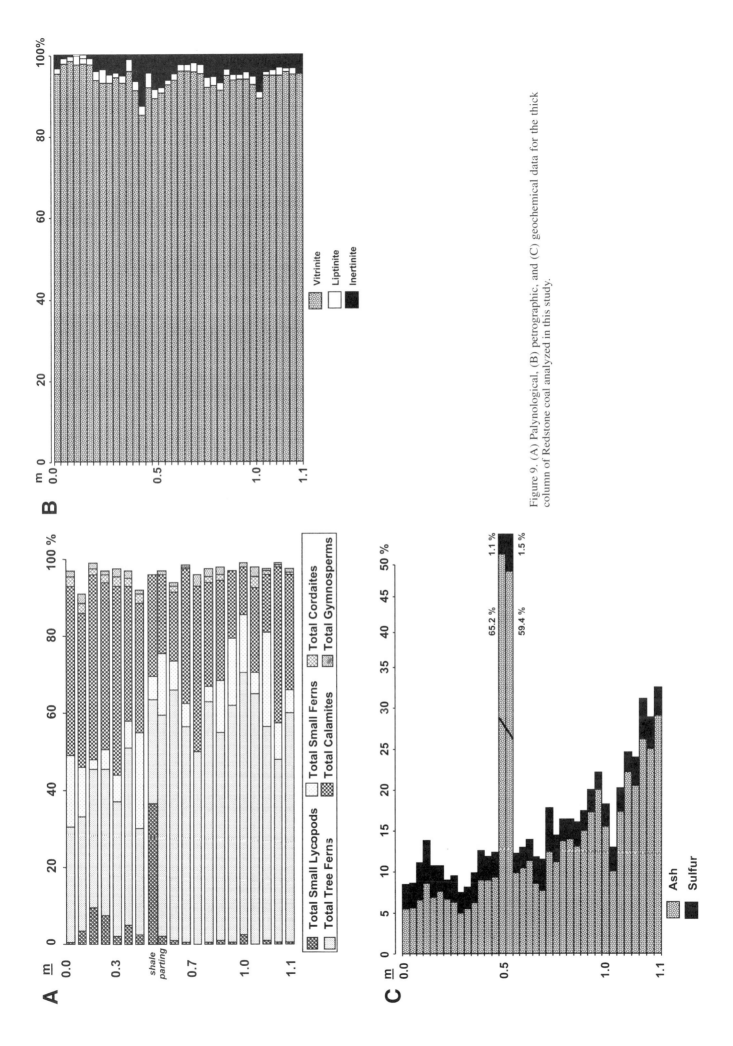

Figure 9. (A) Palynological, (B) petrographic, and (C) geochemical data for the thick column of Redstone coal analyzed in this study.

The upper bench of this column shows a return to a calamite/tree fern spore co-dominance, but with a continued presence of *Endosporites globiformis*. Calamite spores are more abundant in the upper bench (avg 41.9%) than they are in the lower bench (avg 26.0%). Overall, calamite spores are more abundant in the Redstone coal (avg 31.3%) than they are in the Pittsburgh coal (avg 7.8%).

Petrographically, the Redstone coal is strongly dominated by vitrinite (avg 93.9%, mmf), with correspondingly low percentages of liptinite (avg 1.8%, mmf) and inertinite macerals (avg 4.3%, mmf). Vertical distribution of macerals is relatively uniform without any significant difference being seen for the upper and lower benches. Geochemically, the lower bench of this column shows a progressive decrease upward in ash yield, but relatively uniform total sulfur contents. The upper bench ash and sulfur profiles are more uniform in distribution. The Redstone coal has a higher average vitrinite content than the Pittsburgh coal (93.9% versus 81.4%) but also contains higher amounts of ash and sulfur.

Thin Redstone Column

Palynological, petrographic, and geochemical data for a thin (0.3 m) column of Redstone coal are shown in Figure 10. This column contains much higher percentages of other gymnosperm pollen (avg 13.5%), as well as high amounts of *Endosporites* (avg 7.5%). Gymnosperm pollen typically is rare in the Redstone coal, with increases in abundance usually paralleling increases in *Endosporites*. Petrographically, this column contains a reduced amount of vitrinite (avg 80.1%, mmf), and much higher inertinite (avg 18.7%, mmf) than the thick Redstone column. The same is true for ash yield (avg 23.2%) and sulfur content (avg 3.6%). In this respect, it differs from the thin column of Pittsburgh coal, which was high in vitrinite, and low in ash and sulfur.

The high percentages of *Endosporites*, other gymnosperm pollen, inertinite and ash are probably the result of this column being near the edge of the mire. *Endosporites* and other gymnosperm pollen were produced by plants that probably inhabited clastic environments adjacent to mires (DiMichele and Phillips, 1994). The increased amount of inertinite indicates accelerated oxidation of the peat, possibly due to a fluctuating water table or increased frequency of wildfire. The high ash yields are most likely a function of repeated clastic influx, which would affect mire margin areas more than mire interior areas. Some of the inertinite may also have been brought into the mire along with sediment, and thus be of allochthonous origin.

Sewickley Coal Bed

Thick Sewickley Column

Figure 11 shows palynological, petrographic, and geochemical data for a 1.7-m-thick column of Sewickley coal. Like the thick Redstone and Pittsburgh columns discussed previously, this Sewickley column can be subdivided into a lower and upper bench on the basis of a thin (3 cm) bone coal parting near the middle of the bed. The bottom bench of this column starts out with a heterogeneous palynoflora containing relatively high percentages of other gymnosperm (8%) and cordaite (4.8%) pollen and calamite spores (37.6%). Following this is a progressive increase in tree fern spores up to the deposition of the bone coal parting.

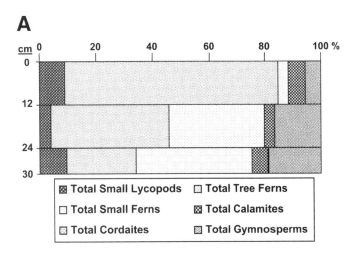
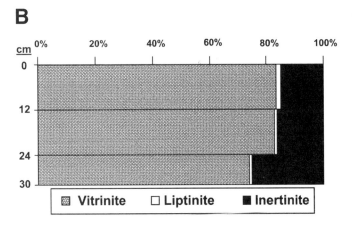
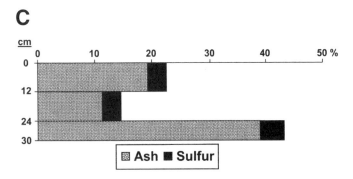

Figure 10. (A) Palynological, (B) petrographic, and (C) geochemical data for the thin column of Redstone coal analyzed in this study.

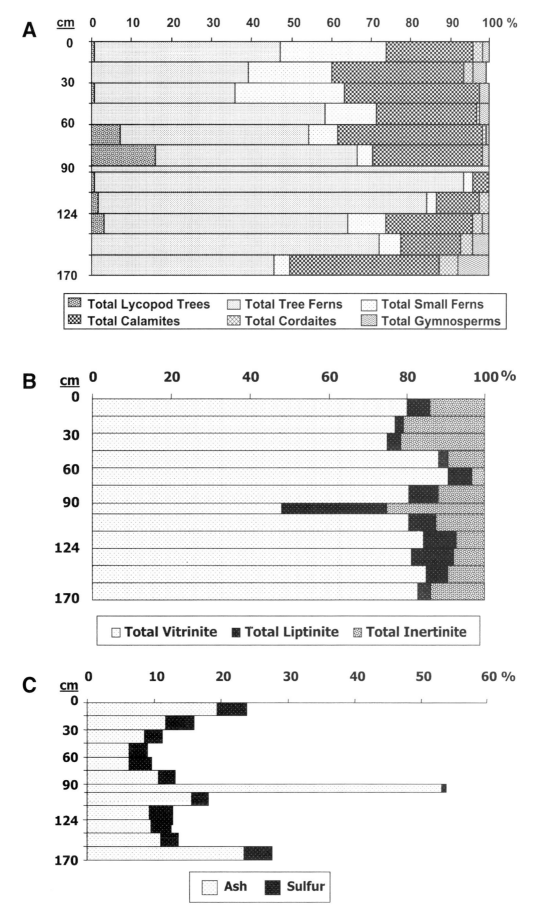

Figure 11. (A) Palynological, (B) petrographic, and (C) geochemical data for the thick column of Sewickley coal analyzed in this study.

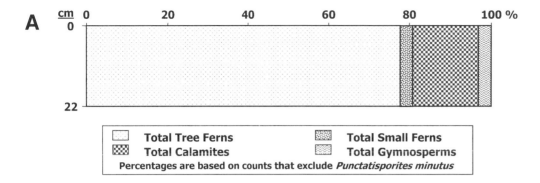

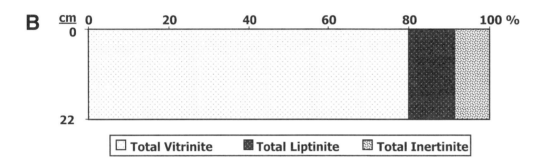

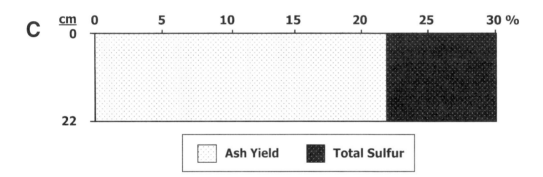

Figure 12. (A) Palynological, (B) petrographic, and (C) geochemical data for the thin column of Sewickley coal analyzed in this study.

The bone parting, which has an ash yield of 53.1%, is essentially half organic matter, half mineral matter. Unlike the palynofloras in partings in the Pittsburgh and Redstone coals, this parting is 100% *Punctatisporites minutus*, the tree fern spore that was excluded from the counts because of its overwhelming abundance. This represents yet another response of palynoflora to a disturbance within the mire. It most resembles the palynoflora of the bearing-in bench of the Pittsburgh coal, although the dominant tree fern species in the bearing-in bench is *Thymospora thiessenii*, not *Punctatisporites minutus*. Both partings also have similar ash yields and elevated amounts of inertinite.

The upper bench of the thick Sewickley column starts out with increased percentages of *Crassispora kosankei*, the spore of *Sigillaria* (avg 11.6%). A comparable increase in *Crassispora* can be seen in the brick coal bench in the thick Pittsburgh column. The remaining portion of the upper bench of the thick Sewickley column displays high percentages of calamite and small fern spores, along with greater inertinite contents and ash yield.

Small ferns account for much of the diversity we see among Pennsylvanian-age ferns, but almost none of the biomass. Small ferns, and herbaceous plants in general, simply weren't large enough to contribute to any significant peat accumulation. What they do potentially indicate is an exposed substrate, at least part of the time, as a deep, continuous water cover would have drowned small ground-cover plants, and also would have prevented them from completing their reproductive cycle (DiMichele and Phillips, 1994).

Alternatively, if the small fern spores recovered from the Sewickley coal bed represent epiphytic plants (Rothwell, 1991) then an exposed peat surface may not have been necessary. The large root zone of *Psaronius* tree fern trunks may have served not only as a protected substrate for juvenile epiphytes, but also as a subaerial platform for adult plants.

Thin Sewickley Column

A thin (0.22m) column of Sewickley coal was also examined (Fig. 12). Palynologically, this column is dominated by tree fern spores. Other gymnosperm pollen is present, but not nearly in the amounts seen in the thin column of Redstone coal. *Crassispora*, which increases just above the bone coal in the thick column of Sewickley coal, is numerically absent in the thin column, as is *Endosporites*. Petrographically, the coal contains abundant vitrinite (80%), which is similar to the average vitrinite content of the thick Sewickley column (82.2%). The ash yield (21. 9%) and sulfur content (8.2%) are both high.

Although this column is thin, and located close to the mire margin (according to regional mapping), no increase in "extra-mire" palynomorphs is seen, relative to the thick Sewickley column. This response (or lack of one) is somewhat analogous to the bearing-in bench of the Pittsburgh coal, which also exhibits no palynological change. The reason for this is unclear, but probably related to the flora growing on the clastic environments adjacent to the paleomire. Although plants like *Chaloneria*, pteridosperms, and conifers were focused outside of mires, tree ferns were much more cosmopolitan in distribution, occupying both peat and clastic environments. Therefore, it is plausible that the paleoflora growing adjacent to the location where the thin Sewickley column was sampled were mainly tree ferns. There simply may not have been enough *Sigillaria*, *Chaloneria*, and gymnospermous plants in the immediate vicinity to show up in the dispersed spore and pollen record.

SUMMARY

The Pittsburgh, Redstone, and Sewickley coal beds occur in the Late Pennsylvanian Pittsburgh Formation of the Monongahela Group. The distributions of these coal beds are highly dissimilar, and probably related to paleotopography. The Pittsburgh coal is underlain by fine-grained sediments over most of its extent, which apparently provided a stable platform for widespread peat accumulation. In contrast, the Redstone coal is underlain by the Pittsburgh sandstone, which limited peat development mainly to interfluve areas. Likewise, the Sewickley coal was influenced by paleotopography (the underlying Lower Sewickley sandstone), but not to the extent of the Redstone coal.

All three coal beds contain palynofloras dominated by tree fern spores, though species dominance varies. In the Pittsburgh coal, *Thymospora thiessenii* is the dominant species. In the Redstone coal, *Laevigatosporites minimus*, *Punctatosporites minutus* are usually the most abundant species, with *Thymospora pseudothiessenii* being locally abundant. *Thymospora thiessenii* is rare or absent in the Redstone coal. In the Sewickley coal, three species of *Thymospora* (*T. thiessenii, T. pseudothiessenii,* and *T. obscura*) occur frequently, as does *Laevigatosporites minimus*.

In all three coals, changes in palynomorph composition are commonly, but not always, associated with the occurrence of high-ash and/or high-inertinite coal layers, and inorganic partings. For example, in the Pittsburgh coal, a thin, high-inertinite layer that marks the boundary between the bottom and brick coal benches contains increased percentages of *Crassispora kosankei* and of calamite spores, as well as higher amounts of other gymnosperm pollen. This represents a pronounced change from the interval dominated by tree fern spores directly below this layer. This was a widespread change in the Pittsburgh paleomire, as this layer can be traced across the extent of the coal, which may have been caused by a regional drop in base level, followed by wildfire.

In contrast to this, the bearing-in bench, which is another high-inertinite, high-ash parting of basinwide extent, shows no change in palynoflora; the bearing-in bench and adjacent coal layers are both dominated by tree fern spores. Why this mire disturbance is not reflected by a change in palynoflora is unknown, but it may be related to the bearing-in bench resulting from a regional rise in base level (i.e., temporary drowning). Partings in the Sewickley coal behave similarly, though most cannot be correlated regionally the way partings in the Pittsburgh coal can. Some partings show distinct palynofloral changes from adjacent coal layers, typically with increased percentages of *Crassispora kosankei*, calamite spores, and other gymnosperm pollen. Others show essentially no change from adjacent coal layers.

Response to mire disturbance in the Redstone coal is somewhat different. Rather than increased percentages of *Crassispora kosankei*, *Endosporites* and other gymnosperm pollen usually are associated with high-ash and/or high-inertinite layers in the Redstone coal. In this regard, *Endosporites* seems to mimic the distribution of *Crassispora kosankei* in the Pittsburgh and Sewickley coal beds. However, it is unclear why *Endosporites* is locally abundant in the Redstone coal, but rare to absent in the Pittsburgh and Sewickley, with the opposite being true for *Crassispora kosankei* (i.e., locally abundant in the Pittsburgh and Sewickley coals, but rare to absent in the Redstone coal). One possible explanation could be the differences in mire distribution. The Pittsburgh and Sewickley are more continuous in distribution and represent broad, blanket-like peats. In contrast, the Redstone coal is much more discontinuous and podlike, forming from localized peat bodies.

The climate must have been sufficiently wet during the times that the three paleomires formed, though probably not so wet as to promote the development of ombrogenous, domed peat. Rather, all three paleomires, which produced coal with moderate to high ash yields and total sulfur contents, appear to have been topogenous and planar (Cecil et al., 1985).

ACKNOWLEDGMENTS

The authors wish to thank Debra Willard and Peter Warwick, both with the U.S. Geological Survey, for reviewing this manuscript. This paper has benefited greatly from their respective comments. The authors also wish to thank Steven Greb for organizing and editing this volume.

APPENDIX 1

APPENDIX 1: THICK PITTSBURGH COAL BED

	rider	coal	zone	top main coal					B Bearing in Bench
	9508	9509	9510	9511	9512	9513	9514	9515	9516
Punctatisporites minutus	57.6	39.6	58	30.4	30	38.8	30.4	49.6	66
Crassispora kosankei		0.8	2	0.8					
Total Lycopod Trees	0	0.8	2	0.8	0	0	0	0	0
Thymospora pseudothiessenii					0.4				
T. thiessenii	84.4	85.6	88.4	90	94	84	89.2	63.6	81.6
Laevigatosporites minimus	4.4	4	1.6	1.2	0.4	0.8	1.2	1.2	2.4
Punctatisporites parvipunctatus	1.6			0.4				0.8	
Punctatosporites minutus	1.6	1.6	0.4	1.6	1.2			1.2	8
Spinosporites exiguus	0.8	2	0.8	0.4	1.2	10	6	20	2.8
Total Tree Ferns	92.8	93.2	91.2	93.6	96.8	94.8	96.4	86.8	94.8
Granulatisporites parvus	0.4	0.4	0.4		0.4				
G. granulatus						0.4			
G. verrucosus									
Cyclogranisporites minutus	0.8	0.8							
C. aureus									
C. microgranus	0.4	0.4							
Leiotriletes subadnatoides	0.4								
L. levis	0.4	0.4							
Acanthotriletes aculeolatus			0.4					0.4	
Gillespieisporites venustus									
Verrucosisporites microtuberosus									
V. verrucosus				0.4	0.8				
Microreticulatisporites sulcatus									
Punctatisporites glaber									
P. punctatus									
P. breviornatus	0.8			0.8		0.4			0.4
P. aerarius									
Total Small Ferns	3.2	2	0.8	1.2	1.2	0.8	0	0.4	0.4
Calamospora flexilis					0.4				
C. pallida									
C. pedata									
Laevigatosporites minor	3.6	3.6	5.2	3.6	0.8	4	3.6	12	4
L. vulgaris									
Total Calamites	3.6	3.6	5.2	3.6	1.2	4	3.6	12	4
Florinites mediapudens									
F. florini	0.4			0.8					
F. pumicosus									
Total Cordaites	0.4	0	0	0.8	0	0	0	0	0
Vesicaspora wilsonii			0.4						
Pityosporites westphalensis			0.4		0.4	0.4		0.8	0.8
Hamiapollenites sp.									
Total Gymnosperms	0	0	0.8	0	0.4	0.4	0	0.8	0.8
Indospora boletus		0.4							
Total Unknown Affinity	0	0.4	0	0	0	0	0	0	0
Thickness (cm)	15.2	12.7	20.3	4.6	15.2	15.2	15.2	12.5	3.1
Ash (dry basis)	94.1	44.9	79.4	40.3	15.5	12.3	17.9	15.2	65.8
Sulfur (dry basis)	0.2	3	0.4	2	3.9	3.3	1.8	2.3	0.8
Total Vitrinite (mmf)	n/d	76.3	n/d	56.3	77.8	73.9	79.6	75.6	19.3
Total Liptinite (mmf)	n/d	3.1	n/d	12.7	8.5	9.1	4.6	7.9	22.7
Total Inertinite (mmf)	n/d	20.6	n/d	31.0	13.7	17.1	15.8	16.5	58.0

(*continued*)

APPENDIX 1: THICK PITTSBURGH COAL BED (continued)

						A Fusain Parting				base main coal
	9517	9518	9519	9520	9521	9522	9523	9524	9525	9526
Punctatisporites minutus	52.8	30.4	61.2	54.8	74.4	73.2	61.2	55.6	57.6	64.8
Crassispora kosankei	1.2		7.6	4.4	2.4		7.6	0.4		
Total Lycopod Trees	**1.2**	**0**	**7.6**	**4.4**	**2.4**	**0**	**7.6**	**0.4**	**0**	**0**
Thymospora pseudothiessenii										
T. thiessenii	70.8	89.2	52	56.8	38.8	33.2	52	86.8	87.6	74
Laevigatosporites minimus	5.2		5.6	4.4	2.4	6.4		1.2	1.2	1.2
Punctatisporites parvipunctatus								0.4		
Punctatosporites minutus	2.4		5.2	1.6	0.8	7.2		0.8	0.8	0.8
Spinosporites exiguus	12.8	6	12.4	20	30.8	4	12.4	3.2	7.6	17.2
Total Tree Ferns	**91.2**	**95.2**	**75.2**	**82.8**	**72.8**	**50.8**	**64.4**	**92.4**	**97.2**	**93.2**
Granulatisporites parvus				0.4	0.4	0.8				
G. granulatus										
G. verrucosus			0.8							
Cyclogranisporites minutus										0.4
C. aureus										
C. microgranus								0.4		
Leiotriletes subadnatoides				0.8	0.8	0.8				
L. levis										
Acanthotriletes aculeolatus										
Gillespieisporites venustus										0.4
Verrucosisporites microtuberosus					0.4					
V. verrucosus										
Microreticulatisporites sulcatus						0.4				0.4
Punctatisporites glaber						0.4				
P. punctatus	0.4									
P. breviornatus	0.4		0.8			1.6	1.6	0.4		
P. aerarius						0.4				
Total Small Ferns	**0.8**	**0**	**1.6**	**1.2**	**1.6**	**4.4**	**1.6**	**0.8**	**0**	**1.2**
Calamospora flexilis								0.4		
C. pallida								0.4		
C. pedata					0.8					
Laevigatosporites minor	6	4.8	14	7.6	16.8	23.6	19.6	4	2.8	3.2
L. vulgaris						0.8		0.4		
Total Calamites	**6**	**4.8**	**14**	**7.6**	**17.6**	**24.4**	**19.6**	**5.2**	**2.8**	**3.2**
Florinites mediapudens					0.4	1.6	1.6			0.4
F. florini	0.4		0.4	0.4	4.4	6.8		1.2		0.4
F. pumicosus										1.2
Total Cordaites	**0.4**	**0**	**0.4**	**0.4**	**4.8**	**8.4**	**1.6**	**1.2**	**0**	**2**
Vesicaspora wilsonii				0.4		1.6				
Pityosporites westphalensis	0.4		1.2	3.2	0.8	9.6	9			0.4
Hamiapollenites sp.						0.8				
Total Gymnosperms	**0.4**	**0**	**1.2**	**3.6**	**0.8**	**12**	**9**	**0**	**0**	**0.4**
Indospora boletus										
Total Unknown Affinity	**0**	**0**	**0**	**0**	**0**	**0**	**0**	**0**	**0**	**0**
Thickness (cm)	15.2	15.2	15.2	15.2	7.6	3.7	15.8	15.2	15.2	18.9
Ash (dry basis)	8.2	7.1	5.5	5.4	5	5.5	5.4	4.3	10	14.5
Sulfur (dry basis)	1.5	2.4	1.9	1.2	1.5	2.2	2.2	1.1	5.3	4.5
Total Vitrinite (mmf)	79.3	81.2	84.6	69.2	83.9	63.0	87.7	89.3	88.1	86.4
Total Liptinite (mmf)	8.9	5.8	2.7	12.4	4.7	2.2	3.2	1.9	1.1	2.1
Total Inertinite (mmf)	11.8	13.1	12.7	18.4	11.4	34.9	9.2	8.8	10.8	11.5

APPENDIX 1: THIN PITTSBURGH COAL BED

	Top		Base
Punctatisporites minutus	32.4	66.4	88.4
Crassispora kosankei	1.2	5.2	5.6
Total Lycopod Trees	1.2	5.2	5.6
Endosporites globiformis			0.4
Total Small Lycopods	0	0	0.4
Thymospora pseudothiessenii	0.4		1.2
T. thiessenii	71.6	56.4	20
Laevigatosporites minimus	3.6	4.4	7.6
L. ovalis	0.4		
Punctatosporites minutus	0.8		3.2
Spinosporites exiguus	1.6	13.6	14.4
Total Tree Ferns	78.4	74.4	46.4
Granulatisporites parvus	0.4	0.4	0.8
G. pannosites			1.2
G. verrucosus			0.4
C. microgranus			0.4
Leiotriletes subadnatoides			0.4
Raistrickia sp.			0.4
Microreticulatisporites sulcatus			0.4
P. breviornatus			2
Total Small Ferns	0.4	0.4	6
Calamospora pallida	0.4		
Laevigatosporites minor	18.4	18.8	34
L. vulgaris	0.8	0.4	2
Total Calamites	19.6	19.2	36
Florinites florini			2.4
Total Cordaites	0	0	2.4
Vesicaspora wilsonii	0.4		0.4
Pityosporites westphalensis		0.8	2
Total Gymnosperms	0.4	0.8	2.4
Tantillus triquetrus			0.4
Diaphanospora parvigracilis			0.4
Total Unknown Affinity	0	0	0.8
Thickness (cm)	10.7	15.2	15.2
Ash (dry basis)	64.8	5.1	5.7
Sulfur (dry basis)	5.6	1.9	2.2
Total Vitrinite (mmf)	34.8	88.6	88.3
Total Liptinite (mmf)	8.2	1.8	1.5
Total Inertinite (mmf)	57	9.6	10.2

APPENDIX 1. THICK REDSTONE COAL BED

Palynologic Sample Increments	Top 1	3	5	7	9	11	13	15	16	17
Punctatisporites minutus	n/d	n/d	n/d	n/d	n/d	n/d	n/d	n/d	n/d	n/d
Endosporites globiformis	2	3.5	9.5	7.5	3.5	5	7.5	36.5	2	1
Total Small Lycopods	2	3.5	9.5	7.5	3.5	5	2.5	36.5	2	1
Thymospora pseudothiessenii	0.5	2				1	2	1.5		
T. thiessenii		1.5			0.5		0.5			
Laevigatosporites minimus	26.5	16		23.5	23.5	14.5	15	6	4	11.5
Punctatisporites parvipunctatus	0.5	1.5	33.5	3.5	4	15	7.5	4	6.5	1.5
Punctatosporites minutus	2.5	10	2.5	11	7	16.5	2.5	15.5	47	52
Total Tree Ferns	30	31	36	38	35	47	27.5	27	57.5	65
Granulatisporites parvus							1			
G. pallidus								1.5		
G. piroformis							1.5	0.5		
Cyclogranisporites minutus		2			0.5	0.5	4.5	1	3.5	2.5
C. microgranus	0.5	0.5	0.5	2	0.5	2	4	0.5	7	1.5
Leiotriletes subadnatoides	3.5	0.5	1	1	2	0.5				
L. gracilis										
L. adnatus		0.5				0.5	1.5	1		
Acanthotriletes aculeolatus	3	0.5		0.5	1	1	0.5			
Lophotriletes insignitus										
L. commissuralis	11.5	7	0.5	1	2	0.5	4.5	0.5	0.5	2.5
Verrucosisporites donarii							0.5			
Raistrickia diversa							0.5			
Apiculatasporites spinulistratus		0.5					0.5			
A. sp.			0.5	0.5	0.5	0.5	2	0.5	0.5	
Convolutispora venusta		0.5						0.5		
C. sp.		0.5								
Punctatisporites breviornatus		0.5	1		0.5	1	4		4.5	1
Total Small Ferns	18.5	13	3.5	5	7	6.5	25	6	16	7.5
Calamospora hartungiana							1	1.5		
C. microrugosa		0.5			0.5		0.5			
Laevigatosporites minor	44	39.5	47.5	43.5	48	35	31.5	20.5	20.5	16.5
L. medius			0.5		0.5		0.5	4.5		1.5
Total Calamites	44	40	48	43.5	49	35	33.5	26.5	20.5	18
Florinites mediapudens										0.5
F. florini	2.5	2.5	1.5	2	2.5	2	2			1.5
F. pumicosus						0.5	0.5			
Total Cordaites	2.5	2.5	1.5	2	2.5	2.5	2.5	0	0	2
Vesicaspora wilsonii		1.5	1.5		1.5	0.5	1.5			0.5
Pityosporites westphalensis										0.5
Protohaploxypinus amplus		0.5		0.5						
Vittatina sp.						0.5				
Alisporites recurvus	1.5	0.5		0.5	0.5	1	1		1	0.5
Total Gymnosperms	1.5	2.5	1.5	1	2	2	2.5	0	1	1.5
Triquitrites minutus	1	7		1.5	0.5	2	2.5	3	2.5	4.5
T. cf. spinosus	0.5			1.5	0.5		1	0.5	0.5	0.5
Planisporites granifer							0.5			
Dictyotriletes danvillensis								0.5		
Complexisporites polymorphus		0.5								
Total Unknown Affinity	1.5	7.5	0	3	1	2	4	4	3	5

(continued)

APPENDIX 1. THICK REDSTONE COAL BED (continued)

Palynologic Sample Increments	19	21	23	25	27	29	31	33	35	Base 37
Punctatisporites minutus	n/d	n/d	n/d	n/d	n/d	n/d	n/d	n/d	n/d	n/d
Endosporites globiformis	0.5		1	1	0.5	2.5	1.5	1	0.5	0.5
Total Small Lycopods	0.5	0	1	1	0.5	2.5	1.5	1	0.5	0.5
Thymospora pseudothiessenii										
T. thiessenii										
Laevigatosporites minimus	15.5	45	27.5	26	16.5	5.5	24.5	8.5	15.5	16
Punctatisporites parvipunctatus		0.5	1	0.5	2	8	9.5	12		1
Punctatosporites minutus	40.5	4.5	34	27.5	43	54.5	31	35	32	42.5
Total Tree Ferns	56	50	62.5	54	61.5	68	65	55.5	47.5	59.5
Granulatisporites parvus										
G. pallidus										
G. piroformis			0.5				2	0.5		0.5
Cyclogranisporites minutus	1		0.5	6.5	5.5	3.5	0.5	3		0.5
C. microgranus	2		0.5	2	5.5	4		10	7	2
Leiotriletes subadnatoides						0.5		0.5		
L. gracilis										
L. adnatus										
Acanthotriletes aculeolatus						0.5		0.5		
Lophotriletes insignitus			2							
L. commissuralis	2.5			4.5	3	2		4	1	2.5
Verrucosisporites donarii										
Raistrickia diversa								0.5		
Apiculatasporites spinulistratus							0.5			
A. sp.	0.5		0.5		0.5			0.5		
Convolutispora venusta										
C. sp.										
Punctatisporites breviornatus				0.5	2.5	5	2	5.5	1.5	0.5
Total Small Ferns	6	0	4	13.5	17.5	15	5.5	24.5	9.5	6
Calamospora hartungiana									0.5	0.5
C. microrugosa										
Laevigatosporites minor	33.5	42	27	26	15	11.5	21	13.5	39	27.5
L. medius	1.5	1			2.5	1	1	1.5	1	2
Total Calamites	35	43	27	26	17.5	12.5	22	15	40.5	30
Florinites mediapudens	0.5									
F. florini			1.5	1.5			3	1	0.5	0.5
F. pumicosus										
Total Cordaites	0.5	0	1.5	1.5	0	0	3	1	0.5	0.5
Vesicaspora wilsonii						0.5			0.5	
Pityosporites westphalensis										
Protohaploxypinus amplus										
Vittatina sp.										
Alisporites recurvus	0.5	3	2	2		0.5	2.5	0.5		1
Total Gymnosperms	0.5	3	2	2	0	1	2.5	0.5	0.5	1
Triquitrites minutus	0.5	3	1	1	3	0.5	0.5	1	0.5	2
T. cf. spinosus	1	1	1	1		0.5		1	0.5	0.5
Planisporites granifer										
Dictyotriletes danvillensis										
Complexisporites polymorphus								0.5		
Total Unknown Affinity	1.5	4	2	2	3	1	0.5	2.5	1	2.5

(continued)

APPENDIX 1. THICK REDSTONE COAL BED (continued)

Thickness (cm)	Palynologic Increment	dry ash	dry sulfur	mmf Vitrinite	mmf Liptinite	mmf Inertinite
3.05	1	5.5	3.0	97.8	1.3	0.9
3.05		5.6	3.0	98.4	1.1	0.5
3.05	3	6.6	4.5	97.6	2.2	0.2
3.05		8.7	5.2	97.9	1.4	0.7
3.05	5	6.9	3.9	97.6	1.6	0.8
3.05		7.6	3.1	93.6	2.3	4.1
3.05	7	6.6	2.4	93.0	3.2	3.8
3.05		6.3	3.2	93.2	2.0	4.8
3.05	9	5.0	2.5	94.3	1.3	4.4
3.05		5.5	2.7	92.9	2.0	5.2
3.05	11	6.2	3.6	96.1	2.8	1.1
3.05		9.0	3.6	91.2	2.2	6.5
3.05	13	9.0	2.9	85.3	2.2	12.5
3.05		9.4	3.0	92.0	3.4	4.5
3.05	15	65.2	1.1	89.3	2.1	8.6
3.05	16	59.4	1.5	90.6	1.3	8.1
3.05	17	9.9	2.3	92.7	1.1	6.2
3.05		10.5	2.5	93.7	1.6	4.6
3.05	19	11.4	2.6	95.9	1.7	2.4
3.05		8.6	3.2	95.8	1.6	2.6
3.05	21	7.7	3.8	95.8	2.2	2.0
3.05		12.4	5.4	95.3	2.1	2.6
3.05	23	11.2	3.3	91.9	2.4	5.7
3.05		13.7	2.7	92.4	2.2	5.4
3.05	25	14.0	2.4	91.2	1.8	6.9
3.05		13.1	2.9	94.8	1.7	3.5
3.05	27	15.0	2.4	93.7	1.4	4.9
3.05		17.3	2.7	94.0	1.0	5.0
3.05	29	20.0	2.2	94.0	1.7	4.3
3.05		15.5	2.7	92.7	1.9	5.4
3.05	31	10.1	2.9	89.3	1.5	9.2
3.05		17.4	2.8	94.8	1.0	4.2
3.05	33	22.2	2.5	94.9	1.3	3.8
3.05		20.5	3.4	94.9	2.1	3.0
3.05	35	26.2	5.0	95.7	0.9	3.4
3.05		25.1	3.8	94.9	1.5	3.6
3.05	37	29.1	3.5	95.2	0.4	4.4

APPENDIX 1. THIN REDSTONE COAL BED

	Top		Base
Punctatisporites minutus	n/d	n/d	n/d
Endosporites globiformis	9.5	4	9
Total Small Lycopods	**9.5**	**4**	**9**
Laevigatosporites minimus	2	2.5	3.5
Punctatisporites parvipunctatus	12	28.5	68
Punctatosporites minutus	10.5	10.5	4
Total Tree Ferns	**24.5**	**41.5**	**75.5**
Cyclogranisporites minutus	2	0.5	0.5
C. microgranus	0.5	2	
Leiotriletes subadnatoides	0	0.5	0.5
Lophotriletes commissuralis	1.5	1	0.5
Verrucosisporites donarii	0.5	0.5	
Microreticulatisporites sulcatus	0.5	0.5	
Apiculatasporites spinulistratus	0.5		
Triquitrites minutus	34	28	2
P. breviornatus	1.5	1	0.5
Total Small Ferns	**41**	**34**	**4**
Calamospora pallida	0	0.5	0.5
Laevigatosporites minor	5.5	3	5
L. vulgaris			0.5
Total Calamites	**5.5**	**3.5**	**6**
Florinites florini	0.5		
Total Cordaites	**0.5**	**0**	**0**
Vesicaspora wilsonii	2	1	1.5
Pityosporites westphalensis		0.5	
Alisporites recurvus	16.5	15	4
Total Gymnosperms	**18.5**	**16.5**	**5.5**
Proprisporites tectus	0.5		
Spackmanites rotundus		0.5	
Total Unknown Affinity	**0.5**	**0.5**	**0**
Thickness (cm)	12.2	12.2	6.1
Ash (dry basis)	39.1	11.2	19.2
Sulfur (dry basis)	4.1	3.3	3.3
Total Vitrinite	74.1	82.7	83.4
Total Liptinite	0.9	1	1.7
Total Inertinite	25	16.3	14.9

APPENDIX 1. THICK SEWICKLEY COAL BED

	Top						Bone Parting					Base
Punctatisporites minutus	48.8	81.6	78.4	74.4	84.8	28.8	100	44	93.6	87.2	88.8	46.4
Crassispora kosankei	0.8		0.8		7.2	16		0.8	1.6	3.2		
Total Lycopod Trees	0.8	0	0.8	0	7.2	16		0.8	1.6	3.2	0	0
Thymospora pseudothiessenii	6.4	8.8	9.6	9.6	6.4	11.2		39.2	22.4	3.2		
T. obscura	11.2	5.6	7.2	5.6	5.6	0.8			8	0.8		
T. thiessenii	12	4	4.8	19.2	8	16.8		47.2	28	2.4		
Laevigatosporites minimus	7.2	9.6	10.4	17.6	26.4	16		4	8.8	17.6	16.8	15.2
L. ovalis		2.4		1.6		1.6						
Punctatisporites parvipunctatus	2.4	3.2		3.2	0.8	1.6			7.2	12	32	28
Punctatosporites minutus	0.8	0.8		0.8		0.8		2.4	8	24	23.2	2.4
Spinosporites exiguus	3.2	2.4	1.6	0.8		0.8				0.8		
Apiculatasporites saetiger	3.2	2.4	1.6			0.8						
Total Tree Ferns	46.4	39.2	35.2	58.4	47.2	50.4	100	92.8	82.4	60.8	72	45.6
Granulatisporites parvus		0.8	0.8							0.8		
G. piroformis	0.8											
Cyclogranisporites microgranus										1.6		
Leiotriletes subadnatoides	4.8	4.8	7.2	8	1.6	2.4			1.6			0.8
L. adnatus				1.6								
L. levis		0.8	1.6							0.8		
L. subintortus	1.6											
Acanthotriletes aculeolatus			2.4		0.8							
Lophotriletes pseudaculeatus						0.8						
L. commissuralis	0.8	1.6	0.8	0.8								
L. microsaetosus		1.6	0.8									0.8
Gillespieisporites venustus	1.6	1.6	4.8					1.6		0.8	0.8	0.8
Verrucosisporites compactus										0.8		
V. sifati			2.4									
V. donarii			2.4									
V. verrucosus			0.8									
V. bacculatus	1.6											
Raistrickia sp.	1.6				0.8							
Microreticulatisporites nobilis	0.8											
Apiculatasporites spinulistratus		1.6							0.8		1.6	
Triquitrites minutus	12	4.8	3.2	2.4	2.4	0.8		0.8			2.4	
Punctatisporites glaber		1.6										
P. pseudolevatus						0.8						
P. punctatus	0.8					0.8				4.8		
P. breviornatus												1.6
P. aerarius		1.6								0.8		
Total Small Ferns	26.4	20.8	27.2	12.8	7.2	4		2.4	2.4	9.6	5.6	4
Calamospora pedata	0.8					0.8						
C. pallida			0.8			0.8				1.6		
C. microrugosa			4	1.6	2.4	3.2			0.8	0.8		
C. breviradiata	3.2				2.4					1.6		4
Laevigatosporites minor	15.2	28.8	28	22.4	30.4	21.6		4	8	20	15.2	33.6
L. vulgaris	2.4	4	1.6	0.8	1.6				0.8			
L. maximus	0.8	0.8		0.8		1.6						
Total Calamites	22.4	33.6	34.4	25.6	36.8	28		4	11.2	22.4	15.2	37.6
Florinites mediapudens												
F. florini	2.4	2.4		0.8	0.8				2.4	2.4	3.2	4.8
Total Cordaites	2.4	2.4	0	0.8	0.8	0		0	2.4	2.4	3.2	4.8
Vesicaspora wilsonii	1.6	1.6	1.6	1.6		0.8					1.6	1.6
Pityosporites westphalensis		1.6	0.8	0.8						0.8	1.6	6.4
Potoneisporites elegans					0.8						0.8	
Schopfipollenites ellipsoides						0.8				0.8		
Total Gymnosperms	1.6	3.2	2.4	2.4	0.8	1.6		0	0	1.6	4	8
Diaphanospora parvigracilis		0.8										
Total Unknown Affinity	0	0.8	0	0	0	0	0	0	0	0	0	0
Thickness (cm)	15.2	15.2	15.2	15.2	15.2	15.2	3.0	15.2	15.2	15.2	15.2	15.2
Ash Yield (dry)	19.3	11.7	8.6	6.3	6.3	10.6	53.1	15.7	9.3	9.6	11.0	23.5
Total Sulfur Content (dry)	4.5	4.2	2.6	2.7	3.4	2.5	0.7	2.5	3.6	3.1	2.7	4.1
Total Vitrinite	80	76.8	74.8	88	90.4	80.4	48	80.4	84	81.2	84.8	82.8
Total Liptinite	5.6	2.4	3.6	2.4	6.4	7.6	26.8	6.8	8.8	10.8	5.6	3.2
Total Inertinite	14.4	20.8	21.6	9.6	3.2	12	25.2	12.8	7.2	8	9.6	14

APPENDIX 1. THIN SEWICKLEY COAL BED

Punctatisporites minutus	85.6
Laevigatosporites minimus	40
Punctatisporites parvipunctatus	13.6
Punctatosporites minutus	23.2
Spinosporites exiguus	0.8
Total Tree Ferns	**77.6**
Lophotriletes commissuralis	0.8
Verrucosisporites verrucosus	0.8
Punctatisporites punctatus	1.6
Total Small Ferns	**3.2**
Calamospora breviradiata	1.6
Laevigatosporites minor	14.4
Total Calamites	**16**
Vesicaspora wilsonii	0.8
Pityosporites westphalensis	2.4
Total Pteridosperms	**3.2**
Thickness	**21.9**
Ash	**21.9**
Sulfur	**8.2**
Total Vitrinite	**80**
Total Liptinite	**11.2**
Total Inertinite	**8.8**

REFERENCES CITED

American Society for Testing and Materials, 1997a, ASTM D 5142-90, Standard test methods for proximate analysis of the analysis sample of coal and coke by instrumental procedures: 1997 Annual book of ASTM standards, v. 05.05, Gaseous fuels; coal and coke, p. 444–448.

American Society for Testing and Materials, 1997b, ASTM D 3177-89, Standard test methods for total sulfur in the analysis sample of coal and coke: 1997 Annual book of ASTM standards, v. 05.05, Gaseous fuels; coal and coke, p. 306–309.

Cecil, C.B., 1990, Paleoclimate controls on stratigraphic repetition of chemical and clastic rocks: Geology, v. 18, p. 533–536, doi: 10.1130/0091-7613(1990)018<0533:PCOSRO>2.3.CO;2.

Cecil, C.B., Stanton, R.W., Dulong, F.T., and Renton, J.J., 1979, Some geologic factors controlling mineral matter in coal, in Donaldson, A.C., Presley, M.W. and Renton, J.J., eds., Carboniferous coal guidebook. West Virginia Geological and Economic Survey Bulletin, v. B-37-3, p. 43–56.

Cecil, C.B., Stanton, R.W., Dulong, F.T., and Renton, J.J., 1982. Geologic factors that control mineral matter in coal, in Filby, R.H., Carpenter, S.B. and Ragaini, R.C., eds., Atomic and nuclear methods in fossil energy research: New York, Plenum Press, p. 323–335.

Cecil, C.B., Stanton, R.W., Neuzil, S.G., Dulong, F.T., Ruppert, L.F., and Pierce, B.S., 1985, Paleoclimatic controls on Late Paleozoic sedimentation and peat formation in the central Appalachian Basin, USA: International Journal of Coal Geology, v. 5, p. 195–230, doi: 10.1016/0166-5162(85)90014-X.

Clayton, G., Coquel, R., Doubinger, J., Gueinn, K.J., Loboziak, S., Owens, B., and Streel, M., 1977, Carboniferous miospores of western Europe: Illustration and zonation: Mededelingen Rijks Geologische Dienst, v. 29, p. 1–71.

Clendening, J.A., 1965, Characteristic small spores of the Redstone coal in West Virginia: Proceedings of the West Virginia Academy of Science, v. 37, p. 183–189.

Clendening, J.A., and Gillespie, W.H., 1963, Characteristic small spores of the Pittsburgh coal in West Virginia and Pennsylvania: Proceedings of the West Virginia Academy of Science, v. 35, p. 141–150.

Cross, A.T., 1947, Spore floras of the Pennsylvanian of West Virginia and Kentucky: Journal of Geology, v. 55, p. 285–308.

Cross, A.T., 1952, The Geology of the Pittsburgh coal: Stratigraphy, petrology, origin and composition, and geologic interpretation of mining problems. Second conference on the origin and constitution of coal, June 18–20, 1952, Nova Scotia Department of Mines, Crystal Cliffs, Nova Scotia, p. 32–111. (Reprinted 1954 as *The Geology of the Pittsburgh Coal*: West Virginia Geological and Economic Survey Publication, v. RI-10, 80 p.)

Cross, A.T., and Schemel, M.P., 1956, Economic resources of the Ohio River Valley in West Virginia: West Virginia Geological and Economic Survey Publication V-22b, part 2, 129 p.

DiMichele, W.A., and Phillips, T.L., 1994, Paleobotanical and paleoecological constraints on models of peat formation in the Late Carboniferous of Euramerica: Palaeogeography, Palaeoclimatology, Palaeoecology, v. 106, p. 39–90, doi: 10.1016/0031-0182(94)90004-3.

Donaldson, A.C., 1974, Pennsylvanian sedimentation of central Appalachians, in Briggs, G. ed., Carboniferous of the southeastern United States: Geological Society of America Special Paper 148, p. 47–78.

Donaldson, A.C., 1979, Depositional environments of the Upper Pennsylvanian Series, in Englund, K.J., Arndt, H.H. and Henry, T.W., eds., Proposed Pennsylvanian System stratotype, Virginia and West Virginia. International Congress of Carboniferous Stratigraphy and Geology, 9th, Guidebook for Field Trip No. 1: Falls Church, Virginia, American Geological Institute Selected Guidebook Series no. 1, p. 123–131.

Donaldson, A.C., and Shumaker, R.C., 1979, Late Paleozoic molasse of the central Appalachians, in Donaldson, A.C., Presley, M.W. and Renton, J.C., eds., Carboniferous coal short course and guidebook. West Virginia Geological and Economic Survey Bulletin B-37-3, p. 1–42.

Donaldson, A.C., and Shumaker, R.C., 1981, Late Paleozoic molasse of the central Appalachians, in Miall, A.D., ed., Sedimentation and tectonics in alluvial basins: Geological Association of Canada Special Paper 23, p. 99–124.

Eble, C.F., 1985, Palynology and paleoecology of the Redstone coal (Upper Pennsylvanian) in West Virginia [M.S. thesis]: Morgantown, West Virginia University, 285 p.

Eble, C.F., Pierce, B.S., and Grady, W.C., 2003, Palynology, petrography and geochemistry of the Sewickley coal bed (Monongahela Group, Late Pennsylvanian), Northern Appalachian Basin, USA: International Journal of Coal Geology, v. 55, p. 187–204, doi: 10.1016/S0166-5162(03)00110-1.

Grady, W.C., and Eble, C.F., 1990, Relationships among macerals, miospores and paleoecology in a column of Redstone coal (Upper Pennsylvanian) from north-central West Virginia: International Journal of Coal Geology, v. 15, p. 1–26, doi: 10.1016/0166-5162(90)90061-3.

Gray, R.J., 1951, Microfossils and general stratigraphy of the Waynesburg coal [M.S. thesis]: Morgantown, West Virginia University, 128 p.

Habib, D., 1968, Spore and pollen paleoecology of the Redstone seam (Upper Pennsylvanian) of West Virginia: Micropaleontology, v. 14, p. 199–220.

Henry, T.W., Lyons, P.C., and Windolph, J.F., Jr., 1979, Upper Pennsylvanian and Lower Permian (?) series in the area of the proposed Pennsylvanian System stratotype, in Englund, K.J., Arndt, H.H., and Henry, T.W., eds., Proposed Pennsylvanian System stratotype, Virginia and West Virginia. International Congress of Carboniferous Stratigraphy and Geology, 9th, Guidebook for Field Trip No. 1: Falls Church, Virginia, American Geological Institute Selected Guidebook Series no. 1, p. 81–86.

Kosanke, R.M., 1943, The characteristic plant microfossils of the Pittsburgh and Pomeroy coals of Ohio: American Midland Naturalist, v. 29, p. 119–132.

Kosanke, R.M., 1988, Palynological analyses of Upper Pennsylvanian coal beds and adjacent strata of the proposed Pennsylvanian System stratotype in West Virginia. U.S. Geological Survey Professional Paper 1486, 24 p., 2 plates.

Peppers, R.A., 1996, Palynological correlation of major Pennsylvanian (Middle and Upper Carboniferous) chronostratigraphic boundaries in the Illinois and other basins. Geological Society of America Memoir 188, 111 p., 1 plate.

Phillips, T.L., Peppers, R.A., Avcin, M.J., and Laughnan, P.F., 1974, Fossil plants and coal: Patterns of change in Pennsylvanian coal swamps of the Illinois Basin: Science, v. 184, p. 1367–1369.

Phillips, T.L., Peppers, R.A., and DiMichele, W.A., 1985, Stratigraphic and interregional changes in Pennsylvanian coal-swamp vegetation: Environmental inferences: International Journal of Coal Geology, v. 5, p. 43–109, doi: 10.1016/0166-5162(85)90010-2.

Presley, M.W., 1979. Facies and depositional system of Upper Mississippian and Pennsylvanian strata in the central Appalachians, in Donaldson, A.C., Presley, M.W. and Renton, J.C., eds., Carboniferous coal short course and guidebook. West Virginia Geological and Economic Survey Bulletin B-37-1, p. 1–50.

Rothwell, G.W., 1991, *Botryopteris forensis* (Botryopteridaceae), a trunk epiphyte of the tree fern *Psaronius*: American Journal of Botany, v. 78, p. 782–788.

Stevenson, J.J., 1877, Report of progress in the Fayette and Westmoreland [Counties] district of the bituminous coal-fields of western Pennsylvania—Part 1, Eastern Allegheny County and Fayette and Westmoreland Counties west from Chestnut Ridge: Pennsylvania Geological Survey, 2nd series, KKK, viii, 437 p.

Stevenson, J.J., 1878. Report of progress in the Fayette and Westmoreland [Counties] district of the bituminous coal-fields of western Pennsylvania—Part 2, The Ligonier Valley: Pennsylvania Geological Survey, 2nd series, KKK, x, 331 p.

Tewalt, S.J., Ruppert, L.F., Bragg, L.J., Carlton, R.W., Brezinski, D.K., Wallack, R.N., and Butler, D.T., 2000, A digital resource model of the Upper Pennsylvanian Pittsburgh coal bed, Monongahela Group, northern Appalachian Basin coal region. U.S. Geological Survey Professional Paper 1625-C, 102 p.

Thiessen, R., 1930, The microscopic structure of coals of the Monongahela Series: West Virginia Academy of Science, v. 3, p. 159–198.

Thiessen, R., and Staud, J.N., 1923, Correlation of coal beds in the Monongahela Formation of Ohio, Pennsylvania and West Virginia: Pittsburgh, Carnegie Institute of Technology, Coal Mining Investigations Bulletin, v. 9, 64 p.

MANUSCRIPT ACCEPTED BY THE SOCIETY 28 JUNE 2005

Geological Society of America
Special Paper 399
2006

From wetlands to wet spots: Environmental tracking and the fate of Carboniferous elements in Early Permian tropical floras

William A. DiMichele*
Department of Paleobiology, National Museum of Natural History, Smithsonian Institution, Washington, D.C. 20560, USA

Neil J. Tabor
Department of Geological Sciences, Southern Methodist University, Dallas, Texas 75275, USA

Dan S. Chaney
Department of Paleobiology, National Museum of Natural History, Smithsonian Institution, Washington, D.C. 20560, USA

W. John Nelson
Illinois State Geological Survey, 615 East Peabody Drive, Champaign, Illinois 61820, USA

ABSTRACT

Diverse wetland vegetation flourished at the margins of the Midland Basin in north-central Texas during the Pennsylvanian Period. Extensive coastal swamps and an ever-wet, tropical climate supported lush growth of pteridosperm, marattialean fern, lycopsid, and calamite trees, and a wide array of ground cover and vines. As the Pennsylvanian passed into the Permian, the climate of the area became drier and more seasonal, the great swamps disappeared regionally, and aridity spread. The climatic inferences are based on changes in sedimentary patterns and paleosols as well as the general paleobotanical trends. The lithological patterns include a change from a diverse array of paleosols, including Histosols (ever-wet waterlogged soils), in the late Pennsylvanian to greatly diminished paleosol diversity with poorly developed Vertisols by the Early–Middle Permian transition. In addition, coal seams were present with wide areal distribution in the late Pennsylvanian whereas beds of evaporates were common by the end of the Early Permian. During this climatic transition, wetland plants were confined to shrinking "wet spots" found along permanent streams where the vegetation they constituted remained distinct if increasingly depauperate in terms of species richness. By Leonardian (late Early Permian) time, most of the landscape was dominated by plants adapted to seasonal drought and a deep water table. Wetland elements were reduced to scattered pockets, dominated primarily by weedy forms and riparian specialists tolerant of flooding and burial. By the Middle Permian, even these small wetland pockets had disappeared from the region.

Keywords: climate change, paleosols, Permian, tropics, wetlands.

---

*E-mail: dimichel@si.edu

DiMichele, W.A., Tabor, N.J., Chaney, D.S., and Nelson, W.J., 2006, From wetlands to wet spots: Environmental tracking and the fate of Carboniferous elements in Early Permian tropical floras, *in* Greb, S.F., and DiMichele, W.A., Wetlands through time: Geological Society of America Special Paper 399, p. 223–248, doi: 10.1130/2006.2399(11). For permission to copy, contact editing@geosociety.org. ©2006 Geological Society of America. All rights reserved.

INTRODUCTION

The fossil plant record of the Late Paleozoic tropics suggests a climate change from ever-wet during the Pennsylvanian to seasonally dry during the Early Permian. The floras characteristic of these different climatic conditions share few species and represent different biomes or species pools (Broutin et al., 1990; DiMichele and Aronson, 1992; Falcon-Lang, 2003): wetland, seasonally dry, or arid with a wet season. The wetland flora was dominated by lycopsids, ferns, and primitive seed plants, and first assembled in the tropics during the late Mississippian (Namurian) (Pfefferkorn et al., 2000), drawing on earlier wetlands of generally similar structure but composition that differed to varying degrees (e.g., Scheckler, 1986; Falcon-Lang, 2004a). It became progressively more areally restricted in most of the tropical belt particularly during the Permian, but in Cathaysia, present-day China representing the eastern-most parts of the tropical belt, climates remained ever-wet (Rees et al., 2002) and wetland floras very similar to those of the middle Late Carboniferous continued to survive well into the Permian (Guo, 1990; Tian et al., 1996; Hilton et al., 2001).

Floras characteristic of habitats inferred from independent sources (such as sedimentology or paleosols) to be seasonally dry existed prior to the Pennsylvanian (see for example, Scott et al., 1984) but were composed of a different flora than the one that appears in tropical seasonally dry habitats of Pennsylvanian age. This latter flora was dominated by evolutionarily derived seed plants and appears in Euramerica beginning in the early part of the Pennsylvanian (Lyons and Darrah, 1989; Falcon-Lang et al., 2004). Evidence of this derived flora occurs initially as rare and isolated deposits in red bed sequences or as fragmentary material transported into basins from nearby upland environments, and occurs only much later as well-preserved macrofossils (Cridland and Morris, 1963; Broutin et al., 1990; DiMichele and Aronson, 1992). This pattern suggests that seasonally dry floras evolved in tropical upland areas or paratropical regions (Zhou, 1994) and migrated into tropical basinal lowlands during times when these lowland basins began to experience seasonal moisture limitation. During the early and middle Pennsylvanian, such times of seasonality may have been largely during the drier periods of glacial-interglacial cycles (Cecil, 1990; Falcon-Lang, 2003, 2004b). Initially, this flora was not as highly divergent as it would ultimately be when seasonal drought became more widespread and presumably more severe in the late Pennsylvanian and Permian. The third biome, that of very dry environments with only a short wet season or seasons, was also present in the tropics and makes its first appearance near the end of the Early Permian, again in what is now western North America, which in the Permian was at the western end of the tropical landmass (DiMichele et al., 2001).

The objective of this paper is to document the final phases of this widespread vegetational change and to interpret it dynamically within an ecological context. We find that opportunistic wetland plants, because of their architecture and reproductive biology, are those most likely to survive in the fragmented wetter parts of dry landscapes. Dynamically direct resource competition among plants from wetland and seasonally dry habitats seems to play a minor part, if any at all, in the overall pattern of floristic change. Rather, environmental tracking of climatic conditions dominates the stratigraphic and spatial patterns of floristic distribution.

The pattern of change is revealed especially well in north-central Texas (Fig. 1), which preserves an excellent record of plants from the Virgilian (late Pennsylvanian/Stephanian) through the end of the Early Permian. The following synopsis is based upon surface and subsurface studies that began in the early 1990s and that now encompass ~41,000 km^2 and 15 counties in north-central Texas. This effort started as a reconnaissance of plant localities worked previously by field parties of the 1940–1941 Clay County Unit of the State-Wide Paleontologic-Mineralogic Survey (Works Projects Administration [WPA] collections, Texas Memorial Museum) and by our colleague S.H. Mamay, as well as other plant sites found by Chaney, R.W. Hook, and the late Nicholas Hotton III in the course of fossil vertebrate prospecting. From the outset, this work was interdisciplinary in nature with particular regard for the depositional context and stratigraphic position of fossil plant occurrences.

GEOLOGIC SETTING AND STRATIGRAPHY

Most of our study area lies on the Eastern Shelf of the Midland Basin. During Permian time, this region lay near the southern margin of the North American craton, inland from the Ouachita Mountains, which had arisen during Late Carboniferous time. Except for minor, intermittent fault movements along the Red River-Matador uplift (Brister et al., 2002), the Eastern Shelf was tectonically stable during the Permian. Primary sources of clastic sediment to the Eastern Shelf were the Ouachita Mountains to the east and the Arbuckle and Wichita Mountains to the northeast (Smith 1974; Hentz 1988). Gradient of the Eastern Shelf was extremely gentle, as shown by great lateral continuity of thin carbonate and gypsum beds.

Facies on the Eastern Shelf grade from terrestrial red beds on the northeast, near the source terranes, to shallow marine on the southwest approaching the Midland Basin. The outcrop belt crosses this facies trend obliquely, so that southern outcrops are generally more marine than northern. Thus, red beds of the Bowie and Wichita Groups of north-central Texas intergrade southward with marine shale and carbonate rocks of the Cisco and Albany Groups, respectively (Hentz 1988). In similar fashion, the Clear Fork Group of north-central Texas is entirely continental, yielding terrestrial plant and vertebrate fossils. Southward, interbeds of marine limestone and dolomite increase in number and thickness (Olson 1958, 1989; Olson and Mead 1982). The Pease River Group comprises two units: the San Angelo Formation of sandstone and conglomerate and the overlying Blaine Formation of interbedded mudstone, carbonate rock, and gypsum representing shallow marine and coastal sabkha settings. The San Angelo comprises an upward-fining clastic wedge derived from the Llano Uplift, which merges with a similar but smaller wedge

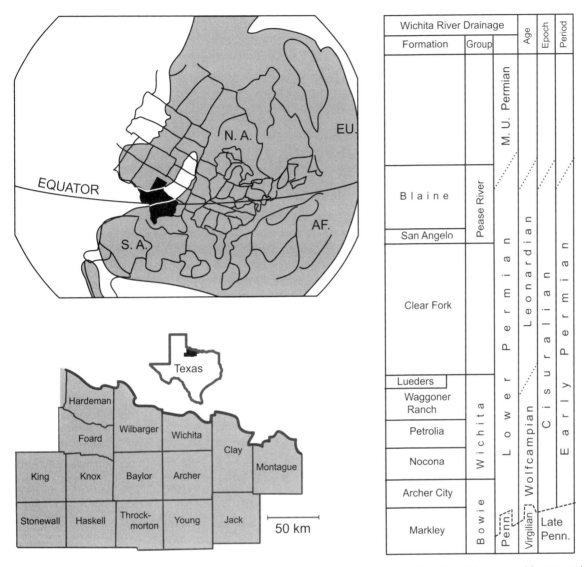

Figure 1. Study area in paleogeographic, modern geographic, and stratigraphic context. Top left: Paleogeographic map projection of Early Permian (from Christopher Scotese, Paleomap Project): N.A.—North America, EU.—Europe, S.A.—South America, AF.—Africa. Land areas shown in shading; oceanic areas in white. Texas is shown in black. Bottom left: Texas with study area in north-central region shown in black; counties shown below. Right: Stratigraphic nomenclature as used in this paper (refer to text).

from the Wichita Mountains. The upper San Angelo intergrades with the Blaine; sandstones pinch out in subsurface a short distance west of the outcrop (Franklin 1951; Mear 1963 and 1984; Smith 1974).

Age determinations on the Eastern Shelf are based on marine fossils, principally fusulinids, ammonoids, and nautiloids. Most authors place the Carboniferous-Permian boundary within the Harpersville Formation of the Cisco Group, correlating with the Markley Formation of the Bowie Group (Roth, 1931; Dunbar and Skinner, 1937; Plummer and Scott, 1937; Henbest, 1938; Miller and Youngquist, 1947). The Wolfcampian-Leonardian boundary apparently lies close to the Elm Creek Limestone in the Albany Group, which equates to the basal Petrolia Formation in the Wichita Group (Bose, 1917; Sellards, 1932; Plummer and Scott, 1937; Miller and Youngquist, 1947; Hentz and Brown, 1987). This horizon corresponds to the first appearance of gigantopterid plants in north-central Texas (Read and Mamay, 1964). The Leonardian-Guadalupian boundary on the Eastern Shelf is rather controversial, being based on scattered ammonoid finds in the Blaine. The weight of evidence favors placing this boundary within the Blaine (DiMichele et al. 2004).

PALEOCLIMATIC TRENDS

Paleoclimatic conditions in the study area have been inferred from four lines of evidence: (1) the nature of the primary sedimentary deposits, (2) the character of the paleosols, (3) the taxonomy and physiognomy of fossil plants, and (4) animal fossils, including trace fossils such as burrows and aestivation assemblages. Sedimentary patterns integrate climate and regional patterns of drainage with tectonics. Paleosols reflect local topographic and drainage signatures overprinted by regional climatic conditions. Plant fossils, especially those that were transported only short distances prior to burial and fossilization, reflect the local conditions proximal to the site of deposition.

Based on all these lines of evidence, we infer a trend toward generally drier conditions from the late Pennsylvanian through the end of the Early Permian in the north-central Texas study area. These changes were neither gradual nor continuous. There are some clear, relatively abrupt, directional changes in the succession and some intervals during which climate appears to have fluctuated repeatedly and, on average, non-directionally.

Lithologic Patterns

Bowie Group

The Bowie Group encompasses the Markley and Archer City Formations. The Markley Formation (Figs. 2A and 2B) is composed mainly of mudstone and sandstone, along with minor deposits of conglomerate, limestone, coal, and underclay. Fluvial sandstones are a prominent part of the Markley Formation, both at the surface and in the shallow subsurface. Braided fluvial channels are a common part of these deposits with a change to meandering fluvial systems up-section and westward (Galloway

Figure 2. Upper Pennsylvanian outcrops showing basic sedimentological features. (A) "Walker" locality (NMNH 40007), Markley Formation, Pennsylvanian-Permian transition. Note shallow U shape of the deposit; paleosol at base. (B) "Maxey Ranch" locality (NMNH 40602), Markley Formation, Pennsylvanian-Permian transition. Details of typical lithological succession outcrops of this age and type. 1—Basal paleosol of Pedotype B. 2—Kaolinite layer at top of Pedotype B typical of lithologies enclosing a seasonally dry plant assemblage. 3—Organic shale layer, Pedotype A, containing a plant assemblage typical of swampy, ever-wet conditions. 4—Mudstones and sandstones deposited in a flood basin setting enclosing a plant assemblage of ever-wet conditions.

and Brown, 1972; Hentz and Brown, 1987; Hentz, 1988; Brown et al., 1987, 1990; McGowen et al., 1967 [1991]).

The Archer City and Markley Formations are very similar lithologically (Hentz and Brown, 1987). In general, fluvial sandstones of the Archer City Formation represent multistoried, high-sinuosity meandering channels (Sander, 1989). As detailed below, most of the Archer City mudstones exhibit paleosol characteristics typical of younger formations. Pedogenic carbonates first occur in this formation, and only a few exposures include organic shales rich in plant fossils, indicative of swamp conditions.

The widespread occurrence of bituminous coal in the Bowie Group is critical to an understanding of paleoclimatic conditions that prevailed during the latest Carboniferous and earliest Permian of this region. Coals within the Markley Formation have been mapped on outcrop from near the Red River in the Colorado River Valley, a distance of ~300 km, and from well data nearly 140 km westward in the subsurface (Mapel, 1967), indicating long periods of time during which climate was wet enough for peat formation to occur over large areas.

The distribution of Bowie/Cisco sandstones in the subsurface, as shown by well-log analysis, is a second point of paleoenvironmental importance. Sandstones exposed on outcrop have little basinward persistence. Sands appear to have become impounded in the vast, low-energy coastal plain that existed along the margin of the Eastern Shelf, which corresponds to sediment-starved conditions in the Midland Basin. This pattern continued through the Wolfcampian and Leonardian, even during periods of elevated clastic influx in the lower Clear Fork and basal Pease River Groups, and presaged the development of coastal evaporite pans in the Leonardian.

Wichita Group and Lueders Formation

The Wichita Group and Lueders Formation, in gross sedimentological character, comprise a large-scale fining-upward sequence that records a shift from alluvial upper coastal plain environments to lower coastal plain settings (Hentz 1988). That is to say, the coastline advanced landward transgressively over time.

Relatively few plant-bearing localities have been found in the Nocona Formation. This impoverished record contrasts with the abundance of vertebrate localities known from these same strata (Hook, 1989) and may be a taphonomic effect, the lack of plant preservation obscuring the true pattern of plant occurrences at that time.

Numerous plant deposits have been found in the overlying Petrolia Formation (Fig. 3B), dating back to some of the earliest collectors in the area (e.g., White, 1911). Additional plant localities occur in equivalent formations of the northern Albany Group in southern Baylor County; foremost among these is the Emily Irish locality (Mamay, 1968). The Petrolia Formation is mudstone dominated, and most sandstones in the southwestern portion of the outcrop represent single storey, suspended-load deposits of high-sinuosity fluvial channels. Local beds of carbonate-pebble and mudstone-clast conglomerate that contain mixed assemblages of fragmentary marine invertebrates and freshwater to terrestrial vertebrates are thought to represent supratidal storm deposits (Parrish, 1978; Hentz, 1988). Channel-fill deposits may contain a variety of animal fossils, some suggestive of brackish-water, especially in the lower and upper parts of the formation, and thus possibly tidal channels. Included are the following seen by us and reported by Hentz (1988): sharks and bony fishes of both freshwater and marine types; spirorbid and serpulid worms, various pelecypods including *Pinna*, and pecenoids; nautiloids and ammonoids; scraps of tetrapod bones—thus, a mixture of nonmarine, brackish, and fully marine life forms, although most of the latter might have been washed or driven ashore during storms. Unequivocal evidence of tidal sedimentation is lacking.

In the subsurface, the middle to upper part of the Petrolia Formation and equivalent rocks of the Albany Group include extensive deposits of dolomite and bedded anhydrite, or, at shallow depths, gypsum (Moore, 1949), which split and thin eastward into mudstones and pinch out in western Throckmorton and central Baylor Counties (Abilene Geological Society, 1949, 1953) and coincide with increased pedogenic carbonates and decreased fluvial sedimentation. This represents the first occurrence of widespread evaporites in the lower coastal plain, a facies pattern that recurs throughout Leonardian time in the region.

The Waggoner Ranch Formation includes brackish to possibly marine invertebrates and vertebrates in restricted channel-form lenses that may represent tidal deposits similar to those of the underlying Petrolia Formation (Read, 1943; Johnson, 1980).

Although it is a thin unit (15–25 m), the Lueders Formation of north-central Baylor County has attracted considerable paleontological interest (Berman, 1970; Dalquest and Kocurko, 1986, 1989). Plant, invertebrate, and vertebrate remains occur in small channel deposits. We tentatively interpret these as tidal channels, although they lack a clear lithological signature of tidal deposition, on the basis of the following evidence: brackish-water invertebrates together with terrestrial plants and tetrapods; channels laterally correlative to or intercalated with carbonate rocks containing marine invertebrates such as cephalopods.

At least two failed vertebrate aestivation assemblages, i.e., colonies of dozens to hundreds of calcified burrows containing skeletal remains of their inhabitants, occur in the Lueders Formation; "failed" aestivation ends in death for the animal, suggesting, especially when found in large numbers, that conditions suitable for reemergence did not reappear in a timely manner. More specifically, prolonged drought probably killed these animals in their burrows. In north-central Texas, such concentrations of lungfish and lysorophoid amphibian skeletons were generally thought to be limited to the Clear Fork Formation, but we verified in the field that the aestivation assemblages reported by Romer and Olson (1954) and Dalquest and Carpenter (1975) occur, respectively, in the uppermost and lowermost Lueders Formation. These older Leonardian aestivation assemblages reiterate the development of seasonal-moisture stress in coastal plain environments as indicated by paleosol data.

Figure 3. Lower Permian outcrops showing basic sedimentological features. (A) "North of Cedar Top" locality (NMNH 40972), middle Clear Fork Formation, Lower Permian. Note U shape of deposit and laminated, claystone fill. (B) "Red Hollow" locality (NMNH 40032), Petrolia Formation, Lower Permian. Paleosol at base truncated by active channel with features include trough cross beds, erosional contacts and multistory architecture. Plant fossils occur in swales of cross beds.

Clear Fork Formation

In this paper we classify the Clear Fork in north-central Texas as a formation that is informally divided into three parts, as outlined by Nelson et al. (2001). The lower 55–70 m in the general area of Lake Kemp includes a thin but widespread (<0.5 m) dolomite bed and two widespread fluvial channel belts. The dolomite resembles intertidal to supratidal dolomites of the underlying Wichita/Albany Groups. The lower sandstone belt consists of multistory, suspended-load deposits of a high-sinuosity, meandering system. Numerous, claystone-rich channel fills (Fig. 3A) within this lithofacies have yielded the majority of lower Clear Fork plant and vertebrate remains. In contrast, the upper sandstone belt is coarser grained, contains no significant fossiliferous channel fills, and is distinguished by large-scale bedforms characteristic of high-energy braided fluvial channels. Lower Clear Fork mudstones generally are reddish-brown and exhibit well-developed paleosol features described below. Other than failed aestivation assemblages (Olson and Bolles, 1975), fossil remains are exceedingly sparse in these well-drained flood-plain facies.

The middle 125–160-m-thick portion of the Clear Fork includes three mappable fluvial channel belts, all of which are similar sedimentologically and paleontologically to the high-sinuosity, suspended-load fluvial channel belt of the lower Clear Fork (Edwards et al. 1983) (Fig. 4). Upward in the middle Clear Fork, pedogenic carbonates gradually diminish and gypsum nodules, veins, and stringers increase in both vertical distribution and lateral persistence. Failed vertebrate aestivation assemblages occur intermittently through the middle Clear Fork section.

Figure 4. Lower Permian outcrop showing basic sedimentological features. "North Fork Pens" locality, middle Clear Fork Formation, Lower Permian, illustrating large scale accretion beds deposited in active channel and containing fossil plants.

The upper 165–180 m of the Clear Fork Group is a monotonous succession of reddish-brown, gypsum-rich mudstones that is nearly devoid of significant sandstone deposits. A few fossil localities are represented by channel-fills and an aestivation assemblage in the lowermost part of the upper section, but the remainder is nonfossiliferous (Olson, 1958; Murry and Johnson, 1987). The combination of dolomite and bedded anhydrite is common in the subsurface and indicates primary evaporites in the lower coastal plain.

Pease River Group

The Pease River Group consists of two formations in north-central Texas, the San Angelo and Blaine. In its lower portions the San Angelo consists of sandstones that Smith (1974) reported to be of a deltaic origin. Examination of these and other exposures indicate that sandstones of the San Angelo Formation are fluvial and have bedding characteristics of broad, shallow, multi-channel systems in some outcrops, and of incised, suspended-load, meandering streams in others. The upper half of the San Angelo Formation is composed mainly of red mudstones, small, discontinuous tabular and channel sandstones, and gypsum nodules. The contact with the overlying Blaine Formation is gradational and placed by convention at the lowest occurrence of bedded gypsum (Hentz and Brown, 1987). The Blaine Formation is composed chiefly of reddish-brown and gray-green siltstones, gypsum beds up to ~10 m thick, and regionally persistent dolomite beds. Local concentrations of marine invertebrates (gastropods, pelecypods, and cephalopods) preserved in some dolomite beds may represent storm-washed deposits (Clifton, 1944; Jones and Hentz, 1988). Rhythmically interbedded mudstones, dolomites, and gypsum in the lower to middle Blaine Formation have been interpreted as a combination of tidal flat and sabkha deposits (Smith, 1974). Bedded salt deposits in excess of 2 m thick occur in the upper part of the Blaine Formation in the subsurface of western King and Stonewall Counties.

An impoverished and generally fragmentary record of almost exclusively terrestrial vertebrates is known from the San Angelo of our study area (Olson, 1962); large herbivores are a significant part of this record. No terrestrial vertebrates have been reported from the Blaine Formation of Texas. As noted below, fossil plants also are rare in the Pease River Group. A series of presumed tidal-channel deposits near the San Angelo-Blaine contact contains a precocious flora that is distinct from all other Permian records in the region (DiMichele et al., 2001). Some 164 m higher in the section, a single upper Blaine deposit that also originated as a coastal plain channel has produced a flora that, again, is unique, though of low diversity (DiMichele et al., 2004). Nearly all the Pease River plant occurrences are accompanied by local concentrations of copper minerals in partially pyritized fossil wood.

Paleosol and Geochemical Evidence and Patterns

The characteristics and stratigraphic distribution of soil types, which we refer to here as "pedotypes" (sensu Retallack 1994), are perhaps the most direct indicator of moisture availability in terrestrial paleoenvironments. Virgilian through Leonardian rocks of north-central Texas include eight major pedotypes (soil types), designated A through H, and defined by paleosol (ancient soil) structure (horizonation, structure, fabric, and color), mineralogy, and chemical characteristics (Figs. 5–7). The following descriptions and interpretations are summarized from the detailed treatment of Tabor and Montañez (2004).

Figure 5. Paleosol types from study area. (A) Pedotype A above Pedotype B at "Cooper" locality (NMNH 39991), Markley Formation, upper Pennsylvanian. (B) Pedotype A above Pedotype B at "Cooper" locality (NMNH 39991), Markley Formation, upper Pennsylvanian. (C) Slickensides typical of Pedotype D and Pedotype G, Nocona Formation stratotype, Lower Permian. (D) Clastic dykes typical of Pedotype D and Pedotype G, Nocona Formation stratotype, Lower Permian.

Paleosols are an important component of Virgilian through Leonardian strata of the Eastern Shelf. Pedotypes A through D occur only in the Virgilian and lower Wolfcampian Bowie Group. These paleosols have morphological, mineralogical, and chemical characteristics indicative of ever-wet to seasonally dry conditions. In particular Pedotypes A–D preserve redoximorphic features, which are recognized by "redox depletions" and "redox concentrations." Specifically, redox depletions refer to drab gray, yellow, and green colors of the paleosol matrix and/or mottles, whereas redoximorphic concentrations are nodular or vermicular concentrations of red, orange, and bright yellow (e.g., hematite, goethite, jarosite; e.g., Vepraskas 1994). In modern soils, redoximorphic features form in seasonally saturated profiles through removal of Fe and Mn from areas of low Eh (redox depletions) and re-precipitation as Fe- and Mn-oxides (redox concentrations) in better-oxidized areas. Gley matrix colors indicate reduced conditions, which are typical of relatively prolonged saturation (25%–50% of the year), whereas yellow-brown to reddish mottles record seasonal soil drying (Daniels et al., 1971; Duchaufour, 1982). By analogy to modern soils, paleosol redoximorphic features are interpreted to have formed in seasonally saturated portions of the profile with sufficient organic content to yield reducing conditions (see Pipujol and Buurman 1994). Preservation of these features likely reflects alternating episodes of soil saturation and aeration (Soil Survey Staff, 1975, 1998).

Pedotypes F through H occur only in Permian (Wolfcampian and lower Leonardian) rocks within the study area. These paleosols have a range of morphological, mineralogical, and chemical characteristics indicative of seasonally dry to seasonally wet moisture budgets that reflect a seasonal climate regime. Both the morphological diversity and stratigraphic occurrence of paleosols are greatly diminished in the late Leonardian San Angelo and Blaine Formations. The vast majority of the floodplain facies in these formations exhibit little or no diagnostic evidence of pedogenesis. The most common profiles at this stratigraphic level are weakly developed, thin (<50 cm) type E paleosols, which have characteristics indicative of a shallow fluctuating water table leading to alternating oxidation and reduction of iron compounds and associated mottling. In addition, these paleosols are observed typically as weakly rooted horizons within channel sandstones that apparently developed along banks of fluvial and/or tidal channels.

Pedotype A (Histosols)

Pedotype A (Figs. 5A, 5B, 8, 9) is characterized by an upper organic layer of weakly degraded plant material and an underlying fine angular blocky kaolinitic claystone that contains abundant, thin root traces that are relatively shallow in the profile (~10 cm) and have a tabular orientation. Type A paleosols occur in the Markley Formation. They are developed within laminated to thin-bedded claystones and silty claystones that, on outcrop appear to be shallow depressions on poorly drained flood plains and coastal plains. Bituminous coal to highly carbonaceous shale forms the organic-rich portion of these paleosols, which may

Figure 6. Paleosol types from study area. (A) Rhizolith in Pedotype E, Wellington Formation (Oklahoma), Lower Permian. (B) Pedotype G at "Grayback" locality (NMNH 40059), Lueders Formation, Lower Permian. (C) Pedotype H at "Wichita River Cutbank" locality, Waggoner Ranch Formation, Lower Permian. (D) Rhizolith in Pedotype H, near "FM 1919 Lungfish" locality, middle Clear Fork Formation, Lower Permian.

Figure 7. Paleosol from study area. Pedotype H with calcic horizon, near Little Moonshine Creek, upper Waggoner Ranch Formation, Lower Permian.

have been similar to modern, peat-bearing Histosols (Soil Survey Staff, 1998). Paleosols of this type are good indicators of regional climate when they are observed to extend over wide geographic areas. Small isolated pockets of this pedotype might indicate a micro-environment on a landscape.

Pedotype B (Ultisols)

Pedotype B (Figs. 5A, 5B, 8, 9) consists of two intergradational units, an upper highly differentiated kaolinitic claystone that has an angular blocky structure with clay coatings, and a lower massive claystone cemented by iron and manganese oxides with redoximorphic features. Type B paleosols are found in the Markley Formation, stratigraphically associated with Pedotypes A and C, and are developed within mudstone and claystone deposits of flood-plain facies. The upper portion of the claystone is interpreted as an argillic (clay-rich) horizon, whereas the lower, iron-cemented claystone is interpreted as a plinthite horizon. The redoximorphic features and plinthite horizon record fluctuation of a relatively shallow water table and saturation of soil horizons at depth (Duchaufour, 1982; Soil Survey Staff, 1998); moreover, the argillic (Bt) horizon of Type B paleosols requires periodically well-drained conditions.

As modern parallels of periodic soil drainage, consider the flood plains along the Mekong, Yellow, Missouri, Mississippi, Nile, Amazon, and every other pre-historical flood plain that did not have a natural dam. During the rainy season, or after large, individual rainfall events, rivers overtop their levees and flood extensive regions of their flood plains, saturating soils. Such flooding leads to soil anoxia with relatively little physical transport of solids within the soils. After flooding has waned and soils have drained, subsequent rainfall may pass through the soil profile by way of percolating, free drainage. Under these conditions, physical transport of solids and chemical leaching occurs and may lead to the formation of argillic horizons. This scenario is common in nature and is explicitly related to dynamics of the groundwater table.

Pedotype B is interpreted as an Ultisol that formed upon relatively better drained local topographic highs (or more permeable substrates) within Permo-Carboniferous flood plains and coastal plains, possibly pedogenesis upon stream levees or stream terraces (see Markewich and Pavich, 1991).

Pedotype C (Inceptisols)

Pedotype C (Figs. 8, 9) is a highly differentiated mudstone with weak, angular, blocky structure and a plinthite horizon at or very near the presumed upper surface of the paleosol. Redoximorphic features are present in the upper portions of these paleosols. Deeper portions are massive and grade downward to weakly laminated, green to brown mudstone. Type C paleosols occur primarily in the Markley Formation and into the basal most parts of the Archer City Formation. They are developed within mudstones and claystones of flood-plain facies and are generally stratigraphically associated with channel sandstones. The presence of redoximorphic features and plinthite horizons near the surface of the paleosol record a shallow, fluctuating paleo-groundwater table and generally poorly drained conditions during soil formation (see Duchaufour, 1982). Pedotype C paleosols are interpreted to have been inceptisols that likely developed upon flood-plain deposits near fluvial channels, but intervals between floods were long enough to permit development of soil structure and horizonation (Kraus, 1987; Wright and Marriott, 1996). Although this pedotype is broadly distributed across the Pennsylvanian flood plains, its morphological characteristics are attributed to local, micro-environmental controls related to a shallow, fluctuating groundwater table.

Pedotype D (Vertisols)

Pedotype D (Figs. 5C, 5D, 8, 9) consists of red smectitic (expandable clays) mudstone that grades downward to gray smectitic mudstone with wedge-shape aggregate structure, slickensides, and sand-filled clastic dikes. Redoximorphic features are present throughout the profile. Type D paleosols occur in the Markley Formation in association with type E paleosols (above Sandstone (ss) 13 [Hentz, 1988] on Figs. 2, 5C, 5D).

Type D paleosols are developed within mudstone intercalated with cross-laminated sandstones of the braid-bar facies, reflecting their restricted distribution in the western portion of the study area (Hentz, 1988). The wedge-shaped structure, slickensides, and clastic dikes are all indicative of strongly contrasted seasonal wetting and drying, typical of Mediterranean or monsoon-type climates (Soil Survey Staff, 1998). However, the presence of redoximorphic features throughout these paleosols indicates moist, aquic (high water table) conditions for extended periods punctuated by only short intervals of soil-moisture deficit (drought, or drying). Type D paleosols are interpreted as aquerts that were strongly influenced by local factors such as a shallow water table. Nevertheless, this characteristic morphology is indicative of strongly contrasted, seasonal changes in moisture availability upon portions of the Pennsylvanian flood plains.

Pedotype E (Entisols)

Pedotype E (Figs. 6A, 8, 9) is composed of poorly differentiated, massive sandstone with several superimposed horizons of vertical and laminar, iron-oxide-rich root traces and hematitic nodules. These paleosols are the most common and widely distributed of those found in the north-central Texas section, occurring throughout the upper Pennsylvanian and Lower Permian.

The most diagnostic evidence of soil formation typically found in these weakly developed paleosols is rooting structures and rhizoliths (e.g., Fig. 6A). They are developed in (1) finely laminated or bioturbated mudstones that cap cross-laminated, fine-grained sandstones of levee, crevasse splay, and proximal lacustrine deposits and (2) sandstones of levee and crevasse splay deposits. The great majority of type E paleosols in the Bowie Group exhibit redoximorphic features. In the Permian section, these paleosols are the only pedotype that may be gleyed, meaning soils that exhibit mottled drab yellow, gray, and green colors due to intermittent water saturation and the associated oxidation and reduction of iron compounds. These characteristics are indicative of generally very wet conditions likely associated with a shallow water table. The development of these paleosols is controlled by the availability of shallow groundwater, even in driest times. Such conditions reflect "humid" micro-environmental conditions upon the paleolandscape. Although this pedotype is observed throughout the Permo-Pennsylvanian strata of north-central Texas, it is particularly important to note that such paleosols are observed in strata with independent lithological and geochemical evidence that is indicative of regionally dry climate (Tabor et al., 2002; Tabor and Montañez, 2004). These humid micro-environments are most likely the paleosols of "wet spots" upon the paleolandscape and therefore may have provided the substrates upon which many of the fossil plants grew in the Lower–Middle Permian strata of north-central Texas.

Pedotype F (Alfisols)

Pedotype F (Figs. 8, 9) is a highly differentiated angular blocky red mudstone with clay skins in the upper horizons and calcium carbonate nodules and vertically oriented calcareous rhizoliths (root casts or traces replaced by calcite) at depth. Type F paleosols are found throughout the Lower Permian strata of the

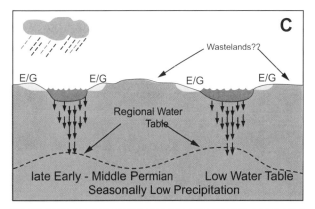

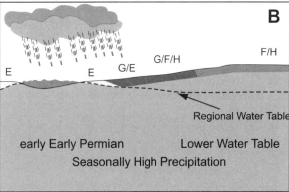

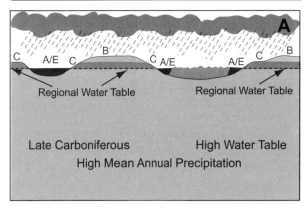

Figure 8. Diagrammatic representation of pedotype development on the landscape in relation to groundwater supply. Pedotypes marked by letters. (A) Late Pennsylvanian ever-wet conditions. (B) Early Permian seasonally dry conditions. (C) Middle Permian with short wet season.

study area. They occur most commonly in the mud-rich strata of the flood-plain facies.

Pedotype F compares most favorably to Modern Alfisols, which form on stable, well-drained portions of the landscape in subhumid to semiarid climates (Franzmeier et al., 1985; Retallack, 1990); Alfisols typically do not occur in climates characterized by exceedingly high rainfall (~1200 mm/yr) and extensive hydrolysis of base-rich parent material and soil (Buol et al., 1997; Soil Survey Staff, 1975). They are a typical component of tropical landscapes with limited rainfall (Wilson, 1999). The

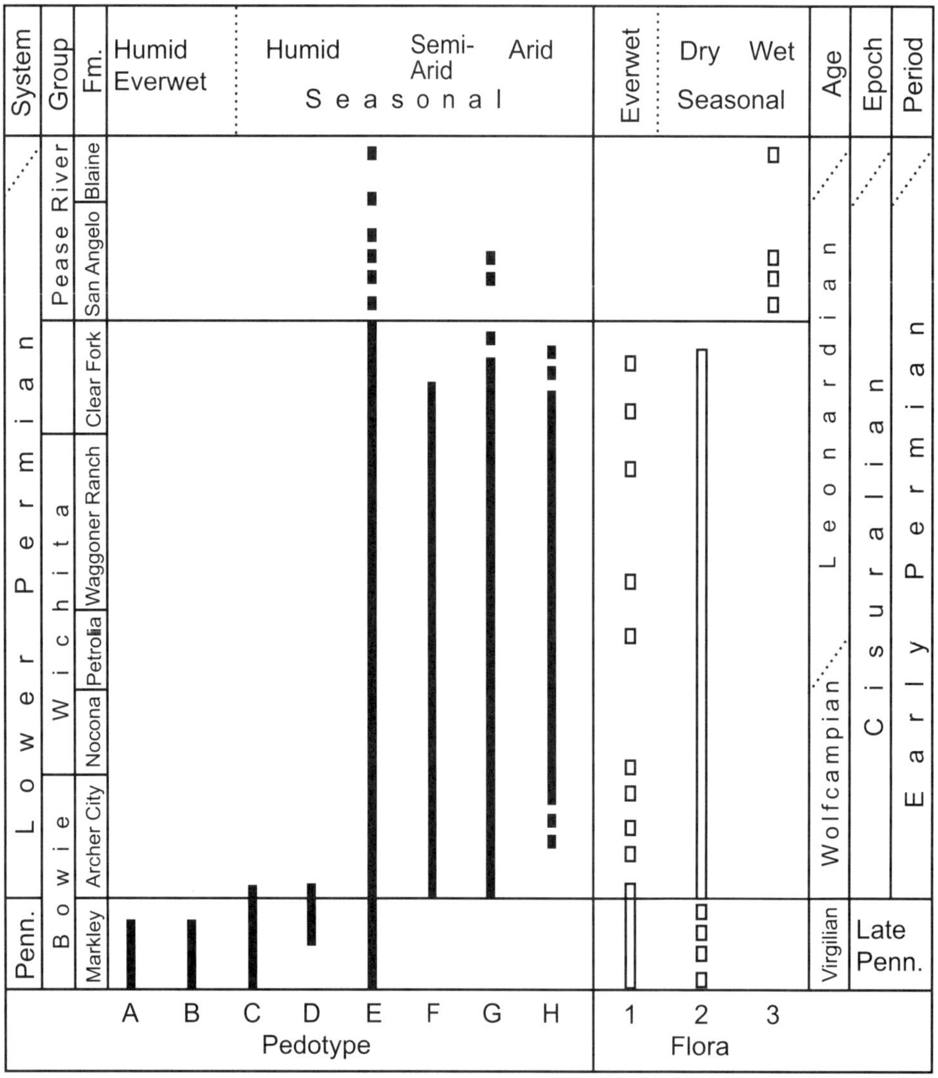

Figure 9. Stratigraphic distributions of pedotypes and paleofloras. Formational thicknesses not to scale.

dominant pedogenic process in the Alfisol order is translocation of expansible 2:1 phyllosilicates (e.g., smectite). Translocation of these clay-size particles is facilitated by leaching of Ca^{2+} from phyllosilicate interlayers. Type F (Alfisols) are typically leached of carbonate above and within the upper layers of argillic horizons, further indicating that these paleosols formed in a manner analogous to modern Alfisols. Pedogenic carbonate beneath noncalcareous argillic horizons in type F (Alfisols) likely indicates formation in a xeric, or semiarid, dry soil moisture regime, characterized by incomplete soil leaching due to seasonal drying (Buol et al., 1997; Soil Survey Staff, 1975).

We interpret type F (Alfisols) to have developed upon stable topographic highs of the interchannel flood-plain facies. This interpretation is based on the fine-grained nature of the sedimentary strata associated with the profiles, and features such as argillic horizons, that formed upon older, more stable portions of the landscape (Soil Survey Staff, 1975; Buol et al., 1997). Profiles were apparently well drained and formed well above the water table, judging from (1) the presence of argillic horizons and (2) oxidized paleosol matrix colors without redoximorphic features.

In contrast to Pedotype E, Pedotype F paleosols can be regionally extensive and are reasonably good indicators of regional climate. Although these paleosols do indicate an abundant flora (e.g., Retallack, 1990) during soil formation, they are also indicative of stable landscapes that are distal from major depocenters. In this regard, the floras associated with these paleosols are not likely preserved as part of the fossil record.

Type G (Vertisols)

Type G paleosols (Figs. 6B, 8, 9) have highly differentiated profiles with red to brown claystone, wedge-shaped aggregate structure, slickensides, and calcium carbonate nodules. Smectite is the dominant mineral in the <2 μm fraction of this profile, with trace amounts of kaolinite and weathered micas. Type G paleosols are found throughout the Lower Permian, from the Archer City through the Clear Fork Formations and sporadically in the San Angelo Formation. They developed in fine-grained muds of the flood-plain facies, reflecting their widespread distribution in the study area during Early Permian time.

Pedotype G is interpreted as a Vertisol, based on evidence of cracking in the form of clastic dikes and slickensides with wedge-shaped aggregates between 25 cm and 1 m depth (Soil Survey Staff 1975; 1998). Type G (Vertisols) exhibit two major differences from type D (Vertisols): (1) accumulation of pedogenic carbonate and (2) lack of redox features (Fig. 6B). Paleosol matrix colors in type G (Vertisols) indicate that these profiles formed under oxidized conditions well above the paleo–water table. The development of pedogenic slickensides records periods of soil saturation that allowed for periodic shrinking and swelling of expandable smectite clays. We interpret type G (Vertisols) to have developed on seasonally flooded, low-lying portions of the lower coastal plain province in positions proximal to stream channels or in internally drained, shallow depressions in the upper coastal plain and piedmont settings.

Type H (Inceptisols)

Type H paleosols (Figs. 6C, 6D, 7–9) are highly differentiated angular blocky red mudstones. They may exhibit an upper layer of massive carbonate and subsurface horizons with weakly expressed wedge-shaped aggregate structure and slickensides. However, all Type H paleosols have horizons of mudstone with common calcareous nodules and rhizoliths (Figs. 6C, 6D) that may cement entire layers with $CaCO_3$ (Fig. 7). Hydroxy-interlayered minerals and mica-like minerals dominate the <2 µm fraction in this pedotype, with trace amounts of kaolinite and chlorite. Type H paleosols are found in the Lower Permian Archer City through Clear Fork Formations in what we infer to be upper and lower coastal-plain environments.

Type H paleosols may have been Aridisols. However, such a classification requires specific knowledge of the soil moisture content and temperature during Permian time. These values are not known. In reality, Pedotype H paleosols may be classified only as Inceptisols, despite their apparent level of morphological maturity, which includes thick (≥15 cm) zones of carbonate accumulation (Soil survey Staff, 1998). Nevertheless, chlorite in the <2 µm fraction of Type H (Inceptisols) suggests arid conditions, given that this mineral is not stable in sub-humid and humid climates (Yemane, et al. 1996).

From estimates for development of modern calcic-horizons and their typical stratigraphic association with fine-grained sediments, we interpret Type H (Inceptisols) to have formed on stable regions of the flood-plain facies, removed from any frequently flooded fluvial systems. However, slickensides in layers coincident with, and beneath, calcic horizons suggest seasonal variations in soil moisture.

Similar to Pedotype F, Pedotype H developed upon stable portions of the landscape that were distal from active sedimentary environments. Its morphological characteristics likely represent regional climate, and it is unlikely that the floral elements associated with this pedotype would have been incorporated into the fossil record.

Summary of Permo-Pennsylvanian Paleosols

The paleo-landscape positions associated with the formation of different Permo-Pennsylvanian paleosols result in a representative suite of eight pedotypes. The morphology of pedotypes that record soil formation in stable, well-drained portions of the landscape, which were distal from major depocenters, likely record the effect of regional paleoclimate. These well-drained pedotypes are B, F, G, and H. Nevertheless, the regional continuity of pedotypes with morphologies that are indicative of poorly drained conditions (A, C, D, and E) through the upper Pennsylvanian strata of north-central Texas probably corresponds to a regionally humid paleoenvironment characterized by a very shallow groundwater table (Tabor and Montañez, 2004). However, pedotypes with morphologies indicative of poor drainage (most notably Pedotype E) are volumetrically very small through the Lower Permian strata and probably represent aberrant moisture availability upon a semiarid to arid regional landscape. In this regard, our most parsimonious explanation of "ever-wet" fossil floras preserved in the dry climate associated with the deposition of the Lower Permian strata is that these aberrant, poorly drained paleosols provided miniature "inliers" of ever-wet refuge, or "wet spots," for these plants.

PATTERNS OF PLANT PRESERVATION AND DISTRIBUTION

The plant fossil record, in general, is strongly biased toward floras from lowland basins, mainly representing plants that grew within a few meters of where their remains are found (Scheihing and Pfefferkorn, 1984; Burnham et al., 1992). There also are other, slightly smaller scale "megabiases" (Behrensmeyer and Hook, 1992). Most importantly, leaves have limited potential for long-distance transport (Spicer, 1981; Gastaldo, 1989). Leaf assemblages generally represent species growing in the wetter parts of the local landscape (pedotypes A, C, D, and E), close to standing bodies of water where burial and subsequent fossilization are most likely. Consequently, plant fossils can provide a significantly "wetter" climatic signature for a region than do paleosols or geochemical indicators. This is of importance when considering plants as indicators of climatic trends. At still smaller scales, it is necessary to consider the nature of the burial processes that preceded fossilization. Parautochthonous plant assemblages are transported but deposited within the habitat of growth (Bateman, 1991); these are most common. Autochthonous assemblages are not transported; they occur rarely, mainly as in situ tree stumps (e.g., Gastaldo, 1986), or where plants are buried in place by floods (Andrews et al., 1977) or volcanic ash (Wing et al., 1993). Allochthonous plant assemblages are transported and deposited in environments different from those in which the parent plants grew; although allochthonous plant debris is common in some lithological sequences, well-preserved allochthonous assemblages are rare. The overly "wet" bias notwithstanding, the north-central Texas section preserves a clear pattern of diminishment of Late Carboniferous dominants and their replacement by seed plants of more xeromorphic aspect.

Plant fossils in rocks of the upper Pennsylvanian and Lower Permian of north-central Texas are mostly parautochthonous and occur almost exclusively in small channel-form deposits composed of claystone, mudstone, and very fine grained sandstone. Such deposits rarely exceed 20 m in width or 2 m in thickness. The bases of these channel-form deposits are in erosional contact with underlying rocks, frequently paleosols, and may include sandstone accretion beds. Within these deposits, plant fossils are most often preserved in finely laminated claystones and mudstones deposited from suspension, probably during late stages of channel abandonment or as oxbow lakes. Less frequently, plant fossils may occur in sandstones or coarse siltstones deposited at the bases of accretion surfaces, probably on bars during slack-water intervals following floods in active channels. Some deposits are isolated stratigraphically and geographically; others recur in multiple, discrete deposits within a restricted stratigraphic interval that extends across several square kilometers. The most extensive plant-bearing deposits are those of the coal-bearing parts of the Virgilian to lower Wolfcampian Bowie Group (DiMichele et al., 1991), which are traceable for tens of kilometers along outcrop. Those parts of the Permian section that lack significant channel facies, such as the Leonardian upper Clear Fork (Nelson et al., 2001), have not yielded plant assemblages. In this respect, and at a somewhat coarse scale of resolution, the north-central Texas plant record can be regarded as "isotaphonomic," meaning that it formed under the same kind of conditions at the scale of analysis.

Physiognomic and Taxonomic Patterns

The most commonly preserved plants indicative of ever-wet conditions include tree ferns, sphenopsids, and lycopsids (Fig. 11), which have architectures and reproductive biologies that are closely tied to wet substrates (for a review, see DiMichele and Phillips, 1994). For example, lycopsids are heterosporous, which is a plant life history closely tied to water. The female reproductive organs of these plants appear to be water dispersed (Phillips, 1979). In addition, their root systems are broadly spreading yet shallow and have growth tips and lateral appendages that are structurally incapable of penetrating hard substrates (Phillips and DiMichele, 1992). They also have limited water-conducting systems that require high levels of moisture availability. Calamitean sphenopsids have underground rhizomes that permit them to recover from burial by sediment; they are generally found in disturbance-prone streamside, bayside, and lakeside deposits and may recover from repeated rapid, shallow burial (e.g., Gastaldo, 1992). In some instances they may form a succession to buried lycopsid forests in Pennsylvanian equatorial wetlands (Calder et al., this volume). Marattialean tree ferns are homosporous, with freely released spores that require moist, but not flooded, conditions to complete their life cycle. They also have highly divided, large fronds indicative of high water transpiration rates and thus high levels of soil moisture. Tree fern rooting systems have numerous air chambers, an anatomy found in extant plants that grow all or part of the year under submerged conditions, and appear to have been shallowly penetrating (DiMichele and Phillips, 2002).

In contrast, plants typical of seasonally dry conditions have small or reduced foliage, sunken stomatal pores, various kinds of water-retention attributes of the leaf surface or subsurface, reduced stature if broad-leaved, and deep, vertical patterns of rooting (Algeo and Scheckler, 1998). Conifers, gigantopterids, peltasperms, taeniopterids, and cordaites are examples of plants that bear some or all of these characteristics, depending on the extent to which they are known (Kerp, 1990).

Composite Floristic Patterns

The Upper Carboniferous "Wet" Flora

Upper Pennsylvanian wetland plant assemblages of north-central Texas (Flora 1, Figs. 9–11) can be divided into two groups, those from organic shales and those from flood-plain mudstones. Although they are compositionally similar, patterns of dominance differ and there are some species that apparently occur exclusively or nearly exclusively in one ecological setting or the other.

Organic shales and coals, deposits formed in forested wetlands (swamps and mires), represent the wettest habitats supporting plants on the landscape. As noted above, these deposits are associated with Pedotype A, as the organic upper horizon, and have a regional extent in parts of the Bowie and Cisco Groups. Organic-rich deposits commonly contain thick lenses of fusain (fossil charcoal), indicating recurring fires during the time of forest growth, and thus probable periods of short-term drying. The plant assemblages of these deposits are dominated by medullosan pteridosperms, particularly the species *Neuropteris scheuchzeri*. Associated elements that are locally abundant include the marattialean foliage *Pecopteris*, which is dominant in the youngest known deposit of this kind. *Sigillaria brardii*, a small arborescent lycopsid, is locally common to abundant, but is not everywhere a member of the flora. Calamitean stems and foliage attributable to *Asterophyllites equisetiformis* are the most common members of this group of plants in organic-rich deposits. Locally common are the reproductive organs *Macrostachya*, although no particular foliage has been found that can be attributed to this cone. In general, these floras contain few species, with the pteridosperm *Neuropteris scheuchzeri* dominating (as high as 90% of the biomass determined by quadrat analysis; see Pfefferkorn et al., 1975, for description of the techniques used).

Mudstone-filled hollows with erosional basal contacts commonly directly overlie organic shales and are closely associated with Pedotype C. These deposits are a mixture of small-scale channels and scours with standing-water mud infillings,

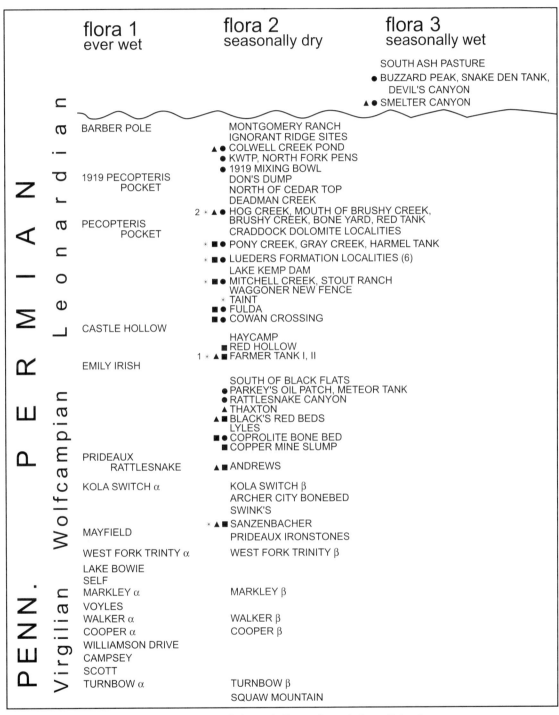

Figure 10. Stratigraphic distributions of the three major floras. Localities are noted by their informal names. Sites joined by a horizontal line and marked with α and β indicate occurrences in the same outcrop of the wet (α) flora and seasonally dry (β) flora (see Fig. 2B for an example). Symbols adjoining locality names in Flora 2 and Flora 3 represent occurrences of wet elements either as part of plant assemblages otherwise compositionally typical of seasonally dry conditions, or as dominant elements of an assemblage.

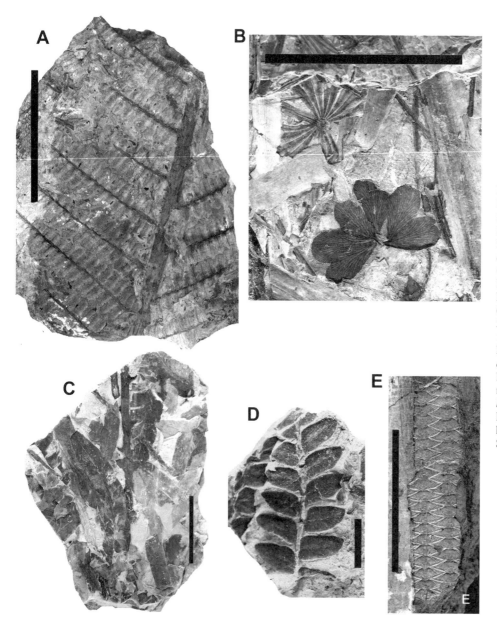

Figure 11. Common plant fossils of Flora 1. (A) *Pecopteris* sp. (USNM 526025), Castle Hollow locality (USGS 8960), Petrolia Formation, Lower Permian. (B) Sphenopsids: top, *Annularia carinata*; bottom, *Lilpopia raciborskii* (USNM 526026); both, Williamson Drive locality (NMNH 40013), Markley Formation, upper Pennsylvanian. (C) *Neuropteris scheuchzeri* (USNM 526027), Self School locality (USGS 10056), Markley Formation, upper Pennsylvanian. (D) *Pseudomariopteris cordata-ovata* (USNM 526028), Cooper locality (NMNH 39992), Markley Formation, upper Pennsylvanian. (E) *Sigillaria brardii* (USNM 526029), Maxey Ranch locality (NMNH 40602), Markley Formation, upper Pennsylvanian. Scale bars, 5 cm in A–C, E; 1 cm in D.

interbedded with thin, poorly developed, lightly rooted paleosols that represent flood-plain deposits (Fig. 2). The plant assemblages characteristic of these environments (Fig. 11) are typical of late Pennsylvanian clastic deposits in coal-bearing intervals from the midcontinent of the United States and from western Europe (see especially Wagner, 1983, Kerp and Fichter, 1985, and Laveine, 1989, for floristic comparisons). Dominant elements vary from location to location in north-central Texas, but common species include the pteridosperms *Neuropteris ovata*, *Neuropteris auriculata*, and *Alethopteris zeilleri*. Many species of marattialean tree fern foliage also are common, including the *Pecopteris cyathea*-group (sensu Zodrow, 1990), *Lobatopteris puertollanensis*, and *Polymorphopteris polymorpha* (Wagner,

1983). The most common calamitean remains are stems and foliage attributable to *Annularia carinata* (Kerp and Fichter, 1985). A wide variety of ground cover is associated with these arborescent taxa including the sphenopsids, *Sphenophyllum oblongifolium* and *Lilpopia raciborskii* (Kerp, 1984), and small fern foliage of the genus *Sphenopteris*. This assemblage is derived from a flora typical of humid atmospheric conditions and wet substrates environments.

The Permian "Seasonally Dry" Flora

During the Early Permian the most common elements of plant assemblages are seed plants (Fig. 12). The first distinctively Permian assemblages identified (Flora 2 of Fig. 10), occur in

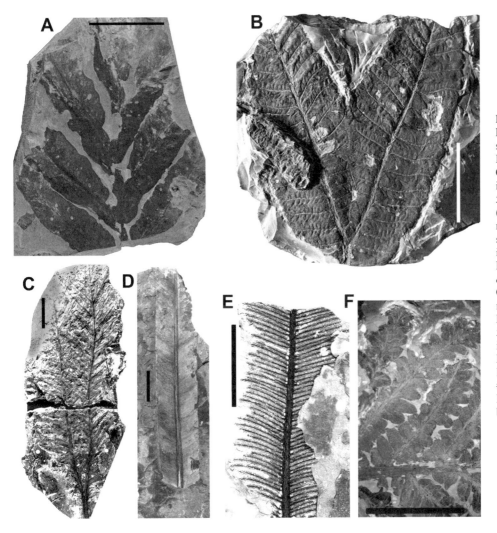

Figure 12. Common plant fossils of Flora 2. (A) Comioid peltasperm, undescribed (USNM 526030), 1919 Mixing Bowl locality (NMNH 40992), middle Clear Fork Formation, Lower Permian. (B) *Zeilleropteris wattii* (USNM 526031), Brushy Creek #2 locality (NMNH 40048), lower Clear Fork Formation, Lower Permian. (C) *Supaia* sp. (USNM526032), KWTP #1 locality (NMNH 40979), middle Clear Fork Formation, Lower Permian. (D) *Taeniopteris* sp. (USNM 526033), Mouth of Colwell Creek locality (NMNH 41005), middle Clear Fork Formation, Lower Permian. (E) Walchian conifer (USNM 526034), Mouth of Colwell Creek locality (NMNH 41006), middle Clear Fork Formation, Lower Permian. (F) Callipterid (USNM 526035), Brushy Creek #2 locality (NMNH 40048), lower Clear Fork Formation, Lower Permian. Scale bars, 5 cm.

association with physical indicators of seasonal dryness. The specific and generic composition of Flora 2 (Figs. 10, 12) changes gradually during the Early Permian. Most widespread and long-ranging are conifers of the genera *Walchia*, *Culmitzschia*, and *Ernestiodendron* (Clement-Westerhoff, 1988). Associated with these plants are a number of peltasperms (seed plants belonging to the order Peltaspermales), most commonly *Autunia* and related callipterids (Kerp, 1988). In the Wolfcampian, but not the Leonardian, part of the section, *Sphenopteridium* is common (Mamay, 1992). In the Leonardian part of the section, the probable peltasperms *Comia*, *Supaia*, *Brongniartites*, and, possibly, *Compsopteris* are prominent (Naugolnykh,1999; Read and Mamay, 1964). Also in the Leonardian, several genera of gigantopterids, which may be peltaspermous seed plants, also become prominent elements of the flora; included, in the approximate stratigraphic order of their appearance and relative abundance, are *Gigantopteridium*, *Cathaysiopteris*, *Zeilleropteris*, *Evolsonia*, and *Delnortea* (Mamay, 1986, 1989; Mamay et al., 1988). Other seed plants locally common include *Russelites* (Mamay, 1968), cordaitean foliage, and *Odontopteris* of the *Mixoneura* form. *Taeniopteris*, a form that may represent seed plants and ferns, is locally common to abundant.

Flora 3 (Figs. 10, 13) contains elements that are precocious, appearing much earlier than previously known from the fossil record. This flora occurs in the San Angelo and Blaine Formations of the Pease River Group. These rocks are upper Lower Permian to lower Middle Permian and are characterized by abundant gypsum and weakly developed paleosols, characteristic of deposition under conditions with a short wet season and long hot, dry intervals. The flora contains elements that previously were known only from the Mesozoic, including such taxa as the putative cycad *Dioonitocarpidium* and some conifers such as *Podozamites*, and others known only from the Late Permian, such as *Ulmannia* and *Pseudovoltzia* (DiMichele et al., 2001, 2004). Flora 3 has been identified at several stratigraphic horizons in the San Angelo Formation at a single locality in the Blaine Formation. However, the composition of the plant assemblages from these two formations is different, indicating much drier conditions in

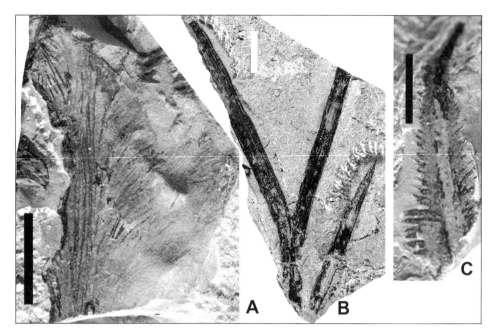

Figure 13. Selected elements of Flora 3. (A) Gigantopterid (USNM 520384), South Ash Pasture locality (NMNH 40968), Blaine Formation, Middle Permian. (B) *Dicranophyllum* sp. (USNM 526036), Buzzard Peak locality (NMNH 41394), contact of San Angelo and Blaine Formations, Lower Permian. (C) *Dioonitocarpidium* sp. (USNM 526037), Buzzard Peak locality (NMNH 41384), contact of San Angelo and Blaine Formations, Lower Permian. Scale bars, 1 cm.

the younger Blaine deposit. In the precociousness of their taxonomic composition, these floras parallel the early appearance of Flora 2 in isolated deposits or as thin beds intercalated within typical wetland deposits containing Flora 1 during the Virgilian in the Late Carboniferous.

Patterns of Change in the Distribution of Wetland Plants

The distribution of deposits bearing wetland plants is illustrated in Figures 9 and 10. In general, plants of the wetland biome (Flora 1) clearly predominate during the latest Carboniferous and earliest Permian. However, there is a shift to dominance by plants of the seasonally dry biome (Flora 2) in the early Early Permian (Wolfcampian), and this pattern persists through the later parts of the Early Permian (Leonardian). In the latest Early Permian or Middle Permian, a third biome appears, characterized by yet more evolutionarily derived seed plants (Flora 3); some of these taxa appear elsewhere in the world in rocks of Late Permian and early Mesozoic age (DiMichele et al., 2001).

Outcrop-Scale Alternation of Wetland and Seasonally Dry Biomes in the Upper Carboniferous and Lower Permian

In the Late Carboniferous deposits of the study area, wetland floras are diverse and found in a variety of lithologies typical of the environmentally heterogeneous, wet tropics. Habitat specialization can be documented, however, by the restriction of species occurrence or of dominance to particular lithologies. For example, there is a distinct pattern of differentiation between the most common forms of calamite foliage: *Annularia carinata* is abundant in flood-plain mudstones, whereas *Asterophyllites equisetiformis* is most common in swamp deposits represented by organic shales. The only macro-trend visible in these deposits is the disappearance of pteridosperm dominance in organic shales/coals and its replacement by tree fern dominance in the youngest organic-shale deposit. Otherwise, the same basic plant taxa are found in all wetland deposits with the same general patterns of lithological distribution. These deposits are typically associated with Pedotypes A and C.

Occurring with these wetland floras in the same Upper Carboniferous and lowermost Permian outcrops is a distinct flora composed of seed plants typical of the seasonally dry biome that is dominant through most of the Lower Permian. Conifers, callipterid peltasperms, *Neuropteris*, and *Sphenopteridium* form the dominant elements, with rare scraps of foliage similar to *Russelites* or *Charliea* (Mamay, 1968, 1990). This flora occurs in distinct, kaolinite-rich beds associated with Pedotype B (Figs. 5A, 5B, 8, 9), which may have formed on topographic highs, during periods of lowered water table. Some of these kaolinite deposits appear to mark the base of channel fills and may have contributed to ponding of water and initial development of subsequently deposited organic-rich mucks that contain plant assemblages representative of the wetland biome (see Gardner et al., 1988).

These two floras are spatially and temporally distinct, inferred from the positions and extents of their host sediments on outcrop. Typically, deposits of this type occur in coal/organic-shale-bearing portions of the Markley Formation but continue into the Lower Permian where such heterogeneous deposits become progressively less common. One of the youngest examples (Kola Switch) occurs ~8 m below the base of informal sandstone member 8 of the Archer City Formation (Hentz and Brown, 1987). These deposits have significant depositional relief and are lithologically heterogeneous. Where well exposed, they are shallow, broadly U-shaped, channel-form deposits, ~6–10 m deep and 50–100 m in width. Some have erosive sandstones and intraformational conglomerates at the base, indicative of proximal crevasse-splay channels. Others lack coarser-grained clastics and may have originated as more distal overbank channels, minor tributaries, or interior flood-plain channels. Floristically, the distinct plant assemblages are preserved in several different lithologies. Flood-plain paleosols commonly contain tree fern roots, generally preserved as casts and molds but occasionally with the distinctive anatomy characteristic of these organs. Rarely occurring kaolinitic lenses associated with the upper boundary of Pedotype C paleosols contain allochthonous floras typical of seasonally dry conditions. Autochthonous to parautochthonous swamp deposits of organic shale to shaly coal occur above the kaolinites and may be up to 1 m in thickness, and preserve a densely packed, low-diversity flora characteristic of the ever-wet biome. Overlying flood-plain sediments typically include small, mud-filled scours intermixed with weakly developed paleosols. These small channels contain a diverse, parautochthonous assemblage also typical of the ever-wet biome.

Patterns in the Lower Permian Wichita, Clear Fork, and Pease River Groups

Typically Carboniferous plants occur in association with the seasonally dry floras of the Lower and Middle Permian in two ways (Fig. 10): Either the Carboniferous wetland elements occur intermixed as rare elements within assemblages heavily dominated by the seasonally dry biome, or the wetland plants dominate (often nearly monospecifically) in assemblages that contain few or no species typical of seasonally dry habitats. There are a few rare and notable exceptions to this pattern, where a strong wetland component dominates with minor occurrences of plants from the seasonally dry biome (see below).

The wetland species most commonly encountered are the marattialean tree fern foliage *Pecopteris* and calamite stems (Fig. 14). Other wet indicators, such as *Sphenopteris,* a true fern, occur in the earliest Permian but are rare to absent in younger rocks. The lycopsid tree *Sigillaria* (Fig. 14), the ground cover sphenopsid *Sphenophyllum* (Fig. 14), and medullosan foliage scraps have been found quite rarely in the Leonardian as components of otherwise seasonally dry floras. Such unusual and unexpected occurrences suggest the existence of unsampled wet areas from which these plants were transported into the seasonally dry assemblages. The Late Carboniferous wetland plants that occur most commonly as holdovers in Permian deposits are largely "weedy," opportunistic forms also tolerant of wet substrates (see also Calder and Falcon-Lang, this volume, for similar ecological interpretations of these same plants in the early Pennsylvanian Joggins section). Tree ferns are most notable in this respect. They are cheaply constructed, with root mantles and, often, stems filled with air spaces, suggesting rapid growth and minimal resource allocation to the vegetative tissues used to attain tree heights. Quantitative analyses of the "cost" of tree fern tissues per unit volume show them to have the lowest biomass among all groups of Late Carboniferous swamp plants (Baker and DiMichele, 1997). In addition, their large fronds produce very large numbers of tiny, highly dispersible spores, permitting them to locate and colonize disturbed open sites rapidly. Similarly, sphenopsids appear to have strongly favored streamside and aggradational flood-basin habitats where their rhizomes allowed rapid recovery from catastrophic burial. Other groups, such as the lycopsid tree *Sigillaria* or ground cover plants such as the sphenopsid *Sphenophyllum*, are much rarer and may have existed only in isolated populations.

Monospecific assemblages of wetland plants in Lower Permian rocks are dominated by tree ferns exclusively. Their range of dispersal, due to their small, wind-dispersed spores, probably gave them access to short-lived and areally limited wet spots that could not be reached by other plants restricted to wet habitats. The youngest Clear Fork locality known is dominated by tree ferns. Calamitean sphenopsids are very common to abundant in rare instances. For example, at the North Fork Pens site, of the Middle Clear Fork Group, large numbers of calamite stems were encountered lying three-dimensionally within red siltstones, crossing bedding planes and flattened unidirectionally, indicating rapid burial by sediment-laden flood waters. Such calamite-dominated deposits are of very limited areal extent.

The Emily Irish locality in the basal Petrolia Formation was collected intensively by Sergius Mamay and colleagues in the 1960s and is perhaps the best sampled Early Permian flora from the United States. It is overwhelmingly dominated by tree fern foliage but contains a diversity of other kinds of plants, many of which are typical of the seasonally dry biome. Emily Irish was sampled in bulk; no effort was made to collect specimens layer by layer. Thus, we do not know if the tree-fern-rich assemblages occurred in the same rock layers as those enriched in evolutionarily derived seed plants. If they did co-occur, then this assemblage would represent one of the only examples in the study area where plants of the seasonally dry biome occurred as minor elements in a wetland-dominated assemblage. The occurrence of rare elements, either wetland or seasonally dry, in assemblages dominated by plants of the other biome, suggests strongly that these two vegetation types coexisted in close proximity on Early Permian landscapes, their distributions controlled by water table level rather than regional climate.

The seasonally wet biome of the Pease River Group is characteristic of environments with a relatively short wet season within a generally xeric climatic background. It is known from small channel fills that bear entirely distinct floras of unique composition compared with other Lower Permian floras of the paleotropical belt (DiMichele et al., 2001, 2004). There are virtually no known wetland plants associated with these floras, which, in fact, also lack species common in the earlier seasonally dry biome.

DISCUSSION

The loss of wetland assemblages during the Pennsylvanian-Permian transition in north-central Texas is mirrored by the increase of seasonally dry assemblages during the later

Figure 14. Ever-wet elements persisting as components of Early Permian seasonally dry landscapes presumably in "wet spots." (A) *Calamites* sp. stem (USNM 526038), Deadman Creek locality (NMNH 40670), lower Clear Fork Formation, Lower Permian. (B) *Pecopteris* sp. (USNM 526039), Pecopteris pocket locality (NMNH 40645), lower Clear Fork Formation, Lower Permian. (C) *Sigillaria brardii* (USNM 526040), Mouth of Brushy Creek locality (NMNH 38907), lower Clear Fork Formation, Lower Permian. (D) *Annularia spicata* (USNM 526041), Brushy Creek West locality (NMNH 38908), lower Clear Fork Formation, Lower Permian. Scale bars, 1 cm for A, C, D; 5 cm for B.

Pennsylvanian and earliest Permian on a global basis. When sampled on a bed-by-bed basis, it appears that few species cross over between assemblages drawn from these two floras (Kerp and Fichter, 1985; Broutin et al., 1990; DiMichele and Aronson, 1992). The flora typical of seasonally dry conditions begins to appear sporadically during the Stephanian (late Pennsylvanian) intercalated within geological sequences otherwise dominated by assemblages of plants typical of ever-wet conditions (McComas, 1988; Cridland and Morris, 1963; Doubinger et al., 1995; Mamay and Mapes, 1992; Rothwell and Mapes, 1988). In some cases plant assemblages have been collected from single, well-documented beds and often these are quite distinct in dominance and diversity patterns from wetland assemblages. In the latest Pennsylvanian of north-central Texas, for example, the seasonally dry and ever-wet floras occur within the same outcrops but in different beds. In some of the other studies the plants are binned by stratigraphic intervals or units, such as formations, and thus the two floras are analytically combined and treated as part of a single flora, rather than being broken down into specific time-space assemblages. This approach limits understanding to broad trends only. Using either approach, ultimately the seasonally dry flora becomes dominant in the early Lower Permian.

The conditions and patterns surrounding the origin of the seasonally dry flora are not quite as clear as its rise to prominence in the Pennsylvanian-Permian transition. Elements such as conifers make their initial appearances as scrappy plant remains, sometimes preserved as charcoal, evidently washed or blown into the depositional lowlands from adjacent "upland" areas (Lyons and Darrah, 1989) beginning in the middle Westphalian (middle Pennsylvanian) (Scott and Chaloner, 1983; Galtier et al., 1992). However, there is considerable evidence of "upland" vegetation prior to or concurrent with the first appearances of the conifer-rich "Permian" flora, which raises questions about the degree to which vegetation was differentiated in extrabasinal areas (sensu Pfefferkorn, 1980). For example, Leary (1975) and Leary and Pfefferkorn (1977) described early Pennsylvanian–aged plant assemblages from the margin of the Illinois basin that are distinct from contemporaneous basinal assemblages associated with coal beds; however, the marginal floras do not contain conifers or any plants that could be seen as precursors to the plants that come to dominate Permian seasonally dry habitats. Similarly, Falcon-Lang and Bashforth (2004) describe early Pennsylvanian forested uplands dominated by cordaitalean trees, but not intermixed with conifers and other elements that might have been parts of the seasonally dry vegetation that subsequently dominates the lowland tropics. Interestingly, some plants characteristic of late Mississippian seasonal climates, such as *Sphenopteridium*, reappear in late Pennsylvanian and Early Permian conifer-rich floras (Mamay, 1992), but not in these other kinds of Pennsylvanian extrabasinal assemblages, suggesting the possibility of biogeographic differentiation controlled largely by climatic factors—the conifer-rich flora may have been primarily subtropical or true upland in evolutionary origin and primary distribution.

Paleosols from the late Pennsylvanian and Early Permian in north-central Texas closely mirror the changes in the patterns of plant distribution. The most striking characteristic of the late Pennsylvanian paleosols is redox features. The combination of (1) redox features in Pedotypes B, C, and D (Ultisols, Inceptisols, and Aquerts) and (2) organic matter accumulation and tabular rooting systems in Pedotype A (Histosols) indicates that poorly drained conditions varied between continuous and episodic, suggesting a regionally shallow water table. In addition, the dominance of kaolinite in late Pennsylvanian profiles is consistent with an environment characterized by high rates of chemical weathering, conditions expected in humid tropical environments (Yemane et al., 1996).

In contrast, Permian paleosols typically do not exhibit redoximorphic features indicative of a regionally extensive shallow groundwater table. Oxidized paleosol matrix colors, vertically extensive rooting structures, and pedogenic carbonate accumulation in Pedotypes F, G, and H (Alfisols, Vertisols, and Inceptisols) indicate that large portions of the Lower Permian landscape were well drained and remained above the water table for extended periods. In addition, the presence of smectite, hydroxy-interlayered phyllosilicate minerals, and chlorite clay minerals suggests well-drained and dry conditions (Tabor and Montañez, 2004).

Another indicator of changing hydrologic conditions is the paucity of pedogenic carbonate in late Pennsylvanian paleosols and its abundance in Early Permian paleosols. Pedogenic carbonate accumulation is a response to net moisture deficit in which annual evaporation exceeds precipitation (Birkeland 1999; Retallack 1990). In most cases carbonate will not be retained in the soil if mean annual precipitation exceeds 760 mm (Royer 1999). This suggests that carbonate-bearing Early Permian paleosols developed under conditions of low precipitation and net moisture deficiency, whereas Late Carboniferous paleosols, which do not contain pedogenic carbonate, probably received higher mean annual precipitation without extended episodes of moisture deficiency. Early Permian paleosol morphologies in the Archer City, Nocona, Petrolia, Waggoner Ranch, and lower half of the Clear Fork Formations suggest development under warm and relatively dry, subhumid to semiarid climate.

Finally, the paucity of soil profiles across the lower coastal plain facies of the San Angelo and Blaine Formations may indicate generally unfavorable conditions for soil formation. The fact that the only diagnostic morphologies (i.e., root traces and soil structure) indicative of soil formation occur along channel sandstones may indicate either (1) regional water table levels generally were very low except along stream galleries, where the regional water table intersects surface topography, or (2) these stream corridors provided the only regional sources of fresh water for plant-growth and soil formation.

Lithological features of the Pease River Group support the second option above. These include thick deposits of bedded gypsum interbedded with marine-invertebrate-bearing limestone and dolomite beds, which record repeated incursions of salt water from the Midland Basin into the study area, suggesting a landscape of hot, arid coastal mudflats and salt pans largely devoid of vegetation. Between marine incursions, it is likely that nothing grew except along the margins of little brackish creeks, such as the one sampled at South Ash Pasture in the Blaine Formation (DiMichele et al., 2004). The water table may have been high (the region frequently was under water), yet the alkali-laden soil and extreme drought conditions precluded most plant growth or development of soil horizons.

Thus, the study of paleosols shows that the climate became drier through time and the regional water table dropped. Late Pennsylvanian paleosol morphologies suggest formation in a warm and moist climate with a generally shallow and seasonally fluctuating groundwater table. Groundwater table levels were more variable during Permian time and apparently followed the regional physiography of the eastern Midland Basin. Morphologies indicative of generally wet soil conditions apparently formed proximal to streams and oxbow ponds, whereas morphologies indicative of drier soil-moisture budgets formed upon the open coastal flood plains, distant from major streams. Finally, beginning in the upper Clear Fork Formation (middle Lower Permian) and continuing through the San Angelo and Blaine Formations (upper Lower Permian/lower Middle Permian), paleosol distribution and diversity of morphologies progressively diminish until the only evident soils formed along stream galleries. These stream galleries may have comprised the only plant-accessible source of fresh water at this time, and indicate either that the regional groundwater was deep or that despite a high water table the coastal sabkhas were hostile to plant life.

Thus, vegetation tolerant of periodic drought became more common and ultimately dominated the landscape (diagrammatically represented in Fig. 15). Floras of ever-wet habitats retreated to isolated wet areas along permanent streams. Initially, these "wet spots" were numerous and hosted a fairly diverse selection of wetland plants persisting from the latest Carboniferous. Ultimately, these little wetlands diminished, both in area and species diversity, until only *Pecopteris* and calamites were common. The

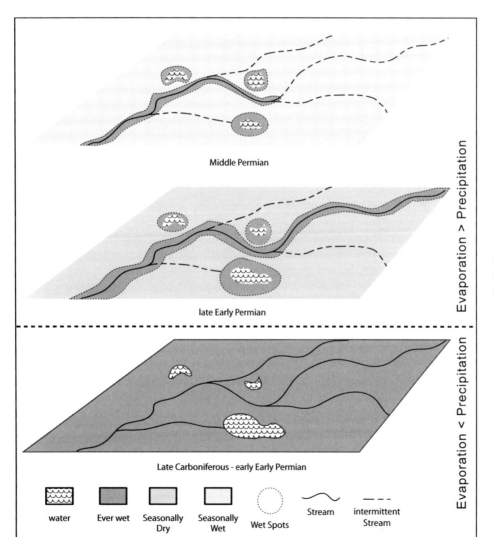

Figure 15. From wetlands to wet spots. Landscape progression and restriction of available habitat area for wetland plants under progressive moisture limitation.

plants that survived in the areally diminished wetlands represent two major ecological strategies. One is that of the sphenopsids, particularly the calamites. Calamites are the only tree group of the late Paleozoic with clonal growth, permitting them to recover from burial by sediment and thus to occupy aggradational, disturbed environments. Calamites occurred and locally flourished throughout the Early Permian and into the early Middle Permian. Some specimens record evidence of burial and recovery from disturbance, in the form of streamside flooding. Calamite foliage is very rare; stems routinely are the only fossil encountered. However, these plants may have resembled modern *Equisetum*, which has highly reduced foliage and photosynthetic stems.

The second "survivor" group is the marattialean tree ferns. These trees were "cheaply constructed." Anatomical studies demonstrate that their root mantles, and sometimes their stems and petioles, were filled with air spaces. In addition, they produced massive numbers of spores on large fronds, the wide broadcast of which permitted them to colonize any suitable landscape patch. Cheap construction and massive reproduction typify opportunistic, "weedy" species. It is likely that this life history permitted these plants to survive in a wide array of habitats, those where sufficient moisture for reproduction was available during part of the year, and where substrates stayed wet enough for the year-round maintenance of large fronds. Opportunism may have encouraged an evolutionary response to changing environments as well, permitting the formation of many populational isolates, some of which may have evolved greater tolerance to limited periods of drought.

More problematic is the very rare occurrence, isolated both stratigraphically and spatially, of such plants as *Sigillaria* and *Sphenophyllum*. These taxa were abundant in the Late Carboniferous in widespread wetlands, although *Sigillaria brardii*, the likely species in question, is heir to a legacy of complex ecologies, including some species that may have been tolerant of disturbance and moisture stress (DiMichele and Phillips, 1994; Calder and Falcon-Lang, this volume). The discovery of these taxa as transported singletons in rocks of the Clear Fork Group, well above previously known occurrences, indicates that they persisted in isolated refuges, unlikely to be sampled. Nonetheless, the low dispersibility of these plants, relative to tree ferns or calamites, requires that wetlands have enough connectivity for populations to survive as wet spaces fluctuated in space and time. These rare occurrences thus may be taken as a barometer of the inadequacy of the fossil record itself (Darwin, 1859). The fossils are much rarer than the weedy wetland forms, but the overall sample size of fossil-bearing deposits themselves is very small compared with the vast size of the landscape being sampled. We thus can infer that small, almost completely unseen, populations must have existed continuously, and that the populations of the more commonly encountered, if still rare, forms were of significant size. That these plants were most typical of the wettest parts of landscapes, but are so rarely found, indicates that suitable habitat patches also were uncommon.

The fossil record of wetlands is relevant to larger questions of ecosystem organization that are central to debates in modern ecology. This is of particular interest when wetlands are examined in the context of a large-scale environmental shift, such as that at the transition from the Pennsylvanian to the Permian. The patterns described here indicate that environmental tracking by plants is a major factor in their spatial distribution. Why should this be surprising? Much of the debate in ecology has centered on the "individualism" of species, suggesting that plant associations or "communities" at any given place and time are simply happenstance, reflecting the momentary co-occurrence of plants with similar resource requirements under currently ambient climatic conditions. As visualized, such communities have no spatio-temporal bounds and, hypothetically at least, any combination of species is possible under this kind of lottery. The alternative is a system with some kind of identifiable limits to possible species combinations, and perhaps even identifiable rules governing species combinations (Weiher and Keddy, 1999). "Individualistic" responses of species to local changes in habitat conditions have been documented in studies of Holocene and Quaternary floras, particularly of North America and Europe (Delcourt and Delcourt, 1991; Bennett, 1997), including "non-analogue" floras (Overpeck et al., 1992). However, spatial patterns indicate that there may be limits to such dynamics, perhaps at the scale of biomes. Individualism in postglacial floras, such as it is, appears to be restricted largely to mixing within particular vegetation types/biomes. The deep fossil record, such as that of the Permian described here, suggests that there are environmental limits, or bounds, to the extent of these responses. The very concept of the biome has embedded within it the recognition that species are not randomly distributed across vast physical spaces in a gradient-like fashion. Rather, species tend to cluster around certain environmental norms, and to track those norms in a haphazard but nonetheless identifiable manner. The fossil record, despite its flaws, may be the best place to search for general patterns of this nature.

ACKNOWLEDGMENTS

Without the assistance of the following people, over more than 15 years, this project would not have been possible: Kenneth Craddock, Louis Todd, Donald Gregg, Walt Dalquest, John Kocurko, Nicholas Hotton III, S.H. Mamay, Todd Thomas, David Hale, Daniel Ortuño, Tucker Hentz, Christopher Durden, and Isabel Montañez. We also thank numerous landowners who graciously permitted access to their property and offered material support when needed. We particularly want to thank Bob Hook for his collaboration and substantial comments on the manuscript. Finally, but not least, we thank John Calder and Howard Falcon-Lang who shared ideas and made numerous constructive comments on the manuscript. This research was supported by several grants from the Scholarly Studies Program, and by the Mary Walcott and Roland W. Brown funds of the Smithsonian Institution. This is a contribution from the Evolution of Terrestrial Ecosystems Program of the Smithsonian Institution, which also provided partial support.

REFERENCES CITED

Abilene Geological Society, 1949, Cross section, Stonewall County to Hood County, Texas: Abilene Geological Society, one sheet.

Abilene Geological Society, 1953, Cross section, Schleicher County to Childress County, Texas: Abilene Geological Society, one sheet.

Algeo, T.J., and Scheckler, S.E., 1998, Terrestrial-marine teleconnections in the Devonian: Links between the evolution of land plants, weathering processes, and marine anoxic events: Royal Society of London Philosophical Transactions, ser. B, v. 353, p. 113–130.

Andrews, H.N., Kasper, A.E., Forbes, W.H., Gensel, P.G., and Chaloner, W.G., 1977, Early Devonian Flora of the Trout Valley Formation of northern Maine: Review of Palaeobotany and Palynology, v. 23, p. 255–285, doi: 10.1016/0034-6667(77)90052-5.

Baker, R., and DiMichele, W.A., 1997, Resource allocation in Late Pennsylvanian coal-swamp plants: Palaios, v. 12, p. 127–132.

Bateman, R.M., 1991, Palaeoecology, in C.J. Cleal, ed., Plant fossils in geological investigation: The Palaeozoic: Chichester, UK, Ellis-Horwood, p. 34–116.

Behrensmeyer, A.K., and Hook, R.W., rapporteurs, 1992, Paleoenvironmental contexts and Taphonomic Modes, in Behrensmeyer, A. K. et al., eds., Terrestrial ecosystems through time: Chicago, University of Chicago Press, p. 14–136.

Bennett, K.D., 1997, Evolution and ecology, the pace of life: Cambridge, UK, Cambridge University Press, 241 p.

Berman, D.S., 1970, Vertebrate fossils from the Lueders Formation, Lower Permian of north-central Texas: University of California Publications in Geological Sciences, v. 86, p. 1–61.

Birkeland, P.W., 1999, Soils and geomorphology (third edition): New York, Oxford University Press, 430 p.

Bose, E., 1917, The Permo-Carboniferous ammonoids of the Glass Mountains, West Texas, and their stratigraphical significance: University of Texas Bulletin, no. 1762, 241 p.

Brister, B.S., Stephens, W.C., and Norman, G.A., 2002, Structure, stratigraphy, and hydrocarbon system of a Pennsylvanian pull-apart basin in north-central Texas: American Association of Petroleum Geologists Bulletin, v. 86, p. 1–20.

Broutin, J., Doubinger, J., Farjanel, G., Freytet, F., Kerp, H., Langiaux, J., Leberton, L., Sebban, S., and Satta, S., 1990, Le renouvellement des flores au passage Carbonifère Permien: Approaches stratigraphique, biologique, sédimentologique: Comptes Rendus de l'Académie des Sciences, Paris, v. 311, p. 1563–1569.

Brown, L.F., Jr., Solis-Iriarte, R.F., and Johns, D.A., 1987, Regional stratigraphic cross sections, Upper Pennsylvanian and Lower Permian strata (Virgilian and Wolfcampian Series), north-central Texas: University of Texas at Austin Bureau of Economic Geology Cross Sections, 27 p.

Brown, L.F., Jr., Solis-Iriarte, R.F., and Johns, D.A., 1990, Regional depositional systems tracts, paleogeography, and sequence stratigraphy, Upper Pennsylvanian and Lower Permian strata, north- and west-central Texas: University of Texas at Austin Bureau of Economic Geology Report of Investigations, v. 197, p. 1–116.

Buol, S.W., Hole, F.D., McCracken, R.J., and Southard, R.J., 1997, Soil genesis and classification: Ames, Iowa, Iowa State University Press, 527 p.

Burnham, R.J., Wing, S.L., and Parker, G.G., 1992, The reflection of deciduous forest communities in leaf litter: Implications for autochthonous litter assemblages from the fossil record: Paleobiology, v. 18, p. 30–49.

Cecil, C.B., 1990, Paleoclimate controls on stratigraphic repetition of chemical and siliciclastic rocks: Geology, v. 18, p. 533–536, doi: 10.1130/0091-7613(1990)018<0533:PCOSRO>2.3.CO;2.

Clement-Westerhoff, J.A., 1988, Morphology and phylogeny of Paleozoic Conifers in Beck, C. B., ed., Origin and evolution of gymnosperms: New York, Columbia University Press, p. 298–337.

Clifton, R.L., 1944, Paleoecology and environments inferred from some marginal middle Permian marine strata: American Association of Petroleum Geologists Bulletin, v. 28, p. 1021–1031.

Cridland, A.A., and Morris, J.E., 1963, *Taeniopteris, Walchia,* and *Dichophyllum* in the Pennsylvanian System of Kansas: University of Kansas Science Bulletin, v. 44, p. 71–85.

Dalquest, W.W., and Carpenter, R.M., 1975, A new discovery of fossil lungfish burrows: Texas Journal of Science, v. 26, p. 611.

Dalquest, W.W., and Kocurko, M.J., 1986, Geology and vertebrate paleontology of a Lower Permian delta margin in Baylor County, Texas: Southwestern Naturalist, v. 31, p. 477–492.

Dalquest, W.W., and Kocurko, M.J., 1989, Notes on Permian fishes from Lake Kemp, Baylor County, Texas, with a synopsis of Texas palaeonisciform fishes: Southwestern Naturalist, v. 33, p. 263–274.

Daniels, R.B., Gamble, E.E., and Nelson, L.A., 1971, Relations between soil morphology and water table levels on a dissected North-central Carolina coastal plain surface: Soil Science Society of America Proceedings, v. 35, p. 157–175.

Darwin, C., 1859, The origin of species by means of natural selection: London, John Murray, 513 p.

Delcourt, H.R., and Delcourt, P.A., 1991, Quaternary ecology, a paleoecological perspective: New York, Chapman and Hall, 242 p.

DiMichele, W.A., and Aronson, R.B., 1992, The Pennsylvanian-Permian vegetational transition: A terrestrial analogue to the onshore-offshore hypothesis: Evolution—International Journal of Organic Evolution, v. 46, p. 807–824.

DiMichele, W.A., and Phillips, T.L., 1994, Paleobotanical and paleoecological constraints on models of peat formation in the Late Carboniferous of Euramerica: Palaeogeography, Palaeoclimatology, Palaeoecology, v. 106, p. 39–90, doi: 10.1016/0031-0182(94)90004-3.

DiMichele, W.A., and Phillips, T.L., 2002, The ecology of Paleozoic ferns: Review of Palaeobotany and Palynology, v. 119, p. 143–159, doi: 10.1016/S0034-6667(01)00134-8.

DiMichele, W.A., Hook, R.W., Mamay, S.H., and Willard, D.A., 1991, Paleoecology of Carboniferous-Permian transitional vegetation in north-central Texas: American Journal of Botany, Abstracts Supplement, v. 78, no. 6, p. 111–112.

DiMichele, W.A., Mamay, S.H., Chaney, D.S., Hook, R.W., and Nelson, W.J., 2001, An Early Permian flora with Late Permian and Mesozoic affinities from north-central Texas: Journal of Paleontology, v. 75, p. 449–460.

DiMichele, W.A., Hook, R.W., Nelson, W.J., and Chaney, D.S., 2004, An unusual Middle Permian flora from the Blaine Formation (Pease River Group, Leonardian-Guadalupian Series) of King County: West Texas: Journal of Paleontology, v. 78, p. 765–782.

Doubinger, J., Vetter, P., Langiaux, J., Galtier, J., and Broutin, J., 1995, La flore fossile du bassin houiller de Saint-Étienne: Mémoires du Muséum National d'Histoire Naturelle, Paléobotanique, v. 164, p. 1–355.

Duchaufour, P., 1982, Pedology: Pedogenesis and classification: London, Allen and Unwin, 448 p.

Dunbar, C.O., and Skinner, J.W., 1937, Permian Fusulinidae of Texas: University of Texas Bulletin, no. 3701, The geology of Texas, v. 3, part 1, p. 516–732.

Edwards, M.B., Eriksson, K.A., and Kier, R.S., 1983, Paleochannel geometry and flow patterns determined from exhumed Permian point bars in north-central Texas: Journal of Sedimentary Petrology, v. 53, p. 1261–1270.

Falcon-Lang, H.J., 2003, Late Carboniferous tropical dryland vegetation in an alluvial-plain setting, Joggins, Nova Scotia, Canada: Palaios, v. 18, p. 197–211.

Falcon-Lang, H.J., 2004a, Early Mississippian lycopsid forests in a delta-plain setting at Norton, near Sussex, New Brunswick, Canada: Journal of the Geological Society [London], v. 161, p. 964–981.

Falcon-Lang, H.J., 2004b, Pennsylvanian tropical rain forests responded to glacial-interglacial rhythms: Geology, v. 32, p. 689–692.

Falcon-Lang, H.J., and Bashforth, A., 2004, Pennsylvanian uplands were forested by giant cordaitalean trees: Geology, v. 32, p. 417–420, doi: 10.1130/G20371.1.

Falcon-Lang, H.J., Rygel, M.C., Calder, J.H., and Gibling, M.R., 2004, An early Pennsylvanian waterhole deposit and its fossil biota in a dryland alluvial plain setting, Joggins, Nova Scotia: Journal of the Geological Society [London], v. 161, p. 209–224.

Franklin, D.W., Chairman, 1951, Fall field excursion: Abilene Geological Society, 24 p.

Franzmeier, F.P., Bryant, R.B., and Steinhardt, G.C., 1985, Characteristics of Wisconsinan glacial tills in Indiana and their influence on Argillic horizon development: Soil Science Society of America Journal, v. 49, p. 1481–1486.

Galloway, W.E., and Brown, L.F., Jr., 1972, Depositional systems and shelf-slope relationships in Upper Pennsylvanian rocks, north-central Texas: University of Texas at Austin Bureau of Economic Geology Report of Investigations, v. 75, p. 1–62.

Galtier, J., Scott, A.C., Powell, J.H., Glover, B.W., and Waters, C.N., 1992, Anatomically preserved conifer-like stems from the Upper Carboniferous of England: Proceedings of the Royal Society of London, ser. B, v. 247, p. 211–214.

Gardner, T.W., Williams, E.G., and Holbrook, P.W., 1988, Pedogenesis of some Pennsylvanian underclays: Ground-water, topographic and tectonic controls, in Reinhardt, J. and Sigleo, W.R., eds., Paleosols and weathering through geological time: Principles and applications: Geological Society of America Special Paper 216, p. 81–102.

Gastaldo, R.A., 1986, Implications on the paleoecology of autochthonous lycopods in clastic sedimentary environments of the Early Pennsylvanian of Alabama: Palaeogeography, Palaeoclimatology, Palaeoecology, v. 53, p. 191–212, doi: 10.1016/0031-0182(86)90044-1.

Gastaldo, R.A., 1989, Processes of incorporation of plant parts in deltaic coastal sedimentary environments: Implications for paleoecological restorations: International Carboniferous Congress of Stratigraphy and Geology, 11th, Beijing, China, Compte Rendu, v. 3, p. 109–120.

Gastaldo, R.A., 1992, Regenerative growth in fossil horsetails following burial by alluvium: Historical Biology, v. 6, p. 203–219.

Guo, Y., 1990, Paleoecology of flora from coal measures of Upper Permian in western Guizhou: Journal China Coal Society, v. 15, p. 48–54.

Henbest, L.G., 1938, Notes on the ranges of Fusulinidae in the Cisco Group (restricted) of the Brazos River region, north-central Texas: The University of Texas, publication 3801, p. 237–247.

Hentz, T.F., 1988, Lithostratigraphy and paleoenvironments of upper Paleozoic continental red beds, north-central Texas: Bowie (new) and Wichita (revised) Groups: The University of Texas at Austin Bureau of Economic Geology Report of Investigations, v. 170, p. 1–55.

Hilton, J., Wang, S.-J., Galtier, J., and Li, C.-S., 2001, An Early Permian plant assemblage from the Taiyuan Formation of northern China with compression/impression and permineralized preservation: Review of Palaeobotany and Palynology, v. 114, p. 175–189, doi: 10.1016/S0034-6667(01)00045-8.

Hentz, T.F., and Brown, L.F., Jr., 1987, Wichita Falls-Lawton Sheet. University of Texas at Austin Bureau of Economic Geology, Geologic Atlas of Texas, scale 1:250 000.

Hook, R.W., 1989, Stratigraphic distribution of tetrapods in the Bowie and Wichita Groups, Permo-Carboniferous of north-central Texas, in Hook, R.W., ed., Permo-Carboniferous vertebrate paleontology, lithostratigraphy, and depositional environments of north-central Texas: Society of Vertebrate Paleontology, 49th Annual Meeting, Austin, Texas, Field Trip Guidebook 2, p. 47–53.

Johnson, G.D., 1980, Early Permian vertebrates from Texas: Actinopterygii (Schaefferichthys), Chondrichthyes (including Pennsylvanian and Triassic Xenacanthodii), and Acanthodii [Ph.D. Dissertation]: Dallas, Southern Methodist University, 653 p.

Jones, J.O., and Hentz, T.F., 1988, Permian strata of north-central Texas, in Hayward, O.T., ed., South-Central Section of the Geological Society of America: Boulder, Colorado, Geological Society of America Centennial Field Guide, v. 4, p. 309–316.

Kerp, J.H.F., 1984, Aspects of Permian palaeobotany and palynology. III. A new reconstruction of Lilpopia raciborskii (Lilpop) Conert et Schaarschmidt (Sphenopsida): Review of Palaeobotany and Palynology, v. 40, p. 237–261, doi: 10.1016/0034-6667(84)90011-3.

Kerp, J.H.F., 1988, Aspects of Permian palaeobotany and palynology. X. The west- and central European species of the genus Autunia Krasser emend. Kerp (Peltaspermaceae) and the form-genus Rhachiphyllum Kerp (callipterid foliage): Review of Palaeobotany and Palynology, v. 54, p. 249–360, doi: 10.1016/0034-6667(88)90017-6.

Kerp, H., 1990, The study of fossil gymnosperms by means of cuticular analysis: Palaios, v. 5, p. 548–569.

Kerp, J.H.F., and Fichter, J., 1985, Die Makrofloren des saarpfälzischen Rotliegenden (?Ober-Karbon-Unter-Perm; SW Deutschland): Mainzer Geowissenschaftliche Mitteilungen, v. 14, p. 159–286.

Kraus, M.J., 1987, Integration of channel and flood-plain suites. II. Vertical relations of alluvial paleosols: Journal of Sedimentary Petrology, v. 57, p. 602–612.

Laveine, J.-P., 1989, Guide Paleobotanique dans le terrain Houiller Sarro-Lorrain: Documents des Houilleres du Bassin de Lorraine, Imprimerie de Houillères du Bassin de Lorraine, 154 p.

Leary, R.L., 1975, Early Pennsylvanian paleoecology of an upland area, western Illinois, USA: Bulletin de la Société Géologique Belgique, v. 84, p. 19–31.

Leary, R.L., and Pfefferkorn, H.W., 1977, An Early Pennsylvanian flora with Megalopteris and Noeggerathiales from west-central Illinois: Illinois State Geological Survey Circular 500, p. 1–77.

Lyons, P.C., and Darrah, W.C., 1989, Earliest conifers in North America: Upland and/or paleoclimatic indicators?: Palaios, v. 4, p. 480–486.

Markewich, H.W., and Pavich, M.J., 1991, Soil chronosequence studies in temperate to subtropical, low-latitude, low relief terrain with data from the eastern United States: Geoderma, v. 51, p. 213–239, doi: 10.1016/0016-7061(91)90072-2.

Mamay, S.H., 1968, Russellites, new genus, a problematical plant from the Lower Permian of Texas: U.S. Geological Survey Professional Paper 593-I, 15 p.

Mamay, S.H., 1986, New species of Gigantopteridaceae from the Lower Permian of Texas: Phytologia, v. 61, p. 311–315.

Mamay, S.H., 1989, Evolsonia, a new genus of Gigantopteridaceae from the Lower Permian Vale Formation, north-central Texas: American Journal of Botany, v. 76, p. 1299–1311.

Mamay, S.H., 1990, Charliea manzanitana, n. sp., and other enigmatic parallel-veined foliar forms from the Upper Pennsylvanian of New Mexico and Texas: American Journal of Botany, v. 77, p. 858–866.

Mamay, S.H., 1992, Sphenopteridium and Telangiopsis in a Diplopteridium-like association from the Virgilian (Upper Pennsylvanian) of New Mexico: American Journal of Botany, v. 79, p. 1092–1101.

Mamay, S.H., and Mapes, G., 1992, Early Virgilian plant megafossils from the Kinney Brick Company Quarry, Manzanita Mountains, New Mexico: New Mexico Bureau of Mines and Mineral Resources Bulletin, v. 138, p. 61–85.

Mamay, S.H., Miller, D.M., Rohr, D.M., and Stein, W.E., 1988, Foliar morphology and anatomy of the gigantopterid plant Delnortea abbottiae, from the Lower Permian of west Texas: American Journal of Botany, v. 75, p. 1409–1433.

Mapel, W.J., 1967, Bituminous coal resources of Texas: U.S. Geological Survey Bulletin, v. 1242-D, p. D1–D28.

McComas, M.A., 1988, Upper Pennsylvanian compression floras of the 7-11 Mine, Columbiana County, northeastern Ohio: Ohio Journal of Science, v. 88, p. 48–52.

McGowen, J.H., Hentz, T.F., Owen, D.E., Pieper, M.K., Shelby, C.A., and Barnes, V.E., 1967 [revised 1991], Sherman Sheet, University of Texas at Austin, Bureau of Economic Geology, Geologic Atlas of Texas, scale 1:250 000, 1 sheet.

Mear, C.E., 1963, Stratigraphy of Permian outcrops, Coke County, Texas: Bulletin of the American Association of Petroleum Geologists, v. 47, p. 1952–1962.

Mear, C.E., 1984, Stratigraphy of Upper Permian rocks, Midland Basin and Eastern Shelf, Texas: Society of Economic Paleontologists and Mineralogists, Permian Basin Section, Publication no. 84–23, p. 89–93.

Miller, A.K., and Youngquist, W., 1947, Lower Permian cephalopods from the Texas Colorado River Valley: University of Kansas Paleontological Contributions, no. 2, Mollusca, Article 1, p. 1–15.

Moore, R.C., 1949, Rocks of Permian(?) age in the Colorado River Valley, north-central Texas: U.S. Geological Survey Oil and Gas Investigation Preliminary Map 80, scale 1:62 500, 2 sheets.

Murry, P.A., and Johnson, G.D., 1987, Clear Fork vertebrates and environments from the Lower Permian of north-central Texas: Texas Journal of Science, v. 39, p. 254–266.

Naugolnykh, S.V., 1999, A new species of Compsopteris Zalessky from the Upper Permian of the Kama River Basin (Perm Region): Palaeontological Journal, v. 33, p. 686–697.

Nelson, W.J., Hook, R.W., and Tabor, N., 2001, Clear Fork Group (Leonardian, Lower Permian) of north-central Texas: Oklahoma Geological Survey Circular 104, p. 167–169.

Olson, E.C., 1958, Fauna of the Vale and Choza, 14: Summary, review, and integration of the geology and the faunas: Fieldiana, Geology, v. 10, p. 397–448.

Olson, E.C., 1962, Late Permian terrestrial vertebrates: USA and USSR: Transactions of the American Philosophical Society, v. 52, p. 3–224.

Olson, E.C., 1989, The Arroyo Formation (Leonardian: Lower Permian) and its vertebrate fossils: Texas Memorial Museum Bulletin, v. 35, 25 p.

Olson, E.C., and Bolles, K., 1975, Permo-Carboniferous fresh water burrows: Fieldiana, Geology, v. 33, p. 271–290.

Olson, E.C., and Mead, J.G., 1982, The Vale Formation (Lower Permian), its vertebrates and paleoecology: Texas Memorial Museum Bulletin, v. 29, 46 p.

Overpeck, J.T., Webb, R.S., and Webb, T., III, 1992, Mapping eastern North American vegetation change of the past 18 ka: No-analogs and the future: Geology, v. 20, p. 1071–1074, doi: 10.1130/0091-7613(1992)020<1071: MENAVC>2.3.CO;2.

Parrish, W.C., 1978, Paleoenvironmental analysis of Lower Permian bonebed and adjacent sediments, Wichita County, Texas: Palaeogeography, Palaeoclimatology, Palaeoecology, v. 24, p. 209–237, doi: 10.1016/0031-0182(78)90043-3.

Pfefferkorn, H.W., 1980, A note on the term "upland flora": Review of Palaeobotany and Palynology, v. 30, p. 157–158, doi: 10.1016/0034-6667(80)90011-1.

Pfefferkorn, H.W., Mustafa, H., and Hass, H., 1975, Quantitative Charakterisierung ober-karboner Abdruckfloren (Quantitative characterization of compression-impression floras of Upper Carboniferous age): Neues Jahrbuch für Geologie und Paläontologie, Abhandlungen, v. 150, p. 253–269.

Pfefferkorn, H.W., Gastaldo, R.A. and DiMichele, W.A. 2000. Ecological stability during the Late Paleozoic cold interval, in Gastaldo, R.A., and DiMichele, W.A., eds., Phanerozoic terrestrial ecosystems: Paleontological Society Special Papers v. 6, p. 63–78.

Phillips, T.L., 1979, Reproduction of heterosporous arborescent lycopods in the Mississippian-Pennsylvanian of Euramerica: Review of Palaeobotany and Palynology, v. 27, p. 239–289, doi: 10.1016/0034-6667(79)90014-9.

Phillips, T.L., and DiMichele, W.A., 1992, Comparative ecology and life-history biology of arborescent lycopods in Late Carboniferous swamps of Euramerica: Annals of the Missouri Botanical Garden, v. 79, p. 560–588.

Pipujol, M.D., and Buurman, P., 1994, The distinction between groundwater gley and surface-water gley phenomena in Tertiary paleosols of the Ebro Basin, NE Spain: Palaeogeography, Palaeoclimatology, Palaeoecology, v. 110, p. 103–110.

Plummer, F.B., and Scott, G., 1937, Upper Paleozoic ammonites in Texas: University of Texas Bulletin, no. 3701, The geology of Texas, v. 3, part 1, p. 1–516.

Read, C.B., and Mamay, S.H., 1964, Upper Paleozoic floral zones and floral provinces of the United States: U.S. Geological Survey Professional Paper 454-K, p. 1–35.

Read, W.F., 1943, Environmental significance of a small deposit in the Texas Permian: Journal of Geology, v. 51, p. 473–487.

Rees, P.M., Ziegler, A.M., Gibbs, M.T., Kutzbach, J.E., Behling, P.J., and Rowley, D.B., 2002, Permian phytogeographic patterns and climate data/model comparisons: Journal of Geology, v. 110, p. 1–31, doi: 10.1086/324203.

Retallack, G.J., 1990, Soils of the past: An introduction to paleopedology: Boston, Unwin-Hyman, 520 p.

Retallack, G.J., 1994, A pedotype approach to latest Cretaceous and earliest Tertiary paleosols in eastern Montana: Geological Society of America Bulletin, v. 106, p. 1377–1397, doi: 10.1130/0016-7606(1994)106<1377: APATLC>2.3.CO;2.

Romer, A.S., and Olson, E.C., 1954, Aestivation in a Permian lungfish: Breviora, v. 30, p. 1–8.

Roth, R., 1931, New information on the base of the Permian in north-central Texas: Journal of Paleontology, v. 5, p. 295.

Rothwell, G.W., and Mapes, G., 1988, Vegetation of a Paleozoic conifer community, in Mapes, G., and Mapes, R.H., eds., Regional geology and paleontology of upper Paleozoic Hamilton Quarry Area in southeastern Kansas: Geological Society of America, South-Central Section, 33rd Annual Meeting, Guidebook, p. 213–223.

Royer, D.L., 1999, Depth to pedogenic carbonate horizon as a paleoprecipitation indicator?: Geology, v. 27, p. 1123–1126, doi: 10.1130/0091-7613(1999)027<1123:DTPCHA>2.3.CO;2.

Sander, P.M., 1989, Early Permian depositional environments and pond bonebeds in central Archer County, Texas: Palaeogeography, Palaeoclimatology, Palaeoecology, v. 69, p. 1–21, doi: 10.1016/0031-0182(89)90153-3.

Scheckler, S.E., 1986, Floras of the Devonian-Mississippian transition, in Gastaldo, R.A., ed., Land plants: Notes for a short course: University of Tennessee, Studies in Geology, v. 15, p. 81–96.

Scheihing, M.H., and Pfefferkorn, H.W., 1984, The taphonomy of land plants in the Orinoco Delta: A model for the incorporation of plant parts in clastic sediments of Late Carboniferous age of Euramerica: Review of Palaeobotany and Palynology, v. 41, p. 205–280, doi: 10.1016/0034-6667(84)90047-2.

Scott, A.C., and Chaloner, W.G., 1983, The earliest fossil conifer from the Westphalian B of Yorkshire: Proceedings of the Royal Society of London, ser. B, v. 220, p. 163–182.

Scott, A.C., Galtier, J., and Clayton, G., 1984, Distribution of anatomically-preserved floras in the Lower Carboniferous in Western Europe: the Royal Society of Edinburgh, Earth Sciences, v. 75, p. 311–340.

Sellards, E.H., 1932, Pre-Paleozoic and Paleozoic systems in Texas, in Sellards, E.H., Adkins, W.S., and Plummer, F.B., eds., University of Texas Bulletin, no. 3232, The Geology of Texas, v. 1, p. 15–238.

Smith, G.E., 1974, Depositional systems, San Angelo Formation (Permian), North Texas—Facies control on red-bed copper mineralization: University of Texas at Austin Bureau of Economic Geology Report of Investigations, v. 80, p. 1–74.

Soil Survey Staff, 1975, Soil taxonomy: Washington, D.C., U.S. Department of Agriculture Handbook 436, 754 p.

Soil Survey Staff, 1998, Keys to soil taxonomy: Washington, D.C., U.S. Department of Agriculture, Natural Resources Conservation Service, 325 p.

Spicer, R.A., 1981, The sorting and deposition of allochthonous plant material in a modern environment at Silwood Lake, Silwood Park, Berkshire, England. U.S. Geological Survey Professional Paper 1143, p. 1–77.

Tabor, N.J., and Montañez, I.P., 2004, Morphology and distribution of fossil soils in the Permo-Pennsylvanian Wichita and Bowie Groups, north-central Texas, USA: Implications for western equatorial Pangean paleoclimate during icehouse-greenhouse transition: Sedimentology, v. 51, p. 851–884, doi: 10.1111/j.1365-3091.2004.00655.x.

Tabor, N.J., Montañez, I.P., and Southard, R.J., 2002, Paleoenvironmental reconstruction from chemical and isotopic compositions of Permo-Pennsylvanian pedogenic minerals: Geochimica et Cosmochimica Acta, v. 66, p. 3093–3107, doi: 10.1016/S0016-7037(02)00879-7.

Tian, B.L., Wang, S.J., Gao, Y.T., Chen, G.R., and Zhao, H., 1996, Flora of Palaeozoic coal balls in China: Palaeobotanist, v. 45, p. 247–254.

Wagner, R.H., 1983, Upper Stephanian stratigraphy and palaeontology of the Puertollano Basin, Ciudad Real, Spain, in Lemos de Sousa, M.J., and Wagner, R.H., eds., Papers on the Carboniferous of the Iberian Peninsula: Annals Facultad de Ciências, Porto, Supplement, v. 64, p. 171–231.

Weiher, E., and Keddy, P., 1999 (2001). Assembly rules as general constraints on community composition, in Weiher, E., and Keddy, P., eds., Ecological assembly rules: Cambridge, UK, Cambridge University Press, p. 251–271.

White, D.W., 1911, in Gordon, C. H., Geology and underground waters of northeastern Texas: U.S. Geological Survey Water-Supply Paper 276 p. 1–78.

Wilson, M.J., 1999, The origin and formation of clay minerals in soils; past, present and future perspectives: Clay Minerals, v. 34, p. 7–25.

Wing, S.L., Hickey, L.J., and Swisher, C.C., 1993, Implications of an exceptional fossil flora for Late Cretaceous vegetation: Nature, v. 363, p. 342–344, doi: 10.1038/363342a0.

Wright, V.P., and Marriott, S.B., 1996, A quantitative approach to soil occurrence in alluvial deposits and its application to the Old Red Sandstone of Britain: Journal of the Geological Society [London], v. 153, p. 907–913.

Yemane, K., Kahr, G., and Kelts, K., 1996, Imprints of post-glacial climates and paleogeography in the detrital clay mineral assemblages of an Upper Permian fluviolacustrine Gondwana deposit from north-central Malawi: Palaeogeography, Palaeoclimatology, Palaeoecology, v. 125, p. 27–49, doi: 10.1016/S0031-0182(96)00023-5.

Zhou, Y.-X., 1994, Earliest pollen-dominated microfloras from the early Late Carboniferous of the Tian Shan, northwest China: Their significance for the origin of conifers and palaeophytogeography: Review of Palaeobotany and Palynology, v. 81, p. 193–211, doi: 10.1016/0034-6667(94)90108-2.

Zodrow, E.L., 1990, Revision and emendation of *Pecopteris arborescens* Group, Permo-Carboniferous: Palaeontographica, ser. B, v. 217, p. 1–49.

MANUSCRIPT ACCEPTED BY THE SOCIETY 28 JUNE 2005

Carbon isotopic evidence for terminal-Permian methane outbursts and their role in extinctions of animals, plants, coral reefs, and peat swamps

Gregory J. Retallack*
Department of Geological Sciences, University of Oregon, Eugene, Oregon 97403-1272, USA

Evelyn S. Krull*
CSIRO Land and Water, Adelaide Laboratories, PMB2, Glen Osmond, South Australia 5064, Australia

Dedicated to the memory of William T. Holser, colleague and friend.

ABSTRACT

A gap in the fossil record of coals and coral reefs during the Early Triassic follows the greatest of mass extinctions at the Permian-Triassic boundary. Catastrophic methane outbursts during terminal Permian global mass extinction are indicated by organic carbon isotopic ($\delta^{13}C_{org}$) values of less than –37‰, and preferential sequestration of ^{13}C-depleted carbon at high latitudes and on land, relative to low latitudes and deep ocean. Methane outbursts massive enough to account for observed carbon isotopic anomalies require unusually efficient release from thermal alteration of coal measures or from methane-bearing permafrost or marine methane-hydrate reservoirs due to bolide impact, volcanic eruption, submarine landslides, or global warming. The terminal Permian carbon isotopic anomaly has been regarded as a consequence of mass extinction, but atmospheric injections of methane and its oxidation to carbon dioxide could have been a cause of extinction for animals, plants, coral reefs and peat swamps, killing by hypoxia, hypercapnia, acidosis, and pulmonary edema. Extinction by hydrocarbon pollution of the atmosphere is compatible with many details of the marine and terrestrial fossil records, and with observed marine and nonmarine facies changes. Multiple methane releases explain not only erratic early Triassic carbon isotopic values, but also protracted (~6 m.y.) global suppression of coral reefs and peat swamps.

Keywords: Permian, Triassic, boundary, methane, mass extinction, coal, reef

INTRODUCTION

Mass extinctions of the past, like a good murder mystery, can be instructive for the living. The Permian-Triassic boundary, for example, not only was the greatest biodiversity crisis of the past 500 m.y. (Erwin, 1993), but it also was followed by

*E-mails: gregr@darkwing.uoregon.edu; Evelyn.Krull@csiro.au.

the longest temporal gap in the fossil record of coals since their Late Devonian appearance (Retallack, et al., 1996) and of coral reefs since their Ordovician appearance (Weidlich et al., 2003). No coals or coral reefs are known for the entire Early Triassic (Fig. 1), which has a currently dated duration of ~6 m.y. (Mundil et al., 2004; Gradstein et al., 2005). The causes of this global ecological crisis are potentially relevant to modern anthropogenic extinctions, wetland pollution (Whiting and Chanton, 2001), and coral reef predation and bleaching (Pandolfi, 1992; Marshall and McCulloch, 2002).

A diagnostic clue to the life-crisis at the Permian-Triassic boundary is a marked carbon isotopic anomaly (negative $\delta^{13}C$ spike) in both carbonate and organic matter (Magaritz et al., 1992; Morante, 1996), which has been instrumental for recognition of this mass extinction at many localities worldwide (Fig. 2). When comprehensively sampled and well preserved, the isotopic anomaly has a characteristic shape of rapid but steady decline to a very low value, then rebound followed by one or more low spikes, before settling on Early Triassic values often a little lower than during the Late Permian (Figs. 3–5). Other abrupt or erratic carbon isotopic excursions at the Permian-Triassic boundary may reflect local disconformities and inadequately dense sampling, but one or more negative spikes and later persistent lower values are common to most carbon isotopic records across the Permian-Triassic boundary. Here we consider the magnitude, shape, and paleogeographic distribution of numerous Permian-Triassic carbon isotopic spikes as evidence for the nature of the carbon-cycle crisis some 251 m.y. ago.

The negative carbon isotopic anomaly at the Permian-Triassic boundary has been variously attributed to curtailed biological productivity in the ocean (Kump, 1991; Wang et al., 1994), curtailed biological productivity on land (Broecker and Peacock, 1999), change from woody to herbaceous vegetation (Gorter et al., 1995; Foster et al., 1998), erosion of coals exposed during low sea level (Holser and Magaritz, 1987; Magaritz et al., 1992), overturn of a stratified ocean carbonated with dissolved CO_2 (Hoffman et al., 1991; Kajiwara et al., 1994; Knoll et al., 1996), introduction of extraterrestrial carbon from comet or asteroid

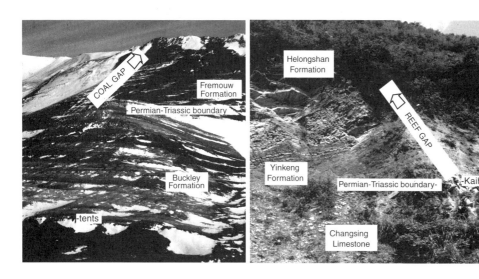

Figure 1. Permian-Triassic boundary sections showing strata (A) within the global coal gap in nonmarine strata of Graphite Peak, Antarctica and (B) within the global reef gap in marine strata of Meishan, China. For scale: at Graphite Peak, small triangular tents; at Meishan, Kunio Kaiho collecting samples.

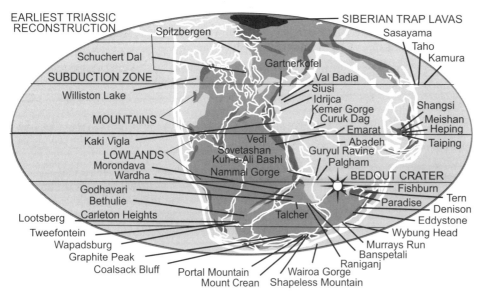

Figure 2. World paleogeographic map at the Permian-Triassic boundary (251 Ma, from Scotese 1994) showing sites analyzed for carbon isotopic anomalies.

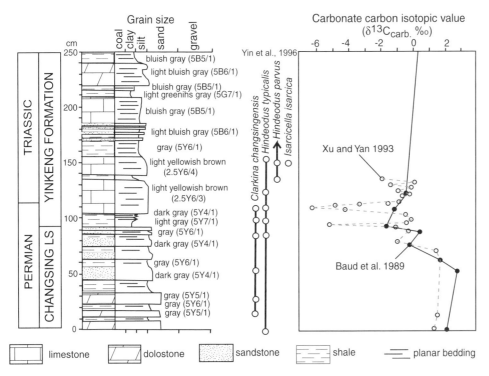

Figure 3. Stratigraphic section at the global Permian-Triassic stratotype in Meishan quarry D, China, measured by Retallack, with conodont ranges from Yin et al. (1996) and carbonate carbon isotopic data after Xu and Yan (1993) and Baud et al. (1989). Additional isotopic data of d'Hondt et al. (2000) show a shift comparable to that in the data of Baud et al. (1989), which were preferred for our compilation (Table 1).

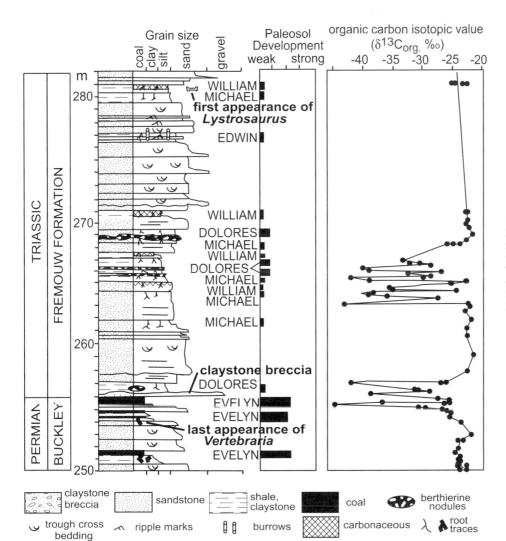

Figure 4. Stratigraphic section of the Permian-Triassic boundary at Graphite Peak, with stratigraphic and fossil range data from Retallack and Krull (1999) and carbonaceous carbon isotopic data from Krull and Retallack (2000) and Retallack et al. (2005).

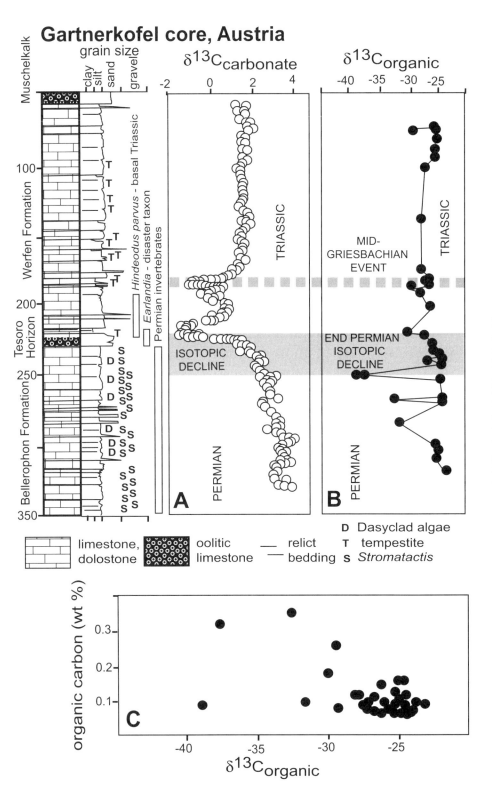

Figure 5. A pronounced negative excursion in carbon isotopic value ($\delta^{13}C$) of carbonate at the Permian-Triassic boundary in the Gartnerkofel-1 core, Austria (A), corresponds to very low isotopic values for organic carbon (B) but no correlation between total organic carbon content and carbon isotopic value of carbon in these samples (C). Samples with low values are mostly high in organic carbon content and so are inferred to be reliable analyses (data from Holser et al. 1991; Magaritz et al. 1992).

impact (Rampino and Haggerty, 1996), degassing associated with flood basalt eruption (Landis et al., 1996; Conaghan et al., 1994; Isozaki, 1997), methane clathrate dissociation (Morante, 1996; Vermeij and Dorritie, 1996; de Wit et al., 2002), and thermogenic methane outbursts from volcanic intrusion of coal measures (suggested here, as in Paleocene case described by Svensen et al., 2004). All of these hypotheses were viable when the isotopic excursion was only thought to be some −3‰ $\delta^{13}C$ and was unknown in nonmarine sections across the Permian-Triassic boundary. Now that numerous isotopic spikes of much larger magnitude have been discovered from a global network of carbon isotopic profiles in both marine and nonmarine facies (Table 1), the most likely explanation for the carbon isotopic shift is atmospheric pollution from massive outbursts of coalbed or clathrate methane. Such methane release and atmospheric pollution events have already been hypothesized from the postglacial (12 ka: Thorpe et al., 1998; Kennett, 2003), Ypresian (55 Ma, Paleocene-Eocene boundary: Sinha et al., 1995; Bains et al., 1999), Danian (65 Ma, Cretaceous-Tertiary boundary: Max et al. 1999), Aptian (117 Ma, Early Cretaceous: Jahren et al. 2001), Toarcian (175 Ma, Early Jurassic: Hesselbo et al. 2000), and Hettangian (200 Ma, earliest Jurassic: Hesselbo et al., 2002).

Productivity and vegetation-change hypotheses all imply that the carbon isotopic excursion was a consequence of mass extinction, but the methane outburst scenario, like the ocean overturn, bolide, and volcanic hypotheses, allows a causative role for CH_4 and CO_2 as kill mechanisms. Neither gas is benign at high concentrations. Methane reacts with atmospheric NO_x to create formaldehyde, and forms explosive mixtures with air when between 5.5 vol% and 14 vol% (Patnaik, 1992). Methane asphyxiates humans at concentrations high enough to displace oxygen (87%–90%) and anesthetizes at concentrations in excess of 1000 ppmV (Clough, 1998). Methane oxidizes to CO_2 in the atmosphere within 7–24 yr (Khalil et al., 2000), and CO_2 is also an asphyxiant, fatal to humans in excess of 15 vol% and inducing headache and breathing difficulty in excess of 2% (Kulkarni and Mehendale, 1998). Acidosis and pulmonary edema from high CO_2 levels can both be fatal (Fiddian-Green, 1995; Engoren, 2004; Retallack 2004c). Key questions for assessing the role of methane in extinction of organisms and ecosystems are timing and dose, which we address before discussing mechanisms of release and kill.

DATA COMPILATION AND METHODS

We have compiled data on the transient $\delta^{13}C$ excursion and long-term offset across the Permian-Triassic boundary at all studied localities, together with estimates of paleolatitude and paleoelevation of those sites (Table 1). Localities in Nammal Gorge (Pakistan), Schuchert Dal (Greenland), and near Meishan (China) have been analyzed for $\delta^{13}C$ several times with different results, but published protocols encourage us to use the most recent studies (Baud et al., 1996; Bowring et al., 1998; D'Hondt et al., 2000; Twitchett et al., 2001). Our estimates of paleolatitude from paleogeographic maps (Scotese, 1994) are limited by propagation of errors in plate restoration transects (±600 km) and uncertainties in position of small terranes (±2000 km). The highest paleoelevation (+300 m) was for the sequence at Graphite Peak, Antarctica, which was deposited in a fluvial basin well inland of an Andean-style volcanic mountain range (Retallack and Krull, 1999). Negative paleoelevations were used for depth below sea level, which for many sequences of limestone with marine algae was within the photic zone. The lowest paleoelevation (−3000 m) was for the cherty deep-marine sequence at Sasayama, Japan, assumed to have been deposited below calcium carbonate compensation depth for its paleolatitude (Ishiga et al., 1993). Uncertainties in depth below sea level reflect the coarseness of facies and biotic depth zonation as well as marine transgression across the Permian-Triassic boundary (Table 1).

New isotopic analyses of Early Triassic fossil plants also are presented here (Table 2) in order to assess natural variability of plant sources for early Triassic organic carbon. These were obtained by one of us (ESK) following protocols outlined by Bestland and Krull (1999) in the Department of Geological Sciences, Indiana University.

ISOTOPIC EVIDENCE FOR COMPLETENESS OF BOUNDARY SEQUENCES

The Permian-Triassic carbon isotopic sequence most comprehensively sampled and analyzed is still, after a decade of subsequent studies, the Gartnerkofel core of the Carnic Alps of Austria (Fig. 5). This sequence has Permian marine faunas to a depth of 230 m, the first appearance of the conodont *Hindeodus parvus* at 225 m, and a crisis fauna of *Earlandia* foraminifera and microgastropods from 224 to 190 m (Holser et al., 1991). The first appearance of *Hindeodus parvus* at the stratotype section of Meishan in China marks the Permian-Triassic boundary (Jin et al., 2000). Late Permian carbonate $\delta^{13}C$ values remain fairly constant at about +3.0‰ until 251 m, which is 26 m below the boundary defined by the newly evolved conodont *H. parvus*, and estimated using cyclostratigraphy to be 260 k.y. before the boundary (Rampino et al., 2000). Carbonate isotopic values then begin a regular decline like that of a mixing curve to the lowest values of −1.3‰ $\delta^{13}C$ at 220 m. The beginning of this decline in carbonate corresponds to anomalously low organic carbon isotopic values in the same core (at 251 m in Fig. 5B). After this point, carbon and carbonate isotopic values track each other more closely. Following the Permian-Triassic boundary, carbonate isotopic values rebound to a maximum of +0.7‰ $\delta^{13}C$ at 206 m, estimated to be some 190 k.y. after the boundary (Rampino et al., 2000). Carbonate isotopic values then fall to −1.1‰ $\delta^{13}C$ at 186 m some 390 k.y. after the boundary, then rebound to an earliest Triassic plateau of about +1.3‰ $\delta^{13}C$ at 162 m, some 630 k.y. after the boundary. The two declines in isotopic values have been attributed to soil formation from two separate horizons (Heydari et al., 2001), but there is no independent evidence of soil formation in the core or nearby outcrop other than pervasive dolomitization, which can

TABLE 1. CARBON ISOTOPIC OFFSET AND EXCURSION ACROSS THE PERMIAN-TRIASSIC BOUNDARY

LOCALITY	Reference	$\delta^{13}C$ offset	$\delta^{13}C$ excursion	Paleolatitude	Paleoelevation (m)
Carbonate					
Abadeh, Iran	Heydari et al., 2000	−1.3	−5.1	−15 ± 10	−50 ± 40
Çürük Dağ, Turkey	Baud et al., 1989	−2.9	−3.3	+5 ± 6	−6 ± 10
Emarat, Iran	Baud et al., 1989	−3.2	−4.9	−7 ± 10	−10 ± 15
Gartnerkofel-1, Austria	Magaritz et al., 1992	−1.7	−4.6	+26 ± 6	−120 ± 100
Guryul Ravine, India	Baud et al., 1996	−3.5	−5.6	−29 ± 6	−40 ± 30
Heping, China	Krull et al., 2004	−0.9	−3.4	+1 ± 6	−10 ± 15
Idrijca, Slovenia	Baud et al., 1989	−3.6	−2.6	+13 ± 6	−10 ± 15
Kaki Vigla, Greece	Baud et al., 1989	−1.3	−2.5	+6 ± 6	−500 ± 400
Kamura, Japan	Musashi et al., 2001	−0.5	−2.3	30 ± 6	−6 ± 12
Kemer Gorge, Turkey	Baud et al., 1989	−2.8	−3.7	+6 ± 6	−10 ± 10
Kuh-e Ali Bashi, Iran	Baud et al., 1989	−2.2	−3.3	−3 ± 10	−20 ± 15
Lootsberg, South Africa	de Wit et al., 2002	−1.2	−3.3	−54 ± 6	+200 ± 150
Meishan-D, China	d'Hondt et al., 2000	−3	−4.6	+2 ± 6	−40 ± 30
Nammal Gorge, Pakistan	Baud et al., 1996	−3.6	−4.4	−26 ± 6	−10 ± 20
Palgham, India	Baud et al., 1996	−4.2	−5.8	−28 ± 6	−40 ± 30
Schuchert Dal, Greenland	Twitchett et al., 2001	−5	−8	+38 ± 6	−100 ± 90
Shangsi, China	Baud et al., 1989	−3.5	−5	+3 ± 10	−100 ± 100
Siusi, Italy	Newton et al., 2004	−3.1	−6.9	+24 ± 6	−120 ± 100
Sovetashan, Armenia	Baud et al., 1989	−1.6	−3.5	−2 ± 10	−6 ± 12
Taho, Japan	Musashi et al., 2001	−0.7	−2.1	0 ± 6	−6 ± 12
Taiping, China	Krull et al., 2004	−2.1	−3.4	10 ± 15	10 ± 15
Vedi, Armenia	Baud et al., 1989	−2.2	−2	+2 ± 10	−8 ± 12
Nonmarine carbonate					
Bethulie, South Africa	MacLeod et al., 2000	−2	−13.1	−53 ± 6	+200 ± 150
Tweefontein, South Africa	Ward et al., 2004	−3.9	−9.6	−56 ± 6	+200 ± 150
Wapadsberg, South Africa	Ward et al., 2004	−5.8	−8.4	−56 ± 6	+200 ± 150
Marine organic					
Abadeh, Iran	Heydari et al., 2000	−0.9	−3.2	−15 ± 10	−50 ± 40
Fishburn-1, Australia	Morante, 1996	−6.6	−8	−40 ± 6	−100 ± 80
Gartnerkofel-1, Austria	Magaritz et al., 1992	−1.8	−5.8	+31 ± 6	−120 ± 100
Heping, China	Krull et al., 2004	−2.7	−5.3	+1 ± 6	−10 ± 15
Kamura, Japan	Musashi et al., 2001	−1.0	−2.2	30 ± 6	−6 ± 12
Paradise 1–6, Australia	Morante, 1996	−7.7	−9.4	−37 ± 6	−50 ± 50
Sasayama, Japan	Ishiga et al., 1995	−0.9	−6.5	+31 ± 6	−3000 ± 1000
Schuchert Dal, Greenland	Twitchett et al., 2001	−6	−9	+38 ± 6	−100 ± 90
Festningen, Spitzbergen	Wignall et al., 1998	−5.3	−5.6	+46 ± 6	−100 ± 90
Taiping, China	Krull et al., 2004	−1.7	−4.4	0 ± 6	−10 ± 15
Taho, Japan	Musashi et al., 2001	−0.3	−2.5	29 ± 6	−6 ± 12
Tern-3, Australia	Morante, 1996	−3.1	−10.6	−35 ± 6	−60 ± 80
Wairoa Gorge, New Zealand	Krull et al, 2000	−5.2	−15	−70 ± 10	−400 ± 300
Williston Lake, Canada	Wang et al., 1994	−1.8	−4.1	+35 ± 30	−2000 ± 1000
Nonmarine organic					
Banspetali, India	Sarkar et al., 2003	−1.9	−9.7	−42 ± 10	+40 ± 40
Carleton Heights, S. Africa	Ward et al., 2004	−0.9	−3.3	−54 ± 6	+200 ± 150
Coalsack Bluff, Antarctica	Retallack et al., 2004	−0.8	−4.4	−69 ± 6	+40 ± 50
Denison, Australia	Morante, 1996	−3.4	−5.5	−48 ± 6	+200 ± 90
Eddystone, Australia	Morante, 1996	−2	−3.2	−52 ± 6	+150 ± 90
Godhavari Coalfield, India	de Wit et al., 2002	−3.0	−11.2	−44 ± 6	+40 ± 40
Lootsberg, South Africa	Ward et al., 2004	−0.8	−2.2	−55 ± 6	+200 ± 150
Morondava, Madagascar	de Wit et al., 2002	?	−7.9	−42 ± 6	+20 ± 20
Graphite Peak, Antarctica	Krull and Retallack, 2000	−4	−22.2	−70 ± 6	+300 ± 100
Mount Crean, Antarctica	Retallack et al., 2004	−1.0	−3.3	−65 ± 6	+40 ± 50
Murrays Run, Australia	Morante, 1996	−1.7	−2.9	−58 ± 6	+40 ± 50
Portal Mountain, Antarctica	Retallack et al., 2004	−1.0	−3.1	−66 ± 6	+40 ± 50
Raniganj Coalfield, India	de Wit et al., 2002	−4.2	−13.8	−42 ± 10	+40 ± 40
Shapeless Mt, Antarctica	Retallack et al., 2004	−3.2	−3.4	−64 ± 6	−40 ± 50
Talcher Coalfield, India	de Wit et al., 2002	−3.1	−8.8	−45 ± 6	−50 ± 40
Wardha Coalfield, India	de Wit et al., 2002	−2.1	−3.7	−41 ± 6	+40 ± 40
Wybung Head, Australia	Retallack and Jahren, unpub.	−1.3	−2.4	−59 ± 6	+40 ± 50

TABLE 2. CARBON ISOTOPIC COMPOSITION OF FOSSIL PLANT LEAVES
FROM THE SYDNEY BASIN, AUSTRALIA

Taxon	Locality	Triassic age	Specimen	C wt%	$\delta^{13}C_{org}$
Tomiostrobus australis	Terrigal	Late Early	P12194C	10.5	−27.5
Tomiostrobus australis	Terrigal	Late Early	P5594C	6.9	−26.8
Cylostrobus sydneyensis	Turimetta Head	Late Early	P4031C	9.8	−22.5
Isoetes beestonii	South Bulli Mine	Earliest	P12200a	8.4	−26.8
Lepidopteris callipteroides	South Bulli Mine	Earliest	P12200a	5.4	−27.9

be explained in other ways (Boeckelman and Magaritz, 1991). In any case, known depression of carbonate isotopic values by soil formation on carbonates is less profound and does not show such strongly curved depth-functions (Allan and Matthews, 1977). Soil formation also fails to explain decoupled carbon isotopic variation in coexisting organic matter (Fig. 5). Instead, the smooth variation of isotopic values in these marine rocks suggest oceanic mixing curves and recovery from at least two massive injections of isotopically light carbon.

Not all carbon isotopic records through the Permian-Triassic boundary show the high sampling density or stratigraphic completeness of the Gartnerkofel Core. For example, several sequences (Abadeh carbonate, Bethulie carbonate, Çürük Dağ, Paradise 1–6, Sovetashan, and Talcher Coalfield of Table 1) go from values within the range of the Late Permian analyses to the lowest of boundary isotopic values from one sample to the next. Such records probably reflect incompleteness of sampling or of stratigraphic record (Heydari et al. 2001). Conversely, there are other sequences, such as Meishan, China (Fig. 3), Graphite Peak, Antarctica (Fig. 4), and Murrays Run bore, southeastern Australia (Morante and Herbert, 1994; Morante, 1996), where disconformities have long been a local stratigraphic issue (Yin et al., 1996; Retallack and Krull, 1999; Retallack, 1999a), but which appear comparable in completeness to the Gartnerkofel core in resolving decline from latest Permian to earliest Triassic isotopic values.

ISOTOPIC EVIDENCE FOR EARLY TRIASSIC ATMOSPHERIC METHANE

A surprising discovery of recent years is that the carbon isotopic anomaly of the Permian-Triassic boundary is not merely a phenomenon of marine carbonates. The carbon isotopic excursion and offset are similar for both marine carbonate and organic carbon in the Gartnerkofel core (Fig. 5B) and elsewhere (Heydari et al., 2000; Twitchett et al., 2001; Musashi et al., 2001; Krull et al., 2004). Isotopic excursions also are seen in paleosol organic matter (Krull and Retallack, 2000), paleosol carbonate, and tusks of therapsid reptiles (MacLeod et al., 2000). The ubiquity and magnitude of this signal are clues that it reflects more than just marine productivity, different plant physiological pathways, or diagenetic artifacts.

Magnitude of Carbon Isotopic Spike

The magnitude of the carbon isotopic anomaly is an important new clue. Some organic carbon isotopic values are so low (−39‰ to −29‰ $\delta^{13}C$) that they were not at first thought credible by Magaritz et al. (1992), but very low isotopic values of organic carbon are now known from 12 Permian-Triassic boundary sections (Ghosh et al., 1998; Krull et al., 2000; Krull and Retallack, 2000; de Wit et al., 2002; Sarkar et al., 2003; Newton et al., 2004; Ward et al., 2005). The very low Gartnerkofel values are from samples at least as high in total organic carbon as other samples analyzed (Fig. 3C), so it is unlikely that these low values are due to contamination in carbon-lean samples.

Carbon isotope values less than −37‰ in organic matter are known from biologically produced methane, which is typically −60‰ and can have values as low as −110‰ (Whiticar, 2000), and from coalbed thermogenic methane which is −35 to −55‰ (Clayton, 1998). The average isotopic excursion for organic carbon in both marine and nonmarine rocks at the Permian-Triassic boundary is −6.4 ± 4.4‰ ($n = 30$; Table 1). The average nonmarine excursion in organic carbon of −6.0 ± 5.1‰ ($n = 16$) is not significantly different from the marine organic carbon excursion of −7.0 ± 3.5‰ ($n = 14$). Such a global average isotopic excursion requires rapid release to the atmosphere of at least 622 ± 589 Gt ($= 10^{15}$g) of carbon in methane (using the mass balance equation of Jahren et al., 2001, assuming methane at −60‰ $\delta^{13}C$, and pCO$_2$ from Retallack, 2001, 2002, recalculated by method of Wynn, 2003). Earliest Triassic methane release and oxidation of 299 ± 285 ppmV CH$_4$ added to Late Permian CO$_2$ inventories of 1385 ± 631 ppmV (Retallack, 2001; Wynn, 2003) would have created a transient atmosphere with 1684 ± 916 ppmV CO$_2$, a value with the same order of magnitude as estimates from the stomatal index of earliest Triassic seed ferns (Retallack, 2001, 2002; Wynn, 2003). This large amount of released methane is much less than proposed for the Permian-Triassic boundary by Ryskin (2003), and no more than 6% of the current inventory of methane hydrates in the world as estimated by Kvenvolden (2000). However, Milkov (2004) estimates global inventories of methane hydrates at only 500–2500 Gt. If Permian reservoirs were similar in size, they would not have yielded methane in amounts sufficient to explain the observed carbon isotope excursion. A more voluminous source is thermogenic methane release by igneous intrusion of carbonaceous sediments (Svensen et al., 2004), because this mechanism can generate 1000 Gt methane from 300–1800 tonnes of coal (Clayton, 1998). High levels of methanogenic CH_4 would not have been sustained in the atmosphere for more than a few tens of years because of oceanic mixing, biotic recycling, and burial with organic matter (Berner, 2002). Claystone breccias, paleosol development, and braided stream paleochannels of the earliest Triassic all indicate an accelerated rate of erosion, sedimentation, and carbon burial following this greatest of all life crises (Retallack et al., 1998, 2003; Retallack 1999a; Retallack and Krull, 1999; Ward et al., 2000).

Our estimates of *minimum* emission volume (n) are based on equilibration of land plant organic carbon isotopic composition with atmospheric carbon isotopic composition, according to the relationship

$$(GtC_{atmosphere} Permian + n\, GtC) \cdot (\delta^{13}C_{atmosphere} Triassic) =$$

$$GtC_{atmosphere} Permian \cdot (\delta^{13}C_{atmosphere} Permian) + n\, GtC \cdot (\delta^{13}C_{emission})$$

This formulation of Jahren et al. (2001) equilibrating organic matter with the atmosphere is comparable to an earlier formulation of Dickens et al. (1995) equilibrating carbonate carbon with oceanic bicarbonate, which is suspect for Permian-Triassic applications because of likely diagenetic alteration of isotopic values of marine carbonate. We are not convinced by the argument of Heydari et al. (2001) that covariance of oxygen and carbon isotopic composition in latest Permian and earliest Triassic carbonates necessarily indicates diagenetic alteration, because plant physiological processes can create such covariance in terrestrial organic matter (Farquhar et al.1993), which contaminates most marine environments. More troubling is the finding of Mii et al. (1997) that Permian-Triassic brachiopod and bivalve shells are variably recrystallized and their carbon isotopic composition altered from original values. Yet another problem is assumed equilibration of the entire ocean with such transient events as massive outbursts of methane from clathrates (Dickens et al., 1995) or intruded carbonaceous sediments (Svensen et al., 2004). Rapid dispersal and oxidation of methane to carbon dioxide, then assimilation by terrestrial vegetation and marine phytoplankton, may keep much methanogenic carbon from equilibration with deep oceanic bicarbonate. For these reasons, our estimates are based entirely on nonmarine organic carbon isotopic values. Nevertheless, dynamic modeling including land and sea data by de Wit et al. (2002) and Berner (2002) gives comparable magnitudes and time scales of Permian-Triassic atmospheric hydrocarbon pollution following massive methane outbursts.

These calculations allow reassessment of past explanations for the Permian-Triassic carbon cycle perturbation involving (1) non-methanogenic organic carbon such as biomass oxidation (Wang et al., 1994; Twitchett et al., 2001), vegetation change (Foster et al., 1998), coal erosion (Magaritz et al., 1992), and oceanic overturn (Knoll et al.,1996); (2) meteoritic carbon from extraterrestrial impacts (Rampino and Haggerty, 1996); or (3) Earth's mantle carbon from general degassing (Isozaki, 1997) or eruptive degassing of the Siberian traps (Conaghan et al., 1994). None of these mechanisms can explain the magnitude of the carbon isotopic excursions by themselves, as outlined below.

Non-methanogenic biomass and fossil fuels have carbon isotopic composition ranging from –35‰ to –15‰. Early Triassic plant values show less range (Table 2). Even if all the oxidized organic matter from biomass, vegetation change, or oceanic overturn had the unrealistically low carbon isotopic value of –35‰, some 4017 GtC would be needed to lower global isotopic organic carbon composition by –6.4‰ (calculations again by formula above). This large mass of carbon is five times the current biomass and more than twice the current soil carbon reserves (including peats) of Earth (Siegenthaler and Sarmiento, 1993), unlikely for the Late Permian (Hotinski et al., 2001; Beerling and Woodward, 2001).

Meteorites have carbon isotopic values of –47‰ to +1100‰ (Grady et al., 1986; Pillinger, 1987), and comets and interplanetary dust particles have carbon isotopic values down to –120‰ (Jessberger, 1999; Messenger, 2000). Carbonaceous chondrite fragments from the Permian-Triassic boundary in Antarctica are much less carbonaceous and more mineral rich than comets or interplanetary dust particles (Basu et al., 2003). Assuming all the carbon of a carbonaceous chondrite had an isotopic value of –47‰, carbon content as high as 7 wt% (Vdyovkin and Moore 1971), and bulk density as low as 2.2 g.cm^{-3} (Wasson, 1974), the observed –6.4‰ isotopic shift would require a 25-km-diameter asteroid or comet to deliver 1262 GtC at the Permian-Triassic boundary. Evidence from size of shocked quartz, magnitude of iridium anomalies, amounts of extraterrestrial-helium-bearing fullerenes, and size of Bedout Crater all indicate a Permian-Trassic impactor no bigger than the Cretaceous-Tertiary impactor of ~10 km (Retallack et al., 1998; Becker et al., 2001, 2004). A 10–60 km bolide has been proposed on the basis of sulfur isotopic anomalies at one site (Kaiho et al., 2001), but this result has been challenged as more likely due to changes in marine bacterial sulfate reduction (Koeberl et al., 2002).

The Permian-Triassic isotopic anomaly is also too large to be explained by carbon from mantle degassing by way of hydrothermal vents or volcanic eruptions. Addition of volcanic gas with carbon isotopic value of –7‰ to –5‰ (Dickens et al. 1995) into an atmosphere already with carbon isotopic composition of –6.4‰ to –9.6‰ (Arens et al., 2000) would not increase $\delta^{13}C$ of organic matter from its usual value of about –27‰ (Erwin, 1993).

Catastrophic methane outbursts near the Permian-Triassic boundary also explain multiple negative isotopic excursions in many sections (Magaritz et al., 1992; Morante, 1996; Ghosh et al., 1998; Krull et al., 2000; Krull and Retallack, 2000; Heydari et al., 2000; de Wit et al., 2002). Continued Triassic releases of methane may have contributed to unusually protracted recovery from the mass extinction extending some 6 m.y. through most of the Early Triassic (Retallack et al., 1996, 2005; Twitchett et al., 2001; time scale of Gradstein et al., 2005). Additional injections of methane also could have contributed to the generally lower carbon isotopic values of early Triassic organic matter and carbonate (long-term offset of Fig. 6B, 6D), but the size of this difference from late Permian values (~ –2‰ $\delta^{13}C$) is not uniquely methanogenic and could be as much due to productivity, oceanic, volcanic, or paleoclimatic effects (Kump, 1991; Broecker and Peacock, 1999). Modeling indicates that methane outbursts at the Permian-Triassic boundary probably did not continue to destabilize all near-surface methane reservoirs as temperatures rose (de Wit et al., 2002). Instead, methane was released in distinct and transient pulses.

Paleogeographic Variation of Isotopic Anomalies

Organic carbon isotopic data show greater overall variation and paleogeographic spread than carbonate carbon isotopic data, although much deep-water organic matter was presumably formed at shallower depths than where it was preserved. Both long-term carbon isotopic offset and the transient excursion across the Permian-Triassic boundary are less at low latitudes than at high ones (Fig. 6A, B). In addition, the largest long-term offsets and transient excursions were on land, and the smallest in the sea (Fig. 6C, D).

Why would high-latitude land plants have been such effective sinks for isotopically light carbon compared with low-latitude plants and phytoplankton? This problem is exacerbated by lack of evidence for peat-forming ecosystems as a carbon sink during the earliest Triassic at any paleolatitude, probably as a consequence of the extinctions in wetland ecosystems (Retallack et al., 1996). Also distinctive of the earliest Triassic is global warming indicated by migration of thermophilic plants and paleosols into high latitudes of both the Northern and Southern Hemispheres (Retallack, 1999a; Retallack and Krull, 1999; Looy, 2000). There were pioneering plant communities of herbaceous quillworts, and woody seed ferns and conifers at high latitudes (Retallack, 1997, 2002; Retallack and Krull, 1999; Twitchett et al., 2001). Tropical earliest Triassic vegetation was similar, with perhaps a greater abundance of herbaceous isoetaleans (Looy, 2000; Wang and Chen, 2001). From these perspectives the latitudinal difference in isotopic composition could be attributed to differences in plant decay, productivity, woodiness, or physiology. Polar versus tropical differences in decay, productivity, water stress, or herbaceous versus woody plants cannot account for the magnitude of the isotopic differences or their absolute values, because these effects do not alter plant carbon isotopic composition by more than 5‰ (Bestland and Krull, 1999; Beerling and Woodward, 2001). Another possibility is that tropical plants had C_4 or CAM photosynthetic pathways and were thus isotopically heavier than high-latitude plants. This does not account for the extremely light absolute values either (de Wit et al., 2002), and in any case is not supported by our C_3 isotopic values of $\delta^{13}C_{org}$ for earliest Triassic fossil *Isoetes*, a genus that today includes aquatic CAM species (Table 2). The C_4 and CAM pathways are adaptations for conservation of CO_2 in the atmosphere or underwater when CO_2 is limiting (Beerling and Woodward, 2001), but neither the Late

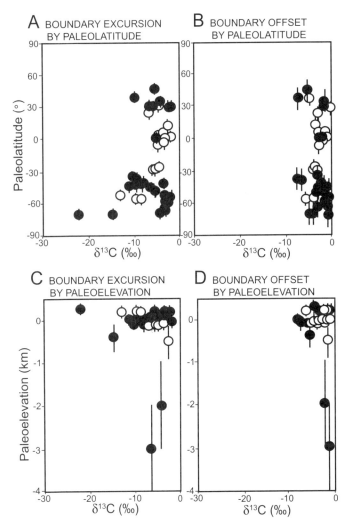

Figure 6. Preferential sequestration of anomalously light carbon at high latitudes and near surface environments is indicated by paleolatitudinal (+ to north, – to south) and paleoelevational (+ on land, – below sea level) distribution of the magnitude of carbon isotopic ($\delta^{13}C$) boundary excursion (A, C) and long-term offset (B, D) of organic matter (solid symbols) and carbonate (open symbols) across the Permian-Triassic boundary (data from Table 1). Carbonate carbon results may be compromised by diagenetic alteration (Mii et al. 1997; Heydari et al. 2001), but comparable patterns are shown by organic carbon results.

Permian nor Early Triassic had such low levels (<500 ppmV) of CO_2 (Retallack, 2001, 2002).

Perhaps high latitudes were a source as well as a sink for isotopically light carbon. High-latitude lakes, peat, permafrost, marine continental shelves, and coal measures are currently the largest near-surface methane reservoirs (Clayton, 1998; Kvenvolden, 2000). There was high-latitude permafrost and peat during the modest Late Permian greenhouse, but not during the Early Triassic postapocalyptic greenhouse (Retallack et al., 1996; Retallack, 1999a, 1999b; Retallack and Krull, 1999). Once released, CH_4 is oxidized to CO_2 after ~7 yr in the tropical atmosphere and 24 yr in temperate atmospheres (Khalil et al., 2000), The CO_2 then loses its $\delta^{13}C$ signature as it mixes with CO_2 from other sources (Whiticar, 2000). Time scales for global atmospheric mixing of isotopically distinct CH_4 and CO_2 are probably less than a century, because Greenland and Antarctic ice cores show rapid postglacial rise of CO_2 and CH_4 coincident within a century (Chapellaz et al., 2000). Global mixing times are probably more than two years because the ^{14}C pulse from nuclear weapons testing peaked in Germany in August 1963 after the test ban treaty, but ^{14}C peaked in January 1965 in South Africa and South America, remote from test sites (Levin et al., 1992). Since that time, the main source of ^{14}C has been from nuclear power plants in the Northern Hemisphere, and cruises in 1986 and 1988 showed that ^{14}C in atmospheric CO_2 and CH_4 declined markedly from northern to southern latitudes (Levin et al., 1992). Differences in carbon isotopic composition of plants proximal and distal to burning coal also indicate that uptake of local carbon dioxide can be more rapid than atmospheric mixing (Gleason and Kryser, 1984). Similarly, the paleolatitudinal and paleoelevational variation of carbon isotopic values of earliest Triassic disaster-recovery communities (Fig. 6), indicates that biotic uptake of methane-derived CO_2 was more rapid than global mixing of this isotopically distinct gas.

The paleogeographic distribution of carbon isotopic anomalies also allows reconsideration of past hypotheses for the isotopic excursion. The more profound excursion on land than in the sea cannot be explained by losses of only marine biomass and productivity (Wang et al., 1994). In scenarios in which isotopically light CO_2 bubbled up out of the deep ocean, in a way comparable to the fatal degassing of Lake Nyos in Cameroon (Knoll et al., 1996; Ryskin, 2003), one would expect that the largest excursions would be closer to marine sources, yet the largest isotopic excursions in South Africa, Australia, and Antarctica were well inland and isolated from the ocean by an Andean-style volcanic and fold-mountain range (Fig. 2). The ocean-overturn hypothesis also requires a positive carbon isotopic excursion in deep water and a negative carbon isotopic excursion in shallow water (Hoffman et al., 1991), yet both deep and shallow marine carbon isotopic excursions are negative. The greater polar than equatorial pattern of the carbon isotopic anomalies is also a difficulty for hypotheses that the carbon isotopic anomalies came from eruptive degassing of the Siberian Traps (Isozaki, 1997), a single bolide impact (Becker et al., 2004), or any other single point source.

MECHANISMS FOR MASSIVE METHANE OUTBURST

The massive amounts of methane indicated by the magnitude of the carbon isotopic anomaly at the Permian-Triassic boundary require an efficient release mechanism, which could overwhelm natural methane sinks. Annual sinks of methane now are ~5 Gt (Khalil et al., 2000). Extraterrestrial impact, large volcanic eruptions, submarine landslides, or a combination of these could have exceeded a comparable threshold, releasing methane from hydrates in permafrost and high latitude continental shelves or from thermal alteration of coal measures. These mechanisms are not mutually exclusive, because flood basalt eruption and submarine slides could have been consequences of bolide impact (Boslough et al., 1996; Max et al., 1999). There is evidence of varying quality for each of these methane-clathrate release mechanisms at the Permian-Triassic boundary, as outlined below.

Volcanic Eruption and Intrusion

Substantial amounts of methane could have come from melting of Angaran permafrost by lavas of the Siberian Traps (Vermeij and Dorritie, 1996), which erupted beginning at the Permian-Triassic boundary (Renne et al., 1995; Reichow et al., 2002). Some of these flows are pillowed, vesiculated, and brecciated (Daragan-Sushchov, 1989; Wooden et al., 1993), but latest Permian paleosols overrun by the Siberian Traps have not yet been studied for evidence of permafrost and lava-ice interaction. There is paleosol evidence of permafrost in *Glossopteris*-dominated Middle and Late Permian peatlands at comparable high paleolatitudes of the Gondwana supercontinent, antipodal to the Siberian Traps (Conaghan et al., 1994; Krull, 1999; Retallack, 1999b; Retallack and Krull, 1999).

Intrusions into organic sediments could also have played a role comparable to that proposed for terminal Paleocene global warming (Svensen et al., 2004). The Siberian Traps are prime suspects because of their earliest Triassic age (Renne et al., 1995; Reichow et al., 2002), and extensive thermal metamorphism of coal measures in the Tunguska Basin by feeder dikes to the lavas (Bogdanova, 1991; Mazor et al., 1979). In addition, the Pourakino Trondjemite and Hekeia Gabbro intrusions into Permian marine rocks of the Longwood Range of the South Island of New Zealand have been radiometrically dated at the Permian-Triassic boundary ($^{40}Ar/^{39}Ar$ plateau ages 251.2 ± 0.4 and 249.0 ± 0.4 respectively; Mortimer et al., 1999).

Also plausible is release of marine methane hydrates destabilized by global greenhouse from release of CO_2 and water vapor from Siberian Traps (Krull et al., 2000; Kidder and Worsely, 2004). Modeling by Berner (2002) indicates that the Siberian traps did not release sufficient greenhouse gases to destabilize marine clathrates, but that independently added CO_2 from flood basalts exacerbated atmospheric greenhouse. Such models need to be revised, because yields of CO_2 from lavas may have been higher than previously thought: as much as 2100 ppm from FTIR studies of glasses (Wallace, 2003) and as much as 8500 metric tons/day from volcano monitoring (Gerlach et al., 2002). Alkaline volcanics, such as the Siberian Trap precursor Miamecha-Kotui carbonatites, are much more prolific sources of CO_2 and could be related to flood basalt eruptive cycles (Morgan et al., 2004).

Bolide Impact

The Bedout High in the offshore, subsurface Canning Basin of Western Australia is the most promising of candidate Permian-

Triassic-boundary craters, although this interpretation remains controversial (Glikson, 2004; Renne et al., 2004; Wignall et al., 2004). Impact breccia from its central uplift has an $^{40}Ar/^{39}Ar$ plateau age of 250.1 ± 4.5 Ma (Harrison for Becker et al., 2004; but see analytical reservations of Mundil et al., 2004). This new age determination supports an earlier K-Ar age of 253 ± 5 Ma (Gorter 1996). Remote sensing by gravity and seismic data suggest that Bedout Crater was ~100 km in diameter (Becker et al., 2004). Most breccia clasts from its central peak are Late Permian volcanics, but there are also clasts of fossiliferous limestone in the breccia (Becker et al., 2004). Late Permian shallow marine sediments are a plausible source of methane clathrates. Late Permian permafrost paleosols have been found along the southeast coast of Australia (Retallack, 1999b), but a long way south of Bedout Crater at 32° south paleolatitude (Fig. 2), where permafrost was unlikely in the modest greenhouse of the Late Permian (Retallack, 2002).

Other plausible craters are inadequately documented or not quite the right age. A possible small (35 km diameter) impact crater has been reported near latest Permian cold temperate to frigid peatlands, in New South Wales, Australia (Tonkin, 1998). Lorne Crater includes fossil pollen of the *Protohaploxypinus samoilovichi* zone and fossil plants of the *Dicroidium zuberi* zone high in its conglomeratic fill (Pratt and Herbert, 1973; Holmes and Ash, 1979; Retallack, 1995; Tonkin, 1998), indicating an age at the base of the sequence close to the Permian-Triassic boundary. The Noril'sk iron ores of Siberia have been proposed as remains of a large (20 km diameter) metallic asteroid, whose enormous (>200 km diameter) crater was obliterated by subsequent eruption of the Siberian Traps (Dietz and McHone, 1992; Jones et al., 2002). Araguainha Crater of Brazil is well documented and dated, but small (40 km), and its age ($^{40}Ar/^{39}Ar$ plateaus of 245.5 ± 3.5 Ma and 243.3 ± 3.0 Ma) is some 5 m.y. after the Permian-Triassic boundary (Hammerschmidt and Engelhardt, 1995). The Woodleigh Crater of Western Australia may be either 40 or 120 km in diameter (Reimold et al., 2000), and though poorly dated is most likely the Early Jurassic age of its lacustrine fill (Mory et al., 2000).

Impact of an asteroid a few kilometers in diameter at the Permian-Triassic boundary also is indicated by iridium anomalies in the range 80–130 ppt (ng.g^{-1}), shocked quartz grains, Fe-Si-Ni-microspherules, fullerenes with extraterrestrial noble gases, and chondritic meteorite fragments (Zhou and Chai, 1991; Retallack et al., 1998; Chijiwa et al., 1999; Becker et al., 2001, 2004; Kaiho et al., 2001; Basu et al., 2003; Poreda and Becker, 2003; Shukla et al., 2003). Some of these supposed shocked quartz grains have subsequently turned out to be metamorphic quartz grains (Langenhorst et al., 2005). An unusual positive europium anomaly in rare earth elements of the Permian-Triassic boundary near Spiti, India, has been attributed to an eucritic bolide impact (Bhandari et al., 1992), but a thick sequence of sandstone with comparable europium anomalies near Banspetali, India, has been interpreted as due to erosion of Archaean gneisses (Sarkar et al., 2003).

Submarine Landslide

Methane also could have come from large submarine slides, as in the Permian-Triassic boundary sequences at Kaki Vigla, Greece (Baud et al., 1989), and Wairoa Gorge, New Zealand (Krull et al., 2000). The Little Ben Sandstone of New Zealand is an enormous marine mass emplacement deposit mappable throughout the South Island, and it contains a profound methane-related isotopic excursion (Krull et al., 2000). Submarine landslide unroofing of continental shelf clathrates can release massive amounts of methane (Thorpe et al., 1998; Kennett, 2003).

DEATH BY METHANE

Many theories of Permian-Triassic boundary events treat the carbon isotopic excursion as a consequence of death and destruction of biomass (Twitchett et al., 2001), but could methane and its oxidation product, CO_2, have been a cause of the extinctions? Two observations support this idea.

First, both marine and terrestrial extinctions were close to the low point of the carbon isotopic excursion in sequences of suitable stratigraphic resolution. Onset of the isotopic excursion before final extinction near the isotopic minimum is well established in the marine sections near Meishan, China (Fig. 3), in the Gartnerkofel core, Austria (Fig. 5), and Bonaparte Basin of Western Australia (Morante 1996), in coal measures of Graphite Peak, Antarctica (Fig. 4), Murrays Run bore, southeastern Australia (Morante and Herbert, 1994), and in the vertebrate-bearing beds of Bethulie, South Africa (MacLeod et al., 2000). A marine sequence at Schuchert Dal in Greenland was claimed to show marine extinction before the isotopic excursion (Twitchett et al., 2001), but sample spacing and temporal resolution of data presented fail to make this case, as is common with mass extinctions (Marshall 1995). The last marine body and trace fossils at Schuchert Dal are only one sample below the sample showing the beginning of the carbon isotopic excursion. Final extinction of Permian pollen and spores at Schuchert Dal is at a second and higher carbon isotopic minimum and fungal spike (Looy et al., 2001). In no case do the extinctions or isotopic minimum coincide exactly with the Permian-Triassic boundary which is defined at the first appearance of the newly evolved Triassic conodont *Hindeodus parvus* at Meishan, China (Fig. 3; Erwin, 1993; Jin et al., 2000). Initiation of isotopic decline, isotopic minima, and extinctions, were all close to modeled times for methane outburst and dispersal (Berner, 2002).

Second, the magnitude of the carbon isotopic excursion when converted to likely amount of methane released into the atmosphere correlates very well with contemporaneous percentage of generic extinction (Fig. 7). Extinction percentages were taken from Sepkoski (1996). Our calculations of methane release (Table 3) use only nonmarine organic carbon and follow an algorithm using the mass of carbon in the atmosphere (Jahren et al., 2001), rather than the mass of carbon in the ocean (Dickens et al., 1995), because of problems of mixing of such transient

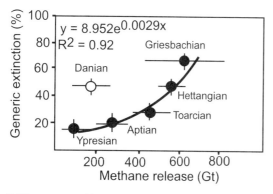

Figure 7. The amount of isotopically light CH_4 released from late Paleocene (Ypresian, 55 Ma), Early Cretaceous (Aptian, 117 Ma), Early Jurassic (Toarcian, 175 Ma), Triassic-Jurassic (Hettangian, 200 Ma), and Permian-Triassic boundary (251 Ma) methane-hydrate dissociation events (Sinha et al. 1995; Hesselbo et al. 2000, 2002; Jahren et al. 2001) correlates well with percentage of generic extinction at those times (Sepkoski 1996). The Cretaceous-Tertiary boundary (65 Ma) isotopic excursion (open circle; after Arens and Jahren 2000, 2002; Beerling et al. 2001) lies off the curve and has overwhelming evidence of asteroid impact. Calculations are for the transient isotopic excursion based on equilibration of land plant and atmospheric carbon, and do not include suspect marine isotopic data or effects of oceanic mixing.

releases (Berner, 2002) and of diagenesis of marine records (Mii et al., 1997; Heydari et al., 2001). Methane isotopic composition is assumed to be −60‰, and estimates of initial atmospheric CO_2 are from stomatal index of fossil plants (Retallack, 2001, 2002; Wynn, 2003). We have not found any comparably significant relationship between extinction magnitude and impact crater diameter, iridium anomalies, shocked quartz abundance, flood basalt volumes, strontium isotopic variation, or sea-level change (see also Wignall, 2001).

Methane could have killed by reacting with atmospheric NO_x to form formaldehyde, by igniting in concentrations greater than 5% by volume, or by anesthetizing and asphyxiating with prolonged exposure in excess of 0.1% by volume (Patnaik, 1992; Clough, 1998). Even release of 622 Gt C in methane would not maintain concentrations greater than 0.1% for more than a few hundred kilometers from individual point sources. Because of methane oxidation to CO_2 in only 7–24 yr (Khalil et al., 2000), the most widespread lethal effects of methane would be similar to those of an unusually severe CO_2 greenhouse. This would include respiratory difficulties and acidosis (Knoll et al., 1996; Retallack et al., 2003), as well as possible effects of acid rain (Retallack 1996, 2004b).

Oceanic Hypercapnia, Hypoxia, Warming, and Reef Gap

In the ocean, excess CO_2 (hypercapnia) and low O_2 (hypoxia) have been suggested to explain preferential extinction of marine invertebrates with heavily calcified skeletons, poor respiratory and circulatory systems, and low basal metabolic rate, such as corals, bryozoans, and brachiopods (Knoll et al. 1996). Hypercapnia also has been invoked to explain unusual abiotic calcite seafloor cements, which occur at the Permian-Triassic boundary, as well as at other stratigraphic levels of negative carbon isotopic excursions within the Early Triassic (Woods et al., 1999). Unlike the model developed from these observations by Knoll et al. (1996), in which CO_2 bubbled out of formerly stratified deep-ocean water, our scenario envisages limited CO_2 from marine oxidation of local rapid release of clathrate or coalbed methane, with the principal global CO_2 enrichment of the ocean being diffusion back from the atmosphere of oxidized CH_4.

Some degree of anoxia (O_2 deficit) also may explain unbioturbated, gray marine shales and cherts of earliest Triassic age (Ishiga et al., 1993; Wignall et al., 1998). It has been tempting to compare Permian-Triassic anoxia to oceanic anoxic events, marked by very pyritic (>3% pyrite) and carbonaceous (>10% TOC) shales, due to high biological productivity and oceanic stagnation (Wignall, 1994; Kuhnt et al., 2002). The biggest problem with the stagnation-anoxia model for the Permian-Triassic extinctions has always been the low pyrite and organic carbon

TABLE 3. ESTIMATED METHANE HYDRATE RELEASE FOR TRANSIENT ORGANIC CARBON ISOTOPIC EXCURSIONS

Age	Ma	Initial CO_2 ppmV	Initial $\delta^{13}C_{org}$	$\Delta^{13}C_{org}$ excursion	Methane release (Gt)	Methane release (ppmV)	Genera extinct (%)
Ypresian	55	333 ± 31	−23.5	−3.6 ± 0.7	75 ± 16	36 ± 7	13 ± 2
Danian	65	748 ± 364	−23.5	−3.4 ± 0.4	159 ± 21	77 ± 10	46 ± 3
Aptian	117	871 ± 706	−22.2	−4.9 ± 1.5	268 ± 92	130 ± 45	17 ± 5
Toarcian	175	871 ± 28	−25.5	−7.0 ± 1.0	459 ± 82	222 ± 40	26 ± 6
Hettangian	200	1935 ± 466	−25.5	−4.2 ± 1.3	570 ± 202	275 ± 98	46 ± 5
Griesbachian	250	1385 ± 631	−24.0	−6.4 ± 5.0	622 ± 589	299 ± 285	66 ± 4

Note: Methane release has been estimated in both GtC (= 10^{15} g) and ppmV. Initial CO_2 is from unsmoothed data of Retallack (2001) recalculated using method of Wynn (2003); isotopic excursions from Sinha et al. (1995), Hesselbo et al. (2000, 2002), Jahren et al. (2001) and herein; extinction values from Sepkoski (1996); methane release calculated by method of Jahren et al. (2001).

content of marine shales of earliest Triassic age (Retallack, 2004a; Wignall and Newton, 2004). Although Isozaki (1997) claimed that shales of the earliest Triassic are carbonaceous, he presented no analytical data. Hundreds of published total organic carbon and inorganic sulfur analyses of marine shales of earliest Triassic age do not exceed 3 wt%, whereas shales of Late Permian and Middle Triassic age in the same sequences are much more carbonaceous (Suzuki et al., 1993; Kajiwara et al., 1994; Wang et al., 1994; Wolbach et al., 1994; Morante, 1996; Wignall et al., 1998; Krull et al., 2000, 2004). By our methane poisoning scenario, anoxia arises not from respiration in a highly productive surface ocean, but from oxidation of methane in depopulated seas, which would result in only local and minor accumulation of organic matter and pyritization.

Postapocalyptic earliest Triassic greenhouse warming is indicated by evidence from fossil soils (Retallack, 1999a), plants (Retallack, 2002), and vertebrates (Retallack et al., 2003). Oceanic warming is a prime culprit in coral bleaching today (Marshall and McCulloch, 2002) and may also enable outbreaks of coral predation (Pandolfi, 1992). A combination of warm tropical sea temperatures and hypercapnia may explain the lack of fossil coral reefs anywhere in the world during the Early Triassic, the longest gap in the geological record of reefs (Weidlich et al., 2003). Trepostome and fenestellid bryozoans and rugose and tabulate corals are prominent among casualties of the mass extinction, and hexacorals did not appear until after the Early Triassic (Erwin, 1993). These simple organisms had limited ventilatory capacity, and oxygen deficiency would have been exacerbated once their oxygenating symbiotic zooxanthellae were expelled during thermal bleaching. Abundant earliest Triassic cyanobacterial stromatolites (Schubert and Bottjer 1992) and algal wrinkle structures (Pruss et al., 2004) recall greenhouse conditions of the Late Cambrian.

Soil Warming, Acidification, Hypoxia, Erosion, and Coal Gap

On land, there are multiple lines of evidence for a postapocalyptic greenhouse supplementary to the already modest Late Permian greenhouse (Retallack 1999a, 2002). A contribution of methane is needed to explain extreme fluctuations in carbon isotopic depth variation within earliest Triassic paleosols (Krull and Retallack, 2000). Abrupt onset of earliest Triassic warmth at high latitudes is indicated by unusually strong chemical weathering compared with physical weathering of Antarctic and Australian paleosols, which include such deeply weathered soils as Ultisols (Retallack, 1999a; Retallack and Krull, 1999). In addition, fossil leaves of *Lepidopteris* immediately above the Permian-Triassic boundary in Australia and India have exceptionally low stomatal index, which, by comparison with comparable modern leaves of *Ginkgo*, indicates atmospheric CO_2 levels of 7876 ± 5693 ppmV or 28 ± 20 PAL (1 PAL is present atmospheric level: Retallack, 2001, 2002, Wynn, 2003). Early Triassic fossil plants are notably herbaceous, succulent, heterophyllous, sclerophyllous, and thickly cutinized, characteristics that can be considered adaptations to hot or dry conditions (Retallack, 1997; Looy, 2000) but also are adaptations to high levels of atmospheric CO_2 (Retallack, 2001, 2002).

Evidence for acid rain at the Permian-Triassic boundary is mixed. Australian terrestrial sequences have strongly leached, kaolinitic Permian-Triassic boundary breccias, but coeval boundary breccias in Antarctica are little leached (Retallack et al., 1998). Nevertheless, paleosol sequences examined chemically in Australia, Antarctica, and South Africa show evidence of increased precipitation and chemical weathering across the Permian-Triassic boundary (Retallack, 1999a; Retallack and Krull, 1999; Retallack, et al., 2003). Strontium-isotopic composition of carbonates provide a global proxy of acidic leaching of continents, but the predicted rise of $^{87}Sr/^{86}Sr$ due to acid rain is difficult to distinguish from a longer-term mid-Permian to mid-Triassic steep rise in $^{87}Sr/^{86}Sr$ values (Martin and MacDougall, 1995). High-resolution $^{87}Sr/^{86}Sr$ analyses across the boundary in the stratotype section at Meishan, China, do show a rise compatible with acidification between the boundary beds and the first appearance of the earliest Triassic conodont *Hindeodus parvus* (Kaiho et al., 2001). Thus there is evidence for increased rainfall, and for chemical weathering due to higher atmospheric CO_2 levels, but not yet clear indications of strong acid rain, such as nitric and sulfuric acid rain created by impact at the Cretaceous-Tertiary boundary (Retallack, 1996, 2004b).

Soil hypoxia in wetlands is indicated by earliest Triassic paleosols directly above coal measures (Retallack, 1999a; Retallack and Krull, 1999). Large berthierine nodules indicating unusually low soil oxidation are found in earliest Triassic Dolores paleosols of Antarctica (Sheldon and Retallack, 2002). Signs of hypoxia are not evident from red, highly oxidized earliest Triassic paleosols, interpreted as originally well drained (Retallack, et al., 2003). Elevated levels of carbon dioxide and methane that are not serious problems for plants of well-drained soils could have harmed plants of wetland soils already marginally aerated. Mass balance modeling suggests a drop in atmospheric O_2 to as low as 12% by volume as methane oxidized to carbon dioxide during the earliest Triassic (Berner, 2002). Such low oxygen levels in the air would be further lowered a short distance down into biologically active, swampy soils, thus restricting root respiration to very shallow levels within soils (Jenik, 1978). When waterlogged soils become completely stagnant, swampland trees die and peats are inundated by lakes and lagoons.

A variety of sedimentary features attest to extensive plant dieback in marginally aerated lowland sedimentary environments. Sedimentary response to terminal Permian plant extinctions may be compared to the response to forest clear-cutting today (Retallack, 2005). Unlike underlying alluvial sequences of latest Permian age, those of earliest Triassic age are dominated by weakly developed paleosols (Retallack and Krull, 1999; Retallack et al., 2003) and extensive claystone breccias from local soil erosion (Retallack et al., 1998; Retallack, 2005). Of similar origin is a 9-m-thick mudflow deposit at the Permian-Triassic boundary in the Raniganj Basin, India (Sarkar et al., 2003). Following wetland plant dieback, mudflows and gully erosion of floodplain soils would have propagated into upland regions and induced the observed replacement of latest Permian meandering streams by earliest Triassic braided streams (Retallack and Krull, 1999; Ward et al., 2000).

Extinction by hypoxia is also an explanation for the Early Triassic coal gap. This was a time when peat formed nowhere in the world for 6 m.y. (following time estimates of Gradstein et al., 2005), and the only long-term hiatus in peat formation documented since the Devonian evolution of swamps (Retallack et al., 1996). The idea of methane-induced soil hypoxia explains not only preferential extinction of wetland plants, but also the protracted duration of the coal gap, prolonged by additional methane outbursts throughout the Early Triassic (Fig. 4; 5; Krull and Retallack, 2000). *Vertebraria*'s aquatically adapted glossopterid roots extend down as much as 80 cm into latest Permian peaty paleosols (Fig. 8B), but it was extinguished at the Permian-Triassic boundary throughout the Gondwana supercontinent (Retallack and Krull, 1999). Similarly, peat-forming

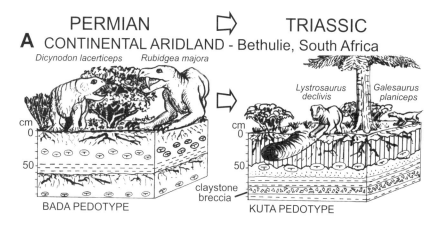

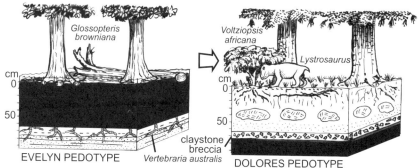

Figure 8. Reconstructed paleoenvironments before and after the Permian-Triassic boundary at (A) Bethulie, South Africa (Retallack et al., 2003); (B) Graphite Peak, Antarctica (Retallack and Krull, 1999); and (C) Meishan, China (Yin et al., 1996).

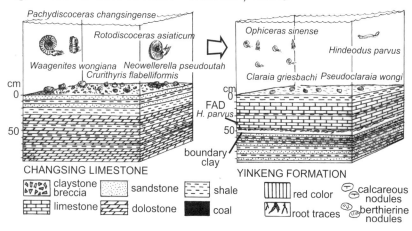

ecosystems of ruflorian cordaites in Siberia (Meyen, 1982), and of lycopsids in China (Wang and Chen, 2001) were destroyed utterly at the end of the Permian. Other regions without Late Permian coal, such as Europe and North America, do not show such profound levels of extinction (Rees, 2002). This is partly a taphonomic artifact, because coal measures preserve most fossil plants, but even within Permian-Triassic coal-measure-terminating sequences, non-peat-forming plants such as conifers and seed ferns survived preferentially (Meyen, 1982; Retallack, 1995; Wang and Chen, 2001). Terminal-Permian lowland woody plant dieback and decay also is indicated by unusually abundant fungal spores, preferential extinction of gymnosperms relative to pteridophytes, and shifting composition of palynodebris (Visscher et al., 1996; Looy, 2000). One problem with this interpretation is that many hyphae previously interpreted as fungal may instead have been zygnematalean algae (Krassilov and Afonin, 1999; Foster, 2001), but some fungal spikes are still evident from fungal spores (Retallack, 1995). When peat-forming ecosystems reappeared during the Middle Triassic (ca. 245 Ma), they were dominated by conifers and cycad-like plants, which were upland plants during the Permian (DiMichele et al., 2001) and unrelated to plants thought to have formed Late Permian coals (Retallack et al., 1996).

Vertebrate Hypoxia, Acidosis, and Pulmonary Edema

Among large animals on land, high methane and carbon dioxide and low oxygen would impair mitochondrial oxidative phosphorylation (Fiddian-Green 1995). Consequent changes in blood pH would be most telling in the lungs, and death by pulmonary edema formerly restricted to high altitudes would have occurred at lower altitudes. Early Triassic oxygen levels of 12 vol% of the atmosphere, estimated by modeling (Berner, 2002), would have made oxygenation near sea level comparable to that in mountains more than 4000 m high (Retallack et al., 2003), where mountaineers are vulnerable to nausea, headache, hypertension, and pulmonary edema, a group of maladies commonly called mountain sickness (Hultgren, 1997). Although the Permian-Triassic hydrocarbon pollution crisis envisaged here would have been normobaric hypoxia with acidosis, and mountain sickness today is hypobaric hypoxia with alkalosis, both elicit similar physiological responses in experiments with rats (Russell and Crook, 1968; Schoene, 1990). These responses include exaggerated (Cheney-Stokes) breathing, pulmonary vasoconstriction, and high red blood cell count (Hultgren, 1997). Fossilizable adaptations to hypoxia of altitude among large mammals include large barrel chests relative to body size and low-birth-weight babies (Beall, 1982; Bouverot, 1985; Schoene, 1990). Earliest Triassic survival of small, barrel-chested *Lystrosaurus* (Fig. 8A, 8B) offers support to this concept of extinction by hypoxia. Evolution of a muscular diaphragm giving greater aerobic scope of survivors may also explain the thickened thoracic but reduced lumbar ribs of Triassic compared with Permian therapsid reptiles, especially *Galesaurus* and *Thrinaxodon*

(Fig. 8A, 8B; Brink, 1956). Another ventilatory advantage of many modern mammals over reptiles is a short snout and long secondary palate, creating a less obstructed airway in *Lystrosaurus* and all subsequent dicynodonts compared with their Permian ancestors (Cruickshank, 1968; King, 1991). Mammals and birds also can acclimate to low oxygen, unlike frogs, which are more closely related to Permian amphibians and reptiles. Acclimatization may also have aided survival of therapsids such as *Lystrosaurus* (Engoren, 2004; Retallack, 2004c). Finally, there is evidence from fossil skeletons in burrows that *Lystrosaurus*, like other Permian therapsids, was a burrower (Smith, 1987; Groenewald, 1991). Burrows are poorly aerated spaces, and burrowing animals adapt to hypercapnia and hypoxia. Large, non-burrowing reptiles were not under such selection pressures and did not survive the Permian-Triassic boundary (Smith and Ward, 2001; Benton et al., 2004). The Permian-Triassic acquisition of near-mammalian nares, palate, and rib-cage was "a defining moment for amniote physiology and evolution" (Graham et al., 1997).

RAGNAROK HYPOTHESIS

We call our hypothesis of terminal-Permian death by methane the Ragnarok hypothesis. In ancient Norse mythology, Ragnarok is the twilight of the gods, a mythic end of the world as we know it, and harbinger of a new age of harmony. "First will come a triple winter .... and the sun impart no gladness.... The earth.... will tremble, the sea leave its basin, the heavens tear asunder, and men perish in great numbers.... The gods and their enemies having fallen in battle, Surtur, having killed Freya, darts fire and flames over the world" (Bullfinch, 1993). The Ragnarok myth captures our concept of impact or volcanic winter with massive methane outbursts and CO_2+CH_4 greenhouse as a cause for the Permian-Triassic mass extinction.

Our Ragnarok hypothesis of terminal-Permian death by methane outbursts is a testable hypothesis that explains more effectively than competing hypotheses (Table 4) a variety of observations on the carbon isotopic excursion across the Permian-Triassic boundary at a global network of localities (Fig. 2, Table 1). It is a disturbing vision of biotic vulnerability to impact, volcanic, or other destabilization of coalbed and clathrate methane. Parallels with historic atmospheric pollution with hydrocarbons and consequent global warming, coral bleaching, and wetland stagnation are especially intriguing.

ACKNOWLEDGMENTS

We dedicate this work to our late colleague William T. Holser, whose wit and wisdom are missed. M. Whiticar, J. Hayes, L. Becker, P. Olsen, and D. Gröcke offered useful discussion. P. Wignall, L. Kump, and E. Grossman offered valuable reviews of earlier drafts. Work was funded by National Science Foundation grant OPP 9315228 and American Chemical Society grant 31270-AC8.

TABLE 4. EVALUATION OF ALTERNATIVE HYPOTHESES FOR SEVERAL OBSERVATIONS OF THE CARBON ISOTOPIC EXCURSION AT THE PERMIAN-TRIASSIC BOUNDARY

Hypotheses	$\Delta^{13}C$ excursion as much as −8‰	Greater $\delta^{13}C$ excursion at high than low latitude	Greater $\delta^{13}C$ excursion on land than in sea	Negative $\delta^{13}C$ excursion in deep sea	$\delta^{13}C$ excursion on land and in sea
Methane outburst	Yes	Yes	Yes	Yes	Yes
Comet/asteroid impact	Maybe (1)	Maybe (2)	Yes	Yes	Yes
Volcanic degassing	No	Maybe (3)	Yes	Yes	Yes
Erosion of coal	No	Yes	Yes	Yes	Yes
Change from trees to herbs	No	No	Yes	Yes	Yes
Curtailed productivity on land	No	No	Yes	Yes	Yes
Curtailed productivity in sea	No	No	No	Yes	Yes
Overturn of seltzer ocean	No	No	No	No	Yes

Note: The "maybe" entries are dependent on (1) chemically rare chondritic bolide, >25 km diameter; (2) bipolar impacts; and (3) bipolar eruptions.

REFERENCES CITED

Allan, J.R., and Matthews, R.K., 1977, Carbon and oxygen isotopes as diagenetic and stratigraphic tools: Surface and subsurface data, Barbados: West Indies: Geology, v. 21, p. 771–775.

Arens, N.C., and Jahren, A.H., 2000, Carbon isotope excursion in atmospheric CO_2 at the Cretaceous-Tertiary boundary: Evidence from terrestrial sediments: Palaios, v. 15, p. 314–322.

Arens, N.C., and Jahren, A.H., 2002, Chemostratigraphic correlation of four fossil-bearing sections in southwestern North Dakota, *in* Hartman, J.H., Johnson, K.R., and Nichols, D.S., eds., The Hell Creek Formation and the Cretaceous-Tertiary boundary in the northern Great Plains: An integrated continental record of the end of the Cretaceous: Geological Society of America Special Paper 301, p. 75–93.

Arens, N.C., Jahren, A.H., and Amundson, R., 2000, Can C3 plants faithfully record the carbon isotopic composition of atmospheric carbon dioxide?: Paleobiology, v. 26, p. 137–164.

Bains, S., Corfield, R.M., and Norris, R.D., 1999, Mechanisms of climate warming at the end of the Paleocene: Science, v. 285, p. 724–727, doi: 10.1126/science.285.5428.724.

Basu, A.R., Petaev, M.I., Poreda, R.J., Jacobsen, S.B., and Becker, L., 2003, Condritic meteorite fragment associated with the Permian-Triassic boundary in Antarctica: Science, v. 302, p. 1388–1392, doi: 10.1126/science.1090852.

Baud, A., Magaritz, M., and Holser, W.T., 1989, Permian-Triassic of the Tethys: Carbon isotope studies: Geologische Rundschau, v. 78, p. 649–677, doi: 10.1007/BF01776196.

Baud, A., Atudori, V., and Sharp, Z., 1996, Late Permian and Early Triassic evolution of the northern Indian margin: Carbon isotope and sequence stratigraphy: Geodinamica Acta, v. 9, p. 57–77.

Beall, C.M.A., 1982, A comparison of chest morphology in high altitude Asian and Andean populations: Human Biology, v. 54, p. 145–163.

Becker, L., Poreda, R., Hunt, H.G., Bunch, T.E., and Rampino, M., 2001, Impact event at the Permian-Triassic boundary: Evidence from extraterrestrial noble gases in fullerenes: Science, v. 291, p. 1530–1533, doi: 10.1126/science.1057243.

Becker, L., Poreda, R.J., Basu, A.R., Pope, K.O., Harrison, T.M., Nicholson, C., and Iasky, I.R., 2004, Bedout: A possible end-Permian impact crater offshore of northwestern Australia: Science, v. 304, p. 1469–1476, doi: 10.1126/science.1093925.

Beerling, D.J., and Woodward, F.I., 2001, Vegetation and the terrestrial carbon cycle: Modelling the first 400 million years: Cambridge, UK, Cambridge University Press, 405 p.

Beerling, D.J., Lomax, B.H., Upchurch, G.R., Nichols, D.J., Pillmore, C.L., Handley, L.C., and Scrimgeour, C.M., 2001, Evidence for the recovery of terrestrial ecosystems ahead of marine primary production following a biotic crisis at the Cretaceous-Tertiary boundary: Journal of the Geological Society [London], v. 158, p. 737–740.

Benton, M.J., Tverdokhlbov, V.P., and Surkhov, M.V., 2004, Ecosystem remodelling among vertebrates at the Permian-Triassic boundary in Russia: Nature, v. 432, p. 97–100, doi: 10.1038/nature02950.

Berner, R.A., 2002, Examination of hypotheses for the Permo-Triassic boundary extinction by carbon cycle modeling: Proceedings of the National Academy of Sciences USA, v. 99, p. 4172–4177, doi: 10.1073/pnas.032095199.

Bestland, E.A., and Krull, E.S., 1999, Palaeoenvironments of early Miocene Kisingiri Volcano *Proconsul* sites: Evidence from carbon isotopes, palaeosols and hydromagmatic deposits: Journal of the Geological Society [London], v. 156, p. 965–976.

Bhandari, N., Shukla, P.N., and Azmi, R.J., 1992, Positive europium anomaly at the Permo-Triassci boundary Spiti, India: Geophysical Research Letters, v. 19, p. 1531–1534.

Boeckelman, K., and Magaritz, M., 1991, The Permian-Triassic of Gartnerkofel-1 core (Carnic Alps, Austria): Dolomitization of the Permian-Triassic sequence, *in* Holser, W.T., and Schönlaub, H.P., eds., The Permian-Triassic boundary in the Carnic Alps of Austria (Gartnerkofel Region): Abhandlungen Bundesanstalt Autriche, Vienna, v. 45, p.61–68.

Bogdanova, L.A., 1971, Metamorficheskii ryd uglei tungusskogo basseina (Metamorphism of coal seams in the Tunguska Basin): Litologiya i Poleznhie Iskopaemie, v. 6, p. 84–98.

Boslough, M.B., Chael, E.P., Trucano, T.B., Crawford, D.A., and Campbell, D.L., 1996, Axial focusing of impact energy in the Earth's interior: A possible link to flood basalts and hotspots, *in* Ryder, G., Fastovsky, D., and Gartner, S., eds., The Cretaceous-Tertiary event and other catastrophes in Earth history: Geological Society of America Special Paper 307, p. 541–550.

Bouverot, P., 1985, Adaptation to altitude hypoxia in vertebrates. Berlin, Springer, 176 p.

Bowring, S.A., Erwin, D.H., Jin, Y.G., Martin, M.W., Davidek, E.K., and Wang, W., 1998, U/Pb zircon geochronology and tempo of the end-Permian mass extinction: Science, v. 280, p. 1039–1045, doi: 10.1126/science.280.5366.1039.

Brink, A.S., 1956, Speculations on some advanced mammalian characteristics in higher mammal-like reptiles: Palaeontographica Africana, v. 4, p. 77–86.

Broecker, W.S., and Peacock, S., 1999, An ecological explanation for the Permo-Triassic carbon and sulfur isotope shifts: Global Geochemical Cycles, v. 13, p. 1167–1172, doi: 10.1029/1999GB900066.

Bullfinch, T., 1993, The golden age of myth and legend: London, Bracken Books, 495 p.

Chappellaz, J., Raynaud, D., Blunier, T., and Stauffer, B., 2000, The ice core record of atmospheric methane, *in* Khalil, M.A.K., ed., 2000, Atmospheric methane: Berlin, Springer, p. 9–24.

Chijiwa, T., Arai, T., Sugai, T., Shinohara, H., Kumazawa, M., Takano, M., and Takani, S., 1999, Fullerenes found in the Permo-Triassic mass extinction period: Geophysical Research Letters, v. 26, p. 767–770, doi: 10.1029/1999GL900050.

Clayton, J.L., 1998, Geochemistry of coalbed gas—a review: International Journal of Coal geology, v. 35, p. 159–173.

Clough, S., 1998, Methane, in Wexler, P., ed., Encyclopedia of toxicology: San Diego, Academic Press, v. 2, p. 294–295.

Conaghan, P.J.G., Shaw, S.E., and Veevers, J.J., 1994, Sedimentary evidence of Permian/Triassic global crisis induced by Siberian hotspot, in Embry, A.F., Beauchamp, B., and Glass, D.J., eds., Pangea: Global resources and environments: Canadian Society of Petroleum Geologists Memoir, v. 17, p. 785–795.

Cruickshank, A.R.I., 1968, A comparison of the palates of Permian and Triassic dicynodonts: Palaeontographica Africana, v. 11, p. 23–31.

Daragan-Sushchov, Y.I., 1989, Development patterns of the Early Triassic lakes of the Tunguska Syneclise: International Geology Review, v. 31, p. 1007–1017.

de Wit, M.J., Ghosh, J.G., de Villiers, S., Rakotosolofo, N., Alexander, J., Tripathi, A., and Looy, C., 2002, Multiple organic carbon isotope reversals across the Permo-Triassic boundary of terrestrial Gondwanan sequences: Clues to extinction patterns and delayed ecosystem recovery: Journal of Geology, v. 110, p. 227–240, doi: 10.1086/338411.

d'Hondt, S., Zachos, J.C., Bowring, S., Hoke, G., Martin, M., Erwin, D., Jin, Y.-G., Wang, W., Cao, C.-Q., and Wang, Y., 2000, Permo/Triassic events and the carbon isotope record of Meishan: Geological Society of America Abstracts with Programs, v. 32, p. A368.

Dickens, G.R., O'Neil, J.R., Rea, D.K., and Owen, R.M., 1995, Dissociation of oceanic methane hydrate as a cause of the carbon isotopic excursion at the end of the Paleocene: Paleoceanography, v. 10, p. 965–971, doi: 10.1029/95PA02087.

Dietz, R.S., and McHone, J.F., 1992, Noril'sk/Siberian Plateau basalts and Bahama hot spot: Impact triggered?: Houston, Lunar and Planetary Institute Contributions, v. 790, p. 22–23.

DiMichele, W.A., Mamay, S.H., Chaney, D.S., Hook, R.W., and Nelson, W.J., 2001, An early Permian flora with late Permian and Mesozoic affinities from north central Texas: Journal of Paleontology, v. 75, p. 449–460.

Engoren, M., 2004, Vertebrate extinction across the Permian-Triassic boundary in Karoo Basin, South Africa: Comment: Geological Society of America Bulletin, v. 116, p. 1294, doi: 10.1130/B25504.1.

Erwin, D.H., 1993, The great Paleozoic crisis: New York, Columbia University Press, 327 p.

Farquhar, G.D., Lloyd, J., Taylor, J.A., Flanagan, L.B., Syvertsen, J.P., Hubrick, K.T., Wong, C.S., and Ehleringer, J.R., 1993, Vegetation effects on the isotopic composition of oxygen in atmospheric CO_2: Nature, v. 363, p. 439–443, doi: 10.1038/363439a0.

Fiddian-Green, R., 1995, Gastric intramucosal pH, tissue oxygenation and acid-base balance: British Journal of Anaesthesiology, v. 74, p. 591–606.

Foster, C.B., 2001, Micro-organism uncertainty casts doubt on extinction theory: Australian Geological News, v. 61, p. 8–9.

Foster, C.B., Logan, G.A., and Summons, R.E., 1998, The Permian-Triassic boundary in Australia: Where is it and how is it expressed?: Royal Society of Victoria Proceedings, v. 110, p. 247–266.

Gerlach, T.M., McGee, K.A., Elias, T., Sutton, A.J., and Doukas, M.P., 2002, Carbon dioxide emission rate of Kilauea Volcano: Implications for primary magma and summit reservoir: Journal of Geophysical Research, v. 109, no. B9, ECV3, p. 1–15, doi: 10.102912015B000407.

Ghosh, J.G., Tripathy, A., Rakotosoloto, N.A., and de Wit, M., 1998, Organic carbon isotope variation of plant remains: Chemostratigraphical markers in terrestrial Gondwanan sequences: Journal of African Earth Sciences, v. 27, p. 83–85.

Gleason, J.D., and Kryser, T.K., 1984, Stable isotope compositions of gases and vegetation near naturally burning coal: Nature, v. 307, p. 254–257, doi: 10.1038/307254a0.

Glikson, A., 2004, Comment on "Bedout: A possible end-Permian impact crater offshore of northwestern Australia": Science, v. 306, p. 613, doi: 10.1126/science.1100404.

Gorter, J., 1996, Speculation on the origin of the Bedout High—A large circular structure of pre-Mesozoic age in the offshore Canning Basin, Western Australia: Petroleum Exploration Society of Australasia News, v. 4, p. 32–34.

Gorter, J.D., Foster, C.B., and Summons, R.E., 1995, Carbon isotopes and the Permian-Triassic boundary in the north Perth, Bonaparte and Carnarvon Basins, Western Australia: Petroleum Exploration Society of Australasia Journal, v. 4, p. 21–38.

Gradstein, F.M., Ogg, J.G., and Smith, A.G., eds., 2005, A geologic time scale 2004. Cambridge, Cambridge University Press, 589 p.

Grady, M.W., Wright, J.P., Carr, L.P., and Pillinger, C.J., 1986, Compositional differences in enstatite chondrites based on carbon and nitrogen stable isotope measurements: Geochimica et Cosmochimica Acta, v. 50, p. 2799–2813, doi: 10.1016/0016-7037(86)90228-0.

Graham, J.B., Aguilar, N., Dudley, R., and Gans, C., 1997, The late Paleozoic atmosphere and the ecological and evolutionary physiology of tetrapods, in Sumida, S.S., and Martin, K.L., eds., Amniote origins: Completing the transition to land: San Diego, Academic Press, p. 141–167.

Groenewald, G.H., 1991, Burrow casts from the Lystrosaurus-Procolophon assemblage zone: Koedoe, v. 34, p. 13–27.

Hammerschmidt, K., and Engelhardt, W.V., 1995, $^{40}Ar/^{39}Ar$ dating of the Araguainha impact structure, Mato Grosso, Brazil: Meteoritics, v. 30, p. 227–233.

Hesselbo, S.P., Gröcke, D.R., Jenkyns, H.C., Bjerrum, C.J., Farrimond, P., Bell, H.S.M., and Green, O.R., 2000, Massive dissociation of gas hydrate during a Jurassic oceanic anoxic event: Nature, v. 406, p. 392–395, doi: 10.1038/35019044.

Hesselbo, S.P., Robinson, S.A., Surlyk, F., and Piasecki, S., 2002, Terrestrial and marine extinction at the Triassic-Jurassic boundary synchronized with major carbon cycle perturbation: A link to initiation of massive volcanism: Geology, v. 30, p. 251–254.

Heydari, E., Hassanzadeh, J., and Wade, W.J., 2000, Geochemistry of central Tethyan Upper Permian and Lower Triassic strata, Abadeh region, Iran: Sedimentary Geology, v. 137, p. 85–99, doi: 10.1016/S0037-0738(00)00138-X.

Heydari, E., Wade, W.J., and Hassandzadeh, J., 2001, Diagenetic origin of carbon and oxygen isotopic composition of Permian-Triassic boundary strata: Sedimentary Geology, v. 143, p. 191–197, doi: 10.1016/S0037-0738(01)00095-1.

Hoffman, A., Gruszcyński, M., and Malkowski, K., 1991, On the interrelationship between temporal trends in $\delta^{13}C$, $\delta^{18}O$ and $\delta^{34}S$ in the world ocean: Journal of Geology, v. 99, p. 355–370.

Holmes, W.B.K., and Ash, S.R., 1979, An Early Triassic megafossil flora from the Lorne Basin, New South Wales: Proceedings of the Linnaean Society of N.S.W., v. 103, p. 47–70.

Holser, W.T., and Magaritz, M., 1987, Events near the Permian-Triassic boundary: Modern Geology, v. 11, p. 155–180.

Holser, W.T., Schönlaub, H.P., and Magaritz, M., 1991, The Permian-Triassic of the Gartnerkofel-1 core (Carnic Alps, Austria): synthesis and conclusions, in Holser, W.T., and Schönlaub, H.P., eds., The Permian-Triassic boundary in the Carnic Alps of Austria (Gartnerkofel Region): Abhandlung Bundesanstalt Autriche, Vienna, v. 45, p. 213–232.

Hotinski, R.M., Bice, K.L., Kump, L.R., Najjar, R.G., and Arthur, M.A., 2001, Ocean stagnation and end-Permian anoxia: Geology, v. 29, p. 7–10, doi: 10.1130/0091-7613(2001)029<0007:OSAEPA>2.0.CO;2.

Hultgren, H., 1997, High altitude medicine: Stanford, California, Hultgren Publications, 550 p.

Ishiga, H., Ishida, K., Sampei, Y., Musashino, M., Yamakita, S., Kajiwara, Y., and Morikiyo, T., 1993, Oceanic pollution at the Permian-Triassic boundary in pelagic condition from carbon and sulfur isotopic excursion, southwest Japan: Japanese Geological Survey Bulletin, v. 44, p. 721–726.

Isozaki, Y., 1997, Permian-Triassic superanoxia and stratified superocean: Records from the lost deep sea: Science, v. 276, p. 235–238, doi: 10.1126/science.276.5310.235.

Jahren, A.H., Arens, N.C., Sarmiento, G., Guerro, J., and Amundson, R., 2001, Terrestrial record of methane hydrate dissociation in the Early Cretaceous: Geology, v. 29, p. 159–162, doi: 10.1130/0091-7613(2001)029<0159:TROMHD>2.0.CO;2.

Jenik, J., 1978, Roots and root systems in tropical trees: Morphologic and ecologic aspects, in Tomlinson, P.B., and Zimmermann, M.H., eds., Tropical trees as living systems: Cambridge, UK, Cambridge University Press, p. 323–349.

Jessberger, E.K., 1999, Rocky cometary particulates: Their elemental, isotopic and mineralogical ingredients: Space Science Reviews, v. 90, p. 91–97, doi: 10.1023/A:1005233727874.

Jin, Y.-G., Wang, Y., Sheng, Q., Cao, C.-Q., and Erwin, D.H., 2000, Pattern of marine mass extinction across the Permian-Triassic boundary in South China: Science, v. 289, p. 432–436, doi: 10.1126/science.289.5478.432.

Jones, A.P., Price, G.R., Price, N.J., De Carli, P.S., and Clegg, R.A., 2002, Impact-induced melting and the development of large igneous provinces:

Earth and Planetary Science Letters, v. 202, p. 551–561, doi: 10.1016/S0012-821X(02)00824-5.

Kaiho, K., Kajiwara, Y., Nakano, T., Miura, Y., Kawahata, H., Tazaki, K., Ueshima, M., Chen, Z.-Q., and Shi, G.R., 2001, End-Permian catastrophe by a bolide impact: Evidence of a gigantic release of sulfur from the mantle: Geology, v. 29, p. 815–819, doi: 10.1130/0091-7613(2001)029<0815:EPCBAB>2.0.CO;2.

Kajiwara, Y., Yamakita, S., Ishida, K., Ishiga, H., and Imai, A., 1994, Development of a largely anoxic stratified ocean and its temporary massive mixing at the Permian/Triassic boundary supported by the sulfur isotopic record: Palaeogeography, Palaeoclimatology, Palaeoecology, v. 111, p. 367–379, doi: 10.1016/0031-0182(94)90072-8.

Kennett, J.P., 2003, Methane hydrates in Quaternary climate change: The clathrate gun hypothesis: Washington, D.C., American Geophysical Union, 216 p.

Khalil, M.A.K., Shearer, M.J., and Rasmussen, R.A., 2000, Methane sinks, distribution and trends, in Khalil, M.A.K., ed., Atmospheric methane: Berlin, Springer, p. 86–97.

Kidder, D.L., and Worsely, T.R., 2004, Causes and consequences of extreme Permo-Triassic warming to globally equable climate and relation to the Permo-Triassic extinction and recovery: Palaeogeography, Palaeoclimatology, Palaeoecology, v. 203, p. 207–238, doi: 10.1016/S0031-0182(03)00667-9.

King, G.W., 1991, The Dicynodonts: A study in paleobiology: London, Chapman and Hall, 233 p.

Knoll, A.H., Bambach, R.K., Canfield, D.E., and Grotzinger, J.P., 1996, Comparative earth history and the Late Permian mass extinction: Science, v. 273, p. 452–457.

Koeberl, C., Gilmour, I., Riemold, W.U., Clayes, P., and Ivanov, B., 2002, End-Permian catastrophe by bolide impact: Evidence of a gigantic release of sulfur from the mantle: Comment: Geology, v. 30, p. 855–856, doi: 10.1130/0091-7613(2002)030<0855:EPCBBI>2.0.CO;2.

Krassilov, V.A., and Afonin, S.A., 1999, *Tympanicysta* and the terminal Permian events: Permophiles, v. 35, p. 16–17.

Krull, E.S., 1999, Permian palsa mires as paleoenvironmental proxies: Palaios, v. 14, p. 530–544.

Krull, E.S., and Retallack, G.J., 2000, $\delta^{13}C_{org}$ depth profiles from paleosols across the Permian-Triassic boundary: Evidence for methane release: Geological Society of America Bulletin, v. 112, p. 1459–1472, doi: 10.1130/0016-7606(2000)112<1459:CDPFPA>2.0.CO;2.

Krull, E.S., Retallack, G.J., Campbell, H.J., and Lyon, G.L., 2000, $\delta^{13}C_{org}$ chemostratigraphy of the Permian-Triassic boundary in the Maitai Group, New Zealand: Evidence for high-latitude methane release: New Zealand Journal of Geology and Geophysics, v. 43, p. 21–32.

Krull, E.S., Lehrmann, D., Druke, D., Kessell, B., Yu, Y.-Y., and Li, R.-X., 2004, Stable carbon isotope stratigraphy across the Permian-Triassic boundary in shallow marine carbonate platforms, Nanpanjiang Basin, South China: Palaeogeography, Palaeoclimatology, Palaeoecology, v. 204, p. 297–315, doi: 10.1016/S0031-0182(03)00732-6.

Kuhnt, W., El Hassane, C., Holbourn, A., Luderer, F., Thurow, J., Wagner, T., El Albani, A., Beckmann, B., Herbin, J.-P., Kawamuna, H., Kolonic, S., Nederbragt, S., Street, C., and Ravilous, K., 2002, Morocco Basin's sedimentary record may provide correlations for Cretaceous paleoceanographic events worldwide: Eos (Transactions, American Geophysical Union), v. 82, p. 361–364.

Kulkarni, S.G., and Mehendale, H.M., 1998, Carbon dioxide, in Wexler, P., ed., Encyclopedia of toxicology: San Diego, Academic Press, v. 1, p. 222–223.

Kump, L.R., 1991, Interpreting carbon isotope excursions: Strangelove oceans: Geology, v. 19, p. 299–302, doi: 10.1130/0091-7613(1991)019<0299:ICIESO>2.3.CO;2.

Kvenvolden, K.A., 2000, Natural gas hydrate: Introduction and history of discovery, in Max, M.D., ed., Natural gas hydrate in oceanic and permafrost environments: Dordrecht, Kluwer, p. 9–16.

Landis, G.P., Rigby, J.K., Sloan, R.E., Hengston, R., and Smee, L.W., 1996, Pelée hypothesis: Ancient atmospheres and geologic-geochemical controls in evolution, survival and extinction, in MacLeod, N., and Keller, G., eds., Cretaceous-Tertiary mass extinction: Biotic and environmental change: New York, W.W. Norton, p. 519–556.

Langenhorst, F., Kyte, F.T., and Retallack, G.J., 2005, Re-examination of quartz grains from the Permian-Triassic boundary section at Graphite Peak, Antarctica [abs.]: 26th Lunar and Planetary Conference, Houston, no. 2358, ftp://www.lpi.usra.edu/pub/outgoing/lpsc2005/full76.pdf (accessed October 6, 2005).

Levin, I., Bösinger, R., Bonani, G., Francey, R.J., Kromer, B., Münnich, K.O., Suter, M., Trivett, N.B.A., and Wölfi, W., 1992, Radiocarbon in atmospheric carbon dioxide and methane: Global distribution and trends, in Taylor, R.E., Long, A., and Kra, R.S., eds., Radiocarbon after four decades: An interdisciplinary perspective: New York, Springer, p. 503–518.

Looy, C.V., 2000, The Permian-Triassic biotic crisis: Collapse and recovery of terrestrial ecosystems: Laboratory for Palaeobotany and Palynology Contributions Utrecht, v. 13, 114 p.

Looy, C.V., Twitchett, R.J., Dilcher, D.L., van Konijnenberg-van Cittert, J.H.A., and Visscher, H., 2001, Life in the end-Permian dead zone: Proceedings of the National Academy of Sciences USA, v. 98, p. 7879–7883, doi: 10.1073/pnas.131218098.

MacLeod, K.G., Smith, R.M.H., Koch, P.L., and Ward, P.D., 2000, Timing of mammal-like reptile extinctions across the Permian-Triassic boundary in South Africa: Geology, v. 28, p. 227–230, doi: 10.1130/0091-7613(2000)028<0227:TOMLRE>2.3.CO;2.

Magaritz, M., Krishnamurthy, R.V., and Holser, W.T., 1992, Parallel trends in organic and inorganic carbon isotopes across the Permian-Triassic boundary: American Journal of Science, v. 292, p. 727–740.

Marshall, C.R., 1995, Distinguishing between sudden and gradual extinctions in the fossil record: Predicting the position of the Cretaceous-Tertiary iridium anomaly using the ammonite fossil record on Seymour Island, Antarctica: Geology, v. 23, p. 731–734, doi: 10.1130/0091-7613(1995)023<0731:DBSAGE>2.3.CO;2.

Marshall, J.T., and McCulloch, M.T., 2002, An assessment of Sr-Ca ratios in shallow water hermatypic corals as a proxy for sea surface temperature: Geochimica et Cosmochimica Acta, v. 66, p. 32, p. 3263–3280.

Martin, E.E., and MacDougall, J.D., 1995, Seawater Sr isotopes at the Permian/Triassic boundary: Chemical Geology, v. 125, p. 73–79, doi: 10.1016/0009-2541(95)00081-V.

Max, M.D., Dillon, W.P., Nishimura, A.C., and Hurdle, B.G., 1999, Sea-floor methane blow-out and global firestorm at the K-T boundary: Geomarine Letters, v. 18, p. 285–291, doi: 10.1007/s003670050081.

Mazor, Y.R., Piskarev, Y.V., and Bocharova, L.V., 1979, Characteristics of the accumulation and transformation of coals of the Tunguska Basin: Vestnik Moskovskogo Universiteta Geologiya, v. 34, p. 32–39.

Messenger, S., 2000, Identification of molecular-cloud material in interplanetary dust particles: Nature, v. 404, p. 968–971, doi: 10.1038/35010053.

Meyen, S.V., 1982, The Carboniferous and Permian floras of Angaraland (a synthesis): Lucknow, International Publishers, 109 p.

Mii, H.-S., Grossman, E.L., and Yancey, T.E., 1997, Stable carbon and oxygen isotope shifts in Permian seas of west Spitzbergen: Global change or diagenetic artifact?: Geology, v. 25, p. 227–230, doi: 10.1130/0091-7613(1997)025<0227:SCAOIS>2.3.CO;2.

Milkov, A., 2004, Global estimates of hydrate-bound gas in marine sediments: how much is really out there?: Earth Science Reviews v. 66, p. 183–197.

Morante, R., 1996, Permian and early Triassic isotopic records of carbon and strontium in Australia and a scenario of events about the Permian-Triassic boundary: Historical Biology, v. 11, p. 289–310.

Morante, R., and Herbert, C., 1994, Carbon isotopes and sequence stratigraphy about the Permian-Triassic boundary in the Sydney Basin, in Diessel, C.F.K., and Boyd, R. L., eds., Proceedings, Symposium on Advances in the Study of the Sydney Basin, Newcastle, Australia, v. 28, p. 102–109.

Mortimer, N., Gans, P., Calvert, A., and Walker, N., 1999, Geology and thermochronometry of the east edge of the Median Batholith (Median Tectonic Zone): A new perspective on Permian to Cretaceous crustal growth of New Zealand: The Island Arc, v. 8, p. 404–425, doi: 10.1046/j.1440-1738.1999.00249.x.

Morgan, J.P., Reston, T.J., and Ranero, C.R., 2004, Contemporaneous mass extinctions, continental flood basalts, and "impact signals": Are mantle plume-induced lithospheric gas explosions the causal link?: Earth and Planetary Science Letters, v. 217, p. 263–284, doi: 10.1016/S0012-821X(03)00602-2.

Mory, A.J., Iasky, R.P., Glikson, A.Y., and Pirajno, F., 2000, Woodleigh, Carnarvon Basin, Western Australia: A new 120 km diameter impact structure: Earth and Planetary Science Letters, v. 177, p. 119–128, doi: 10.1016/S0012-821X(00)00031-5.

Mundil, R., Ludwig, K.R., Metcalfe, I., and Renne, P.R., 2004, Age and timing of the Permian mass extinctions: U/Pb dating of closed system zircons: Science, v. 305, p. 1760–1763, doi: 10.1126/science.1101012.

Musashi, M., Isozaki, Y., Koike, T., and Kreulon, R., 2001, Stable carbon isotope signature in mid-Panthalassa shallow water carbonates across the

Permo-Triassic boundary: Evidence for ^{13}C depleted superocean: Earth and Planetary Science Letters, v. 191, p. 9–20, doi: 10.1016/S0012-821X(01)00398-3.

Newton, R.J., Pevitt, E.L., Wignall, P.B., and Bottrell, S.H., 2004, Large shifts in the isotopic composition of seawater sulphate across the Permo-Triassic boundary in northern Italy: Earth and Planetary Science Letters, v. 218, p. 331–345, doi: 10.1016/S0012-821X(03)00676-9.

Pandolfi, J.M., 1992, A paleobiological examination of the geological evidence for recurring outbreaks of the crown of thorns starfish, *Acanthaster planci* (L.): Coral Reefs, v. 11, p. 87–93, doi: 10.1007/BF00357427.

Patnaik, P., 1992, A comprehensive guide to the hazardous properties of chemical substances: New York, Van Nostrand-Reinhold, 763 p.

Pillinger, C.T., 1987, Stable isotope measurements of meteorites and cosmic dust grains: Royal Society of London Philosophical Transactions, ser. A, v. 323, p. 313–322.

Poreda, R.J., and Becker, L., 2003, Fullerenes and interplanetary dust at the Permian-Triassic boundary: Astrobiology, v. 3, p. 75–90, doi: 10.1089/153110703321632435.

Pratt, G.W., and Herbert, C., 1973, A reappraisal of the Lorne Basin: Geological Survey of New South Wales Records, v. 15, p. 205–212.

Pruss, S., Fraiser, M., and Bottjer, D.J., 2004, Proliferation of Early Triassic wrinkle structures: Implications for environmental stress following the end-Permian mass extinction: Geology, v. 32, p. 461–464, doi: 10.1130/G20354.1.

Rampino, M.R., and Haggerty, B.M., 1996, The "Shiva hypothesis": Impacts, mass extinctions, and the galaxy: Earth, Moon, and Planets, v. 72, p. 441–460, doi: 10.1007/BF00117548.

Rampino, M.R., Prokoph, A., and Adler, A., 2000, Tempo of the end-Permian event: High-resolution cyclostratigraphy at the Permian-Triassic boundary: Geology, v. 28, p. 643–646, doi: 10.1130/0091-7613(2000)028<0643:TOTEPE>2.3.CO;2.

Rees, P.M., 2002, Land-plant diversity and the end-Permian mass extinction: Geology, v. 30, p. 827–830, doi: 10.1130/0091-7613(2002)030<0827:LPDATE>2.0.CO;2.

Reichow, M.K., Saunders, A.D., White, R.V., Pringle, M.S., Al'Mukhamedov, A.A., Medvedev, A.I., and Kirda, N.P., 2002, ^{40}Ar/^{39}Ar dates from West Siberian Basin: Siberian Flood Basalt Province doubled: Science, v. 296, p. 1846–1849, doi: 10.1126/science.1071671.

Reimold, W.U., Koeberl, C., Mory, A.J., Iasky, R.P., Glikson, A.Y., and Pirajno, F., 2000, Woodleigh, Carnarvon Basin, Western Australia: A new 120 km diameter impact structure: discussion and reply: Earth and Planetary Science Letters, v. 184, p. 353–365, doi: 10.1016/S0012-821X(00)00282-X.

Renne, P.R., Zhang, Z., Richards, M.A., Black, M.T., and Basu, A.R., 1995, Synchrony and causal relations between Permian-Triassic boundary crisis and Siberian flood volcanism: Science, v. 269, p. 1413–1416.

Renne, R.R., Melosh, H.J., Farley, K.A., Riemold, W.U., Koeberl, C., Kelly, S.P., and Ivanov, B.A., 2004, Is Bedout an impact crater? Take 2: Science, v. 306, p. 610–611, doi: 10.1126/science.306.5696.610.

Retallack, G.J., 1995, Permian-Triassic life crisis on land: Science, v. 266, p. 77–80.

Retallack, G.J., 1996, Acid trauma at the Cretaceous-Tertiary boundary in eastern Montana: GSA Today, v. 6, no. 5, p. 1–5.

Retallack, G.J., 1997, Earliest Triassic origin of *Isoetes* and quillwort evolutionary radiation: Journal of Paleontology, v. 71, p. 500–521.

Retallack, G.J., 1999a, Post-apocalyptic greenhouse paleoclimate revealed by earliest Triassic paleosols in the Sydney Basin, Australia: Geological Society of America Bulletin, v. 111, p. 52–70, doi: 10.1130/0016-7606(1999)111<0052:PGPRBE>2.3.CO;2.

Retallack, G.J., 1999b, Permafrost palaeoclimate of Permian palaeosols in the Gerringong volcanics of New South Wales: Australian Journal of Earth Sciences, v. 46, p. 11–22, doi: 10.1046/j.1440-0952.1999.00683.x.

Retallack, G.J., 2001, A 300-million-year record of atmospheric carbon dioxide from fossil plant cuticles: Nature, v. 411, p. 287–290, doi: 10.1038/35077041.

Retallack, G.J., 2002, *Lepidopteris callipteroides*, an earliest Triassic seed fern of the Sydney Basin, southeastern Australia: Alcheringa, v. 26, p. 475–500.

Retallack, G.J., 2004a, Contrasting deep-water records from the Upper Permian and Lower Triassic of South Tibet and British Columbia: Evidence for a diachronous mass extinction (Wignall and Newton 2003): Comment: Palaios, v. 19, p. 101–102.

Retallack, G.J., 2004b, End-Cretaceous acid rain as a selective extinction mechanism between birds and dinosaurs, *in* Currie, P., Koppelhus, E., Shugar, M., and Wright, J., eds., Feathered dragons: Studies in the transition from dinosaurs to birds: Bloomington, Indiana University Press, p. 35–64.

Retallack, G.J., 2004c, Vertebrate extinction across the Permian-Triassic boundary in Karoo Basin, South Africa: Reply: Geological Society of America Bulletin, v. 116, p. 1295–1296, doi: 10.1130/B25504.1.

Retallack, G.J., 2005, Earliest Triassic claystone breccia and soil erosion crisis: Journal of Sedimentary Petrology, v.75, p. 663–679, doi:10.2110/jsr.2005.055.

Retallack, G.J., and Krull, E.S., 1999, Landscape ecological shift at the Permian-Triassic boundary in Antarctica: Australian Journal of Earth Sciences, v. 46, p. 785–812, doi: 10.1046/j.1440-0952.1999.00745.x.

Retallack, G.J., Veevers, J.J., and Morante, R., 1996, Global early Triassic coal gap between Permo-Triassic extinction and middle Triassic recovery of swamp floras: Geological Society of America Bulletin, v. 108, p. 195–207, doi: 10.1130/0016-7606(1996)108<0195:GCGBPT>2.3.CO;2.

Retallack, G.J., Seyedolali, A., Krull, E.S., Holser, W.T., Ambers, C.P., and Kyte, F.T., 1998, Search for evidence of impact at the Permian-Triassic boundary in Antarctica and Australia: Geology, v. 26, p. 979–982, doi: 10.1130/0091-7613(1998)026<0979:SFEOIA>2.3.CO;2.

Retallack, G.J., Smith, R.M.H., and Ward, P.D., 2003, Vertebrate extinction across the Permian-Triassic boundary in the Karoo Basin of South Africa: Geological Society of America Bulletin, v. 115, p. 1133–1152, doi: 10.1130/B25215.1.

Retallack, G.J., Jahren, A.H., Sheldon, N.D., Charkrabarti, R., Metzger, C.A., and Smith, R.M.H., 2005, The Permian-Triassic boundary in Antarctica: Antarctic Science v.17, p. 241-258, doi: 10.1017/S0954102005002658.

Russell, J.A., and Crook, L., 1968, Comparison of metabolic responses of rats produced by two methods: American Journal of Physiology, v. 214, p. 111–116.

Ryskin, G., 2003, Methane driven oceanic eruptions and mass extinctions: Geology, v. 31, p. 741–744, doi: 10.1130/G19518.1.

Sarkar, A., Yoshioka, H., Edihara, M., and Naraoka, H., 2003, Geochemical and organic isotope studies across the continental Permo-Triassic boundary of the Raniganj Basin, eastern India: Palaeogeography, Palaeoclimatology, Palaeoecology, v. 191, p. 1–14, doi: 10.1016/S0031-0182(02)00636-3.

Schubert, J.K., and Bottjer, D.J., 1992, Early Triassic stromatolites as post-mass-extinction disaster forms: Geology, v. 20, p. 883–886, doi: 10.1130/0091-7613(1992)020<0883:ETSAPM>2.3.CO;2.

Schoene, R.B., 1990, High altitude pulmonary edema: search for a mechanism, *in* Sutton, J.P., Coates, G., and Remmers, J.E., eds., Hypoxia: The adaptations: Toronto, B.C. Decker, p. 246–259.

Scotese, C.R., 1994, Early Triassic paleogeographic map, *in* Klein, G. de V., ed., Pangaea: Paleoclimate, tectonics, and sedimentation during accretion, zenith and breakup of a supercontinent: Geological Society of America Special Paper 288, p. 7.

Sepkoski, J.J., 1996, Patterns of Phanerozoic extinction: A perspective from global data bases, *in* Walliser, O.H., ed., Global events and event stratigraphy in the Phanerozoic: Berlin, Springer, p. 35–51.

Sheldon, N.D., and Retallack, G.J., 2002, Low oxygen levels in earliest Triassic soils: Geology, v. 30, p. 919–922, doi: 10.1130/0091-7613(2002)030<0919:LOLIET>2.0.CO;2.

Shukla, A.D., Bhandari, N., and Shukla, P.N., 2003, Shocked quartz at the Permian-Triassic boundary (P/T) in Spiti Valley, Himalaya, India: Lunar and Planetary Science Proceedings, v. 34, p. 1490.

Siegenthaler, U., and Sarmiento, J.L., 1993, Atmospheric carbon dioxide and the ocean: Nature, v. 365, p. 119–125, doi: 10.1038/365119a0.

Sinha, A., Aubry, M.-P., Stottt, L.D., Thiry, M., and Berggren, W.A., 1995, Chemostratigraphy of the lower Sparnacian deposits (Argiles Plastiques Bariolées) of the Paris Basin: Israel Journal of Earth Sciences, v. 44, p. 223–237.

Smith, R.M.H., 1987, Helical burrow casts of therapsid origin from the Beaufort Group (Permian) of South Africa: Palaeogeography, Palaeoclimatology, Palaeogeography, v. 60, p. 155–170, doi: 10.1016/0031-0182(87)90030-7.

Smith, R.M.H., and Ward, P.D., 2001, Pattern of vertebrate extinction across an event bed at the Permian-Triassic boundary in the Karoo Basin of South Africa: Geology, v. 29, p. 1147–1150, doi: 10.1130/0091-7613(2001)029<1147:POVEAA>2.0.CO;2.

Suzuki, N., Ishida, K., and Ishida, H., 1993, Organic geochemical implications of black shales relative to the Permian-Triassic boundary, Tamba Belt, southwest Japan: Japanese Geological Survey Bulletin, v. 44, p. 707–720.

Svensen, H., Planke, S., Malthe-Sørensen, A., Tamtwell, B., Myklehurst, R., Eldem, T.R., and Rey, S.S., 2004, Release of methane from a volcanic basin as a mechanism for initial Eocene global warming: Nature, v. 429, p. 542–545, doi: 10.1038/nature02566.

Thorpe, R.B., Pyle, J.A., and Nisbet, E.G., 1998, What does the ice-core record imply concerning the maximum climatic impact of possible gas hydrate release at termination 1A? in Henriet, J.-P., and Meinert, J., eds., Gas hydrates: Relevance to world margin stability and climate change: Geological Society [London] Special Publication 137, p. 319–326.

Tonkin, P.C., 1998, Lorne Basin, New South Wales: Evidence for a possible impact origin?: Australian Journal of Earth Sciences, v. 45, p. 669–671.

Twitchett, R.J., Looy, C.V., Morante, R., Visscher, H., and Wignall, P.B., 2001, Rapid and synchronous collapse of marine and terrestrial ecosystems during the end-Permian biotic crisis: Geology, v. 29, p. 351–354, doi: 10.1130/0091-7613(2001)029<0351:RASCOM>2.0.CO;2.

Vdyovkin, G.P., and Moore, C.B., 1971. Carbon, in Mason, B., ed., Handbook of elemental abundance in meteorites: New York, Gordon and Breach, p. 81–91.

Vermeij, G.J., and Dorritie, D., 1996, Late Permian extinctions: Science, v. 274, p. 1550.

Visscher, H., Brinkhius, H., Dilcher, D.L., Elsik, W.C., Eshet, Y., Looy, C.V., Rampino, M.R., and Traverse, A., 1996, The terminal Permian fungal event: Evidence of terrestrial ecosystem destabilization and collapse: Proceedings of the National Academy of Sciences USA, v. 93, p. 2155–2158, doi: 10.1073/pnas.93.5.2155.

Wallace, P.J., 2003, From mantle to atmosphere: magma degassing, explosive eruptions and volcanic volatile budgets, in De Vivo, B., and Bodnar, R.J., eds., Melt inclusions in volcanic systems: Methods, applications and problems: Amsterdam, Elsevier, 258 p.

Wang, K., Geldsetzer, H.H.J., and Krouse, H.R., 1994, Permian-Triassic extinction: Organic $\delta^{13}C$ evidence from British Columbia, Canada: Geology, v. 22, p. 580–584, doi: 10.1130/0091-7613(1994)022<0580:PTEOCE>2.3.CO;2.

Wang, Z.-Q., and Chen, A.-S., 2001, Traces of arborescent lycopsids and dieback of the forest vegetation in relation to the terminal Permian mass extinction in North China: Review of Palaeobotany and Palynology, v. 117, p. 217–243, doi: 10.1016/S0034-6667(01)00094-X.

Ward, P.D., Montgomery, D.R., and Smith, R., 2000, Altered river morphology in South Africa related to the Permian-Triassic extinction: Science, v. 289, p. 1740–1743, doi: 10.1126/science.289.5485.1740.

Ward, P.D., Botha, J., Buick, R., de Kock, M.O., Erwin, D.H., Garrison, G.H., Kirschvink, J.C., and Smith, R., 2005, Abrupt and gradual extinction among Late Permian land vertebrates in the Karoo Basin, South Africa: Science, v. 307, p. 709–714, doi: 10,1126/sciencev1107068.

Wasson, J.T., 1974, Meteorites: Classification and properties: Berlin, Springer, 316 p.

Weidlich, O., Kiessling, W., and Flügel, E., 2003, Permian-Triassic boundary interval as a model for forcing marine collapse by long-term atmospheric oxygen drop: Geology, v. 31, p. 961–964, doi: 10.1130/G19891.1.

Whiticar, M.J., 2000, Can stable isotopes and global budgets be used to constrain atmospheric methane budgets? in Khalil, M.A.K., ed., Atmospheric methane: Berlin, Springer, p. 63–85.

Whiting, G.J., and Chanton, J.P., 2001, Greenhouse carbon balance of wetlands: Methane emission versus carbon sequestration: Tellus, ser. B, Chemical and Physical Meteorology, v. 53, p. 521–528.

Wignall, P.B., 1994, Black shales: Oxford, Oxford University Press, 127 p.

Wignall, P.B., 2001, Large igneous provinces and mass extinctions: Earth Science Reviews, v. 53, p. 1–33, doi: 10.1016/S0012-8252(00)00037-4.

Wignall, P.B., and Newton, R.J., 2004, Contrasting deep-water records from the Upper Permian and Lower Triassic of South Tibet and British Columbia: Evidence for a diachronous mass extinction (Wignall and Newton 2003): Reply: Palaios, v. 19, p. 102-104.

Wignall, P.B., Morante, R., and Newton, R., 1998, The Permo-Triassic transition in Spitzbergen: $\delta^{13}C_{org}$ chemostratigraphy, Fe and S geochemistry, facies, fauna and trace fossils: Geological Magazine, v. 135, p. 47–62, doi: 10.1017/S0016756897008121.

Wignall, P.B., Thomas, B., Willink, R., and Watling, J., 2004, Is Bedout an impact crater? Take 1: Science, v. 306, p. 609, doi: 10.1126/science.306.5696.609d.

Wolbach, W.S., Roegge, D.R., and Gilmour, I., 1994, The Permian-Triassic of the Gartnerkofel-1 core (Carnic Alps, Austria): Organic carbon isotope variation: Houston, Lunar and Planetary Institute Contributions, v. 825, p. 133–134.

Wooden, J.L., Czamanske, G.K., Fedorenko, V.A., Arndt, N.T., Chauvel, C., Bouse, R.M., King, B.W., Knight, R.J., and Siems, D.F., 1993, Isotopic and trace element constraints on mantle and crustal contributions to Siberian continental flood basalts, Noril'sk area, Siberia: Geochimica et Cosmochimica Acta, v. 57, p. 3677–3704, doi: 10.1016/0016-7037(93)90149-Q.

Woods, A.D., Bottjer, D.J., Mutti, M., and Morrison, J., 1999, Lower Triassic large sea-floor carbonate cements: Their origin and a mechanism for the prolonged biotic recovery from the end-Permian mass extinction: Geology, v. 27, p. 645–648, doi: 10.1130/0091-7613(1999)027<0645:LTLSFC>2.3.CO;2.

Wynn, J.G., 2003, Towards a physically based model of CO_2-induced stomatal frequency response: The New Phytologist, v. 157, p. 391–398.

Xu, D.-Y., and Yan, Z., 1993, Carbon isotopes and iridium event markers near Permian-Triassic boundary in the Meishan section: Zhejiang: Palaeogeography, Palaeoclimatology, Palaeoecology, v. 104, p. 171–176, doi: 10.1016/0031-0182(93)90128-6.

Yin, H.-F., Wu, S.-B., Ding, M.-H., Zhang, K.-X., Tong, J.N., Yang, F.-Q., and Lai, X.-L., 1996, The Meishan section candidate of the global stratotype section and point of the Permian-Triassic boundary, in Yin, H.-F., ed., The Paleozoic-Mesozoic boundary candidates of global stratotype section and point of the Permian-Triassic boundary, China: Wuhan, China, University of Geosciences, p. 31–48.

Zhou, Y.-Q., and Chai, C.-F., 1991, The discovery of shocked quartz and stishovite in the Permian-Triassic boundary clay of Huangshi, China: Meteoritics, v. 26, p. 413.

MANUSCRIPT ACCEPTED BY THE SOCIETY 28 JUNE 2005

Controls on the formation of an anomalously thick Cretaceous-age coal mire

T.A. Moore*
Solid Energy NZ Ltd., P.O. Box 1303, Christchurch, New Zealand, and Department of Geological Sciences, University of Canterbury, Christchurch, New Zealand

Z. Li
School of Biological, Earth and Environmental Sciences, University of New South Wales, Sydney, New South Wales 2052, Australia

N.A. Moore
Newman Energy Research Ltd., 2 Rose Street, Christchurch, New Zealand

ABSTRACT

The Main Seam in the Greymouth coalfield (Upper Cretaceous Paparoa Coal Measures) is exceptionally thick (>25m) and occurs in three locally thick pods, termed north, middle, and south. These pods are separated by areas of thin or absent ("barren") coal. The barren zone between the north and middle coal pods is characterized by a sequence that is 60 m thick comprising relatively thin (1–2.5 m thick) but laterally extensive (up to 500 m) sandstone units. The orientation of both the thin and the barren coal zones is approximately east to west. This is coincident with basement fault systems that occur in the region. Therefore, the stacked nature of the sandstones within this narrow zone may be a result of differential subsidence across basement fault blocks.

The Main Seam, like the sandstone units in the "barren" zone, is inferred to represent a stacked sequence. Two zones of thin partings (<20 cm in thickness) occur in the coal, and even where these zones do not occur, an interval of abundant vitrain bands is present. As has been suggested for other coal beds, intervals with high vitrain content may represent a demarcation between different paleomire systems, or, as in the case of the Main Seam, periods where the paleomire was rejuvenated with plant nutrients, allowing continued aggradation of the mire. The low ash yield (<5% dry basis) indicates that the Main Seam was rarely affected by flood incursions. This may have been the result of both doming of the peat surface as well as restriction of the dominant sediment flow by syn-sedimentary faulting.

Palynological analyses indicate that the Main Seam mire throughout most of its time was dominated by gymnosperms, particularly a relative of the Huron pine (*Lagarostrobus franklinii*). However, a distinct floral change to a *Gleichenia*-dominated mire occurs in the upper few meters of the Main Seam. This vegetation change may have resulted from basinwide environmental or climatic change. *Gleichenia* does not produce much biomass, and if it was the dominant mire plant it may not have been

---

*E-mail: tim.moore@solidenergy.co.nz

Moore, T.A., Li, Z., and Moore, N.A., 2006, Controls on the formation of an anomalously thick Cretaceous-age coal mire, *in* Greb, S.F., and DiMichele, W.A., Wetlands through time: Geological Society of America Special Paper 399, p. 269–290, doi: 10.1130/2006.2399(13). For permission to copy, contact editing@geosociety.org. ©2006 Geological Society of America.

able to keep peat accumulation rates higher than subsidence. Whether the cause was a decrease in peat accumulation or a drying of mire, the result would have been lowering of the surface to a degree that flooding and final termination would be likely.

Keywords: New Zealand, Cretaceous, Paparoa coal measures, *Lagarostrobus franklinii*, *Gleichenia*, coal lithofacies, Greymouth Coalfield, *Phyllocladdidites mawsonii*, Huon Pine, paleomire, syndepositional fault, paleochannels, stacked mires.

INTRODUCTION

The New Zealand landmass split from the Australian continent in the Middle to Late Cretaceous and has been isolated since (Laird, 1993, 1994). During the rifting process, peat mires were an early depositional component of New Zealand basins. Major peat mires are represented not only in Cretaceous sediments of New Zealand, but also within its Paleocene through to Holocene deposits. Because of its isolation for the last 70 million years, plant evolutionary trends, as represented by mire vegetation, afford a good opportunity to understand island ecologic systems. From a coal utilization perspective, New Zealand also offers a chance to understand the specific role of vegetation on resultant coal properties (Newman, 1989; Li et al., 2001). This paper will focus on the beginning of mire formation during the Cretaceous of New Zealand. We will examine one Cretaceous-age paleomire in an attempt to understand major contributing factors that resulted in the unique coal bed it later formed. It is hoped that the results and conclusions from this investigation can be used in broader studies of New Zealand plant evolution. We targeted the Cretaceous age "Main Seam" in the Greymouth coalfield on the West Coast of the South Island (Fig. 1) for this study because it represents one of the first significant paleomires to have formed within New Zealand.

Some of the least known Cretaceous sequences are found in the relatively small and confined basins of New Zealand. Most of these basins occur over an area of a few tens of square kilometers but contain very thick sedimentary sequences of mostly terrestrial deposits, often greater than 1.5 km in thickness (e.g., Shearer, 1995). A prominent feature in these sequences is the presence of coal measures that constitute much of the Greymouth Basin. It is estimated that these Cretaceous coal deposits account for over one billion metric tons of coal resources in New Zealand (Barry et al., 1994). Most of the Cretaceous coal deposits occur in the western or southern South Island. The variability of the sediment types is high and is, for the most part, thought to be controlled by syn-sedimentary tectonism (Gage, 1952; Newman and Newman, 1992; Sherwood et al., 1992; Laird, 1993; Shearer, 1995). However, what makes Cretaceous deposits in New Zealand of special interest is the exceptional thickness of some of the coal beds. Most coals from the Cretaceous elsewhere form seams that rarely exceed 20 m (see Shearer et al., 1995; Retallack et al., 1996), but in New Zealand seams of this thickness are common.

Previous work along the west coast of the South Island has shown thick, laterally discontinuous, but high-quality coal deposits of Upper Cretaceous age (Newman, 1987; Newman and Newman, 1992; Ward et al., 1995; Li et al., 1999; Li, 2002). The thick, economically mineable seams occur mainly within the Rewanui Member of the Paparoa Coal Measures (Fig. 2), and these subbituminous to bituminous rank seams have been mined since the 1930s.

Many coal beds in New Zealand have an interesting geochemistry, largely because the mineral matter in these coals is unusually low, less than 1% (dry basis). The Cretaceous Main Seam in particular has intervals that contain abnormally low ash yield values (<0.1%; Li, 2002). These values are less than what would be found in most plant material and it has been postulated that leaching of inorganic material, beginning within the paleomire, may have helped to lower an already low mineral matter content (Li et al., 2001). Select aspects of the geochemistry therefore can be used to help understand mire processes in these Late Cretaceous deposits.

Palynological studies of coal seams are extremely important in mire reconstruction. The New Zealand palynological record is fairly well known (Mildenhall, 1980; Edbrooke et al., 1994; and many others).The Cretaceous mires in New Zealand were dominated by gymnosperms, principally podocarps (Warnes, 1988; Shearer and Moore, 1994; Ward et al., 1995). Notably, angiosperms are virtually absent from Cretaceous mires but are present in riparian communities in adjacent rock types. However, few investigations have probed the details of palynological variations within individual seams, information that would improve our understanding of Cretaceous seams in New Zealand.

The Cretaceous Main Seam has been previously identified as one of the thickest Cretaceous coal seams in the world (Shearer et al., 1995). To help understand the processes of its formation, we set out three goals for this paper: (1) to delineate geometry and geochemistry of the Main Seam in relation to the associated sediments, to understand sedimentological processes involved; (2) to determine the ecological evolution within Late Cretaceous mires; and (3) to synthesize these data to form a model that explains the Main Seam's development and properties.

DATA AND METHODS

Lithological Analyses

The Main Seam occurs primarily in the subsurface, and thus the three-dimensional distribution of lithofacies has to be inferred mostly from drill core and geophysical logs. Some drill core data have been assessed previously (Newman, 1987) but recent drilling programs have supplemented geological data and allowed more detailed models to be tested. In this study 56 drill cores were used to reconstruct the geometry of the Main Seam (Fig. 3). However, only a small subset of these data was used in a more detailed study encompassing an area of ~4.5 km^2. In this area, 35 drill cores, all with geophysical logs were examined and 16 of the rock and coal cores were described on the basis of grain size, color, and sedimentary features.

Description of the rock core served two purposes. First, it allowed us to measure the frequency of occurrence of rock types. Second, it allowed us to calibrate geophysical logs to actual rock types, and thus we could interpret other logs as to lithology without having to describe each core in detail—a time-consuming and tedious effort. This methodology has been used elsewhere for maximizing the amount of information from geophysical logs (Flores et al., 1982). In our rock descriptions we noted the same characteristics for all sediments; although not exhaustive, this approach provided consistency between cores. For each core description, we marked boundaries between visually different rock types (with a minimum interval thickness greater than 15 cm) and then described each unit on the basis of color (e.g., black, dark-brown, olive-green, gray), grain size (e.g., conglomerate, sandstone, mudstone, claystone, limestone, coal), and sedimentary/biogenic features (e.g., roots, ripples, cross-beds). The use of the term "mudstone" is meant to denote a rock type containing both silt- and clay-size clastic sedimentary particles (Ferm et al., 1985); in contrast, "claystone" is applied to sediments without any noticeable particles of silt size fraction. To facilitate stratigraphic analyses, we then transferred all descriptions of the cores into a three-digit code for computer plotting using a system based on that of Ferm et al. (1979; 1985).

Two attributes can be used to describe the macroscopic characteristics of coal: (1) vitrain band thickness and abundance and (2) luster of the areas between the vitrain bands (referred to as the attrital matrix). Although luster is a subjective measurement, most workers can distinguish between "dull" and "bright," and we used these two qualifiers in this investigation (see also Cameron, 1978; Davis, 1978; Moore et al., 1992). In contrast, vitrain band abundance and thickness are easily quantifiable (see e.g., Moore et al., 1993; Shearer and Moore, 1994). In this study we used both descriptions of macroscopic coal types and measurements of thickness, because the relative abundance of coal types has been shown to control certain physical and chemical properties of coal (Esterle et al., 1994; 1995; Moore, 1995).

Geochemical Analyses

To understand the geochemical composition of the Main Seam, we analyzed major element compositions of 95 samples from four boreholes (drill holes 703, 755 in the north part of the Main Seam deposit, drill holes 819 and 836 in the south) and one underground site (MU). These samples consisted of coal and inorganic partings as well as roof and floor sediments.

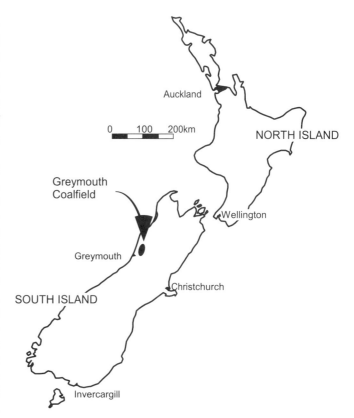

Figure 1. Index map of New Zealand showing the location of the Greymouth Coalfield.

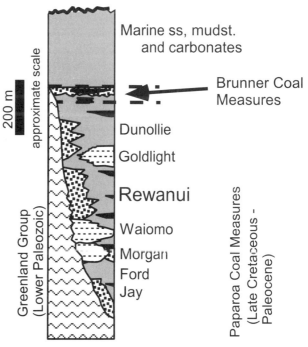

Figure 2. Generalized stratigraphic column for the Greymouth Coalfield.

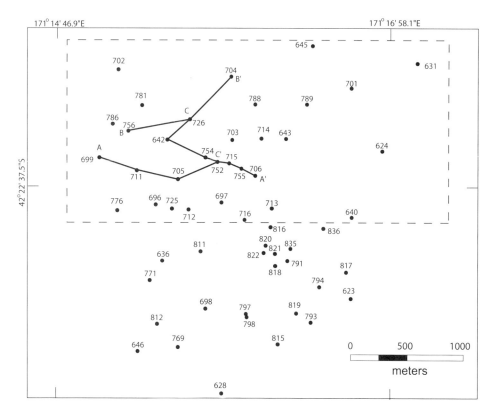

Figure 3. Location of drill holes used in the study. Dotted box shows the area studied in detail. Note position of cross section lines A–A′, B–B′, and C–C′. Cross section lines are shown in Figures 13 and 14.

A Philips PW-2400 wavelength dispersive X-ray fluorescence (WD-XRF) spectrometer in the Department of Geological Sciences, University of Canterbury, was used to analyze bulk rock samples for the major elements. In this study, XRF has been used routinely to analyze major elements (Si, Al, Ti, Fe, Ca, Mg, Mn, K, Na, P, S) of both high-temperature ash (HTA) of coal and associated sedimentary rocks (in-seam inorganic partings, roof and floor sediments). Samples were prepared as fused glass discs following a procedure similar to that of Norrish and Hutton (1969) and Leake et al. (1969). Detailed sample locations and major element results for all the above-mentioned samples were previously reported by Li et al. (2001) and Li (2002).

Palynological Analyses

Coal samples for palynological analyses were taken from seam intersections in 12 drill holes (640, 697, 699, 703, 704, 705, 706, 711, 712, 714, 716, and 726). Where ply intervals exceeded ~1.5 m, the ply was divided into sub-plies, each with a thickness of no more than 1 m. Initial sampling compared the assemblages of single "point" samples with continuous, whole-ply samples (Moore, 1996b). The results demonstrated considerable variation between point and continuous samples. Subsequent sampling, therefore, collected material from the entire thickness of each ply, in order for results to be representative of the whole ply.

The palynological samples were crushed to 1.5 ϕ (0.35 mm), small enough to allow rapid chemical oxidation of the coal, yet large enough to minimize damage to individual pollen grains. Palynomorphs were extracted from the coal using the wet Schultze process, whereby reactive organic materials in the coal, such as vitrinite and liptinite, are removed by oxidation with concentrated nitric acid/potassium chlorate ($HNO_3/KClO_3$) solution.

Residues were neutralized with potassium hydroxide, and remaining coarse material was removed by sieving through a 100 μm screen. Unwanted fine material, such as colloids, was removed by addition of 10% bleach solution for 10–20 min, followed by "short centrifuging" (accelerating samples from 0 to 2000 rpm over 30 seconds). Any mineral matter present was removed by heavy liquid separation in potassium iodide solution (specific gravity 1.3–1.5 g/mL). Processed samples were mounted in glycerine jelly on glass slides.

The pollen assemblages were examined with a Zeiss KM microscope, using a magnification of 300× or greater for the identification of small palynomorphs. The proportions of pollen groups were estimated using a 250 grain count.

RESULTS

Clastic Lithofacies

Documentation of rock types in the 16 cores from the Rewanui Member allows some generalizations to be made about typical lithologies. Examples of the major inorganic rock types are shown in Figures 4 and 5, and for convenience they can be subdivided by

Figure 4. Representative photographs of coarse-grained sediments encountered in cores of the Rewanui Member. (A) Greywacke pebble conglomerate. (B) Quartz pebble conglomerate. (C) Mixed greywacke and quartz pebble conglomerate. (D) Coal band/spar conglomerate. (E) Rippled sandstone with mudstone lenses. (F) Massive sandstone.

Figure 5. Representative photographs of fine-grained sediments encountered in cores of the Rewanui Member. (A) Light olive-gray rippled mudstone with sandstone lenses. (B) Burrowed mudstone with sandstone lenses. (C) Light olive-gray mudstone with carbonaceous lenses. (D) Light olive-gray mudstone, root penetrated. (E) Olive-gray massively bedded mudstone. (F) Dark-gray to black, massive claystone.

grain size into conglomerate, sandstone, mudstone, and claystone (Table 1; see also Ferm and Moore, 1997). Although none of these rock types are rare in the Main Seam interval, mudstone is by far the most common and claystone is the least frequent.

The conglomerates in the study area vary considerably in clast size and composition. The majority of clasts are derived from pre-Tertiary basement (i.e., well-indurated olive-gray mudstone), generally rounded but flat in shape, and poorly to moderately sorted (Fig. 6). Thicknesses of individual greywacke conglomerate beds vary from a few tens of centimeters to tens of meters. A few intervals in the cores studied also contain conglomerates composed of quartz pebbles, which tend to be subangular to subrounded as well as poorly sorted. A third clast type comprises spars and bands of coaly material in a matrix of usually fine- to coarse-grained quartz sandstone. The coal spars/bands are considered clasts (and thus a conglomerate rock type; Ferm et al. 1985) because they represent large (0.5–2 cm in thickness) woody material that would settle even in relatively high water velocities.

Gradation from conglomerates into sandstones can be gradual to abrupt; i.e., from 3 to 15 cm (Fig. 6). Sandstones are mostly olive-gray in color, a result probably from their mixed quartz and clay composition. Although some massive (i.e., no visible bedding surfaces) sandstone intervals were observed, most contained mudstone interbedding on the order of 0.5–2 cm in thickness. Rippled and less commonly cross-bedded sandstones could also be noted in the cores. Other structures such as burrows and root penetration are readily identifiable in core as well as in nearby equivalent-age sediments (Fig. 7).

Mudstones commonly grade into and out of sandstones over varying intervals from less than 1 cm to over 1 m. Usually in this gradation interval mudstone interbeds would become more dominant; when over 50% of the interval was mudstone interbeds, it was termed mudstone with sandstone. A common feature in mudstones is the presence of carbonaceous material either as subhorizontal layers (Figs. 5C, 8) or as vertical features representing root penetration (Fig. 5D). Burrowing structures occur locally.

Unlike the other grain size types, claystone units are relatively rare in the Main Seam interval. Where claystones occur, they are commonly black to medium gray in color and are either

TABLE 1. MAJOR INORGANIC ROCK TYPES OCCURRING IN THE MAIL SEAM INTERVAL

Grain Size	Color	Sedimentary features
Conglomerate	light grey to olive grey (highly dependant on clast type	clasts <0.5 to 7 cm three clast types: greywacke pebbles quartz pebble (<1 cm) coal spars greywacke pebbles well rounded quartz pebbles subangular to well rounded coal spars lenticular in shape
Sandstone	light grey to olive grey	mostly fine to medium grain but some very fine and coarse material sandstone types: massive (no visible bedding) rippled root penetrated
Mudstone (siltstone)	black to light olive grey	four types of features common: massive (no visible features) sandstone interbeds carbonaceous material rooted and/or burrowed
Claystone	black to medium grey	mostly massive bedding but some root penetration

Figure 7. Outcrop photograph of Rewanui root-penetrated sandstone. Arrows point to root traces; scale bar is 4 cm.

Figure 8. Outcrop photograph of Rewanui carbonaceous mudstone. Arrows point to carbonaceous material; hammer handle for scale is 22 cm.

Figure 6. Outcrop photograph of Rewanui greywacke conglomerate interbedded with massively bedded sandstone. Hammer for scale is 33 cm.

massive or have bedding disturbed by roots. Claystones tend to occur only directly beneath coal or impure coal and are usually less than 25 cm thick.

Coal Lithofacies

Five macroscopically recognizable coal types (bright luster, nonbanded; bright with 10%–20% vitrain bands; bright with 20%–40% banding; steely-gray luster with 10%–20% banding; and dull luster with 20%–40% banding) were identified in the coal cores of the Main Seam (Fig. 9). Of the five coal types, three were uniformly bright in their matrix luster but varied in vitrain band content whereas the other two types had a dull or steely-gray matrix luster (Table 2). All coal types were also identified by qualitatively estimating matrix brightness, and for the banded types, quantitatively point-counting seam intervals to determine

Controls on the formation of an anomalously thick Cretaceous-age coal mire 275

Figure 9. Macroscopic coal types of the Cretaceous Main Seam. (A) Bright luster, nonbanded. (B) Bright luster, banded 10%–20%. (C) Bright luster, banded 20%–40%. (D) Steely-gray luster, banded 10%–20%. (E) Dull luster, banded 20%–40%.

TABLE 2. MACROSCOPIC COAL TYPES IN THE MAIN SEAM

Matrix lustre	Vitrain abundance (%)
Bright	non banded
	10–20
	20–40
Steely-grey	10–20
Dull	20–40

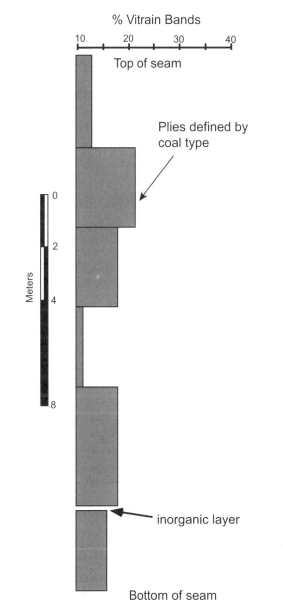

Figure 10. Representative vertical profile of quantitative vitrain band counts (drill core 697; 23 m in thickness). Note that coal is all of bright luster.

vitrain band abundance and brightness. The quantification of the exact proportion of vitrain bands (and their average thickness) for all intervals allowed trends in the seam to be delineated. The coal types found in the Main Seam are very similar to ones reported for the "D" seam (Ward et al., 1995), which is adjacent to the study area in a stratigraphically similar zone.

Most intervals of the Main Seam are composed of the three bright coal types (Table 2). The nonbanded coal type, although relatively common, occurs less frequently than banded coal and generally only in areas where the seam is the thickest (e.g., core hole 752). The two non-bright luster coal types are thin and rare. Preliminary chemical and petrographic analysis of the steely-gray luster coal type indicates that it is high in inertinite as well as in CaO. The dull-luster coal type occurs more frequently than the steely-gray type but was common only near the base or top of the seam.

A representative vertical profile of banding character for a coal core from the Main Seam is given in Figure 10; in addition, a size histogram for vitrain band thickness for the three major coal types for all cores is shown in Figure 11. A phi (ø) scale is used because, as with other sediments, this transformation normalizes the size distribution of organic particles (Lorente, 1990; Moore and Hilbert, 1992). The size histogram illustrates that although the proportion of vitrain bands may vary, the mean size of the bands remains the same regardless of coal type.

Spatial Distribution of Coal and Clastic Lithofacies

After an initial survey of all available geophysical logs in the study area we found that lithologic variability was quite high. We constructed several seam correlation possibilities using a "best fit" and circular correlation techniques. We then selected the most probable correlation scenario, and from this a total coal thickness isopach was produced (Fig. 12). Although faulting does occur in the region, there was no evidence of tectonic thickening in the coals examined.

Three features of the coal thickness isopach are of note: First, a significantly large part of the study area is composed of coal greater than 20 m. Second, the thickest coal occurs in three separate oval or disc-shaped structures (referred to as "pods"). Third, the three pods are separated by a zone of thin or absent coal oriented approximately east-west.

The seam correlations used in this study can be observed in the cross sections shown in Figures 13 and 14. In addition, the relationship between coal and inorganic rock types can be noted in these sections. One of the keys to correlation of the coal body was dividing the areas into a "north," a "middle," and a "south" zone, which were demarcated by a thin coal area. In the middle, cross section A–A′ transects the thickest part of the coal (drill hole 752). The general west-east orientation of this section shows that the coal body splits at both ends into at least two beds. On the eastern side of the deposit the relatively thick coal found in drill hole 715 (23 m) splits rapidly over a distance of less than 300 m into two beds of 15 m and 7 m thickness and then only the upper bed retains any appreciable thickness farther eastward in drill hole 706. On the western margin of the deposit, the coal attains a thickness of 20 m, but then farther westward splits into

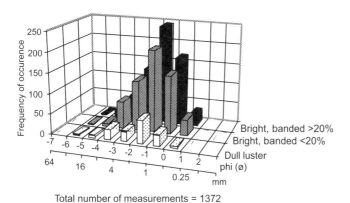

Figure 11. Size histogram for all counts of vitrain bands on all cores of the Main Seam. Mean (X), standard deviation (s), and number of counts (n) for the three coal types are as follows: banded 20%–40%, X = 1.62ø, s = 1.12ø, n = 651; banded <20%, X = 1.60ø, s = 1.14ø, n = 579; dull luster, X = 1.91ø, s = 1.25ø, n = 142.

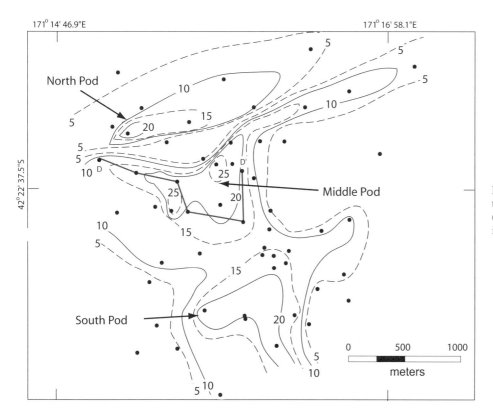

Figure 12. Isopach map for total coal thickness of the Cretaceous Main Seam (in meters). Cross section D–D′ is given in Figure 15A.

at least three seams; the thickest and central seam is ~12 m, whereas the upper and lower seams are less than 3 m (drill hole 699). Although the coal in this southern pod is thick and does not contain thick clastic partings, several small partings can be observed in the geophysical logs (e.g., drill hole 705) and in the coal core.

Coarse-grained sediments (sandstones and conglomerates) are common both above and below the Main Seam. However, in general conglomerates occur mostly below and on the western edge of the Main Seam interval. Examples of exceptions to this are the coal spar conglomerates, which interfinger with the splits of the Main Seam in drill core 699, and the mudstone pebble conglomerates, which occur in abundance between the splits of the coal in drill hole 704. Previous studies have indicated that the mudstone pebble conglomerates have been derived from the west from greywacke Greenland Group basement; in contrast the predominantly quartz pebble conglomerates are derived from the eastern granitic basement areas (Newman and Newman, 1992).

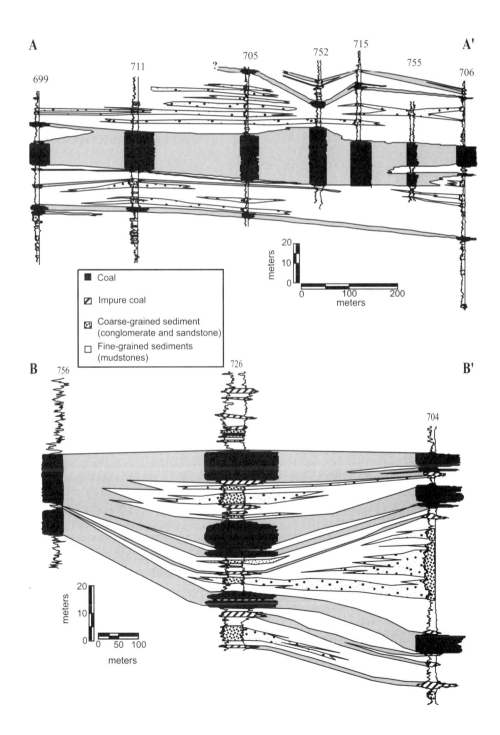

Figure 13. Interpretive cross sections using gamma and density geophysical downhole logs; A–A′ (Main Seam middle pod) and B–B′ (Main Seam north pod). See Figure 3 for location of cross sections.

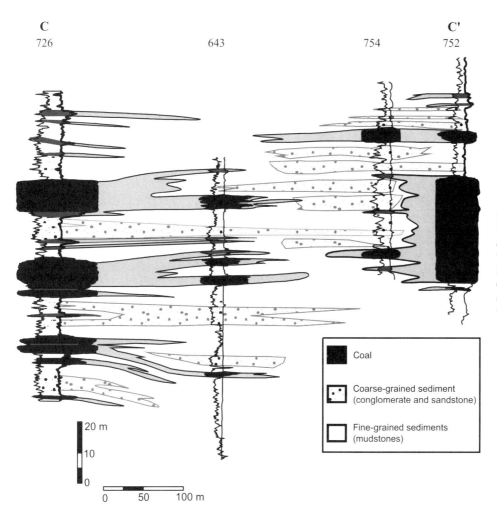

Figure 14. Interpretive cross section using gamma and density geophysical downhole logs; C–C' through the "want" area and connecting the Main Seam north and middle pods. Note the common point of drill hole 752 in this and the A–A' cross sections. See Figure 3 for location of cross sections.

However, recent cathodoluminescence data indicate that there may be no distinction between sources (Ettmuller, 2003). Sandstones, however, are abundant, overlying the seam in beds ranging from a few tens of centimeters to a few meters. In particular, these intervals occur in drill cores 711, 705, 715, 755, and 706. In all cases, the sandstones have a "sheet" geometry without erosional bases. In addition, they are rippled or cross-bedded, usually coarsen upward, and are interbedded with mudstone or carbonaceous mudstone. Although relatively thin (~1 m), some of these sheet sandstones can be correlated over 500 m.

The thickest sets of sandstone beds occur in the area where there is the least coal. It is important to note, however, that individual sandstone beds, although most abundant in the "want" area (see drill cores 643 and 754; Fig. 14), rarely exceed 2.0–2.5 m in thickness. Drill core indicates that these sandstones generally are planar but in rare circumstances have erosive bases. In addition, these sandstones almost always fine upward and are penetrated by roots and/or bioturbated throughout their whole thickness. Some sandstone units are weakly rippled and the predominant interbeds are light olive-gray carbonaceous mudstones with roots. The sandstone units observed in core 754 are laterally correlated to the south to be equivalent to the Main Seam coal. If this correlation is correct, then it is significant not only that the Main Seam is thickest adjacent to the "want" area (>30 m) but also that clastic partings are absent and that the seam is extremely low in ash yield (2%–3%, dry basis).

The sandstone units in the "want" area between the north and middle pods can be correlated laterally northward and occur as splits between plies of the Main Seam north "pod." The sandstones, where they were described in drill cores (e.g., 726; Figs. 13, 14), have planar bases, are rippled and/or cross-bedded, and are locally penetrated by roots throughout. The sandstones are interbedded both with light olive-gray mudstones as well and with thin beds of impure coal and coaly mudstone. Erosional surfaces do occur but were found most commonly within sandstone units rather than at the bases.

Coal Type Distribution

In general, the most-banded coal types were found to occur at the top and bottom of the coal seam and at the margins of the deposit (Fig. 15). This latter relationship is illustrated in

Figure 15B where vitrain band occurrence can be shown to vary in direct relation with seam thickness. A higher proportion of banding is also noted in many of the described cores in intervals that are laterally correlatable to inorganic partings. Specifically within the Main Seam middle "pod" of coal, the seam can be vertically segregated into three units based on the banded/less banded nature of the coal.

Geochemical Variability

As a measure of lateral variability of both inorganic matter and geochemistry in the coal, ash contents and major elements of the Main Seam have been investigated and modeled (Figs. 16–20). Expectedly, the lowest ash yields for the Main Seam correlate with the thickest parts of the seam (Figs. 12 and 16). Toward all margins, especially the east and northwest, ash yields tend to increase. The two highest ash zones, located in the northwest and eastern parts of the study area, are probably related to syndepositional detrital inorganic sources (Li et al., 2001; Li, 2002). Bowman (1984) divided the Rewanui Coal Measures broadly into two distinctive zones in terms of the provenance of the sediment. In the northwest and western parts of the basin the sediments are characterized by high-energy-transported bedded greywacke conglomerate, and fining-upward cycles with sandstone, mudstone, and coal, which are thought to be derived from the Greenland Group greywackes and argillites. In contrast, in the eastern and southern parts of the basin the sediments are characterized by quartzofeldspathic sandstone and fining-upward cycles with mudstone and coal, which are thought to be derived from a granitic source that probably came from at least 20 km north of the present position of the coalfield. Therefore, the northwestern parts of the basin were prone to being inundated by clastic detritus derived from the erosion of the Greenland Group rock. The higher ash zones along the eastern margins (Fig. 16) are thought to be related to fluvial paleochannels, which periodically flooded the peat margins and supplied granitic-derived sediments (Bowman, 1984; Newman and Newman, 1992).

To further understand the spatial distribution of ash yield, we investigated four vertical profiles for the Main Seam in detail. For example, in the ash profile drill hole 703 with a coal thickness of 21m (Fig. 21), the ash yields are around 2%–3% (as-received basis) at both the top and the bottom of the seam. However, the ash yield decreases dramatically toward the center, where it is as low as 0.58%. The ash yield profile of drill hole 755 is very similar to that of drill hole 703, perhaps because these two sample locations are only ~200 m apart from each other. The ash content increase from point A to point B (Fig. 21) in the profile drill hole 703 is probably correlated to the inorganic parting of drill hole 755 due to the same flooding event as seen in drill hole 755. Moore (1996c) had also correlated this interval on the basis of lithotype data.

In the study area of the Main Seam, ash yield and the major elements Si, Al, K, and Ti vary laterally in a manner very similar to each other. This is not surprising, because Si, Al, and K (as

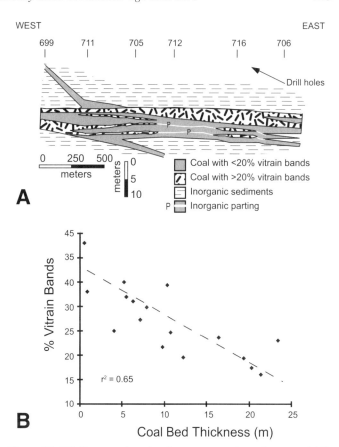

Figure 15. (A) Cross section showing macroscopic coal type variability. (B) Relationship between coal bed thickness and the average % vitrain band abundance within the complete bed. See Figure 12 for location of drill holes. r^2 is a Pearson correlation coefficient

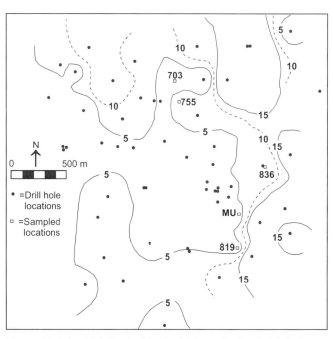

Figure 16. Ash yield (%, air dried basis) isopachs for the Main Seam.

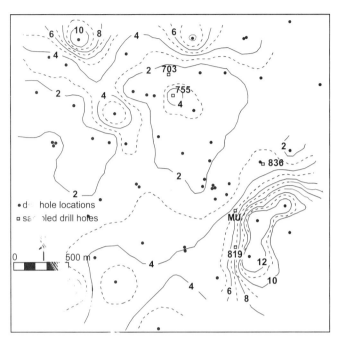

Figure 17. SiO$_2$ (%) isopach contours for the Main Seam.

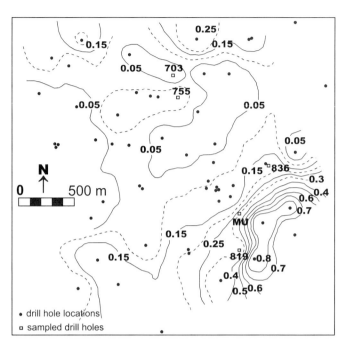

Figure 19. K$_2$O (%) isopach contours for the Main Seam

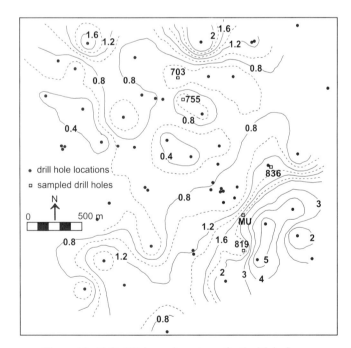

Figure 18. Al$_2$O$_3$ (%) isopach contours for the Main Seam.

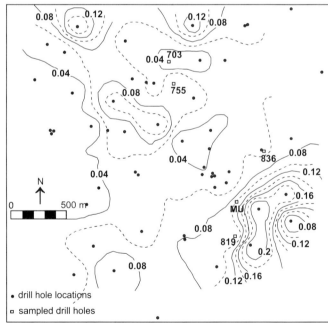

Figure 20. TiO$_2$ (%) isopach contours for the Main Seam

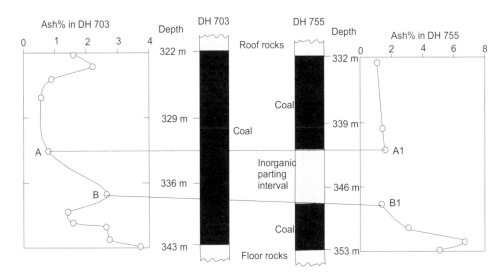

Figure 21. Ash yield (air-dried basis) profiles for drill holes 703 and 755 of the Main Seam. The ash yield increases from point A to B in profile of 703 is probably the lateral equivalent to the inorganic parting occurring in the 755 profile.

clays and quartz) are the controlling factors in the ash yield as a result of clays and quartz being the dominant minerals in the coal (Li, 2002). In addition, Ti is also primarily associated with clay minerals as discrete oxide (rutile or anatase) inclusions in clays.

In the Main Seam, ash content tends to be systematically high in the lowermost bench (in some cases in the uppermost bench as well) and the lowest ash contents are always in the middle bench (Li et al., 2001; Li, 2002). Major elements Si, Al, K, and Ti also vary vertically concordantly with ash yield. A profile of major elements observed in drill hole 703 is shown in Figure 22. SiO_2 and Al_2O_3 follow the same vertical variation trend as ash yield. These trends are consistent with the mineralogical observations that show predominance of clays and quartz in both the Main and E seams. Fe_2O_3 and S show a similar trend across the profile but this trend is different from that of the ash yield and SiO_2. Similar variations in major elements were also observed in other profiles from the Main Seam (Li, 2002).

In the Main Seam, the K contents are elevated more in the basal layers than in the top portions in all locations investigated. This is interesting because potassium is one of the essential nutrients for plant growth and potassium enrichment in the top layers is usually expected owing to plant bioaccumulation mechanisms.

Among 88 seam-composite samples from the Main Seam coal quality database of Solid Energy Ltd. (R. Boyd, 2000, personal commun.), 63 samples (72% of total population) have total S of <0.4% (Li, 2002). Sulfur in the Main Seam is primarily organic sulfur with minor amounts of pyritic and sulfate sulfur. This suggests a freshwater depositional environment in the peat swamp of the Main Seam.

Si, Al, K, and Ti have a linear correlation (r = 0.95–0.98) with the ash yield. Al, K, and Ti also have linear correlations (r = 0.95–0.96) with Si in the Main Seam (Li, 2002). These correlations confirm, from a geochemistry perspective, that clay minerals and quartz are the controlling factors on ash composition and concentrations. There is also a general, though not strong, correlation between Fe_2O_3, and S. Ward et al. (1999) reported a similar observation in the New South Wales coals of Australia.

Vertical and Lateral Palynological Variability

Palynological analyses have been simplified to show variations in abundance of spore types. By far the dominant palynological constituents were *Gleichenia* and *Phyllocladidites mawsonii* and these are shown in Figure 23. Angiosperms were present in most samples, but only in small concentrations (Moore, 1996b). A brief description for the major palynological constituents is given below.

Phyllocladidites mawsonii

The most abundant pollen type in most of the samples analyzed is the gymnosperm *P. mawsonii*. *P. mawsonii* pollen are distinctive, characterized by small proximal tubercles. Although *P. mawsonii* has sacci for air transport, these are smaller than those of many other gymnosperms. The pollen type first appeared in the mid-Cretaceous, and was widespread in New Zealand and Australia in the Late Cretaceous and early Tertiary (Martin, 1984).

P. mawsonii pollen is homologous to that produced by modern *Lagarostrobus franklinii*, the Huon Pine (Playford and Dettman, 1978). Today restricted to Tasmania, Australia, *L. franklinii* is a tree 20–30 m in height, with spreading branches and a dense, pyramidal crown. The extant Huon Pine is locally dominant in temperate rain forests along riverbanks and lake shores, and on swamp flats. The plant can propagate vegetatively along waterways (Gibson and Brown, 1991). *L. franklinii* is sensitive to fire and can be eliminated from areas by a single fire event (Gibson, 1988). Most stands are therefore restricted to within ten meters of channels.

Remaining Gymnosperms

Remaining gymnosperms are far less abundant than *P. mawsonii* in all the samples examined. Because many of the

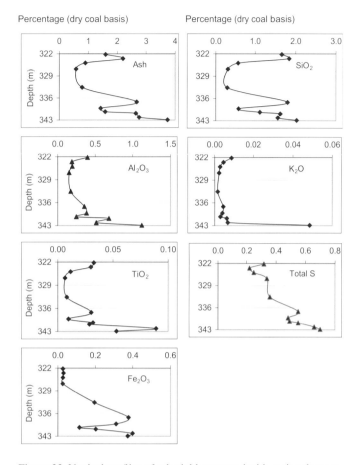

Figure 22. Vertical profiles of ash yield compared with major elements in drill hole 703.

gymnosperm species are podocarps with large sacci, implying wind transport, it is probable that some of the non-*P. mawsonii* gymnosperms are regional in origin and derived from the adjacent flood plains and higher ground.

Gleichenia

Gleicheniaceae is a family of primitive ferns found throughout Australasia. Some species of *Gleicheniaceae* spread along the ground, whereas others form thick tangles supported by trees and shrubs. Most extant species of Gleicheniaceae prefer at least seasonally wet conditions, and many are found in mires and wetlands (Fig. 24). *Gleichenia* macrofossils closely resemble modern specimens and are presumed to have preferred similar environments (Kennedy, 1993).

Other Spores

A variety of other spore form species are also present in the coal samples, often outnumbering *Gleichenia*. These include *Clavifera*, *Cyatheadities*, *Laevigatosporites*, *Peromonolites*, *Stereisporites*, and various *Triletes* form species. These spores are produced by tree and ground ferns, algae and fungi.

Angiosperms

Angiosperm pollen are uncommon in Cretaceous coals, although they form the majority of pollen species in many Tertiary coals. The highest abundance recorded in this study was 12.1%; in comparison, Pocknall (1985) reported angiosperm proportions in excess of 90% in the late Eocene–late Oligocene Waikato Coal Measures of the North Island, New Zealand.

A summary for all samples analyzed showing the proportion of spores, gymnosperms, and angiosperms is given in Figure 25.

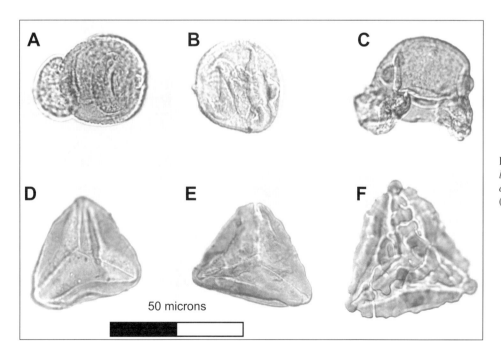

Figure 23. Typical views of (A–C) *Phyllocladidites mawsonii*, (D) *Gleichenia circinidites*, (E) *Clavifera triplex*, and (F) *Clavifera rudis*.

Most samples contain less than 10% angiosperms, the primary variation being between gymnosperms and spores. The samples shown as circles in Figure 25 are in the eastern part of the study area (from drill hole 640) and is thought to be a marginal area of the mire. The clustering of data from drill hole 640 indicates a less gymnosperm-dominated character than in western sites (the remainder of the data), with higher proportion of spore palynomorphs. In addition, drill hole 640 also contains a slightly higher angiosperm population than in the western parts of the study area.

Ply-by-ply analysis of Main Seam intersections has demonstrated that at each site the proportion of *Gleichenia* increases several meters below the roof of the seam. Detailed vertical variation of palynological components, as well as ash yield, is given in Figure 26 for drill hole 711 and demonstrates this trend. Cross sections showing major palynomorph concentrations also confirm that *Gleichenia* consistently increases at the top of Main Seam over most of the study area (Fig. 27). Analyses of other Rewanui Member seams in drill hole 640, and the underlying Sub-main Seam in drill holes 704 and 726 has shown that lower Rewanui Member seams generally contain very little *Gleichenia* (Moore, 1996b).

DISCUSSION

Controls on Sediment Type Distribution

From the available data, we constructed a model for the lithologic distribution of the Main Seam interval (Fig. 28). Three salient features within this model need to be addressed: (i) paleochannel type, (ii) paleochannel orientation, and (iii) coal type distribution. First, the thin coal area (often referred to as a "want" area by coal miners) separating the two main deposits of the Main Seam is based on only a few drill holes. However, the sedimentary features described within the cores (e.g., cross-bedding, graded bedding, root penetration) as well as the lateral relationships with other sediments suggest that the "want" area is a paleochannel–flood plain complex. It is important to note that this zone did not contain a single large river channel but consisted of narrow fluvial tracts (which vertically aggraded through time) with small channels separated from peat by restricted flood plain. Drill cores in the "want" area show sandstones that are only 2–2.5 m thick and these were difficult to correlate with each other. Thus it is inferred that the channels themselves would have been relatively shallow (<2 m) and small in width (<100 m) but the flood plain could have been as wide as a few hundred meters.

The second feature to note in the model shown in Figure 28 is the orientation of the paleochannel separating the two pods of the Main Seam. The approximately east-west direction is one which has been documented in other seams of different stratigraphic intervals in the Greymouth area (P. Caffyn, 1996, personal commun.) including the "D" seam (Ward et al., 1995). It is evident from the cross sections that the relatively small channels aggraded vertically while affecting the coal only in a restricted area. To the north of the channel tracts the seam is

Figure 24. Modern *Gleichenia* in a New Zealand peat mire (see also Moore and Shearer, 2003).

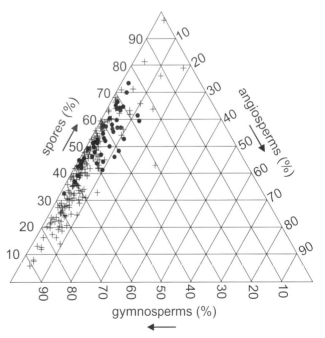

Figure 25. Palynomorph type proportions from all samples analyzed. Points depicted as circles are from drill hole 640 and are believed to represent marginal mire environments.

split into several plies, whereas to the south the coal contains fewer clastic splits (Moore, 1995). These features suggest that the location and confinement of the paleochannels could have been controlled by structural components, namely an active fault, oriented approximately east-west with its down-thrown side to the north. An active fault system would allow greater localized subsidence and thus be the locus of stream activity. A basement fault has also been hypothesized (Laird 1994; Ward 1996) in this same area, on the basis of sedimentation patterns of other members of the Paparoa Coal Measures. Furthermore, from seismic data Bishop (1992) delineated a WNW fault system in a graben feature offshore of the west coast of the South Island in-filled with Pororari Group (Motuan, middle Cretaceous) sediments. Although not conclusive, these data represent consistent circumstantial evidence that the orientation of the paleochannels found in the Paparoa Coal Measures (Haumurian–Maastrichtian) in the Greymouth area may have been controlled by active basement fault systems. Finally, the pattern of coal bed splitting, as well as the hypothesized fault placement, is highly reminiscent of known basement fault controls on coal bed geometry inferred by Weisenfluh and Ferm (1991) for some Carboniferous coal measures in the Appalachian Basin. Coincidence of faults and seam splits has also been observed in the Eocene Brunner coals of New Zealand (Titheridge, 1993; Flores and Sykes, 1996).

The final feature to note in the model is that of coal type distribution. The most-banded coal types occur at the edge as well as the top and bottom of the deposit. Previous studies have suggested that higher concentrations of banding at the margins of coal seams may be related to nutrient supply during peat accumulation (Esterle, 1984; Warwick and Stanton, 1988; Esterle and Ferm, 1994). Periodic flooding during mire initiation and subsequently along the margins has been shown in modern studies to allow larger, more robust vegetation to grow and that peat tends to have more wood (roots) pieces in it (Anderson, 1964; Bruenig, 1970; Esterle and Ferm, 1994). These woody pieces are equivalent to vitrain bands. It is

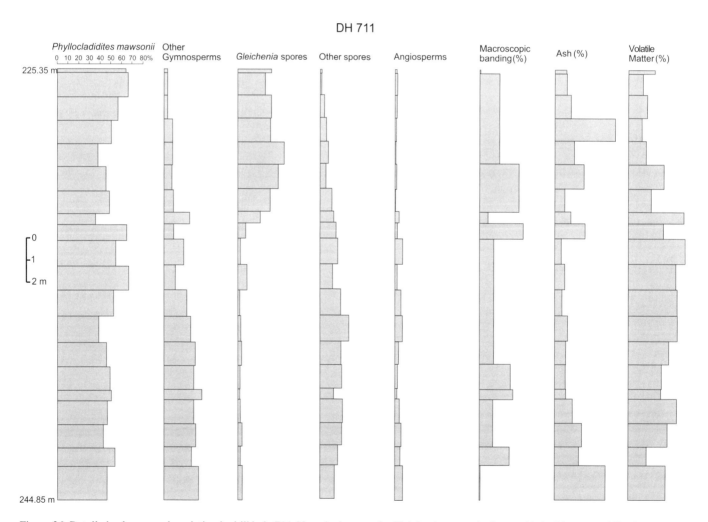

Figure 26. Detailed palynomorph variation in drill hole 711. Note the increase in *Gleichenia* spores in the top third of the seam. All palynomorph values represent percentage of total counts and follow the same scale.

interesting to note that plant type appears not to have changed across the thin zones of banding in the Main Seam and changes significantly only in the upper third of the bed (Moore, 1996a). This may indicate that the periodic flooding helped to maintain the nutrient levels and thus the flora over a significant amount of time. It is conceivable, too, that these thin partings and highly banded zones may roughly correspond to the two thick inorganic intervals separating the three major coal splits observed in the Main Seam north "pod" (Fig. 14). If this is true, then this might suggest some larger, allogenic control on the inundations of the Main Seam.

The variability in vitrain band proportion while the size distribution of those bands remains constant is curious. It is known that there is an optimal size distribution for roots in mires (~1 - 3 mm; Puustjärvi, 1982), and thus in the Main Seam (and possibly other seams as well) variability in vitrain band proportion is likely a reflection of degradational processes without a size-class bias.

Controls on Geochemistry

Ash yield variability of the Main Seam is similar to the trends of Si, Al, K, and Ti. This is probably related to the geochemical controls on the original detrital clastic sources of the Greymouth coal basin. Rare earth element signature of both the Main Seam coals and its associated inorganic partings (Li, 2002) and sedimentary rocks (e.g., roof and floor rocks) also strongly supports this conclusion. Moreover, the two high ash content zones in the northwest and southeast of the study area (Fig. 16) are coincidently correlated to the two inorganic sources proposed in previous studies (Bowman, 1984; Newman and Newman, 1992). These two groups are the Greenland Group greywacke and argillite rocks derived from the northwest of the basin as well as the granite and granitoids derived sediments from the eastern margin of the basin. These two sediment sources have apparently supplied the bulk of the inorganic matter into the Main Seam basin and in return controlled the geochemical composition of the Main Seam

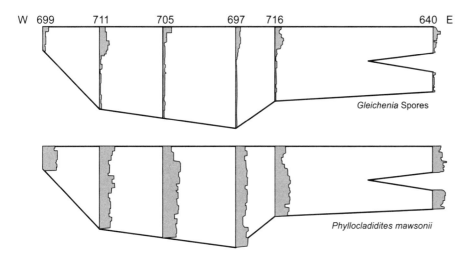

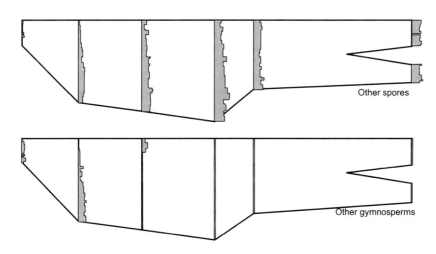

Figure 27. Comparison of palynomorph abundances in six drill holes in a section running from west to east (see Figure 3 for position of drill holes).

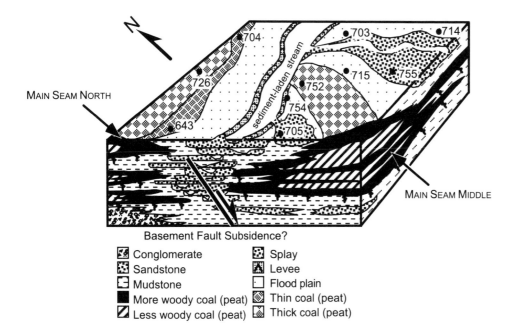

Figure 28. Schematic model for inorganic and organic lithologic variability within the Main Seam. Note that the drill hole locations are meant only to give a perspective of how the model was derived. The inferred basement fault is explained in the text

coal (Li et al., 2001; Li, 2002). Therefore, the amount of the ash content in the Main Seam is partially dependent on the availability of and proximity to those two original sediment sources. Li et al. (2001) also suggested that in-seam leaching may also have controlled the ash content and geochemical compositions of the Main Seam coal. The general low sulfur (mainly organic) content supports that the Main Seam was deposited and developed in a freshwater environment.

Ecological Evolution of the Main Seam

Although pollen in peat may have been transported by water prior to deposition, the very low mineral matter content of the Main Seam suggests that drainage was limited, and that much of the pollen observed was probably deposited near the parent plants. Palynomorph abundances cannot be related directly to plant populations, as different plant species and groups produce different amounts of pollen or spores. However, differences in pollen populations and proportions between samples can provide information on paleofloral changes both laterally and over time, indicating changes in growing conditions.

The potential environmental range of plants that would correspond to *P. mawsonii* is unknown. Modern communities of the closest relative, the Huron pine (*Lagarostrobus franklinii*), occur in a variety of environmental settings, from temperate lowland forests to highlands of up to 700 m altitude. Mean annual temperatures of the various communities range from 6–12 °C, with annual rainfall of 1000–3000 mm. The non-riverine communities in Tasmania today are thought to be glacial refugia, still expanding after the last glaciation 10,000–12,000 yr ago.

Given that the Huon pine thrives in channels and areas with a high water table, it might easily have dominated the centers of ombrotrophic mires, or topogenous mires with restricted drainage and sediment supply. If the host mire contained bodies of standing water and internal drainage channels, *L. franklinii* could have dominated wide areas.

In a mire with limited water circulation, it is probable that most *P. mawsonii* pollen would be deposited near the parent plant. Coal samples with a high spore and pollen yield but low diversity, such as many Main Seam samples, may be interpreted as representing a forest setting. A dense forest canopy would tend to inhibit the circulation of air beneath the canopy, restricting the range of pollen air transport.

In the geological record, colonization of mires by *Gleichenia* is believed to represent seasonally drier or better-drained conditions than those represented by *P. mawsonii* (Luly et al. 1980; Ward et al.1995). Such conditions might occur in mire centers without sag ponds, away from fluvial channels, and with an unstable water table. Luly et al. (1980) concluded that, with limited water for transport, spores would tend to be deposited within 100 m of the parent plant.

In contrast to *L. franklinii*, *Gleichenia* is thought to be very tolerant of fire (Ward et al., 1995). After a mire fire, *Gleichenia* might be locally dominant until competing species had recovered. The replacement of a protective arboreal canopy of flora such as *L. franklinii* with *Gleichenia* might result in increased evapotranspiration of groundwater from the peat. A consistently drier peat might then have been unsuitable for *L. franklinii* regeneration, and possibly susceptible to recurrent fires. However, it should be noted that *Gleichenia* occurs in modern mires in New Zealand and appears to be able to tolerate wet or dry conditions (Moore and Shearer, 2003).

Previous studies (e.g., Diessel, 1985) have indicated that peat accumulates more slowly in environments where the peat surface may dry out. There is evidence that the top of the Main Seam was more susceptible to drying and/or fire. Moore (1996a, 1996b) reports of increased inertinite (both fusinite and semifusinite) in drill hole 711. The proportion of *Gleichenia* was found to increase in the upper plies of the Main Seam at ten sites, separated by up to 2.5 km. Such a widespread change is interpreted as indicating a significant, basinwide change in the mire flora. If, as it suggests, the increase in the proportion of *Gleichenia* indicates a drying of the mire, a slowing of the peat accumulation rate would result in higher concentrations of mineral matter. A concomitant increase in ash yield is seen in the same drill core (Fig. 26).

Mire drying can result from changes in climate, mire drainage, or loss of a protective canopy resulting in increased evapotranspiration. Possible climate change scenarios include a reduction in mean annual rainfall, an increase in mean annual temperature resulting in increased evapotranspiration, or greater seasonal variability in rainfall or temperature, leading to periodic drying of the peat. Foliar physiognomic analysis of the macrofossil leaf record of the surrounding basin sediments might characterize changes in temperature or rainfall.

At most sites, the Main Seam coals are characterized by low mineral matter content, indicating that the mire was raised and isolated from the basin's main fluvial channels, and that the water table was maintained by rainfall onto the mire (i.e., ombrotrophic). Increased mire drainage/evapotranspiration and the loss of bodies of standing water could favor an increase in *Gleichenia*.

If drier mire conditions were the result of sudden loss of a protective canopy of *L. franklinii*, for example by an incidental fire, the loss would be unlikely to affect such a wide area or produce such a long-lasting change. At some sites the *Gleichenia*-rich plies are several meters thick, indicating that the vegetative change persisted for probably thousands of years. Given its prolific pollen production and ability to reproduce vegetatively, *L. franklinii* would almost certainly have been able to revegetate its former habitats if climatic conditions were still suitable, spreading from wetter mire drainage channels toward the drier raised centers. The permanent change in vegetation is therefore interpreted as indicating a change in regional environment, rather than a change in the mire environment due to the loss of vegetation. The increase in *Gleichenia* might therefore reasonably be considered a chronostratigraphic horizon.

The increase in *Gleichenia* occurs 5–7 m below the top of the Main Seam at most of the sites examined. Variations in the overall thickness of the Main Seam both above and below the *Gleichenia* increase indicate that peat accumulation finished at different times in different parts of the basin, or that accumulation rates varied from site to site.

Ward et al. (1995) examined palynological and petrographic trends in the approximately coeval Upper Rewanui "D" seam, using the same palynological indicator species as the Main Seam study. The analyses demonstrated an increase in *Gleichenia* in the lower plies of the seam at the center of the mire, before returning to a *P. mawsonii*-dominated assemblage. This trend suggested that peat at the mire center was initially drier or better drained than at the margin. The increased *P. mawsonii* abundance suggested wetter peats, due to reduced drainage, and possibly the development of bodies of standing water.

A Main Seam-like *Gleichenia* increase in the upper plies of the "D" seam near the mire margin corresponded to abundant oxidized tissues and fragments, indicating that the mire margin had become drier or at least better drained. The trend was reversed in the uppermost ply, indicating a return to wetter, less aerobic conditions before peat accumulation ceased.

Thick Coal Formation

One of the most unusual properties of the Main Seam is its extraordinary thickness. As pointed out by Shearer et al. (1994), modern peats are reported to be no thicker than 20 m (a constraint placed upon the system by the balance between plant growth and degradation; Clymo, 1983), although coal beds, like the Main Seam, are significantly thicker even without compaction being taken into account. However, Shearer et al. (1994) believed that most coal beds are "stacked" sequences of paleomires, rather than a single episode of uninterrupted peat accumulation. In most cases even a precursory look at thick coal seams reveals that they can contain at least a few thin, inorganic partings (e.g., see Chao et al., 1984; Moore and Shearer, 1993). Even where inorganic partings cannot be found, zones rich in oxidized components or liptinites and often appearing duller in luster than the surrounding coal can be distinguished macroscopically; in either type of parting (i.e., inorganic or organic) these intervals probably represent hiatuses in mire aggradation and demarcations between paleomire systems (Shearer et al., 1994). These stacking sequences may even be an outcome of larger basin wide sequences (Holdgate et al., 1995; Banerjee et al., 1996; Holdgate and Clarke, 2000).

No doubt the inorganic sediment separating the splits in the north pod of the Main Seam represents hiatuses in peat accumulation; in a more subtle way the thin partings observed within the middle pod and the associated highly banded coal types at the same intervals in other locations probably also demarcate cessation in peat accumulation. In a study of a Powder River Basin coal bed, Moore (1991; 1994) showed that a half-meter interval of low-ash coal was laterally equivalent to 90 m clastic split sediment. Therefore, in the case of the Main Seam the cessation between splits or plies based on the high-vitrain layers may have been only brief. Palynological data indicate that floral type did not change significantly until the upper third of the seam, that is, above the upper thin parting/highly banded zone. Therefore, one of the factors contributing to the great thickness of the Main Seam was probably the delicate balance between peat accumulation and minor periodic incursions of nutrient-rich waters allowing rejuvenation and "stacking" of mire systems.

A second factor in the unusual thickness of the Main Seam may have been the presence of an active fault system in the area. Differential subsidence would have, for the most part, preferentially attracted drainage and thus most of the inorganic sediments. Entrenchment of the paleochannel would have compounded any interruptions in peat accumulation, allowing thicker than normal coal to develop. Indeed, the Main Seam north pod probably represents a "more normal" accumulation in which the mire was interrupted by periodic extensive flooding which deposited significant inorganic sediment. On the up-thrown side of the fault, however, the slightly positive relief would have further inhibited influx of sediment. Another possibly allied mechanism for preventing sediment deposition in the paleomire is that most thick, low-ash peats form with a topographically domed surface (Anderson, 1964; Esterle and Ferm, 1994). It is thus likely that only extreme flooding events would have affected the Main Seam middle pod.

CONCLUSIONS

Examination of sediment types and their distribution, as well as the geochemistry and palynology of the Cretaceous Main Seam interval, has allowed us to construct a model for their formation. This model helps to explain, among other things, how this seam could have formed to such an extraordinary thickness relative to other Cretaceous coal seams worldwide. Three predominant factors apparently enabled such thick peat formation:

1. Vertical stacking of individual paleomire systems as a result of periodic flooding and rejuvenation of the peat floral community;
2. A possible syn-depositional fault system in the area attracting and restricting inorganic sediments to a narrow "want" zone (~100 m); and
3. A seamwide and possibly basinwide floral change.

Regarding the first factor, seam geometry, geochemistry, and coal types all indicate that there are at least three separate stacked paleomire systems forming the thick Main Seam middle pod. Although not fully understood, this process of mire stacking is probably indicative of a fairly stable tectonic environment but one periodically punctuated by small floods—events sufficiently large enough to cause cessation of peat accumulation but bringing in enough essential plant nutrients for rejuvenation of mire communities, thus staving off normal limitations on mire growth as a result of plant community type and degradation.

With regard to the second factor, the sedimentary units indicate that the paleochannels were quite small (less than 100 m) and shallow (less than a few meters in thickness) but vertically stacked. Sediment flow was inhibited into the adjacent paleomire both by topographic effects originating from a lower base level as a result of differential subsidence and by a higher surface level in the peat due to ombrogeny.

The cause of final termination of the Main Seam paleomire system is not known but may be related to the third factor. A change in flora to *Gleichenia* could be symptomatic of a changing environment. This change in environmental conditions may have put the peat-accumulating machine out of balance. *Gleichenia* does not produce much biomass, and if it was the dominant mire plant it may not have been able to keep peat accumulation rates higher than subsidence. Either a decrease in peat accumulation or a drying of the mire would have resulted in lowering of the surface to a degree where flooding and final termination would be likely.

ACKNOWLEDGMENTS

This research was partially funded through a grant by the Foundation for Research, Science and Technology, New Zealand (CRA301). Solid Energy NZ Ltd. and the former Greymouth Coal Ltd. are also thanked for access to drill core as well as geophysical downhole logs. The project has benefited greatly from discussions with P. Caffyn, J.C. Shearer, J. Newman, J. McNee, A. Nicol, F. Taylor, R. Boyd, S.D. Weaver, and S.D. Ward. Thanks also to Dave Kelly for a simple, but highly astute, observation on plants, stress, and time. M.G. Laird, D.G. Titheridge, and G.A. Weisenfluh all made extensive comments on earlier versions of the manuscript. Finally the reviews of Guy Holdgate and Maria Mastalerz have further improved the paper.

REFERENCES CITED

Anderson, J.A.R., 1964, The structure and development of the peat swamps of Sarawak and Brunei: Journal of Tropical Geography, v. 18, p. 7–16.

Banerjee, I., Kalkreuth, W., and Davies, E.H., 1996, Coal seam splits and transgressive-regressive coal couplets: A key to stratigraphy of high-frequency sequences: Geology, v. 24, no. 11, p. 1001–1004, doi: 10.1130/0091-7613(1996)024<1001:CSSATR>2.3.CO;2.

Barry, J.M., Duff, S.W., and MacFarlan, D.A.B., 1994, Coal resources of New Zealand: Ministry of Commerce, Resource Information Report 16, 73 p.

Bishop, D., 1992, Extensional tectonism and magmatism during the middle Cretaceous to Paleocene, North Westland, New Zealand: New Zealand Journal of Geology and Geophysics, v. 35, p. 81–91.

Bowman, R.G., 1984, Greymouth Coalfield Report, Part 1: Wellington, New Zealand, Mines Division, Ministry of Energy, New Zealand Coal Resources Survey (Coalfield Geology and Coal Resources), 212 p.

Bruenig, E.F., 1970, Stand structure, physiognomy and environmental factors in some lowland forests in Sarawak: Tropical Ecology, v. 11, no. 1, p. 26–43.

Cameron, A.R., 1978, Megascopic description of coal with particular reference to seams in southern Illinois, in Dutcher, R.R., ed., Field description of coal, ASTM STP 661, American Society for Testing Materials, p. 9–32.

Chao, E.C.T., Minkin, J.A., Back, J.M., and Pierce, F.W., 1984, Petrographic characteristics and depositional environment of the Paleocene 6-m-thick subbituminous Big George coal bed, Powder River Basin, Wyoming, in 1984 Symposium on the Geology of Rocky Mountain Geology, Bismarck, North Dakota, p. 41–60.

Clymo, R.S., 1983, Peat, in Gore, A.J.P., ed., Mires: Swamp, bog, fen and moor: New York, Elsevier, Ecosystems of the World, v. 4A, p. 159–224.

Davis, A., 1978, Compromise in coal seam description, in Dutcher, R.R., ed., Field description of coal, ASTM STP 661, American Society for Testing Materials, p. 33–40.

Diessel, C.F.K., 1985, Australian Mineral Foundation Workshop Course 282/84.

Edbrooke, S.W., Sykes, R., and Pocknall, D.T., 1994, The geology of the Waikato Coal Measures, Waikato Coal Region, New Zealand: Institute of Geological and Nuclear Sciences Monograph 6, 236 p.

Esterle, J.S., 1984, The Upper Hance coal seam in southeastern Kentucky: A model for petrographic and chemical variation in coal seams [M.S. thesis]: Lexington, University of Kentucky, 105 p.

Esterle, J.S., and Ferm, J.C., 1994, Spatial variability in modern tropical peat deposits from Sarawak, Malaysia and Sumatra, Indonesia: Analogues for coal: International Journal of Coal Geology, v. 26, p. 1–41, doi: 10.1016/0166-5162(94)90030-2.

Esterle, J.S., O'Brien, G., and Kojovic, T., 1994, Influence of coal texture and rank on breakage energy and resulting size distributions in Australian coals, in Proceedings, Sixth Australian Coal Science Conference, Newcastle, p. 175–181.

Esterle, J.S., O'Brien, G.O., and Moore, T.A., 1995, A comparison of breakage behaviour for New Zealand and Australian coals, in Proceedings of the 6th New Zealand Coal Conference, p. 373–383.

Ettmuller, F., 2003, The integrated CL method and the Paparoa Trough: A new approach to provenance analysis [B.S. honours thesis]: Christchurch, NZ, University of Canterbury, text 61 pp.; figures and tables 70 p.

Ferm, J.C., and Moore, T.A., 1997, Guide to cored rocks in the Greymouth Coalfield: Coal Research Ltd., CRL Report no. 97-11189, 41 p.

Ferm, J.C., Berger, J.T., and Hedge, S.S., 1979 A revised system for plotting core hole logs, in Ferm, J.C., and Horne, J.C., eds., Carboniferous depositional environments in the Appalachian Region: Columbia, S.C., University of South Carolina, p. 581–583.

Ferm, J.C., Smith, G.C., Weisenfluh, G.A., and DuBois, S.B., 1985, Cored rocks in the Rocky Mountain and High Plains coal fields: Lexington, University of Kentucky, Department of Geology, 90 p.

Flores, R.M., and Sykes, R., 1996, Depositional controls on coal distribution and quality in the Eocene Brunner Coal Measures, Buller Coalfield, South Island, New Zealand: International Journal of Coal Geology, v. 29, p. 291–336, doi: 10.1016/0166-5162(95)00028-3.

Flores, R.M., Toth, J.C., and Moore, T.A., 1982, Use of geophysical logs in recognizing depositional environments in the Tongue River Member of the Fort Union Formation, Powder River Area, Wyoming and Montana: U.S. Geological Survey Open-File Report 82-576.

Gage, M., 1952, The Greymouth Coalfield: New Zealand Geological Survey Bulletin, v. 45, 232 p.

Gibson, N., 1988, A description of the Huon Pine (*Lagarostrobus franklinii* (Hook. F.) C.J. Quinn) forests of the Prince of Wales and King Billy Ranges: Papers and Proceedings of the Royal Society of Tasmania, v. 122, p. 127–133.

Gibson, N., and Brown, M.J., 1991, The ecology of *Lagarostrbus franklinii* (Hook. f.) Quinn (Podocarpaceae) in Tasmania. 2. Population structure and spatial pattern: Australian Journal of Ecology, v. 16, p. 223–229.

Holdgate, G.R., and Clarke, D.A., 2000, A review of Tertiary Brown coal deposits in Australia—Their depositional factors and eustatic correlations: American Association of Petroleum Geologists Bulletin, v. 84, no. 8, p. 1129–1151.

Holdgate, G.R., Kershaw, A.P., and Sluiter, I.R.K., 1995, Sequence stratigraphic analysis and the origins of Tertiary brown coal lithotypes, Latrobe Valley, Gippsland Basin, Australia: International Journal of Coal Geology, v. 28, no. 2–4, p. 249–276, doi: 10.1016/0166-5162(95)00020-8.

Kennedy, E.M., 1993, Palaeoenvironment of an Haumurian plant fossil locality within the Pakawau Group, North West Nelson, New Zealand [M.Sc. thesis]: Christchurch, NZ, University of Canterbury.

Laird, M.G., 1993, Cretaceous continental rifts: New Zealand region, in Ballance, P.F., ed., South Pacific sedimentary basins: Sedimentary basins of the world: Amsterdam, Elsevier, p. 37–49.

Laird, M.G., 1994, Geological aspects of the opening of the Tasman Sea, in van der Lingen, G., Swanson, K.M., and Muir, R., eds., Evolution of the Tasman Sea Basin: Rotterdam, A.A. Balkema, p. 11–17.

Leake, B.E., Hendry, G.L., Kemp, A., Plant, A.G., Harbey, P.K., Wilsom, J.R., Coats, J.S., Aucot, J.W., Lunel, T., and Howarth, R.J., 1969, The chemical analysis of rock powders by automated X-ray florescence: Chemical Geology, v. 5, p. 7–86, doi: 10.1016/0009-2541(69)90002-3.

Li, Z., 2002, Mineralogy and trace elements of the Cretaceous Greymouth coals and their combustion products [Ph.D. thesis]: Christchurch, NZ, University of Canterbury, 340 p.

Li, Z., Moore, T.A., and Weaver, S.D., 1999, The mineralogy and geochemistry of "Main Bed" coal (Cretaceous), Greymouth, New Zealand, in Proceedings, Eighth New Zealand Coal Conference, Wellington, New Zealand, p. 183–199.

Li, Z., Moore, T.A., and Weaver, S.D., 2001, Leaching of inorganics in the Cretaceous Greymouth coal beds, South Island, New Zealand: International Journal of Coal Geology, v. 47, p. 235–253, doi: 10.1016/S0166-5162(01)00044-1.

Lorente, M.A., 1990, Textural characteristics of organic matter in several subenvironments of the Orinoco Upper Delta: Geologie en Mijnbouw, v. 69, p. 263–278.

Luly, J., Sluiter, I.R., and Kershaw, A.P., 1980, Pollen studies of Tertiary brown coals; preliminary analyses of lithotypes within the Latrobe Valley, Victoria: Monash Publications in Geography 23, Monash University, 77 pp.

Martin, H.A., 1984, The use of quantitative relationships and paleoecology in stratigraphic palynology of the Murray Basin of New South Wales: Alcheringa, v. 8, p. 253–272.

Mildenhall, D.C., 1980, New Zealand Late Cretaceous and Cenozoic plant biogeography: A contribution: Palaeogeography, Palaeoclimatology, Palaeoecology, v. 31, p. 197–233, doi: 10.1016/0031-0182(80)90019-X.

Moore, N.A., 1996a, Seam identification, correlation and coal quality prediction using in-seam variations in key palynomorph abundances, in Proceedings, 29th Annual Conference of the Australasian Institute of Mining and Metallurgy, Greymouth, New Zealand, p. 228–246.

Moore, N.A., 1996b, Palynology and coal petrography of Rewanui Member seams in the Rapahoe sector, Greymouth coalfield [M.Sc. thesis]: Christchurch, NZ, University of Canterbury, 130 p.

Moore, T.A., 1991, The effects of clastic sedimentation on organic facies development within a Tertiary subbituminous coal bed, Powder River Basin, Montana, U.S.A.: International Journal of Coal Geology, v. 18, p. 187–209, doi: 10.1016/0166-5162(91)90050-S.

Moore, T.A., 1994, Organic compositional clues to a stacked mire sequence in the Anderson-Dietz #1 coal bed (Paleocene), Montana, U.S.A., in Flores, R.M., Mehring, K.M., Jones, R.W., and Beck, T.L., eds., Field Trip Guide to Tertiary Basins in Northcentral Wyoming: Laramie, Geological Survey of Wyoming Public Information Circular 33, p. 165–174.

Moore, T.A., 1995, Developing models for spatial prediction of mining and utilisation potentials in coal seams: An example from the Greymouth Coalfield, in Proceedings, Sixth New Zealand Coal Conference, p. 385–402.

Moore, T.A., 1996c, Rock and coal type distribution in the Greymouth area: Applications for mining, in Proceedings, 29th Annual Conference of the Australasian Institute of Mining and Metallurgy, Greymouth, New Zealand, p. 200–227.

Moore, T.A., and Hilbert, R.E., 1992, Petrographic and anatomical characteristics of plant material from two peat deposits of Holocene and Miocene age, Kalimantan, Indonesia: Review of Palaeobotany and Palynology, v. 72, p. 199–227, doi: 10.1016/0034-6667(92)90027-E.

Moore, T.A., and Shearer, J.C., 1993, Processes and possible analogues in the formation of Wyoming's coal deposits, in Snoke, A.W., Steidtmann, J.R., and Roberts, S.M., eds., Geology of Wyoming: Geological Survey of Wyoming Memoir 5, p. 874–896.

Moore, T.A., and Shearer, J.C., 2003, Coal/peat type and depositional environment—Are they related?: International Journal of Coal Geology, v. 56, p. 233–252, doi: 10.1016/S0166-5162(03)00114-9.

Moore, T.A., Ferm, J.C., and Weisenfluh, G.A., 1992, Guide to Eocene coal types in Kalimantan Selatan, Indonesia: Lexington, University of Kentucky, Department of Geological Sciences, 74 p.

Moore, T.A., Shearer, J.C., and Esterle, J.S., 1993, Quantitative macroscopic textural analysis: Society of Organic Petrology Newsletter, v. 9, no. 4, p. 13–16.

Newman, J., 1987, Coal type and paleoenvironments in the Rapahoe Sector Greymouth Coalfield: Wellington, New Zealand, Ministry of Commerce, Resource Management and Mining, Coal Geology Technical Report 3, 28 p.

Newman, J., 1989, Why are some high rank Tertiary coals more peculiar than others? Some thoughts on climate and floral assemble, in New Zealand Coal Research Association Conference, 3rd, Wellington, New Zealand, p. 182–196.

Newman, J., and Newman, N.A., 1992, Tectonic and paleoenvironmental controls on the distribution and properties of Upper Cretaceous coals on the West Coast of the South Island, New Zealand, in McCabe, P.J., and Parrish, J.T., eds., Controls on the distribution and quality of Cretaceous Coals: Geological Society of America Special Paper 267, p. 347–368.

Norrish, K., and Hutton, J.T., 1969, An accurate X-ray spectrographic method for the analysis of a wide range of geological samples: Geochimica et Cosmochimica. Acta, v. 33, p. 431–453, doi: 10.1016/0016-7037(69)90126-4.

Playford, G., and Dettman, M.E., 1978, Pollen of *Dacrydium franklinii* Hook. F. and comparable Early Tertiary microfossils: Pollen and Spores, v. 20, p. 513–534.

Pocknall, D.T., 1985, Palynology of Waikato Coal Measures (Late Eocene–Late Oligocene) from the Raglan area, North Island, New Zealand: New Zealand Journal of Geology and Geophysics, v. 28, p. 329–349.

Puustjärvi, V., 1982, The structure of peat substrate: Helsinki, Finland, Association of Finnish Peat Industries, Peat and Plant Yearbook 1981–1982, p. 21–27.

Retallack, G.J., Veevers, J.J., and Morante, R., 1996, Global coal gap between Permian-Triassic extinction and Middle Triassic recovery of peat-forming plants: Geological Society of America Bulletin, v. 108, no. 2, p. 195–207, doi: 10.1130/0016-7606(1996)108<0195:GCGBPT>2.3.CO;2.

Shearer, J.C., 1995, Tectonic controls on styles of sediment accumulation in Late Cretaceous coal measures of the Ohai coalfield, New Zealand: Cretaceous Research, v. 16, p. 367–384, doi: 10.1006/cres.1995.1027.

Shearer, J.C., and Moore, T.A., 1994, Grain-size and botanical analysis of two coal beds from the South Island of New Zealand: Review of Palaeobotany and Palynology, v. 80, no. 1-2, p. 85–114, doi: 10.1016/0034-6667(94)90095-7.

Shearer, J.C., Staub, J.R., and Moore, T.A., 1994, The conundrum of coal bed thickness: A theory for stacked mire sequences: Journal of Geology, v. 102, p. 611–617.

Shearer, J.C., Moore, T.A., and Demchuk, T.D., 1995, Delineation of the distinctive nature of Tertiary coal beds, in Demchuk, T.D., Shearer, J.C., and Moore, T.A., eds., Controls on the character of Tertiary coal beds: International Journal of Coal Geology, v. 28, p. 71–98.

Sherwood, A.M., Lindqvist, J.K., Newman, J., and Sykes, R., 1992, Cretaceous coals in New Zealand, in McCabe, P. J., and Parrish, J. T., eds., Controls on the distribution and quality of Cretaceous Coals: Geological Society of America Special Paper 267, p. 325–346.

Titheridge, D.G., 1993, The influence of half-graben syn-depositional tilting on thickness variation and seam splitting in the Brunner Coal Measures, New Zealand: Sedimentary Geology, v. 87, p. 195–213, doi: 10.1016/0037-0738(93)90004-O.

Ward, C.R., Spears, D.A., Booth, C.A., Staton, I., and Gurba, L.W., 1999, Mineral matter and trace elements in coals of the Gunnedah Basin, New South Wales, Australia: International Journal of Coal Geology, v. 40, p. 281–308, doi: 10.1016/S0166-5162(99)00006-3.

Ward, S.D., 1996, Application of lithostratigraphic and chronostratigraphic analysis to seam modelling in the Rapahoe Sector, western Greymouth Coalfield, in Proceedings, 29th Annual Conference of the Australasian Institute of Mining and Metallurgy, Greymouth, New Zealand, p. 173–199.

Ward, S.D., Moore, T.A., and Newman, J., 1995, Floral assemblage of the "D"-coal seam (Cretaceous): Implications for banding characteristics in New Zealand coal seams: New Zealand Journal of Geology and Geophysics, v. 38, no. 3, p. 283–297.

Warnes, M.D., 1988, The palynology of the Ohai Coalfield, Southland [M.Sc. thesis]: Christchurch, NZ, University of Canterbury.

Warwick, P.D., and Stanton, R.W., 1988, Depositional models for two Tertiary coal-bearing sequences in the Powder River Basin, Wyoming, USA: Journal of the Geological Society [London], v. 145, p. 613–620.

Weisenfluh, G.A., and Ferm, J.C., 1991, Application of depositional models to mining problems, in Peters, D.C., ed., Geology in coal resource utilization: Fairfax, Virginia, TechBooks, p. 189–202.

MANUSCRIPT ACCEPTED BY THE SOCIETY 28 JUNE 2005

Paleoecology of a late Pleistocene wetland and associated mastodon remains in the Hudson Valley, southeastern New York State

Norton G. Miller
Biological Survey, New York State Museum, Albany, New York 12230-0001, USA

Peter L. Nester
Paleontological Research Institution/Museum of the Earth, 1259 Trumansburg Road, Ithaca, New York 14850, USA

ABSTRACT

Late Quaternary history and paleoecology of a small oxbow wetland on glaciated terrain were investigated using sediment lithology (cores, bulk samples, backhoe-dug trenches), ground-penetrating radar, vascular plant and moss macrofossil stratigraphies, and accelerator mass spectrometric radiocarbon dating. A nearly complete mastodon skeleton was recovered from late Pleistocene detrital peat and peaty marl near the top of the sediment sequence. Sedimentation in the basin began with silt and clay over dense cobble outwash transported southward from the nearby Hyde Park Moraine. Overbank sediment deposition occurred between ~13,000 and 12,220 yr B.P. during a period of tundra vegetation, which ended with a sharp rise in spruce needle abundance and a shift to autochthonous marl and finally peat deposition. Fossils of aquatic and wetland plants began to accumulate before the tundra-spruce transition and increased after it. Rich fen wetland began to infill the pond with peat, while the upland supported open white spruce and later white spruce–balsam fir–tamarack forest. The mastodon, 11,480 ± 40 radiocarbon years old, was contemporaneous with spruce–balsam fir–tamarack forest and rich fen wetland. Many mastodon bones were articulated or nearly so, indicating that the animal died in the basin and that post-mortem bone dispersal was slight.

Keywords: plant macrofossils, calcareous wetland, late Pleistocene, mastodon, New York, paleoecology.

INTRODUCTION

Freshwater lakes of various size, depth, and origin were produced in glacial drift as the continental ice sheet disappeared from northern North America at the close of the Pleistocene. Sediments of such natural lakes and the wetlands now associated with them contain evidence of postglacial environments and environmental change both within and beyond the water-holding basin. The record usually starts soon after local retreat of the ice, and sedimentation may continue without interruption to the present, unless the basin is emptied by erosion or fills to capacity. The first sediments were sand, silt, and clay. As climate warmed and lake productivity increased, organic-rich lake mud accumulated, starting at or near the Pleistocene/Holocene boundary or early in the Holocene. Peat accumulation began in southern parts of glaciated North America in the early Holocene. However, the extensive peatlands now characteristic of northern United States and Canada formed and, in some places, expanded over the last four or five thousand years.

Wetland basins contain stratigraphically deposited fossils and fossil assemblages that have been subject to Quaternary paleoenvironmental research. In North America, for example, pollen analysis began with studies of sediments taken from peat-accumulating wetlands (e.g., Auer, 1927; Potzger, 1933; Sears, 1935). Although sediment in lakes is now preferred for research into regional patterns of vegetation and climate change, wetland sediments remain essential for investigations of basin infilling processes and wetland ecosystem development.

Here we present results of investigations of a fen wetland that developed in a small basin, the Hyde Park site, created by a stream draining southward from a recessional moraine during the late Pleistocene deglaciation of the Hudson River Valley. We use plant macrofossils and accelerator mass spectrometric (AMS) radiocarbon dating to characterize wetland and associated upland vegetation types and chronological relationships between them. A nearly complete and largely unaltered skeleton of a mastodon (*Mammut americanum* (Kerr)) was recovered from peaty fen sediment in the basin. We interpret the taphonomy of these remains with reference to the plant macrofossil record, and to the sedimentary history of the basin as revealed by ground-penetrating radar and other evidence.

Northern peat-accumulating wetlands are separated into fen and bog on the basis of water chemistry, plant indicator species, hydrology, and landform type (Glaser, 1992). Bogs are domed, whereas fens are concave landforms. In bogs, the water table is raised above that in adjacent non-bog terrain. Fens have acidic to alkaline waters (pH 4.2–7.2; Ca^{2+} concentration 2–50 mg l^{-1}), mineral input from ground or surface water flow (versus the peat mass in bogs isolated above ground and surface water), and a distinctive set of bryophyte and tracheophyte indicator species, including different conifers, sedges, and mosses, than those typical of bogs. Fens range from acidic (poor) to calcareous (rich). Indicator plants and water chemistry distinguish between poor and rich fens. Stratigraphic studies of peatland sediment have shown replacement upward of calcareous fen plant communities by plants typical of poor fens or bogs (Janssens et al., 1993; Futyma and Miller, 1986).

Mastodon cheek teeth, tusks, bones, and more or less complete skeletons have been found throughout the Great Lakes states and southern Ontario (Dreimanis, 1968; Holman, 2001; McAndrews and Jackson, 1988) in sediment of shallow wetland depressions. Most mastodon fossils in New York State have come from western and central regions and the Hudson Valley and adjacent uplands (Dreimanis, 1968; Drumm, 1963; Thompson, 1994). Although the Hiscock site in western New York has continued to yield large numbers of mastodon fossils and a diverse associated fauna and flora (Laub, 2003), most other mastodon finds have been of single animals.

For most New York and many other mastodons, detailed information about the paleoenvironmental setting of the animal was never gathered. The Hyde Park wetland yielded a nearly complete mastodon skeleton dated at 11,480 ± 50 radiocarbon years (Beta-141061; G. Robinson, 2001, written comm.), and fen sediments and plant fossils of the same age. The proximity of many elements of the Hyde Park skeleton when it was excavated indicates minimal postmortem bone displacement. Therefore, discovery of the Hyde Park mastodon provided an uncommon opportunity to reconstruct the paleoenvironment in which the proboscidean lived and the postmortem history of the skeleton.

SITE DESCRIPTION

The Hyde Park site wetland developed in a small (40 m north to south, 28 m east to west) oxbow initiated during deglaciation by the Fall Kill (Miller, 2006), a stream that drains southward from and beyond the Hyde Park Moraine (Connally and Sirkin, 1986), which is located 1.5 km north of the oxbow basin. The wetland is in Dutchess County, New York, 3 km east of Hyde Park at 40°46'45" N, 73°53'40" W, 68 m asl, and ~145 km north of the Wisconsinan terminal moraine, which extends along the length of Long Island, New York, and across Staten Island and northern New Jersey. Relief near the site is low (~75 m).

The Hyde Park Moraine is one of several east-west trending moraines that were deposited as the continental ice sheet retreated northward up the Hudson River valley. A glacial lake was ponded in the valley between the ice front and a dam located to the south. Glacial varves accumulated in the lake, and large deltas were deposited along the valley rim by rivers draining into the lake from melting glacier ice. The chronology of deglaciation and glacial lake history in the Hudson River valley is poorly constrained.

The age of the Hyde Park Moraine is not known precisely. On the basis of the radiocarbon age of a composite bulk sediment sample associated with the onset of organic deposition at Eagle Hill Camp Bog, 25 km north of the Hyde Park Moraine, Connally and Sirkin (1986) calculated that the Eagle Hill Camp Bog basin began to fill at 16,070 yr B.P. However, none of the AMS ages of tundra plant fossils from the basal inorganic sediment in the Hyde Park wetland are older than 13,000 yr B.P., indicating that moraines in the mid-Hudson Valley are probably a few thousand years younger than previously thought (Miller, 2006).

When mastodon bones were discovered in 1999 at the Hyde Park site, the basin (hereafter referred to as Lozier Pond) had already been enlarged and deepened during excavations begun in 1966. As a result, it was not possible to study the contemporary wetland vegetation, which in any case was secondary, having become established after this earlier episode of pond construction. The property owner indicated that before disturbance a swamp, which exists several hundred meters northward beyond the pond margin, also covered the area now occupied by the pond. Extant upland, but mainly secondary, forests near the basin are dominated by oak (*Quercus* spp.) and hickory (*Carya* spp.) on drier sites, and hemlock (*Tsuga canadensis* (L.) Carr), sugar maple (*Acer saccharum* Marsh.), and other broad-leaved deciduous trees on mesic slopes and in shaded ravines. As a result of agricultural practices and settlement patterns, forest cover is discontinuous.

Pollen stratigraphic studies by Maenza-Gmelch (1997a, 1997b) at lakes 48 and 71 km south-southwest of the Hyde Park site show that forests of the middle and lower Hudson Valley were dominated by oak from middle to late Holocene. These were preceded by an early Holocene period of pine (*Pinus*) and oak abundance, and by late Pleistocene forests initially with abundant spruce (*Picea*) and later with reduced spruce and an increased amount of balsam fir (*Abies balsamea* (L.) Mill.).

METHODS

Large-volume sediment samples for plant macrofossil analysis were taken from a sidewall of trench HPM-2, dug by workers during the mastodon excavation 1 m southeast of the area in which mastodon bones occurred (Fig. 1). Sampling began at an arbitrary 0 m datum placed at the bottom of a zone of sediment disturbed by pond building and continued to a depth of 1 m. Additional samples were obtained by bucket auger or posthole digger from 1 to 1.9 m. Cobbles at 1.9 m stopped further penetration. Two cores were taken with a 4-cm-diameter Eijkelkamp piston sampler (6987 EM Giesbeek, The Netherlands) from a wooden platform positioned where the least amount of sediment had been removed during the August 1999 excavation. Core LP3A was at the northern end of the pond, and 2.27 m of sediment was obtained. Core LP2A passed directly through the ribcage of the mastodon, and 2.25 m of sediment was collected. Although sediment depth at the time of our studies exceeded 2 m, the total thickness of sediment in the basin before disturbance can only be estimated. Lithology of the ~1–2 m of sediment removed during pond constructions is not known. Water depth of Lozier Pond in June 2000 before in situ mastodon remains were discovered was 1 m at the location of core LP2A. All measurements

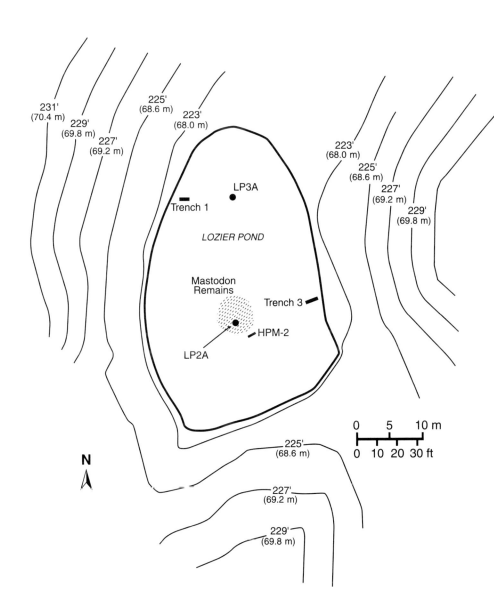

Figure 1. Topography of the Hyde Park Site basin, showing the abandoned meander channel of the paleo–Fall Kill. The mastodon bone bed is indicated by the hatched pattern.

of pre-disturbance sediment thickness assume that the top of the pond sediment before excavation was at the altitude of the floor of the swamp north of the pond. Because this surface is 1 m higher than the minimum water depth, sediment thickness has been increased by 1 m at the location of core LP2A to correspond to the pre-disturbance level.

Sediment lithology was described in the laboratory after microscopic examination of subsamples, testing with dilute hydrochloric acid, and other procedures. Sample volume was measured by displacement in water. Samples were disaggregated by hand in warm deionized water and washed with a low-velocity jet of tap water through a 250-μm-mesh sieve until clean. Identifiable plant macrofossils (leaves and needles, twigs, seeds and seedlike fruits, seed and pollen cones of conifers, megaspores, leafy moss plant fragments and individual moss leaves) were picked from the residues under low magnification and identified. Moss fragments and leaves were cleaned of adhering sediment with 00 sable hair brushes, dissected and/or sectioned by hand, and mounted in Hoyer's solution for examination by light microscopy. A large reference collection of plant parts taken from herbarium specimens was used when needed. Counts were made of fossils in each recognized category, and these data were standardized to fossil number per 300 ml of sediment for all samples and all macrofossil types identified, except mosses, for which frequency and abundance were described by a different method (Table 2). Plant macrofossils have been placed in the Quaternary Paleobotanical Collection of the New York State Museum.

To determine the nature and distribution of sediment elsewhere in the pond, we dug two backhoe trenches, one on the northwest side of the pond (Trench 1) and one on the southeast side (Trench 3), and the trench faces were described and sampled (Fig. 1). We also studied the wall of the mastodon excavation pit, and removed samples every 5 cm for macro- and microfossil analysis under the direction of David Burney (Department of Biological Sciences, Fordham University), using techniques described by Burney and Robinson (2006). Exploratory augering north and south of the pond established the presence or absence of sediments similar to those in the oxbow basin. As further confirmation of stratigraphic continuity, ground-penetrating radar (GPR) transects were run perpendicular to the axis of the sedimentary basin using a Pulse Ekko 100 GPR unit. After test lines were run at 50, 100, and 200 MHz, we determined that the unconsolidated sediments were best resolved with 100 MHz source emissions. Eight parallel transects, spaced at 1-m intervals, were run across the northern end of Lozier Pond. With the emitter set to pulse every meter, the device was placed in a plastic sled and pulled along the ground and across the water. The returning signals were recorded in the field on a laptop computer. Post-acquisition processing of the raw data was conducted using Win EKKO Pro software, and graphical displays were optimized using EKKO Mapper. Resolvable horizons were observed at depths of 4 m or greater (including standing water). The GPR profiles in conjunction with field observations of lithology permitted thickness and area of the sedimentary units to be determined throughout the basin. A full report of this work is presented by Nester et al. (2006).

A wooden post was driven into the sediment near where the first in situ mastodon bone was found during the 2000 excavation. Using a compass and tape measure, we established a 2 × 2 m grid, with stakes marking corners of the squares. All larger bones were tagged with an identification number while still in the pond sediment. Each bone was mapped in two-dimensional space by recording distances along two 1-m-long Jacob staffs positioned along the north-south and east-west sides of the squares. Depth was measured from the same 0 datum established for sediment samples taken within the mastodon excavation zone, namely a point just above the highest exposed bone. To maintain the same zero-point across the bone field, we extended string across the grid from the zero-datum point, and used a spirit level to ensure a uniform horizontal reference line above the squares. The distance between the zero datum and a bone was measured along a plumb-bob line. Because of possible disturbance and breakage, we removed smaller bones before mapping them, but not before we had recorded their positions within a square. Data for individual bones were then placed in virtual space using a graphics software package (Discreet's 3ds max 4), which allowed three-dimensional relationships to be studied. All original maps of the disarticulated skeleton created in the field and the computer graphics files are in the Collections Department, Paleontological Research Institution, Ithaca, New York.

Miller was responsible for the paleobotanical data and its interpretation; Nester collected and analyzed data pertaining to the distribution of sediments in the basin and the mastodon skeleton.

RESULTS

Sediment Packages

We recognized five sediment types. The bottommost was (1) cobbles in a matrix of silt and clay. Above this was (2) clayey silt with some coarse gravel and cobbles. Marly, clayey silt (3) occurred next higher up. Above this, marl and degraded plant matter (peaty marl [4]) were mixed with gastropod, bivalve, and ostracod shells, along with fine sand and silt. Capping these sediments was a layer of fine-grained, non-marly detrital peat (5). Between units (2) and (3) and units (3) and (4) the contact was gradational, whereas between units (1) and (2) and units (4) and (5) it was abrupt. Nowhere in the pond did the nonsystematic backhoe excavation penetrate deeper than unit (4). Sedimentary units are continuous within the study area (Fig. 2). Exploratory augering revealed that the depth to the cobble horizon, the lowest unit encountered stratigraphically, decreased toward the north and south ends of the pond, with the deepest part of the basin 8–25 m northeast of the mastodon bone field. Ground penetrating radar transects at the north end of the pond indicated a more or less symmetrical basin, with the pond axis oriented roughly north to south (Fig. 2). These profiles also revealed truncation of the peaty marl (unit 4) by the overlying peat (unit 5). Insufficient

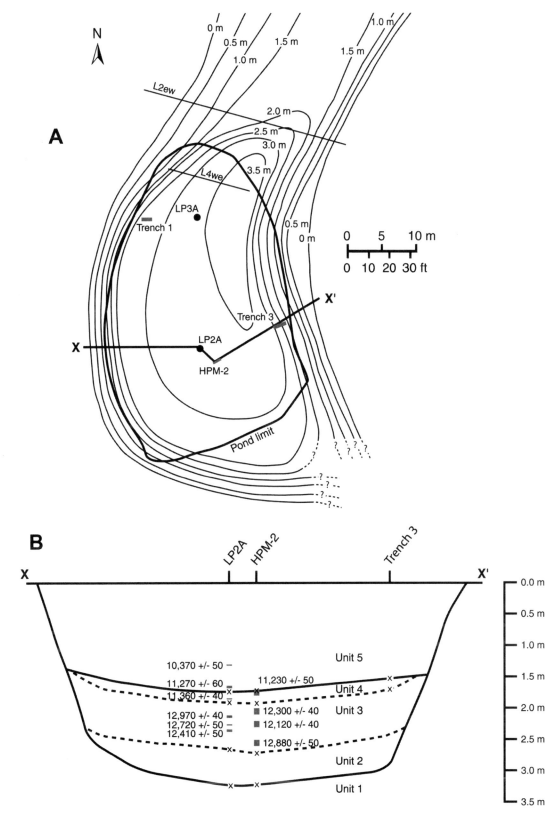

Figure 2. Lithology and stratigraphic thicknesses of the sediments of Lozier Pond. (A) Depth mapped to cobbles (unit 1). Locations of two key ground penetrating radar transects are indicated by L2ew and L4we; other trenches and cores used to construct this diagram are noted. (B) Cross section of the stratigraphy of the basin. AMS ages mentioned in the text are shown in their proper stratigraphic relationships. Abrupt contacts are represented by solid lines, gradational ones by dashed lines. "**X–X'**" indicates where direct observations of unit thicknesses were made. Vertical exaggeration is 5×.

data were collected to the southeast of Lozier Pond to extrapolate beyond the limit of the existing water body.

Plant Macrofossils

The sediment samples produced a robust vascular plant macrofossil stratigraphy (Fig. 3; Table 1), with Tundra (lower) and Spruce (upper) zones clearly demarcated by plant macrofossils. The transition between Tundra and Spruce zones was abrupt (Fig. 3). As expected, given the history of digging at the wetland, the record was truncated at the top, and no Holocene sediment and macrofossils were present in the studied sample series. The Tundra zone yielded fossils of 21 species and eight additional genera, whereas the Spruce zone contained fossils of 22 species and 11 additional genera. Fossils identifiable only to genus had no species-diagnostic features. The full record is presented and interpreted by Miller (2006), but here we focus mainly on fossil wetland plants.

Macrofossils of wetland plants occurred from the upper Tundra zone through the Spruce zone, with only a few recovered lower in the Tundra zone. Rooted aquatics (*Myriophyllum sibiricum* Kom., *Najas flexilis* (Willd.) Rostk. & Schmidt, *Potamogeton* spp., *Sparganium* sp.) had uninterrupted records in the Spruce zone, whereas others (*Carex vesicaria* L., *Hippuris vulgaris* L., *Lycopus americanus* Muhl., *Ranunculus* sp., *Sagittaria latifolia* Willd., *Selaginella selaginoides* (L.) Link, *Viola* sp.) were present discontinuously. Fossils of emergent wetland plants (*Carex aquatilis* Wahlenb., *Eleocharis palustris* (L.) R. & S., *Hypericum virginicum* L., *Scirpus tabernaemontani* Gmelin, *Typha* sp.) were restricted to the topmost Spruce zone samples. Fossil sedge achenes (*Carex* spp.; nine types in addition to those with well-preserved perigynia, which for this reason could be identified to species) were recovered from samples throughout the Spruce zone, and also from the top three Tundra zone samples. Although not the subject of this paper, it is important to note that aquatic

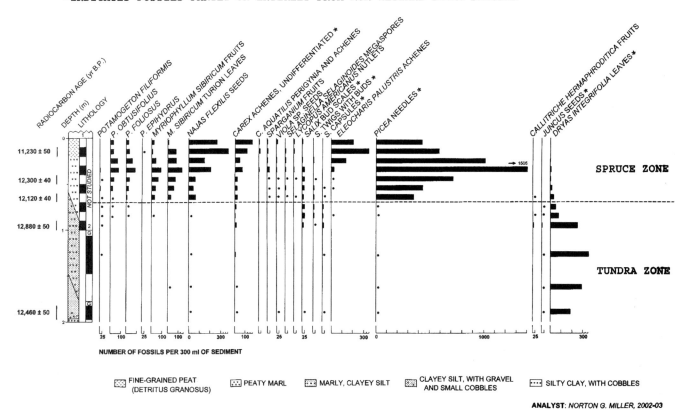

Figure 3. Stratigraphic distribution of vascular plant macrofossils, mainly wetland taxa, at sample position HPM-2, Hyde Park mastodon site. The bars show numbers of fossils per 300 ml of sediment. Solid black circles indicate fossil numbers of 1–5. Radiocarbon ages and sediment types are given on the left side of the diagram. Two adjacent sediment sample series (1, 2) were studied. "G" indicates gaps in these sequences.

and wetland plants occurred with fossils of upland (non-wetland) species—leaves of mountain avens (*Dryas integrifolia* Vahl) and other arctic-alpine plants in the Tundra zone and needles and cones of white spruce (*Picea glauca* (Moench) Voss), balsam fir, and tamarack (*Larix laricina* (DuRoi) K. Koch) in the Spruce zone (Fig. 3).

Moss fossils, including fragments of leafy stems, leaves, calyptrae, opercula (with leafy stems most common), were present in all samples of both zones (Tables 1, 2). Wetland species were most frequent and abundant in Spruce zone sediment. Eighty-four slides were prepared, and from these, 30 moss taxa (identifications to species, genus, or family) were recognized, including 22 species, and fossils of plants representing five genera, four of which were genera other than those of the 22 species. One taxon, *Polytrichum juniperinum* Hedw./*P. strictum* Brid. type, includes two possible species, neither of which can be conclusively identified from incomplete fossil material. Nine slides contained unidentified fossils, seven of which were poorly preserved, whereas the other two were well-preserved fossils of unrecognized species. Wetland mosses included the following: *Calliergon giganteum* (Schimp.) Kindb., *Calliergonella cuspidata* (Hedw.) Loeske, *Campylium stellatum* (Hedw.) C. Jens., *Drepanocladus aduncus* (Hedw.) Warnst., *D. sordidus* (C. Müll.) Hedenäs, *Meesia triquetra* (Richt.) Ångstr., and *Warnstorfia exannulata* (Schimp. in B.S.G.) Schimp. Other species of wet habitats, including seeps over rock and soil, were *Philonotis fontana* (Hedw.) Brid. and *Dicranella schreberiana* (Hedw.) Hilf. ex

TABLE 1. STRATIGRAPHIC OCCURRENCE OF SPRUCE ZONE WETLAND VASCULAR PLANT MACROFOSSILS NOT GRAPHED IN FIGURE 3 AND MOSSES NOT LISTED IN TABLE 2

VASCULAR PLANTS
Carex vesicaria: 0–10(4), 10–20(2), 30–40(1), 40–50(2), 50–60(1).
Hypericum virginicum: 0–10(1), 80–90(?1).
Typha sp.: 0–10(2), 10–20(1).
Sagittaria latifolia: 1–10(3), 20–30(1), 50–60(1), 80–90(1).
Ranunculus sp.: 0–10(1), 20–30(1).
Scirpus tabernaemontani: 0–10(3), 10–20(2).
Hippuris vulgaris: 0–10(1), 10–20(2), 50–60(1).

MOSSES
Bryum sp.: 0–10(t), 20–30(t), 50–60(+), 60–70(+), 70–80(+), 80–90(+), 90–100(+), 145–175(+).
Campylium chrysophyllum (Brid.) J. Lange: 80–90(t).
Encalypta sp.: 105–145(t).
Mniaceae (cf. *Cinclidium* sp.): 50–60(t).
Pohlia sp.: 145–175(t).
Polytrichaceae undifferentiated: 70–80(t), 90–100(t).
Polytrichum juniperinum/*P. strictum* type: 50–60(++).

Note: For both, number of fossils are given in parentheses following sample depth (in cm); for mosses, abundance designations are explained in Table 2.

TABLE 2. STRATIGRAPHIC OCCURRENCE OF FOSSIL MOSSES, SAMPLE SERIES HPM-2, HYDE PARK SITE, DUTCHESS COUNTY, NEW YORK

DEPTH (cm)	Calliergon giganteum*	Drepanocladus sordidus*	Drepanocladus aduncus*	Scorpidium scorpioides*	Calliergonella cuspidata*	Meesia triquetra*	Dicranella schreberiana*	Philonotis fontana*	Tortula sp.	Campylium stellatum*	Polytrichastrum alpinum	Ditrichum flexicaule	Distichium sp.	Hypnum vaucheri	Warnstorfia exannulata*	Mnium thomsonii	Tortella fragilis	Timmia norvegica	Hypnum revolutum	Abietinella abietina	Ceratodon purpureus	Tortula norvegica	Bryoerythrophyllum recurvirostre	Σ wetland spp. : Σ upland spp. ratio
0–10	++	+++		+																				3
10–20	+	+++	+	+																				4
20–30				+																				3
30–40	t	++	++	+	t																			5
40–50		+				t			t															1
50–60	+		t			t	+	+	t		+													3
60–70			+	+					t			+	+	t										1.5
70–80				+				t		t		+	+	t	++	t	t	t	t	t				1
80–90				++					t	t		+	+		+									0.4
90–100	+			++					t			+	+	t	+				+		t			0.6
105–145										t		++	++	+	+++							t	t	0.7
145–175			+++									+++	+++		+++									0.5
180–195			+++																					0.6

Key: + (2–5 plants or fragments); ++ (6–10); +++ (>10); t = one leaf, plant, or plant fragment.
Note: Species marked with an asterisk (*) are wetland and wet seep mosses; the others are non-wetland mesic or xeric species. The bold line between samples 60–70 and 70–80 cm separates the Tundra (lower) and Spruce (upper) zones recognized on the basis of vascular plant macrofossils (Miller, 2006).

Crum and Anders. Non-wetland mosses were mostly restricted to sediments below 70 cm, i.e., the Tundra zone (Table 2). Species of dry to moist soil and rock were present, including *Abietinella abietina* (Hedw.) Fleisch., *Bryoerythrophyllum recurvirostre* (Hedw.) Chen, *Ceratodon purpureus* (Hedw.) Brid., *Distichium* sp., *Ditrichum flexicaule* (Schwaegr.) Hampe, *Hypnum revolutum* (Mitt.) Lindb., *H. vaucheri* Lesq., *Mnium thomsonii* Schimp., *Polytrichastrum alpinum* (Hedw.) G. L. Sm., *Timmia norvegica* (Web.) Wahlenb. *ex* Lindb., *Tortella fragilis* (Hook. & Tayl. *in* Drumm.) Limpr., and *Tortula norvegica* (Web.) Wahlenb. *ex* Lindb.

Mastodon Remains

More than 80% of the mastodon skeleton was recovered, including all long bones, ribs, vertebrae anterior to the sixteenth caudal, the skull, mandible, and both tusks (Table 3). Possible missing bones include several of the most distal caudal vertebrae and a few foot bones. All bones were pristine, and the only postmortem damage to the skeleton was caused by the backhoe used in the excavation of the pond by the landowner (Fisher, 2006). Original positions of the larger bones of the Hyde Park Mastodon skeleton when it was excavated in August 2000 are shown in Figure 4. Missing from the figure are the caudal vertebrae and ribs, toe, wrist/ankle, and other small bones (e.g., hyoids). These were found but have been omitted from the diagram for clarity. Prior to the discovery of mastodon bones by the landowner in 1999, the excavation had disturbed and removed several bones, including the skull, left humerus, right radius and ulna, the thirteenth caudal vertebra, and several toe bones. Therefore, their original position in the sediment is unknown. One exception was the skull, which was found in a spoil bank on the east side of the pond. The proximal quarter of the left tusk was still lodged within the tusk cavity, with ~50 cm projecting beyond the alveolar margin. The complementary section of tusk was found in situ in the pond sediment, thereby allowing the original location of the skull to be determined. The majority of the ribs were found immediately north and east of the reference post shown in Figure 4. Figure 5 shows most of the bone field as it was being excavated.

Large sections of the skeleton were articulated or nearly so. The proximal end of the right femur was in a state of near-articulation with the corresponding socket of the right innominate (pelvis). Similarly, the distal end of this femur was in near-articulation with the articulated right tibia and right fibula. The distal portion of these two skeletal elements was just above the nearly complete and well-articulated right pes (hind foot). Only the most distal phalanx of the first toe was missing from the right pes. Several complete sections of the vertebral column were also found, with many of the cervical, thoracic, and lumbar vertebrae articulated. A segment from the sixth cervical vertebra to the ninth thoracic vertebra was uncovered at the southern end of the bone field. A segment of spine from the twelfth thoracic vertebra to the third lumbar vertebra was found in a state of articulation to near-articulation. Bones were present in a zone extending from the lower 50 cm of detrital peat (unit 5) down to the lower portion of the marly, clayey silt (unit 3). An even greater concentration of skeletal remains occurred over a 40-cm interval that brackets the peaty marl (unit 4). The lowest bones recovered, both stratigraphically and in absolute depth, were foot bones from the articulated right pes, which were planted in marly, clayey silt (unit 3) near the bottom of the shallow pond by the animal before it died.

DISCUSSION

Sediments and Sedimentary History

Two depositional processes operated to fill the Hyde Park oxbow basin at different times in its history. During the early phase, which probably began as glacier ice withdrew from the Hyde Park Moraine or another position closer to the ice front,

TABLE 3. LARGER BONES RECOVERED FROM THE HYDE PARK SITE

No.	Names	Identification Numbers
1	left tusk and skull	HP1002, 1000
2	right tusk	HP1001
3	mandible	HP060
4	right humerus	HP322
5	right scapula	HP506
6	left radius	HP308
7	left ulna	HP366
8	left scapula	HP321
9	right fibula	HP303
10	right tibia	HP602
11	right femur	HP368
12	left fibula	HP148
13	left tibia	HP320
14	left femur	HP328
15	left innominate (pelvis)	HP300
16	right innominate (pelvis)	HP309
17	atlas and axis	HP523, 524
18	cervical vertebrae 3,4,5	HP317, 507. 335
19	cervical vertebrae 6 (north), 7; thoracic vertebrae 1-8 (south)	HP509, 510, 511, 512, 513, 145, 136, 147, 307, 064
20	thoracic vertebrae 9 (west), 10, 11 (east)	HP069, 324, 325
21	thoracic vertebrae 12–20 (south); lumbar vertebrae 1,2,3 (north)	HP521, 520, 517, 519, 518, 516, 350, 351, 352, 353, 354, 355
22	right pes (hind foot)	various

Note: Numbers in the column at the left refer to numbered bones illustrated in Figure 5. The field identification numbers were used in cataloging parts of the skeleton.

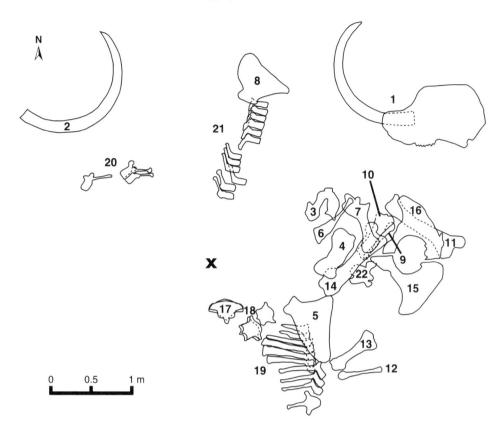

Figure 4. Map view of the in situ skeleton of the Hyde Park mastodon. Excluded from this figure are the ribs, toe bones, and other minor components of the skeleton (e.g., hyoids, patella, sternum). Bone identifications are listed in Table 3. "**X**" indicates position of the reference post used to map bone locations.

periodic overbank flooding by the paleo-Fall Kill delivered silt, clay, plant matter, and sometimes gravel and cobbles. In the second phase, the basin became separated from the Fall Kill, and autochthonous marl deposition began. The shallow pond of phase two supported rooted aquatic and emergent plants. Later, intermixed marl and detrital peat accumulated as peat-generating plant communities became established along the margin of the pond. Finally, the deposition of marl ceased and detrital peat accumulated. The onset of detrital peat deposition was abrupt, beginning ca. 11,230 ± 50 yr B.P. on the basis of the age of a balsam fir (*Abies balsamea*) cone scale deposited in the wetland from a tree growing on the adjacent upland. The transition period between marly peat and non-calcareous detrital peat deposition appears to have been one of brief erosion or non-deposition, as indicated by the truncated reflectors as well as the sharp contact seen in field observation. The early phase of deposition was estimated to begin ca. 13,000 and continue to ca. 12,200 yr B.P. (Miller, 2006).

Menking et al. (2006) studied core LP2A in detail and obtained late Pleistocene ages for peat (unit 5), peaty marl (unit 4), and marly, clayey silt (unit 3). Peat from 1.64 to 1.68 m (unit 5) was 10,370 ± 50 yr B.P. (AMS; Beta-169810). Peaty marl (unit 4) at 1.84–1.85 m was 11,360 ± 40 yr B.P. (AMS; Beta-159335). Three ages were obtained from marly, clayey silt (unit 3): 12,410 ± 50 yr B.P. (AMS; Beta-159338) from 1.33 to 1.375 m, 12,720 ± 50 yr B.P. (AMS; Beta-159337) from 1.25 to 1.27 m, and 12,970 ± 40 yr B.P. (AMS; Beta-159336) from 1.125 to 1.165 m.

Figure 5. Bones of the Hyde Park mastodon as they appeared in the pond on 16 September 2000, facing northeast. The pond was drained 2 m to allow excavation and sediment sampling. White arrow shows the position of the reference post, from which all bone locations were determined. Ribs of the mastodon dominate the left-central portion of this photograph, with the right scapula, pelvis, and other long bones visible to the right. University of Michigan's Dan Fisher is standing behind the mandible.

Ages from sample series HPM-2 are also late Pleistocene, from 11,230 ± 50 yr B.P. (AMS; Beta-175554) obtained from a single balsam fir cone scale at the top of unit 4 to 12,880 ± 50 yr B.P. (AMS; Beta-175557) obtained from six *Dryas integrifolia* (Rosaceae) leaves near the bottom of unit 3. This date and 12,970 ± 40 yr B.P. (Menking et al., 2006) are the oldest reported from the pond. Two other dates obtained in this sample set from spruce (*Picea*) needles (12,300 ± 40 yr B.P. [AMS; Beta-175555] at 1.40–1.50 m and 12,120 ± 40 yr B.P. [AMS; Beta-175556] at 1.60–1.70 m) record an inversion in unit 3 at roughly the same levels as the inversion reported above from core LP2A. This is possibly the result of reworked sediment during overbank deposition or perhaps bioturbation by either aquatic organisms living in the oxbow basin or a terrestrial mammal such as a mastodon in search of food or water. The Hyde Park mastodon, 11,480 ± 50 yr B.P., is younger than these ages. Vascular plant macrofossil and pollen stratigraphies from the face of trench HPM-2 (Miller, 2006; Robinson and Burney, 2006) revealed a tundra environment at the site from the deposition of the basal cobbles (unit 1) to an abrupt increase in spruce needle and pollen deposition at 0.7 m (adjusted depth 2.3 m, i.e., from the original projected surface prior to excavation disturbance). A Holocene record was not present in the HPM-2 sample series.

Vegetation Reconstruction

The plant macrofossil assemblage and stratigraphic changes in it reflect the floristic composition of the wetland plant community through time. Open, shallow, standing pond water occurred in the oxbow during the Spruce zone on the basis of the continuous presence of pondweed and water milfoil fossils, which are produced by obligately aquatic plants. However, at the wet margin of the pond grew sedges, bulrushes, spike-rushes, and other emergent herbaceous plants. Pollen assemblages from the same interval of pond sediment (Robinson and Burney, 2006) support this determination but provide less taxonomic resolution than the macrofossils. Sedge (Cyperaceae) pollen (5%–13% of the sum of terrestrial herbs, shrubs, and trees) was registered consistently in all Spruce zone samples. The stratigraphic distribution of sedge pollen parallels the sedge macrofossil record. Bulrush, spike-rush, and *Carex aquatilis* macrofossils occurred in the uppermost Spruce zone samples, indicating an expansion of the area supporting these rooted, emergent plants as the basin shallowed along the pond margin from infilling.

The pond periphery was a calcareous fen associated with circumneutral pond water of moderately high cation content and conductivity. Water chemistry (pH and alkalinity) measurements from numerous wetlands throughout New England where plants of the Hyde Park pondweed and water milfoil species grow at present (Crow and Hellquist, 1983; Hellquist, 1980; Hellquist and Crow, 1980) indicate that the Hyde Park pond water fell within a pH range of 6.5–9.8, with alkalinity between 15 and 170 mg HCO_3^- liter^{-1} (Miller, 2006). These conditions are sufficient to support a fen of the floristic composition indicated by the macrofossils.

The type of wetland is further specified by the fossil moss assemblage that was recovered from sediments of the Spruce zone (Table 2). Although mosses in a wetland form the ground layer beneath the emergent vascular plants (and therefore mosses are part of a broadly defined vegetation), a fossil moss record is largely independent of a paleoenvironmental assessment on the basis of fossils of vascular plants alone. Fossil mosses can be used, therefore, to test and interpret inferences derived from other proxy data sources. Miller (1980) has reviewed the application of moss fossils in Quaternary paleoecology. Information about Pleistocene and Holocene fossil bryophytes from wetland and other types of Quaternary sediment has increased greatly in recent years, particularly in connection with investigations into the development of peat-generating systems. The moss assemblage of the Hyde Park Spruce zone indicates rich fen wetland (Miller, 1980). Peatland of this sort develops in association with base-rich water that is replenished by ground- or stream-water flow. Rich fens are found in areas of calcareous bedrock or glacial drift, and they and/or calcareous seeps appear to have been common in the late Pleistocene. Some late Pleistocene rich fens persisted through the Holocene (Futyma and Miller, 2001), and organisms in them can be of special conservation concern. *Scorpidium scorpioides* (Miller, 1980) and *Meesia triquetra* (Montagnes, 1990) are considered to be rich-fen indicator species. These mosses and others in the Hyde Park Spruce zone, *Calliergon giganteum*, *Calliergonella cuspidata*, and *Campylium stellatum*, are common associates in extant rich fens. Janssens (1983) considered *Drepanocladus sordidus* (as *D. sendtneri*; see Hedenäs 1998 for a taxonomic revision of these and related species) to be a moss of rich fens and reported that its associate in the Hyde Park Spruce zone assemblage, *D. aduncus*, occurred in contemporary peatlands having the following water chemistry: pH 6.6–7.5, conductivity 150– 575 μS, and Ca^{2+} concentration 23–61 ppm in his study area, which extended between British Columbia and Yukon Territory eastward to Minnesota and Michigan.

There is no paleobotanical evidence from the samples studied that fen plants were replaced upward stratigraphically by fossils of oligotrophic fen or bog plants. Heath (Ericaceae) pollen from bog and oligotrophic fen plant communities occurred (Robinson and Burney, 2006), but only sporadically and in low frequency in Spruce zone sediment. However, because the basin did not capture much of a Holocene record, the wetland type occurring in later stages of basin infilling cannot be specified. Wood from the top of an unexcavated island of sediment located above the mastodon remains was Holocene (5420 ± 50 yr B.P.; Menking et al., 2006), indicating that some Holocene sediment may have been present in the basin prior to the two episodes of pond excavation. But because the basin is shallow and most of the sediment filling is late Pleistocene, water was never more than about one to two meters deep. Therefore, it seems likely that a long Holocene record never accumulated. Alternatively, if one did, the record was lost as organic sediment in the shallow basin oxidized and disappeared with changes in the water table position caused by variation in precipitation and groundwater input. Thus, the vascular plant and moss macrofossils were from a wetland that existed in the basin over a relatively short period beginning just before the Pleistocene/Holocene boundary and perhaps continuing some hundreds of years into the early Holocene.

The radiocarbon (AMS) age of collagen extracted from a sample of one of the Hyde Park mastodon tusks was 11,480 ± 50 yr B.P. Thus, the time of death of the animal was ~200 radiocarbon years before the deposition of a cone scale of balsam fir, which occurred with other macrofossils of balsam fir and with those of white spruce and tamarack. These macrofossils, and spruce- and balsam-fir-containing pollen assemblages from the same interval, indicate that the upland forest vegetation near the basin consisted of these trees, rather than of white spruce alone, macrofossils of which occurred in low numbers and sporadically in samples from the lower Spruce zone. It is likely that pine (perhaps jack pine, *Pinus banksiana* Lamb.) and tree birches (*Betula* sp.) grew near the basin throughout the Spruce zone, but there is no macrofossil evidence of this. These two pollen types were, however, registered during pollen analysis of the interval (Robinson and Burney, 2006).

The Mastodon and Its Paleoenvironment

Haynes (1991) noted that the number of bones in the skeleton of a healthy adult American mastodon might have been variable. A review of the published literature (Osborn, 1936; Olsen, 1972; Haynes, 1991) indicated the number of bones to be 235–255, with the number of bones in the feet and tail in greatest contention. Two hundred and fourteen bones of the Hyde Park mastodon have been accounted for, with only possibly the distal caudal vertebrae and 34 of 128 foot bones, most of them the smaller phalanges and sesamoids, not recovered. The Hyde Park mastodon skeleton is therefore 84%–91% complete.

The rearticulated Warren mastodon from near Newburgh, New York (Osborn, 1936), offers some useful comparisons. Like the Hyde Park mastodon, this skeleton is essentially complete, with only some toe bones and caudal vertebrae unaccounted for. The skeleton measures 4.55 m (14' 11") from the most distal caudal vertebra to the tusk base, and 2.78 m (9' 2") from the floor to the highest spinous process of the thoracic vertebrae. Bones of the Warren and Hyde Park mastodons have similar measured dimensions. These animals, both males, were probably nearly the same size at death.

Taphonomic information about the skeleton and the position of the bones in the pond sediment contribute information about the sedimentary history of the basin and paleoenvironment at the time of and immediately following the death of the animal. Considering that all of the bones found in situ originated from an area ~5 × 5 m, and that many elements of the skeleton remained articulated, there was very little postmortem scattering of the bones. Discovery of the lower thoracic and lumbar vertebrae (the smallest bones of the spinal column anterior to the pelvis) in a state of semi-articulation indicates settling and burial of this part of the mastodon carcass in quiet water. Although some of the skeletal elements are disorganized relative to one another, the articulated nature of others suggests little transport and disruption of the decaying carcass. The plant macrofossils also show that the Hyde Park oxbow had become isolated from the Fall Kill and contained a rich fen and associated shallow-water pond at the time of death of the mastodon, while land above the basin supported (Miller, 2006) white spruce and/or white spruce-balsam fir-tamarack forest. Bones of the Hyde Park mastodon's right rear leg was found in what appears to be life position, with the right pes planted in the clayey silt. Therefore, it is likely that the mastodon, in one of its final actions before dying, pushed its leg deep into the unconsolidated pond sediment in which it became mired. The Hyde Park mastodon was not transported by stream-water flow to the oxbow basin but instead died and was buried and preserved near or at its place of death. The mastodon died in the basin during the transition from peaty marl to detrital peat deposition (Fig. 2).

Water levels must have been high enough that the decaying animal was never exposed subaerially, but shallow enough that the animal was able to walk to this location and plant its foot into the sediment. We estimate this depth to have been 1–2 m. In addition, the excellent condition of the bones indicates that the water table remained high enough for the remains to become completely covered by sediment. The skeleton was minimally disturbed during the nearly 11,500 radiocarbon years following its death. The movement that took place appears to have been the result of random floating of individual segments of the carcass in a closed basin, because there is no preferential orientation of the remains. Because of the pristine condition of the bones and the amount of articulation observed, there is a strong possibility that the unaccounted for bones are still buried in the sediments of Lozier Pond. The complete disarticulation history of the skeleton is presented by Fisher (2006).

Mastodon remains have been found in a variety of depositional basins in the Great Lakes region of North America. Small kettle holes seem to be the most productive sites for mastodon fossils, but remains have been found in spring deposits (Laub, 2003), potholes (Hall, 1868), sinkholes (Harington et al., 1993), kames (Hodgson et al., 2006), swales of uncertain origin (McAndrews and Jackson, 1988), and on the submerged Atlantic Coastal Plain (Oldale et al., 1987). Most discoveries have been made fortuitously, and often as a result of draining or modifying land for agriculture or recreation. When contextual paleoenvironmental data were obtained in conjunction with recovery at sites in the Great Lakes region (Table 4), mastodons are often associated with calcareous, peaty, rich-fen sediment and peat consisting largely of sedge remains. Calcareous sedge fens are an initial successional stage in kettle hole basin infilling (terrestrialization). They can persist through the Holocene to the present or are replaced by oligotrophic fen communities and sometimes bogs, depending on changes in groundwater hydrology and chemistry, ecosystem productivity, and other autogenic and allogenic factors that may be specific to a basin or carry a regional message related to climate and other large-scale environmental factors. In contrast to the wetland vegetation, the upland near the Hyde

TABLE 4. DATA PERTAINING TO WELL-DOCUMENTED MASTODON FINDS AND ASSOCIATED VEGETATION RECORDS IN THE GREAT LAKES REGION OF NORTH AMERICA (MICHIGAN, INDIANA, OHIO, NEW YORK)

Site name	Location	Basin type	Sediment type surrounding bones	Radiocarbon age (yr B.P.); Material Dated	Inferred upland vegetation	Paleovegetation data source	Reference
Powers	Van Buren County, Michigan; 42°7'N, 85°54'W	Swale on sorted sand in lake bed plain between moraines	Gray clay and peat	11,220 ± 310, bone	Spruce and balsam fir forest	Pollen spectrum	Garland and Cogswell (1985)
Eldridge	Montcalm County, Michigan; 43°15'N, 85°55'W	Shallow lake	Marl-peat transition	10,190 ± 80, 12,020 ± 120; wood above and below bones	Open spruce parkland	Pollen profile, wood	Kapp et al. (1990)
Rappuhn	Lapeer County, Michigan; 43°11'N, 83°10'W	Sallow kettle hole	Resting on clay below peat	10,400 ± 400, 10,750 ± 400; wood below bones	Spruce forest with balsam fir, tamarack, pine, and poplar	Pollen profile, plant macrofossils	Kapp (1986), Wittry (1965)
Heisler	Calhoun County, Michigan; 42°23'N, 84°44'W	Sall depression	Clay-marl transition below peat	11,160 ± 110, wood associated with bones	Spruce-balsam fir forest	Pollen profile, wood	Bearss and Kapp (1987)
Shelton	Oakland County, Michigan; 42°51'N, 82°20'W	Sallow lake on outwash plain	Sandy matrix with pebbles, cobbles, and wood above lake clay and below interbedded sand, silt, and clay indicating lake margin environment	12,320 ± 110, wood; 11,700 ± 110, wood	White spruce forest	Pollen profile, plant macrofossils, incl. abundant conifer cones	Shoshani et al. (1989)
Christensen	Hancock County, Indiana; 39°52'N, 85°50'W	Sall kettle hole	Marl and peat	13,220 ± 100, wood in bone layer; 12,060 ± 100, wood in peat above bone layer	Open white spruce forest	Pollen profile, plant macrofossils	Whitehead et al. (1982)
Kolarik	Starke County, Indiana; 41°11'N, 86°35'W	Dpression in outwash	Sandy muck	10,115 ± 205 & 11,185 ± 270, bone; 11,760 ± 280, associated wood	Open spruce-dominated forest	Pollen profile, plant macrofossils	Jackson et al. (1986)
Wells	Fulton County, Indiana; 41°N, 86°13'W	Dpression in outwash	Marl; wind-blown sand and peat above	12,000 ± 450, wood beneath bones	Spruce, balsam fir, birch, and ash forest	Pollen profile	Gooding and Ogden (1965)
Burning Tree	Licking County, Ohio; 39°59'N, 82°27'W	Sall wetland on end moraine	Fibric and humic peat	10,860 ± 70, bone collagen; 11,660 ± 120 & 11,450 ± 70, gut contents	Spruce, balsam fir, pine forest	Pollen spectrum from tooth cementum, pollen profile, plant macrofossils from digesta and in adjacent peat	Lepper et al. (1991)
Hiscock	Genesee County, New York; 43°05'N, 78°5'N	Shallow, 0.8 hectare, spring-associated wetland	Gray, sandy, silty clay, some gravel and cobbles	11,250 ± 140 and 10,930 ± 90, associated wood	Open spruce-jack pine forest	Pollen profile, plant macrofossils	Laub et al. (1988), Miller (1988)
Hyde Park	Dutchess County, New York; 40°46'N, 73°54'W	Wetland in oxbow	Dtrital peat; peaty marl below	11,480 ± 50, tusk collagen; 11,230 ± 50, *Abies balsamea* cone scale	Open white spruce-balsam fir-tamarack forest	Pollen profile, plant macrofossils, incl. wood	Miller (2006), Robinson and Burney (2006)

Park basin supported open white spruce forest initially and later open forest of white spruce, balsam fir, and tamarack. Dreimanis (1967, 1968) observed that a spruce forest–mastodon association was common in the Great Lakes region and suggested mastodon extinction was in response to a sharp decrease in the area occupied by this kind of conifer forest. It is by no means certain that his hypothesis explains the loss of this megavertebrate, but the association between mastodons and late-glacial conifer forest has remained well established for the Great Lakes region.

ACKNOWLEDGMENTS

The authors wish to thank the property owners, Larry and Sheryl Lozier, for donating their discovery to science. David Burney and Guy Robinson provided much expertise in the field and aided in the interpretation. Daniel Fisher and James Sherpa carried out mastodon bone identification, and Sherpa also provided helpful comments on a draft of the manuscript. Fieldwork was performed with the assistance of Warren Allmon, Paul Harnik, Robert Ross, and Elizabeth Humbert. Three-dimensional placement of the bones was conducted with the assistance of Elizabeth Humbert and Brian Dunston. Financial support for the project was provided by a grant from the Eppley Foundation. We also thank Daniel Fisher and John McAndrews for comments and suggestions on the submitted manuscript.

REFERENCES CITED

Auer, V., 1927, Stratigraphical and morphological investigations of peat bogs in southeastern Canada: Communicationes ex Instituto Quaestionum Forestalium Findlandiae, v. 12(5) p. 1–62 + 11 unpaged pl.

Bearss, R.E., and Kapp, R.O., 1987, Vegetation associated with the Heisler Mastodon site, Calhoun County, Michigan: Michigan Academician, v. 19, p. 133–140.

Burney, D., and Robinson, G., 2006, Excavating and interpreting flooded megafaunal sites, in Allmon, W.D., Nester, P.L., and Chiment, J.C., eds., Mastodon paleobiology, taphonomy, and paleoenvironments in the late Pleistocene of New York State: Studies on the Hyde Park, Chemung, and Java sites: Bulletins of American Paleontology (in press).

Connally, G.G., and Sirkin, L., 1986, Woodfordian ice margins, recessional events, and pollen stratigraphy of the mid-Hudson valley, in Cadwell, D.H., ed., The Wisconsinan Stage of the First Geological District, eastern New York: New York State Museum Bulletin 455, p. 50–72.

Crow, G.E., and Hellquist, C.B., 1983, Aquatic vascular plants of New England, Part 6, Trapaceae, Haloragaceae, Hippuridaceae: New Hampshire Agriculture Experiment Station Bulletin 524, 26 p.

Dreimanis, A., 1967, Mastodons, their geologic age and extinction in Ontario, Canada: Canadian Journal of Earth Sciences, v. 4, p. 663–675.

Dreimanis, A., 1968, Extinction of mastodons in eastern North America: testing a new climatic-environmental hypothesis: Ohio Journal of Science, v. 68, p. 257–272.

Drumm, J., 1963, Mammoths and mastodons, Ice Age elephants of New York: New York State Museum and Science Service Educational Leaflet 13, 31 p.

Fisher, D.C., 2006, Taphonomy of the Hyde Park mastodon, in Allmon, W.D., Nester, P.L., and Chiment, J.C., eds., Mastodon paleobiology, taphonomy, and paleoenvironments in the late Pleistocene of New York State: Studies on the Hyde Park, Chemung, and Java sites: Bulletins of American Paleontology (in press).

Futyma, R.P., and Miller, N.G., 1986, Stratigraphy and genesis of the Lake Sixteen peatland, northern Michigan: Canadian Journal of Botany, v. 64, p. 3008–3019.

Futyma, R.P., and Miller, N.G., 2001, Postglacial history of a marl fen: Vegetational stability at Byron-Bergen Swamp, New York: Canadian Journal of Botany, v. 79, p. 1425–1438, doi: 10.1139/cjb-79-12-1425.

Garland, E.B., and Cogswell, J.W., 1985, The Powers Mastodon site: Michigan Archaeologist, v. 31, p. 3–39.

Glaser, P.H., 1992, Vegetation and water chemistry, in Wright, H.E., Jr., Coffin, B.A., and Aaseng, N.E., eds., The patterned peatlands of Minnesota: Minneapolis and London, University of Minnesota Press, p. 15–26.

Gooding, A.M., and Ogden, J.G., III, 1965, A radiocarbon dated pollen sequence from the Wells Mastodon site near Rochester, Indiana: Ohio Journal of Science, v. 65, p. 1–11.

Hall, J., 1868 (pub. 1871), Notes and observations on the Cohoes mastodon: Report of the Regents of the University of the State of New York on the Condition of the State Cabinet of Natural History, v. 21, p. 99–148 + 7 unpaged pls.

Harington, C.R., Grant, D.R., and Mott, R.J., 1993, The Hillsborough, New Brunswick, mastodon and comments on other Pleistocene mastodon fossils from Nova Scotia: Canadian Journal of Earth Sciences, v. 30, p. 1242–1253.

Haynes, G., 1991, Mammoths, mastodonts, and elephants: Biology, behavior, and the fossil record: New York, Cambridge University Press, 413 p.

Hedenäs, L., 1998, An overview of the *Drepanocladus sendtneri* complex: Journal of Bryology, v. 20, p. 83–102.

Hellquist, C.B., 1980, Correlation of alkalinity and the distribution of *Potamogeton* in New England: Rhodora, v. 82, p. 331–344.

Hellquist, C.B., and Crow, G.E., 1980, Aquatic vascular plants of New England, Part 1, Zosteraceae, Potamogetonaceae, Zannichelliaceae, Najadaceae: New Hampshire Agriculture Experiment Station Bulletin 515, 68 p.

Hodgson, J.A., Allmon, W.D., Sherpa, J.M., and Nester, P.L., 2006, Taphonomy and geology of the North Java mastodon site, Wyoming County, New York, in Allmon, W.D., Nester, P.L., and Chiment, J.C., eds., Mastodon paleobiology, taphonomy, and paleoenvironments in the late Pleistocene of New York State: Studies on the Hyde Park, Chemung, and Java sites: Bulletins of American Paleontology (in press).

Holman, J.A., 2001, In quest of Great Lakes Ice Age vertebrates: East Lansing, Michigan, Michigan State University Press, 230 p.

Jackson, S.T., Whitehead, D.R., and Ellis, G.D., 1986, Late-glacial and early Holocene vegetational history at the Kolarik Mastodon site, northwestern Indiana: American Midland Naturalist, v. 115, p. 361–373.

Janssens, J.A., 1983, Past and extant distribution of *Drepanocladus* in North America, with notes on the differentiation of fossil fragments: Journal of the Hattori Botanical Laboratory, v. 54, p. 251–298.

Janssens, J.A., Hansen, B.C.S., Glaser, P.H., and Whitlock, C., 1993, Development of a raised-bog complex, in Wright, H.E., Jr., Coffin, B.A., and Aaseng, N.E., eds., The patterned peatlands of Minnesota: Minneapolis and London, University of Minnesota Press, p. 189–221.

Kapp, R.O., 1986, Late-glacial pollen and macrofossils associated with the Rappuhn Mastodont (Lapeer County, Michigan): American Midland Naturalist, v. 116, p. 368–377.

Kapp, R.O., Cleary, D.L., Snyder, G.G., and Fisher, D.C., 1990, Vegetational and climatic history of the Crystal Lake area and the Eldridge Mastodont site, Montcalm County, Michigan: American Midland Naturalist, v. 123, p. 47–63.

Laub, R.S., ed., 2003, The Hiscock site: Late Pleistocene and Holocene paleoecology and archaeology of western New York State: Bulletin of the Buffalo Society of Natural Sciences, v. 37, 327 p.

Laub, R.S., DeRemer, M.F., Dufort, C.A., and Parsons, W.L., 1988, The Hiscock site: A rich late Quaternary locality in western New York State, in Laub, R.S., Miller, N.G., and Steadman, D.W., eds., Late Pleistocene and early Holocene paleoecology of the eastern Great Lakes region: Bulletin of the Buffalo Society of Natural Sciences, v. 33, p. 67–81.

Lepper, B.T., Frolking, T.A., Fisher, D.C., Goldstein, G., Sanger, J.E., Wymer, D.A., and Ogden, J.G., III, 1991, Intestinal contents of a late Pleistocene mastodont from midcontinental North America: Quaternary Research, v. 36, p. 120–125.

Maenza-Gmelch, T.A., 1997a, Late-glacial–early Holocene vegetation, climate, and fire at Sutherland Pond, Hudson Highlands, southeastern New York, U.S.A.: Canadian Journal of Botany, v. 75, p. 431–439.

Maenza-Gmelch, T.A., 1997b, Vegetation, climate, and fire during the late-glacial–Holocene transition at Spruce Pond, Hudson Highlands, southeastern New York, USA: Journal of Quaternary Science, v. 12, p. 14–24.

McAndrews, J.H., and Jackson, L.J., 1988, Age and environment of late Pleistocene mastodont and mammoth in southern Ontario, in Laub, R.S., Miller,

N.G., and Steadman, D.W., eds., Late Pleistocene and early Holocene paleoecology of the eastern Great Lakes region: Bulletin of the Buffalo Society of Natural Sciences, v. 33, p. 161–172.

Menking, K.M., Schneiderman, J., Nester, P.L., Bedient, K.D., Collins, B.C., and Feingold, B.J., 2006, Lithology and stratigraphy of core LP2A from the Hyde Park mastodon site, southeastern New York, *in* Allmon, W.D., Nester, P.L., and Chiment, J.C., eds., Mastodon paleobiology, taphonomy, and paleoenvironments in the late Pleistocene of New York State: Studies on the Hyde Park, Chemung, and Java sites: Bulletins of American Paleontology (in press).

Miller, N.G., 1980, Mosses as paleoecological indicators of late-glacial terrestrial environments: Some North American studies: Bulletin of the Torrey Botanical Club, v. 107, p. 373–391.

Miller, N.G., 1988, The late Quaternary Hiscock site, Genesee County, New York: Paleoecological studies based on pollen and plant macrofossils, *in* Laub, R.S., Miller, N.G., and Steadman, D.W., eds., Late Pleistocene and early Holocene paleoecology and archeology of the eastern Great Lakes region: Bulletin of the Buffalo Society of Natural Sciences, v. 33, p. 83–93 + unpaged foldout.

Miller, N.G., 2006, Contemporary and prior environments of the Hyde Park, New York, mastodon on the basis of plant macrofossils, *in* Allmon, W.D., Nester, P.L., and Chiment, J.C., eds., Mastodon paleobiology, taphonomy, and paleoenvironment in the late Pleistocene of New York State: Studies on the Hyde Park, Chemung, and Java sites: Bulletins of American Paleontology (in press).

Montagnes, R.J.S., 1990, The habitat and distribution of *Meesia triquetra* in North America and Greenland: The Bryologist, v. 93, p. 349–352.

Nester, P.L., Brown, L.D., and Miller, N.G., 2006, Stratigraphy of mastodon sites in New York State using ground penetrating radar, *in* Allmon, W.D., Nester, P.L., and Chiment, J.C., eds., Mastodon paleobiology, taphonomy, and paleoenvironments in the late Pleistocene of New York State: Studies on the Hyde Park, Chemung, and Java sites: Bulletins of American Paleontology (in press).

Oldale, F.N., Whitmore, F.C., Jr., and Grimes, J.R., 1987, Elephant teeth from the western Gulf of Maine, and their implications: National Geographic Research, v. 3, p. 439–446.

Olsen, S.J., 1972, Osteology for the archeologist. 3. The American mastodon and the woolly mammoth: Papers of the Peabody Museum of Archeology and Ethnology Harvard University, v. 56, no. 3, p. 1–47.

Osborn, H.F., 1936, Proboscidea: A monograph of the discovery, evolution, migration and extinction of the mastodonts and elephants of the world, Vol. I, Moeritherioidea, Deinotherioidea, Mastodontoidea: New York, American Museum Press, 802 p.

Potzger, J.E., 1933, Succession of forests as indicated by fossil pollen from a northern Michigan bog: Science, v. 75, p. 366–368.

Robinson, G., and Burney, D. 2006, The Hyde Park mastodon and palynological clues to megafaunal extinction, *in* Allmon, W.D., Nester, P.L., and Chiment, J.C., eds., Mastodon paleobiology, taphonomy, and paleoenvironment in the late Pleistocene of New York State: Studies on the Hyde Park, Chemung, and Java sites: Bulletins of American Paleontology (in press).

Sears, P.B., 1935, Types of North American pollen profiles: Ecology, v. 16, p. 488–499.

Shoshani, J., Fisher, D.C., Zawiskie, J.M., Thurlow, S.J., Shoshani, S.L., Benninghoff, W.S., and Zoch, F.H., 1989, The Shelton mastodon site: Multidisciplinary study of a late Pleistocene (Twocreekan) locality in southeastern Michigan: Contributions from the Museum of Paleontology University of Michigan, v. 27, p. 393–436.

Thompson, L.M., 1994, Discoveries of the American mastodon (*Mammut americanum*) in New York State: 1921–1994 [Master's essay]: Rochester, New York, University of Rochester, 66 p.

Whitehead, D.R., Jackson, S.T., Sheehan, M.C., and Leyden, B.W., 1982, Late-glacial vegetation associated with caribou and mastodon in central Indiana: Quaternary Research, v. 17, p. 241–257.

Wittry, W.L., 1965, The Institute digs a mastodon: Cranbrook Institute of Science News Letter, v. 35, p. 14–19.

MANUSCRIPT ACCEPTED BY THE SOCIETY 28 JUNE 2005

Index

A

Abadeh: **250**, **254**, 255
Abies balsamea: **29**, 293, 299
Abietinella abietina: **297**, 298
Abietites: 20
aborescent plants: 5, 7
Acadian orogen: 58, 81, 104
Acanthodes: 108
acanthodians
    Acanthodes: 108
    in Devonian: 87, 99, 108
    Gyracanthus: 108, **159**, **162**, 163
    habitats of
        Albert Formation: 108
        Catskill Formation: 87, 99
        characteristics of: 123
        in Devonian: 87, 99, 108
        East Kirkton: 9
        in Mississippian: 9, 108, 164
    in Mississippian: 9, 108, 164
Acanthostega: 9, 156
Acanthotriletes aculeolatus: **214–15**, **217–18**, **220**
Acer: 25
Acer saccharum: 292
acid rain: 261
acritarchs: 44, 99
Acrostichum: 21, **24**, **26**
actinopterygian fish: 94–95
Adiantites adiantoides: 180, **183**
Adiantites tenuifolius: **106**
Aegyptopithecus: 26
Aeronian: **42**, 44–55
aestivation, "failed:" 227
Africa
    Bahariya: 21
    Cameroon: 258
    coal in: **15**, **23**
    in Devonian: 99
    Egypt: 26–27, 30, 232
    Lake Nyos: 258
    lungfish in: 9
    maps of: **15**, **23**, **225**
    in Miocene: **23**
    Niger Delta: 178, 190
    primates in: 26
    South Africa. *see* South Africa
    Uganda: 26
Agathis: 22
ageleodid sharks: 99
Aglaophyton: **4**, **6**
Aglosperma: 98
Aglosperma quadripartita: **81**, 89, 98
aïstopods: 165
Alabama
    Bangor Limestone: **141**
    barrier islands off: 151–52
    Birmingham anticlinorium: 140
    Cumberland Plateau: 140
    Dauphin Island: **152**
    Hartselle Sandstone: 140–52
    Homewood: 140–52
    Jefferson County: 140–52
    map of Homewood: **140**
    Mobile Delta: 122
    Monteagle Limestone: **141**
    Pride Mountain Formation: 140–41, 150
    Tuscumbia Limestone: **141**
    Walker County: 140–52
Alaska: 22, 29
Albany Group: 224–25, 227
Albert Formation: 104–6, 108, 118–23
Alberta: 99
Alces alces: 16
Alcicornepteris: 98
alders: 23–25
Aldrovanda: 22
Alethopteris decurrens: 180, **183**
Alethopteris discrepans: 180, **183**
Alethopteris u rophylla: 238
Alethopteris zeilleri: 180
Alfisol, formation of: 233–34
algae
    charophycean: 51
    coal and: 8
    Courvoisiella ctenomorpha: 95
    dasyclad: 252
    in Devonian: 95
    habitats of
        Bellerophon Formation: **252**
        Catskill Formation: 95
        in Devonian: 95
        fossilization and: 44
        Massanutten Sandstone: 51–52
        Passage Creek: 51–52
        in Permian: **252**
        Rhynie Chert: 5
        in Siegenian: 5
        in Silurian: 51–52
    nematophytes and: 43
    in Permian: **252**
    in Siegenian: 5
    in Silurian: 51–52
    in Triassic: 261
    zygnematalean: 263
Alisporites: 208
Alisporites recurvus: **217–19**
Alken-an-der-Mosel: 5
Alleghenian orogeny: 150
alligators: 14, 17, 25, **27**
allochthonous, definition of: 58, 235–36
alluvial fans, characteristics of: 62
Alnus: 23–25
Alportian: **157**
Alps: **250**, 252–55, 259
Amazon Basin: 21, 25, 232
Ambelodon: 27
amber: 22, 27
Ambrona: 30
Ames Limestone: **157**, **199**
ammonoids: 225, 227
amniotes
    acclimatization by: 263
    diapsids and: 13
    habitats of: 13, 156, **186**
    in Pennsylvanian: 13, 156, **186**
    protorothyridids: 13, **14**, 171
    reptiles. *see* reptiles
    reptiliomorphs and: 9
amphibians
    Dendrerpeton acadianum: 186, **188**
    in Devonian: 9, 108
    in Eocene: 25
    habitats of
        characteristics of: 123
        in Devonian: 9, 108
        in Eocene: 25
        "failed" aestivation: 227
        in Holocene: 9
        Horton Bluff Formation: 108
        in Jurassic: 18
        in Mississippian: 108
        vs. reptiles: 13
    in Holocene: 9
    in Jurassic: 18
    lysorophoid: 227
    in Mississippian: 108
    reptiliomorphs and: 9
    in Triassic: 18
amphipods: 22
Amynodon: 25–26
analodesmatids: **141**, **144**, 150–51
Anatidae: 27–29
Andreyevka: **157**
Aneimites acadica: **106**
aneurophytes: 80
Angaran wetlands: 11
angiosperms
    Archaefructus: **19**, 20
    birches: 29, 301
    birds and: 27
    in Carboniferous: 23
    carnivorous: 22
    coal and: 23
    in Cretaceous: 19–23, 270, 281–84
    in Eocene: 24–25
    grasses: 6, 28, 151
    habitats of
        Amazon Basin: 25
        in Cretaceous: 20, 23, 270, 281–84
        in Eocene: 24–25
        Gulf Coast: 25
        in Holocene: 6, 25
        latitude of: 25
        London Clay: 25
        in Miocene: 23–24
    in Holocene: 6
    in Jurassic: **19**
    K-T event and: 23
    in Miocene: 23–24
    origin of: 19–20
    palms: 20–21, 23–27
    sedges. *see* sedges

in Tertiary: 19, 23
Angulisporites splendidus: **199**
Ankyropteris brongniartii: 131, 134
Annularia: 180, **237**
Annularia carinata: 238, 240
Annularia spicata: **242**
Antarctica
  acid rain in: 261
  Buckley Formation: **250**, **251**
  carbon cycle in: 258
  chondrites in: 256
  coal in
    maps of: **15**, **17**
    Permian: 14, **250**
    Triassic: 16
  Coalsack Bluff: **250**, **254**
  in Cretaceous: 20, 25
  in Devonian: 99, 163
  Fremouw Formation: **250**, **251**
  Graphic Peak: **250–51**, 253–55, 259
    maps of: **15**, **17**, **250**
  Mount Crean: **250**, **254**
  paleosols: 261, **262**
  in Permian
    acid rain in: 261
    chondrites from: 256
    coal in: 14
    isotopic analysis of: 258
    map of: **15**, **250**
  photograph of Granite Peak: **250**
  Portal Mountain: **250**, **254**
  Shapeless Mountain: **250**, **254**
  in Triassic: 16, **17**
anthracosaurs
  Callignethelon: 186
  in Devonian: 108
  habitats of
    in Devonian: 108
    in East Kirkton: 9
    Horton Bluff Formation: 108
    Joggins Formation: 186
    in Mississippian: 9, 108, 163, 165
    in Pennsylvanian: 186
  in Mississippian: 9, 108, 163, 165
  morphology of: 165
  in Pennsylvanian: 186
  reptiles and: 108
  seymouriamorphs: 108, 163, 165
  terrestrialization of: 165
anthracotheres: 26, 27
anthraxylon, definition of: 200
antiarch fish: 108
ants: 22
Apiculatasporites: **217–18**
Apiculatasporites saetiger: 203, **220**
Apiculatasporites spinulistratus: **217–20**
Appalachians
  in Carboniferous: 156, 284
  Casselman Formation: **199**
  Catskill Formation: 80–99, 163
  coal in: 14
  Conemaugh Group: **199**
  in Devonian: 96
  faults in: 284
  Glenshaw Formation: **199**

Hartselle Sandstone: 140–52
lepidodendrids in: 107
Maritimes Basin and: 171–72
Massanutten Sandstone: 44–55
in Mississippian: 150
Monongahela Group: 198–221
in Pennsylvanian: 205
Pittsburgh Formation: 198–221
Tuscarora Formation: 52–53
Uniontown Formation: 198, **199**, 202–3
Aptian: 253, **260**
"aquacarps:" 10
aquatic wetlands, definition of: 2
Araceae: 20
arachnids
  in Devonian: 99
  in Emsian: 5
  Graeophonus carbonarius: 186, **187**
  habitats of
    Alken-an-der-Mosel: 5
    Catskill Formation: 99
    in Devonian: 99
    in Emsian: 5
    Joggins Formation: **184**, 186
    in Pennsylvanian: **184**, 186
    in Pridolian: 44
    in Siegenian: 5–6
  mites: 5
  in Pennsylvanian: **184**, 186
  in Pridolian: 44
  scorpions. *see* scorpions
  spiders: 5, 186, **187**
  trigonotarbids: 5–6, 99
Araguainha Crater: 259
Araucariaceae: 18
araucariacean trees: 18, 22
araucarian swamps: 22
Araucarioxylon: 14, 17
Arbuckle Mountains: 224
Archaefructus: **19**, 20
Archaeocalamites
  Equisetum and: 134
  habitats of
    Fayetteville Formation: 130–35
    Horton Bluff Formation: 111–13, 122
    in Mississippian
      Blue Beach: 107, 111–13, 122
      characteristics of: 10
      Lincoln Dam area: 130–35
  in Mississippian
    habitats of
      Blue Beach: 107, 111–13, 122
      characteristics of: 10
      Lincoln Dam area: 130–35
  photograph of sample: **135**
Archaeocalamites esnostensis: 130–31
Archaeocalamites radiatus: 130–33
Archaeocalamites scrobiculatus: **106**
Archaeocetes: 27
Archaeopteris
  Callixylon and: 80, 98
  characteristics of: 80
  in Devonian
    drawing of: **7**, **9**
    extinction: 80, 104

    habitats of
      Africa: 99
      Alberta: 99
      Antarctica: 99
      Australia: 99
      Bear Island: 99
      Belgium: 98
      China: 99
      fish and: 99
      Hampshire Formation: 98
      New York (state): 99
      Pennsylvania: 98
      Red Hill: 80–81, 85, 88, 90–91, 95–97
      riparian wetlands: 9
      Russia: 99
      Siberia: 99
      South America: 99
      swamps: 7
    height of: 7, 80
  habitats of
    Catskill Formation: 80–81, 85, 88, 90–91, 95–97, 99
    in Devonian
      Africa: 99
      Alberta: 99
      Antarctica: 99
      Australia: 99
      Bear Island: 99
      Belgium: 98
      China: 99
      fish and: 99
      Hampshire Formation: 98
      New York (state): 99
      Pennsylvania: 98
      Red Hill: 80–81, 85, 88, 90–91, 95–97
      riparian wetlands: 9
      Russia: 99
      Siberia: 99
      South America: 99
      swamps: 7
    Evieux Formation: 98
    Oswayo Formation: 98
    Yahatinda Formation: 99
  height of: **4**
  Rhacophyton and: 80
  root system of: 97
  seasons, response to: 94–97
Archaeopteris halliana: **81**, 88, **89**, 98
Archaeopteris hibernica: **81**, 88, 98, 99
Archaeopteris macilenta: **81**, 88, 98
Archaeopteris obtusa: **81**, 88, 98
Archaeopteris sphenophyllifolia: 98
Archaeosperma arnoldii: 80, 98
Archanodon: **173**, 186–87, 189
Archanodon catskillensis: 186
Archanodon westoni: **186**
archeological sites, wetlands and: 30
Archer City Formation
  composition of: 227
  depositional environment of: 227, 232, 234–35, 240–41, 243
  fossils in: 240–41
  geologic time period of: **225**
  Markley Formation and: 227
  sedimentary structures in: 232, 234–35

Archer County: **225**
Archeria: 163
archosaurs: 17, 18
Arecaceae: 21, 25–26
Arenigian: **42**
Arizona
    Battleship Wash Formation: 10, 151
    Chinle Formation: 16–17
    in Mississippian: 151
    Petrified Forest: 16–17
Arkansas
    Fayetteville Formation: 127–35
    maps of: **128**
Armenia: **250**, **254**, 255
Arnsbergian. *See also* Namurian
    Greer: **157**, **159**, **162**, 164
    Hancock County (Kentucky): 156–65
aroids: 20
Arran: 190
arsinöitheres: 26
Artemia: 6
Arthrophycus: 53
Arthropleura: 5, 13
arthropods
    centipedes: 5, **6**, 22, 44
    Chelicerata: 5
    crustaceans. *see* crustaceans
    in Devonian: 5, 79, 99
    diplopods. *see* diplopods
    flight, development of: 13
    habitats of: 99
    insects. *see* insects
    in Ordovician: 5, 43
    in Pennsylvanian: 13
    terrestrialization of: 5
Arundian: **157**
Asbian: **157**, 165
ash trees: 189, **302**
Ashgillian: **42**, 52–53
Asia
    China. *see* China
    coal in
        Carboniferous: **11**
        Cretaceous: 20
        Devonian: 8
        Jurassic: 18, **19**, 29
        maps of: **11**, **15**, **17**, **19**, 20
        palms in: 24
        Pennsylvanian: 10
        Permian: 14, **15**, **250**
        Sphagnum and: 29
        Triassic: **17**
    Japan: **250**, 253, **254**
    maps of
        in Carboniferous: **11**
        coal deposits: **11**, **15**, **17**, **19**–20
        in Cretaceous: **20**
        in Jurassic: **19**
        in Miocene: **23**
        in Permian: **15**
        at Permian-Triassic boundary: **250**
        in Triassic: **17**
    Mekong River: 232
    Mongolia: 18
    Vietnam: 232

Asterocalamites scrobiculatus: **106**
asteroids. *See* meteorites
Asterophyllites equisetiformis: 236, 240
Asteroxylon: **4**, **6**
Asteroxylon mackiei: 6
Asthenodonta: 186. See also *Archanodon*
Atchafalaya Bay: 178
Athol Syncline
    geologic setting of: 171
    Joggins Formation. *see* Joggins Formation
Athrotacites: 20
Atlantic Coastal Plain: 301
Atoka Formation: **129**
Atokan: **129**
attritus, definition of: 202
Auchenreoch Glen: **157**
Australia
    Baragwanathia in: 5
    Bedout Crater: **250**, 256, 258–59
    Bonaparte Basin: 259
    Canning Basin: **250**, 256, 258–59
    in Carboniferous: 156
    coal in
        beech in: 25
        Carboniferous: **11**
        Cretaceous: 20
        geochemical analysis of: 281
        Jurassic: **19**
        maps of: **11**, **15**, **17**, **19**
        Miocene: **23**
        Permian: 14, **15**
        Triassic: **17**
    in Cretaceous: **20**, 20
    Denison: **250**, **254**
    in Devonian: 99, 108, 163
    Eddystone: **250**, **254**
    Fishburn: **250**, **254**
    in Holocene: 9
    in Jurassic: **19**
    Lagarostrobus franklinii in: 281, 286
    Lorne Crater: 259
    maps of
        Carboniferous: **11**
        coal deposits: **11**, **15**, **17**, **19**, **20**, **23**
        Cretaceous: **20**
        Jurassic: **19**
        Miocene: **23**
        Permian: **15**
        at Permian-Triassic boundary: **250**
        Triassic: **17**
    Middle Paddock: 156
    in Miocene: **23**, 25
    Murrays Run: **254**, 255, 259
    New Zealand and: 270
    in Oligocene: 25
    Paradise: **250**, **254**, 255
    in Pennsylvanian: **11**
    in Permian
        acid rain in: 261
        coal from: 14, **15**
        isotopic analysis of: **254**, 255, 258
        location of: **254**, 259
        maps of: **15**, **250**
        meteor craters: **250**, 256, 258–59
        permafrost in: 259

    South Bulli Mine: 255
    Sydney Basin: 255
    Tern: **250**, **254**
    Terrigal: 255
    tetrapod trackways in: 108
    in Triassic: **17**
    Woodleigh Crater: 259
    Wybung Head: **250**, **254**
Austria: **250**, 252–55, 259
autochthonous, definition of: 58, 235
Autunia: 239
Avalon Terrane: 81, 171
avens: 297
Avicennia: 26
Avon Gorge assemblage: 98–99
Avonport: 104–5, **105**, 107, 109–17
Axel Heiberg Island: 24

B

Babalus bubalis: 16
bacteria: 44, 51–52, 261
Baggy Beds: 99
Bahariya: 21
Bairdia: 107
Bakerstown coal bed: **199**
bald cypress: 24
Ballyheigue deposits: 99
balsam firs: **29**, 293, 297, 299–303
Baltic: 22, 25
Bangor Limestone: **141**
Banspetali: **250**, **254**, 259
Baptemys: **27**
Baragwanathia: 5
Barataria Bay: 178
Barinophyton: 98–99
Barinophyton citrulliforme: 89
Barinophyton obscurum: **81**, 89, 98
Barinophyton sibericum: **81**, 89–90, 98
Baronyx: 21
barrier islands: 151–52, 178
Barsostrobus: 98
Bashkirian: **129**. See also Morrowan; Namurian; Westphalian
basilosaurs: 26
Batesville Formation: **129**
Battery Point Formation: 42–43, 63, 74, 76
Battleship Wash Formation: 10, 151
Baxter State Park: 6, 58–76
Bay of Fundy: **105**, **171**, 190
bayberry: 23, **24**
Baylor County: **225**, 227
Bear Island: 99
"bearing-in bench:" 205–6
beavers: 29–30
Bedout Crater: **250**, 256, 258–59
beech: 20, 25
bees: 22
beetles: 22
Belarus: 10
Belgium: 20, 88, 98
Bellerophon Formation: **252**
bennetites: 18
Benwood Limestone: **199**
Bernissart: 20
berthierine nodules: 261

Bold numbers indicate material in figures and tables.

Bethulie
    isotopic analysis of: **254**, 255, 259
    map of: **250**
    at Permian-Triassic boundary: **254**, **262**
    sedimentary structures in: **262**
Betula: 29, 301
Betulaceae: 20, 25
bicarbonate: 22
Big Bone Lick: 30
Big Sewickley Creek: 200
Bijou Creek: 190
billabongs: 10
birches: 29, 301
birds: 27–28, 263
Birmingham anticlinorium: 140
bivalves
    Archanodon: **173**, 186, 189
    coal and: 191
    in Cretaceous: 21
    Edmondia: **141**, **144**, 150–51
    habitats of
        barrier islands: 151
        in Cretaceous: 21
        endobyssate: 151
        euryhaline: 151
        Forty Brine Seam: 172
        Hartselle Sandstone: 142, 144, 150–51
        Joggins Formation: 172, 174, 186
        Lozier Pond: 294
        in Mississippian: 142, 144, 150–51
        in Pennsylvanian: 172, 174, 186
        St. Catherines Island: 151–52
        Trout Valley Formation: **60–61**, 67
    in Mississippian: 142, 144, 150–51
    Naiadites: **186**
    in Pennsylvanian: 172, 174, 186
    Permian-Triassic boundary and: 256
    Streblochondria: **144**, 150–51
    Streblopteria: 150–51
    Weichselia reticulata and: 21
black flies: 22
black gum: 25
black spruce: 29
Black Warrior Basin: 140–52, 190, 192
bladderworts: 22
Blaine Formation
    composition of: 224, 229
    depositional environment of: 229–30, **234**, 239–40, 243–44
    fossils in: 225, 229, 239
    geologic time period of: **225**
    photographs of samples: **240**
    San Angelo Formation and: 224–25, 229
    sedimentary structures in: 230, **234**
Bloomsburg Formation: 43, 47
Bloyd Formation: **129**
Blue Beach: 104–5, 107, 109–17, 122
Bluefield Formation: 132
"boghead" coal: 8
bogs. *See also* forest mires
    archeological sites in: 30
    carnivorous plants in: 22
    in Cretaceous: 20
    definition of: 5, 9
    in Devonian: 5, 9

    vs. fens: 9, 292
    in Jurassic: 18
    latitude of: 14–15
    in Miocene: 23
    in Ordovician: 3
    origin of: 9
    in Silurian: 3
Bonaparte Basin: 259
Boone Formation: **129**
Borden Formation: 151
Boss Point Formation: 176
Bothrodendron: 180, 187, **189**
Bothrodendron punctatum: **182**
Bothrodendrostrobus: 149, 150
Bowie Group
    Albany Group and: 224
    Archer City Formation. *see* Archer City Formation
    Cisco Group and: 224
    depositional environment of: 224, 230–34, 240–41
    fossils in: 236–38, 240–41
    geologic time period of: 225
    Markley Formation: 225–27, 230–34, **238**, 240–41
    sedimentary structures in: 230–34
brachiopods: 151, 256, 260
branchiopods: 6
Brazil
    Amazon Basin: 21, 25, 232
    Araguainha Crater: 259
    Pantanal River: 21
    tetrapod trackways in: 108
Brigantian
    Delta: **157**, **159**, **162**, 164
    East Kirkton: 9–10, 156–**57**, **159**, 164–65
    Gilmerton: **157**
bristletail: 5
Bristol: 98–99
British Columbia: 300
Broiliellus: 161
Brongniarites: 239
Broxburn: **157**
Bruguiera: 26
Brunner Coal Measures: **271**, 284
Brush Creek coal bed: **199**
Brush Creek Limestone: **199**
Bryoerythrophyllum recurvirostre: **297**, 298
bryophytes
    in Devonian: **68**
    habitats of
        bogs vs. fens: 292
        in Devonian: **68**
        Massanutten Sandstone: 51
        in Ordovician: 3
        Passage Creek: 51
        peats and: 300
        in Permian: 14
        in Silurian: 3, 51
        Trout Valley Formation: **68**
    in Holocene: 14
    hornworts: 20, 43, 51
    liverworts: 43, 51
    in Ordovician: 3
    origin of: 3, 43

    in Permian: 14
    polysporangiophytes and: 43
    in Silurian: 3, 51
    Sporongonites: **68**
bryozoans: 260–61
Bryum: **297**
Buckley Formation: **250**, **251**
buffalo: 16
Buffalo Wallow Formation: 158–60
bulrushes: 28, 300
Burma: 22
Burning Tree site: **302**
bushes: 6
butterworts: 22

C

Cabombaceae: **19**, **20**, **26**, **29**
Cacops: 161
caddis flies: 22
calamiteans. *See also* neocalamites
    Annularia: 180, **237**
    Asterophyllites equisetiformis: 236, 240
    Calamites. *see Calamites*
    in Devonian: 6
    drawing of: **12**
    growth strategies of: 245
    habitats of
        Black Warrior Basin: 192
        Bowie Group: 236–38
        Cisco Group: 236–38
        Clear Fork Formation: 241
        in Devonian: 6
        Joggins Formation: 171–80, 189, 191–92
        in Mississippian: 9, 161
        in Namurian: 192
        in Pennsylvanian
            characteristics of: 10–11
            Nova Scotia: 171–80, 189, 191–92
            Texas: 236–38, 240
            West Virginia: 203–13, **214–15**, **217–21**
        in Permian: 244–45
        Pittsburgh Formation: 203–13, **214–15**, **217–21**
        Ruhr Basin: 192
    Laevigatosporites: 282
    Mesocalamites: 10
    in Mississippian: 9, 161
    Montrichardia aborescens and: 189
    morphology of: 189
    in Namurian: 192
    in Pennsylvanian
        habitats of
            characteristics of: 10–11
            Nova Scotia: 171–80, 189, 191–92
            Texas: 236–38, 240
            West Virginia: 203–13, **214–15**, **217–21**
        height of: 11
        reproductive strategies of: 11
    in Permian: 244–45
    photograph of in situ: **176**
    photographs of samples: **242**
    regeneration by: 75
    reproductive strategies of: 11, 161, 236
Calamites
    drawing of: **12**

Equisetum and: 245
  habitats of: 161, 180
  in Mississippian: 161
  in Pennsylvanian: 180
  photographs of samples: **242**
  regeneration by: 75
  reproductive strategies of: 161
Calamites carinatus: 180
Calamites suckowi: 180
Calamospora: 160–61, 204
Calamospora breviradiata: **220–21**
Calamospora flexlis: **214–15**
Calamospora hartungiana: **217–18**
Calamospora microrugosa: **217–18**, 220
Calamospora pallida: **214–16, 219–20**
Calamospora pedata: **214–15**, 220
Calamus: 25
Calathiops: 80
Caledonia Highlands: 171
Calhoun County: **302**
California: 44
Calliergon giganteum: 297, 300
Calliergonella cuspidata: 297, 300
Calligenethlon: 163
Callignethelon: 186
callipterids: 239–40
Callitriche hermaphroditica: **296**
Callixylon: 80, 98
Callixylon erianum: 98
Cambrian: **42**, 43
Cameroon: 258
Camishaella: 107
Campylium chrysophyllum: **297**
Campylium stellatum: 297, 300
Canada
  Alberta: 99
  amber deposits in: 22
  Axel Heiberg Island: 24
  Bay of Fundy: **105, 171**, 190
  British Columbia: 300
  Chignecto Bay: **171**
  coal in: 10
  Cobequid Bay: **171**
  Ellesmere Island: 99
  Horton Group: 103–23, **157**
  Hudson Bay area: 14, 29
  Maritimes Basin. *see* Maritimes Basin
  Melville Island: 98, 99
  in Mississippian: 10
  New Brunswick. *see* New Brunswick
  Newfoundland. *see* Newfoundland
  Nova Scotia. *see* Nova Scotia
  Ontario: 292
  in Pennsylvanian: 10
  Quebec. *see* Quebec
  wetland system: 2
  Williston Lake: **250, 254**
  Yukon Territory: 300
Canning Basin: **250**, 256, 258–59
Canobius: 108
Cap-aux-Os: 76
capybaras: 29
Caradocian: **42**
carbon cycle
  in Devonian: 8–9

  in Jurassic: 18
  latitude and: 257–58
  mires and: 9
  nuclear emissions and: 258
  Permian-Triassic boundary and: 250–58, 260–61
  photosynthesis and: 28, 257
  root systems and: 8
  in salt marshes: 28
  sedges and: 28
  Triassic-Jurassic mass extinction and: 18
carbon dioxide
  coal and: 258
  humans and: 253
  from Lake Nyos: 258
  methane and: 253–61
  nuclear emissions and: 258
  Permian-Triassic boundary and: 57, 250–61, 263
  plants and
    at Permian-Triassic boundary: 57, 261
    photosynthesis and: 28, 257
    root systems and: 8
  Siberian Traps and: 258
  from volcanoes: 258
  wetlands and: 261
carbonate: 243, 250–57, 260–61
Carbondale Formation: 147–48
Carboniferous. *See also* Mississippian; Pennsylvanian
  atmospheric oxygen levels in: 190
  coal from
    Forty Brine Seam: **171**, 172
    Joggins Formation: 170–84, 186–87, 189–92
    Lycospora and: 160
    map of: **11**
    Springhill Mines Formation: 172, 176, 191
  Cumberland Basin. *see* Cumberland Basin
  East Kirkton: 9–10, 156–**57**, 159, 164–65
  herbivory in: 80
  Horton Group: 103–23, **157**
  "Icehouse:" 190
  Joggins Formation. *see* Joggins Formation
  maps of: **11, 158**
  Romer's Gap: 104, **157**, 164
  tetrapods in: 156
  wetland evolution in: 6–11, 13–17
Carbonita: 107
carbonitacean ostracods: 107
Cardiopteridium: 132, **133**
Cardiopteridium hirta: 134
Carex: 28, **29**
Carex achenes: **296**
Carex aquatilis: 296, 300
Carex vesicaria: 296, **297**
Carleton Heights: **250, 254**
carnivorous plants: 22
Carolina Beach: **152**
Carpolithus tenellus: **106**
cartilaginous fish. *See* sharks
Carya: 292
Caseyville Formation: **159**
Casineria: 165
Casselman Formation: **199**
Castoridae: 29–30

Castoroides ohioensis: 30
Castracollis: 6
catenas: 21
Cathaysia: 224
Cathaysiopteris: 239
Catskill Delta Complex: 81, 97
Catskill Formation: 80–99, 163
cattails: **27**, 28, 296, **297**
centipedes: 5, **6**, 22, 44
Central America
  coal in: **19**, 20
  K-T event and: 23
  maps of: **19, 20**
  Mexico: 18, 22
cephalopods: 128, 227, 229
Cephalopteris mirabilis: 99
Ceratodon purpureus: **297**
Ceratodus: 17
Ceratophyllaceae: 20, 43, 51
Ceratophyllales: **19**, 20
Ceriops: 26
Chadian: **157**
Chaloneria
  Endosporites globiformis and: 208–10, 213, **216–19**
  habitats of: 10, 208–10, 213, **216–19**
  Hartsellea and: 148–50
  in Mississippian: 10
  in Pennsylvanian: 208–10, **216–19**
Chaloneria cormosa: 148–50
Chaloneria periodica: 149
Chamovnicheskian: **157**, 199
Changsing Limestone: **250**, 251, **262**
charadriiform shorebirds: 27
Charcharodontosaurus: 21
charcoal
  in Bowie Group: 236
  in Catskill Formation: 90, **93**, 94–98
  in Cisco Group: 236
  in Joggins Formation: 176, 179, 182, **184**, 189–90
  in Pittsburgh Formation: 205
Charliea: 240
charophyceans: 51
Cheese Bay: **157**, 165
Cheiledonites major: **199**
Cheirolepidiaceae: 18
Chelicerata: 5
Cheremshanskian: **157**
Chesterian. *See also* Elvirian; Gasperian; Hombergian; Namurian; St. Genevieve time period; Serpukhovian
  Batesville Formation: **129**
  Buffalo Wallow Formation: 158–60
  climate in Kentucky: 160
  Fayetteville Formation: 127–35
  Greenbrier Formation: 140, 146–48, 150–51
  Hancock County (Kentucky): 156–65
  Hartselle Sandstone: 140–52
  Hindsville Formation: **129**
  Menard Limestone: 158–60
  Palestine Sandstone/Shale: 158–61
  Pitkin Formation: **129**
chestnut trees: 25
Cheverie Formation: **105–6**

Bold numbers indicate material in figures and tables.

Chignecto Bay: **171**
Chile: 30
Chilopoda: 5, **6**, 22, 44
China
    Changsing Limestone: **250, 251, 262**
    coal in
        Carboniferous: **11**
        Cretaceous: 20
        Devonian: 8
        Jurassic: 18, **19**, 29
        maps of: **11, 15, 17, 19, 20**
        Pennsylvanian: 10
        Permian: 14, **15**, 250
        Sphagnum and: 29
        Triassic: **17**
    in Devonian: 99
    in Eocene: 27
    Helongshan Formation: **250**
    Heping: **250**, 254
    maps of
        Carboniferous: **11**
        coal deposits: **11, 15, 17, 19, 20**
        Cretaceous: **20**
        Jurassic: **19**
        Permian: **15**
        at Permian-Triassic boundary: **250**
        Triassic: **17**
    Meishan: **250–51**, 253–55, 259, 261–62
    in Mississippian: 10
    in Pennsylvanian: 10, 15, 224
    in Permian: 15, 224
    at Permian-Triassic boundary: **250–51**, 253–55, 259, 261–62
    Shangsi: **250**, 254
    Taiping: **250**, 254
    Yellow River: 232
    Yinkeng Formation: **250, 251, 262**
Chinle Formation: 16–17
chitinozoans: 44
Chlidanophyton dublinensis: 98
Chokierian. *See* Hancock County (Kentucky)
chondrichthyans: 9
chondrites: 256
Christensen site: **302**
Chunerpeton: 18
Ciconiidae: 26, 27
Cimmeria: **15, 19**
Cinclidium: **297**
Cisco Group: 224, **225**, 227, 236–38
Cisuralian
    Archer City Formation. *see* Archer City Formation
    Blaine Formation. *see* Blaine Formation
    Bowie Group. *see* Bowie Group
    Clear Fork Formation. *see* Clear Fork Formation
    Lueders Formation: **225**, 227, **231**
    Markley Formation: 225–27, 230–34, **238**, 240–41
    Nocona Formation. *see* Nocona Formation
    Pease River Group. *see* Pease River Group
    Petrolia Formation. *see* Petrolia Formation
    San Angelo Formation. *see* San Angelo Formation
    Waggoner Ranch Formation. *see* Waggoner Ranch Formation

cladoxylaleans: 80
"clam coals:" 172
Claraia griesbachi: **262**
Clarkina changsingensis: **251**
Classopollis-type pollen: 18
clastic swamps: 11, 108, 179
Clavifera: 282
Clay County, map of: **225**
claystone, definition of: 271
Clear Fork Formation
    composition of: 228–29
    depositional environment of: 224, 227–29, 234–35, 243–44
    fossils in: 227–29, 236, 239, 241
    geologic time period of: **225**
    photographs of samples: **239, 242**
    photographs of sites: **228–29, 231**
    sedimentary structures in: 234–35
Clementstone deposits: 99
Clevelandodendron: 149
Clinton County, map of: **81**
Clore Limestone: 158, 160
club mosses: 26
coal
    amber and: 22
    ash
        Joggins Formation: 191
        "Main Seam:" 270, 278–82, **284**, 285–86
        Pittsburgh Formation: 198, 204–13, **214–21**
        Powder River Basin: 287
    attributes of: 271
    bivalves and: 191
    "boghead:" 8
    carbon dioxide from: 258
    Carboniferous
        Forty Brine Seam: **171**, 172
        Joggins Formation: 170–84, 186–87, 189–92
        Lycospora and: 160
        map of: **11**
        Springhill Mines Formation: 172, 176, 191
    "clam coals:" 172
    "coal balls:" 160–61
    conifers and: 263
    Cretaceous: 20, 29, 270–88
    cycads and: 263
    Devonian: 8, 97, 98
    Eocene: 23, 26
    Famennian: 98
    formation of
        algae and: 8
        angiosperms and: 23
        "boghead:" 8
        faults and: 287
        flooding and: 284, 287–88
        iridium and: 23
        peat and: 8, 11–13, 287
    Frasnian: 8, 98
    geochemical analysis of: 279–81
    Givetian: 98
    Jurassic: 18
    K-T event and: 23
    Lycospora and: 160
    maps of: **11, 15, 17, 23**
    methane from: 255–57
    Miocene: **23**

    Mississippian: 10, 107, 159–61
    Pennsylvanian: 10–14
        in Bowie Group: 227, 230, 236
        in Cisco Group: 236
        in Illinois Basin: 243
        in Pittsburgh Formation: 198–213
    Permian: 14–15
        in Bowie Group: 227
        Triassic boundary: 16, 249–56, 258–59, 261–63
    rheotrophic: 190–92
    rider zone in: 197
    root systems and: 160–61
    sapropelic: 8
    successions in: 11–13, 23
    sulfur in: 191, 198, 204–13, **214–21**
    thickness of: 23
    Triassic
        map of: **17**
        Permian boundary: 16, 249–56, 258–59, 261–63
        in Tunguska Basin: 258
"coal balls:" 160, 161
Coal Mine Point: 170–71
Coalsack Bluff: **250**, 254
coastal wetlands
    amber deposits in: 22
    Battleship Wash Formation: 10, 151
    in Cretaceous: 21–22
    crocodilians in: 17, 27
    in Devonian: 6, 7, 58–76
    in Eocene: 22, **24**, 25
    Greenbrier Formation: 140, 146–48, 150–51
    Hancock County (Kentucky): 156–65
    Hartselle Sandstone: 140–52
    Joggins Formation. *see* Joggins Formation
    in Jurassic: 18–19
    in Llandoverian: 52–53
    Menard Limestone: 158–60
    in Mississippian: 10, 140–52, 151, 158–60
    Newman Limestone: 151
    Pittsburgh Formation: 198–221
    plant morphology: 6–7, 19
    ravinement of: 140
    vs. tidal flats: 139–40
    Trout Valley Formation: 6, 58–76
    Tuscarora Formation: 52–53
    whales in: 27
Cobequid Bay: **171**
Cobequid Fault zone: 171
Cobequid Highlands: 171–72, 191
cockroaches: 13
Coelophysis: 17
Colorado
    Bijou Creek: 190
    Green River Formation: 27
    South Platte River: 53
Colorado River Valley: 227
colosteids: **159**, 161–62, 164–65
Colosteus: 161
comets. *See* meteorites
Comia: 239
Complexisporites polymorphus: **217–18**
Compsopteris: 239
Condrusia: 98

Conemaugh Group: **199**
Conemaughian. *See also* Chamovnicheskian; Dorogomilovskian; Krevyakinskian
    Ames Limestone: **157**, **199**
    Garnett: **157**
    Howard: **157**
    Sangre de Cristo: **157**
conifers
    Abietites: 20
    amber and: 22
    Araucariaceae: 18
    Araucarioxylon: 14, 17
    Athrotacites: 20
    black spruce: 29
    Cheirolepidiaceae: 18
    coal and: 263
    cordaites and: 11
    in Cretaceous: 20, 22–23
    Culmitzschia: 239
    Ernestiodendron: 239
    glaciation cycles and: 29
    habitats of
        bogs vs. fens: 292
        in Cretaceous: 20, 22–23
        in Holocene: 29
        in Jurassic: 190
        Lozier Pond: 294
        in Miocene: 23
        Navajo Sandstone: 190
        Pease River Group: 239
        in Pennsylvanian: 213, 243
        in Permian: 239–40
        swamps: 22
        in Triassic: 16–17, 257
        in Westphalian: 243
    in Jurassic: 18, 190
    K-T event and: 23
    mastodons and: 303
    Metasequoia: 20, 24
    in Miocene: 23
    Moriconea: 20
    origin of: 243
    in Pennsylvanian: 213, 243
    in Permian: 239–40, 263
    photograph of sample: **239**
    Pityoxylon: 22
    in Pliocene: 29
    podocarpaceous: 18, 20, 25
    Podozamites: 20, 239
    Protophyllocladus: 20
    seasons, response to: 236
    Sequoia: 20, 23, **24**
    tamarack: 29, 297, 301–3
    taxodiaceous: 18, 20, 24
    tree ring analysis of: 190
    in Triassic: 16–17, 257, 263
    Walchia: 239
    in Westphalian: 243
Coniopteris: 18
conodonts
    Clarkina changsingensis: **251**
    habitats of: 128, **251**, 253, 259
    Hindeodus parvus: **251–52**, 253, 259, **262**
    Hindeodus typicalis: **251**
    Isarcicella isarcica: **251**
    in Mississippian: 128
    in Permian: 253
    in Triassic: 259
Convolutispora: **217–18**
Cooksonia: 4, **42**, 43
Copelandella: 107
copper: 229
coral reefs: 22, 249, 260–61
cordaites
    Cordaites palmaeformis: 180
    Cordaites principalis: 180
    drawing of: **12**
    Florinites and: 204, 208
    habitats of
        conifers and: 11
        Joggins Formation: **189**, 190
        in Pennsylvanian
            China: 15
            conifers and: 243
            mires: 11
            Nova Scotia: 190
            West Virginia: 203–6, 208–11, **214–20**
        in Permian: 15, 239, 263
        Pittsburgh Formation: 203–6, 208–11, **214–20**
    mangroves and: 11
    in Pennsylvanian
        habitats of
            China: 15
            conifers and: 243
            mires: 11
            Nova Scotia: 190
            West Virginia: 203–6, 208–11, **214–20**
        root system of: 11
    in Permian: 15, 239, 263
    root system of: 11
    ruflorian: 15, 263
    seasons, response to: 236
    tree ring analysis of: 190
    voynovskyalean: 15
cordgrass: 28, 151
Cormophyton mazonensis: 147–48
cormose lycopsids
    Chaloneria. see *Chaloneria*
    classification of: 149
    Cormophyton mazonensis: 147–48
    in Devonian
        habitats of: 80
            Red Hill: 87, 90, 95, 97
        root system of: 87
    habitats of
        Carbondale Formation: 147–48
        Catskill Formation: 87, 90, 95, 97
        in Devonian: 80
            Red Hill: 87, 90, 95, 97
        Greenbrier Formation: 140, 146–47
        Hartselle Sandstone: 140–45, 147–50
        Illinois Basin: 147–48
        Mazon Creek: 148
        in Mississippian: 10, 140–50
    Hartsellea: **141**, 143–45, 148–51
    in Mississippian: 10, 140–50
    origin of: 149
    in Pennsylvanian: 147–48
    photographs of samples: **86**, **143**

Cornus: 23
Coryphodon: 26
corystosperms: 16
cottonwoods: 25
County Kerry: 99
County Kilkenny: 99
County Wexford: 99
Courvoisiella ctenomorpha: 95
Cowardin classification: 3
Cowdenbeath: **157**
Crassigyrinus: **162**, 164
Crassispora kosankei: 204–5, 207–8, 212–16, **220**
Crean, Mount: **250**, **254**
Cretaceous. *See also* Haumurian; Maastrichtian; Motuan
    amber deposits: 22
    carnivorous plants in: 22
    Greymouth Basin: 270–88
    maps of: **20**
    methane release during: 253, **260**
    Pororari Group: 284
    Tertiary boundary event. *see* K-T event
    wetland evolution in: 19–22
crocodilians
    in Cretaceous: 21, 23
    in Eocene: 26
    habitats of: 14, 17–18, 21, 26
    in Jurassic: 18
    labyrinthodonts and: 15, 17
    Machimosaurus: 18
    phytosaurs and: 17–18
    rhinesuchids and: 15
    suchians and: 17–18
    in Tertiary: 23
    Weichselia reticulata and: 21
    whales and: 27
crossopterygians: 108
Crurithyris flabelliformis: **262**
crustaceans: 5–6, 22
Cruziana: 107, 108
cryptospores: **42**, 44
ctenacanthid sharks: 99
Ctenodus: 108
Culicidae: 22
Culmitzschia: 239
Cumberland Basin
    faults in: 171
    fossils in: 180
    geologic setting of: 171–72
    Joggins Formation. *see* Joggins Formation
    Little River Formation: **171**, 172, 176
    seasonal changes in: 190
    Springhill Mines Formation: 172, 176, 191
    stratigraphy of: 172
    subsidence of: 171, 178–79
Cumberland Plateau: 140
Cunninghamia: **24**
Çürük Dağ: **250**, **254**, 255
cuticle-like fragments: **42**, 44, 52
cyanobacteria: 44, 51–52, 261
Cyatheadities: 282
cycadeoids: 16, 18
cycadophytes: 16, 19, 20
cycads: 15–18, 239, **240**, 263
Cyclogranisporites aureus: **214–15**

Bold numbers indicate material in figures and tables.

Cyclogranisporites microgranus: 214–20
Cyclogranisporites minutus: 214–15, **217–19**
Cyclogranisporites obliquus: **199**
Cyclostigma: 8, 87, 99, 149
Cyclostigma kiltorkense: 99
Cyclostrobus: 16
Cyclostrobus sydneyensis: **255**
Cymodocea: 22
Cyperaceae
    Carex: 28, **29**
    Cyperacites: 28
    Cyperus: 28
    in Eocene: 26
    habitats of: 26, 28
    Scirpus: 28
    sedges. *see* sedges
Cyperacites: 28
Cyperites: 180
Cyperus: 28
Cyperus papyrus: 28, 30
cypress: 24, 26, 108
cypress swamps: 108
cyptospores: 42
Cyrilliceae: **24**
Cystosporites: 182
Cystosporites diabolicus: **184, 185**

D

Dadoxylon: 14
Danian: 253, **260**
dasyclad algae: **252**
Dauphin Island: **152**
Dawson, William: 177
decapod crustaceans: 22
Deinosuchus: 18
Delnortea: 239
Delta: **157, 159, 162**, 164
Deltoidospora: 18
Dendrerpeton acadianum: 186, **188**
dengue fever: 22
Denison: **250, 254**
Densignathus rowei: 99
densopores: 172
desmocollinite: 202
Devonian. *See also* Eifelian; Emsian; Famennian; Frasnian; Gedinnian; Givetian; Lochkovian; Pragian; Siegenian
    Avon Gorge assemblage: 98–99
    Baggy Beds: 99
    Ballyheigue deposits: 99
    Battery Point Formation: 42–43, 63, 74, 76
    carbon cycle in: 8–9
    Carboniferous boundary mass extinction: 8, 80, 104
    Catskill Formation: 80–99, 163
    coal from: 8, 97, 98
    Evieux Formation: 98
    fossils from: **42**
    glaciation cycles: 8
    herbivory in: 79
    Hook Head deposits: 99
    Horton Bluff Formation: 103–17, 120–23, **157**
    Kiltorcan deposits: 99
    maps of: **158**
    marine-bottom anoxic events in: 58
    mass extinction: 8
    Memramcook Formation: **105**
    miospore zones in: **105**
    nitrogen cycle in: 8
    Oswayo Formation: 98
    plant partitioning in: 7, 9, 80
    Red Hill: 80–99, **157**, 163
    Rhynie Chert: 5–6, 42–43
    Taffs Well assemblage: 98
    Trout Valley Formation: 6, 58–76
    wetland evolution in: **4**, 5–9
diadectomorphs: 165
Diaphanospora parvigracilis: **216, 220**
Diaphorodendron
    Cystosporites diabolicus and: **185**
    habitats of: 180, 183, 187, **189**
diapsids: 13
Dicksoniaceae: 18
dicots: 23
dicotyledons: 29
Dicranella schreberiana: 297–98
Dicranophyllum: **240**
Dicroidium: 16
Dicroidium zuberi: 259
Dictyotriletes danvillensis: **217**
Dicynodon: 16
Dicynodon lacerticeps: **262**
dicynodonts: 15–17, 263
dinoflagellates: 44
dinosaurs: 17–18, 20–21, 23
Dioonitocarpidium: 239, **240**
Diplocraterion: **60–61**, 65–66
diplopods: 5, 13, 22, 186, **187**
Diplotmema patentissimum: **106**
Diploxylon: 179, **180**
dipnoans
    Ceratodus: **17**
    Ctenodus: 108
    in Devonian: 163
    Gnathorhiza: 163
    habitats of: **159**, 160–64
    in Mississippian: **159**, 160–64
    Neoceratodos forsteri: 9
    in Pennsylvanian: 163
    Protopterus: 9
    Tranodis: **162**, 163–64
Dipoides: 29
Diptera: 22
Dissorophus: 161
Distichium: **297**, 298
Distichlis: 28
Ditrichum flexicaule: **297**, 298
Docodonta: 18
dogwood: 23
Dominican Republic: 22
Dorinnotheca streelii: 89
Dorogomilovskian: **157**
douglastownese-eurypterota spore assemblage zone: 58
dragonflies: 13, 22
Drepanocladus aduncus: 297, 300
Drepanocladus sendtneri: 300
Drepanocladus sordidus: 297, 300
Drepanophycus: **68**, 70
Drepanophycus gaspianus: **68**
Drosera: 22
Droseraceae: 22
Dry Brook, map of: **59**
Dryas integrifolia: **296**, 297, 300
Duckmantian. *See* Cumberland Basin
ducks: 27–29
duckweed: **19**, 20, 27
dugongs: 27
Dunollie: **271**
Dutchess County: 292–303

E

Eagle Hill Camp Bog: 292
Earlandia: **252**, 253
earthquake, New Madrid: 190
East Greenland: **157**
East Kirkton: 9–10, 156–**57**, **159**, 164–65
Eastern Interior Basin: 156–65. *See also* Illinois Basin
ecotones: 7
Eddystone: **250, 254**
Edmondia: **141, 144**, 150–51
"egg crate" texture: **118**
Egypt
    Fayum: 26–27
    Nile River Valley: 30, 232
Eichhornia: **19**
Eifelian
    fossils from: **42**
    height of plants in: **4**
    morphology in, plant: 6
    Trout Valley Formation: 6, 58–76
elasmobranch sharks: 108, 123
Elatides: 18
Eldeceeon: 165
Eldridge site: **302**
Eleocharis palustris: **296**
elephants: 27, 30. *See also* mammoths; mastodons
Elk Lick coal bed: **199**
Elkins: 80, 90, 97–98
Elkinsia polymorpha: 88, 98
Ellesmere Island: 99
Elm Creek Limestone: 225
Elonichthys: 108
Elvirian
    Buffalo Wallow Formation: 158–60
    Goreville: **157, 159, 162**, 164
    Hancock County (Kentucky): 156–65
    Menard Limestone: 158–60
    Palestine Sandstone/Shale: 158–61
Emarat: **250, 254**
embolomeres: **159**, 162–65
embryophytes: 3, 43, 52, **68**
Emsian
    Alken-an-der-Mosel: 5
    arthropods in Canada: 5
    Battery Point Formation: 42–43, 63, 74, 76
    fossils from: **42**
    height of plants in: **4**, 6
    Trout Valley Formation: 6, 58–76
Encalypta: **297**
encephalitis: 22
Endosporites globiformis: 208–10, 213, **216–19**
Endosporites globosus: 150
England

amber deposits in: 22
Bristol: 98–99
coal in: 10
East Kirkton: 9–10, 156–**57**, **159**, 164–65
in Eocene: 26
London Clay: 25, 26
North Devon: 99
in Pennsylvanian: 10
Eocene. *See also* Ypresian
   Brunner Coal Measures: **271**, 284
   coal from: 23, 26
   global warming in: 24, 26, 258
   Paleocene boundary: 253, 258
   wetland evolution in: **19**, 21–27
Eoherpeton: 164
Eohostimella: 43
Eospermatopteris: **4**, 7
Ephemeroptera
   flight process of: 13
   mayflies: **13**, 22
Epihippus: 25
epiphytic plants: 212
equisetales
   Archaeocalamites scrobiculatus: **106**
   habitats of: **106**
   horsetails. *see* horsetails
   Nematophyllum: **106**
Equisetum
   Archaeocalamites and: 134
   Calamites and: 245
   in Cretaceous: 20
   habitats of: 10, 17, 18, 20
   in Holocene: 10
   in Jurassic: 18
   in Pennsylvanian: 10
   in Triassic: **17**, 17
Ericaceae: **24**
Erieopterus: 67
Ernestiodendron: 239
Eryops: 15
estuarine wetlands
   in Cretaceous: 22
   crocodilians in: 17–18
   in Devonian: 63, 65–68, **71**–76, 99
   in Eocene: 22, **24**, 25
   Hancock County (Kentucky): 156–65
   in Jurassic: 18
   mangroves and: 21
   in Mississippian: 140–52, 156–65
   Palestine Shale: 158–61
   in Pennsylvanian: 13
Etapteris: 131
Eucalamites: 180
Eucritta: 165
Euphrates River: 30
Europe
   Acadian orogen: 58, 81, 104
   Alps: **250**, 252–55, 259
   archeological sites in: 30
   Austria: **250**, 252–55, 259
   Belarus: 10
   Belgium: 20, 88, 98
   coal in
      Carboniferous: **11**
      Cretaceous: **20**

Jurassic: **19**
maps of: **11**, **23**
Miocene: **23**
palms in: 24
Pennsylvanian: 10
Permian: **15**
Sphagnum and: 29
Triassic: **17**
in Cretaceous: **20**
France: 13
Germany. *see* Germany
Griesbachian event: **252**, **260**
Hungary: 26
Ireland: 99
Isle of Wight: 21
Italy: **250**, **254**
in Jurassic: **19**
Latvia: 108
maps of
   Carboniferous: **11**
   coal deposits: **11**, **15**, **17**, **19**, **20**, **23**
   Cretaceous: **20**
   Jurassic: **19**
   Miocene: **23**
   Permian: **15**, **225**
   at Permian-Triassic boundary: **250**
   Triassic: **17**
in Miocene: **23**, 24
in Mississippian: 10
Norway: 99
in Oligocene: 24
in Pennsylvanian: 10, **11**
in Permian: **15**
Portugal: 18
Slovenia: **250**, **254**
Spain: 30, 190
in Triassic: **17**
United Kingdom. *see* United Kingdom
europium: 259
eurypterids: 5, 44, **60–61**, 67
euthycarcinoids: **6**
Everglades: 17
Evieux Formation: 98
Eviostachya: 98
Evolsonia: 239

F

Fabasporites. See *Punctatisporites minutus*
Fagaceae: **24**, 25, **29**, 292–93
Fagus: 25
fairy shrimp: 6
Fall Kill stream: 292, **293**, 299, 301
Famennian: **157**. *See also* Strunian
   Andreyevka: **157**
   Catskill Formation: 81–99, 163
   coal from: 98
   Forbes: **157**
   Hampshire Formation: 80, 90, 97–98
   height of plants in: **4**
   Horton Bluff Formation: 103–17, 120–23, **157**
   Ketleri: **157**
   maps of: **158**
   Memramcook Formation: 105
   miospore zones in: **105**
   Ningxia Hui: **157**

Pavari: **157**
Red Hill: 80–99, **157**, 163
Fayetteville Formation: 127–35
Fayum: 26–27
fenestellid bryozoans: 261
fens
   vs. bogs: 9, 292
   calcium levels in: 300
   carbon cycle and: 9
   carnivorous plants in: 22
   coal from: 8
   conductivity of: 300
   in Cretaceous: 20
   definition of: 2, 8
   in Devonian: 8
   in Eocene: 24
   groundwater and: 292, 300
   in Holocene: 13
   Hyde Park Basin: 292–303
   indicator species: 300
   in Miocene: 23, 24
   in Mississippian: **4**
   in Ordovician: 3
   pH level in: 292, 300
   in Silurian: 3
ferns
   Acrostichum: 21, **24**, 26
   Ankyropteris brongniartii: 131, 134
   biomass of: 241
   coal and: 14
   Coniopteris: 18
   in Cretaceous: 19–20, **21**, 23, 281–82
   depositional environment of: 212
   in Devonian: 80, 87–88
   in Eocene: 25, 26
   Etapteris: 131
   filicalean: 15, **189**
   filicopsids. *see* filicopsids
   Gillespiea randolphensis: **81**, 85–90, **92**, 95–98
   Gleichenia: 281–88
   growth strategies of: 11, 241
   habitats of
      Bowie Group: 238
      in Carboniferous: 14
      Catskill Formation: 87–88
      Cisco Group: 238
      Clear Fork Formation: 241
      in Cretaceous: 19–20, 281–82
      in Devonian: 80, 87–88
      in Eocene: 25, 26
      Fayetteville Formation: 130–35
      Joggins Formation: 180, 190
      in Jurassic: 18–19
      Messel: 26
      in Miocene: 25
      in Mississippian: 10, 130–35
      in Namurian: 10
      in Oligocene: 25
      in Pennsylvanian
         Archaeopteris and: 80
         coal and: 14
         mires: 13–14
      Nova Scotia: 180, 190
      Pennsylvania: 80
      Texas: 236–38, 240

Bold numbers indicate material in figures and tables.

West Virginia: 203–13, **214–21**
in Permian: 14, 236, 240–41
Petrolia Formation: 241
Pittsburgh Formation: 203–13, **214–21**
Silesian Basin: 10
in Triassic: 16–17, 257
Hausmannia: 19
in Jurassic: 18–19
K-T event and: 23
lyginopterids. *see* lyginopterids
manoxylic wood production by: 131
Marsileaceae: 19
Medullosa steinii: 131
medullosan: 190
in Miocene: 25
in Mississippian: 10, 130–35
in Namurian: 10
in Oligocene: 25
in Pennsylvanian
  drawing of: **12**
  growth strategies of: 11
  habitats of
    mires: 13–14
    Nova Scotia: 180, 190
    Pennsylvania: 80
    Texas: 236–38, 240
    West Virginia: 203–13, **214–21**
in Permian
  drawing of: **15**
  habitats of: 14, 236, 240–41
  seed plants and: 80
  Triassic boundary: 263
Psaronius: 10–11, 212–13
pteridosperms. *see* pteridosperms
reproductive strategies of: 241
Rhacophyton. *see* *Rhacophyton*
Rhetinangium arberi: 131
spermatopsids: 19, 130–31
Sphenopteris. *see* *Sphenopteris*
tree. *see* tree ferns
in Triassic: 16–17, 255, 257
Trivena arkansana: 131
zygopterid. *see* zygopterid ferns
Festningen: **254**
filicalean ferns: 15, **189**
filicopsids
  Adiantites tenuifolius: **106**
  in Devonian: **81**
  habitats of: **81**, **106**, 180
  in Pennsylvanian: 180
  preservation of: 109
  stauropterids: **81**, 85–90, **92**, 95–98
  zygopterid ferns. *see* zygopterid ferns
fish
  acanthodians. *see* acanthodians
  actinopterygian: 94–95
  antiarch: 108
  cartilaginous. *see* sharks
  chondrichthyans: 9
  in Cretaceous: 21, 27–28
  crossopterygians: 108
  crustaceans and: 6
  in Devonian, habitats of: 9
    Blue Beach: 107–8
    Red Hill: 80, 86–87, 94–95, 99

Sussex: 108
in Eocene: 25
gars: **27**
habitats of: 123
  Albert Formation: 108
  Catskill Formation: 80, 86–87, 94–95, 99
  in Cretaceous: 21, 27–28
  in Devonian: 9
    Archaeopteris and: 99
    Blue Beach: 107–8
    Red Hill: 80, 86–87, 94–95, 99
    Sussex: 108
  in Eocene: 25
  Hancock County (Kentucky): 161, 164
  Horton Bluff Formation: 107–8
  in Mississippian: 107–8, 108, 161, 164
  in Permian: 227
  Petrolia Formation: 227
Knightia: **27**
lobe-finned. *see* lobe-finned fish
lungfish. *see* lungfish
megalichthyids: 87, 99
in Mississippian
  habitats of
    Blue Beach: 107–8
    Kentucky: 161, 164
    Sussex: 108
palaeoniscoids. *see* palaeoniscoids
in Permian: 227
rhizodonts: 99, 108, **159**, 161–64
sea grass and: 22
Semionotus: **17**
traces of, in geologic record: 108
tristichopterids: 99
Undichna: 108
Weichselia reticulata and: 21
Fishburn: **250**, **254**
Fishpot Limestone: **199**
flamingoes: 27
flooding
  in Amazon Basin: 232
  Archaeopteris and: 7
  biodiversity and: 20
  biota burial and: 190
  coal and: 284, 287–88
  Cooksonia and: 4
  floral partitioning and: 6
  in Nile River Valley: 232
  of oxbow lakes: 161
  reeds, rushes, and: 7
  root systems and: 7–8
  sand deposition in: 190
  sedimentary structures and: 117
  soil profiles and: 232
  sphenopsids and: 132
Florida
  archeological sites in: 30
  in Eocene: 22
  Everglades: 17
  Windover: 30
Florinites: 204, 208
Florinites florini: **214–20**
Florinites mediapudens: **214–15**, **217–18**, **220**
Florinites pumicosus: **214–15**, **217–18**
flowers: 22

Floyd Formation: **141**
Foard County, map of: **225**
foraminifera
  in Devonian: 104–5, 107, 122
  Earlandia: **252**, 253
  habitats of
    in Devonian: 104–5, 107, 122
    Fayetteville Formation: 128
    Horton Bluff Formation: 104–5, 107, 122
    in Mississippian: 104–5, 107, 122, 128
    in Permian: 253
  in Mississippian: 104–5, 107, 122, 128
  in Permian: 253
  sea grass and: 22
  Trochammina: 107
Forbes: **157**
Ford: **271**
forest mires
  Bowie Group. *see* Bowie Group
  carbon cycle and: 9
  Cisco Group: 224, **225**, 227, 236–38
  clastic swamps and: 11
  in Cretaceous: 20, **21**
  definition of: 2
  in Devonian: 8–9
  in Eocene: 23–25
  in Holocene: 13
  Horton Group: 108
  in Jurassic: 18
  in Miocene: 23, **24**
  in Mississippian: **4**
  in Pennsylvanian: 10, **12**, 13
    Bowie Group. *see* Bowie Group
    Cisco Group: 224, **225**, 227, 236–38
  in Permian: 14–15
forest tundra: 29
forests
  amber from: 22
  burial of: 190
  carbon cycle and: 8
  in Devonian: 80
  inventorying of: 120, 179
  nitrogen cycle and: 8
formaldehyde: 253, 260
Fort Payne Formation: 151
Forty Brine Seam: **171**, 172
Fountain Lake Group: 104
France: 13
Francis Creek Shale: 147–48
Frasnian
  coal from: 8, 98
  height of plants in: **4**
  progymnosperms in: 7
  Scat Craig: **157**
  Yahatinda Formation: 99
Fraxinus: **189**
Fremouw Formation: **250**, **251**
frenelopsids: 18, 19
frogs: 18, 23, 263
fullerenes: 256, 259
Fulton County: **302**
Fundy seam: 170, **173**, 174
fungi
  in amber: 22
  in Cretaceous: 22

habitats of
    Bloomsburg Formation: 43
    in Cretaceous: 22
    in Devonian: **68**, 76
    fossilization and: 44
    in Ludlovian: 43
    Massanutten Sandstone: 51–53
    Passage Creek: 51–53
    Rhynie Chert: 5
    riparian wetlands: 42
    in Siegenian: 5
    in Silurian: 51–53
    Trout Valley Formation: **68**, 76
    in Wenlockian: 43
mushrooms: 22
nematophytes and: 43
Prototaxites: 42–43, **68**, 76
in Siegenian: 5
in Silurian: 51–53
fusain. *See* charcoal
fusulinds: 225

G

Galesaurus: 263
Galesaurus planiceps: **262**
gametophytes: 5
Gangamopteris: 14
Garnett: **157**
gars: **27**
Gartnerkofel: **250**, 252–55, 259
Gaspé Bay: 42–43, 63, 74, 76
Gasperian: **157**, **159**, **162**, 164
Gastrioceras subcrenatum: 172
gastropods. *See also* microgastropods
    in Cretaceous: 21
    in Eocene: 26
    habitats of
        Blaine Formation: 229
        in Cretaceous: 21
        in Eocene: 26
        Hartselle Sandstone: 142
        in Jurassic: 18
        Lozier Pond: 294
        in Mississippian: 142
        in Permian: 229
        Thailand: 26
        Trout Valley Formation: **60–61**, 67
    in Jurassic: 18
    in Mississippian: 142
    in Permian: 229
    Weichselia reticulata and: 21
gavials: 14, 17
Gedinnian: **4**, 5. *See also* Lochkovian
Geisina: 107
Geissel: 26
Genesee County: 292, **302**
Genselia: 146
Georgia
    barrier islands off: 151
    St. Catherines Island: 151–52
    in Westphalian: 132
geothermal wetlands
    in Devonian: 5–6, 42–43
    East Kirkton: 9–10, 156–**57**, **159**, 164–65
    in Mississippian: 9–10, 156–**57**, **159**, 164–65

Permian-Triassic boundary and: 256
preservation in: 42
Rhynie Chert: 5–6, 42–43
in Siegenian: 5–6, 42–43
Germany
    Alken-an-der-Mosel: 5
    archeological sites in: 30
    carbon cycle in: 258
    coal in: 23, 26
    in Eocene: 25
    Geissel: 26
    Messel: 26
    Schöningen: 30
    in Triassic: 18
Gifford Brook, map of: **59**
Gigantophis: 26
Gigantopteridium: 239
gigantopterids: 225, 236, 239, **240**
Gilboa: 5, 7
Gillespiea randolphensis: **81**, 85–90, **92**, 95–98
Gillespieisporites venustus: **214–15**, **220**
Gilmerton: **157**
Ginkgophyllum kiltokense: 99
ginkgophytes: 15, 18
Givetian
    coal from: 98
    fossils from: **42**
    wetland evolution in: **4**, 6
glaciation cycles
    conifers and: 29
    in Devonian: 8
    lacustrine wetlands and: 291
    in Pennsylvanian: 224
    in Permian: 15
    root systems and: 8
    Sphagnum and: 29
    vascular plants and: 58
Gleichenia: 281–88
Glenshaw Formation: **199**
gleyed pedotypes, definition of: 233
global cooling: 8
global warming
    in Eocene: 24, 26
    Paleocene-Eocene boundary and: 258
    Permian-Triassic boundary and: 257, 261
    Siberian Traps and: 258
    Triassic-Jurassic boundary and: 18
glossopterids: 18
Glossopteris: 14–16, 258
Glossopteris browniana: **262**
Glumiflorae
    in Miocene: **24**
    reeds. *see* reeds
Glyptolepis: 108
Glyptostrobus: 24
Gnathorhiza: 163
gnetaleans: 16
Godhavari: **250**, **254**
Goldlight: **271**
Gondwana
    coal in: 14
    in Devonian: 163
    in Permian: 14, **15**
    at Permian-Triassic boundary: 16, **250**, 258
    Siberian Traps: **250**, 256, 258

in Tertiary: 23
goniatites: 172
Goreville: **157**, **159**, **162**, 164
Gorstian: **42**
Graeophonus carbonarius: 186, **187**
Grandispora cornuta: 83
Granulatisporites: 203
Granulatisporites granulatus: **214–15**
Granulatisporites pallidus: **217–18**
Granulatisporites pannosites: **216**
Granulatisporites parvus: **214–18**, **220**
Granulatisporites piroformis: **217–18**, **220**
Granulatisporites verrucosus: **214–15**
Graphic Peak: **250–51**, 253–55, 259, **262**
grasses: 6, 28, 151
Great Britian
    England. *see* England
    Scotland. *see* Scotland
    Wales: 98
Great Lakes region
    Indiana. *see* Indiana
    mastodons in: 292, 301–3
    Michigan: 300, **302**
    New York (state). *see* New York (state)
    Ohio. *see* Ohio
    Pennsylvania. *see* Pennsylvania
    Wisconsin: 52
Greece: **250**, **254**, 259
Green Mountain: 45
Green River Formation: 27
Greenbrier Formation: 140, 146–48, 150–51
Greenland
    carbon cycle in: 258
    in Devonian: 108
    Schuchert Dal: **250**, 253, **254**, 259
Greenland Group: **271**, 277, 279, 285
Greer: **157**, **159**, **162**, 164
Greererpeton: 161, **162**
Grenville Terrane: 171
Greymouth Basin: 270–88
Griesbachian: **252**, **260**
groenlandaspids: 87, 99
ground sloths: 30
groundwater
    bogs vs. fens: 292
    in Chesterian: 161
    in moors: 13
    paleosol formation and: 230–35
    stilting of: 8
    in topogenous mires: 13
    tree ring analysis and: 190
Guadalupian. *See* Blaine Formation
Guimarota mine: 18
Gulf Coast: 23–25, 151–52, 178
Guryul Ravine: **250**, **254**
Guttenberg Formation: 52
gymnosperms
    Aglosperma: 98
    Alisporites: 208
    coal and: 14
    cordaites. *see* cordaites
    in Cretaceous: 23, 270, 281–85
    in Devonian
        habitats of
            Belgium: 88, 98

Bold numbers indicate material in figures and tables.

Elkins: 97–98
England: 99
Ireland: 99
Red Hill: 80–81, 85, 88–98
Scotland: 99
West Virginia: 88
reproductive strategies of: 88–89, 98
size of: 88
Dicroidium: 16
Dicroidium zuberi: 259
Dorinnotheca streelii: 89
Elkinsia polymorpha: 88, 98
Florinites. see *Florinites*
Gangamopteris: 14
Glossopteris: 14–16, 258
Glossopteris browniana: **262**
habitats of
Albert Formation: **106**
in Antarctica: 16
Baggy Beds: 99
Ballyheigue deposits: 99
Catskill Formation: **81**, 85, 88–98
Cheverie Formation: **106**
in Cretaceous: 23, 270, 281–85
in Devonian
Belgium: 88, 98
Elkins: 97–98
England: 99
Ireland: 99
Red Hill: 80–81, 85, 88–98
Scotland: 99
West Virginia: 88
Evieux Formation: 98
Hampshire Formation: 97–98
Horton Bluff Formation: **106**
Joggins Formation: 180, 189
in Pennsylvanian: 11, 80
Nova Scotia: 180, 189
West Virginia: 203–6, 208–13, **214–20**
in Permian: 14
Pittsburgh Formation: 203–6, 208–13, **214–20**
in Triassic: 16
in Westphalian: 11
incertae sedis: **106**
Moresnetia: 88
in Pennsylvanian
habitats of: 11, 80
Nova Scotia: 180, 189
West Virginia: **203**, 203–6, 208–13, **214–20**
in Permian: 14
at Permian-Triassic boundary: 263
photograph of sample: **89**
Phyllocladidites mawsonii: 281–82, **284**–87
pteridosperms. see pteridosperms
Rhacophyton and: 97
Sphenopteridium: 99, **106**, 239–40, 243
Sphenopteris. see *Sphenopteris*
Telangiopsis: 98–99, 146, **147**
in Triassic: 16
in Westphalian: 11
wildfires and: 97–98
Xenotheca devonica: 99
Gyracanthus: 108, **159**, **162**, 163

H

Hale Formation: **129**
Halogragaceae: 20
halophytes. *See* sea grass
Hamamelidaceae: 24
Hamiapollenites: **214**–15
Hampshire Formation: 80, 90, 97–98
Hancock County (Indiana): **302**
Hancock County (Kentucky): 156–65
Hardeman County, map of: **225**
Harlem coal bed: **199**
Harpersville Formation: 225
Hartselle Sandstone: 140–52
Hartsellea: **141**, 143–45, 148–51
Hartsellea dowensis: **143**, 148–49, 151
Haskell County, map of: **225**
Hastarian
Albert Formation: 104–6, 108, 118–23
Horton Bluff Formation: 103–17, 120–23, **157**
Memramcook Formation: **105**
miospore zones in: **105**
Romer's Gap: **157**
Haumurian. *See* Paparoa Coal Measures
Hausmannia: 19
heaths: 29, 300
Hebeia Gabbro: 258
Heisler site: **302**
Heliminthopsis: **60**–**61**, 64–66
Helongshan Formation: **250**
hemlock: 292
Heping: **250**, 254
herbivory
in Battery Point Formation: 43
in Carboniferous: 80
definition of: 43
in Devonian: 43, 79
in Rhynie Chert: 43
herons: 26, 27
Hettangian: 253, **260**
hexacorals: 261
Hexapoda: 5. *See also* insects
hickory: 292
Hierogramma: 98
Hindeodus parvus: **251**–52, 253, 259, **262**
Hindeodus typicalis: **251**
Hindsville Formation: **129**
Hinton Formation: 132
Hippopotamus: 26–27
Hippopotamus amphibius: 16
Hippuris vulgaris: 296, **297**
Hiscock site: 292, **302**
Holkerian
Middle Paddock: 156
Wardie: **157**
Holocephali: **159**
Holotheria: 18
Hombergian
Greenbrier Formation: 140, 146–48, 150–51
Greer: **157**, **159**, **162**, 164
Hartselle Sandstone: 140–52
Homerian: **42**
Homewood: 140–52
homosporous plants: 5
Hook Head deposits: 99
Horneophyton: **4**, 6

hornworts: 20, 43, 51
horses: 25, 30
horsetails
calamiteans and: 10
Equisetites: 18
Equisetum. see *Equisetum*
habitats of: 14, 17
in Jurassic: 18
in Permian: 14, **15**
in Triassic: 17
Horton Bluff Formation: 103–17, 120–23, **157**
Horton Group: 103–23, **157**
Howard: **157**
Hostimella crispa: 80
Hudson Bay area: 14, 29
Hudson River Valley: 292–303
humans: 30, 253, 263
Hungary: 26
Huon pines: 281, 286
hurricanes: 151–52
Hyde Park Basin: 292–303
Hyde Park Moraine: 292
Hydrocaritaceae: 22
Hylonomus: 13, **14**, 171
Hylonomus lyelli: 171
Hylopus: 108
Hyneria: 87
Hynerpeton bassetti: 99
Hypericum virginicum: 296, **297**
hyphomycetes: 52
Hypnum revolutum: **297**, 298
Hypnum vaucheri: **297**, 298

I

Iapetus Ocean: 81
Ibyka: **4**
Ichthyostega: 156
Idrijca: **250**, 254
Iguanodons: 20–21
Illinois
Goreville: **157**, **159**, **162**, 164
Illinois Basin. *see* Illinois Basin
map of: **158**
Illinois Basin
Buffalo Wallow Formation: 158–60
Carbondale Formation: 147–48
in Carboniferous: 156
Clore Limestone: 158, 160
Francis Creek Shale: 147–48
Goreville: **157**, **159**, **162**, 164
Hancock County (Kentucky): 156–65
Iowa and: **162**
Kentucky and: **162**
lithology of: 158
map of: **158**
Mazon Creek: 13, 148, **157**
Menard Limestone: 158–60
Nova Scotia and: **162**
Palestine Sandstone/Shale: 158–61
in Pennsylvanian: 243
Vienna Limestone: 158
West Virginia and: **162**
India
Banspetali: **250**, 254, 259
coal in: 14, **15**, **19**, 27

Godhavari: **250**, **254**
Guryul Ravine: **250**, **254**
maps of: **15**, **19**
Palgham: **250**, **254**
Raniganj Basin: **250**, **254**, 261–62
Spiti: 259
Talcher: **250**, **254**, 255
Wardha: **250**, **254**
Indiana
    Fulton County: **302**
    Hancock County: **302**
    Illinois Basin. *see* Illinois Basin
    map of: **158**
    Starke County: **302**
Indospora boletus: 214–15
insects
    in amber: 22
    ants: 22
    bees: 22
    beetles: 22
    black flies: 22
    bristletail: 5
    caddis flies: 22
    cockroaches: 13
    in Cretaceous: 22
    in Devonian: 22
    as disease carriers: 22
    dragonflies: 13, 22
    in Emsian: 5
    flight, development of: 13
    in Givettian: 5
    habitats of: 5, 22, 186
    mayflies: **13**, 22
    Meganeura: 13
    megasecopterans: 13, 186, **187**
    metamorphosis of: 13
    mosquitoes: 22
    paleodictyopterans: 13
    in Pennsylvanian: 13, 186
    potter wasps: 22
    in Pragian: 5
    praying mantids: 22
    Rhyniognatha: 5
    roaches: 13, 22
    stone flies: 13
    wasps: 22
    waterbugs: 22
Iowa
    Delta: **157**, **159**, **162**, 164
    Illinois Basin and: **162**
    Kentucky and: **162**
    Nova Scotia and: **162**
    West Virginia and: **162**, 164
Iran
    Abadeh: **250**, **254**, 255
    coal in: 18
    Emarat: **250**, **254**
    in Jurassic: 18
    Kuh-e-Ali Bashi: **250**, **254**
Ireland: 99
iridium
    K-T event and: 23
    Permian-Triassic boundary and: 259–60
Iridopterids: 6
Isarcicella isarcica: **251**

Isle of Wight: 21
isoetaleans. *See also* cormose lycopsids
    anatomy of: 142
    habitats of: 16, **81**, **106**, 257
    lepidodendraleans and: 149
    origin of: 149
    in Triassic: 16, 257
Isoetes: 16, 257
Isoetes beestonii: **255**
Isopodichnus: 107
isopods: 22, 107
Italy: **250**, **254**
Ivorian
    Albert Formation: 104–6, 108, 118–23
    Auchenreoch Glen: **157**
    Cheverie Formation: **105–6**
    Horton Bluff Formation: 103–17, 120–23, **157**
    miospore zones in: **105**
    Romer's Gap: **157**
    Weldon Formation: **105**

J

Jack County, map of: **225**
Japan
    Kamura: **250**, **254**
    Sasayama: **250**, 253, **254**
    Taho: **250**, **254**
Jarrow: **157**
Jay: **271**
Jefferson County: 140–52
Joggins Formation
    Atchafalaya Bay and: 178
    Barataria Bay and: 178
    "clam coals" in: 172
    Dawson on: 177
    depositional environment of: 123, 156, 165, 172, 174, 177–79, 186–91
    fossils in: 13–14, 156, 165, 170–90
    geologic setting of: 171–72
    Horton Bluff Formation and: 123
    Little River Formation and: 172
    Lyell on: 170, 191
    map of: **171**
    Niger Delta and: 178
    paleoflow patterns in: 176–78
    palynological analyses of: 172, 180–85
    photographs of samples: **181–85**, **187–88**
    photographs of site: **175–76**
    Polly Brook Formation and: 172
    rose diagrams of: **177**
    schematic transect of: **189**
    seasonal changes in: 190
    sedimentary structures in: 108, 123, 172–79
    Springhill Mines Formation and: 172
    stratigraphy of: **172–73**
    Western European Carboniferous Basin and: 186
    wildfires in: 189–90
    woodcuts of: **171**, **180**
Juglandaceae: 23–25
Juncaceae
    habitats of: 28
    Juncus: 28, **296**
    rushes. *see* rushes
Juncus: 28, **296**

Jurassic. *See also* Hettangian; Toarcian
    carbon cycle in: 18
    coal from: 18
    maps of: **19**
    methane release during: **260**
    Triassic boundary: 18, 253, **260**
    wetland evolution in: 18–19
    Woodleigh Crater: 259

K

Kaki Vigla: **250**, **254**, 259
kames: 301
kampecarids: 5
Kamura: **250**, **254**
Kansas: 108
kaolinite: 23, 243
Karinopteris tennesseana: 180
Karinopteris dernonrcourtii: 180, **184**
Karoo Basin: 15–16
Kashirskian: **157**
Kaulangiophyton: **68**
Kaulangiophyton akantha: 65, **68**, 72
Kauri pines: 22
Kemer Gorge: **250**, **254**
Kemp, Lake: 228
Kennebecasis Formation: **106**
Kentucky
    Big Bone Lick: 30
    Buffalo Wallow Formation: 158–60
    Clore Limestone: 158, 160
    Cumberland Plateau: 140
    Fort Payne Formation: 151
    Hancock County (Kentucky): 156–65
    Hartselle Sandstone: 140–52
    Illinois Basin. *see* Illinois Basin
    Iowa and: **162**, 164
    map of: **158**
    Newman Limestone: 151
    Nova Scotia and: **162**
    Vienna Limestone: 158
    West Virginia and: **162**, 164
Ketleri: **157**
kettle holes: 301, **302**
Kiltorcan deposits: 99
Kinderscoutian: **157**
King County: **225**, 229
Klazminskian: **157**
Knightia: 27
Knox County, map of: **225**
Kolarik site: **302**
Kounova: **157**
Krassilov taphonomic analysis methods: 59
Krevyakinskian: **157**
K-T event
    acid rain from: 261
    birds and: 27
    coal and: 23
    impactor size: 256
    iridium and: 23
    methane release during: 253, **260**
    Permian-Triassic boundary and: 253, 256
    wetland evolution and: 23
Kuh-e-Ali Bashi: **250**, **254**
Kutchicetus minimus: 27

Bold numbers indicate material in figures and tables.

L

labyrinthodonts: 15, 17, 18
lacustrine wetlands
　archeological sites in: 30
　Bowie Group. *see* Bowie Group
　in Cretaceous: 19–20
　crocodilians in: 27
　in Devonian: 6, **7**, 103–23, **157**
　East Kirkton: 9–10, 156–**57**, **159**, 164–65
　in Eocene: 24, 26
　Geissel: 26
　glaciation cycles and: 291
　Hancock County (Kentucky): 156–65
　in Holocene: 6–7
　Horton Group: 103–23, **157**
　in Jurassic: 18
　mastodons in: 30, **302**
　Messel: 26
　in Mississippian
　　East Kirkton: 9–10, 156–**57**, **159**, 164–65
　　Hancock County (Kentucky): 156–65
　　sphenopsids and: 132
　in Pennsylvanian
　　Bowie Group. *see* Bowie Group
　　evolution of: 13
　in Permian: 15
　Pittsburgh Formation: 198–221
　Rhynie Chert: 5–6, 42–43
　in Siegenian: 5–6, 42–43
　in Triassic: 18
　whales in: 27
Laevigatosporites: 282
Laevigatosporites maximus: 204, **220**
Laevigatosporites medius: 204, **217–18**
Laevigatosporites minimus: 203, 213, **214–21**
Laevigatosporites minor: 204, **214–21**
Laevigatosporites ovalis: 203, **216**, **220**
Laevigatosporites punctatus: 203
Laevigatosporites vulgaris: 204, **214–16**, **219**, **220**
Lagarostrobus franklinii: 281, 286–87
Lagenicula: 182, **184**
Lagenicula horrida: **184**, **185**
Lagenoisporites: 182
Lagenoisporites rugosus: **184**, **185**
landslides: 258–59
Langsettian
　Joggins Formation. *see* Joggins Formation
　zonation in: 10
Lapeer County: **302**
Larix: 22
Larix laricina: 29, 297
Latensia trileta: **199**
Latosporites minutus: **199**
Latvia: 108
Lauraceae: **24**
Leclercqia: **4**, **68**
Leclercqia complexa: **68**
Leiotriletes: 203
Leiotriletes adnatus: **217–18**, **220**
Leiotriletes gracilis: **217–18**
Leiotriletes levis: **214–15**, **220**
Leiotriletes subadnatoides: **214–20**
Leiotriletes subintorus: **220**
Lemna: 27
Lemnaceae: **19**, 20, 27

Lentibulariaceae: 22
Leonardian. *See also* Cisuralian
　Blaine Formation. *see* Blaine Formation
　Clear Fork Formation. *see* Clear Fork Formation
　Lueders Formation: **225**, **227**, **231**
　Pease River Group. *see* Pease River Group
　San Angelo Formation. *see* San Angelo Formation
　Waggoner Ranch Formation. *see* Waggoner Ranch Formation
　Wichita Group. *see* Wichita Group
Lepidocaris: 6
lepidodendraleans
　definition of: 105
　in Devonian: 105–7
　habitats of: 10, 105–7, 120–21
　isoetaleans and: 149
　Lepidodendron. see *Lepidodendron*
　in Mississippian: 10, 105–7
　root system of: 149
lepidodendrids
　habitats of: 174, 177–79, 187–92
　life span of: 190
　morphology of: 190
　in Namurian: 192
　in Pennsylvanian: 174, 177–79, 187–92
　photographs of in situ: **175**
　reproductive strategies of: 190
　structural support for: 190
Lepidodendron
　drawing of: **12**
　habitats of
　　Horton Bluff Formation: **106**
　　Joggins Formation: 183, 187–89
　　in Mississippian: 161
　　in Pennsylvanian: 183, 187–89
　　in Westphalian: 130–32
　Lagenicula horrida and: **185**
　Lycospora pusilla and: 161
　in Mississippian: 161
　in Pennsylvanian: 183, 187–89
　in Westphalian: 130–32
Lepidodendron aculeatum: 179–80, **182**
Lepidodendron lycopodioides: 180
Lepidodendron veltheimii: 130–34
Lepidodendron volkmannianum: 130–34
Lepidodendropsis
　in Carboniferous: 97
　in Devonian, habitats of
　　Blue Beach: 103–16, 120–22
　　China: 8
　　Ireland: 99
　　Red Hill: **81**, 87, 97
　　Sussex: 118–22
　habitats of
　　Albert Formation: **106**, 118–22
　　in Carboniferous: 97
　　Catskill Formation: **81**, 87, 97
　　in Devonian: 8
　　　Blue Beach: 103–16, 120–22
　　　Ireland: 99
　　　Red Hill: **81**, 87, 97
　　　Sussex: 118–22
　　Horton Bluff Formation: 103–16, 120–22

　　Kiltorcan deposits and: 99
　　in Mississippian: 103–16, 118–22
　　Price Formation: 122
　height of: **4**
　in Mississippian
　　habitats of: 103–16, 118–22
　　spatial analysis of: 120–22
　photograph of samples: **106**, **110**, **116**, **118**
　Protostigmaria and: 103, 107, 111, 122
　root system of
　　Horton Group: 103, 107, 111, 122
　　Price Formation: 111
　Stigmaria and: 122
Lepidodendropsis corrugatum: **106**
Lepidophloios
　in Carboniferous: 190
　drawing of: **12**
　habitats of: 10, 130–32, 180, 187–90
　in Mississippian: 10
　in Pennsylvanian: 180, 187–89
　in Westphalian: 130–32
Lepidophloios laricinus: 180, **182**
Lepidophyllum fibriatum: **106**
Lepidopteris: 261
Lepidopteris callipteroides: **255**
Lepidosigillaria: **4**, 8
lepidosigillarioid lycopsids: **4**, 8, 80
Lepidostrobophyllum: 105–6
Lepidostrobophyllum fibriatum: **106**
Lepidostrobophyllum majus: 180
Lepidostrobus gallowayi: 80
Lepisosteus: 27
lepospondyls: 165
Leptophloeum: 104
liana growth strategy: 11
lichen: 3, 51
Licking County: **302**
lignites. *See* coal
Lilpopia raciborskii: 238
limnetic zones: 20, 23
Limnobiophyllum: **19**
Limnopus vagus: **186**, **188**
Lincoln Dam area: 127–35
Linton: **157**
lissamphibians: 156
Little Ben Sandstone: 259
Little Clarksburg coal bed: **199**
Little Pittsburgh Formation: **199**
Little River Formation: **171**, 172, 176
Little Waynesburg coal bed: **199**
littoral zones: 20, 23
liverworts: 43, 51
lizards: 18
Llandoverian
　fossils from: **42**
　Massanutten Sandstone: 44–55
　Passage Creek: 44–55
　Tuscarora Formation: 52–53
　wetland evolution in: **4**
Llano Uplift: 224–25
Llanvirnian: **42**
Loanhead: **157**
Lobatopteris puertollanensis: 238
lobe-finned fish
　in Devonian: 9, 108

Glyptolepis: 108
  habitats of: 9, 108
  in Mississippian: 108
  rhizodonts: 99, 108, **159**, 161–64
Lochkovian. *See also* Gedinnian
  fossils from: **42**
  height of plants in: **4**
Lock Haven Formation: 81
Lockeia: 107, 150
London Clay: 25, 26
Long Island: 292
Longwood Range: 258
Lootsberg: **250**, **254**
Lophotriletes: 203
Lophotriletes commissuralis: **217–21**
Lophotriletes insignitus: **217–18**
Lophotriletes microsaetosus: **220**
Lophotriletes pseudaculeatus: **220**
Lorne Crater: 259
lotus: **19**, 20
Lower Sewickley Sandstone: 200, 213
loxommatids: 164, 165
Lozier Pond: 292–303
Ludfordian: **42**
Ludlovian. *See also* Gorstian; Ludfordian
  Bloomsburg Formation: 43
  fossils from: **42**
  height of plants in: **4**
  lycopsids in: 5
  Massanutten Sandstone: 44–55
Lueders Formation: **225**, 227, **231**
lungfish
  Ctenodus: 108
  in Devonian: 9, 108
  habitats of: 123
    Albert Formation: 108
    Clear Fork Formation: 227
    in Devonian: 9, 108
    in East Kirkton: 9
    "failed" aestivation: 227
    in Holocene: 9
    Lueders Formation: 227
    in Mississippian: 9, 108, 161, 163
    in Permian: 227
  in Mississippian: 9, 108, 161, 163
  Neoceratodos forsteri: 9
  in Permian: 227
  photographs of samples: **162**
  Protopterus: 9
  Tranodis: **162**, 163–64
Luronne coal seam: 20
lycophytes: **4**–6, **68**, 70
lycopods
  Arthropleura and: 13
  Chaloneria. see *Chaloneria*
  coal and: 14
  in Devonian: 8–9
  habitats of
    in Devonian: 8–9
    East Kirkton: 10
    in Mississippian: 10, 161
    in Pennsylvanian: 14, 203–11, **214–20**
    Pittsburgh Formation: 203–11, **214–20**
    Trout Valley Formation: 68
  in Mississippian: 10, 161

Paralycopodites. see *Paralycopodites*
  in Pennsylvanian: 14, 203–11, **214–20**
  in Permian: **15**
lycopsids
  arthropods and: 5
  Baragwanathia: 5
  Barsostrobus: 98
  Chaloneria. see *Chaloneria*
  coal and: 8
  cormose. see cormose lycopsids
  Cyclostigma: 8, 87, 99, 149
  in Devonian
    arborescence and: 7–8
    habitats of: 5, 7–9, 80
      Bear Island: 99
      Belgium: 98
      Blue Beach: 103–16, 120–22
      Elkins: 98
      Melville Island: 98
      Rawley Springs: 98
      Red Hill: 80–81, 85, 87, 90, **92**, 95, 97
      Sussex: 118–22
    height of: 7
    root system of: 7–9, 87
  diameter of: 120–21, 142
  in Emsian: 5
  in Eocene: 25
  fissures in: 178
  in Givettian: 5
  habitats of
    Alken-an-der-Mosel: 5
    Australia: 5
    Battleship Wash Formation: 10, 151
    Black Warrior Basin: 190, 192
    Bowie Group: 236–37
    Carbondale Formation: 147–48
    Catskill Formation: **81**, 85, 87, 90, **92**, 95, 97
    Cisco Group: 236–37
    in Devonian: 5, 7–9
      Bear Island: 99
      Belgium: 98
      Blue Beach: 103–16, 120–22
      Elkins: 98
      Melville Island: 98
      Rawley Springs: 98
      Red Hill: 80–81, 85, 87, 90, **92**, 95, 97
      Sussex: 118–22
    in Emsian: 5
    in Eocene: 25
    Evieux Formation: 98
    Fayetteville Formation: 130–35
    Francis Creek Shale: 147–48
    Gilboa: 5
    in Givettian: 5
    Greenbrier Formation: 140, 146–47
    Hampshire Formation: 98
    Hartselle Sandstone: 140–45, 147–50
    Horton Bluff Formation: 103–16, 120–22
    Joggins Formation: 14, 156, 170–83, 187–92
    Mazon Creek: 148
    in Mississippian: 10
      Arizona: 151
      Blue Beach: 103–16, 120–22
      Jefferson County: 140–45, 147–50
      Lincoln Dam area: 130–35

      Sussex: 118–22
      West Virginia: 140, 146–47
    in Namurian: 10, 192
    in Pennsylvanian: 10–11, 15
      Illinois: 147–48
      Nova Scotia: 14, 156, 170–83, 187–92
      Texas: 236–37
    in Permian: 14–16, 236, 241, 263
      Ruhr Basin: 192
    in Tournaisian: 105–6
    in Triassic: 16–17
    in Westphalian: 130–32
  Hartsellea: **141**, 143–45, 148–51
  isoetaleans. see isoetaleans
  lepidodendraleans. see lepidodendraleans
  Lepidodendron. see *Lepidodendron*
  Lepidodendropsis. see *Lepidodendropsis*
  Lepidophloios. see *Lepidophloios*
  Lepidosigillaria: **4**, 8
  lepidosigillarioid: **4**, 8, 80
  Lepidostrobus gallowayi: 80
  in Ludlovian: 5
  Megaloxylon wheelerae and: 134
  in Mississippian
    habitats of: 10
      Arizona: 151
      Blue Beach: 103–16, 120–22
      Jefferson County: 140–45, 147–50
      Lincoln Dam area: 130–35
      Sussex: 118–22
      West Virginia: 140, 146–47
    height of: 145
  in Namurian: 10, 192
  Omphalophloios: **12**, 13, 190
  Paralycopodites. see *Paralycopodites*
  in Pennsylvanian
    growth strategies of: 11
    habitats of: 10–11, 15
      Illinois: 147–48
      Nova Scotia: 14, 156, 170–83, 187–92
      Texas: 236–37
    in ombrogenous mires: 13
    reptiles and: 14
    root systems of: 10
    sphenopsids and: 236
    types of: 80
  in Permian: 14–16, 236, 241, 263
  at Permian-Triassic boundary: 16, 263
  photographs of in situ: **175**
  photographs of samples: **86**, **130**, **143**, **181–82**
  polycarpic. see polycarpic lycopsids
  Prolepidodendron breviinternodium: 80
  reproductive strategies of: 10, 236
  reptiles and: 14
  root system of: 107, 145–46, 149, 236
  in Siegenian: 5
  Stigmaria. see *Stigmaria*
  in Tournaisian: 105–6
  in Triassic: 16–17
  in Westphalian: 130–32
  Wexfordia: 99
Lycopus americanus: 296
Lycospora: 204
Lycospora micropapillata: 160
Lycospora orbicula: 160

Bold numbers indicate material in figures and tables.

Lycospora pusilla: 160–61
Lyell, Charles: 170
lyginopterids
    habitats of: 130–32, 134–35, 160, 190
    Lyginopteris: 160–61
    in Mississippian: 130–32, 134–35, 160
    in Pennsylvanian: 190
    Sphenopteris. see *Sphenopteris*
Lyginopteris: 160–61
Lyginopteris royalii: 131
Lymnobiophyllum: 20
lysorophoid amphibians: 227
Lystrosaurus: 16, **251**, **262**, 263
Lystrosaurus declivis: **262**

M

Maastrichtian. *See* Paparoa Coal Measures
Machimosaurus: 18
Macrostachya: 236
Madagascar: **250**, **254**
Magnoliaceae: **24**
"Main Seam:" 270–71, 273–87
Maine
    Baxter State Park: 58–76
    Dry Brook: **59**
    Gifford Brook: **59**
    in Llandoverian: 43
    map of Trout Brook: **59**
    South Branch Ponds Brook: **59**
    Trout Brook: 58–72
    Trout Valley Formation: 6, 58–76
malaria: 22
Malerisaurus: **17**
mammals
    acclimatization by: 263
    buffalo: 16
    Docodonta: 18
    habitats of: 18, 26
    Holotheria: 18
    in Jurassic: 18
    manatees: 16, 22, 27
    moose: 16
    Multituburculata: 18
    primates. see primates
    proboscideans. see proboscideans
    rodents: 18, 29
    water buffalo: 16
    wooly rhinoceroses: 30
mammoths: 30
Mammut americanum: 292–94, 298–301
manatees: 16, 22, 27
Mandibulata. *See* Uniramia
mangals: 10, 18, 21, 26
mangroves
    Acrostichum: 21, **24**, **26**
    Avicennia: 26
    birds and: 27
    Bruguiera: 26
    Ceriops: 26
    coral reefs and: 22
    Cordaites and: 11
    in Cretaceous: 21, **26**
    crocodilians in: 17
    definition of: 21
    in Eocene: 23–26
    functions of: 21
    habitats of
        Amazon Basin: 25
        Bahariya: 21
        in Cretaceous: 21
        in Eocene: 21, 26
        latitude of: 15
        New Zealand: 26
        in Niger Delta: 178
        Tasmania: 26
        Tethys Ocean and: 25–26
    Nypa: 21, 25–26
    origin of: 21
    in Paleocene: **26**
    reproductive strategies of: 26
    Rhizophora: **15**, 25, 26
    root system of: 26
    sea grass meadows and: 22
    vs. swamps: 26
    viviparity in: 26
Manicaria: 25
Manning Canyon: **157**
manoxylic wood: 131
Maracaibo Basin: 26
marattialean tree ferns
    habitats of: 10–11, 236–38, 241, 245
    Lobatopteris puertollanensis: 238
    Pecopteris: 160, 236–38, 241–**42**, 244–45
    in Pennsylvanian: 10–11, 236–38
    in Permian: 241, 245
    Polymorphopteris polymorpha: 238
    reproductive strategies of: 236, 245
    root system of: 236, 245
marcasite nodules: 160–61
Marcoduria: 23
Maritimes Basin
    Appalachians and: 171–72
    in Carboniferous: 171
    Cumberland Basin. *see* Cumberland Basin
    description of: 171
    in Devonian: 104
    Joggins Formation. *see* Joggins Formation
    Kansas and: 108
    in Permian: 171
    petrogenesis of: 104
    in Tournaisian: 107
Maritime-West European Province: 171–72
Markley Formation: 225–27, 230–34, **238**, 240–41
Marsdenian: **157**
marshes
    archeological sites in: 30
    barrier islands and: 151
    carbon cycle and: 8
    Chinle Formation: 16–17
    classification of: 139–40
    definition of: 2, 6, 108
    in Devonian: **4**, 5–9
    East Kirkton: 9–10, 156–**57**, **159**, 164–65
    in Eocene: **24**, 26
    Hancock County (Kentucky): 156–65
    in Holocene: 4, 6–7, 9, 15
    inundation of: 6
    in Jurassic: 18
    malaria and: 22
    methane in: 151
    in Miocene: 23, **24**
    in Mississippian: **4**, 9–10, 140–52, 156–65
    nitrogen cycle and: 8
    in Pennsylvanian: 11–13
    in Permian: 15
    in Quebec: 5
    root systems in: 151
    salt. *see* salt marshes
    in Silurian: 4
    sulfur in: 151
    vs. swamps: 7, 26, 108, 140
    in Triassic: 16–18
Marsilea: 19
Martinsburg Formation: 53
Maryland
    map of: **199**
    Monongahela Group: 198–221
Massanutten Mountain: 53
Massanutten Sandstone: 44–55
mastodons
    conifers and: 303
    drawing of: **29**
    habitats of: 30, 292–94, 298–303
    humans and: 30
    in Pleistocene: **29**, 30, 292–94, 298–303
Matthewichnus velox: **186**
Mauch Chunk Formation: 150
Mauritia: 25
Mauritia flexuosa: 25
mayflies: **13**, 22
Mazon Creek: 13, 148, **157**
McAdams Lake Formations: 104
Medullosa steinii: 131
medullosan ferns: 190
medullosan pteridosperms: 11, 236–38
Meesia triquetra: 297, 300
megalichthyids: 87, 99
Megaloxylon wheelerae: 130–31, 134–35
Meganeura: 13
megasecopterans: 13, 186, **187**
Meguma Terrane: 171
Meishan: **250–51**, 253–55, 259, 261–62
Mekong River: 232
Melekesskian. *See* Joggins Formation
Melville Island: 98, 99
Memramcook Formation: **105**
Menard Limestone: 158–60
Meramecian: **159**. *See also* St. Louis time period; Salem time period; Warsaw time period
Mesocalamites: 10
Messel: 26
Metamynodon: 26
Metasequoia: 20, 24
meteorites
    Araguainha Crater: 259
    Bedout Crater: **250**, 256, 258–59
    bulk density of: 256
    carbon content of: 256
    europium and: 259
    iridium and: 23, 259–60
    isotopic analysis of: 256
    K-T event and. *see* K-T event
    Lorne Crater: 259
    Permian-Triassic boundary and: 250–53, 256, 258–60

Woodleigh Crater: 259
methane: 151, **252**–63
metoposaurs: 17
Metroxylon: 25
Mexico: 18, 22
Miamecha-Kotui: 258
Michigan: 300, **302**
Michigan Basin: 163
microgastropods: 253
Microrecticulatisporites nobilis: **220**
Microrecticulatisporites sulcatus: **214**–**16**, **219**
microsaurs: **162**, 165
Middle Paddock: 156
Midland Basin: 224–45
Midland Valley: 156
Milankovitch band: 191
millipedes: 5, 13, 22, 186, **187**
Minas Basin: 180
"minimum species concept:" 129–30
Minnesota: 300
Minudie Anticline: 171
Miocene: 22–28
mires
    carnivorous plants in: 22
    in Cretaceous: 19–20, 23
    in Devonian: 8–9
    in Eocene: 25
    forest. *see* forest mires
    in Holocene: 3
    in Jurassic: 18
    K-T event and: 23
    in Mississippian: 10
    nonforested. *see* fens
    ombrogenous: 13
    ombrotrophic: 23, 286
    origin of: 8
    palsa: 14
    in Pennsylvanian: 10–14
    in Permian: 14–16
    pines in: 286
    raised: 13, 20, 23
    root systems in: 285
    shrub: **4**, 8
    successions in: 11–13, 20
    vs. swamps: 11, 108
    topogenous: 13, 286
    in Triassic: 16–17
Mississippi River: 108, 178, 190, 232
Mississippian. *See also* Chesterian; Hombergian; Meramecian; Namurian; Osagean; Serpukhovian; Tournaisian; Viséan
    Albert Formation: 104–6, 108, 118–23
    Batesville Formation: **129**
    Battleship Wash Formation: 10, 151
    Boone Formation: **129**
    Borden Formation: 151
    Buffalo Wallow Formation: 158–60
    Cheveric Formation: **105**–**6**
    coal from: 10, 107, 159–61
    East Kirkton: 9–10, 156–**57**, **159**, 164–65
    Fayetteville Formation: 127–35
    Hancock County (Kentucky): 156–65
    Hartselle Sandstone: 140–52
    Hindsville Formation: **129**
    Horton Bluff Formation: 103–17, 120–23, **157**

maps of tetrapod distribution: **158**
Memramcook Formation: **105**
Menard Limestone: 158–60
miospore zones in: **105**
Palestine Sandstone/Shale: 158–61
Pennington Formation: 151
Pitkin Formation: **129**
Tetrapod Province: 164
tetrapods in
    East Kirkton: 9–10, 156
    Romer's Gap: 104, **157**, 164
Weldon Formation: **105**
wetland evolution in: **4**
wildfires in: 10
Missouri River: 232
Missourian: **199**
mites: 5
Mnium thomsonii: **297**, 298
Mobile Delta: 122
Moeritherium: 26–27
molluscs: 18, 22, 142
Moncton Basin: 104–6, 108, 118–23
Mongolia: 18
monocotyledons: 29
Monongahela Group: 198–221
Monongahela River: 198
monsoons: 96
Montague County, map of: **225**
Montcalm County: **302**
Montceau-les-Mines: 13
Monte Verde: 30
Monteagle Limestone: **141**
Montrichardia aborescens: 189
moors: 13, **24**
moose: 16
Moresnetia: 88
Morgan: **271**
Moriconea: 20
Morondava: **250**, **254**
Morrowan: **129**
Moscovian: **129**. *See also* Morrowan; Westphalian
mosquitoes: 22
moss
    Bryum: **297**
    Calliergon giganteum: **297**, 300
    Calliergonella cuspidata: **297**, 300
    Campylium chrysophyllum: **297**
    Campylium stellatum: **297**, 300
    Cinclidium: **297**
    club: 26
    Cowardin classification of: 3
    in Cretaceous: 20
    in Devonian: 9
    Drepanocladus aduncus: **297**, 300
    Drepanocladus sordidus: **297**, 300
    Encalypta: **297**
    habitats of
        bogs vs. fens: 292
        in Cretaceous: 20
        in Devonian: 9
        in Holocene: 3, 9
        Lozier Pond: 294, 297, 300–301
        in Ordovician: 3
        in Pleistocene: 297, 300–301
        in Silurian: 3

    in Holocene: 3, 9
    K-T event and: 23
    Meesia triquetra: 297, 300
    in Ordovician: 3
    origin of: 3
    in Pleistocene: 297, 300–301
    Pohlia: **297**
    polysporangiophytes and: 43
    Polytrichum juniperinum: 297
    Polytrichum strictum: 297
    Scorpidium scorpioides: **297**, 300
    in Silurian: 3
    Sphagnum: 3, 14–15, 29
    Warnstorfia exannulata: 297
Motuan: 284
mudflows: 190, 261–62
mudstone: 271
mukkara structures: 96
Multituburculata: 18
Murphy Brook Formation: 104
Murrays Run: **250**, **254**, 255, 259
Muschelkalk Horizon: **252**
mushrooms: 22
Myachkovskian: **157**
myriapods: 5, 99
Myrica: 23, **24**
Myriophyllum sibiricum: 296
Myrtaceae: 23–25
myrtles: 24

N

Naiadites: **186**
Najas flexilis: 296
Nammal Gorge: **250**, 253, **254**
Namurian. *See also* Alportian; Arnsbergian; Chesterian; Chokierian; Kinderscoutian; Marsdenian; Pendleian
    Bluefield Formation: 132
    Cumberland Basin. *see* Cumberland Basin
    Fayetteville Formation: 127–35
    Gastrioceras subcrenatum band and: 172
    Hale Formation: **129**
    Hancock County (Kentucky): 156–65
    Hinton Formation: 132
    Joggins Formation. *see* Joggins Formation
    Pitkin Formation: **129**
    tetrapods in: **157**
    Westphalian boundary: 172
nautiloids: 225, 227
Navajo Sandstone: 190
Nebraska: 53
nectridians: 165
Nelumbonaceae: **19**, 20
Nematophyllum: **106**
nematophytes: 43
Nematothallus: **42**–43, 47, 52
neocalamites: 17
Neoceratodos forsteri: 9
Neowellerella pseudoutah: **262**
Neuropteris: 133, 180, **184**, 240
Neuropteris auriculata: 238
Neuropteris hollandica: 180, **184**
Neuropteris ovata: 238
Neuropteris scheuchzeri: 236–38
New Brunswick

Bold numbers indicate material in figures and tables.

Albert Formation: 104–6, 108, 118–23
Cobequid Fault zone: 171
Cumberland Basin. *see* Cumberland Basin
in Devonian: 104
Horton Group: 103–23, **157**
maps of: **105, 171**
in Mississippian: 104
Moncton Basin: 104–6, 108, 118–23
Sussex: 105, 108, 118–22
New Jersey
amber deposits in: 22
barrier islands off: 151
in Cretaceous: 22
Wisconsinan moraine: 292
New Madrid earthquakes: 190
New Mexico: 163
New York (state)
in Devonian: 7, 99
Dutchess County: 292–303
Eagle Hill Camp Bog: 292
Genesee County: 292, **302**
Gilboa: 5, 7
Hiscock site: 292
Hudson River Valley: 292–303
Hyde Park Basin: 292–303
Hyde Park Moraine: 292
Long Island: 292
Lozier Pond: 292–303
mastodons in: 292–94, 298–301
Newburgh: 301
pH level of ponds: 300
Schoharie Valley: 63
Staten Island: 292
Wisconsinan moraine: 292
New Zealand
Australia and: 270
Brunner Coal Measures: **271**, 284
coal in
beech in: 25
Cretaceous: 20, 270–88
maps of: **20**
palms in: 24
in Cretaceous: 20, 270–88
in Eocene: 26
Greymouth Basin: 270–88
K-T event and: 23
Little Ben Sandstone: 259
Longwood Range: 258
maps of: **20, 271**
in Miocene: 25
in Oligocene: 25
Paparoa Coal Measures: 270–88
in Permian: 258
Pororari Group: 284
seismic studies of: 284
Waikato Coal Measures: 282
Wairoa Gorge: **250, 254**, 259
Newburgh mastodon: 301
Newfoundland: 103–23, **157**
Newman Limestone: 151
Newsham: **157**
Niger Delta: 178, 190
Nile River Valley: 30, 232
Ningxia Hui: **157**
nitrogen cycle: 8, 28

nitrous oxide: 253, 260
Nocona Formation
Archer City Formation and: **225**
depositional environment of: **234**, 243
fossils in: 227
Petrolia Formation and: **225**
photographs of site: **230**
Noginskian: **157**
nonforested mire. *See* fens
Noril'sk iron ores: 259
North America
Acadian orogen: 58, 81, 104
Appalachians. *see* Appalachians
Canada. *see* Canada
coal in
Carboniferous: **11**
Cretaceous: 20
maps of: **11, 20, 23**
Miocene: **23**
palms in: 24
Pennsylvanian: 10–14
Permian: 14
Sphagnum and: 29
Tertiary: 23
in Cretaceous: 21
Eastern Interior Basin: 156–65
Great Lakes region. *see* Great Lakes region
Grenville Terrane: 171
Illinois Basin. *see* Illinois Basin
K-T event and: 23
maps of
Carboniferous: **11**
coal deposits: **11, 20, 23**
Cretaceous: **20**
Miocene: **23**
Permian: **225**
at Permian-Triassic boundary: **250**
Michigan Basin: 163
in Miocene: **23**, 24
in Pennsylvanian: 10
in Permian: **225**
Taconic Highlands: 53
USA. *see* United States of America (USA)
Wisconsinan moraine: 292
North Carolina: **152**
North Devon: 99
North Sea: 65
Norway: 99
Notalacerta: 186, **188**
Nothofagus: 20, 25
Nova Scotia
Athol Syncline: 171
Avonport: 104–5, 107, 109–17
Blue Beach: 104–5, 107, 109–17, 122
Caledonia Highlands: 171
in Carboniferous: 156, 171
Coal Mine Point: 170–71
Cobequid Highlands: 171–72, 191
Cumberland Basin. *see* Cumberland Basin
Forty Brine Seam: **171**, 172
Horton Bluff Formation: 103–17, 120–23, **157**
Iowa and: **162**
Joggins Formation. *see* Joggins Formation
Little River Formation: **171**, 172, 176
maps of: **105, 171**

Minudie Anticline: 171
in Permian: 171
Point Edward: **157, 162**
Polly Brook Formation: 172
Springhill Mines Formation: 172, 176, 191
stratigraphy of: **172**
Windsor Basin: 103–17, 120–23, **157**
nuclear emissions: 258
Nymphaeaceae: **19**, 20, 26, **29**
Nyos, Lake: 258
Nypa: 21, 25–26
Nypa fructicans: 21
Nýřany: **157**
Nyssa: 23–25

O

oak: 25, **29**, 292–93
Oakland County: **302**
obstacle marks: 117–19, 123
Odontopteris: 239
Ohio
Licking County: **302**
map of Pittsburgh Formation: **199**
Monongahela Group in: 198–221
Pomeroy coal bed: 200, 202
Ohio River: 198
Oklahoma
Fayetteville Formation: 127–35
map of: **129**
Ozark Dome: 128
in Permian: 163
Wellington Formation: **231**
Oligocene: 22, 24–28
ombrogenous mires: 13
ombrotrophic mires: 23, 286
Omphalophloios: **12**, 13, 190
Ontario: 292
oolites: 151
Ophiceras sinense: **262**
Ordovician
fossils from: 5, **42**, 52
Guttenberg Formation: 52
wetland evolution in: 3
Orinoco Delta: 189
ornithischian dinosaurs: 21
Ornithoides trifudus: **186**
ornithopods: 20–21
Osagean: **129**
ostracods
Bairdia: 107
Camishaella: 107
Carbonita: 107
carbonitacean: 107
Copelandella: 107
in Devonian: 104–5, 107
Geisina: 107
habitats of
in Devonian: 104–5, 107
Horton Bluff Formation: 104–5, 107
Joggins Formation: **173**, 174, 186
in Jurassic: 18
Lozier Pond: 294
in Mississippian: 104–5, 107
in Pennsylvanian: **173**, 174, 186
Trout Valley Formation: **60**, 67

in Jurassic: 18
in Mississippian: 104–5, 107
paraparchitacean: 107
in Pennsylvanian: **173**, 174, 186
Shemonaella: 107
Youngiella: 107
Oswayo Formation: 98
Otzinachsonia: 149
Ouachita Mountains: 224
oxbows
in Devonian: 98
Hyde Park Basin: 292–303
in Mississippian: 10, 160–61
organic detritus in: 94
in Pennsylvanian: 236
Oxroadia: 149
oxygen levels
berthierine nodules and: 261
in Carboniferous: 190
methane and: 253, 261–63
in Pennsylvanian: 10
at Permian-Triassic boundary: 261–63
shale formation and: 260–61
Stigmaria and: 10
Streblopteria and: 151
wildfires and: 190
oysters: 152
Ozark Dome: 128

P

Pachydiscoceras changsingense: **262**
Pachytheca: **42**, 43
Pakistan: **250**, 253
Palaeocharinus: **6**
palaeoniscoids: 87, 107–8, **162**, 164. *See also* rays
Palaeophycus: 107
Palaeostachya: 180
Paleoadrovanda splendus: 22
Paleocene
Eocene boundary: 253, 258
global warming in: 258
methane release during: 253, **260**
wetland evolution in: **19**
paleodictyopterans: 13
paleokarst: 44
Palestine Sandstone: 158–60
Palestine Shale: 158–61
Palgham: **250**, **254**
palms: 20–21, 23–27
palsa mires: 14
paludal wetlands. *See also* riverine wetlands
classification of: 3
definition of: 5, 207–8
in Devonian: 5, 7
in Miocene: 29
in Pennsylvanian: **12**
in Pleistocene: 30
in Triassic: 16–18
Pangaea: **11**, **15**, 81
Pantanal River: 21
pantodonts: 26
Paparoa Coal Measures: 270–88
Paradise: **250**, **254**, 255
Paralititan stromeri: 21
Paralycopodites

drawing of: **12**
habitats of: 10, 180, 187
Lagenoisporites rugosus and: **185**
Lycospora and: 160
Lyginopteris and: 161
in Mississippian: 10
in Pennsylvanian: 180, 187
schematic transect of: **189**
paraparchitacean ostracods: 107
Parasuchia: 17–18
parautochthonous, definition of: 58, 235
Parka: **42**, 43, 52
Parkwood Formation: **141**
Passage Creek: **42**, 44–55
Pavari: **157**
Pease River Group
Blaine Formation. *see* Blaine Formation
composition of: 224, 229
depositional environment of: 224, 227, 229–30, **234**, 239–41, 243–44
fossils in: 239
geologic time period of: **225**
San Angelo Formation. *see* San Angelo Formation
sedimentary structures in: **234**
peats
bogs vs. fens: 9, 292
carbon cycle and: 9
in Carboniferous: 9, 14, 16
coal and: 8, 11–13, 287
composition of: 8
cordaites and: 11
in Cretaceous: 20–21
in Devonian: 8–9
dinosaurs and: 21
doming of: 13
in Eocene: 24
in Holocene: 11–14
inertinite and: 210
in Jurassic: 18
lycopsids and: 10
marshes vs. swamps: 7, 26, 108
methane in: 257
in Miocene: 24–25
in Mississippian: **4**, 10
in Oligocene: 25
in Pennsylvanian: 10–11, 14, 200, 205–8
permafrost and: 14, 258
in Permian: 14–16
at Permian-Triassic boundary: 257, 262
pH level in: 8
Rhynie Chert: 5–6, **42**–43
Sphagnum and: 3, 14–15, 29
subsidence and: 205
successions in: 11–13
thickness of: 23, 287
in Triassic: 16, 262
water table and: 13
pecenoids: 227
Pecopteris: 160, 236–38, 241–**42**, 244–45
Pecopteris cyathea: 238
Pederpes: 163, 164
pelecypods: 26, **173**, 227, 229
peltasperms: 15–16, 236, 239–40
Peltobatrachus: 161

Pendleian: **157**
Pennington Formation: 151
Pennsylvania
Big Sewickley Creek: 200
Bloomsburg Formation: 43, 47
Catskill Formation: 80–99, 163
Clinton County: **81**
Lock Haven Formation: 81
maps of: **81**, **199**
Monongahela Group: 198–221
Oswayo Formation: 98
Pittsburgh Formation: 198–221
Potter County: 89
Red Hill: 80–99, **157**, 163
Redstone Creek: 200
Trimmers Rock Formation: 81
Tuscarora Formation: 52–53
Waynesburg coal bed: 198, **199**, 202–3
Pennsylvanian. *See also* Namurian; Stephanian; Virgilian; Westphalian
Ames Limestone: **157**, 199
Atoka Formation: **129**
Bloyd Formation: **129**
Bowie Group. *see* Bowie Group
Carbondale Formation: 147–48
Caseyville Formation: **159**
Catskill Formation: 80–99, 163
Cisco Group: 224, 225, 227, 236–38
coal from
in Bowie Group: 227, 230, 236
in Cisco Group: 236
formation of: 10–14
in Illinois Basin: 243
in Pittsburgh Formation: 198–213
extinctions in: 80
Francis Creek Shale: 147–48
Garnett: **157**
glaciation cycles in: 224
Hale Formation: **129**
Howard: **157**
Joggins Formation. *see* Joggins Formation
maps of: **11**, **158**
Markley Formation: 225–27, 230–34, **238**, 240–41
Mazon Creek: 13, 148, **157**
miospore zones in: **199**
Monongahela Group: 198–221
Pittsburgh Formation: 198–221
plant partitioning in: 80
Sangre de Cristo: **157**
tetrapods in: 156
Uniontown Formation: 198, **199**, 202–3
wetland evolution in: 10–14
zonation in: 10
pentoxylaleans: 16
perissodactyls: 25–26
permafrost: 14, 29, 257–59
Permian. *See also* Leonardian; Wolfcampian
Archer City Formation. *see* Archer City Formation
Blaine Formation. *see* Blaine Formation
Bowie Group. *see* Bowie Group
Buckley Formation: **250**, **251**
Changsing Limestone: **250**, **251**, **262**

Bold numbers indicate material in figures and tables.

Clear Fork Formation. *see* Clear Fork Formation
coal from
　in Bowie Group: 227
　characteristics of: 14–15
　map of: **15**
　Triassic boundary: 16, 249–56, 258–59, 261–63
glaciation cycles in: 15
Lueders Formation: **225**, 227, **231**
Markley Formation: 225–27, 230–34, **238**, 240–41
Nocona Formation. *see* Nocona Formation
Pease River Group. *see* Pease River Group
Petrolia Formation. *see* Petrolia Formation
San Angelo Formation. *see* San Angelo Formation
Triassic boundary: 16, 249–64
Waggoner Ranch Formation. *see* Waggoner Ranch Formation
wetland evolution in: 14–16
Wichita Group. *see* Wichita Group
Yinkeng Formation: **250**, **251**, **262**
Peromonolites: 282, **284**
Pertica: 67–70
Pertica quadrifaria: 6, 68–75
Petrified Forest: 16–17
Petrolia Formation
　composition of: 227
　depositional environment of: 227, **234**, 243
　Elm Creek Limestone and: 225
　fossils in: 227, 241
　geologic time period of: 225
　photograph of sample: **238**
　photographs of site: **228**
　sedimentary structures in: **234**
Pfefferkorn hand sample method: 129
Philonotis fontana: 297
Phoberomys pattersoni: 29
Pholiderpeton: 163
photoautotrophs: 53
photosynthesis: 22, 28, 257
Phragmites: 28
Phthonia sectifrons: 67
Phyllocladidites mawsonii: 281–82, **284–87**
phyllolepidid placoderms: 99
Phyllotheca: 14
phytoplankton: 256
phytosaurs: 17–18
Picea: 22, 293, **296**, 300
Picea glauca: 297
Picea mariana: 29
piedmont: 172, 176, 191
Pinaceae: 22
pines
　amber and: 22
　habitats of: 286, 301–2
　hemlock: 292
　Huon: 281, 286
　Kauri: 22
　Lagarostrobus franklinii: 281, 286
　Pinus: 22, **24**, 293
　in Pleistocene: 301–2
Pinguicula: 22
Pinna: 227

Pinnularia: **160**
Pinus: 22, **24**, 293
Pinus banksiana: 301
pitcher plants: 29
Pitkin Formation: **129**
Pittsburgh coal bed: 198–216
Pittsburgh Formation: 198–221
Pitus: 99
Pityosporites westphalensis: 204, **214–21**
Pityoxylon: 22
Placerias: 17
placoderms: 87, 99, 108
Planisporites granifer: **217–18**
Planolites: 107, 150
plants
　aborescent: 5, 7
　bryophytes. *see* bryophytes
　carbon cycle and: 28, 257
　carbon dioxide, adaptations to: 261
　carnivorous: 22
　community structure of: 128, 245
　embryophytes: 3, 43, 52, **68**
　epiphytic: 212
　homosporous: 5
　origin of land: 3
　photosynthesis by: 22, 28, 257
　pneumatophores: 4
　preservation of: 123
　pre-tracheophyte: 3
　prevascular: 3
　response to burial: 74
　sedimentary structures and: 108
　"turfing in:" 58, 75
　vascular. *see* vascular plants
　zonation of
　　in Devonian: 7, 9, 80
　　in Langsettian: 10
　　in Pennsylvanian: 80
　　in Permian: 14
Platanus: 20, 25
Platte River: 53
Platybelodon: 27
Platyphyllum: 98
Plecoptera: 13
Pleuromeia: 16
Pliocene: 29
pneumatophores: 4
Pneumodesmus: 5
Poaceae: 6, 28, 151
podocarpaceous conifers: 18, 20, 25
podocarps
　in Cretaceous: 20, 270, 282
　habitats of: 25, 270
　in Jurassic: 18
　in Miocene: 25
　in Oligocene: 25
Podolskian: 13, 148, **157**
Podozamites: 20, 239
Pohlia: **297**
Point Edward: **157**, **162**
Polly Brook Formation: 172
polycarpic lycopsids
　Diaphorodendron. *see* *Diaphorodendron*
　habitats of: 183
　in Pennsylvanian: 183

Sigillaria. see *Sigillaria*
Polymorphopteris polymorpha: 238
polysporangiophytes: 5, 43
　Cooksonia: 4, **42**
Polytrichastrum alpinum: **297**, 298
Polytrichum juniperinum: 297
Polytrichum strictum: 297
Pomeroy coal bed: 200, 202
pond cypress: 24
pond weed: **19**, 23–**24**, **29**, 296
poplars: 25, **302**
Populus: 25
Populus deltoides: 25
Pororari Group: 284
Portal Mountain: **250**, **254**
Portugal: 18
Posidonia: **19**, 22
Potamogeton: **19**, 23, **24**, **29**, 296
Potamogeton epihydrus: **296**
Potamogeton filiformis: **296**
Potamogeton foliosus: **296**
Potamogeton obtusifolius: **296**
Potamogetonaceae
　in Cretaceous: 22
　habitats of: 22
　pond weed: **19**, 23–**24**, **29**, 296
　Posidonia: **19**, 22
　sea grass: **19**, 22, 27
　Thalasssodendron: 22
Potomac River: 189, 190
Potonieisporites bharadwaji: **199**
Potonieisporites elegans: 204
Potonieisporites novicus: **199**
Potter County: 89
potter wasps: 22
Pourakino Trondjemite: 258
Powder River Basin: 287
Powers site: **302**
Pragian
　fossils from: **42**
　Rhynie Chert: 5–6, 42–43
　wetland evolution in: **4**, 5, 43
prasinophyceans: 44
praying mantids: 22
precipitation: 8, 18
Presbyornis: 27
pre-tracheophyte plants: 3
prevascular wetlands: 3
Price Formation: 111, **113**, 122, 123
Pride Mountain Formation: 140–41, 150
Pridolian
　arthropods in: 44
　fossils from: **42**
　wetland evolution in: **4**
primates
　Aegyptopithecus: 26
　in Eocene: 26
　habitats of: 26
　humans: 30, 253, 263
　in Oligocene: 26
　Propliopithecus: 26
　Siamopithecus: 26
proboscideans
　elephants: 27, 30
　habitats of: 26, 27

mammoths: 30
mastodons. *see* mastodons
Procaimanoidea: **27**
Proganochelys: 18
progymnosperms
   aneurophytalean: 80
   arthropods and: 5
   Callixylon erianum: 98
   Cardiopteridium and: 132
   in Devonian: 7, 80–81, 88
   in Frasnian: 7
   in Givettian: 5
   habitats of: 5, 7, 80–81, 88, 180
   in Pennsylvanian: 180
   in Permian: **15**
   root system of: 122
   schematic transect of: **189**
Prolepidodendron breviinternodium: 80
Propliopithecus: 26
Proprisporites tectus: **219**
proterogyrinids: **162**, 163–64
Proterogyrinus: **162**
Protobarinophyton pennsylvanicum: 89
Protodonata: 13
Protohaploxypinus amplus: **217–18**
Protohaploxypinus samoilovichi: 259
Protophyllocladus: 20
Protopterus: 9
protorothyridids: 13, **14**, 171
Protosalvinia: 52
protosaurs: **17**
Protosphagnales: 29
Protostigmaria: 103, **106–7**, **110–13**, 120–22
Protostigmaria eggertiana: 111
Prototaxites: 42–43, **68**, 76
Proto-Tethys sea: 96
Psaronius: 10–11, 212–13
Pseudadiantites rhomboideus: 180, 189
Pseudobornia ursina: 99
Pseudobradypus: **186**
Pseudoclaraia wongi: **262**
Pseudofrenelopsis: 19
Pseudomariopteris cordata-ovata: **238**
Pseudosporochnus: **4**
Pseudovoltzia: 239
Psilophyton: **4**, 64, 67–75
Psilophyton dapsile: **68**, 71–72
Psilophyton forbesii: 65, 68–69, 71–72, **73**
Psilophyton microspinosum: **68**, 71, 74
Psilophyton princeps: 65, 68–69, 71–72
pteridophytes: 18, 23
pteridosperms
   Aglosperma: 98
   Alethopteris decurrens: 180, **183**
   Alethopteris discrepans: 180, **183**
   Alethopteris zeilleri: 238
   Alethopteris cf. *u rophylla*: 180
   Aneimites acadica: **106**
   Cardiopteridium and: 132
   in Devonian: 81, **106**, 122
   Dicroidium: 16
   Dicroidium zuberi: 259
   Diplotmema patentissimum: **106**
   Gensella: 146
   Glossopteris: 14–16, 258

Glossopteris browniana: **262**
habitats of
   Albert Formation: **106**
   Bowie Group: 236–38
   Catskill Formation: **81**
   Cheverie Formation: **106**
   Cisco Group: 236–38
   in Devonian: 81, **106**, 122
   Hartselle Sandstone: **141**, 142, 146
   Horton Bluff Formation: **106**
   Horton Group: **106**, 122
   Joggins Formation: 180, **184**, 189
   in Mississippian
     East Kirkton: 10
     Horton Group: **106**, 122
     Jefferson County: **141**, 142, 146
   in Namurian: 10
   in Pennsylvanian
     characteristics of: 11, 213
     Nova Scotia: 180, **184**, 189
     Texas: 236–38, 240
     West Virginia: 204, **221**
   Pittsburgh Formation: 204, **221**
   Silesian Basin: 10
   in Triassic: 16
lyginopterids. *see* lyginopterids
medullosan: 11, 236–38
in Mississippian, habitats of
   characteristics of: 10
   East Kirkton: 10
   Horton Group: **106**, 122
   Jefferson County: **141**, 142, 146
in Namurian: 10
Neuropteris auriculata: 238
Neuropteris ovata: 238
Neuropteris scheuchzeri: 236–38
in Pennsylvanian
   growth strategies of: 11
   habitats of
     characteristics of: 11
     Nova Scotia: 180, **184**, 189
     Texas: 236–38, 240
     West Virginia: 204, **221**
at Permian-Triassic boundary: 16, 263
photographs of samples: **147**
preservation of: 109
schematic transect of: **189**
Schopfipolenites: 203
Schulzospora and: 160
Sphenopteridium: 99, **106**, 239–40, 243
Sphenopteris strigosa: **106**
Telangiopsis: 98–99, 146, **147**
in Triassic: 16
Triphyllopteris minor: **106**
Triphyllopteris virginiana: **106**
Pteroplax: 163
Pteroshonus: 25
Ptilophyllum: 18
Puerto Rico: 27
Puertollano: 190
Pulaski: 111, **113**, 122, 123
Punctatisporites: 203
Punctatisporites aerarius: **214–15**, 220
Punctatisporites breviornatus: **214–20**
Punctatisporites glaber: **214–15**, 220

Punctatisporites minutus
   habitats of: 203, 212, **214–19**, 221
   miospore zone: **199**
Punctatisporites parvipunctatus: 203, 208, **214–15**, **217–21**
Punctatisporites pseudolevatus: **220**
Punctatisporites punctatus: **214–15**, **220–21**
Punctatoporites minutus: 203, 213, **214–21**
Purbeck beds: 19
pyrite: 260–61

Q

Quaestora amplecta: 130–31, 134–35
Quebec
   Battery Point Formation: 42–43, 63, 74, 76
   in Emsian: 5, 42
Quercus: 25, **29**, 292
quillworts: 16, 257

R

Ragnarok hypothesis: 263
raised mires: 13, 20, 23
Raistrickia: **216**, **220**
Raistrickia diversa: **217–18**
Ramsar Convention: 2, 6, 22
Raniganj Basin: **250**, **254**, 261–62
Ranunculus: 296, **297**
Raphia: 25
Rappuhn site: **302**
rattan palm: 25
Rawley Springs: 80, 90, 97–98
rays: 25, 99, 108
Red Hill: 80–99, **157**, 163
red mangrove: **15**, 25, 26
Red River: 227
Red River-Matador uplift: 224
Redoak Hollow Formation: 150
redoximorphic features: 230
Redstone coal bed: **199**–205, 208–10, 212–13, **217–19**
Redstone Creek: 200
Redstone Limestone: **199**
reed grass: 28
reeds
   functions of: 7
   habitats of: 23, 28
   in Miocene: 23, **24**
   Typha: **27**, 28, 296, **297**
Renalia: 42
Renaultia footneri: 180, **183**
Renaultia schatzlarensis: 180, **183**
reptiles
   alligators: 14, 17, 25, **27**
   anthracosaurs and: 108
   crocodilians. *see* crocodilians
   dinosaurs: 17–18, 20–21, 23
   habitats of: 13–14, 165, 171
   in Holocene: 13–14
   lizards: 18
   in Pennsylvanian: 14, 165, 171
   Permian-Triassic boundary and: 255
   suchians: 17–18
   therapsids: 16, 255, 263
   tortoises: 26

Bold numbers indicate material in figures and tables.

in Triassic: 18, 263
turtles: 18, 21, 23, 25–27
reptiliomorphs: 9, 186
Rewanui: 270–72
Rhacophyton
   Archaeopteris and: 80
   Callixylon and: 80
   Cephalopteris mirabilis and: 99
   coal and: 8
   in Devonian
      drawing of: **7**
      habitats of
         characteristics of: 7–9, 98
         Red Hill: 80–81, 85, 90–91, 95–97
      size of: 87
      tetrapods and: 9
   gymnosperms and: 97
   habitats of
      Catskill Formation: 80–81, 85, 90–91, 95–97
      in Devonian
         characteristics of: 7–9, 98
         Red Hill: 80–81, 85, 90–91, 95–97
         tetrapods and: 9
   height of: **4**
   reproductive strategies of: 97
   root system of: 97
   wildfires and: 96–97
Rhacophyton ceratangium: **81**, 87, 98–99
Rhadinichthys: 108
rheotrophic coal: 190–92
Rhetinangium arberi: 131
rhinesuchids: 15
rhinoceroses, wooly: 30
rhizodonts: 99, 108, **159**, 161–64
Rhizodus hardingi: 108
rhizoliths: 233
Rhizophora: **15**, 25, 26
Rhodea subpetioleata: 133
Rhuddanian: **42**
Rhynia: **6**
Rhynia gwynne: **4**
Rhynie Chert: 5–6, 42–43
Rhyniognatha: 5
rhyniophytes
   arthropods and: 5
   Cooksonia: 4, **42**, 43
   definition of: 4
   in Devonian: 5–6, 68, 76
   in Emsian: 5
   habitats of: 4–6, 68, 76
   height of: 4
   origin of: 4
   reproductive strategies of: 5
   in Siegenian: 5
   in Silurian: 4
   Taeniocrada: 59, 65, 68–69, 71–72
rice: 30
rider coal zone: 197
riparian wetlands
   archeological sites in: 30
   calamiteans in: 205
   in Cretaceous: 21
   definition of: 4
   in Devonian: 6–9
   dinosaur trackways in: 21
   in Eocene: 24
   Hancock County (Kentucky): 156–65
   in Holocene: 4, 9, 17
   in Jurassic: 18
   in Miocene: 23
   in Mississippian: 132, 156–65
   in Pennsylvanian: 11–13
   in Permian: 15–16
   in Silurian: 4
   in Triassic: 15–18
river flats, definition of: 54
riverine wetlands. *See also* paludal wetlands
   Battery Point Formation: 42–43, 63, 74, 76
   bedload transportation in: 54
   braided: 54, 63
   characteristics of: 53
   in Cretaceous: 20
   crocodilians in: 27
   definition of: 5
   in Devonian
      aborescence and: 7
      amphibians in: 9
      Battery Point Formation: 42–43, 63, 74, 76
      progymnosperms in: 7
      rhyniophytes in: 5
      tetrapods in: 9
      trimerophytes in: 5
      Trout Valley Formation: 6, 58–76
      zosterophylls in: 5
   in Eocene: 26
   Geissel: 26
   in Holocene: 6–7, 9
   in Llandoverian: 44–55
   Massanutten Sandstone: 44–55
   Messel: 26
   in Mississippian: 158–60
   Palestine Sandstone: 158–60
   Passage Creek: **42**, 44–55
   in Pennsylvanian: **12**
   in Pragian: 43
   Rhynie Chert: 5–6, 42–43
   in Siegenian: 5–6, 42–43
   in Silurian: 4, **42**, 44–55
   Springhill Mines Formation: 172, 176, 191
   in Triassic: **17**
   Trout Valley Formation: 6, 58–76
   Tuscarora Formation: 52–53
   whales in: 27
roaches: 13, 22
rodents: 18, 29
Rogersville Shale: 43
Romer's Gap: 104, **157**, 164
Rubidgea majora: **262**
ruflorian cordaites: 15, 263
Rugispora flexuosa: 83
rugose corals: 261
Ruhr Basin: 192
rushes
   calamiteans and: 10
   flooding and: 7
   habitats of: 28, 300
   Juncus: 28, **296**
   in Miocene: 28
   origin of: 28
   in Pleistocene: 300
   reeds and: 7
Rusophycus: 107
Russelites: 239–40
Russia
   coal in: 18, 20
   in Cretaceous: 20
   in Devonian: 99, 108
   in Jurassic: 18
   in Mississippian: 10
   in Permian: 29
   Siberia. *see* Siberia
   in Triassic: **250**, 256, 258
   Tunguska Basin: 258

S

sabkha deposits: 229, 244
Sagenodus: **162**
Sagittaria latifolia: 296, **297**
St. Catherines Island: 151–52
St. Genevieve time period. *See* East Kirkton
St. Helens, Mount: 190
St. Louis time period. *See* Delta
salamanders: 18
Salem time period: **159**
Salix: 25, **29**, **296**
salt: 6, 229
salt grass: 28
salt marshes
   biodiversity in: 4
   carbon cycle in: 28
   definition of: 4
   in Devonian: 75, 122
   frenelopsids in: 19
   grasses in: 28
   lycopsids in: 122
   in Mississippian: 10, 122
   plant morphology: 4
   ravinement of: 140
   root systems in: 151
   vs. tidal flats: 139–40
   in Triassic: 16
Salvinia: 19
San Angelo Formation
   Blaine Formation and: 224–25, 229
   composition of: 224, 229
   depositional environment of: 229–30, 234–35, 239, 243–44
   fossils in: 229, 239
   geologic time period of: **225**
   Llano Uplift and: 224–25
   photographs of samples: **240**
   sedimentary structures in: 230, 234–35
Sangre de Cristo: **157**
sapropelic coal: 8
Sasayama: **250**, 253, **254**
Sawdonia: **4**
Scat Craig: **157**
Schoharie Valley: 63
Schöningen: 30
Schopfipolenites: 203
Schopfipolenites ellipsoides: 204, **220**
Schuchert Dal: **250**, 253, **254**, 259
Schulzospora: 160
Sciadophtyon: **68**
Sciatophyllum: 16

Sciodopitys: **24**
Scirpus: 28
Scirpus tabernaemontani: 296, **297**
Scorpidium scorpidoides: **297**, 300
scorpions
  in Devonian: 99
  in Emsian: 5
  habitats of: 5, 44, 99, **184**, 186
  in Pennsylvanian: **184**, 186
  in Pridolian: 44
Scotland
  Arran: 190
  in Carboniferous: 156
  Cheese Bay: **157**, 165
  Clementstone deposits: 99
  in Devonian: 108
  East Kirkton: 9–10, 156–**57**, **159**, 164–65
  Midland Valley: 156
  in Mississippian: 156, 164–65
  Rhynie Chert: 5–6, 42–43
  in Silurian: 5
scour: 118–19
scratch circles: 115–17
sea grass: **19**, 22, 27
seat earths, rooted: 11–13
sedges
  bulrushes: 28, 300
  carbon cycle and: 28
  Carex: 28, **29**
  Cyperus: 28
  in Eocene: 26
  habitats of
    Australia: 28
    bogs vs. fens: 292
    in Eocene: 26
    in Holocene: 6, 28–29
    kettle holes: 301
    Lozier Pond: **296**, 300
    Messel: 26
    in Miocene: 28
    in Oligocene: 28
    in Pleistocene: **29**, **296**, 300
  in Holocene: 6
  in Miocene: 28
  origin of: 28
  photosynthesis by: 28
  Phragmites: 28
  in Pleistocene: **29**, **296**, 300
  Scirpus: 28
Selaginella selaginoides: 296
Semionotus: **17**
Senftenbergia plumosa: **184**
Sequoia: 20, 23, **24**
Serpukhovian. *See also* Chesterian; Namurian
  Fayetteville Formation: 127–35
  Pitkin Formation: **129**
serpulids: 107, 174, 227
Setosisporites: 149–50
Setosisporites praetextus: 150
Sewickley coal bed: **199**–205, 210–13, **220**–21
seymouriamorphs: 108, 163, 165
Shangsi: **250**, 254
Shapeless Mountain: **250**, 254
sharks
  ageleodid: 99
  in Cretaceous: 21
  ctenacanthid: 99
  in Devonian: 99, 108
  elasmobranch: 108, 123
  in Eocene: 25
  habitats of
    Albert Formation: 108
    Catskill Formation: 99
    in Cretaceous: 21
    in Devonian: 99, 108
    in Eocene: 25
    in Jurassic: 18
    in Mississippian: 108
    in Permian: 227
    Petrolia Formation: 227
  in Jurassic: 18
  in Mississippian: 108
  in Permian: 227
  Stethacanthus: 108
  Weichselia reticulata and: 21
Sheinwoodian: **42**
Shelton site: **302**
Shemonaella: 107
shrimp: 6
shrub mires: **4**, 8
shrubs: 6–7
Siamopithecus: 26
Siberia
  coal in: **11**, 15
  in Devonian: 99
  maps of: **11**, **15**, **250**
  Noril'sk iron ores in: 259
  in Permian: 15
  Sphagnum in: 14, 29
Siberian Traps: **250**, 256, 258
sickle cell disease: 22
Siegenian. *See also* Pragian
  height of plants in: **4**, 5
  Rhynie Chert: 5–6, 42–43
Sigillaria
  Crassispora kosankei and: 204–5, 207–8, 212–16, **220**
  drawing of: **12**
  habitats of
    Joggins Formation: 156, 179–80, 183, 187
    in Pennsylvanian
      Nova Scotia: 156, 179–80, 183, 187
      West Virginia: 204–8, 212, **214**–**16**, **220**
    in Permian: 241
    Pittsburgh Formation: 204–8, 212, **214**–**16**, **220**
    Verdeña: 179
    in Westphalian: 130–32
  in Pennsylvanian, habitats of
    characteristics of: 11
    Nova Scotia: 156, 179–80, 183, 187
    West Virginia: 204–8, 212, **214**–**16**, **220**
  in Permian: 241
  schematic transect of: **189**
  Tuberculatisporites and: 182
  in Westphalian: 130–32
Sigillaria brardii
  habitats of: 236–38, 245
  in Pennsylvanian: 236–38
  in Permian: 245
  photographs of samples: **238**, **242**
Sigillaria laevigata: 180
Sigillaria mamillaris: 180, **181**
Sigillaria rugosa: 180
Sigillaria scutellata: 180, **181**
Sigillariostrobus: 180, **181**
Silesian Basin: 192
Silurian: 3–5, **42**, 44–55. *See also* Llandoverian; Ludlovian; Pridolian; Wenlockian
Silvanerpeton: 165
sinkholes: 301
sirenians: 16, 22, 25–27
Siusi: **250**, 254
Skolithos: **60**–**61**, 64–67, 108
Skolithus: 53
slickensides: 160, 233
Slovenia: **250**, 254
snakes: 25–26
South Africa
  Bethulie. *see* Bethulie
  carbon cycle in: 258
  Carleton Heights: **250**, 254
  coal in: 14
  Karoo Basin: 15–16
  Lootsberg: **250**, 254
  in Permian: 258, 261
  Tweefontein: **250**, 254
  Wapadsburg: **250**, 254
South America
  Amazon Basin: 21, 25, 232
  Brazil. *see* Brazil
  carbon cycle in: 258
  in Carboniferous: **11**
  Chile: 30
  coal in: **11**, **19**–**20**, **23**
  in Cretaceous: 20, **20**, 25
  in Devonian: 99
  in Jurassic: **19**
  maps of
    Carboniferous: **11**
    coal deposits: **11**, **19**, **20**, **23**
    Cretaceous: **20**
    Jurassic: **19**
    Miocene: **23**
    Permian: **225**
    at Permian-Triassic boundary: **250**
  in Miocene: **23**
South Branch Ponds Brook: **59**
South Bulli Mine: **255**
South Platte River: 53
Sovetashan: **250**, 254, 255
Spackmanites rotundus: **219**
Spain
  Ambrona: 30
  Puertollano: 190
  Torralba: 30
Sparganium: 296
Spartina: 28, 151
Spathicephalus: **162**
spermatopsids: 19, 130–31
Spermolithus devonicus: 99
Sphagnum: 3, 14–15, 29
Sphenophylls: **237**
Sphenophyllum: 241, 245
Sphenophyllum oblongifolium: 238

Bold numbers indicate material in figures and tables.

Sphenophyllum subtenerrimmum: 98
sphenopsids
   Annularia carinata: 238, 240
   Archaeocalamites. see *Archaeocalamites*
   calamitean. see calamiteans
   classification of: 130
   in Devonian: 6–7, 98
   equisetales. see equisetales
   flooding and: 132
   growth strategies of: 7
   habitats of
     Catskill Formation: 98
     in Devonian: 6–7, 98
     Fayetteville Formation: 130–35
     Joggins Formation: 180, 187, 190
     Markley Formation: **238**
     in Mississippian: 10, 130–35, 161
     in Namurian: 10
     in Pennsylvanian
       characteristics of: 80, 132
       Nova Scotia: 180, 187, 190
       Texas: 236–38
     in Permian: 236, 241, 244–45
     Silesian Basin: 10
   Lilpopia raciborskii: 238
   in Mississippian: 10, 130–35, 161
   in Namurian: 10
   in Pennsylvanian
     growth strategies of: 11
     habitats of
       characteristics of: 80, 132
       Nova Scotia: 180, 187, 190
       Texas: 236–38
     lycopsids and: 236
   in Permian: 14, 236, 241, 244–45
   Phyllotheca: 14
   Pseudobornia ursina: 99
   reproductive strategies of: 236
   Sphenophyllum oblongifolium: 238
   in Vicean: 10
Sphenopteridium: 99, **106**, 239–40, 243
Sphenopteridium macconochiei: **106**
Sphenopteris
   in Devonian: 90, 98, 99
   habitats of
     Baggy Beds: 99
     Bowie Group: 238
     Cisco Group: 238
     in Devonian: 90, 98, 99
     Fayetteville Formation: 131–35
     Hampshire Formation: 98
     in Mississippian: 9, 131–35
     in Pennsylvanian: 238
     in Permian: 239
   Lyginopteris and: 160
   in Mississippian: 9, 131–35
   in Pennsylvanian: 238
   in Permian: 239
   photographs of samples: **133**
Sphenopteris effusa: 180, **183**
Sphenopteris hookeri: 99
Sphenopteris mississippiana: 131, 133, 135
Sphenopteris schimperiana: 133–34
Sphenopteris strigosa: **106**
spiders: 5, 186, **187**

Spinosaurus: 21
Spinosporites exiguus: **199**, 203, **214–16**, **220–21**
Spirobis: **173**
spirorbid worms: 227
Spiti: 259
Spitzbergen: **250**, **254**
Spongiophyton: 42, 52
Sporongonites: **68**
Springhill Mines Formation: 172, 176, 191
springtail: **6**
spruces: 22, **296**–97, 300–303
Starke County: **302**
Staten Island: 292
stauropterids: **81**, 85–90, **92**, 95–98
Stephanian. See also Conemaughian; Klazminski-
   an; Missourian; Noginskian; Virgilian
   Ames Limestone: **157**, **199**
   Garnett: **157**
   Howard: **157**
   Kounova: **157**
   miospore zones in: **199**
   mires in: 11
   Pittsburgh Formation: 198–221
   Sangre de Cristo: **157**
   swamps in: 11
Stereisporites: 282
stereospondyls: 165
Stethacanthus: 108
Stigmaria
   "coal balls" and: 160
   habitats of
     Battleship Wash Formation: 151
     Horton Group: 107
     Joggins Formation: **173**
     in Mississippian: 107, 151, 160
     in Pennsylvanian: **173**
     Wedington Shale: 132
   Lepidodendropsis and: 122
   in Mississippian: 107, 151, 160
   origin of: 149
   in Pennsylvanian: 10, **173**
Stigmaria ficoides: **130**, 180
stone flies: 13
Stonewall County: **225**, 229
storks: 26, 27
strap-shaped macrofossils: 46–51
Streblochondria: **144**, 150–51
Streblopteria: 150–51
Strepsodus: **162**
strip-cruse method: 120
Stromatactis: **252**
stromatolites: 261
strontium: 260–61
Strunian
   Horton Bluff Formation: 103–17, 120–23, **157**
   Memramcook Formation: **105**
   miospore zones in: **105**
suchians: 17–18
sugar maple: 292
sulfur
   in coal
     Joggins Formation: 191
     "Main Seam:" 281, 286
     Pittsburgh Formation: 198, **214–21**
   Permian-Triassic boundary and: 256

root systems and: 151
sundews: 22
Supaia: 239
Sussex: 105, 108, 118–19
swales: 301, **302**
swamps
   amber deposits in: 22
   araucarian: 22
   avulsion deposits in: 178
   on Axel Heiberg Island: 24
   Bowie Group. see Bowie Group
   carbon cycle and: 8
   carnivorous plants in: 22
   Cisco Group: 224, 225, 227, 236–38
   clastic: 11, 108, 179
   in Cretaceous: 20, 22, 23
   cypress: 108
   definition of: 2, 5, 7, 108
   in Devonian: **4**, 5, 7–8
   in Eocene: 24–26
   Horton Group: 103–23, **157**
   Joggins Formation. see Joggins Formation
   in Jurassic: 18
   K-T event and: 23
   vs. mangroves: 26
   vs. marshes: 7, 26, 108, 140
   vs. mires: 11, 108
   in Mississippian: 10
   nitrogen cycle and: 8
   origin of: 7–8
   in Pennsylvanian
     Bowie Group. see Bowie Group
     characteristics of: 10–14
     Cisco Group: 224, 225, 227, 236–38
     Pittsburgh Formation: 198–221
   in Permian: 14–15
   pines in: 281
   Pittsburgh Formation: 198–221
   in Triassic: 16–17
sweet gum: **24**
Swiss Helm Mountains: **157**
sycamore: 20, 25
sycamores: **27**
Sydney Basin: **255**
synapsids: 20
Syncerus caffer: 16

T

tabulate corals: 261
Taconic Highlands: 53
tadpole shrimp: 6
Taeniocrada: 59, 65, 68–69, 71–72
taenioid shape: 47
taeniopterids: 236, 239
Taeniopteris: 239
Taffs Well: 98
Taho: **250**, **254**
Taiping: **250**, **254**
Talcher: **250**, **254**, 255
tamarack: 29, **297**, 301–3
Tantillus triquetrus: **216**
Tasmania: 26, 281, 286
Taxodiaceae: 24, 26, 108
taxodiaceous conifers: 18, 20, 24
Taxodium: 23, 24

Telangiopsis: 98–99, 146, **147**
telinite: 200
telocollinite: 200
Telychian: **42**
temnospondyls
    Broiliellus: 161
    Cacops: 161
    depositional environment of: **159**
    Dissorophus: 161
    habitats of: 9, 15, 163
    in Mississippian: 9, 163
    morphology of: 161, 165
    Peltobatrachus: 161
    in Permian: 15
tempestite: **252**
Tennessee
    Cumberland Plateau: 140
    Hartselle Sandstone: 140–52
    Rogersville Shale: 43
Tern: **250**, **254**
Terrigal: **255**
Tertiary methane release: 253, **260**
Tesoro Horizon: **252**
Tethys Ocean: 25–26. *See also* Proto-Tethys sea
tetrahedral tetrads: 44, 52
tetrapods: 156
    acanthodians. *see* acanthodians
    Acanthostega: 9, 156
    aïstopods: 165
    amphibians. *see* amphibians
    anthracosaurs. *see* anthracosaurs
    in Carboniferous: 156, 161–64, 171, **173**
    colosteids: **159**, 161–62, 164–65
    Densignathus rowei: 99
    in Devonian, habitats of
        Australia: 108
        Blue Beach: 108
        characteristics of: 9, 156
        Greenland: 108
        Latvia: 108
        Red Hill: 79–80, 80, 99
        Russia: 108
        Scotland: 108
    diadectomorphs: 165
    dipnoans. *see* dipnoans
    embolomeres: **159**, **162**–65
    fish. *see* fish
    in Frasnian: 155
    habitats of
        in Carboniferous: 161–64, 171, **173**
        Catskill Formation: 80, 99
        in Devonian
            Australia: 108
            Blue Beach: 108
            characteristics of: 9, 156
            Greenland: 108
            Latvia: 108
            Red Hill: 80, 99
            Russia: 108
            Scotland: 108
        East Kirkton: 9–10, 156, 164–65
        Horton Bluff Formation: 108, 123
        Joggins Formation: 156, 171, **173**, **186**
        in Mississippian
            Blue Beach: 108, 123
            Kentucky: 161–64
            Milner's province: 164
            Scotland: 9–10, 156, 164–65
        in Pennsylvanian: 156, 171, **173**, **186**
        in Permian: 227
        Petrolia Formation: 227
    Hynerpeton bassetti: 99
    lepospondyls: 165
    lissamphibians: 156
    loxommatids: 164, 165
    microsaurs: **162**, 165
    in Mississippian
        vs. Devonian: 156
        habitats of
            Blue Beach: 108, 123
            Kentucky: 161–64
            Milner's province: 164
            Scotland: 9–10, 156, 164–65
            Romer's Gap: 104, **157**, 164
        morphology of: 163–65
    nectridians: 165
    palaeoniscoids. *see* palaeoniscoids
    in Pennsylvanian: 156, 171, **173**, **186**
    in Permian: 227
    proterogyrinids: **162**, 163–64
    reptiliomorphs: 9, 186
    rhizodonts: 99, 108, **159**, 161–64
    seymouriamorphs: 108, 163, 165
    stereospondyls: 165
    temnospondyls. *see* temnospondyls
    terrestrialization of: 9–10, 156, 163–65
    in Tournaisian: 104, 108
    trackways of
        in Horton Bluff Formation: 108, 123
        in Joggins Formation: **173**, 178, 186
    xenacanthids: **159**, **162**–64
Texas
    Albany Group: 224–25, 227
    Arbuckle Mountains: 224
    Archer County: **225**
    Baylor County: **225**, 227
    Bowie Group. *see* Bowie Group
    Cisco Group: 224, 225, 227, 236–38
    Clay County: **225**
    Clear Fork Formation. *see* Clear Fork Formation
    Foard County: **225**
    Hardeman County: **225**
    Haskell County: **225**
    Jack County: **225**
    King County: **225**, 229
    Knox County: **225**
    Lake Kemp: 228
    Llano Uplift: 224–25
    maps of: **225**
    Midland Basin: 224–45
    Montague County: **225**
    Ouachita Mountains: 224
    Pease River Group. *see* Pease River Group
    in Pennsylvanian: 225–28, 229–44
    in Permian: 163, 224–45
    Stonewall County: **225**, 229
    Throckmorton County: **225**, 227
    Wichita County: **225**
    Wichita Group. *see* Wichita Group
    Wichita Mountains: 224–25
    Wichita River: **223**
    Wilbarger County: **225**
Thailand: 26
Thalassia: 22
Thalassocharis: **19**
Thalasssodendron: 22
thalloid microfossils: **4**, 45–52
thallophytes: 42–43, **68**, 76
therapsids: 16, 255, 263
theropods: 17, 21
Thescelosaurus: 21
Thrinaxodon: 263
Throckmorton County: **225**, 227
Thymospora obscura: **199**, 203, 213, **220**
Thymospora pseudothiessenii: 203, 213, **214–18**, **220**
Thymospora thiessenii
    habitats of
        Pittsburgh Formation: 202–3, 208, 212–13, **214–18**, **220**
        Pomeroy coal bed: 202
    miospore zone: **199**
tidal flats
    Blaine Formation: 229
    Hartselle Sandstone: 150
    vs. marshes: 139–40
    in Niger Delta: 178
Tigris River: 30
Timmia norvegica: **297**, 298
Toarcian: 253, **260**
Tomiostrobus australis: **255**
topogenous mires: 13, 286
Torralba: 30
Tortella fragilis: **297**, 298
tortoises: 26
Torttula norvegica: **297**, 298
Tortula: **297**
Tournaisian. *See also* Hastarian; Ivorian
    Albert Formation: 104–6, 108, 118–23
    Cheverie Formation: **105–6**
    height of plants in: **4**
    Horton Bluff Formation: 103–17, 120–23, **157**
    Memramcook Formation: **105**
    miospore zones in: **105**
    Weldon Formation: **105**
tracheophytes
    filicopsids. *see* filicopsids
        habitats of: **81**, 292
    lycopsids. *see* lycopsids
    zosterophylls. *see* zosterophylls
Tranodis: **162**, 163–64
Transgressive Systems Tract: 150
Traveler Rhyolite: 58, 59
tree ferns
    habitats of: 10–11, 14–16
    in Jurassic: 18
    K-T event and: 23
    marattialean. *see* marattialean tree ferns
    in Pennsylvanian: 10–**12**, 14–15
    in Permian: 15
    in Triassic: 16
Weichselia reticulata: 21
trees
    alders: 23–25
    amber from: 22

Bold numbers indicate material in figures and tables.

araucariacean: 18, 22
ash trees: 189, **302**
balsam firs: **29**, 293, 297, 299–303
beech: 20, 25
birches: 29, 301
black spruce: 29
Bothrodendron: 180, 187, **189**
Chaloneria. see *Chaloneria*
chestnut: 25
conifers. *see* conifers
cottonwoods: 25
Cyclostigma: 8, 87, 99, 149
cypress: 24, 26, 108
in Devonian: 7–9
Diaphorodendron. see *Diaphorodendron*
Eospermatopteris: **4**, 7
Fraxinus: 189
growth rings in: 190
hemlock: 292
hickory: 292
in Holocene: 13
lycopsids. *see* lycopsids
lyginopterids. *see* lyginopterids
mangals: 10, 18, 21, 26
Megaloxylon wheelerae: 130–31, 134–35
oak: 25, **29**, 292–93
palm: 20–21, 23–27
Paralycopodites. see *Paralycopodites*
pines. *see* pines
poplars: 25, **302**
Quaestora amplecta: 130–31, 134–35
reproductive strategies of: 10, 11
root system of: 7–10
sphenopsids. *see* sphenopsids
spruces: 22, **296**–97, 300–303
sugar maple: 292
sweet gum: **24**
sycamores: 27
tamarack: 29, 297, 301–3
white spruce: 297, 301–3
willows: 25, **29**, **296**
Tremadocian: **42**
Třemošná: **157**
trepostome bryozoans: 261
Triadobatrachus: 18
Triassic
Araguainha Crater: 259
carbon cycle in: 18
coal from
map of: **17**
Permian boundary: 16, 249–56, 258–59, 261–63
in Tunguska Basin: 258
Fremouw Formation: **250**, **251**
Jurassic boundary: 18, 253, **260**
map of: **17**
Permian boundary: 16, 249–64
wetland evolution in: 15–18
Yinkeng Formation: **250**, **251**, **262**
Triassochelys: 18
Trichecus sp.: 16
trigonotarbids: 5–6, 99
Triletes
in Cambrian: **42**, 43
in Cretaceous: 282

in Devonian: **42**
habitats of
Battery Point Formation: **42**
in Cretaceous: 282
Greenbrier Formation: **148**
Hartselle Sandstone: **145**, 146, 149
Horton Group: 105–6
Massanutten Sandstone: 44, 52
in Mississippian: **145**, 146, **148**, 149
Passage Creek: **42**, 44, 52
Rhynie Chert: **42**
Rogersville Shale: 43
in Tournaisian: 105–6
in Mississippian: **145**, 146, **148**, 149
in Ordovician: **42**
in Silurian: **42**
in Tournaisian: 105–6
Triletes cheveriensis: **106**
Triletes glaber: **106**
trilobites: 107
trimerophytes: **4**–6, **42**, 64, 67–76
Trimmers Rock Formation: 81
Triops: 6
Triphyllopteris minor: **106**
Triphyllopteris virginiana: **106**
Triquitrites minutus: **217**–20
Triquitrites spinosus: **217**–18
tristichopterids: 99
Trivena arkansana: 131
Trochammina: 107
trophic chain, elements of: 42
Trout Brook
faults along: 58
graphic logs of: **61**
lithology along: 58–72
map of: **59**
photographs of sample sites: **62**, **64–67**, **69**
Trout Valley Formation: 6, 58–76
Tsuga canadensis: 292
Tuberculatisporites: 181–82
Tuberculatisporites mammilarius: **184**, **185**
tundra: 29, 290, 294–96, 298
Tunguska Basin: 258
tupelo: 23–25
Turimetta Head: **255**
Turkey
Curuk Dag: **250**, **254**, 255
Kemer Gorge: **250**, **254**
turtles: 18, 21, 23, 25–27
Tuscarora Formation: 52–53
Tuscumbia Limestone: **141**
Tweefontein: **250**, **254**
Typha: **27**, 28, **296**, **297**
Typhaceae
habitats of: 28
reeds. *see* reeds
Typha: **27**, 28, **296**, **297**

U

Uganda: 26
Ukraine: 10
Ulmannia: 239
Ulodendron: 180
Ultisol, formation of: 232
Undichna: 108

ungulates: 25–27
Uniontown coal bed: **199**
Uniontown Formation: 198, **199**, 202–3
Uniramia
centipedes: 5, **6**, 22, 44
diplopods. *see* diplopods
insects. *see* insects
United Kingdom
England. *see* England
Isle of Wight: 21
Purbeck beds: 19
Scotland. *see* Scotland
Wales: 98
Wealden strata: 18
Wessex Formation: 21
United States Fish and Wildlife classification: 2
United States of America (USA)
Alabama. *see* Alabama
Alaska: 22, 29
Appalachians. *see* Appalachians
Arizona: 10, 16–17, 151
Arkansas: 127–35
California: 44
coal in: 10
Colorado: 27, 53, 190
Colorado River Valley: 227
Florida: 17, 22, 30
Georgia: 132, 151–52
Gulf Coast: 23–25, 151–52, 178
Illinois. *see* Illinois
Indiana. *see* Indiana
Iowa: **157**, **159**, **162**, 164
Kansas: 108
Kentucky. *see* Kentucky
Maine: 6, 43, 58–76
Maryland: 198–221
Michigan: 300, **302**
Minnesota: 300
Mississippi River: 108, 178, 190, 232
Missouri River: 232
Nebraska: 53
New Jersey. *see* New Jersey
New Mexico: 163
New York (state). *see* New York (state)
North Carolina: **152**
Ohio. *see* Ohio
Oklahoma: 127–35, 163, **231**
Pennsylvania. *see* Pennsylvania
Potomac River: 189, 190
Puerto Rico: 27
Tennessee: 43, 140–52
Texas. *see* Texas
Utah: 190
Virginia. *see* Virginia
West Virginia. *see* West Virginia
Wisconsin: 52
Utah: 190
Utricularia: 22

V

Val Badia: **250**
Valvisisporites auritus: 150
Van Buren County: **302**
vascular plants
arthropods and: 5

carbon dioxide and: 57
in Cretaceous: **27**, 28
glaciation cycles and: 58
habitats of: 5, **297**, 300–301
origin of: 4
in Pleistocene: 296, **297**, 300–301
in Siegenian: 5
VCo miospore zone: 83
Vedi: **250**, **254**
Venezuela
fossils in: 29
Maracaibo Basin: 26
Orinoco Delta: 189
Verdeña: 179
Vereiskian: **157**
vernal ponds. *See* wet meadows
Verrucosisporites bacculatus: **220**
Verrucosisporites compactus: **220**
Verrucosisporites donarii: **217**, **219–20**
Verrucosisporites microtuberosus: **214–15**
Verrucosisporites sifati: **220**
Verrucosisporites verrucosus: **214–15**, **220–21**
Vertebraria: 14, **251**, 262
Vertebraria australis: **262**
Vertisols, formation of: 233, 235
Vesicaspora wilsonii: 204, **214–21**
Viburnum: **24**
Vienna Limestone: 158
Vietnam: 232
Viola: 296
Virgilian
Bowie Group. *see* Bowie Group
Markley Formation: 225–27, 230–34, **238**, 240–41
miospore zones in: **199**
Virginia
Green Mountain: 45
Hampshire Formation: 80, 90, 97–98
Martinsburg Formation: 53
Massanutten Sandstone: 44–55
Passage Creek: **42**, 44–55
Price Formation: 111, **113**, 122, 123
Pulaski: 111, **113**, 122, 123
in Triassic: 17
Tuscarora Formation: 52–53
Viséan. *See also* Arundian; Brigantian; Chadian; Chesterian; Holkerian; Meramecian; Osagean
Austria. *see also* Asbian
Batesville Formation: **129**
Boone Formation: **129**
Broxburn: **157**
Cheese Bay: **157**, 165
Delta: **157**, **159**, **162**, 164
East Kirkton: 9–10, 156–**57**, **159**, 164–65
Fayetteville Formation: 127–35
Gilmerton: **157**
Hindsville Formation: **129**
tetrapods in: **157**, **158**
vitrodetrinite: 202
Vittatina: **217–18**
viviparity: 26
volcanoes: 253–61
Voltziopsis africana: **262**
voynovskyalean cordaites: 15

W

Waagenites wongiana: **262**
Waggoner Ranch Formation
depositional environment of: 227, **234**, 243
fossils in: 227
geologic time period of: **225**
photographs of site: **231–32**
sedimentary structures in: **234**
Waikato Coal Measures: 282
Waiomo: **271**
Wairoa Gorge: **250**, **254**, 259
Walchia: 239
Wales: 98
Walker County: 140–52
Wapadsburg: **250**, **254**
Wardha: **250**, **254**
Wardie: **157**
Warnstorfia exannulata: 297
Warren mastodon: 301
Warsaw time period: **159**
wasps: 22
water buffalo: 16
water clover: 19
water hyacinth: **19**
water lilies: **19**, 20, 26, **29**
water milfoils: 20
waterbugs: 22
Waynesburg coal bed: 198, **199**, 202–3
Waynesburg Limestone: **199**
Wealden strata: 18
Wedington Sandstone/Shale: 128–32
Weichselia reticulata: 21
Weldon Formation: **105**
Wellington Formation: **231**
Wells site: **302**
Wenlockian. *See also* Homerian; Sheinwoodian
Bloomsburg Formation: 43, 47
fossils from: **42**
rhyniophytes in: 4
thalloid plants in: **4**
Werfen Formation: **252**
Wessex Formation: 21
West Virginia
Bluefield Formation: 132
in Devonian: 88, 97
Elkins: 80, 90, 97–98
Greenbrier Formation: 140, 146–48, 150–51
Greer: **157**, **159**, **162**, 164
Hampshire Formation: 80, 90, 97–98
Hinton Formation: 132
Illinois Basin and: **162**
Iowa and: **162**
Kentucky and: **162**, 164
maps of Pittsburgh Formation: **199–201**
Nova Scotia and: **162**
Pittsburgh Formation: 198–221
Tuscarora Formation: 52–53
Western European Carboniferous Basin: 186
Westphalian. *See also* Cheremshanskian; Duckmantian; Kashirskian; Langsettian; Melekesskian; Myachkovskian; Podolskian; Vereiskian
Atoka Formation: **129**
Bloyd Formation: **129**
Carbondale Formation: 147–48
cordaites in: 11
extinction: 11
Francis Creek Shale: 147–48
Gastrioceras subcrenatum band and: 172
Jarrow: **157**
Joggins Formation. *see* Joggins Formation
Linton: **157**
map of tetrapod distribution: **158**
Mazon Creek: 13, 148, **157**
Namurian boundary: 172
Newsham: **157**
Nýrany: **157**
Swiss Helm Mountains: **157**
tetrapods in: 156
Tremošná: **157**
wet meadows: 6, 17
wetlands
characteristics of: 2
classification of: 2, 3
definition of: 3
estuarine. *see* estuarine wetlands
functions of: 2–3, 30
human use of: 30
successions in: 11–13
types of
bogs. *see* bogs
coastal. *see* coastal wetlands
fens. *see* fens
forest mire. *see* forest mires
geothermal. *see* geothermal wetlands
lacustrine. *see* lacustrine wetlands
marsh. *see* marshes
nonforested mire. *see* fens
paludal. *see* paludal wetlands
prevascular: 3
riparian. *see* riparian wetlands
riverine. *see* riverine wetlands
salt marshes. *see* salt marshes
swamps. *see* swamps
wet meadows: 6, 17
Wexfordia: 99
whales: 27
Whatcheeria: **159**, **162**–65
white spruce: 297, 301–3
Wichita County, map of: **225**
Wichita Group
Albany Group and: 224
Cisco Group and: 224
depositional environment of: 224, 227, **234**
Lueders Formation: **225**, 227, **231**
Nocona Formation. *see* Nocona Formation
Petrolia Formation. *see* Petrolia Formation
sedimentary structures in: **234**
Waggoner Ranch Formation. *see* Waggoner Ranch Formation
Wichita Mountains: 224–25
Wichita River: **223**
Wilbarger County, map of: **225**
wildfires
atmospheric oxygen levels and: 190
in Carboniferous: 190
charcoal from. *see* charcoal
in Devonian: 96
Gleichenia and: 286–87
in Holocene: 10, 14

Bold numbers indicate material in figures and tables.

Lagarostrobus franklinii and: 281, 287
in Mississippian: 10
in Pennsylvanian: 14, 189–90, 236
Rhacophyton and: 97
Williston Lake: **250**, **254**
willows: **29**, **296**
Windover: 30
Windsor Basin: 103–17, 120–23, **157**
Wisconsin: 52
Wisconsinan moraine: 292
Wolfcampian. *See also* Cisuralian
    Archer City Formation. *see* Archer City Formation
    Blaine Formation. *see* Blaine Formation
    Bowie Group. *see* Bowie Group
    Clear Fork Formation. *see* Clear Fork Formation
    Lueders Formation: **225**, 227, **231**
    Markley Formation: 225–27, 230–34, **238**, 240–41
    Nocona Formation. *see* Nocona Formation
    Pease River Group. *see* Pease River Group
    Petrolia Formation. *see* Petrolia Formation
    San Angelo Formation. *see* San Angelo Formation
    Waggoner Ranch Formation. *see* Waggoner Ranch Formation
    Wichita Group. *see* Wichita Group
Woodleigh Crater: 259
wooly rhinoceroses: 30
worms
    serpulid: 107, 174, 227
    spirorbid: 227
Wybung Head: **250**, **254**

X

xenacanthids: **159**, **162**–64
Xenotheca devonica: 99
Xyloiulus sigillariae: 186, **187**

Y

Yahatinda Formation: 99
Yeadonian: **157**
yellow fever: 22
Yellow River: 232
Yinkeng Formation: **250**, **251**, **262**
Youngiella: 107
Ypresian: 253, **260**
Yukon Territory: 300

Z

Zeilleropteris: 239
Zeilleropteris wattii: **239**
zosterophylls
    Barinophyton obscurum: **81**, 89, 98
    Barinophyton sibericum: **81**, 89–90, 98
    in Devonian: 5, 42, 76
    habitats of: 5, 42, 76, **81**
    height of: **4**, 5
    reproductive strategies of: 5
Zosterophyllum: **4**
zygnematalean algae: 263
zygopterid ferns
    in Carboniferous: 80
    in Devonian: 80–81, 87
    growth architecture of: 134
        habitats of: 80–81, 87, 131, 134
    in Mississippian: 131, 134
    Rhacophyton. *see Rhacophyton*
    Zygopteris: 131, 134
Zygopteris: 131, 134